PHYSICAL AND BIOLOGICAL HAZARDS OF THE WORKPLACE

Second Edition

PHYSICAL AND BIOLOGICAL HAZARDS OF THE WORKPLACE

Second Edition

EDITED BY

Peter H. Wald, M.D., M.P.H.

Principal and Medical Director, WorkCare Inc., Orange, California;
Associate Clinical Professor-Occupational Medicine, University of Southern California;
Assistant Clinical Professor-Occupational Medicine, University of California, Los Angeles and
University of California, Irvine

Gregg M. Stave, M.D., J.D., M.P.H.

Director, Strategic Health Planning, GlaxoSmithKline Research, Triangle Park, NC;
Consulting Assistant Professor-Occupational and Environmental Medicine,
Duke University Medical Center

WILEY-INTERSCIENCE

A JOHN WILEY & SONS, INC., PUBLICATION

The authors and the publishers have exerted every effort to ensure that drug selection and dosage set forth in this text are in accord with current recommendations and practice at the time of publication. However, in view of ongoing research, changes in government regulations, and the constant flow of information relating to drug therapy and drug reactions, the reader is urged to check the package insert for each drug for any change in indications and dosage and for added warnings and precautions. This is particularly important when the recommended agent is a new or infrequently employed drug.

This book is printed on acid-free paper. ∞

Published simultaneously in Canada.

For ordering and customer service, call 1-800-CALL-WILEY.

Library of Congress Cataloging-in-Publication Data:

Physical and biological hazards of the workplace / edited by Peter H. Wald and Gregg M. Stave.- - 2nd ed.

p. cm.

Includes index.

ISBN 0-471-38647-2 (cloth)

1. Medicine, Industrial. 2. Industrial hygiene. 3. Occupational diseases. I. Wald, Peter H., 1955-II. Stave, Gregg M.

RC963 .P48 2001

616.9′803- -dc21 2001046816

Printed in the United States in America.

10 9 8 7 6 5 4 3 2

To our wives—Isabel and Chris
And to our children—Sam, Ben and Elise
Who supported us and missed us while we were working

CONTENTS

FOREWORD

It has been fifteen years since the original publication of *Chemical Hazards of the Workplace.* That work was conceived as a handbook—which would serve as an authoritative guide to current concepts and practices aimed at protecting workerrs from chemical hazards. Over the intervening years, *Chemical Hazards* has been updated twice, but there has been a need for a similar work on the prevention and management of hazardous exposures from physical and biological agents.

I am happy to report that this void has now been filled expertly in the present volume prepared by Drs. Peter Wald and Gregg Stave. This exciting companion piece to *Chemical Hazards* is an important contribution to the practice of Occupational and Environmental Health. It is arranged to function both as an introduction to, and a review of, physical and biological hazards. It provides practical information not previously available in a single source on emerging, reemerging, and classical hazards due to these agents. Topics range from electromagnetic fields, ionizing radiation, and ergonomics, to occupational exposures to tuberculosis, HIV, and hantavirus. The reader will find helpful current information on a broad array of hazardous agents with a selection of timely literature citations for follow-up review.

All health professionals involved in protecting worker health will find this work a valuable addition to their basic reference library.

James P. Hughes, M.D.

PREFACE

Seven years have passed since the first edition of this book appeared. During this time, physical and biological hazards have been increasingly recognized as important hazards of the workplace. We have seen major revisions of government standards and guidelines for physical agents such as ergonomics, shift work and electric power, and biological agents such as tuberculosis, blood-borne pathogens and latex. In addition, we have seen the emergence or spread of biological hazards. Finally, a new third book in the "Workplace" series, *Reproductive Hazards of the Workplace*, was published in 1997.

The reception of the first edition has been very gratifying to us. Many of our colleagues have written to us with suggestions for new topics and agents. We have tried to preserve the style and format of the original edition, while updating and expanding existing content, and adding new agents that we felt have become important over the intervening years. In addition, we are deeply indebted to our contributors. Many of them have returned to update their original chapters, and many colleagues are new contributors for the second edition.

The primary focus of the book continues to be as a practical "how to" reference containing basic information about the physical and biological hazards for occupational health and safety professionals from an occupational health perspective. We are pleased that readers have told us that this is the book that they pull off the shelf when they need a quick introduction or refresher to a topic in physical and biological hazards, just before they go to talk to employees or patients. This is not meant to be a definitive reference book, but rather a first reference which gives a practical overview for the primary health practitioner. It is also intended to be useful for health professionals who have no formal occupational medicine training.

Our goal continues to be to bring you an introduction to the fascinating world of physical and biological hazards. We hope that the second edition will continue to assist all health professionals who are responsible for protecting the health and safety of workers.

Peter H. Wald, M.D., M.P.H.
Gregg M. Stave, M.D., J.D., M.P.H.

CONTRIBUTORS

Brian A. Boehlecke, M.D., M.S.P.H., Associate Professor of Medicine, University of North Carolina School of Medicine, Chapel Hill, North Carolina

Bryce D. Breitenstein, Jr., M.D., M.P.H., Clinical Professor, Division of Occupational Medicine, Department of Preventive Medicine, SUNY-Stoney Brook Medical School, Stoney Brook, New York; and Director, Occupational Medicine, Brookhaven National Laboratory, Upton, New York

Richard Cohen, M.D., M.P.H., Clinical Professor, Division of Occupational and Environmental Medicine, University of California School of Medicine, San Francisco, California

Dennis J. Darcey, M.D., M.S.P.H., Assistant Clinical Professor, Division of Occupational and Environmental Medicine, Department of Community and Family Medicine, Duke University Medical Center, Durham, North Carolina

Robert A. Dobie, M.D., Director, Division of Extramural Research, National Institute on Deafness and Other Communication Disorders, National Institutes of Health, Bethesda, Maryland

Carol A. Epling, M.D., M.S.P.H., Assistant Clinical Professor, Division of Occupational and Environmental Medicine, Department of Community and Family Medicine, Duke University Medical Center, Durham, North Carolina

Eric T. Evenson, M.D., M.P.H., Colonel, MC, USA, Director, Occupational and Environmental Medicine Residency, Uniformed Services University of the Health Sciences, Bethesda, Maryland

Linda M. Frazier, M.D., M.P.H., Assistant Professor, Department of Preventive Medicine, University of Kansas School of Medicine—Wichita, Wichita, Kansas. Associate Consulting Professor, Division of Occupational and Environmental Medicine, Department of Community and Family Medicine, Duke University Medical Center, Durham, North Carolina

Ed Fritsch, Ph.D., Vice-President, Biologic Process Development, Genetics Institute, Cambridge, Massachusetts

Gary N. Greenberg, M.D., M.P.H., Assistant Clinical Professor, Division of Occupational and Environmental Medicine, Department of Community and Family Medicine, Duke University Medical Center, Durham, North Carolina

Craig S. Glazer, M.D., M.S.P.H., Senior Fellow, Division of Pulmonary Sciences and Critical Care Medicine, University of Colorado Health Sciences Center, Denver, Colorado

Charles C. Goodno, M.D., M.P.H., Director, Occupational Medicine Services, Independent Medical Evaluations (IME), Durham, North Carolina

Gail M. Gullickson, M.D., M.P.H., Division of Geriatrics, University of California, San Francisco, San Francisco, California

Thomas R. Hales, M.D., M.P.H., Chief, Medical Section, Hazards Evaluation and Technical Assistance Branch, National Institute for Occupational Safety and Health, Cincinnati, Ohio

James A. Hathaway, M.D., M.P.H., Director, Medical and Product Safety Services, Rhodia Inc., Cranbury, New Jersey

Jessica Herzstein, M.D., M.P.H., Corporate Medical Director, Air Products and Chemicals Inc., Allentown, Pennsylvania

Harold R. Imbus, M.D., M.Sc.D., Health and Hygiene, Inc., Greensboro, North Carolina

George W. Jackson, M.D., Associate Clinical Professor, Director, Employee Occupational Health Service, Department of Community and Family Medicine, Duke University Medical Center, Durham, North Carolina

Robert Jacobs, Ph.D., Associate Professor, Department of Environmental Health Sciences, The University of Alabama at Birmingham, Birmingham, Alabama

Jeffrey R. Jones, M.P.H., M.S., C.I.H., Industrial Hygienist, Environmental Health and Safety Compliance Department, Port of Oakland, Oakland, California

James Kubalik, M.S., C.S.P., Director of Risk Management & Loss Control Services, Keenan and Associates, Torrance, California

Ricky L. Langley, M.D., M.P.H., Public Health Physician, North Carolina Department of Health and Human Services, Raleigh, North Carolina

Christopher J. Martin, M.D., M.Sc., Assistant Professor, Institute for Occupational and Environmental Health, West Virginia University School of Medicine, Morgantown, West Virginia

Glenn Merchant, M.D., M.P.H., & T.M., Assistant Professor of Aerospace and Preventive Medicine, Uniformed Services University of Health Sciences, Bethesda, Maryland

John D. Meyer, M. D., M.P.H., Assistant Professor, Institute for Occupational and Environmental Health, West Virginia University School of Medicine, Morgantown, West Virginia; and Lecturer in Occupational Medicine Centre for Occupational Health, University of Manchester, Manchester, UK

Aubrey K. Miller, M.D., M.P.H., Medical Officer, Technical Assistance Branch, National Institute for Occupational Safety and Health, Englewood, Colorado

Samuel D. Moon, M.D., M.P.H., Chief, Division of Occupational and Environmental Medicine, Division of Occupational and Environmental Medicine, Department of Community and Family Medicine, Duke University Medical Center, Durham, North Carolina

Catherine E. Palmier, M.D., Regional Medical Director, Southeast Region, Aetna US Healthcare, Atlanta, Georgia

James A. Palmier, M.D., M.P.H., M.B.A., Vice President, Pharmaceutical Sales, GlaxoSmithKline, Duluth, Georgia

Gwendolyn S. Powell, M.D., M.P.H., Consulting Associate, Division of Occupational and Environmental Medicine, Department of Community and Family Medicine, Duke University Medical Center, Durham, North Carolina. Medical Director, Product Surveillance, GlaxoSmithKline, Research Triangle Park, NC

Cecile S. Rose, M.D., M.P.H., Associate Professor, Departments of Medicine (Pulmonary Division), and Preventive Medicine and Biometrics,, University of Colorado School of Medicine; and Director, Occupational Medicine Clinical Program, Division of Environmental and Occupational Health Sciences, Department of Medicine, National Jewish Center for Immunology and Respiratory Medicine, Denver, Colorado

John L. Ryan, Ph.D., M.D., Senior Vice President, Experimental Medicine, Wyeth Research/Genetics Institute, Cambridge, Massachusetts

Allene J. Scott, M.D., M.P.H., M.T. (ASCP), Assistant Professor, Graduate School of Public Health, Environmental and Occupational Health, University of Pittsburgh, Pittsburgh, Pennsylvania. Medical Director, Armstrong Occupational HealthCare, Kittanning, Pennsylvania

James P. Seward, M.D., M.P.P., Director, Health Services Department, Lawrence Livermore National Laboratory, Livermore, California

David P. Siebens, M.D., M.P.H., Medical Advisor, GlaxoSmithKline, Research Triangle Park, North Carolina

David H. Sliney, Ph.D., Program Manager, Laser/Optical Radiation Program, US Army Center for Health Promotion and Preventive Medicine, Aberdeen Proving Grounds, MD

David J. Smith, M.D., M.S., Executive Officer, US Naval Hospital, Rota, Spain

Gregg M. Stave, M.D., J.D., M.P.H., Assistant Consulting Professor, Division of Occupational and Environmental Medicine, Department of Community and Family Medicine, Duke University Medical Center, Durham, North Carolina. Director, Strategic Health Planning, GlaxoSmithKline, Research Triangle Park, North Carolina

Woodhall Stopford, M.D., M.S.P.H., Assistant Clinical Professor, Division of Occupational and Environmental Medicine, Department of Community and Family Medicine, Duke University Medical Center, Durham, North Carolina

Jerry J. Tulis, Ph.D., Research Assistant Professor, Division of Occupational and Environmental Medicine, Department of Community and Family Medicine, Duke University Medical Center, Durham, North Carolina

Peter H. Wald, M.D., M.P.H., Principal and Medical Director, WorkCare Inc., Orange, California,, Assistant Clinical Professor- Occupational Medicine, University of California, Los Angeles and, University of California, Irvine, California

Donald E. Wasserman, M.S.E.E., M.B.A., DE Wasserman & Associates, Inc., Human Vibration & Biomedical Engineering Consulting, Cincinnati, Ohio; and Institute for the Study of Human Vibration, University of Tennessee, College of Engineering, Knoxville, Tennessee

Jack Wasserman, Ph.D., P.E., Institute for the Study of Human Vibration, University of Tennessee, College of Engineering, Knoxville, Tennessee

Thomas R. Waters, Ph.D., Chief, Human Factors and Ergonomics Research Section, National Institute for Occupational Safety and Health, Division of Biomedical and Behavioral Science, Applied Psychology and Ergonomics Branch, 4676 Columbia Parkway, Cincinnati, Ohio

David G. Wilder, Ph.D., P.E., C.P.E., Jolt/Vibration/Seating Lab Director, Iowa Spine Research Center, Biomedical Engineering Department, College of Engineering, Occupational and Environmental Health Department, College of Public Health, Orthopaedic Surgery Department, College of Medicine, The University of Iowa, Iowa City, Iowa

William N. Yang, M.D., M.P.H., Occupational Medicine Physician, Coca-Cola Company, Atlanta, Georgia

I

PHYSICAL HAZARDS

1

INTRODUCTION TO PHYSICAL HAZARDS

Peter H. Wald, M.D., M.P.H.

Physical hazards are hazards that result from energy and matter, and the interrelationships between the two. Conceptually, physical hazards in the workplace can be subdivided into worker–material interfaces, the physical work environment, and energy and electromagnetic radiation. The consequences of exposure to these hazards can be modified by worker protection and a variety of human factors. This chapter will review the general principles of basic physics and worker protection.

Physics is the science of energy and matter and of the interrelationships between the two, grouped in traditional fields such as acoustics, optics, mechanics, thermodynamics, and electromagnetism. Quantum physics deals with very small energy forces; relativity deals with objects traveling at very high speeds (which causes time effects). Thus, physical hazards can be thought of as primarily hazards of energy, temperature, pressure, or time. This broad definition allows for the investigation of many hazards that are otherwise hard to classify but nevertheless represent important issues in the workplaces of the 1990s. An understanding of these physical hazards requires familiarity with the two basic concepts of physics: classical mechanics, with its derivatives of thermodynamics and fluid dynamics, and electromagnetic radiation. For measurements, we have used Standard International (SI) units throughout this book, but we have included conversions to other units where they are in common usage. Table 1.1 reviews the standard unit prefixes for mathematics that are used in the physical hazards section. The mathematical equations for principles discussed in this section are included in tables that accompany the text. Although they are not necessary to understand the material, they are presented for those readers who wish to review them.

Physical and Biological Hazards of the Workplace, Second Edition, Edited by Peter H. Wald and Gregg M. Stave
ISBN 0-471-38647-2

Table 1.1 Mathematical unit prefixes.

Prefix	Symbol	Multiplier
Tetra-	T	10^{12}
Giga-	G	10^{9}
Mega-	M	10^{6}
Kilo-	K	10^{3}
Deci-	d	10^{-1}
Centi-	c	10^{-2}
Milli-	m	10^{-3}
Micro-	μ	10^{-6}
Nano-	n	10^{-9}
Pico-	p	10^{-12}

MECHANICS

Mechanics deals with the effects of forces on bodies or fluids at rest or in motion (Table 1.2). From mechanics, we can get to the study of sound, which is a result of the mechanical vibration of air molecules. The behavior of heat arises from the vibration of molecules. Temperature is proportional to the average random vibrational (in solids) or translational (in liquids and gases) kinetic energy. The physics of pressure arises from the laws of motion and temperature. The laws that govern electricity can be derived from special cases of mechanics (see below), and electromagnetic energy and waves are a direct result of the laws that govern electricity.

Classical mechanics is the foundation of all physics. Galileo (1564–1642) first described the study of kinematics. Kinematics is primarily concerned with uniform straight-line motion and motion where there is uniform acceleration. As a practical example, Galileo used these insights to predict the flight of projectiles. In uniform straight-line motion, velocity (v) is equal to the change in displacement (Δs) divided by the change in time (Δt). Acceleration (a) is the instantaneous change of velocity with respect to time, which is calculated by taking the derivative of velocity with respect to time. Where there is uniform acceleration, the new velocity is equal to the original velocity (v_0) plus acceleration times time. The distance traveled under acceleration is described by a combination of the component traveled at the original velocity plus the component traveled under acceleration. The mathematical equations for these forces are summarized in Table 1.3.

Sir Isaac Newton (1642–1727) originally described the study of mechanics in his 1687 *Principia*. He formulated three laws that serve as the foundation of classical mechanics (Table 1.4).

The first law is known as the law of inertia. It states that all matter resists being accelerated and will continue to resist until it is acted upon by an outside force. The second law states that the acceleration of this outside force will be related to the size of that net force (F) but inversely related to the mass (m) of the object. This relationship is described mathematically by the following expression:

$$a \propto \text{net } F/m$$

The third law states that when two bodies exert a force on each other, they do so with an action and reaction pair. The force between two bodies is always an interaction.

Table 1.2 The disciplines of mechanics.

Solid mechanics	
Statics	The study of bodies at rest or equilibrium
Dynamics or kinetics	The study of forces or the change of motion that forces cause
Kinematics	The study of pure motion without reference to forces
Fluid dynamics	
Hydrostatics	The study of still liquids
Hydraulics	The study of the mechanics of moving liquids
Aerodynamics	A special subset of hydraulics that deals with the movement of air around objects

Table 1.3 Mathematical expressions of Galileo's description of kinematics.

$v = \Delta s/\Delta t$	Average straight-line velocity
$s = vt$	Distance traveled at constant velocity
$a = dv/dt$	Acceleration (derivative of velocity with respect to time)
$v = v_0 + at$	Velocity at straight-line acceleration
$s = v_0 t + \frac{1}{2}at^2$	Distance traveled at uniform acceleration

Variables

v = velocity, s = distance, t = time, a = acceleration,
Δs = change of distance, Δt = change of time, v_0 = original velocity

A good example of how all three laws operate can be seen at the bowling alley. When a bowling ball is sitting on the rack, the force of the ball pressing down on the rack (gravity) is equal and opposite to the force of the rack pressing up on the ball to resist gravity (the third law). The speed of the bowling ball at the end of the alley is dependent on the amount of acceleration imparted to it. An adult can apply more force to the ball than a child, so the adult's ball will go faster. However, if smaller balls (i.e. of less mass) are used, less force is required; therefore, a child can accelerate the ball to the same speed (the second law). Once the ball leaves your hand, no more net force is applied to the ball (if we ignore friction), and it travels down the alley at a constant speed (the first law).

Mechanics has been central to the advancement of physics. Two mechanical concepts are central to understanding what strategies to adopt in order to prevent injury and illness from physical hazards: kinetic energy and potential energy. In order for physical hazards to affect humans, they must possess energy to impart to the biological system. Energy is commonly described in terms of either force (F) or work (W). Force equals mass times the acceleration, and the result is a vector. $F = ma$ is the mathematical representation of Newton's second law. The work done on an object equals the amount of displacement times the force component acting along that displacement. In the special case of the force acting parallel to the displacement, work equals force times displacement. These two relationships are described mathematically in Table 1.5, equations 1 and 2.

Kinetic energy (KE) is the energy of a mass that is in motion relative to some fixed (inertial) frame. KE is related to the mass of the object, and the speed at which it is traveling (Table 1.5, equation 3). Potential energy (PE) is stored energy that can do work when it is released as kinetic energy.

Since mass and energy are conserved in all interactions, the sums of potential and kinetic energy from before and after an encounter are equal. The equation for kinetic energy is also important for electromagnetic radiation. An electric system can store electric energy in a magnetic field in an induction coil. The kinetic energy of the electric charges equals the amount of work done to set up the field in the coil, which is stored as potential energy. Work, kinetic energy and potential energy in

Table 1.4 Newton's laws of motion.

Newton's First Law of Motion: A body remains at rest, or if in motion it remains in uniform motion with a constant speed in a straight line, unless it is acted on by an unbalanced external force

Newton's Second Law of Motion: The acceleration produced by an unbalanced force acting on a body is proportional to the magnitude of the net force, in the same direction as the force, and inversely proportional to the mass of the body

Newton's Third Law of Motion: Whenever one body exerts a force upon a second body, the second body exerts a force upon the first body; these forces are equal in magnitude, and oppositely directed

Table 1.5 Mathematical expressions of force, work and energy.

(1) $F = \kappa ma$ $= ma$ (if units = kg-m/s^2 = Newtons)	Force
(2) $W = (F \cos \theta)s = F \cdot s$ $= Fs$ (if F and s are parallel)	Work
(3) $\text{KE} = \frac{1}{2}mv^2$	Kinetic energy—mechanical system
(4) $\text{PE} = \frac{1}{2}LI^2$	Potential energy—electrical system

Variables

κ = a constant, m = mass, a = acceleration, s = displacement, m = mass, v = velocity, L = the inductance of the coil (in henries), I = the current (in amperes)

this system are related to the inductance of the coil and the current (Table 1.5, equation 4). Potential energy is the potential to do work, and theoretically all this work can be turned into kinetic energy. The expressions for kinetic energy in the mechanical system and potential energy in the electric system have an identical form. This form shows the similarity between kinetic and potential energy in mechanics and electromagnetic radiation and lays the groundwork for examining the electromagnetic wave.

ELECTROMAGNETIC RADIATION

By far the most complicated concept related to the understanding of physical hazards is that of electromagnetic radiation (EMR). Energy can be transmitted directly by collision between two objects, or it can be transmitted by EMR. We see direct examples of energy transfer by EMR when we are warmed by the infrared rays of the sun or burned by its ultraviolet rays. EMR is a continuum of energies with different wavelengths and frequencies. Two similarities of all types of EMR are that they all move at the same speed, and they are all produced by the acceleration or deceleration of electric charge. EMR has a dual, particle–wave nature: its energy transfer is best described by a particle, but the behavior of the radiation is best described as a wave. All EMR travels at a constant speed, $c = 3 \times 10^8$ m/s (the speed of light). Each particle of energy, called a photon, is accompanied by an electric field (E-field) and a magnetic field (H-field); these fields are perpendicular to each other and perpendicular to the direction of travel of the wave (Figure 1.1).

It is important to remember that EMR is only produced when an electric charge is moving. Coulomb forces are forces between stationary charges, whereas magnetic forces are due to the motion of charges relative to each other. A moving electric charge (or electric field) induces a magnetic field, and a moving or changing magnetic field induces an electric field. In 1873, James Maxwell linked together these electric and magnetic phenomena into a unified field theory of EMR. As an electric charge moves, it induces a magnetic field, which in turn induces an electric field. The mutual interaction of these two fields is what allows the electromagnetic wave to propagate, and what dictates its physical form in Figure 1.1.

The energy (E) in each photon in the wave can be calculated in joules (J) and is related to the frequency of the radiation in hertz (Hz). Energy is calculated by multiplying the frequency by Planck's constant (6.626176 J/Hz). The mathematical representation of this is shown inTable 1.6, equation 1.

Since the velocity at which the wave travels equals the frequency times the wavelength (Table 1.6, equation 2), we can discover the wavelength (λ) for each frequency by dividing

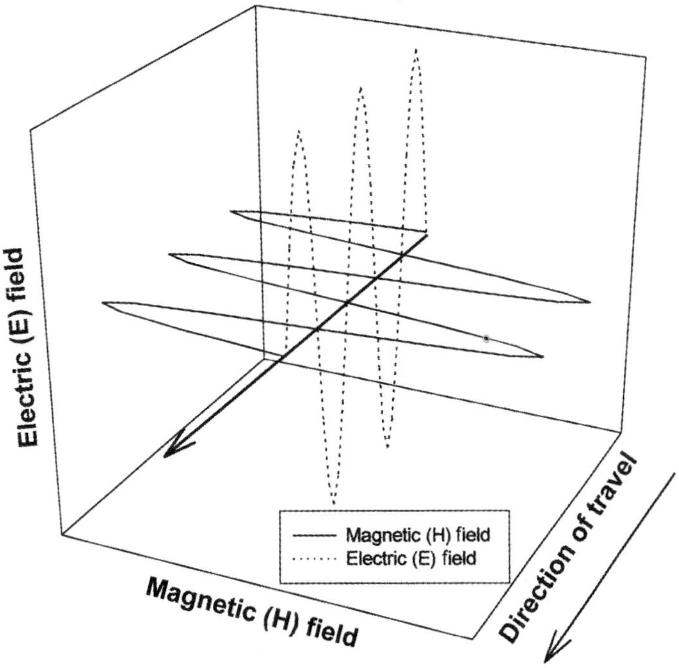

Figure 1.1 Stylized representation of an electromagnetic wave.

3×10^8 m/s (the speed of light, or c) by the frequency (Table 1.6, equation 3). The energy of the wave can also be calculated in terms of the wavelength by substituting the speed of light divided by wavelength for frequency (Table 1.6, equation 4). In biological systems, it is useful to determine photon energy in electron volts from the wavelength. This can be calculated from the wavelength in angstroms (Å) according to Table 1.6, equation 5. The electron volt is a convenient unit to use with biological systems, because it takes greater than roughly 10 electron volts (eV) to cause ionization in tissue.

We can also see from equations 1 and 2 in Table 1.6 that the energy of a given type of EMR varies directly with its frequency and inversely with its wavelength. Figure 1.2 shows a representative cross-section of the electromagnetic spectrum, with the major classes noted. Notice that there are not strict divisions between the different classes of

Table 1.6 Mathematical equations for electromagnetic radiation.

(1) $E = h\nu$	Energy in joules
(2) $c = \nu\lambda$	Wavelength and frequency related to the speed of light
(3) $\lambda = c/\nu,\ \nu = c/\lambda$	Rearrangements of equation 2
(4) $E = hc/\lambda$	Energy in joules, by substituting equation 3 into equation 1
(5) $E = 12400/\lambda$	Energy in electron volts, where λ is in Ångströms (1 Å= 10^{-10} m)
$E = 1.24 \times 10^{-6}/\lambda$	Energy in electron volts, where λ is in meters

Variables

h = Planck's constant (6.626176 J/Hz), ν = frequency, c = speed of light (3×10^8 m/s), λ = wavelength

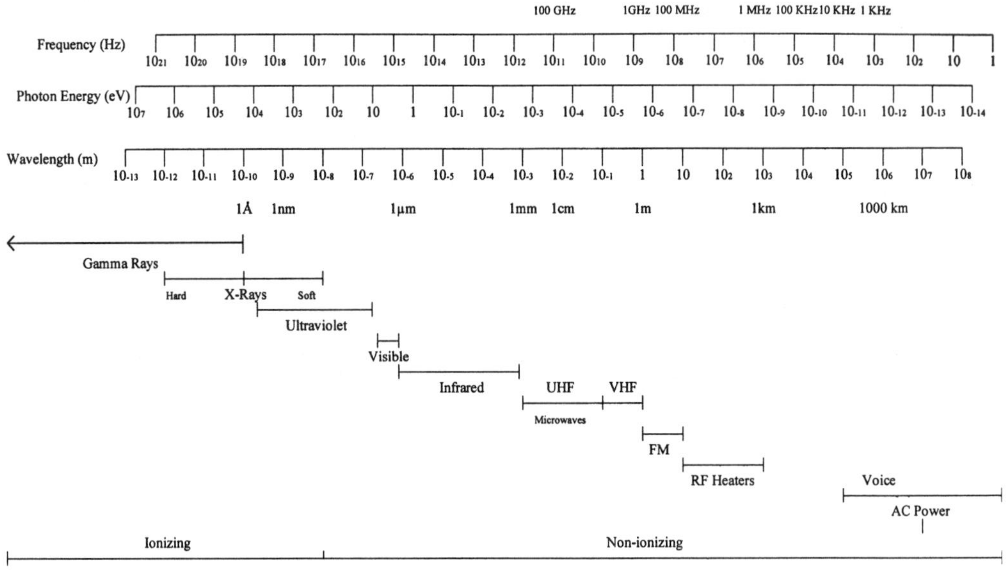

Figure 1.2 The electromagnetic spectrum.

EMR. An important division in the EMR spectrum relates to the ability to ionize chemical bonds in biological tissue. As frequency increases from the radio bands, so does energy, until ionization potential is reached in the "hard" ultraviolet or "soft" X-ray bands.

A final important point about EMR involves the ways in which it can interact with objects. EMR interacts with biological tissues in one of the following three ways: (1) transmission, where the radiation passes through the tissue without any interaction; (2) reflection, where the radiation is unable to pass through the air–tissue interface (also called the boundary layer) and is reflected back into space; and (3) absorption, where the radiation is able to pass through the boundary layer and deposit its energy in the tissue. The frequency of the EMR determines what energy is released in the tissues (heat, electric potential, bond-breaking, etc.). These interactions are summarized in Figure 1.3.

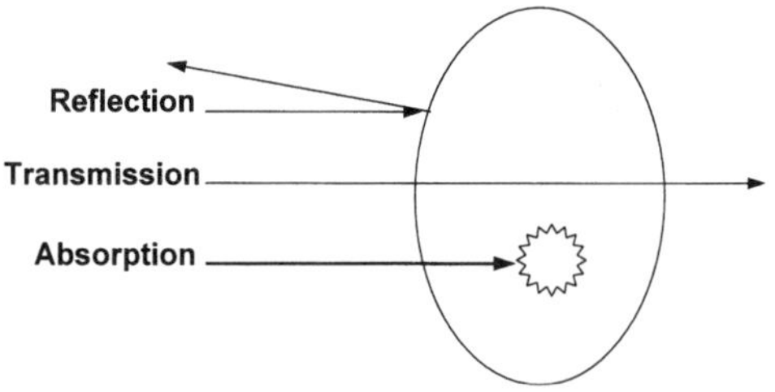

Figure 1.3 Interactions of electromagnetic radiation and biological tissue.

WORKER PROTECTION

Potential energy can also be called a potential hazard. The key to avoiding injuries and illnesses is to prevent the individuals in the workplace from being overexposed to the kinetic energy in the hazards. The major characteristics of the physical hazards covered in this text are reviewed in Table 1.7. Each of the following chapters will deal with the most appropriate method to prevent overexposure. However, there are certain recurring themes.

Since we are trying to prevent exposure, the first step is to educate the workforce. A good training program includes: education about the

Table 1.7 Major characteristics of the physical hazards.

Hazards	Occupational settings	Measurement	Exposure guidelines	Effects of exposure	Surveillance
Worker–material interfaces					
Repetitive ergonomic hazards—extremities	Service and industrial operations	Repetition, force, posture	OSHA (pending)	Musculo-skeletal strain, tunnel syndromes	Survey workers, observe tasks, measure physical parameters of the job
Manual materials handling—backs	Service and industrial operations	Repetition, force, posture	NIOSH lifting guide	Musculo-skeletal strain, disc herniation	Survey workers, observe tasks, measure physical parameters of the job
Vibration	Whole body—vehicle/heavy equipment/industrial equipment operator, hand–arm-powered hand tool users	Frequency, motion	ISO 2631, ANSI S3 NIOSH criteria document, ACGIH	Whole body-low back pain, Hand-arm-hand arm vibration syndrome (HAVS)	Survey workers, observe tasks, measure physical parameters of the job
Mechanical energy—direct injuries	Service and industrial operations	Velocity, distance, acceleration, force, weight, pressure, friction	None	Direct injury	Epidemiology of workplace injuries
The physical work environment					
Hot environments	Hot indoor or outdoor environments	Wet globe bulb temperature, core temperature	ISO 7243, NIOSH criteria document	Heat strain, heat stroke	Heart rate, core temperature, worker selection
Cold environments	Cold indoor or outdoor environments	Wind chill, core temperature	ACGIH-TLV	Frostbite, trench/immersion foot, hypothermia	Worker selection
High-pressure environments	Divers, caisson workers	Pressure in atmospheres absolute (ATA), changes in pressure	OSHA, marine occupational health safety standards	Barotrauma, decompression sickness, indirect effects secondary to pressure acting on other gases (O_2, N_2)	Worker selection
Low-pressure environments	Aircraft crews, private pilots, astronauts	Pressure in atmospheres absolute (ATA), changes in pressure	None	Barotrauma, decompression sickness, hypoxia	Worker selection

(continued)

Table 1.7 *(continued)*

Hazards	Occupational settings	Measurement	Exposure guidelines	Effects of exposure	Surveillance
Shift work	Service and industrial operations	Rotation, duration and hour changes of shift schedule	None	Sleep disturbance, gastrointestinal upset	Worker selection
Energy and electromagnetic radiation					
Ionizing radiation	Pilots, underground miners, radiographers, medial and dental X-ray personnel, operators of high-voltage equipment, nuclear power and fuel cycle workers, medical and scientific researchers, some commercial products manufacturing	Personal dosimetry of radiation exposures	NCRP, ICRP, NRC, UNSCEAR	Acute radiation injury, carcinogenesis	Personal dosimetry. Lung and whole body scanning, biological monitoring as appropriate
Ultraviolet radiation	Outdoor workers, welders, printers	Wavelength, intensity	ACGIH-TLV	Corneal photokeratitis, skin erythema, cataracts	None
Visible light and infrared radiation	Outdoor workers, welders, printers, glass blowers	Visible duration, wavelength, intensity. Infrared wavelength, intensity	ACGIH-TLV	Visible-scotoma, thermal burn, photosensitivity, urticaria. Infrared-thermal burns, cataracts	None
Laser radiation	Service and industrial operations, researchers, medical personal, maintenance personal	Wavelength, power, energy, duration	ACGIH-TLV, (ANSI Z136.1)	Retinal and skin burns	Required for Class 3b and 4 lasers. Baseline and post-exposure retinal exams

Microwave, radiofrequency (MW/RF) and extremely low-frequency (ELF) radiation	MW/RF—communication workers, industrial heating and RF welding operations. ELF—electricians and electrical workers, telephone and cable workers, electric arc welders, movie projectionists	Frequency, electric field, magnetic field, power density, operating mode	ACGIH-TLV, OSH	MW/RF-Thermal effects ELF-No proven effects	None
Noise	Service and industrial operations	Time-weighted dBA	NIOSH, OSHA	Noise induced hearing loss	Hearing conservation program for all exposed workers
Electric power and electrocution injuries	Service and industrial operations	Current, voltage	OSHA	Electrical burns, electrocution	None

Table 1.8 Engineering and administrative controls for physical hazards.

Hazards	Engineering controls	Administrative controls
Worker–material interfaces		
Repetitive ergonomic hazards—extremities	Repetition—mechanical aids, automation, distribution of tasks across the shift and the workforce	More frequent or longer rest breaks, limit overtime, varying work tasks, rotation of workers between less and more ergonomically stressful jobs
	Force—decrease weight of tools/ containers, optimize handles, torque control devices Postures—locate work for mechanical advantage	
Manual materials handling—backs	Same as above	Same as above
Vibration	Whole body—relocate worker away from vibration, mechanically isolate vibration, use vibration-isolating seats in vehicles Hand–arm—use anti-vibration tools	Hand–arm—removal from work for significantly affected workers
Mechanical energy—direct injuries	Guards, interlocks, proper lighting, non-skid floors	None
The physical work environment		
Hot environments	Air conditioning, increase air movement, insulate and shield hot surfaces, decrease air humidity, shade work area, mechanize heavy work	Use recommended work/rest cycles, work during cool hours of the day, provide cool rest areas, use more workers for a given job, rotation of workers between less and more physically stressful jobs, provide fluids for cooling and hydration
Cold environments	Enclose and heat work area	Use recommended work/rest cycles, provide appropriate clothing, provide shelter for break, provide fluids for warming and hydration
High-pressure environments	Engineer a "shirtsleeve" environment which avoids high-pressure work	Work under no decompression guidelines/tables. Adhere to recommended decompression guidelines
Low-pressure environments	Work remotely at low altitude	Wait 12–48 hours after diving to fly, schedule time for acclimation when working at altitude
Shift work	Automate processes to reduce the number of workers/shift	Rotate shifts forward, get worker input for desires of time off and shift design

(*continued*)

Table 1.8 *(continued)*

Hazards	Engineering controls	Administrative controls
Energy and electromagnetic radiation		
Ionizing radiation	Shielding, interlocks, increase worker distance to source, warning signs, enclose radionuclides	Worker removal if dose limit reached, minimize exposure times, use radionuclides only in designated areas using safe handling techniques, limited personnel access
Ultraviolet radiation	Enclosure, opaque shielding and/or tinted viewing windows, interlocks, increase worker distance to source, non-reflective surfaces, warning signs	Minimize exposure times, limited personnel access
Visible light and infrared radiation	Enclosure, shielding, interlocks, increase worker distance to source, non-reflective surfaces, warning signs	Limited personnel access
Laser radiation	Enclosure, interlocks, non-reflective surfaces, warning signs	Limited personnel access
Microwave, radiofrequency (MW/RF) and extremely low-frequency (ELF) radiation	MW/RF—Wire mesh enclosure, interlocks, increase worker distance to source, warning signs ELF—Increase worker distance to source	MW/RF—Limited personnel access
Noise	Enclose sources, warning signs	Limited personnel access
Electric power and electrocution injuries	Interlocks, warning signs	Limited personnel access

potential hazards; the safest procedures to follow for each manufacturing or maintenance operation; correct tool selection and use for each job; use and care of personal protective equipment; and procedures to follow in emergency situations, including fire and loss control, shutdown, rescue, and evacuation.

Substitution of less dangerous equipment or agents is the best protection from hazards, because it totally removes any chance of exposure. However, substitution is often not possible; therefore, worker protection from physical hazards generally focuses on engineering controls. Engineering and administrative controls for physical hazards are summarized in Table 1.8. Often, these controls involve isolation or shielding from the hazard. The most effective isolation involves physically restricting an individual from a hazard area by fencing off the area whenever the hazard is present. Interlocks that inactivate the equipment when the exclusion area is entered are often used to further enhance physical barriers. Alternatively, the hazard can be "locked out" when a worker is present in an area that would become hazardous if the equipment were energized (Chapter 5). This process of excluding maintenance workers from hazardous areas has been institutionalized in the OSHA Lock-Out, Tag-Out (LOTO) Standard (Code of Federal Regulations [CFR] 1910.147).

Another way to protect workers is to specifically shield them from the hazard. In some cases, an individual piece of equipment can be shielded to prevent exposure. With some higher-energy hazards, such as ionizing radiation, shielding may be needed in addition to isolation of the equipment. In special cases where it is not practical to shield the hazard (e.g. cold, low pressure), individual workers can be shielded with personal protective equipment, such as jackets or environment suits. In addition, it is sometimes possible to alter the process so as to decrease exposure. This is often the case with hazards affecting the worker–material interface, where engineering design is often inadequate. Personal protective equipment can also be used as an adjunct to engineering controls. Table 1.9 contains a summary of the most common personal protective equipment used for physical hazards.

The final strategy for hazard control is the use of administrative controls. These controls are implemented when exposures cannot be controlled to acceptable levels with substitution, engineering controls, or personal protective equipment. Administrative measures can be instituted to either rotate workers through

Table 1.9 Commonly used personal protective equipment for physical hazards.

Equipment type	Hazard category	Specific hazard
Helmet	Direct injuries	(1) Falling objects (2) Low clearances/"bump hazards"
Safety glasses	(1) Direct injuries	(1) Flying objects (2) Sparks
	(2) Lasers	Retinal burns
Face shield	Direct injury	(1) Flying objects (2) Molten metal, sparks
Welding helmet/goggles	(1) Direct injury	(1) Flying objects (2) Molten metal, sparks
	(2) Ultraviolet radiation	Skin/conjunctival burns
Earplugs/earmuffs	Noise	Noise
Fall protection systems—safety belt, body harness, lines and/or other hardware	Direct injury	Falls
Respirators	Ionizing radiation	α-Emitters: internal contamination
Clothing		
Leather	Heat	Burns
Aluminized	Heat	Heat stroke, burns
Lead	Ionizing radiation	γ-Emitter, X-rays
Fire-resistant	Direct injury	Burns
Insulating	Cold	Hypothermia
Disposable	Ionizing radiation	α-Emitter: external contamination
Gloves		
Leather	Direct injury	Abrasions, lacerations
Rubber	Electric injury	Electrocution
Metal mesh	Direct injury	Lacerations
Anti-vibration	Vibration	Vibration
Footwear		
Steel toe	Direct injury	Falling objects
"Traction sole"	Direct injury	Slips, trips, falls
Rubber	Electric energy	Electrocution

different jobs to prevent repetitive motion injuries, or to remove workers from ionizing radiation exposure once a predetermined exposure level is reached. Although this is not the preferred method of hazard control, it can be effective in some circumstances. Administrative controls are also reviewed in Table 1.8.

The best way to determine what hazards are present in a specific workplace is to go to the site and walk through the manufacturing or service process. There are a number of excellent texts available on evaluating workplaces from both an industrial hygiene and a safety perspective; they are included in the list of further reading at the end of this chapter. An additional point that will become obvious as you read through the text is that there are some significant measurement issues that need to be addressed by an appropriate health professional. Although larger employers will undoubtedly have such a person on staff, at the majority of smaller work sites, no such person will be available.

If you are unfamiliar with the measurement technology, make sure that you (or the employer) retain someone who knows how to do an exposure assessment. Inaccurate measurements will invalidate the entire process of a prevention program. There are, of course, a number of physical hazards that do not require special measuring and can be handled quite nicely with relatively low-cost safety programs. Several excellent texts describing how to set up general safety programs are included in the further reading list at the end of this chapter.

Finally, remember that the human being is a biological system. For a given exposure, different people will respond differently because of interindividual variation. Most workplace standards are designed with a safety factor to protect against overexposure related to this variation (and to account for any knowledge gaps). In addition, however, a worker's perception of the hazard must also be taken into account. Some workers may have an exaggerated response to a non-existent or low-threat hazard, whereas others may not respond appropriately to a series hazard with which they have "grown comfortable". The challenge in assessing and communicating the relative danger entailed by the hazard is to strike the right balance between these two competing tendencies.

The goal of the first section of this volume is to acquaint the reader with the types of physical hazard that may be present in the workplace. Once these hazards are identified at the site, he or she can refer to the specific chapter that addresses the salient measurement issues, or offers general strategies to control exposures and monitor effects.

FURTHER READING

Balge MZ, Krieger GR. *Occupational health and safety*, 3rd edn. Chicago: National Safety Council Press, 2000.

Burgess WA. *Recognition of health hazards in industry: a review of materials processes*, 2nd edn. New York: JB Wiley & Sons, 1995.

Clayton GD, Clayton FE. *Patty's industrial hygiene and toxicology*, Vol. 1, Parts A and B, 4th edn. New York: Wiley, 1991.

Hagan PE, Montgomery JF, O'Reilly JT. *Accident prevention manual for business and industry: administration and programs*, 12th edn. Chicago: National Safety Council Press, 2001.

Krieger GR, Montgomery JF. *Accident prevention manual for business and industry: engineering and technology*, 11th edn. Chicago: National Safety Council Press, 1997.

Plog B, ed. *Fundamentals of industrial hygiene*, 4th edn. Chicago: National Safety Council Press, 1996.

Serway RA, Faughn JS. *College physics*, 5th edn. Philadelphia: WB Saunders, 1998.

Section I

Worker-Material Interfaces

2

ERGONOMICS AND UPPER EXTREMITY MUSCULOSKELETAL DISORDERS

Thomas R. Hales, M.D., M.P.H.

Ergonomics has been defined as the science of fitting the job to the worker,[1] or the art of matching job demands with worker capabilities. Upper extremity (UE) musculoskeletal disorders (MSDs) are soft tissue disorders of the muscles, tendons, ligaments, peripheral nerves, joints, cartilage, or supporting blood vessels in the neck, shoulder, arm, elbow, forearm, hand, or wrist. Examples of specific disorders include tension neck syndrome, rotator cuff tendinitis, epicondylitis, peritendinitis, and carpal tunnel syndrome (CTS). When job demands overwhelm an employee's mental and/or physical capacity, employee health, comfort and productivity can be adversely affected.[2] While comfort and productivity levels are important outcomes to consider, this chapter will focus upon the effect of workplace physical stressors (repetition, force, posture, and vibration) on the musculoskeletal system of the upper extremities. The chapter will not only review the literature, but will also provide practical tools for healthcare providers to (1) assess physical stressors in the workplace, and (2) recognize, treat, and prevent UE MSDs.

OCCUPATIONAL SETTING

Magnitude of the problem

The Bureau of Labor Statistics (BLS) conducts an annual survey of employer-maintained OSHA 200 Logs for about 165 000 private sector establishments in the USA. In 1998, the BLS estimated that nearly 593 000 employees suffered a work-related MSD serious

Physical and Biological Hazards of the Workplace, Second Edition, Edited by Peter H. Wald and Gregg M. Stave
ISBN 0-471-38647-2

enough to result in days away from work (DAW).[3] The median number of DAW was seven, with 120 000 cases (20%) missing more than 30 days. Sprains and strains represented the most common diagnosis (81%), followed by carpal tunnel syndrome (5%), soreness/pain (4%), tendinitis (3%), and others (7%).[3] The most common responsible event or exposure was "overexertion" (77%), followed by "bending, climbing, crawling, reaching, twisting" (13%), and "repetitive motion" (10%).[3]

In 1989, Webster and Snook identified 6067 UE MSD workers' compensation claims for policy holders in 45 states with an average (mean) cost of $8070 per claim.[4] They estimated the total direct US workers' compensation costs for UE MSDs to be $563 million in 1989.[4] For this same year, Silverstein et al reported an average Washington State UE MSD claim to range from $7093 to $8250, and estimated the total direct workers' compensation costs for the US to be $6.1 billion.[5] Neither of these estimates include the indirect costs, such as administrative costs for claims processing, lost productivity, and the cost of recruiting and training replacements. It has been suggested that indirect costs are two to three times direct compensation costs.[6]

While these numbers are impressive, they do not take into account those workers who suffer a UE MSD but are never recorded onto the OSHA 200 Log and do not file a claim.[7,8] Rossenman et al reported that in 1996 only 25% of workers with work-related MSD filed for workers' compensation.[9] Factors associated with filing a claim included increased length of employment, lower annual income, dissatisfaction with coworkers, physician restriction on activities, type of physician providing treatment, being off work for at least 7 days, decreased current health status, and increased severity of illness.[9]

Industries at risk

According to the BLS annual survey, the manufacturing and service industries had the most MSDs involving DAW.[3] Looking at a category of illnesses labeled by the BLS as "disorders due to repeated trauma" (DRT), meat-packing plants have had the highest DRT rates since the BLS began collecting industry-specific data in 1984. In 1998, the DRT rate for meat-packing plants was 994 cases per 10 000 full-time workers.[3] The motor vehicles/car bodies and the poultry-slaughtering/processing industries have always had some of the highest DRT rates, ranking second and third, respectively, in 1998.[3] Data on CTS from individual state workers' compensation programs and National Health Interview Survey (NHIS) data have identified the same high-risk industries.[5,10–12] Although these industries have higher rates of DRT, the BLS reports substantial numbers of MSD involving lost work days in every industry. This finding is consistent with data from the National Occupational Exposure Survey (NOES), which found physical stressors in most industrial and service operations.[13] The NOES survey estimated that 4.6 million workers are exposed to "arm transports", 2.8 million workers are exposed to "shoulder transports", 4.9 million workers are exposed to "hand/wrist manipulations", and 3.6 million workers are exposed to "finger manipulations".[13]

Occupations at risk

Case reports have given rise to a number of disorders named for the patient's occupation

Table 2.1 Work-related MSD named by occupation.

Bricklayer's shoulder
Carpenter's elbow
Golfer's elbow
Tennis elbow
Janitor's elbow
Stitcher's wrist
Cotton twister's hand
Telegraphist's cramp
Writer's cramp
Bowler's thumb
Jeweler's thumb
Cherry pitter's thumb
Gamekeeper's thumb
Carpetlayer's knee

(Table 2.1). This does not mean, however, that these disorders are unique to their occupations. The BLS reported over 100 occupations with large numbers of cases (over 600 cases of MSD involving DAW per occupation).[3] Looking specifically at MSDs involving DAW resulting from repetitive motion, assemblers had the largest number of cases (Table 2.2). The NHIS described cases of self-reported CTS to be highest among mail/message distributors (prevalence 3.2%), health assessment and treatment occupations (2.7%), and construction trades (2.5%).[12] The Wisconsin workers' compensation program reported wrist injury to be highest among dental hygienists, data-entry keyers, and hand grinding and polishing occupations.[11] Although the various occupations have different rates of MSD, the BLS and workers' compensation data point to almost all occupations reporting at least one case of work-related MSD.

Epidemiology

One of the main purposes of epidemiologic studies is to identify factors that are associated (positively or negatively) with the development or recurrence of adverse medical conditions. No single epidemiologic study determines causality. Rather, results from epidemiologic studies can contribute to the evidence of causality. Over the past decade, several publications have reviewed the medical and ergonomic literature to determine whether scientific evidence supports a relationship between workplace physical factors and MSDs. The most comprehensive review was completed by Bernard at the National Institute for Occupational Safety and Health (NIOSH).[14] This review focused on disorders that affected the neck (tension neck syndrome), upper extremities (shoulder tendinitis, epicondylitis, CTS, hand–wrist tendinitis, hand–arm vibration syndrome), and the low back (work-related low back pain). A database search strategy initially identified 2000 studies, but laboratory, biomechanical, clinical treatment and other non-epidemiologic studies were excluded, leaving 600 for systematic review. The review process consisted of three steps.

The first step gave the increased emphasis, or weight, to studies that had high participation rates (> 70%), physical examinations, blinded assessment of health and exposure, and objective exposure assessment. The second step assessed for any other selection bias and any uncontrolled potential confounders. The final step summarized studies with regard to strength of the associations, consistency in the associations, temporal associations, and exposure–response (dose–response) relationships.

Bernard concluded that a substantial body of credible epidemiologic research provides strong evidence of an association between MSDs and certain work-related physical factors. This is particularly true when there are high levels of exposure or exposure to more than one physical factor (e.g. repetition and forceful exertion). The strength of the associations for specific physical stressors varies from insufficient to strong (Table 2.3). The consistently positive findings from a large number of cross-sectional studies, strengthened by the available prospective studies, provides strong evidence for an increased risk of work-related MSDs for the neck, elbow, and hand–wrist. This conclusion was supported by another exhaustive review conducted by the National Academy of Sciences, National Academy of Engineering, Institute of Medicine, and the National Research Council.[15]

MEASUREMENT—ASSESSMENT

Physical stressors can be grouped into the following categories: repetition, force, posture, and vibration. They arise from excessive job demands, improperly designed workstations, tools, or equipment, or inappropriate work techniques. Recently, Radwin and Lavender have proposed a different model of "external loading factors".[16] They identified motion, force, vibration and temperature as physical stressors that are modified by the properties of

Table 2.2 MSDs involving days away from work resulting from repetitive motion; occupations with 1% or more of total cases, 1998.

	Repetitive motion		Repetitive typing or key entry		Repetitive use of tools		Repetitive placing, grasping, or moving objects, except tools	
Occupation	Number	%	Number	%	Number	%	Number	%
All occupations	65 866	100.0	9784	100.0	9364	100.0	21 164	100.0
Assemblers	6683	10.4	65	0.7	1909	20.4	2405	11.4
Miscellaneous machine operators, n.e.c.	3932	6.0	–	–	533	5.7	1735	8.2
Laborers, non-construction	2483	3.8	–	–	284	3.0	988	4.7
Textile sewing machine operators	1782	2.7	–	–	287	3.1	755	3.6
Secretaries	1557	2.4	1111	11.4	–	–	283	1.3
Cashiers	1467	2.2	241	2.5	–	–	602	2.8
Packaging and filling machine operators	1444	2.2	–	–	79	0.8	930	4.4
Electrical and electronic equipment assemblers	1359	2.1	–	–	268	2.9	479	2.3
Data-entry keyers	1279	1.9	1158	11.8	–	–	44	0.2
Truck drivers	1177	1.8	–	–	63	0.7	639	3.0
Hand packers and packagers	1007	1.5	–	–	–	–	619	2.9
Welders and cutters	973	1.5	–	–	309	3.3	315	1.5
Butchers and meat cutters	881	1.3	–	–	412	4.4	196	0.9
Investigators and adjusters, excluding insurance	870	1.3	456	4.7	19	0.2	45	0.2
Book-keepers, accounting, and auditing clerks	859	1.3	399	4.1	203	2.2	–	–
Freight, stock, and material handlers, n.e.c.	835	1.3	–	–	–	–	576	2.7
General office clerks	819	1.2	317	3.2	–	–	137	0.6
Machine operators, not specified	808	1.2	–	–	136	1.5	304	1.4
Production inspectors, checkers, and examiners	802	1.2	–	–	60	0.6	485	2.3
Carpenters	790	1.2	–	–	215	2.3	335	1.6
Supervisors and proprietors, sales occupations	706	1.1	113	1.1	44	0.5	141	0.7
Stock handlers and baggers	701	1.1	64	0.7	–	–	332	1.6
Maids and housemen	678	1.0	–	–	48	0.5	212	1.0

Source: BLS.[3] n.e.c. = not elsewhere classified.

magnitude, repetition, and duration (Table 2.4). Whatever physical stressor model is used, a number of methods are available to measure/estimate these stressors. The method selected should be based on the purpose of the evaluation. The following grouping provides several options.

Survey methods

Employee or supervisor interviews, employee diaries and employee-completed questionnaires are useful because of their low cost, rapid availability, and, for some, the ability to obtain historical data about previous

Table 2.3 Evidence for causal relationship between physical work factors and MSDs.

Body part and risk factor	Strong evidence (+ + +)	Evidence (++)	Insufficient evidence (+/0)	Evidence of no effect (−)
Neck and neck/shoulder				
Repetition		✓		
Force		✓		
Posture	✓			
Vibration			✓	
Shoulder				
Posture		✓		
Force			✓	
Repetition		✓		
Vibration			✓	
Elbow				
Repetition			✓	
Force		✓		
Posture			✓	
Combination	✓			
Hand/wrist				
Carpal tunnel syndrome				
Repetition		✓		
Force		✓		
Posture			✓	
Vibration		✓		
Combination	✓			
Tendinitis				
Repetition		✓		
Force		✓		
Posture		✓		
Combination	✓			
Hand–arm vibration syndrome				
Vibration	✓			
Back				
Lifting/forceful movement	✓			
Awkward posture		✓		
Heavy physical		✓		
Whole-body vibration		✓		
Static work posture			✓	

Source: Bernard.[14]

exposures.[17–19] One of the most commonly used survey tools is the so-called Borg, or rating of perceived exertion, scale.[20] A 15-point scale (6–20) was created to reflect the linear relationship between physical workload and heart rate divided by 10 (for example, a heart rate of 60 beats/min corresponds to 6 on the scale). The scale is presented to the subject before the start of a job or job task with anchors of "no exertion at all = (6)" to "maximal exertion = (20)". The subject is then asked to rate his or her exertion level

Table 2.4 Relationship between external physical stress factors and their properties.

	Property		
Physical stress	Magnitude	Repetition rate	Duration
Force	Force generated or applied	Frequency with which force is applied	Time for which force is applied
Motion	Joint angle, velocity, acceleration	Frequency of motion	Time to complete motion
Vibration	Acceleration	Frequency with which vibration occurs	Duration of vibration exposure
Cold	Temperature	Frequency of cold exposure	Duration of cold exposure

Source: Adapted from Radwin and Lavender.[16]

after completing the job and/or job task. A 10-point Borg scale was also created (Table 2.5) to account for large muscle group exertion, rather than heart rate or total body exertion.[21,22]

The accuracy of self-assessment surveys has been questioned because of the potential for the worker to either under-report or over-report exposures. For example, highly motivated subjects might underestimate their exertion, while unmotivated subjects might overestimate their exertion. This potential problem has led many to utilize observational checklists (described below).

Table 2.5 Rated perceived exertion scale of Borg.

0	Nothing at all
0.5	Extremely weak (just noticeable)
1	Very weak
2	Weak
3	Moderate
4	Somewhat strong
5	Strong (heavy)
6	
7	Very strong
8	
9	
10	Extremely strong (almost maximal)

Source: Adapted from Borg.[21,22]

Observational methods

Observational methods are commonly employed to objectively assess the workplace for physical stressors. Such observations should arguably follow a checklist to ensure that subtle risk factors are not overlooked.[23] Some checklists can be used by healthcare providers with limited expertise,[24–26] others require some training,[27,28] while others require a considerable amount of experience and training.[29–31] Figures 2.1 to 2.4 provide the reader with four tools for the assessment of workplace physical stressors.

Measuring workers

If the first two methods suggest that physical stressors exist in the workplace, quantitative measurement of those risk factors could be considered. However, quantitative measurement of ergonomic hazards typically utilizes specialized equipment and requires expertise in its use and interpretation of the results. Such measurements should currently be reserved for research settings in which the goal is to evaluate the link between an exposure and a disease outcome, or to demonstrate that ergonomic hazards have been reduced or eliminated through job or workstation redesign.

Methods used to generate quantitative information on physical stressors include electrogoniometers (dynamic measurements of

Risk factors	No	Yes
1. Physical stress:		
1.1 Can the job be done without hand/wrist contact with sharp edges?		
1.2 Is the tool operating without vibration?		
1.3 Are the worker's hands exposed to temperature > 70 degrees F		
1.4 Can the job be done without using gloves?		
2. Force:		
2.1 Does the job require exerting less than 10 lb of force?		
2.2 Can the job be done without using finger-pinch grip?		
3. Posture:		
3.1 Can the job be done without flexion of extension of the wrist?		
3.2 Can the tool be used without flexion or extension of the wrist?		
3.3 Can the job be done without deviating the wrist side to side?		
3.4 Can the tool be used without deviating the wrist side to side?		
3.5 Can the worker be seated while performing the job?		
3.6 Can the job be done without "clothes-wringing"motion?		
4. Workstation hardware:		
4.1 Can the orientation of the work surface be adjusted?		
4.2 Can the height of the work surface be adjusted?		
4.3 Can the location of the tool be adjusted?		
5. Repetitiveness:		
5.1 Is the cycle time longer than 30 s?		
6. Tool design:		
6.1 Are the thumb and finger slightly overlapped in a closed grip?		
6.2 Is the span of the tool's handle between 5 and 7 cm?		
6.3 Is the handle of the tool made from material other than metal?		
6.4 Is the weight of the tool below 4 kg (note exceptions to rule)?		
6.5 Is the tool suspended?		

Figure 2.1 University of Michigan's checklist for Physical Stressors. Adapted from Lifshitz and Armstrong.[24]

posture), accelerometers, and imaging techniques (electronic and laser optical recordings). Two devices that may be useful outside of research settings are spring scales or gauges to estimate force requirements, and simple goniometers to measure static postures. Both of these tools have been used successfully in workplaces, due to their simplicity.

Internal forces can be measured using surface electromyography (EMG), but currently available equipment is expensive, and its use requires expertise that is not widely available. Video and imaging systems as a means to measure posture have been used primarily in the laboratory setting, where the camera's line of sight is perpendicular to the planes of the measured body segments. But given the dynamic nature of most job activities, their use in the workplace seems limited unless multiple cameras can be used from a variety of viewing angles. Goniometer use for measuring static postures is well established, but few jobs require continuous static postures. Electrogoniometers can measure dynamic postures, but their accuracy and associated analytic methods are not well established.

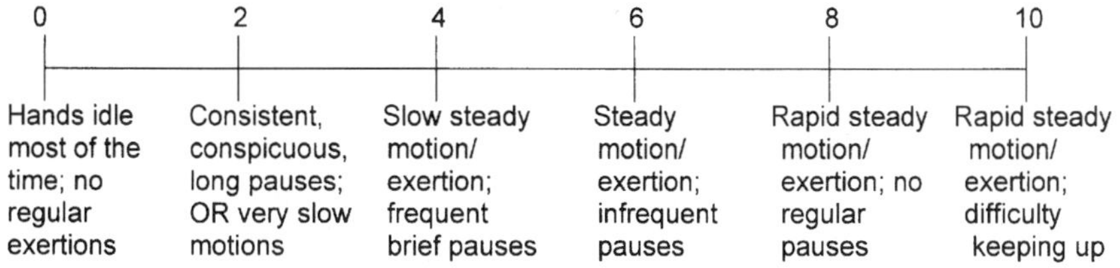

Figure 2.2 Visual-analog scale for rating repetition/hand activity, with verbal anchors.[27]

Body Part	Effort Level	Continuous Effort Time	Efforts/ Min	Priority
Neck/Shoulders	R______	______	______	______
	L ______	______	______	______
Back	______	______	______	______
Arms/Elbows	R______	______	______	______
	L ______	______	______	______
Wrists/Hands/ Fingers	R______	______	______	______
	L ______	______	______	______
Legs/Knees	R______	______	______	______
	L ______	______	______	______
Ankles/Feet/Toes	R______	______	______	______
	L ______	______	______	______

Effort Categories
1 = Light
2 = Moderate
3 = Heavy

Continuous Effort Time Categories
1 = <6 s
2 = 6-20 s
3 = >20 s

Efforts/Minutes Categories
1 = <1/min
2 = 1-5 min
3 = >5/min

Priority for Change

Moderate =	123
	132
	213
	222
	231
	232
	312
High =	223
	313
	322
Very High =	323
	331
	332

Job Title:______________________________

Specific Task:____________________________

Job Number:_______________________________

Department:_______________________________

Location:_________________________________

Contact Person(s):__________________________

Phone:___________________________________

Analyst:__________________________________

Phone:___________________________________

Date of analysis:____________________________

Figure 2.3 Ergonomic job analysis checklist.[28]

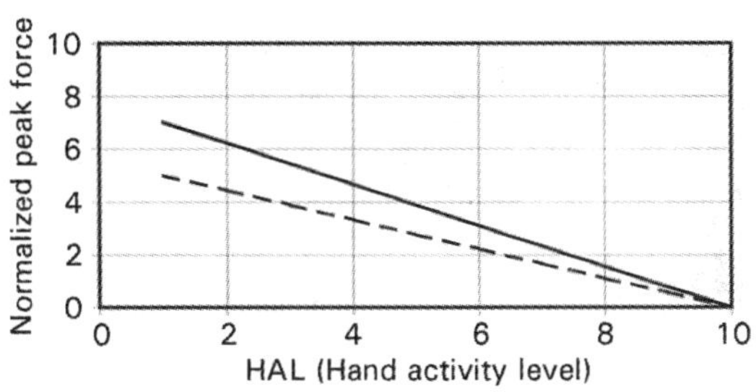

Figure 2.4 The threshold limit value (TLV) for reduction of work-related MDS based on hand activity and force. The top line is the TLV, and the bottom line the action limit. Reprinted with permission from American Conference of Governmental Industrial Hygienist (ACGIH®) TLV® Hand Activity Level Draft Document Copyright 2001.[32]

EXPOSURE GUIDELINES

American Conference of Governmental Industrial Hygienists

In 1999, the American Conference of Governmental Industrial Hygienists (ACGIH) proposed an exposure threshold for UE MSDs based on repetition and peak normal force for mono-task jobs.[32] While MSDs can involve any area of the upper extremity, the focus of the ACGIH threshold limit value (TLV) is the hand, wrist, and forearm. It is based on "mono-task" jobs, defined as jobs that required repeatedly performing a similar set of motions or exertions for four or more hours per day.

Repetition, or average hand activity level (HAL), is determined by a trained observer using the rating scale validated by Latko et al (Figure 2.2),[27] or calculated using information on the frequency of exertion and the work/recovery ratio (Table 2.6).[32] Peak hand force is normalized on a scale of 0–10, where 0 corresponds to no effort and 10 corresponds to 100% maximal effort. Normalized peak hand force is determined for a given task by: (1) determining hand forces and corresponding postures; (2) obtaining strength data for that posture and that worker or work population—in most cases, strength values can be obtained directly or extrapolated from the literature; and (3) calculating normalized peak hand force by dividing required force by strength.

Hand force can be determined by worker ratings, observer ratings, biomechanical analyses, force gauges and EMG. Since the latter three methods require considerable skill and instrumentation, the following discussion will focus on worker and observer ratings. Worker ratings utilize the same Borg scale

Table 2.6 Hand activity level (0–10) is related to exertion, frequency, and duty cycle (% of work cycle where force is greater than 5% of maximum).

Frequency (exertion/s)	Period (s/exertion)	Duty cycle (%) 0–20	20–40	40–60	60–80	80–100
0.125	8.0	1	1	–	–	–
0.25	4.0	2	2	3	–	–
0.5	2.0	3	4	5	5	6
1	1.0	4	5	5	6	7
2	0.5	–	5	6	7	8

Reprinted with permission from American Conference of Governmental Industrial Hygienist (ACGIH®) TLV® Hand Activity Level Draft Document Copyright 2001.[32]

described earlier (Table 2.5). Suppose, for example, a male worker rates his job's grip strength requirements as four (somewhat strong). To normalize this force, we measure the worker's grip strength (300 newtons (N)) and compare this to the average male strength (500 N). Therefore, the normalized peak force = 4 × 300 N/500 N = 2.4. Observer ratings of force utilize the same Latko et al scale described earlier (Figure 2.2).[27] Factors that the observer should consider include the weight, shape and friction of the work object, posture, glove fit and friction, mechanical assists, torque specification of power tools, quality control, and equipment maintenance. The precision of both the worker and observer ratings are improved by having multiple workers/observers rate the same job.

The HAL and the normalized force estimates can now be plotted and compared to the TLV (Figure 2.4). Employees performing job tasks above the solid top line will be at significant risk of acquiring a UE MSD, and specific control measures should be utilized so that the force for a given level of hand activity is below this line. The dotted lower line represents an "action limit", the point at which general controls, including surveillance (discussed below), are recommended. Although professional judgment is used to account for awkward or extreme postures, contact stresses, low temperatures, and vibration, the TLV does not specifically account for these potential stressors. If these stressors are present on jobs, the TLV and the action limit will be lower.

Others

Since ACGIH TLV does not account for all potential physical stressors, the reader is encouraged to review other proposed UE exposure guidelines, such as the strain index proposed by Moore and Garg,[29,33] and the Rapid Upper Limb Assessment (RULA) proposed by McAtamney and Corlett.[30]

Posture is an important potential physical stressor. Posture can be defined as the position of a part of the body relative to an adjacent part, as measured by the angle of the connecting joint. Standard posture definitions (neutral and non-neutral) and normal ranges of motion have been developed by the American Academy of Orthopedic Surgeons.[34] Postural stress develops as a joint reaches its maximal deviation; therefore, postures should be maintained as close to neutral as possible. In addition to postures at the extreme end of a joint's range, tasks that require finger-pinching postures have been associated with UE musculoskeletal disorders.

Kodak has proposed the following posture guidelines:[35]

- Keep the work surface height low enough to permit employees to work with their elbows at their sides, and wrists near their neutral position.
- Keep reaches within 20 inches in front of the work surface so that the elbow is not fully extended when forces are applied.
- Keep motions within 20–30° of the wrist's neutral point.
- Avoid operations that require more than 90° of rotation around the wrist.
- Avoid gripping requirements in repetitive operations that spread the fingers and thumb apart more than 2.5 inches. Cylindrical grips should not exceed 2 inches in diameter, with 1.5 inches being the preferable size.

Federal Ergonomics Standard

Over the past decade the Occupational Safety and Health Administration (OSHA) worked to develop an ergonomics standard. On November 14, 2000 a final Ergonomics Program Standard was issued to take effect January 16, 2001. In March 2001, under the Congressional Review Act, Congress passed and President Bush signed a resolution of disapproval. Accordingly, OSHA removed the standard form the Code of Federal Regulations on April 23, 2001. In its place the Bush Administration pledged to find solutions to ergonomic-related problems affecting the nation's workforce. In addition, OSHA will continue to conduct ergonomic investigations, and, if appropriate, cite and fine companies with ergonomic problems under the authority of the General Duty Clause. The repealed standard is still available from OSHA's web site.[36]

California Ergonomics Standard

In 1993, the California State Legislature required its Occupational Safety and Health Standards Board to develop an ergonomics standard on or before January 1, 1995. Subsequent legal challenges shaped its content and start date. The current standard, adopted in 1999, applies to a job, process, or operation where a repetitive motion injury (RMI) has occurred to more than one employee under the following conditions:

1. A licensed physician objectively identified and diagnosed the RMI; and
2. The RMI was work-related (>50% caused by a repetitive job, process, or operation); and
3. The employees with RMIs were performing a job process, or operation of identical work activity (performing same repetitive motion task); and
4. The employee reported the RMI to the employer in the last 12 months.

If the above conditions are met, the employer is required to develop an ergonomics program with the following three components: worksite evaluation, control of workplace exposures, and employee training. The worksite evaluation requires that each job, process, or operation of identical work activities be evaluated for exposures causing RMIs. If these exposures are found, they must be corrected in a timely manner, or, if not capable of being corrected, have the exposures minimized to the extent feasible. In addition, employees must receive training on:

- The employer's ergonomic program;
- The exposures which have been associated with RMIs;
- The symptoms and consequences of injuries caused by repetitive motion;
- The importance of reporting symptoms and injuries to the employer; and
- Methods used by the employer to minimize RMIs.

Finally, the employer is obligated to make an alternative intervention if the alternative has proven to reduce RMI and would not impose unreasonable employer costs.[37]

Washington Ergonomics Standard

In May 2000, the State of Washington issued its ergonomics rule. The rule is exposure based, and does not include requirements for handling of insurance claims, or for medical management. Employers are covered by the rule if their workplace contains "caution zone jobs" which may be hazardous and require further evaluation. The rule defines caution zone jobs as exposure to physical risk factors (awkward posture, high hand force, highly repetitive motion, repeated impact, heavy/frequent/awkward lifting, and moderate/high hard-arm vibration) for a significant duration, typically for more than two hours per day. Jobs that meet this definition must be analyzed for work-related MSD hazards and, if found, those exposure must be reduced below the hazardous level or to the extent technologically and economically feasible. The rule also requires employees working or supervising in caution zone jobs to receive ergonomics awareness education, and be given the opportunity to participate in the analysis for work-related MSD hazards and selecting measures to reduce these hazards. Compliance with the rule will be phased-in by type of industry and size of the workforce.[38]

NORMAL PHYSIOLOGY AND ANATOMY

Muscles

Muscle consists of muscle fibers (muscle cells), nerve elements (motor neurons, afferent neurons, receptors of different types), connective tissue, and blood vessels. Muscle fibers are classified into two types: type I fibers, also known as slow-twitch or red muscle fibers, and type II fibers, also known as fast-twitch or white muscle fibers. In muscle fibers, the smallest morphologic contractile unit is the sarcomere, built of actin and myosin filaments. The smallest functional unit is the motor unit, which consists of a motor neuron cell and the muscle fibers that its branches supply. The muscles of the body are the generators of internal force that convert chemically stored energy into mechanical work. A muscle contracts its thread-like fibers, which shortens the length of the muscle, thereby generating a contractile force.

Myalgia is the medical term for the symptom of muscle pain. The most common type of myalgia, delayed-onset muscle soreness (DOMS), is a contraction-induced injury after vigorous or unaccustomed exercise. DOMS is a self-limiting condition that typically appears within the first 24 hours after exercise, peaks at 48–72 hours, and resolves within 1 week. Histologic and chemical changes are found in affected muscles, but these changes are not permanent and lead to a conditioning effect when they take place in a graduated manner.[39–41]

Armstrong proposed the following theory for the pathogenesis of DOMS:[39–41]

- High mechanical forces, particularly those associated with eccentric exertions, cause structural damage of the muscle fibers and associated connective tissue structures.
- This structural damage alters the sarcolemma's permeability, producing a net influx of calcium into the cell. This calcium inhibits mitochondrial production of ATP, and activates proteolytic enzymes that degrade Z-discs, troponin, and tropomyosin.
- The progressive degeneration of the sarcolemma is accompanied by diffusion of intracellular products into the interstitium and plasma; this attracts inflammatory cells that release lysosomal proteases, which further degrade the muscle proteins.
- Active phagocytosis and cellular necrosis lead to accumulation of histamine, kinin, and potassium, which stimulate regional nociceptors, resulting in the sensation of DOMS.

Eccentric contractions (muscle activation while the muscle is stretched), rather than isometric contractions, are felt to lead to DOMS.[42] Eccentric contractions can occur when muscles are exposed to either a single rapid stretch or a series of repetitive contractions.[41] Both models are consistent with DOMS requiring a temporary reduction in physical loading because of pain or discomfort. This is followed by a gradual increase in physical loading to stimulate healing and subsequent tissue-remodeling processes.

Muscle also undergoes a number of age-related changes, such as a 20% decrease in muscle mass, a 20% reduction in maximal isometric force, and a 35% decrease in the maximal rate of developing force and power.[43] This latter reduction is not due to differences in muscle recruitment strategies, but rather due to a change in the contractility of the muscle itself.[44] This translates into a marked decrease in the ability to sustain power over repeated contractions in older individuals. In addition, animal experiments have also demonstrated that older muscle damages more easily and heals more slowly.[45,46] These effects may help explain why older athletes seem to require greater rest intervals between training sessions, and why workers in physically demanding jobs tend to change to less demanding jobs with age.[47]

Tendons

As a general rule, tendons transmit the contractile force generated by muscles to bone. Tendons are composed of collagen fibrils grouped into fibers that are collected together into fiber bundles that are united into fascicles.[48] A large number of fascicles form the tendon. The fiber bundles and fascicles are enclosed in thin films of loose connective tissue called the *endotenon*. This connective tissue contains blood vessels, lymphatic vessels, nerves, and elastic fibers, and allows the fascicles to slide relative to one another. The whole tendon is wrapped in connective tissue called the *epitenon*. In some tendons, a further sheath, the *paratenon*, surrounds the tendon. The paratenon is merely a specialization of the areolar connective tissue through which many tendons run. A number of structures associated with tendons control and facilitate their movement. Where tendons wrap around bony pulleys or pass over joints, they are held in place by retaining ligaments (retinaculae, or fibrous sheaths that prevent bowstringing). Tendons glide beneath these retaining structures due to the lubrication provided by the synovial sheath.[49] In some regions, tendons are prevented from rubbing against adjacent structures by bursae. Although tendons generally have a good blood and nerve supply, regions of tendon subjected to friction, compression or torsion are hypovascular or avascular. The general structure of tendons is modified in two regions: the sites where they attach to bone (enthesis) and the region where they are compressed against neighboring structures (around bony pulleys).[50] Fibrocartilage formation at the site of this compression loading is considered a normal/adaptive response. In summary, tendons have the capacity to change their structure and composition in response to mechanical stimulation. In most cases, this mechanical stress is beneficial and adaptive for maintaining cell activity and tissue function.

Peripheral nerves

Peripheral nerves carry signals to and from the central nervous system. A nerve fiber (neuron) consists of the nerve body, which is located in the anterior horn of the spinal cord (motor neuron) or in the dorsal root ganglia (sensory neuron), and a process extending into the periphery—the axon.[51] The axon is surrounded by Schwann cells. In myelinated fibers, a Schwann cell is wrapped around only one axon, in contrast to non-myelinated fibers, where the Schwann cell wraps around several axons. Myelinated and non-myelinated nerve fibers are organized in bundles, called *fascicles*, which are bound by supportive connective tissue, the *perineurium*. The bundles are usually organized in groups, held together by loose connective tissue called the *epineurium*. In between the nerve fibers and their basal membrane is located intrafascicular connective tissue—the *endoneurium*. The amount of connective tissue components varies between nerves, and between various levels along the same nerve. The myelin insulation divides the axon into short, uninsulated regions (nodes of Ranvier) and longer, insulated regions (internodes). Conduction of nerve impulses proceeds by sequential activation of successive nodes without depolarization of the intervening internode (saltatory conduction).

PATHOPHYSIOLOGY AND PATHOGENESIS

Muscles

Myopathy is the medical term for measurable pathologic changes in a muscle, with or without symptoms. Myopathies can be due to a variety of congenital (e.g. muscular dystrophy) or acquired (e.g. inflammatory, metabolic, endocrine, or toxic) disorders. These diseases are not typically work-related and will not be discussed further.

Muscle pain syndromes of unknown etiology can be classified into two categories: general and regional. General muscle pain involving all four quadrants of the body is called primary fibromyalgia. Primary fibromyalgia is not work-related because, by definition, trauma-induced myalgia is excluded by the specific diagnostic criterial set by the American College of Rheumatology.[52] Regional muscle pain syndromes, not involving the whole body, often fall under the term myofascial syndrome. This has been defined as a painful condition of skeletal muscle characterized by the presence of one or more discrete areas (trigger points) that are tender when pressure is applied.[53] These muscle-related syndromes, a common example of which is tension neck syndrome, could be associated with work exposures. A variety of mechanisms have been proposed to account for this syndrome. A few are listed below.

Work and Eccentric Contractions DOMS is a result of eccentric contractions that could occur on or off the job. DOMS has objective histologic and chemical changes, but these changes are part of the normal physiologic response. The pain or discomfort associated with DOMS, however, typically results in a temporary reduction in physical loading due to pain or discomfort. This is followed by a gradual increase in physical loading to stimulate healing and subsequent tissue remodeling. But if workers with physically demanding jobs have little control over the magnitude and duration of loading, the work can aggravate and hinder the healing process, thereby increasing the risk of developing a more chronic condition. Work-hardening programs are specifically designed to minimize this risk by prescribing graduated physical training regimens

Work and gamma motor neurons This theory starts with evidence that muscle pain, inflammation, ischemia or sustained static muscle contractions are known to lead to the release of potassium chloride, lactic acid, arachidonic acid, bradykinin, serotonin and histamine in the affected muscle.[54] These substances, in turn, are known to excite chemosensitive group II and IV afferents, which have a potent effect on gamma-muscle spindle systems and heighten the response of those spindles to stretch. Increased activity in the primary muscle spindle afferents may cause muscle stiffness, leading to further production of metabolites, more stiffness, and repetition of the cycle.

Work and the overload of type I fibers Another hypothesis for the pathogenesis of tension neck syndrome is that prolonged static contractions of the trapezius muscle result in an overload of type I muscle fibers. Type I muscle fibers are used for low static contractions. Support for this hypothesis comes from findings on biopsy. When compared to healthy controls, type I fibers in patients with chronic trapezius muscle pain: (1) were larger, (2) had a lower capillary-to-fiber ratio, (3) had a more "ragged" appearance, and (4) had reduced ATD and ADP levels.[55–57] Whether these findings are due to inadequate muscle recruitment[58] or inadequate tissue oxygenation is unknown.[59]

Work and muscle fatigue Finally, much work has been done on the mechanisms of fatigue relating to muscle disorders. A complete review of these mechanisms can be found in Gandevai et al.[60]

Tendons

Physicians in sports medicine have suggested that tendon disorders fall into four main categories: paratendonitis, paratendonitis with tendinosis, tendinosis, and tendinitis.[61] These categories are based on clinical and histologic findings. It can be difficult to distinguish between these specific conditions on clinical evaluation alone. Because most conditions can be treated conservatively, histologic changes have been documented in only a subset of patients whose cases proceeded to surgery. Thus, many cases are defined simply as "tendinitis" based on history, examination, and impaired function.

Proposed mechanisms for work-related tendon disorders include:[62] (1) ischemia in hypovascular tissues; (2) microinjuries incurred at a rate that exceeds repair potential; (3) thermal denaturation; (4) dysregulation of paratenon–tendon function; and (5) inflammatory processes secondary to some, or all, of these other factors.

Shoulder disorders provide evidence for the ischemic theory. Work above one's head can have two effects: compression (impingement) and reduced local bloodflow. Impingement comes from the narrow space between the humeral head and the tight coracoacromial arch. As the arm is raised in abduction, the rotator cuff tendons and the insertions on the greater tuberosity are forced under the coracoacromial arch.[63] Reduced local bloodflow occurs when the supraspinatus muscle is statically contracted, increasing the intramuscular pressure higher than the arterial pressure of the vessel traversing the supraspinatus muscle belly and supplying it with oxygen.[64] These work factors, in combination with the fact that the entheses of the three tendons comprising the rotator cuff are hypovascular, lead to tendon degeneration manifested by microruptures and calcium deposits. Once the tendons are degenerated, exertion may trigger an inflammatory response resulting in active tendinitis.[65]

The pathogenesis and pathophysiology of lateral epicondylitis is less worked out. Histologic evaluation of tennis elbow tendinosis identifies a non-inflammatory response in the tendon. This histopathology reveals disorganized immature collagen formation in association with immature fibroblastic and vascular elements. This has been named angiofibroblastic tendinosis and is thought to be the result of an avascular degenerative process.[66] This pathology is located at the enthesis of the extensor carpi radialis brevis (ECRB) tendon. Factors associated with its development include age, systemic factors, direct trauma, and repetitive overuse from sports, occupation, and performing arts. It is theorized that multiple repetitive eccentric loading of the ECRB results in tension loading, microruptures of the peritendon and secondary anoxia and degenerative consequences.

Tendon sheaths

Tendons can become trapped in their synovial sheaths due to a narrowing of their fibro-osseous canal. Tendons passing through stenotic canals frequently have a nodular or fusiform swelling and can be covered with granulation tissue.[67] Whether these tendon changes are a cause or an effect of the narrowing is unclear. If the narrowing occurs in the first dorsal compartment, the disorder is known as De Quervain's tenosynovitis.[67] If it occurs in the flexor digits or thumb (A-1 pulley), it is known as trigger finger or trigger thumb.[68]

While the term tenosynovitis implies inflammation of the tendon sheath, inflammatory cells are rarely found on histology. The lack of inflammatory cells could represent a sampling bias, since typically only chronic severe cases are biopsied, when the inflammatory process could have already run its course.[69] The fundamental pathologic change is hypertrophy and/or fibrocartilaginous metaplasia; however, recent studies suggest that the fibrocartilaginous metaplasia represents an adaptive response to compressive forces.[50]

The pathogenesis of tendon entrapment disorders involves static compression, repeated compression, and acute trauma. The static compression model is based on clinicians' observations that De Quervain's tenosynovitis is related to repeated, prolonged, or unaccustomed, exertions that involve the thumb in combination with non-neutral wrist or thumb postures. Tensile loading of the abductor pollicis longus or extensor pollicis brevis, in combination with their turning a corner at the extensor retinaculum, creates a compressive force. The retinaculum responds with functional hypertrophy or fibrocartilaginous metaplasia. The duration of compression is more important than the number (repetition) of compressions. The repeated compression theory relies on the same biomechanical argument, except that the number of episodes of loading (repetition) is more critical than the accumulated duration of loading.

Peripheral nerves

Although a number of peripheral nerve entrapment disorders exist, CTS is the most common, most studied, and will be the only nerve entrapment disorder discussed in this chapter.

CTS is the entrapment of the median nerve within the carpal canal at the wrist. Rempel[70] has summarized three possible mechanisms:

(1) friction associated with repetitive tendon motions, leading to flexor tendon sheath irritation and swelling; (2) repeated direct mechanical trauma to the median nerve by structures within the carpal tunnel; and (3) prolonged elevated pressure within the carpal tunnel, leading to ischemia, tissue swelling, and epineural fibrosis. The last mechanism is supported by the following: (1) carpal tunnel pressure (CTP) is almost always higher in patients with CTS than in normal subjects;[71] (2) surgical decompression (carpal tunnel release surgery) seems to be effective at reducing the elevated CTP and improving symptoms;[72] (3) histologic studies of the flexor tendon sheaths biopsied during carpal tunnel release show edema and vascular changes consistent with long-standing ischemia;[73] (4) animal models of acute and chronic nerve compression show physiologic and histologic findings consistent with nerve ischemia;[71] (5) in human studies, acute elevation of CTP results in acute nerve dysfunction, with the critical threshold varying according to the subject's diastolic blood pressure;[74] and (6) human studies have found a dose–response relationship between CTP and wrist posture,[75] fingertip loading,[76] and repetitive hand activity.[77] These findings are consistent with the static and dynamic biomechanical models of CTS.[78]

Psychosocial

Numerous studies have documented the association between psychological stress and health complaints. However, controversy exists as to whether these health complaints represent an actual increase in disease, an increase in reporting, or somatization. Although several studies have linked psychological stress and medical diseases,[79–89] few have reported a relationship between psychosocial stress and objective signs of UE musculoskeletal disorders.[90–93] The mechanism for this effect is unclear, but some authors have postulated that psychosocial stress can lead to muscle tension, thereby overloading some specific muscle fibers.[94,95] Bongers et al[96] and Moon and Sauter[97] have provided excellent reviews on the subject.

DIAGNOSIS AND TREATMENT

The successful treatment of UE MSDs relies on prompt evaluation, specific diagnosis, and appropriate intervention.

Clinical evaluation

Like all conditions with a potential occupational etiology, the occupational health history is a fundamental component of the evaluation.[98–101] The history should (1) characterize the symptoms, (2) provide an employee description of work activities, and (3) identify predisposing conditions or factors. This information should be documented in the employee's medical record. Figure 2.5 contains a form to assist with the collection of this information. If the employee description of work activities is unclear, or if further information is needed to understand employee job tasks and workplace conditions, this can be ascertained by visiting the workplace or viewing jobs tasks recorded on videotape. Simple reviewing a written description of job tasks may not provide an adequate understanding of the ergonomic stresses involved in the job.

After the history is taken and information obtained about workplace conditions and job tasks, the neck and UE should be examined.[102–104] A comprehensive examination would include inspection, palpation, assessment of the ranges of motion, evaluation of sensory and motor function, and applicable provocative maneuvers. Employees with underlying systemic disease (e.g. diabetes mellitus) may require a more complete examination involving other organ systems. Results of the examination findings, both positive and negative, should be documented in the employee's medical record. Figure 2.6 contains

Date: ____________
Name: ____________________ Age: ______years Gender: female male
Company: ____________________ Dept: ____________________ Job Title: ____________

Symptom Characterization
Onset:
Quality: (let employee describe; check all that apply)

pain	tenderness	weakness	soreness	numbness
tingling	burning	swelling	cramping	throbbing

Location: (R = right, L = left) ________ neck ________ shoulder ________ upper arm
________ elbow ________ lower arm ________ hand/wrist

Radiation: (R = right, L = left) ________ neck ________ shoulder ________ upper arm
________ elbow ________ lower arm ________ hand/wrist

Exacerbating/Relieving Activities:

Prior treatment modalities:

Description of Work Activities
Characterize the required job tasks (particularly with respect to physical stressors [force, repetition, posture, vibration]):
____
____
____

Changes in work patterns:	longer hours	overtime	other
Changes in frequency of tasks:	no	yes	
Changes in duration of tasks:	no	yes	
Unaccustomed work:	no	yes	
Changes in work equipment:	no	yes	

Length of time at various job tasks: ____________________
Frequency and duration of rest breaks: ____________________

Nature of work:	salary	hourly	piecework
Job rotation program:	no	yes	

If yes, list rotation schedule: ____________________

Predisposing Conditions or Factors for CTD

Prior trauma to the symptomatic area:	no	yes
Prior musculoskeletal symptoms or diagnoses:	no	yes
If yes, were these successfully treated?	no	yes

Length of time until complete recovery: ____________________
List hobbies: ____________________
____
List recreational activities: ____________________
____
List underlying diseases: ____________________
____
Other Comments: ____________________
____

Figure 2.5 Musculoskeletal occupational health history.

a form to assist with the collection of this information.

Clinical diagnosis

Using the information from the clinical evaluation, an assessment or diagnosis should be made. Diagnoses should be consistent with the International Classification of Diseases, Ninth Revision (ICD-9) (Table 2.7). Terms such as repetitive motion disorder (RMD), repetitive strain injury (RSI), overuse syndrome, and cumulative trauma disorder (CTD) may be useful for surveillance purposes or epidemiologic investigations, but they are not ICD-9 diagnoses and should not be used as individual medical diagnoses.

Once a diagnosis is made, an opinion is usually rendered regarding whether occupational factors caused, aggravated or contributed to the condition. This tends to be the most difficult portion of the assessment. Unlike the classic occupational diseases, such as asbestosis or silicosis, most occupational MSDs do not have a pathognomonic finding or test specific to the exposure. Therefore, the impor-

Name: ________________ Current job: ________________
Examiner: ________________ Date: ____/____/____
Discomfort scale: 0 = no discomfort, 1 = minimal, 2 = mild, 3 = moderate, 4 = severe, 5 = worst ever
Note: This form does not provide space for extensive neurologic assessment.

Neck
Inspection: Inflammation (red, swollen, warm) ____ yes ____ no
Palpation: *Right* *Left*
Trapezius trigger point ____ ____
Trapezius spasm ____ ____
Maneuvers:
Resisted flexion ____ ____
Resisted extension ____ ____
Resisted rotation ____ ____

Shoulder
Inspection: Acromium inflammation? ____ yes (R or L)____ no
Maneuvers: *Right* *Left*
Passive abduction ____ ____
Active abduction ____ ____
Resisted abduction ____ ____
Deltoid ____ ____

Elbow
Inspection: Olecranon inflammation ____ yes (R or L)____ no
Palpation: *Right* *Left*
Medial epicondyle ____ ____
Lateral epicondyle ____ ____

Forearm
Inspection: Forearm inflammation: ____ yes (R or L)____ no
Maneuvers *Right* *Left*
Passive wrist flexion ____ ____
Passive wrist extension ____ ____
Resisted wrist flexion ____ ____
Resisted wrist extension ____ ____
Resisted finger flexion ____ ____
Resisted finger extension ____ ____
3rd Digit resisted extension ____ ____

Wrist
Inspection: Inflammation ____ yes (R or L)____ no
Extensor ganglion cyst ____ yes (R or L)____ no
Flexor ganglion cyst ____ yes (R or L)____ no
Maneuvers: *Right* *Left*
Guyon Tinel's ____ ____
Carpal Tinel's ____ ____
Phalen's ____ ____

Hands and Fingers
Inspection: Inflammation ____ yes (R or L)____ no
Maneuvers: *Right* *Left*
Trigger finger ____ ____
Finkelstein's ____ ____

Figure 2.6 Physical examination recording form for the neck and upper extremity.

tance of a thorough exposure assessment cannot be overemphasized.[105] In 1979, NIOSH published a guide for state agencies and physicians on the process of determining work-relatedness of disease.[106] The process outlined in this guide, modified for MSD, are still relevant today:

- Has a disease condition been established by accepted clinical criteria?
- Does the literature support that the disease can result from the suspected agent?
- Has exposure to the agent been demonstrated?
- Has the exposure been of sufficient degree and duration to result in the diseased condition?
- Have non-occupational factors been considered?

Table 2.7 Specific ICD-9 diagnosis referred to as cumulative trauma disorders (CTD) by ICD-9 codes.

ICD-9 Code	Diagnosis
353	Nerve root and plexus disorders
	353.0 Brachial plexus lesions (cervical rib syndrome, costoclavicular syndrome, scalenus Anticus syndrome, thoracic outlet syndrome)
	353.9 Unspecified nerve root and plexus disorder
354	Mononeuritis of upper limb and mononeuritis multiplex
	354.0 Carpal tunnel syndrome (median nerve entrapment)
	354.2 Lesions of the ulnar nerve (cubital tunnel syndrome, tardy ulnar nerve palsy)
	354.3 Lesions of the radial nerve
	354.9 Mononeuritis of upper limbs, unspecified
443	Other peripheral vascular disease
	443.0 Raynaud's syndrome
	Raynaud's phenomenon (hand–arm vibration syndrome)
444	Arterial embolism and thrombosis
	444.2 Arteries of the extremities (ulnar artery thrombosis)
723	Other disorders of cervical region
	723.3 Cervicobrachial syndrome (diffuse)
	723.9 Unspecified musculoskeletal disorders and symptoms referable to neck (cervical disorder, NOS)
726	Peripheral enthesopathies and allied syndromes
	(Enthesopathies are disorders of peripheral ligamentous or muscular attachments)
	726.1 Disorders of bursae and tendons in the shoulder region (rotar cuffs syndrome, supraspinatus syndrome, bicipital tenosynovitis)
	726.3 Enthesopathy of elbow region (medical and lateral epicondylitis)
727	Other disorders of synovium, tendon, and bursa
	727.0 Synovitis and tenosynovitis
	727.03 Trigger finger (acquire)
	727.04 Radial styloid enosynovitis (de Quervain's)
	727.05 Other tenosynovitis of hand and wrist
	727.2 Specified bursitides, often of occupational origin
	727.9 Unspecified disorder of synovium, tendon, and bursa
728	Disorders of muscle, ligament, and fascia
	728.9 Unspecified disorder of muscle, ligament, and fascia
729	Other disorders of soft tissue
	729.1 Myalgia and myositis, unspecified (fibromyositis)
	729.8 Other musculoskeletal symptoms referable to limbs (swelling, cramping)
	729.9 Other and unspecified disorders of soft tissue

Nos = Not otherwise specified.

- Have other special circumstances been considered?

Clinical interventions

Clinical interventions should be tailored to the specific diagnosis. However, because soft tissue disorders represent the overwhelming majority of work-related musculoskeletal conditions,[107] resting the symptomatic area and reduction of soft tissue inflammation are the mainstays of conservative treatment.[108–114] The expected duration of treatment, dates for follow-up evaluations and time frames for improvement or resolution of symptoms should be specified at the initial evaluation.

Resting the symptomatic area Reducing or eliminating employee exposure to musculoskeletal risk factors by changing the job conditions (forceful exertions, repetitive activities, extreme or prolonged static postures, vibration, direct trauma) is the most effective way to rest the symptomatic area. This allows employees to remain productive members of the workforce and is best accomplished by engineering and work practice controls in the workplace (see Prevention section).

Until effective controls are installed, employee exposure to biomechanical stressors can be reduced through restricted duty and/or temporary job transfer. The principle of restricted duty and temporary job transfer is to reduce or eliminate the total amount of time spent exposed to the same or similar musculoskeletal risk factors.[115,116] A variety of factors (e.g. symptom type, duration, and severity; response to treatment; and biomechanical stressors associated with work) must be considered when determining the length of time for which an employee is assigned to restricted duty. When trying to determine the length of time assigned to restricted work, the following principle applies: the degree of restriction should be proportional to symptom severity and intensity of the job's biomechanical stressors. In addition, caution must be used in deciding which jobs are suitable for job transfer, because different jobs may pose similar biomechanical demands on the same muscles and tendons.[117]

Complete removal from the work environment should be reserved for severe disorders, workplaces where the only available jobs involve significant biomechanical stressors for the symptomatic area, or workplaces where significant modifications to the current or available jobs are not feasible. For purposes of removal from the work environment, severe disorders can be defined as those that negatively affect that employee's activities of daily living (e.g. difficulty in buttoning clothes, opening jars, brushing hair, etc.)

Immobilization devices, such as splints or supports, can help rest the symptomatic area.[109–114,118] These devices are especially effective off the job, particularly during sleep. Wrist splints, typically worn by patients with possible CTS, should not be worn at work unless the employee's job tasks do not require wrist deviation or bending.[119] Employees who struggle to perform a task requiring wrist deviation with a splint designed to prevent wrist deviation can exacerbate symptoms in the wrist due to the increased force needed to attempt to overcome the splint. This effect may also cause other joint areas (elbows or shoulders) to become symptomatic as work technique is altered.[119,120] Recommended periods of immobilization vary from several weeks to months, depending on the nature and severity of the disorder. Immobilization should be prescribed judiciously and monitored carefully to prevent iatrogenic complications (e.g. disuse muscle atrophy).[120,121] These recommendations do not preclude the use of immobilization devices for patients with special needs due to underlying medical conditions.

Finally, employees with MSDs should be advised about the potential risk posed by hobbies, recreational activities, and other personal habits that involve certain biomechanical stressors.[112–114] Employees should modify their behaviors to reduce such stress.

Treatment for soft tissue/tendon disorders Most clinicians consider cold therapy on the affected area useful to reduce the swelling and inflammation associated with acute tendon-related disorders.[112–114,122] The effectiveness of cold therapy in chronic disorders is not well established, despite several articles discussing its mechanism.[123–126] Cold therapy is inappropriate for employees with neurovascular conditions, such as hand–arm vibration syndrome, thoracic outlet syndrome, and CTS. In addition, prolonged, direct contact with chemical cold packs or ice baths should be avoided to prevent frostbite-type injuries.[127]

Oral anti-inflammatory agents (aspirin or other non-steroidal anti-inflammatory agents) are considered useful to reduce the severity of symptoms, through either their analgesic or anti-inflammatory properties.[109–114,128] Their gastrointestinal and renal side-effects, however, make their chronic prophylactic use among asymptomatic employees inappropriate and may limit their usefulness among employees with chronic symptoms.[128]

Physical and occupational therapy is a valuable adjunct for treatment of MSDs. Useful modalities may include (1) stretching and strengthening programs, (2) individualized training on proper body mechanics, and (3) teaching of appropriate self-care.[112–114,121,129]

Many, if not most, work-related MSDs improve with the above conservative measures. Symptomatic employees should be followed to document symptom improvement. Employees who do not improve within the expected time frames should be re-evaluated, or a second opinion should be obtained.

For some disorders resistant to conservative treatment, local injection of a corticosteroid may be indicated.[109–114,130] The addition of a local anesthetic agent to the injection can provide valuable diagnostic information. For example, to confirm a case of CTS, many clinicians look for rapid symptom improvement following the injection of anesthetic agents into the carpal canal. While some symptoms may indicate the need for immediate surgical intervention, these situations are rare. Surgical options should be reserved for severe, chronic cases that, after an adequate trial of conservative therapy (described above), prevent return to work or show objective signs of disease progression. The length of time needed for an "adequate" trial of conservative therapy depends on many variables, including an employee's ability to remain productive without jeopardizing his or her long-term health.

SURVEILLANCE

Surveillance is "the ongoing systematic collection, analysis, and interpretation of health and exposure data in the process of describing and monitoring a health event".[131] The goal of both hazard and health surveillance systems is the identification of hazardous exposures. Once hazardous exposures are recognized, intervention efforts can be targeted at those exposures with the purpose of preventing future health problems in (as yet) unaffected individuals.

Surveillance should not be confused with screening. Screening is the application of a clinical test to asymptomatic individual at increased risk for a particular disease.[132] The goal of screening tests is the identification of individuals who need further medical evaluation or other intervention. Although clinical tests can identify individuals with MSDs early in their development, there are currently no known screening tests to predict which asymptomatic individuals will develop symptoms and disease. Exposure and health surveillance can be divided into two types: passive and active.

Passive surveillance

Passive health surveillance systems utilize existing databases to identify high-risk jobs. Examples of these databases in occupational medicine include the OSHA 200 Logs, workers' compensation records, medical department logs, clinical laboratory data, hospital discharge records, and accident reports. These databases can be used to identify industries, occupations, jobs or tasks that are associated with disease or injury.

Figure 2.7 illustrates how data can be transcribed onto a spreadsheet and used to calculate disease incidence rates. These incidence rates, usually expressed as cases per 100 full-time workers (ftw) per year, can then be compared to the incidence rates for different industries, plants, departments, or jobs. Those jobs or departments found to have an increased

Department or job (a)	Hours worked (b)	Injuries and illness (c)	Incidence rate (e)	Incidence rate in previous year (e)	% change (f)

(c) Injuries or illness of the upper extremity not caused by accidents.

$$\text{Incidence Rate (d)} = \frac{\text{Number of Injuries + Illnesses (c)} \times 200\,000}{\text{Hours worked (b)}}$$

$$\text{\% Change (f)} = \frac{\text{Current (d)} - \text{Previous (e)}}{\text{Previous (e)}}$$

Figure 2.7 Disease incident rate calculations.

musculoskeletal injury rate can be targeted for further ergonomic and/or medical evaluation. In addition, jobs or departments can be followed over time, to monitor trends in these rates.

Although attractive because of their low cost, passive surveillance databases are sometimes developed for purposes other than surveillance and may have significant limitations.[132,133] These limitations include underreporting, disease misclassification, and exposure misclassification. For example, underreporting in the OSHA 200 Logs can occur for any of the following reasons: symptomatic employees not seeking medical care ("macho" attitude, ignorance that the condition could be work-related, or fear of employer retaliation); restricted or no access to employee health facilities; or misunderstanding about when a case is to be recorded on the OSHA 200 Log. Disease misclassification occurs, for example, when a disorder is recorded as an injury rather than a "disorder due to repeated trauma". Exposure misclassification can occur when employees use a general term to describe their job title; for example, an employee in the poultry industry may report his or her job title as "cutter" in a plant with five distinct cutting positions. Each one of these cutting jobs may be associated with a different ergonomic hazard, and the identification of high-risk jobs requires specific knowledge of the employees' cutting position.

Active surveillance

Active surveillance systems generate more accurate databases to identify high-risk positions. Direct symptom surveys are good examples of active disease surveillance tools in occupational medicine,[134] and have been developed for both pulmonary[135] and musculoskeletal disorders.[136] Symptom surveys collect more accurate information, and can serve a triage function if performed in a confidential manner.[137] Active surveillance systems can also collect information on exposures. This information is typically established by plant personnel or non-medical consultants; however, healthcare providers may be called upon to participate in a comprehensive ergonomics program. If exposure surveillance is a component of that program, the survey instruments in Table 2.5 and Figures 2.1–2.4 could be used to establish an exposure surveillance database.

PREVENTION

The control of identified ergonomic hazards is the most effective means of preventing work-related UE MSDs, and is the primary focus of any ergonomics program. Intervention strategies should follow a three-tiered approach: engineering controls, administrative controls, and medical treatment.

Engineering controls

Numerous studies have shown that engineering interventions can reduce ergonomic hazards.[138–146] In addition, several studies have shown this reduced exposure to result in a reduction in the rates of MSDs.[147–161] In some situations, these ergonomic solutions are obvious and consistent with common sense. On the other hand, work sites frequently require a more comprehensive approach to control of ergonomic hazards. This comprehensive approach should address the following risk factors: repetition, force, posture, and vibration.

To reduce repetitiveness, the following interventions could be used: (1) enlarged work content; (2) automation of some job tasks; (3) uniform spreading of work across the workshift; (4) job restructuring. To reduce force or mechanical stressors, the following interventions should be considered: (1) decreasing the weight of tools, containers, and parts; (2) optimizing the size, shape and friction of handles; and (3) using torque control devices. Reduction of awkward or extreme postures could be achieved by (1) locating the work more appropriately, and (2) selecting tool design and location based upon workstation characteristics. Engineering controls for the reduction of vibration are reviewed in Chapter 4.

In some instances, however, engineering controls are not currently available. Until engineering controls become available, other aspects of an ergonomics program—administrative and medical treatment controls—can be implemented.

Administrative

Administrative controls can be defined as work practices or training used to reduce employee exposure to ergonomic stressors. Examples of work practice controls include: (1) more frequent and longer rest breaks;[162,163] (2) limiting overtime; (3) varying work tasks, or broadening job responsibilities; and (4) periodic rotation of workers between stressful and less stressful jobs. Since job rotation exposes more workers to the more stressful job, it is suitable only where short-term performance of the stressful job poses no appreciable ergonomic hazard. Otherwise, other control methods must be utilized. Training programs range from fundamental instruction on the proper use of tools and materials to instruction on the use of protective devices. Improper work technique has been associated with the development of UE MSDs.[164,165]

Medical treatment

The goal of medical treatment as a component of an ergonomics program is to provide prompt evaluation and treatment to limit the severity, disability and costs associated with these disorders. It should also serve to initiate a re-evaluation of the ergonomic stresses associated with the affected worker's job and institution of appropriate control measures. Medical treatment in this sense is a secondary and tertiary prevention mechanism. Medical management programs are an important component of any successful ergonomic programs.[156,166–168] They should always be used with engineering and administrative controls during the implementation of a complete ergonomics program.[169]

REFERENCES

1. Chaffin DB, Andersson GBJ. *Occupational biomechanics*. New York: Wiley, 1984.
2. Eastman Kodak Company. *Ergonomic design for people at work*, Vol. 1. New York: Van Nostrand Reinhold Company, 1983:3.
3. Bureau of Labor Statistics. *Occupational injuries and illnesses in the United States by industry, 1998*. Washington, DC: BLS, 2000.
4. Webster BS, Snook SH. The cost of compensable upper extremity cumulative trauma disorders. *J Occup Med* 1994; 36:713–27.

5. Silverstein B, Welp E, Nelson N, Kalat J. Claims incidence of work-related disorders of the upper extremities: Washington State, 1987 through 1995. *Am J Public Health* 1998; 88:1827–33.
6. Hagberg M, Silverstein B, Wells R, et al. Introduction. In: Kuorinka I, Forcier L, eds. *Work related musculoskeletal disorders (WMSDs): a reference book for prevention.* London: Taylor & Francis, 1995:1.
7. Pansky G, Synder T, Dembe A, Himmelstein J. Under-reporting of work-related disorders in the workplace: a case study and review of the literature. *Ergonomics* 1999; 42:171–82.
8. Biddle J, Roberts K, Rosenman KD, Welch EM. What percentage of workers with work-related illnesses receive workers' compensation benefits? *JOEM* 1998; 40:325–31.
9. Rosenman KD, Gardiner JC, Wang J, et al. Why most workers with occupational repetitive trauma do not file workers' compensation. *JOEM* 2000; 42:25–34.
10. Franklin GM, Haug J, Heyer N, Checkoway H, Peck N. Occupational carpal tunnel syndrome in Washington state, 1984–1988. *Am J Public Health* 1991; 81(6):741–6.
11. Hanrahan LP, Moll MB. Injury surveillance. *Am J Public Health* 1989; 9(suppl):38–45.
12. Tanaka S, Wild DK, Seligman P, et al. Prevalence and work-relatedness of self-reported carpal tunnel syndrome among US workers: analysis of the Occupational Health Supplement Data of the 1988 National Health Interview Survey. *AJIM* 1995; 27:451–70.
13. NIOSH. *National Occupational Exposure Survey, 1981–1983.* Unpublished provisional data as of 1 July 1990.
14. Bernard B, ed. *Musculoskeletal disorders and workplace factors: a critical review of epidemiologic evidence for work-related musculoskeletal disorders of the neck, upper extremity, and low back.* DHHS (NIOSH) Publication No. 97-141. Cincinnati, OH: Department of Health and Human Services, National Institute for Occupational Safety and Health, 1997.
15. Panel on Musculoskeletal Disorders and the Workplace, Commission on Behavioral and Social Sciences and Education, National Resource Council, and Institute of Medicine. Musculoskeletal Disorders and the Workplace: Low Back and Upper Extremities. Washington DC, National Academy Press, 2001.
16. Radwin RG, Lavender SA. Work factors, personal factors and internal loads: biomechanics of work stressors. In: National Resource Council eds. *Work-related musculoskeletal disorders.* Washington DC: National Academy Press, 1999: 116–151.
17. Baty D, Buckle PW, Stubbs D. Posture recording by direct observation, questionnaire assessment, and instrumentation: a comparison based on a recent field study. In: Corlett N, Wilson J, Manenica I, eds. *The ergonomics of working postures: models, methods, and cases: Proceedings of the First International Occupational Ergonomics Symposium*, Zader, Yugoslavia. London: Taylor & Francis, 1985:283–92.
18. Wiktorin C, Karlqvist PT, Winkel J, et al. Validity of self-reported exposures to work postures and manual materials handling. *Scand J Work Environ Health* 1993; 19:208–14.
19. Viikari-Juntura E, et al. Validity of self-report physical work load in epidemiologic studies on musculoskeletal disorders. *Scand J Work Environ Health* 1996; 22:251–9.
20. Borg G. *An introduction to Borg's RPE-Scale.* Ithaca, NY: Movement Publications, 1985.
21. Borg GAV. Psychological bases of perceived exertion. *Med Sci Sports Exercise* 1982; 4:377–81.
22. Borg GAV. *Borg's perceived exertion and pain scale.* Champaign, IL: Human Kinetics, 1998.
23. Keyserling WM, Stetson DS, Silverstein BA, Brouwer ML. A checklist for evaluating ergonomic risk factors associated with upper extremity cumulative trauma disorders. *Ergonomics* 1993; 36:807–31.
24. Lifshitz Y, Armstrong T. A design checklist for control and prediction of cumulative trauma disorders in hand intensive manual jobs. In: *Proceedings of the 30th Annual Meeting of Human Factors Society,* Santa Monica: Human Factors Society, 1986: 837–41.
25. Kemmert K. A method assigned for the identification of ergonomic hazards—PLIBEL. *Appl Ergonomics* 1995; 26:199–201.

26. International Labour Office. *Ergonomic checkpoints: Practical and easy-to-implement solutions for improving safety, health and working conditions.* Geneva: International Labour Office, 1996.
27. Latko WA, Armstrong TJ, Foulke JA, et al. Development of evaluation of an observational method for assessing repetition in hand tasks. *AIHAJ* 1997; 58:278–85.
28. Rodgers S. Job evaluation in worker fitness determination. In: Himmelstein JS, Pransky GS, eds. *Occupational medicine, state of the art reviews: Worker fitness and risk evaluations.* Philadelphia, PA: Handley & Belfus, Inc., 1988:219–40.
29. Moore JS, Garg A. The strain index: a proposed method to analyze jobs for risk of distal upper extremity disorders. *AIHAJ* 1995; 56:443–58.
30. McAtamney L, Corlett EN. RULA: a survey method for the investigation of work-related upper limb disorders. *Appl Ergonomics* 1993; 24:91–9.
31. Louhevaara V, Suurnakki T. *OWAS: A method for the evaluation of postural load during work.* Training Publication No. 11 Helsinki: Institute of Occupational Health, 1992.
32. American Conference of Governmental Industrial Hygienists. *2000 Threshold limit values for chemical substances and physical agents and biologic exposure indices.* Cincinnati, OH: ACGIH, 2000:119–21.
33. Knox K, Moore SJ. Predictive validity of the strain index in turkey processing. *JOEM* 2001; 43:451–462.
34. American Academy of Orthopedic Surgeons. *Joint motion—method of measuring and recording.* Edinburgh: Churchill Livingstone, 1965.
35. Eastman Kodak Company. *Ergonomic design for people at work*, Vol. 2. New York: Van Nostrand Reinhold Company, 1983:255.
36. Occupational Saftety and Health Administration. Ergonomics Standard and Proposal Archive. Available from URL: http://www.osha-slc.gov/ergonomics-standard/archive-index.html
37. California Department of Industrial Relations. General Industry Safety Orders-Ergonomics. Available from URL: http://www.dir.ca.gov/title8/5110.html
38. Washington Labor and Industries. Ergonomics. Available from URL:http://www.Ini.wa.gov/wisha/ergo.html.
39. Armstrong RB. Mechanisms of exercise-induced delayed onset muscular soreness: a brief review. *Med Sci Sports Exerc* 1984; 16:529–36.
40. Armstrong RB. Initial events in exercise-induced muscular injury. *Med Sci Sports Exerc* 1990; 22:429–35.
41. Armstrong RB, Warren GL, Warren JA. Mechanisms of exercise-induced muscle fiber injury. *J Sports Med* 1991; 12:184–207.
42. Ashton-Miller JA. Soft Tissue Responses to Physical Stressors: Muscles, Tendons, and Ligaments. In: National Resource Council, eds. *Work-related musculoskeletal disorders.* Washington DC: National Academy Press, 1999:39–41.
43. Faulkner JA, Brooks SV. Muscle fatigue in old animals: unique aspects of fatigue in elderly humans. In: Gandevai S, Enoka R, McComas AJ, et al, eds. *Fatigue, neural and muscular mechanisms.* New York: Plenum Press, 1995:471–80.
44. Thelen DG, Aston-Miller JA, Schultz AB, Alexander NB, et al. Do neural factors underlie age differences in rapid ankle torque development? *J Am Geriatr Soc* 1996; 44:804–8.
45. Brooks SV, Faulkner JA. The magnitude of the initial injury induced by stretches of maximally activated muscle fibers of mice and rats increases in old age. *J Physiol* 1996; 497:573–80.
46. Brooks SV, Faulkner JA. Contraction-induced injury: recovery of skeletal muscles in young and old mice. *Am J Physiol* 1990; 258:C436–42.
47. Ashton-Miller JA. Response of muscle and tendon to injury and overuse. In: National Resource Council, eds. *Work-related musculoskeletal disorders.* Washington DC: National Academy Press, 1999:73–97.
48. Gelberman R, Goldberg V, An K-N, et al. Tendon. In: Woo SL-Y, Buckwater JA, eds. *Injury and repair of the musculoskeletal soft tissues.* Park Ridge, IL: American Academy of Orthopedic Surgeons, 1988:5–40.

49. Schumacher HR Jr. Morphology and physiology of normal synovium and the effects of mechanical stimulation. In: Gordon SL, Blair SJ, Fine LJ, eds. *Repetitive motion disorders of the upper extremity.* Rosemont IL: American Academy of Orthopedic Surgeons, 1995:263–76.
50. Vogel KG. Fibrocartilage in tendon: a response to compressive load. In: Gordon SL, Blair SJ, Fine LJ, eds. *Repetitive motion disorders of the upper extremity.* Rosemont IL: American Academy of Orthopedic Surgeons, 1995:205–15.
51. Terzis JK, Smith KL, eds. *The peripheral nerve: structure, function, and reconstruction.* New York: Raven Press, 1990.
52. Wolfe F, Smythe HA, Yunus MB, et al. The American College of Rheumatology 1990: criteria for the classification of fibromyalgia. *Arthritis Rheum* 1990; 33:160–72.
53. Grosshandler S, Burney R. The myofascial syndrome. *North Car Med J* 1979; 40:562–5.
54. Johansson H, Sojka P. Pathophysiological mechanisms involved in genesis and spread of muscle tension in occupational muscle pain and in chronic musculoskeletal pain syndromes: a hypothesis. *Med Hypothesis* 1991; 35:196–203.
55. Lindman R, Eriksson A, Thornell LE. Fiber type composition of the human female trapezius muscle. *Am J Anat* 1991; 190:385–92.
56. Lindman R, Hagberg M, Angquist KA, et al. Changes in the muscle morphology in chronic trapezius myalgia. *Scand J Work Environ Health* 1991; 17:347–55.
57. Larsson SE, Bengtsson A, Bodegard L, et al. Muscle changes in work related chronic myalgia. *Acta Orthop Scand* 1988; 59:552–6.
58. Hagberg M, Angquist KA, Eriksson HE, et al. EMG-relationship in patients with occupational shoulder–neck myofascial pain. In: deGroot, Huijing PA, van Ingen Schenau GJ, eds. *Biomechanics XI-A.* Amsterdam: Free University Press, 1988:450–4.
59. Murthy G, Kahan NH, Gargens AR, et al. Forearm muscle oxygenation decreases with low levels of voluntary contraction. *J Orthop Res* 1997; 15:507–11.
60. Gandevai S, Enoka R, McComas AJ, et al, eds. *Fatigue, neural and muscular mechanisms.* New York: Plenum Press, 1995.
61. Clancy WG Jr. Tendon trauma and overuse injuries. In: Leadbetter W, Buckwater JA, Gordon SL, eds. *Sports-induced inflammation: clinical and basic science concepts.* Park Ridge, IL: American Academy of Orthopedic Surgeons, 1990:609–18.
62. Hart DA, Frank CB, Bray RC. Inflammatory processes in repetitive motion and overuse syndromes: potential role of neurogenic mechanisms in tendon and ligaments. In: Gordon SL, Blair SJ, Fine LJ, eds. *Repetitive motion disorders of the upper extremity.* Rosemont IL: American Academy of Orthopedic Surgeons, 1995:249.
63. Fu FH, Harner CD, Klein AH. Shoulder impingement syndrome: a critical review. *Clin Orthop* 1991; 269:162–73.
64. Jarvholm U, Palmerud G, Styf J, et al. Intramuscular pressure in the supraspinatus muscle. *J Orthop Res* 1988; 6:230–8.
65. Hagberg M, Silverstein B, Wells R, et al. Evidence of the assocciation between work and selected tendon disorders: shoulder tendinitis, epicondylitis, de Quervain's tendinitis, Dupuytren's contracture, Achilles tendinitis. In: Kuorinka I, Forcier L, eds. *Work related musculoskeletal disorders (WMSDs): a reference book for prevention.* London: Taylor & Francis, 1995:55–6.
66. Nirschl RP. Elbow tendinosis: tennis elbow. *Clin Sports Med* 1992; 11:851–70.
67. Moore JS. De Quervain's tenosynovitis. Stenosing tenosynovitis of the first dorsal compartment. *JOEM* 1997; 39:990–1002.
68. Moore JS. Flexor tendon entrapment of the digits (trigger finger and trigger thumb). *JOEM* 2000; 42:526–45.
69. Leadbetter WB. Cell–matrix response in tendon injury. *Clin Sports Med* 1992; 11:533–78.
70. Rempel D. Musculoskeletal loading and carpal tunnel syndrome. In: Gordon SL, Blair SJ, Fine LJ, eds. *Repetitive motion disorders of the upper extremity.* Rosemont IL: American Academy of Orthopedic Surgeons, 1995:123–32.
71. Rempel D, Dahlin L, Lundborg G. Biological response of peripheral nerves to loading:

pathophysiology of nerve compression syndromes and vibration induced neuropathy. In: National Resource Council, eds. *Work-related musculoskeletal disorders*. Washington DC: National Academy Press, 1999:98–115.

72. Gelbermann RH, Rydevik BL, Pess GM, et al. Carpal tunnel syndrome: a scientific basis for clinical care. *Orthop Clin North Am* 1988; 19:115–24.
73. Fuchs PC, Nathan PA, Meyers LD. Synovial histology in carpal tunnel syndrome. *J Hand Surg* 1991; 16A:753–8.
74. Gelbermann RH, Szabo RM, Williamson RV, et al. Tissue pressure threshold for peripheral nerve viability. *Clin Orthop* 1983; 178:285–91.
75. Weiss N, Gordon L, Bloom T, et al. Wrist position of lowest carpal tunnel pressure. Implication for splint design. *J Bone Joint Surg* 1995; 77:1695–9.
76. Rempel D, Smutz WP, So Y, et al. Effect of fingertip loading on carpal tunnel pressure. *Trans Orthop Res Soc* 1994; 19:698.
77. Rempel D, Manojlovic R, Levinsohn D, et al. The effect of wearing a flexible wrist splint on carpal tunnel pressure during repetitive hand activity. *J Hand Surg* 1994; 19A:106–10.
78. Keyserling WM. Workplace risk factors and occupational musculoskeletal disorders, part 2: a review of biomechanical and psychophysical research on risk factors associated with upper extremity disorders. *AIHAJ* 2000; 61:31–43.
79. Feldman M, Walker P, Green JL, et al. Life events stress and psychosocial factors in men with peptic ulcer disease. *Gastroentrology* 1986; 91:1370.
80. Ruberman W. Psychosocial influences on mortality of patients with coronary heart disease. *JAMA* 1992; 267:559–60.
81. Ruberman W, Weinblatt E, Goldberg JD, Chaudhary BS. Psychosocial influences on mortality after myocardial infarction. *N Engl J Med* 1984; 311:552–9.
82. Schnall PL, Pieper C, Schwartz JE, et al. The relationship between job strain, workplace diastolic blood pressure, and left ventricular mass: results of a case–control study. *JAMA* 1990; 263:1929–35.
83. Karasek RA, Baker D, Marxer F, et al. Job decision latitude, job demands, and cardiovascular disease: a prospective study of Swedish men. *Am J Public Health* 1981; 71:694–705.
84. Karasek RA, Theorell T, Schwartz JF, et al. Job characteristics in relation to the prevalence of myocardial infarction in the U.S. Health Examination Survey (HANES). *Am J Public Health* 1988; 78:912–19.
85. Mathews KA, Cottington E, Talbott ED, Kuller LH, Siegel JM. Stressful work conditions and diastolic blood pressure among blue-collar factory workers. *Am J Epidemiol* 1987; 126:280–90.
86. Cohen S, Pyrrell DAJ, Smith AP. Psychological stress and susceptibility to the common cold. *N Engl J Med* 1991; 325:606–12.
87. Graham NM, Douglas RM, Ryan P. Stress and acute respiratory infection. *Am J Epidemiol* 1986; 124:389–401.
88. Calabrese JR, Kling MA, Gold PW. Alterations in immunocompetence during stress, bereavement, and depression: focus on neuroendocrine regulation. *Am J Psychiatry* 1987; 114:1123–34.
89. Shavit Y, Lewis JW, Terman GS, Gale RP, Liebeskind JC. Opioid peptides mediate the suppressive effect of stress on natural killer cell cytotoxicity. *Science* 1984; 223:188–90.
90. Leino P. Symptoms job stress predict musculoskeletal disorders. *J Epidemiol Community Health* 1989; 43:293–300.
91. Dimberg L, Olafsson A, Stefansson E, et al. The correlation between work environment and the occurrence of cervicobrachial symptoms. *J Occup Med* 1989; 31(5):447–53.
92. Hales TR, Sauter SL, Peterson MR, et al. Musculoskeletal disorders among visual display terminal users in a telecommunications company. *Ergonomics* 1994; 37:1603–21.
93. Bernard BP, Sauter S, Fine L, et al. Job task and psychosocial risk factors for work-related musculoskeletal disorders among newspaper employees. *Scand J Work Environ Health* 1994; 20:417–26.
94. Westgaard RH, Bjorklund R. Generation of muscle tension additional to postural muscle load. *Ergonomics* 1987; 30(6):911–23.
95. Edwards RHT. Hypothesis of peripheral and

central mechanisms underlying occupational muscle pain and injury. *Eur J Appl Physiol* 1988; 57:275–81.

96. Bongers PM, deWeinter CR, Krompier MAJ, et al. Psychosocial factors at work and musculoskeletal disorders. *Scand J Work Environ Health* 1993; 19:297–312.

97. Moon SD, Sauter SL, eds. *Psychosocial aspects of musculoskeletal disorders in office work.* London: Taylor & Francis, 1996.

98. Goldman RH, Peters JM. The occupational and environmental health history. *JAMA* 1981; 246:2831–6.

99. Occupational and Environmental Health Committee of the American Lung Association of San Diego and Imperial Counties. Taking the occupational history. *Ann Intern Med* 1983; 99:641–51.

100. Rosenstock L, Logerfo J, Heyer NJ, Carter WB. Development and validation of a self-administered occupational health history questionnaire. *J Occup Med* 1984; 26:50–4.

101. Felton JS. The occupational history: a neglected area in clinical history. *J Fam Prac* 1980; 11:33–9.

102. Piligian G, Herbert R, Hearns M, et al. Evaluation and management of chronic work-related musculoskeletal disorders of the distal upper extremity. *Am J Ind Med* 2000; 37:75–93.

103. Herbert R, Gerr F, Dropkin J. Clinical evaluation and management of work-related carpal tunnel syndrome. *Am J Ind Med* 2000; 37:62–74.

104. Harris JS, ed. *Occupational medicine practice guidelines.* Beverly, MA: American College of Occupational and Environmental Medicine, Occupational and Environmental Medicine Press, 1997.

105. Moore JS. Clinical determination of work-relatedness in carpal tunnel syndrome. *J Occup Rehab* 1991; 1:145–58.

106. Kusnetz S, Hutchison MK, eds. *A guide to the work-relatedness of disease.* DHHS (NIOSH) Publication No. 79-116. Cincinnati, OH: Department of Health and Human Services, National Institute for Occupational Safety and Health, 1979.

107. Moore JS. Carpal tunnel syndrome. *Occupational Med: State of the Art Rev.* 1992; 7:741–63.

108. Moore JS. Function, structure, and responses of components of the muscle–tendon unit. In: Moore JS, Garg A, eds. *Occupational medicine, state of the art reviews: ergonomics: low-back pain, carpal tunnel syndrome, and upper extremity disorders in the workplace.* Philadelphia: Handley & Belfus, Inc., 1992:713–40.

109. Howard NJ. Peritendinitis crepitans: a muscle-effort syndrome. *J Bone Joint Surg* 1937; 19:447–59.

110. Howard NJ. A new concept of tenosynovitis and the pathology of physiologic effort. *Am J Surg* 1938; 42:723–30.

111. Thompson AR, Plewes LW, Shaw EG. Peritendinitis crepitans and simple tenosynovitis: a clinical study of 544 cases in industry. *Br J Ind Med* 1957; 8:150–60.

112. Thorson EP, Szabo RM. Tendinitis of the wrist and elbow. In: Kasdan ML, ed. *Occupational Med: State of the Art Rev* 1989; 4:419–31.

113. Chipman JR, Kasdan ML, Camacho DG. Tendinitis of the upper extremity. In: Kasdan ML, ed. *Occupational hand and upper extremity injuries and diseases.* Philadelphia: Hanley & Belfus Inc., 1991:4030–21.

114. Rempel DM, Harrison RJ, Barnhardt S. Work-related cumulative trauma disorders of the upper extremity. *JAMA* 1992; 267:838–42.

115. McKenzie F, Storment J, Van Hook P, Armstrong TJ. A program for control of repetitive trauma disorders associated with hand tool operations in a telecommunications manufacturing facility. *Am Ind Hyg Assoc J* 1985; 46(11):674–8.

116. Lederman RJ, Calabrese LH. Overuse syndromes in instrumentalists. *Med Probl Perform Art* 1986; 1:7–11.

117. Occupational Safety and Health Administration. *Ergonomic Program Management Guidelines for Meatpacking Plants.* Washington DC: Occupational Safety and Health Administration, US Department of Labor, 1990.

118. Gelberman RH, Aronson D, Wisman MH. Carpal tunnel syndrome—results of a prospective trial of steroid injection and splinting. *J Bone Joint Surg* 1980; 62A:1181.

119. Kessler FB. Complications of the management of carpal tunnel syndrome. *Hand Clinics* 1986; 2:401–6.

120. Hales TR, Bertsche PA. Management of upper extremity cumulative trauma disorders. *Am Assoc Occup Health Nurses J* 1992; 40:118–28.

121. Curwin S, Stanish WD. *Tendinitis: its etiology and treatment*. Lexington, MA: DC Heath, 1984.

122. Simon HB. Current topics in medicine: sports medicine. In: Rubenstein E, Federman DD, eds. *Scientific American Medicine*. New York: Scientific American Inc., 1991:24–5.

123. Olson JE, Stravino VD. A review of cryotherapy. *Physical Ther* 1972; 52:840–53.

124. Thorson O, Lilja B, Ahlgren L, Hemdal B, Westlin N. The effect of local cold application on intramuscular blood flow at rest and after running. *Med Sci Sports Exercise* 1985; 17:710–13.

125. Yackzan L, Adams C, Francis K. The effects of ice massage on delayed muscle soreness. *Am J Sports Med* 1984; 12:159–65.

126. Kaplan PE, Tanner ED. Tendinitis, bursitis, and fibrositis. In: Kaplan PE, Tanner ED, eds. *Musculoskeletal pain and disability*. Norwald, CT: Appleton & Lange, 1989:1–24.

127. Hankin FM. Contact injuries to the hand. *Occupational Med: State of the Art Rev* 1989; 4:478.

128. Simon LS, Mills JA. Drug therapy: nonsteroidal antiinflammatory drugs (pts 1 and 2). *N Engl J Med* 1980; 302:1179–237.

129. Lane C. Hand therapy of occupational upper extremity disorders. In: Kasdan ML, ed. *Occupational hand and upper extremity injuries and diseases*. Philadelphia: Hanley & Belfus Inc. 1991:469–77.

130. Green DP. Diagnostic and therapeutic value of carpal tunnel injection. *J Hand Surg* 1984; 9:850.

131. Klauke DN, Buehler JW, Thacker SB, et al. Guidelines for evaluation of surveillance systems *MMWR* 1988; 17(suppl 5):1–18.

132. Halperin WE, Ratcliffe J, Frazier TM, et al. Medical screening in the workplace: proposed principles. *J Occup Med* 1986; 28:547–52.

133. Baker EL, Melius JM, Millar JD. Surveillance of occupational illness and injury in the United States: current perspectives and future directions. *J Public Health Policy* 1988; 9:188–221.

134. Ehrenberg RL. Use of direct survey in the surveillance of occupational illness and injury. *Am J Public Health* 1989; 79(suppl):11–14.

135. Ferris BG. Epidemiologic Standardization Project (American Thoracic Society). *Am Rev Respir Dis* 1978; 118(No. 6, Part 2):1–120.

136. Kuorinka I, Jonsson B, Kilbom A, et al. Standardized Nordic questionnaires for the analysis of musculoskeletal symptoms. *Appl Ergonomics* 1987; 18:233–7.

137. Hales TR, Bertsche PA. Management of upper extremity cumulative trauma disorders. *Am Assoc Occup Health Nurses J* 1992; 40:118–28.

138. Andersson GBJ. Design and testing of a vibration attenuating handle. *Int J Ind Ergonomics* 1990; 6(2):119–26.

139. Armstrong TJ, Kreutzberg KL, Foulke JA. *Laboratory evaluation of knife handles for thigh boning*. NIOSH Procurement No. 81-2637. Ann Arbor, MI: University of Michigan, 1982.

140. Erisman J, Wick J. Ergonomic and productivity improvements in an assembly clamping fixture. In: Kumar S, ed. *Advances in industrial ergonomics and safety IV.* Philadelphia, PA: Taylor & Francis, 1992:463–8.

141. Knowlton RG, Gilbert JC. Ulnar deviation and short-term strength reduction as affected by a curve-handled ripping hammer and a conventional claw hammer. *Ergonomics* 1983; 26(2):173–9.

142. Little RM. Redesign of a hand tool: a case study. *Semin Occup Med* 1987; 2(1):71–2.

143. Luttman AS, Jager M. Reduction in muscular strain by work design: electromyographical field studies in a weaving mill. In: Kumar S, ed. *Advances in industrial ergonomics and safety IV.* Philadelphia, PA: Taylor & Francis, 1992:553–60.

144. Miller M, Ransohoff J, Tichauer ER. Ergonomic evaluation of a redesigned surgical instrument. *Appl Ergonomics* 1971; 2(4): 194–7.

145. Powers JR, Hedge A, Martin MG. Effects of full motion forearm supports and a negative slope keyboard system on hand–wrist posture while keyboarding. In: *Proceedings of the Human Factors Society 36th Annual Meeting*. Santa Monica, CA: Human Factors Society, 1992:796–800.

146. Wick JL. Workplace design changes to reduce repetitive motion injuries in an assembly task: a case study. *Semin Occup Med* 1987; 2(1):75–8.

147. McKenzie F, Storment J, Van Hook P, Armstrong TJ. A program for control of repetitive trauma disorders associated with hand tool operations in a telecommunications manufacturing facility. *Am Ind Hyg Assoc J* 1985; 46(11):674–8.

148. Echard M, Smolenski S, Zamiska M. Ergonomic considerations: engineering controls at Volkswagen of America. In: ACGIH, ed. *Ergonomic interventions to prevent musculoskeletal injuries in industry.* Industrial Hygiene Science Series. ACGIH, Lewis Publishers, 1987:117–31.

149. Geras DT, Pepper CD, Rodgers SH. *An integrated ergonomics program at the Goodyear Tire & Rubber Company.* Unpublished, 1988.

150. Kilbom A. Intervention programmes for work-related neck and upper limb disorders: strategies and evaluation. *Ergonomics* 1988; 31(5):735–47.

151. LaBar G. A battle plan for back injury prevention. *Occupational Hazards* 1992; 29–33.

152. Luopajarvi T, Kuorinka I, Kukkonen R. The effects of ergonomic measures on the health of the neck and upper extremities of assembly-line packers—a four year follow-up study. In: Noro K, ed. *Proceedings of the 8th Congress of the International Ergonomics Association*, London: Taylor & Francis, 1982, 160–1.

153. Lutz G, Hansford T. Cumulative trauma disorder controls: the ergonomics program at Ethicon, Inc. *J Hand Surg* 1987; 12A(5 Part 2):863–6.

154. Orgel DL, Milliron MJ, Frederick LJ. Musculoskeletal discomfort in grocery express checkstand workers: an ergonomic intervention study. *J Occup Med* 1992; 34(8):815–8.

155. Westgaard RH, Aaras A. The effect of improved workplace design on the development of work-related musculo-skeletal illnesses. *Appl Ergonomics* 1985; 16(2): 91–7.

156. Bernacki EJ, Guidera JA, Schaefer JA, et al. An ergonomics program designed to reduce the incidence of upper extremity work related musculoskeletal disorders. *JOEM* 1999; 41:1032–41.

157. Aarås A. The impact of ergonomic intervention on individual health and corporate prosperity in a telecommunication environment. *Ergonomics* 1994; 37:1679–96.

158. Grant KA, Habes DJ. Summary of studies on the effectiveness of ergonomic interventions, applied occupational and environmental hygiene. *Ergonomics* 1995; 10(16):523–30.

159. Garg A. *Long-term effectiveness of "zero-lift program" in seven nursing homes and one hospital.* Contract No. U60/CCU512089-02 to University of Wisconsin, Milwaukee. Cincinnati, OH: US Department of Health and Human Services, Public Health Service, Centers for Disease Control and Prevention, National Institute for Occupational Safety and Health, 1999.

160. Gjessing C. Ergonomics: Effective workplace practices and programs. In: Transcripts of presentations from the conference held January 8 and 9, 1997, Chicago, Illinois. www.cdc.gov/niosh/ecagenda.html.

161. National Academy of Sciences. *Work-related musculoskeletal disorders*. Report, workshop summary, and workshop papers. Washington, DC: NAS, National Research Council, 1999.

162. Sauter S, Swanson NG. The Effects of Exercise on the Health and Performance of Data Entry Operators. In: Luczak H, Cakir AE, Cakir G, eds. *Work With Display Units 92.* Amsterdam: Elsevier Science Publishers B.B. 1993:288–91.

163. Galinsky TL, Swanson NG, Sauter SL, et al. A field study of supplementary rest breaks for data-entry operators. *Ergonomics* 2000; 43:622–38.

164. Feuerstein M, Fitzgerald TE. Biomechanical factors affecting upper extremity cumulative trauma disorders in sign language interpreters. *J Occup Med* 1992; 34:257–64.

165. Kilbom A, Persson J. Work technique and its consequences for task. *Ergonomics* 1987; 30:273–9.

166. Mehorn MJ, Wilkinson L, Gardner P, et al. An outcomes study of an occupational medicine intervention program for the reduction of musculoskeletal disorders and cumulative trauma disorders in the workplace. *JOEM* 1999; 41:833–46.

167. Government Accounting Office. *Report to congressional requestors: Worker protection, private sector, ergonomics programs yield positive results*. Washington, DC: United States Government Accounting Office/ Health, Education, and Human Services Division Report No. GAO/HEHS-97-163, 1997.

168. Lutz G, Hansford T. Cumulative trauma disorder controls: the ergonomics program at Ethicon, Inc. Part 2. *J Hand Surg* 1987; 12A(5):863–6.

169. National Institute for Occupational Safety and Health. *Elements of ergonomics programs: a primer based on workplace evaluations of musculoskeletal disorders.* DHHS (NIOSH) Publication No. 97-117. Cincinnati, OH: US Department of Health and Human Services, Public Health Service, Centers for Disease Control and Prevention 1997.

3

MANUAL MATERIALS HANDLING

Thomas R. Waters, Ph.D.*

Nearly half the cases of back pain reported by workers each year are work-related. These injuries accounted for more than 543 million lost working days: ~12% of the respondents stopped working or changed jobs due to back pain.[1] This chapter provides an overview of the hazards associated with manual materials handling (MMH); it will focus on muscular strains and sprains, primarily of the lower back. Overuse and overexertion injuries of the upper extremities are covered in Chapter 2. The prevention of these injuries requires sufficient knowledge to both identify workplace hazards and implement changes in the job or process that will reduce or eliminate these hazards.

OCCUPATIONAL SETTING

MMH poses a risk of injury to many workers; injury is more likely to occur when workers perform tasks that exceed their physical capacities. In addition, the physical capacities of individual workers vary substantially. Because MMH hazards are present in many industrial and service operations, workers in a wide variety of industries are potentially at risk. Although the data are not current, the National Occupational Exposure Survey (NOES), conducted by the National Institute for Occupational Safety and Health (NIOSH) in 1982–83, estimated that ~30% of the American workforce is routinely engaged in jobs that expose the worker to the physical hazards associated with manual handling.[2] This figure agrees with the 1988 National Health Interview Survey (NHIS) conducted by the National Center for Health Statistics.[1] In this study, 16% of respondents reported spending 4 h daily in repeated bending, twisting, or reaching. The industries with the greatest estimated total employees exposed to the hazards of lifting were health services, special trade and general building contracting, food and kindred products, and trucking and warehousing.[2]

*Vern Putz-Anderson contributed to the previous edition of this chapter.

Physical and Biological Hazards of the Workplace, Second Edition, Edited by Peter H. Wald and Gregg M. Stave
ISBN 0-471-38647-2

The Bureau of Labor Statistics (BLS)[3] also reported that the majority of manual-handling injuries are associated with lifting activities; however, common activities such as bending, pushing, and pulling, and awkward postures, also contribute to overexertion injuries. Nearly 20% of all injuries and illnesses in the workplace and nearly 34% of the annual workers' compensation payments are attributable to occupational low back pain (LBP).[4] A more recent report by the National Safety Council indicated that overexertion was the most common cause of occupational injury, accounting for 31% of all injuries. The back was the body part most frequently injured (i.e. 22% of 1.7 million injuries) and the most costly to the workers' compensation system.[5]

NORMAL ANATOMY AND PHYSIOLOGY OF THE SPINE

The spine is a complex structure made up of bony, muscular and ligamentous components. The spine can be divided into two major subsystems—the anterior and posterior spine. The **anterior spine** is mainly composed of the large bony vertebral bodies. These vertebral bodies rest atop one another and are separated by the cartilaginous invertebral discs, which act as "shock absorbers". The vertebral bodies and discs are held together by two sets of ligaments. The **posterior spine** is made up of the additional bony structures of the vertebral peduncles and laminae, which together form the spinal canal. The facet joints, which join two adjacent vertebrae, and the lateral and posterior spinous processes also form part of the posterior spine. The spinous processes are the attachment points for muscles that move and support the spine.

The spine is dependent on both bony and non-bony support for stability. Bony support is provided by the interdisc and the facet joints. Non-bony support comes from the ligaments and the attached musculature. Because the bony structures and ligaments do not have enough strength to resist the forces generated during movement and lifting, the spine is dependent on the muscles of the back, abdomen, hip and pelvis for stability. This principle explains why muscular fatigue is so important in the pathophysiology of back injury. The parts of the spine with the greatest degrees of movement are at highest risk. Because the thoracic and sacral vertebrae are fixed in place by the ribs and the pelvis, the lumbar vertebrae are the most common sites of injury.

PATHOPHYSIOLOGY OF INJURY AND RISK FACTORS

The interpretation of the research linking work-related musculoskeletal disorders (MSDs) and manual handling is problematic because of the high prevalence of certain disorders in the general population, such as LBP, and their frequent association with non-occupational factors. In addition, the relationship is further obscured by the wide range of disorders, the non-specific nature of the condition, and the general lack of objective data relating different risk factors to overexertion injury. In 1997, the NIOSH published an extensive review of the epidemiologic literature that assessed the strength of the association between specific work factors and certain upper extremity and lower back MSDs. The NIOSH identified more than 2000 studies, examined more than 600 epidemiologic studies, and published a comprehensive review of the epidemiologic studies of back and upper extremity MSDs and occupational exposures.[6] In addition to the NIOSH's work in this field, there have been other extensive reviews that have also critically evaluated the epidemiologic literature[7–13] and demonstrated the relationship between work-related factors and MSDs. These reviews examined the relationships between workplace exposures and outcomes such as symptoms, physical examination findings, specific diagnoses, or disability, and the effect of potential confounders and effect modifiers such as gender, age, injury and medical history. In addition, population surveys and broad government agency reports have been used to assess the prevalence, incidence and distribution of MSDs across industries and occupations.[1] All of these reviews concluded that there is significant evidence that physical factors are associated with development of MSDs.

It is generally recognized that musculoskeletal injuries are a function of a complex set of variables, including aspects of job design, work environment, and personal factors. Moreover, work-related MSDs may result from direct trauma, a single exertion (overexertion), or multiple exertions (repetitive trauma); typically, it is difficult to determine the specific nature of the causal mechanism. A variety of manual-handling activities increase a worker's risk of developing an MSD, including jobs that involve a significant amount of manual lifting, pushing, pulling, or carrying, and jobs requiring awkward postures, prolonged sitting, or exposure to cyclic loading (whole-body vibration). In addition to these frequent patterns of usage, a variety of personal and environmental factors may affect the risk of developing an MSD. Risk factors increase the probability of occurrence of a disease or disorder, though they are not necessarily causal factors. For the problem of low back injuries involving MMH, three categories of risk factors have been identified by epidemiologic studies—personal, environmental, and job-related.

- **Personal risk factors** are conditions or characteristics of the worker that affect the probability that an overexertion injury may occur. Personal risk factors include attributes such as age, level of physical conditioning, strength, and medical history.
- **Environmental risk factors** are conditions or characteristics of the external surroundings that affect the probability that an overexertion injury may occur. Environmental risk factors include attributes such as temperature, lighting, noise, vibration, and friction at the floor.
- **Job-related risk factors** are conditions or characteristics of the MMH job that affect the probability that an overexertion injury may occur. Job-related risk factors include attributes such as the weight of the load being moved, the location of the load relative to the worker when it is being moved, the size and shape of the object moved, and the frequency of handling.

A summary of significant risk factors is provided in Table 3.1.

Occupationally related MSDs attributed to MMH can result from a direct trauma, a single overexertion, or repetitive loading. The internal tissue response is dependent on tissue strength. It is related to such factors as age, fatigue, and concomitant diseases. Therefore, an injury can occur at different loading levels for different workers. Even for an individual

Table 3.1 Risk factors associated with manual material handling injuries.

Personal factors	Environmental factors	Job-related factors
Gender	Humidity	Location of load relative to the worker
Anthropometry (body weight and height?)	Light	Distance object is moved
Physical fitness and training	Noise	Frequency and duration of handling
Lumbar mobility	Vibration	Bending and twisting
Strength	Foot traction	Weight of object or force required
Medical history		Stability of the load
Years of employment		Postural requirements
Smoking		
Psychosocial factors		
Anatomic abnormality		

worker, a load may be tolerable one day and excessive on another day.[14]

There have been many studies of the multiple factors that contribute to MSDs, some of which have yielded conflicting results. Direct comparisons between epidemiologic studies are often impossible on methodological grounds, and a comparative review is difficult because the variables under study are rarely similar.[15] For example, populations, methods of sampling and data collection, and task and time scales often vary across studies.[16] In addition, studies are confounded by the high prevalence of MSDs in the general population, the wide range of disorders, non-specific symptoms, and frequent association with non-occupational factors. Relationships between MSDs and risk factors would be more apparent if precise and accurate outcome measures were available.[17]

Our knowledge of job-related, environmental and personal risk factors is far from complete. There is a need for long-term, prospective and controlled studies of large samples of workers to identify and quantify workplace factors that contribute to MSDs. Prospective studies should be designed to differentiate factors involved in the development of MSDs from factors that are the result of MSDs.

There have been many reports suggesting that there is a significant relationship between psychosocial factors, such as monotonous work, high perceived workload, low job control and low job satisfaction, and risk of MSD.[18] Bongers and de Winters[19] have published a detailed review of the current literature on psychosocial factors and musculoskeletal injuries. They concluded that studies such as those by Linton[20] and Bigos et al[21] support the contention that the role of psychosocial factors may be as important in affecting the risk of injury as the actual physical demands of the job. The data indicate that as the perceived demands of the job increase and the worker's control over those demands decreases, the rate of injury increases. It is believed that psychosocial factors may increase the effect of other physical factors such as muscle tension and poor body mechanics during MMH. Until more data are available to quantify these relationships, however, it is difficult to develop a comprehensive control strategy for psychosocial factors.

MEASUREMENT ISSUES

A variety of measurement tools are available for the ergonomic evaluation of manual-handling tasks, especially manual lifting tasks. These tools range in complexity from simple checklists, which are designed to provide a general indication of the physical stress associated with a particular MMH job, to complicated computer models that provide detailed information about specific risk factors. Although not exhaustive, Table 3.2 summarizes a variety of ergonomic assessment tools and offers a brief description of their advantages and disadvantages. NIOSH researchers reviewed a number of these MMH assessment methods and presented a case study from warehousing for comparison purposes.[22]

These tools provide objective information about the physical demands of manual-handling tasks that will help the user develop an effective prevention strategy. These tools are generally based on scientific studies that relate physical stress to the risk of musculoskeletal injury, particularly when those stressors exceed the physical capacity of the worker.[23–25] Any assessment of physical stress or human capacity is complicated by the influence of a variety of psychosocial factors, including work performance, motivation, expectation, and fatigue tolerance.

Checklists

A checklist is often the first choice for a rapid ergonomic assessment of a particular workplace. Checklists are designed to provide a general evaluation of the extent of a specific

Table 3.2 Ergonomic assessment tools.

Assessment tool	Advantages	Disadvantages
Checklists	• Simple to use • Best suited for use as a preliminary assessment tool • Applicable to a wide range of manual-handling jobs	• Do not provide detailed information about the specific risk factors • Do not quantify the extent of exposure to the risk factors
Biomechanical models	• Provide detailed estimates of mechanical forces on muscoloskeletal components • Can identify specific body structures exposed to high physical stress	• Not applicable for estimating effects of repetitive activities • Difficult to verify accuracy of estimates • Rely on a number of simplifying assumptions
Psychophysical tables	• Provide population estimates of worker capacities that integrate biomechanical and physiologic stressors • Applicable to a wide range of manual-handling activities	• Reflect more about what a worker will accept than is safe • May over- or underestimate demands for infrequent or highly repetitive activities
Physiologic models	• Provide detailed estimates of physiologic demands for repetitive work as a function of duration • Applicable to a wide range of manual-handling activities	• Not applicable for estimating effects of infrequent activities • Lack of strong link between physiologic fatigue and risk of injury
Integrated assessment models	• Simple to use • Use the most appropriate criterion for the specified task	• Require a significant number of assumptions • Limited range of applications
Videotape assessment	• Economic method of measuring postural kinematics • Can be used to analyze a large number of samples	• Labor-intensive analysis required • Limited to two-dimensional analysis
Exposure monitors	• Provide direct measures of posture and kinematics during manual handling • Applicable to a wide range of manual-handling activities	• Require the worker to wear a device on the body • Lack of data linking monitor output and risk of injury (upper extremity)

hazard that may be associated with a manual-handling task or job. A checklist usually consists of a series of questions about physical stressors such as frequent bending, heavy lifting, awkward or constrained postures, poor couplings at the hands or feet, and hazardous environmental conditions. Some checklists use a yes/no format; others use a numerical rating format. Checklists are easy to use, but they lack specificity and they are imprecise. An example of a manual-handling checklist is presented in Figure 3.1. For more information on checklists, see Grandjean,[26] Eastman Kodak,[27] Alexander and Pulat,[28] or National Occupational Health and Safety Commission.[29]

Biomechanical models

Biomechanics involves the systematic application of engineering concepts to the functioning human body to predict the distribution of internal musculoskeletal forces resulting from the application of externally applied forces. Biomechanical modeling provides a method for predicting the pattern and magnitude of these internal forces during manual handling. These predicted forces can then be compared to predetermined tissue tolerance limits to assess the biomechanical stress associated with specific loading conditions. When a

		YES	NO
1. General			
1.1	Does the load handled exceed 50 lbs?	[ ]	[ ]
1.2	Is the object difficult to bring close to the body because of its size, bulk, or shape?	[ ]	[ ]
1.3	Is the load hard to handle because it lacks handles or cutouts for handles, or does it have slippery surfaces or sharp edges?	[ ]	[ ]
1.4	Is the footing unsafe? For example, are the floors slippery, inclined, or uneven?	[ ]	[ ]
1.5	Does the task require fast movement, such as throwing, swinging, or rapid walking?	[ ]	[ ]
1.6	Does the task require stressful body postures, such as stooping to the floor, twisting, reaching overhead, or excessive lateral bending?	[ ]	[ ]
1.7	Is most of the load handled by only one hand, arm, or shoulder?	[ ]	[ ]
1.8	Does the task require working in environmental hazards, such as extreme temperatures, noise, vibration, lighting, or airborne contaminants?	[ ]	[ ]
1.9	Does the task require working in a confined area?	[ ]	[ ]
2. Specific			
2.1	Does lifting frequency exceed 5 lifts per minute?	[ ]	[ ]
2.2	Does the vertical lifting distance exceed 3 feet?	[ ]	[ ]
2.3	Do carries last longer than 1 minute?	[ ]	[ ]
2.4	Do tasks which require large sustained pushing or pulling forces exceed 30 seconds in duration?	[ ]	[ ]
2.5	Do extended reach static holding tasks exceed 1 minute?	[ ]	[ ]

Comment: "Yes" responses are indicative of conditions that pose a risk of developing low back pain. The larger the percentage of "yes" responses, the greater the possible risk.

Figure 3.1 Manual Material-Handling Checklist.

worker is performing a manual lifting task, for example, the internal reaction forces that are needed to provide equilibrium between the body segments and the external forces are supplied by muscle contractions, tendons, and ligaments at the body joints; in effect, the human body acts as a lever system.

Specifically, the external forces at the hands and the body segments create rotational moments or torques at various body joints, especially the lower back. The skeletal muscles exert forces that result in moments about the joints so as to counteract the moments due to external load and body segment weights. Since the moment arms of the muscles (and ligaments) are much smaller than the moment arms of the external forces and body segment weights, small external forces can produce large muscle, tendon, ligament, and joint reaction forces. On the other hand, the muscles can produce large motions with small degrees of shortening. The concept of muscles loading skeletal structures is extremely important in the biomechanics of low back injuries, because handling light loads in certain postures can create large mechanical loads on the lumbar spine structures.

Biomechanical models vary in complexity, depending upon which factors are included in the model. The simplest biomechanical models are based on two-dimensional static analysis and use only one muscle to model the flexor or extensor muscle groups. The complexity of the model may be increased, however, by adding modeling components that consider the effects of dynamic activity, multiple muscles, intra-abdominal pressure (IAP), muscular co-contraction, and posterior ligamentous structures. Although complex models are more difficult to use, they provide insight into the relative importance of the pattern of load distribution (e.g. compression, shear, or torsion).

Static models are simple and easy to use, but they do not consider the inertial effects of the moving body masses and external loads. **Dynamic models**, on the other hand, more closely simulate the loading of a dynamic system, but they are more difficult to use due to the type of data needed to predict the motion parameters. Two-dimensional models are limited to assessing manual-handling activities with movements restricted to one plane. Current research is focusing on the development of biomechanical models that integrate electromyographic data as input to estimate internal loading.[30] Chaffin and Andersson[31] have published a detailed discussion of biomechanical modeling that describes various two- and three-dimensional models.

Although biomechanical models are typically used to help in the design of infrequent, stressful activities requiring high levels of exertion, recent studies have focused on the assessment of repetitive lifting tasks. In studies of spinal compression tolerance, for example, Brinckmann et al[32] showed that repeated compression loading of the spinal motion segments causes them to fail at lower forces than those required in a single loading cycle.

For more information on biomechanical models, see Chaffin and Andersson[31], Kroemer[33], or Garg.[34]

Psychophysical tables

Psychophysics is a branch of psychology that examines the relationship between the perception of human sensations and physical stimuli. The worker's subjective determination is used to assess the synergistic effects of combined physiologic and biomechanical stress created by various manual-handling factors. Stevens[35] and Snook[25] contend that a worker's actual level of physical stress can be assessed by his or her subjective perception of the physical stress.

Although the vast majority of psychophysical research involving MMH activities has emphasized lifting tasks, the use of psychophysical techniques is not restricted to lifting. Psychophysics is also applicable to lowering, pushing, pulling, holding and carrying activities. The use of psychophysical data to assess the physical demands of manual handling is most appropriate for repetitive activities that are performed more often than once a minute. Databases are available that provide acceptable levels of manual handling for various segments of the population. Acceptable levels of work are typically presented as acceptable weights of lift or carry or acceptable forces of pushing and pulling.[36] A psychophysical table for acceptable weights of carry for industrial workers is shown in Table 3.3. For more information on psychophysical databases, see Snook,[25] Snook and Ciriello,[36] or Ayoub and Mital.[37]

Other psychophysical assessment methods have been developed to assess various MMH activities. For example, self-report measures such as rating of perceived exertion[38] (RPE) and body part discomfort[39] (BPD) have been used to assess a variety of lifting jobs. These assessment measures provide useful information about the worker's perception of the physical demands of the job. Moreover, RPE and BPD compare favorably with measures of physical demand.

Databases containing whole-body and segmental strength measures have also been developed for the design of manual-handling tasks. These include isometric, isokinetic and isoinertial strength databases for whole-body activities, such as lifting, and various databases for the arms, legs, and back. For more information on strength measurement, see Ayoub and Mital[37] and Chaffin and Andersson,[31]

It should be noted that psychophysics relies on self-reports from subjects; consequently, the perceived "acceptable" limit may differ from the actual "safe" limit. In addition, the psychophysical approach may not be valid for all tasks, such as high-frequency lifting.[40]

Physiologic models

One of the goals in designing a manual-handling task is to avoid the accumulation of physical fatigue, which may contribute to an overexertion injury. This fatigue can affect specific muscles or groups of muscles, or it can affect the whole body by reducing the aerobic capacities available to sustain work. Two physiologic factors that affect the suitability of a manual-handling task at the local muscle effort level include the **duration of force exertion** and the **frequency of exertions**. Local muscular fatigue will develop if a heavy effort is sustained for a long period. With heavy loads, the muscles need a substantially longer recovery period to return to their previous state. Small changes in workplace layout or handling heights, however, can often solve a local muscle fatigue problem through a reduction in holding duration. In addition, local muscle fatigue associated with maintaining awkward postures or constant bending can reduce the capacity of the muscles needed for lifting and therefore increase the potential for an overexertion injury to occur.

Table 3.3 Portion of a psychophysical table for maximum acceptable weights of carry.

Height from floor to hands (cm)	Percentage of industrial males	Maximum acceptable weight of carry (kg): 2.1-m carry One carry every:						
		6 s	12 s	1 min	2 min	5 min	30 min	8 h
111	90	10	14	17	17	19	21	25
	75	14	19	23	23	26	29	34
	50	**19**	25	30	30	33	38	44
	25	**23**	30	37	37	41	46	54
	10	**27**	**35**	43	43	48	54	63

Values are available for different heights, distances and sex. Values in bold exceed 8-h physiologic criteria. Adapted from Snook and Ciriello.[36]

Local muscle fatigue may limit the acceptable workloads for manual-handling tasks that are performed for short but intensive periods during a workshift. However, it is the energy expenditure demands of repetitive tasks that have the most profound effect on what a worker is able to do over a longer period of time. Energy expenditure demands are dependent upon the extent of muscular exertion, frequency of activity, and duration of continuous work. A worker's limit for physiologic fatigue is often affected by a combination of discomfort in local muscle groups and more centralized (systemic) fatigue associated with oxygen demand and cardiovascular strain.[27,41]

To assess the cardiovascular demands of manual-handling tasks, physiologic parameters such as heart rate (HR), oxygen consumption and ventilatory rate may be used. In addition, electromyographic (EMG) assessments and blood lactate provide a relative measure of the instantaneous level of physiologic status and muscular fatigue. Assuming that fatigued workers are at a higher risk of overexertion injury, these measures can be used to help prevent overexertion injuries by predicting the limits of fatigue for repetitive handling tasks.

Physiologic models provide a method for estimating the cardiovascular demands associated with a specific manual-handling activity. One such model, developed by Garg,[42] allows the analyst to estimate the energy expenditure demands associated with a complex manual-handling job. The first step is to separate the job into distinct elements or subtasks for which individual energy expenditure values can be predicted, such as standing and bending, walking, carrying, vertical lifting or lowering, and horizontal arm movement. The total energy expenditure requirements for the job are then estimated by summing the incremental expenditures of all of the subtasks.

HR is also useful in predicting physiologic demand, but it is less reliable than direct oxygen consumption measures, due to individual differences in the relationship between HR and energy expenditure. Portable monitors can be used to measure HR and oxygen consumption during manual-handling tasks. For information on assessing physiologic demands, see Astrand and Rodahl[43] or Eastman Kodak.[27]

Integrated assessment models

An integrated assessment model involves a unique approach that considers all three of the primary lifting criteria—biomechanics, physiology, and psychophysics. The integrated approach provides a measure of the relative magnitude of physical demand for a specific manual-handling task which relies on the most appropriate stress measure for that task. The result of the assessment is typically represented as a weight or force limit or as an index of relative severity. An integrated model considers the synergistic effects of the various task factors and uses the most appropriate stress measure to estimate the magnitude of hazard associated with each task factor. Examples of integrated assessment models include Ayoub's Job Severity lndex[44] (JSl) and the NlOSH recommended weight limit (RWL) equation.[40] Details on the revised NIOSH lifting equation are presented later in this chapter.

Videotape assessment

Most ergonomic assessments include the use of videotape analysis, where a video camera is used to record the work activity for later analysis. Videotape recordings make it easy to stop or freeze the action so that body posture or workplace layout can be evaluated. Videotape analysis often consists of general observation by the analyst that results in subjective estimates of physical hazards. More detailed video assessment can also be used to quantify the extent of the hazard objectively.

In fact, complex computerized video analysis systems are available that can capture and analyze individual frames from videotape recordings of workers performing manual-handling activities. These video frames can then be used to make more detailed assessments of spatial or dynamic biomechanical hazards that may not be apparent from the observational approach. These systems may include automatic digitizing capabilities for marking body joint locations and automatic scoring for assessing joint or body segment positions. These systems are generally easy to use, and the output data are presented in a form that is easy to understand and apply. They are limited, however, in their capability to analyze activities that occur outside of the camera focal plane (i.e. the plane parallel to the face of the camera lens). For example, when a body segment or group of segments move outside the camera focal plane, the joint angles and positions measured from the digitized frame are distorted. The amount of distortion depends on the degree of displacement of the joint or segment from the focal plane. Nevertheless, these systems are uniquely suited for certain types of assessments in which large amounts of videotape must be analyzed to determine the extent of the physical hazards.

Guidelines for videotape job analysis, which have been developed by NIOSH researchers, are provided in Table 3.4. For more information on videotape analysis and motion analysis, see Chaffin and Andersson[31] or Eastman Kodak.[27]

Exposure monitors

Monitoring devices have been developed to measure various aspects of physical activity, such as position, velocity, and acceleration of movement. Some monitors can even measure three-dimensional joint angles in real time. These systems consist of mechanical sensors that are attached to various parts of the worker's body, such as the wrist, back, or knees. The mechanical sensors convert angular displacement (rotation) into voltage changes

Table 3.4 Guidelines for recording work activities on videotape.

- If the video camera has the ability to record the time and date on the videotape, use these features to document when each job was observed and filmed. Recording time on videotape can be especially helpful if a detailed motion study will be performed at a later date (time should be recorded in seconds). Make sure that the time and date are set properly before videotaping begins.
- If the video camera cannot record time directly on the film, it may be useful to position a clock or a stopwatch in the field of view.
- At the beginning of each recording session, announce the name and location of the job being filmed so that it is recorded on the film's audio track. Restrict subsequent commentary to facts about the job or workstation.
- For best accuracy, try to remain inobtrusive, i.e. disturb the work process as little as possible while filming. Workers should not alter their work methods because of the videotaping process.
- If the job is repetitive or cyclic in nature, film at least 10–15 cycles of the primary job task. If several workers perform the same job, film at least two or three different workers performing the job to capture differences in work method.
- If necessary, film the worker from several angles or positions to capture all relevant postures and the activity of both hands. Initially, the worker's whole-body posture should be recorded (as well as the work surface or chair on which the worker is standing or sitting). Later, close-up shots of the hands should also be recorded if the work is manually intensive or extremely repetitive.
- If possible, film jobs in the order in which they appear in the process. For example, if several jobs on an assembly line are being evaluated, begin by recording the first job on the line, followed by the second, third, etc.
- Avoid making jerky or fast movements with the camera while recording. Mounting the camera on a tripod may be useful for filming work activities at a fixed workstation where the worker does not move around much.

that can be displayed in real time or-saved to a computer for later analysis. The position measures acquired from the sensors can then be differentiated to obtain rotational velocities and acceleration components. These movement characteristics may be used to estimate the extent of risk associated with a particular task and help to identify potential ergonomic solutions. Examples of positional monitoring equipment include potentiometer-based lumbar and wrist motion monitors, such as those developed at The Ohio State University,[45,46] as well as strain gauge-based strip goniometers, such as those developed by Penny and Giles (Santa Monica, CA, USA).

Another device that has been used to assess the extent of exposure to repetitive movement is an accelerometer-based motion-recording system, or activity monitor. An activity monitor consists of one or more accelerometers mounted within a small aluminum case that is connected with a Velcro strap to a worker's wrist, leg, or trunk. The accelerometers are sensitive to the movements of the body. They are capable of counting and recording rapid movements inherent in a specific task or activity. These measures are important because highly dynamic movements that occur over an extended period of time, such as an 8-h workshift, are believed to increase a worker's risk of musculoskeletal injury. The data acquired from the activity monitor are typically plotted as a series of temporal histogram plots showing the extent of dynamic movement as a function or time. The greater the total dynamic activity, the greater the height of the sequential histogram bars and the greater the potential for injury.[47]

It is important to note that the output from exposure monitors alone cannot provide all the information needed to assess the extent of physical demand required by a manual-handling task. It is also important to know the weight of the load and its position, velocity, and acceleration relative to the body during the task. This approach is best suited for repetitive or high-speed manual-handling tasks where the internal forces on the body may be affected more by extreme postures or rapid movements than by the weight or position of the external load.

GUIDELINES AND STANDARDS

Early attempts to prevent overexertion injuries associated with MMH focused on adopting arbitrary weight limits for lifting loads, hiring strong workers, or using training procedures that emphasized correct (but not necessarily safe) lifting techniques. None of these approaches, however, has proven to be effective in significantly reducing overexertion injuries.[48] Recently, industry leaders have started to recognize the risks associated with MMH. To reduce costs and increase productivity, these companies have implemented ergonomic programs or practices aimed at preventing these injuries. In many cases, these ergonomic programs rely on exposure guidelines or standards recommended by the federal government.

California Occupational Safety and Health Administration (CAL/OSHA)

The state of California has several rules that relate to MMH. According to the California Occupational Safety and Health Act of 1973, every employer has a legal obligation to provide and maintain a safe and healthful workplace for the employees. As of 1991, one of the most important worker protection requirements is that every California employer is responsible for establishing, implementing and maintaining a written Injury and Illness Prevention (IIP) Program. A guide to developing the IIP Program with checklists for self-inspection is available at the CAL/OSHA web site (www.dir.ca.gov/DOSH/dosh_publications/iipp.html).

In addition to the IIP, the CAL/OSHA also passed a standard for preventing repetitive motion injuries (RMIs) in 1997. The standard (CAL/OSHA Title 8—Section 5110, Repetitive Motion Injuries) can also be accessed at the

CAL/OSHA web site: (www.dir.ca.gov/Title8/5110.html). The Cal/OSHA rule includes the following requirements.

(a) This section shall apply to a job, process, operation where a repetitive motion injury (RMI) has occurred to more than one employee under the following conditions:

(1) Work related causation. The repetitive motion injuries (RMIs) were predominantly caused (i.e. 50% or more) by a repetitive job, process, or operation;

(2) Relationship between RMIs at the workplace. The employees incurring the RMIs were performing a job process, or operation of identical work activity. Identical work activity means that the employees were performing the same repetitive motion task, such as but not limited to word processing, assembly or, loading;

(3) Medical requirements. The RMIs were musculoskeletal injuries that a licensed physician objectively identified and diagnosed; and

(4) Time requirements. The RMIs were reported by the employees to the employer in the last 12 months but not before July 3, 1997.

(b) Every employer subject to this section shall establish and implement a program designed to minimize RMIs. The program shall include a worksite evaluation, control of exposures which have caused RMIs and training of employees.

(1) Worksite evaluation. Each job, process, or operation of identical work activity covered by this section or a representative number of such jobs, processes, or operation of identical work activities shall be evaluated for exposures which have caused RMIs.

(2) Control of exposures which have caused RMIs. Any exposures that have caused RMIs shall, in a timely manner, be corrected or if not capable of being corrected have the exposures minimized to the extent feasible. The employer shall consider engineering controls, such as work station redesign, adjustable fixtures or tool redesign, and administrative controls, such as job rotation, work pacing or work breaks.

(3) Training. Employees shall be provided training that includes an explanation of:

(A) The employer's program;

(B) The exposures which have been associated with RMIs;

(C) The symptoms and consequences of injuries caused by repetitive motion;

(D) The importance of reporting symptoms and injuries to the employer; and

(E) Methods used by the employer to minimize RMIs.

(c) Satisfaction of an employer's obligation. Measures implemented by an employer under subsection (b)(1), (b)(2), or (b)(3) shall satisfy the employer's obligations under that respective subsection, unless it is shown that a measure known to but not taken by the employer is substantially certain to cause a greater reduction in such injuries and that this alternative measure would not impose additional unreasonable costs.

To assist employees and employers in reducing exposure to the high-risk factors, the CAL/OSHA developed a document entitled "Easy Ergonomics: A Practical Approach for Improving the Workplace". The document provides information about how ergonomics can be used to change jobs and make them safer (www.dir.ca.gov/DOSH/dosh_publications/EasErg2.pdf). The CAL/OSHA indicates that the information in the booklet is intended to provide general guidance, and there may be instances in which workplace issues are more complex than those presented in the booklet. In those cases, they suggest that you may need to seek the advice of an outside expert.

Washington State Ergonomics Standard

In May 2000, the State of Washington Department of Labor and Industries (WL&I) adopted an Ergonomics Standard to prevent MSDs in the workplace. The Washington State ergonomics rule (Chapter 296-62 WAC General Occupational Health Standards—Part A-1 Ergonomics) has eight key elements:

1. The rule applies only to employers with "caution zone jobs", those where any employee's typical work includes awkward postures, high hand forces, highly repetitive motion, repeated impact, heavy lifting, frequent lifting, awkward lifting, or moderate to high hand–arm vibration.
2. Employers with caution zone jobs must ensure that employees working in or supervising these jobs receive ergonomics awareness education. These employers must also analyze the caution zone jobs to determine if they have hazards.
3. Employers may choose their own method and criteria for identifying and reducing MSD hazards or may use the criteria specified by WL&I.
4. If jobs have hazards, the employer must reduce exposures below hazardous levels or to the extent technologically and economically feasible.
5. Employers must provide for and encourage employee participation.
6. Implementation is delayed to allow time for all employers to comply.
7. WL&I will assist in implementing the rule by providing educational materials, identifying industry best practices, establishing inspection policies and procedures, and conducting demonstration projects.
8. Employers may continue to use effective methods of reducing MSD hazards that were in place before the rule adoption date as long as the methods, taken as a whole, are as effective as the requirements of the rule.

Illustrations of physical risk factors are listed in Appendix A of the rule, and criteria for analyzing and reducing MSD hazards for employers who choose the Specific Performance Approach are listed in Appendix B. The ergonomics rule and the associated appendices can be accessed on the web at: www.lni.wa.gov/wisha/regs/ergo2000/

For lifting, WL&I developed criteria and a five-step process for reducing heavy, frequent, or awkward lifting that are shown in Figure 3.2. To assist employers in identifying and fixing jobs with musculoskeletal hazards, the WL&I has developed a number of publications and programs, listed in Table 3.5.

Proposed Occupational Safety and Health Administration Ergonomics Standard (OSHA)

In November 1999, the Occupational Safety and Health Administration (OSHA) also issued a proposed standard to address the significant risk of work-related MSDs confronting employees in various jobs in general industry workplaces.[49] This standard builds on the ergonomics program for "red meat" packing plants issued in 1990.[50] General industry employers covered by the standard would be required to establish an ergonomics program containing some or all of the elements typical of successful ergonomics programs: (1) management leadership and employee participation; (2) job hazard analysis and control; (3) hazard information and reporting; (4) training; (5) MSD management; and (6) program evaluation. The need to establish a program would depend upon the types of jobs in their workplace and whether a MSD covered by the standard has occurred.

The proposed standard would have required all general industry employers whose employees perform manufacturing or manual-handling jobs to implement a basic ergonomics program in those jobs. The basic program included: management leadership, employee participation, and hazard information and reporting. If an employee in a manufacturing or manual-handling job experienced an OSHA-recordable MSD and it was addition-

Step 1 - Find out the actual weight of the objects the employee lifts.
Actual Weight = ______ lbs.

Step 2 - Determine the Unadjusted Weight Limit using the diagram below.
Unadjusted Weight Limit: ______ lbs.

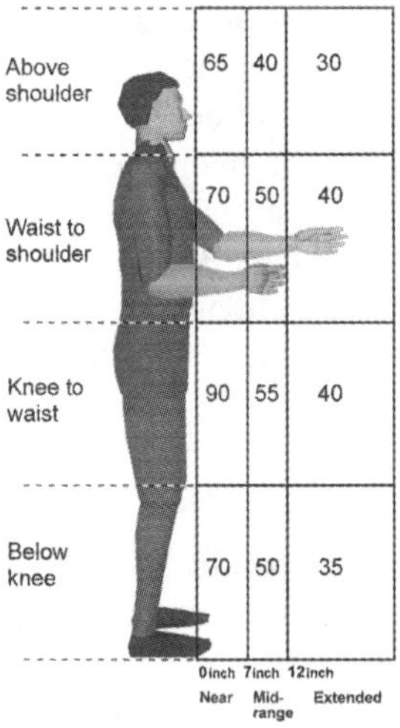

Step 3 - Find the Limit Reduction Modifier from the Table below.

How many lifts per minute?	For how many hours per day?		
	1 hr or less	1 hr to 2 hrs	2 hrs or more
1 lift everry 2–5 mins	1.0	0.95	0.85
1 lift every min	0.95	0.9	0.75
2–3 lifts every min	0.9	0.85	0.65
4–5 lifts every min	0.85	0.7	0.45
6–7 lifts every min	0.75	0.5	0.25
8–9 lifts every min	0.6	0.35	0.15
10+ lifts every min	0.3	0.2	0.0

Limit Reduction Mofifier: ______.

Step 4 - Calculate the Weight Limit.
Start by copying the Unadjusted Weight Limit from Step 2.
Unadjusted Weight Limit: ______. lbs.
If the employee twists more than 45 degrees while lifting, reduce the Unadjusted Weight Limit by multiplying by 0.85. Otherwise, use the Unadjusted Weight Limit
Twisting Adjustment: ______.
Adjusted Weight Limit = ______.
Multiply the Adjusted Weight Limit by the Percentage Modifier from Step 3 to get the Weight Limit.
Weight Limit: = ______ lbs.

Step 5 - **Is this a hazard?** Compare the Weight Limit calculated in Step 4 with the Actual Weight lifted from Step 1. If the Actual Weight lifted is greater than the Weight Limit calculated, then the lifting is a MSD hazard and must be reduced below the hazard level or to the degree technologically and economically feasible.

Note: If the job involves lifts of objects with a number of different weights and/or from a number of different locations, use Steps 1 through 5 above to:
1. Analyze the two worst case lifts—the heaviest object lifted and the lift done in the most awkward posture.
2. Analyze the most commonly performed lift. In Step 3, use the frequency and duration for all of the lifting done in a typical workday.

Figure 3.2 Washington State five-step lifting analysis.

Table 3.5 Resources from Washington State Labor and Industry.

Booklets

Fitting the Job to the Worker: An ergonomics program guideline
Lessons for Lifting and Moving Materials
Office Ergonomics: Practical solutions for a safer workplace
Work Related Musculoskeletal Disorders: Washington State Summary 1992–1994
Cumulative Trauma Disorders and Your Job, Carpal Tunnel Syndrome: A Preventable Disease
Work-Related Disorders of the Back and Upper Extremity in Washington State, 1990–1997 (113 KB PDF file—summary only; please contact SHARP for the entire Technical Report)
Employer Survey of Musculoskeletal Injuries and Illnesses, Risk Factors and Prevention Steps in Washington State Workplaces (42 KB PDF file—summary only; please contact SHARP for the entire Technical Report)
Non-Traumatic Soft Tissue MSDs, 1990–1997 (197 KB PDF file)

WL&I's Safety & Health Assessment & Research for Prevention (SHARP) program has a number of additional research reports on ergonomics and musculoskeletal disorders available, along with a wealth of other safety and health topics

Fact sheets

Commonly Asked Questions about Ergonomics
Quick Tips for Lifting
The Backbelt Fact Sheet
Musculoskeletal Disorders in the Workplace: A Summary of WL&Is' Prevention Efforts 1980s–1999

These publications can be accessed at the L&I web page:
http://www.lni.wa.gov/wisha/ergo/Default.htm

ally determined by the employer to be covered by the proposed standard, the employer would be required to implement the full ergonomics program for that job and all other jobs in the establishment involving the same physical work activities. The full program included, in addition to the elements in the basic program: a hazard analysis of the job; the implementation of engineering, work practice, or administrative controls to eliminate or materially reduce the hazards identified in that job; training the employees in that job and their supervisors; and the provision of MSD management, including, where appropriate, temporary work restrictions and access to a healthcare provider or other professional if a covered MSD occurs. General industry employers whose employees work in jobs other than manual handling or manufacturing and experienced an MSD that was determined by the employer to be covered by the standard would also have been required by the proposed rule to implement an ergonomics program for those jobs. According to OSHA, the proposed standard would affect approximately 1.9 million employers and 27.3 million employees in general industry workplaces.

The proposed ergonomics standard was subsequently overturned by the United States Congress in March 2001, and OSHA is now beginning the process of devloping new ergonomic guidelines or standards.

The National Institute for Occupational Safety and Health (NIOSH)

The NIOSH has not recommended formal exposure limits for general MMH activities. In 1981, however, the NIOSH published its *Work practices guide for manual lifting* (WPG).[51] The 1981 WPG contained an equation for assessing certain manual lifting tasks. The 1981 NIOSH lifting equation (NLE) provided a unique method for determining weight limits for selected two-handed manual lifts, but it was limited in its scope of application. It only applied to lifting tasks that occurred directly in front of the body (sagittal plane lifts) and had optimal hand-to-object couplings (i.e. handles). Responding to the need for a guideline with a brooder application and more flexibility, the NIOSH revised the lifting equation. In addition to the four risk factors addressed by the 1981 equation (i.e. horizontal location, vertical height, vertical distance traveled, and frequency), the Revised NLE includes weight reduction factors for the assessment of asymmetric or non-sagittal plane lifts that begin or end to the side of the body, and for lifting objects with less than optimal hand-to-object couplings (i.e. no handles). The original 1981 NLE was based on the concept that the overall physical stress for a specific lifting task is a function of the various task-related factors that define the lift, such as task geometry, load weight, and lifting frequency. This concept also forms the basis for the recently revised NLE, which provides a practical method for determining the overall physical stress attributable to the various task-related factors. The revised NIOSH equation yields a unique set of evaluation parameters that include (1) intermediate task-related multipliers that define the extent of the physical stress associated with individual task factors; (2) the NIOSH RWL, a task-specific value that defines the load weight that is considered safe for nearly all healthy workers; and (3) the NIOSH Lifting Index (LI), which provides a relative estimate of the overall physical stress associated with a specific manual lifting task.

The RWL is defined by the following equation:

$$\text{RWL} = \text{LC} \times \text{HM} \times \text{VM} \times \text{DM} \times \text{AM} \times \text{CM} \times \text{FM}$$

where the load constant (LC) is equal to 51 lb and the terms HM, VM, DM, AM, CM and FM are the task-specific multipliers within the equation that serve to reduce the recommended weight limit according to the specific task factor to which each multiplier applies. The magnitude of each multiplier will range in value between zero and one, depending on the value of the task factor to which the multiplier applies. The multipliers are defined as follows.

Component	Metric	US Customary
HM (horizontal multiplier)	$25/H$	$10/H$
VM (vertical multiplier)	$1 - (0.003\|V - 75\|)$	$1 - (0.0075\|V - 30\|)$
DM (distance multiplier)	$0.82 + (4.5/D)$	$0.82 + (1.8/D)$
AM (asymmetric multiplier)	$1 - (0.0032A)$	$1 - (0.0032A)$
FM (frequency multiplier)	(Determined from Table 3.6)	
CM (coupling multiplier)	(Determined from Table 3.7)	

In order to use the revised NIOSH lifting equation, you need to make the following measurements:

L = Weight of the load being lifted (lb or kg)

H = Horizontal location of the hands from mid-point between the ankles. Measure at the origin and the destination of the lift (cm or in).

V = Vertical location of the hands above the floor. Measure at the origin and destination of the lift (cm or in).

D = Vertical travel distance between the origin and the destination of the lift (cm or in).

A = Angle of asymmetry, defined as the angular displacement of the load from the sagittal plane when lifts are made to the side of the body. Measure at the origin and destination of the lift (degrees).

F = Average frequency rate of lifting measured in lifts/min. Duration is defined as < 1 hour, 1–2 hours, or 2–8 hours. Specific recovery allowances are required for each duration category.

The LI provides a relative estimate of the level of physical stress associated with a particular manual lifting task. The estimate of the level of physical stress is defined by the

Table 3.6 Frequency multiplier (FM).

	Work duration					
	≤1 hour		≤2 hours		≤8 hours	
Frequency lifts/min	$V < 75$	$V \geq 75$	$V < 75$	$V \geq 75$	$V < 75$	$V \geq 75$
0.2	1.00	1.00	0.95	0.95	0.85	0.85
0.5	0.97	0.97	0.92	0.92	0.81	0.81
1	0.94	0.94	0.88	0.88	0.75	0.75
2	0.91	0.91	0.84	0.84	0.65	0.65
3	0.88	0.88	0.79	0.79	0.55	0.55
4	0.84	0.84	0.72	0.72	0.45	0.45
5	0.80	0.80	0.60	0.60	0.35	0.35
6	0.75	0.75	0.50	0.50	0.27	0.27
7	0.70	0.70	0.42	0.42	0.22	0.22
8	0.60	0.60	0.35	0.35	0.18	0.18
9	0.52	0.52	0.30	0.30	0.00	0.15
10	0.45	0.45	0.26	0.26	0.00	0.13
11	0.41	0.41	0.00	0.23	0.00	0.00
12	0.37	0.37	0.00	0.21	0.00	0.00
13	0.00	0.34	0.00	0.00	0.00	0.00
14	0.00	0.31	0.00	0.00	0.00	0.00
15	0.00	0.28	0.00	0.00	0.00	0.00
>15	0.00	0.00	0.00	0.00	0.00	0.00

Values of V are in cm; 75 cm = 30 in.

Table 3.7 Coupling multiplier (CM).

	Coupling multipliers	
Couplings	$V < 75$ cm (30 in)	$V \geq 75$ cm (30 in)
Good	1.00	1.00
Fair	0.95	1.00
Poor	0.90	0.90

relationship of the weight of the load lifted to the recommended weight limit. The LI is defined by the following equation:

$$\text{LI} = \frac{\text{Load Weight}}{\text{Recommended Weight Limit}} = \frac{\text{L}}{\text{RWL}}$$

A detailed explanation of the use of the revised NIOSH lifting equation, including definitions of terms and procedures, is available in an applications manual for the revised NIOSH lifting equation.[52] The document can be downloaded from the NIOSH web site (www.cdc.gov/niosh/94-110.html).

The developers of the Revised NLE indicated that studies were needed to determine the effectiveness of the NLE in identifying jobs with increased risk of lifting-related LBP for workers. Recently, NIOSH researchers published the results of a cross-sectional epidemiologic study designed to evaluate the effectiveness of the equation in identifying jobs with elevated risk of causing LBP.[53] In the NIOSH study, 50 jobs were evaluated using the Revised NLE. The LI values of the jobs were compared to information obtained about the LBP symptoms in people who worked in those jobs. Using logistic regression modeling, the odds ratio (OR) for LBP was determined for various categories of LI values compared to the unexposed control group. LBP was assessed with a symptom and occupational history questionnaire that was administered to 204 workers employed in lifting jobs and 80 workers employed in non-lifting jobs. Participation was 89–95% among the exposed workers and 82–100% among unexposed workers at four facilities. The authors found that as the LI increased from 1.0 to 3.0, the odds of LBP increased, with a peak and statistically significant OR occurring in the two to three LI category (unadjusted OR = 2.45; CI 1.29–4.85). For jobs with an LI higher than 3.0, however, the OR was lower (1.63; CI 0.66–3.95). The decrease in the OR for these highly exposed jobs is likely to result from a combination of worker selection and a survivor effect. This study also examined several confounding variables such as age, gender, body mass index and psychosocial factors which were included in the multiple logistic models. The highest OR was in the two to three LI category (2.2; CI 1.01–4.96).

DIAGNOSIS AND TREATMENT

The primary objectives of the diagnosis and treatment of occupationally related musculoskeletal injuries are to: (1) assist the recovery of workers and allow for a rapid return to work; (2) ensure that proper diagnostic tools are used so that an accurate assessment is made of the magnitude of the injury; and (3) provide appropriate cost-effective treatments that avoid unnecessary surgery. The system should be based on an organized approach to evaluation, diagnosis, and treatment, which is essential for an early return to work and reduction in costs and lost work time. For example, a standardized diagnostic and treatment protocol has been shown to be effective in significantly and continuously reducing the number of incidents, days lost from work, low back surgery cases, and financial costs of LBP.[54]

Although a variety of MSDs result from MMH, the single most costly medical condition is work-related LBP, which affects millions of Americans. Experts have indicated that there is significant variation in assessment and treatment of LBP that results in inappropriate or at least less than optimal care for many patients with low back disorders. This issue was addressed in 1994 with the publication of the Clinical Practice Guideline for Acute Low Back Pain in Adults by the Agency for Health Care Policy and Research (AHCPR). These guidelines focused on returning the patient to normal activity, and are now in widespread use.[55] Copies of the guidelines are available on the Agency's (now renamed the Agency for Healthcare Research and Quality) web site at www.ahcpr.com

Complicating the diagnosis and treatment of occupationally related LBP are the legal issues of disability and compensation and how they relate to pain and impairment. Pain and impairment, which are direct measures of the extent of injury, primarily depend on the severity of the injury. Disability and compensation, however, may depend more on the nature of the compensation system and laws than on the severity of the injury.[56] Unfortunately, there is often an incentive to magnify pain symptoms to remain off work as well as to increase the size of an eventual disability settlement. However, clinical experience indicates that only ~5% of workers misuse their workers' compensation benefits for secondary gain.

Diagnosis

More than 90% of all episodes of back pain are probably attributable to mechanical causes, but it is rarely possible to precisely identify the specific cause of the pain.[57] Thus, the early diagnostic evaluation of back pain is designed to rule out systemic disease, grossly identify neurologic or anatomic abnormalities that may eventually require surgery such as fracture, tumor, infection, or cauda equina syndrome, and identify characteristic indicators of injury that influence the selection of treatment.

The first step in diagnosing LBP, and perhaps the most important, is the medical history. The medical history should include an assessment of current symptoms, individual history of injury, functional status, and injury documentation.[58–61] Based on the medical history, the clinician should be able to make a differential diagnosis, identify likely problems that may be encountered during the physical examination, and make a preliminary determination of the potential for disability.

The second step in the diagnostic process is a complete physical examination. At minimum, this examination would include a check of the deep tendon reflexes at the Achilles and quadriceps tendons, bilateral straight-leg raising, palpation of the paraspinal muscles for spasm, and a screening motor and sensory neurologic examination. The neurologic examination should give special attention to the L5 and S1 nerve roots, since the vast majority of disc herniations occur at either the L4–L5 or L5–S1 interspaces. It has been suggested that a system should be developed to ensure that all parts of the examination are completed.[60]

Current clinical tests and imaging procedures usually offer few clues to the precise source of pain, except in the minority of cases where a herniated disc is clearly causing nerve root irritation. Experts have concluded that plain radiography is of limited value in distinguishing mechanical causes of pain, because X-ray findings are poorly correlated with symptoms. Nevertheless, radiographs may be necessary to rule out other pathologic conditions, such as infections or tumours, or for making the diagnosis of spinal stenosis, spondylolisthesis, or fracture.[54]

Imaging may be needed (e.g. myelography, computed tomography, or magnetic resonance imaging) to identify an abnormality corresponding to a neurologic deficit from a herniated disc.[62] However, these sophisticated imaging tests should not be ordered too early in the course of back pain, especially when there is an absence of clinical findings that suggest a need for surgical intervention. In general, these tests should be limited to workers who have continued neurologic findings after 4–6 weeks of conservative therapy. It is also important to remember that an abnormal imaging study and back pain are not necessarily related. The clinical course of each worker must be considered before this causal link is made and surgery is considered.

Treatment

Epidemiologic data suggest that the vast majority of low back injuries are not serious and that most workers return to work after a short time with only conservative therapy. In general, only a minority of all affected persons have back pain >2 weeks in duration. By contrast, 90% of patients return to work within 6 weeks of onset of the back pain.[60,63] Conservative therapy typically consists of some combination of ice, bed rest (for up to 2 days if there are no neurologic findings), drug therapy (generally non-steroidal anti-inflammatory drugs), and mild exercise. There is little evidence, however, to indicate which treatment is most appropriate.[60,62,63] Treatment with electric stimulation (TENS) is controversial and unproven, but some workers have shown very good results.

If conservative therapy fails and 4–6 weeks have passed, further diagnostic tests should be scheduled. These cases generally fall into one of the following four categories, by location of the patient's predominant complaints: (1) LBP, which is typically diagnosed as back sprain; (2) leg pain radiating below the knee, which is commonly referred to as sciatica; (3) posterior thigh pain; and (4) anterior thigh pain. Each of these conditions warrants a unique treatment regimen.[54] It may also be appropriate to thoroughly investigate and address psychological and psychosocial issues at this time. Failure to deal with a significant psychological or psychosocial issue will delay the patient's return to work and may ultimately result in treatment failure and total disability.

A few back patients have symptoms suggesting sciatica, which is usually the first clue to a herniated disc. It is estimated that only 5–10% of patients with persistent sciatica require surgery. In general, only patients diagnosed with cauda equina compression (CEC) or a similar mechanoanatomic problem require immediate surgical intervention. Other indications for referral include severe or progressive neurologic deficits, or persistent neurologic finding after 4–6 weeks of conservative therapy. In the overwhelming number of cases of LBP, the appropriate treatment includes a course of conservative therapy.[54,62]

There is general agreement among clinicians that treatment for LBP must be given according to a strict timetable so that patients do not develop a dependence that would prolong symptoms and functional limitations. Also, patient reassurance and education is an important aspect of therapy. The clinician must avoid labeling the patient. The patient should be reassured that the natural history of their condition is favorable and that he or she should be able to return to work in a short time.[62]

MEDICAL SURVEILLANCE

Medical surveillance for MMH is similar to what was discussed for upper extremity MSDs. Active and passive surveillance of the workplace are necessary to establish a database of potential exposures in each workplace. Medical management of injuries and early reporting should be incorporated within the general medical treatment program at each work site. Since there is usually a short latency period between exposure and the onset of symptoms, medical surveillance per se is not helpful in eliminating MMH injuries. It may be useful to screen employees in order to match their physical abilities to the job. However, this is more of an exercise in prevention of injuries, and is thus covered in the next section on prevention.

PREVENTION

Attempts at ergonomic control of MMH problems have included both worker-directed and workplace-directed programs.

Workplace-directed approaches are based on changes in the design of the work to eliminate or minimize the problem; they rely on substitution and engineering controls. Consequently, these approaches are more generalized and less dependent on knowing the detailed capacity of the worker performing the task or job. Workplace-directed approaches include elimination of manual handling from the process through automation, reduction of the amount of physical exertion required to perform the MMH activity by using mechanical aids to assist the worker, or modification of the layout of the workplace to eliminate the MMH problem. In comparison, **worker-directed-approaches** primarily deal with attempts to maintain a balance between the worker's capacity and the demands of the job. This is usually attempted through worker screening, increases in the worker's physical capacity, or protection of the worker from the physical hazard. These controls are based on personal protection and administrative controls. Consequently, they require individualization on a worker-to-worker and job-to-job basis. More important, this approach requires detailed information about the capacity of each worker, as well as the demands of the job. Remember, however, that worker-directed approaches fail to directly reduce the extent of the physical hazard associated with the task or job.

Substitution

Automation Workplace automation should be a top priority when the job has high physical demands, is highly repetitive, or is performed in a hazardous environment. Automation may consist of one or more machines or machine systems such as conveyors, automated handling lines, automated storage and retrieval systems, or robots. This approach is best suited for the design of new work processes or activities or for the redesign of highly stressful tasks. Because automation often requires large capital expenditures, this approach may be prohibitive for small companies with only a few workers.

Mechanical aids In cases where the physical demands are high and automation is not practical, mechanical aids can be used to ameliorate the extent of those demands. Mechanical handling aids include machines or simple devices that provide a mechanical advantage during the MMH task, such as hand trucks, cranes, hoists, lift tables, powered mobile equipment and lift trucks, overhead handling and lifting equipment, and vacuum lift devices. As mentioned earlier in this chapter, the State of California OSHA has developed a document entitled "Easy Ergonomics" that includes a number of mechanical aids that would be useful in reducing exposure to MMH.

Engineering controls

Engineering controls of MMH hazards are preferred to other methods, such as worker selection and testing, training in safe Work practices, or the use of personal protective equipment, which are less reliable and often less effective. The engineering procedure involves modifying tasks and tools using ergonomic principles to reduce the effects of biomechanical stress. For MMH jobs, ergonomic design/redesign may be accomplished by modifying the job layout and procedures to reduce bending, twisting, horizontal extensions, heavy lifting, forceful exertions, and repetitive motions. The ergonomic approach is largely based on the assumption that work activities involving less weight, repetition, awkward postures and applied force are less likely to cause injuries and disorders. The ergonomic approach is desirable because it seeks to eliminate potential sources of problems. Ergonomics also seeks to make safe work practices a natural result of the tool and worksite design.

There are at least four advantages to adopting an ergonomic design/redesign strategy. First, an ergonomic approach does not depend on specific worker capabilities or learned behaviors, such as training. Second, human biological factors and their variations are accounted for in ergonomic approaches, using design data that accommodate large segments of the populations. Third, ergonomic intervention is relatively permanent, since the workplace hazard is eliminated. Fourth, to the extent that sources of biomechanical stress at the work-site are eliminated or significantly reduced, the difficult issues involving potential worker discrimination, lifestyle modifications or attempts at changing behavioral patterns of workers will be of lesser practical significance. Many MMH jobs entail a variety of specialized tasks with overlapping sources of physical stress. Each MMH task may contribute in some unknown manner to the onset of an overexertion injury. Often, numerous adjustments may be required for each activity to minimize the hazard of overexertion injury.

Training and education

The term training has been used to describe two distinctly different approaches to injury prevention and control—**instructional training** in safe materials handling, and **fitness training** (e.g. conditioning, strengthening, or work hardening).

The basic premise of instructional training in MMH is that people can more safely handle greater loads when they perform the task correctly than if they perform the task incorrectly. Although this approach is fundamentally sound, there are some potential problems in its application.

First, it is difficult to clearly distinguish between correct and incorrect materials-handling practices. For example, in manual lifting, there is little agreement on what constitutes a safe lifting style.[48] Perhaps one of the most important things to understand about lifting is that there is no single, correct way to lift. For example, the old adage of lifting with a straight back and bent legs may be inappropriate in many instances. For this reason, the

NIOSH does not recommend a correct lifting style; instead, it suggests that a free-style lift is appropriate in most instances.[51]

Second, the instructional approach relies on the worker's ability to comply with a set of recommended practices, which may be forgotten or changed from time to time. Regardless of the potential associated with MMH training, all workers who perform MMH activities should receive basic instructional training in the recognition of hazardous tasks and should have a thorough knowledge of what to do when a hazardous task is identified. Furthermore, the instructional training should provide information to the worker on how he or she can become involved in the process of preventing and controlling injuries on the job.

Unlike instructional training, the basic premise of **fitness training** is that a worker's risk of injury would decrease if his or her strength or fitness increased. Although this seems intuitive, it is not clear whether an individual's strength may affect his or her risk of injury. Certainly, a worker's capacity to perform heavy work might be increased, but there is some controversy about the relationship between worker strength and risk of injury.[64] Moreover, it is not known how the soft tissues of the body respond to increased loads associated with stronger muscles, especially if the worker performs tasks requiring greater strength demands.

Although both types of training programs have been used to prevent MMH injuries, the effectiveness of training in preventing or controlling injuries is unclear at the present time. Therefore, training programs should be used as a supplement to workplace-directed approaches. Additional information on training programs is provided by Ayoub and Mital[37] and Eastman Kodak.[27]

Employee screening

Some ergonomic experts advocate the use of screening methodologies, which rely on the assessment of one or more physical characteristics of the worker, to select specific workers for certain MMH jobs. In general, screening approaches are designed to identify workers with a high-risk of overexertion injury, or screen workers according to some preselected set of strength or endurance criteria in an attempt to match the capacity of the worker to the demands of the job.

Risk assessment screening Attempts to identify workers with a high risk of overexertion injury have included such activities as spinal radiographs, psychological testing, and medical examinations, which are designed to provide an objective basis for excluding certain individuals from stressful MMH jobs. None of these methodologies, however, have been shown to be reliable in predicting an individual's risk of overexertion injury. It is now widely accepted that the medical risks from radiation associated with radiography far outweigh any potential benefit derived from routine spinal X-ray screening.[51] Although no psychological tests have been found to quantify a worker's risk of overexertion injury, psychological testing can provide an indication of how a worker might respond to a severe injury. Similarly, medical examinations have also failed to reliably identify workers who may have an above-average risk of overexertion injury.

Physical capacity screening Another type of screening approach that has been used to select workers for MMH tasks includes individual testing of physical characteristics, such as strength, aerobic capacity, or functional capability. The underlying basis for using tests such as these to screen workers for MMH jobs is the belief that the risk of injury is dependent on the relationship between the capacity of the worker and the demands of the job. When the physical demands of the job exceed the capacity of the worker, the worker is at risk for developing an MSD. Thus, the idea of this approach is that workers should be matched to jobs according to the demands of the work.

A number of studies have been conducted to develop databases of maximum strength capacities (i.e. population averages) that could be used to design manual-handling tasks and workstations. These studies, however, disagree as to which of the three principal testing methods—isometric, isokinetic, or isoinertial—is most useful for determining strength capacity guidelines. Some researchers argue that traditional isometric lifting strength measurements, by which thousands of workers have been tested, are limited in assessing what workers can do under dynamic task conditions. These researchers suggest that dynamic strength testing is more appropriate than static strength testing for determining strength capacity.[65] This assertion is based on how well the test replicates the job requirements. Other researchers, however, claim that isokinetic lifting strength measurements probably have no greater inferential power to predict risk injury or job performance than any other form of testing.[66] Kroemer[67] claims that isoinertial strength testing is the most appropriate lifting strength testing method because it matches actual lifting conditions. lsoinertial methods, however, have not been generally validated in terms of their ability to predict risk of injury.[68]

Maximum isometric lifting strength (MILS) has been studied and reported extensively. It has a well-established testing procedure[69] and has been reported in field tests to predict risk of injury.[23,70] Extensive measures of MILS have been made for various work postures and activities. One study, for example, measured the isometric strength of 1239 workers in rubber, aluminum and electronic component industries.[71] In another study that measured the standardized isometric strengths (i.e. arm lift, torso lift, and leg lift) of 2178 aircraft manufacturing workers,[64] the employees were followed for >4 years to document back-pain complaints. The investigators found that worker height, weight, age and gender are poor predictors of standardized isometric strength, a finding that agrees with other studies.[71] The investigators also found, however, that standardized isometric strength is a poor predictor of reported back pain, a finding that conflicts with the results of some other studies,[72]

Marras et al[73] published an ergonomics guide for assessing dynamic measures of low back performance. This guide provides information on elements of dynamic performance, techniques to assess dynamic performance, and relationships between testing techniques and internal forces.

Personal protective equipment

Personal protective equipment (PPE) is defined as a device or item used by the worker as protection from recognized hazards, such as heat, cold, vibration, and other physical hazards. Ergonomic stressors should be considered when selecting PPE, and the use of the protective device should not contribute to extreme postures or excessive force. Some devices that have been advertised as PPE, such as braces, splints, back belts, and other similar devices, are not PPE according to the OSHA.[50] There is little evidence that these devices provide any realistic protection from injury for healthy workers performing MMH activities.[74] In fact, there is some evidence that prolonged use of back belts by healthy workers may actually increase the risk of low back injury.[75]

Finally, in an attempt to protect workers from the ergonomic stressors associated with manual handling, certain devices have been developed and marketed that force workers to use good body mechanics. For example, a back inclinometer alarm, which sounds an alarm when the worker's back exceeds a certain flexion angle, was developed to prevent excessive back flexion. These devices, however, have not been shown to reduce the incidence or severity of manual-handling injuries and cannot replace sound ergonomic job design.

In summary, MMH activities, such as excessive lifting, pushing, pulling, and carrying, represent a serious hazard for MSD for many workers. There are analytic tools to identify ergonomic hazards that may result in overexertion injury, and prevention tools that are effective in reducing the potential for risk of work-related overexertion injury. Successful ergonomic programs require the full cooperation of management, labor organizations, government, and the workers themselves. The solution requires a team effort with a commitment to identify and eliminate hazardous material-handling tasks from the workplace.

REFERENCES

1. Park CH, Wagener DK, Winn DM, Pierce JP. Health conditions among the currently employed: United States, 1988. Department of Health and Human Services, National Center for Health Statistics. *Vital Health Stat* 1993; 10(186).
2. National Institute for Occupational Safety and Health. *The National Occupational Exposure Survey*. Publication no. 89-103. Cincinnati: Department of Health and Human Services, National Institute for Occupational Safety and Health, 1989.
3. Bureau of Labor Statistics. *Back injuries associated with lifting*. Bulletin no. 2144. Washington, DC: Department of Labor, Bureau of Labor Statistics, 1982.
4. Leigh JP, Markowitz SB, Fahs M, Shin C, Landrigan PJ. Occupational injury and illness in the United States: estimates of costs, morbidity, and mortality. *Arch Intern Med* 1997; 157:1557.
5. National Safety Council. *Accident facts* 1990 edn. Chicago: National Safety Council, 1990.
6. NIOSH. *Musculoskeletal disorders and workplace factors: a critical review of epidemiological evidence for work-related musculoskeletal disorders of the neck, upper extremity, and low back*. Washington, DC: US Department of Health and Human Services, 1997.
7. Hoogendoorn WE, van Poppel MNM, Bongers PM, Koes BW, Bouter LM. Physical load during work and leisure time as risk factors for back pain. *Scand J Work Environ Health* 1999; 25(5):387–403.
8. National Research Council and the Institute of Medicine. Musculoskeletal disorders and the workplace: low back and upper extremities. Panel on musculoskeletal disorders and the workplace. Washington D.C. National Academy Press, 2001.
9. Viikari-Juntura E, Silversteiri BA. Role of physical load factors in carpal tunnel syndrome. *Scand J Work Environ Health* 1999; 25:163–85.
10. Punnett L. Bergqvist U. *Visual display unit work and upper extremity musculoskeletal disorders. A review of epidemiological findings*. Solna: Arbete Och Hälsa, 1997.
11. Burdorf A, Sorock G. Positive and negative evidence of risk factors for back disorders. *Scand J Work Environ Health* 1997; 23:243–56.
12. Hagberg M, Silverstein B, Wells R, et al. *Work related musculoskeletal disorders (WMSDs,): a reference book for prevention*. London: Taylor and Francis 1995.
13. Hoogendoorn WE, van Poppel MNM, Bongers PM, Koes BW, Bouter LM. Systematic review of psychosocial factors at work and in the personal situation as risk factors for back pain. *Spine* 2000; 25:2114–25.
14. Pope MH, Frymoyer JW, Andersson G. *Occupational low back pain*. New York: Praeger, 1984.
15. Jensen RC. Disabling back injuries among nursing personnel: research needs and justification. *Res Nurs Health* 1987; 10:29–38.
16. Troup JDG, Edwards FC. *Manual handling and lifting: an information and literature review with special reference to the back*. London: HMSO, 1985.
17. Frymoyer JW, Pope MN, Clements JH, Wilder DG, et al. Risk factors in low back pain. *J Bone Joint Surg* 1983; 65A:213–78.
18. Bigos S, Spengler DM, Martin NA, Zeh J, et al. Back injuries in industry: a retrospective study, III. Employee-related factors. *Spine* 1986; 11:252–6.
19. Bongers PM, de Winters CR. *Psychosocial factors and musculoskeletal disease: a review of the literature*. NIPG publication no. 92 .028. Nederlands Instituut voor Praventive-Gezondheidsonderzoek (NIPG-TNO). Amsterdam, Netherlands, 1992.

20. Linton SJ. Risk factors for neck and back pain in a working population in Sweden. *Work Stress* 1990; 4:41–9.

21. Bigos SJ, Battie MC, Spengler DM, Fisher LD et al. A prospective study of work perceptions and psychosocial factors affecting the report of back pain. *Spine* 1991; 16:1–6.

22. Waters TR, Putz-Anderson V, Baron S. Methods for assessing physical demands of manual lifting: A review and case study from warehousing. *Am Ind Hyg Assoc J* 1998; 59:871–81.

23. Chaffin DB, Park KS. A longitudinal study of low-back pain as associated with occupational weight lifting factors. *Am Ind Hyg Assoc J* 1973; 34:513–25.

24. Frymoyer JW, Pope MH Costanza MC, Rosen JC, et al. Epidemiologic studies of low back pain. *Spine* 1980; 5:419–23.

25. Snook SH. The design of manual handling tasks. *Ergonomics* 1978; 21:963–85.

26. Grandjean E. *Fitting the task to the man*. London: Taylor & Francis, 1982.

27. Eastman Kodak Company, Ergonomics Group. *Ergonomic design for people at work*, Vol. 2. New York: Van Nostrand Reinhold, 1986.

28. Alexander DC, Pulat BM. *Industrial ergonomics: a practitioner's guide*. Norcross, GA: Industrial Engineering and Management Press, Institute of Industrial Engineers, 1985.

29. National Occupational Health and Safety Commission. *Safe manual handling: discussion paper and draft code of practice*. Canberra: Australian Government Publishing Service, 1986.

30. Marras WS, Sommerich CM. A three dimensional model of loads on the lumbar spine. Part I, model structure. *Hum Factors* 1991; 33:123–37.

31. Chaffin DB, Andersson GBJ. *Occupational Biomechanics*, 3rd edn. New York: Wiley, 1999.

32. Brinckman P, Biggemann M, Hilweg D. Fatigue fracture of human lumbar vertebrae. *Clin Biomechanics* 1988; 3(suppl 1).

33. Kroemer KHE, Snook ST, Meadows SK, Deutsch S, eds. *Ergonomic models of anthropometry, human biomechanics, and operator–equipment interfaces; proceedings of a workshop*. Washington DC: National Research Council, 1986.

34. Garg A. *The biomechanical basis for manual lifting guidelines*. Report no. 91-222-711. Springfield, VA: National Technical Information Service, 1991.

35. Stevens SS. The psychophysics of sensory function. *Am Sci* 1960; 48:226–53.

36. Snook SH, Ciriello VM. The design of manual handling tasks: revised tables of maximum acceptable weights and forces. *Ergonomics* 1991; 34:1197–213.

37. Ayoub MM, Mital A. *Manual materials handling*. London: Taylor & Francis, 1989.

38. Borg G. Psychophysical scaling with applications in physical work and the perception of exertion. *Scand J Work Environ Health* 1990; 16(suppl 1):55–8.

39. Corlett EN, Bishop RP. A technique for assessing postural discomfort. *Ergonomics* 1976; 19:175–82.

40. Waters TR, Putz-Anderson V, Garg A, Fine LJ. Revised NIOSH equation for the design and evaluation of manual lifting tasks. *Ergonomics* 1993; 36:749–76.

41. Rodgers SH, Yates JW, Garg A. *The physiological basis for manual lifting guidelines*. Report no. 91-227-330. Springfield, VA: National Technical Information Service, 1991.

42. Garg A. *A metabolic rate prediction model for manual materials handling jobs*. Ann Arbor, MI: University of Michigan, 1976.

43. Astrand PO, Rodahl K. *Textbook of work physiology* 3rd edn. New York: McGraw-Hill, 1986.

44. Ayoub MM, Selan JL, Liles DH. *An ergonomics approach for the design of manual materials handling tasks*. Human Factors 1983;25:507–515.

45. Marras WS, Fattalah F. Accuracy of a three dimensional lumbar motion monitor for recording dynamic trunk motion characteristics. *Int J md Ergonomics* 1992; 9:75–87.

46. Marras WS, Schoenmarklin RW. Wrist motions in industry. *Ergonomics* 1993; 36: 341–51.

47. Grant KA, Galinsky TL, Johnson PW. Use of the actigraph for objective quantification of hand/wrist activity in repetitive work. In: *Proceedings of the 37th Annual Human Factors*

Meeting, Seattle, WA. Santa Monica, CA: Human Factors and Ergonomics Sociaty, 1993.

48. Garg A. What basis exists for training workers in correct lifting technique? In: Marras WS, Karwowski W, Smith JC, Pacholski L, eds. *The ergonomics of manual work*. London: Taylor & Francis, 1993.
49. Occupational Safety and Health Administration. *Proposed Ergonomics Program Rule*. Federal Register 64, No. 25, 23 November. Washington, DC: Department of Labor, Safety and Health Standards, 65767–6078.
50. Occupational Safety and Health Administration. *Ergonomics program management guidelines for meatpacking plants*. OSHA document no. 3123. Washington, DC: Department of Labor, Occupational Safety and Health Administration, 1990.
51. National Institute for Occupational Safety and Health. *Work practices guide for manual lifting*. NIOSH technical report no. 81-122. Cincinnati: Department of Health and Human Services, National Institute for Occupational Safety and Health, 1981.
52. Waters TR, Putz-Anderson V, Garg A. *Applications manual for the revised NIOSH lifting equation* DHHS (NIOSH) Pub. No. 94-110. Cincinnati, OH: National Institute for Occupational Safety and Health, 1994.
53. Waters TR, Baron SL, Piacitelli LA, et al. Evaluation of the Revised NIOSH Lifting Equation: a cross sectional epidemiological study. *Spine* 1999; 24:386–94.
54. Boden SD, Wiesel SW. Standardized approaches to the diagnosis and treatment of low back pain and multiply operated low back patients. In: Wiesel SW, eds. *Industrial low back pain: a comprehensive approach*, 2nd edn. Charlottesville, VA: Michie, 1989.
55. Agency for Healthcare Quality and Research. *Acute low back problems in adults*, Clinical Practice Guideline Number 14. Publication No. 95-0642: Rockville, MD: Department of Health and Human Services, December 1994.
56. Andersson GBJ, Pope MH, Frymoyer JW, Snook S. Epidemiology and cost. In: Pope MH, Andersson GBJ, Frymoyer JW, Chaffin DB, eds. *Occupational low back pain: assessment, treatment, and prevention*. St Louis: Mosby-Year Book, 1991, Chapter 5:95–117.
57. White AA, Gordon SL. Synopsis: workshop on idiopathic low back pain. *Spine* 1982; 7:141–9.
58. Silverstein BA, Fine LJ. *Evaluation of upper extremity and low back cumulative trauma disorders—a screening manual*. Ann Arbor, MI: University of Michigan, School of Public Health, Occupational Health Program, 1984.
59. Frymoyer JW, Haldermans S. Evaluation of the worker with low back pain. In: Pope MH, et al, eds. *Occupational low back pain: assessment, treatment, and prevention*. St. Louis: Mosby-Year Book, 1991, Chapter 8:151–83.
60. Andersson GBJ, Frymoyer JW. Treatment of the acutely injured worker. In: Pope MH, Andersson GBJ, Frymoyer JW, Chaffin DB, eds. *Occupational low back pain: assessment, treatment, and prevention*. St Louis: Mosby-Year Book, 1991, Chapter 9:183–94.
61. Putz-Anderson V. *Cumulative trauma disorders: a manual for musculoskeletal disorders of the upper limbs*. London: Taylor & Francis, 1988.
62. Deyo RA, Loeser JD, Bigos SJ. Herniated lumbar intervertebral disc. *Ann Intern Med* 1990; 112:598–603.
63. Deyo RA. Non-operative treatment of low back disorders. In: Frymoyer JW, ed. *The adult spine: principles and practice*. New York: Raven, 1991.
64. Battíe MC, Bigos SJ, Fisher LD, Hansson TH, et al. Isometric lifting strength as a predictor of industrial back pain reports. *Spine* 1989; 14:851–6.
65. Kroemer KHE. Testing individual capability to lift material: repeatability of a dynamic test compared with static testing. *J Safety Res* 1985; 3:4–7.
66. Rothstein JM, Lamb RL, Mayhew TP. Clinical uses of isokinetic measurements. *Phys Ther* 1987; 67:1840–4.
67. Kroemer KHE. An isoinertial technique to assess individual lifting capacity. *Hum Factors* 1983; 25:493–506.
68. Kroemer KHE. Matching individuals to the job can reduce manual labor injuries. *Occup Safety Health News Dig* 1987; 3:4–7.
69. Chaffin DB. Ergonomics guide for the assessment of human static strength. *Am Ind Hyg Assoc J* 1975; 36:505–11.

70. Herein GD, Garret M, Anderson CK. Prediction of overexertion injuries using biomechanical and psychophysical models *Am Ind Hyg Assoc J* 1986; 47:322–30.

71. Keyserling WM, Herein GD Chaffin DB. An analysis of selected work muscle strength. In: *Proceedings of the 22nd Annual Meeting of the Human Factors Society*, Detroit, MI, 1978. Santa Monica, CA: Human Factors and Ergonomics Society.

72. Keyserling WM, Herein GD, Chaffin DB. Isometric strength testing as a means of controlling medical incidents on strenuous jobs. *J Occup Med* 1980; 22:332–6.

73. Marras WS, McGlothlin JD, McIntyre DR, Nordin M, et al. *Dynamic measures of low back performance: an ergonomics guide*. Fairfax, VA: American Industrial Hygiene Association, 1993.

74. Wassell JT, Gardner LI, Landsittel DP, Johnston JJ, et al. A prospective study of back belts for prevention of back pain and injury. *JAMA* 2000; 284(21):2727–32.

75. Redell CR, Congelton JJ, Hutchingson RD, Montgomery JF, et al. An evaluation of a weightlifting belt and back injury prevention training class for airline baggage handlers. *Appl Ergonomics* 1992; 23:319–29.

FURTHER READING

Ayoub MM, Mital A. *Manual materials handling*. London: Taylor & Francis, 1989.

Chaffin DB, Andersson GBJ. *Occupational biomechanics*. New York: Wiley, 1991.

Eastman Kodak Company, Ergonomics Group. *Ergonomic design for people at work*, Vol. 2. New York: Van Nostrand Reinhold, 1986.

Pope MH, Andersson GBJ, Frymoyer JW, Chaffin DB, eds. *Occupational low back pain: assessment, treatment, and prevention*. St Louis: Mosby Year Book, 1991.

4

OCCUPATIONAL VIBRATION EXPOSURE

David G. Wilder, Ph.D., P.E., C.P.E., Donald E. Wasserman, M.S.E.E., M.B.A., and Jack Wasserman, Ph.D., P.E.*

Human beings have been exposed for thousands of years to non-human-generated cyclic or repetitive loading, since the beginning of the use of tools, boats, airplanes, or other transport methods using animals or platforms that could be dragged or placed on runners/skids, rollers or wheels. Impact and vibration (acceleration) has been of increasing interest since the mid-1900s. The issue of human response to impact and vibration has been addressed in books, conference proceedings, theses, dissertations, reports, refereed papers, domestic guidelines/standards in the USA by the American National Standards Institute (ANSI) and the American Conference of Governmental Industrial Hygienists (ACGI H), guidelines/standards by individual countries' appropriate organizations, and international consensus guidelines/standards by the International Standards Organization (ISO).[1,2]

Past work has concentrated primarily on the effect of impact and vibration exposure on men. More and more women are entering the workplace and assuming some of the duties that were previously male-dominated (e.g. truck and bus driving, heavy equipment operation). This has caused the emergence of interest not only in the exposure of female spines to whole-body vibration (WBV), but also to the concurrent medical effects of WBV exposure on the female reproductive organs. Laboratory research is underway,[3–5] prompted by reported cases of miscarriage and/or other gynecologic disorders in WBV-exposed female drivers. This societal change is just one example that demonstrates the need for health and safety surveillance programs in the workplace.

Several new issues have arisen since the previous publication of this chapter: (1) There is a new ISO guideline for human exposure to

*Malcolm Pope, Peter Pelmear and the late William Taylor contributed to the previous edition of this chapter.

Physical and Biological Hazards of the Workplace, Second Edition, Edited by Peter H. Wald and Gregg M. Stave
ISBN 0-471-38647-2

WBV, whereas in the USA the ANSI S3.18 is still the accepted limit; (2) the ACGIH promulgated limits in 1995 regarding WBV; (3) the effect of repeated impact, while not well understood, has received much more attention at the basic and applied research level; (4) it has become clear that the posture held and the responsiveness of the musculoskeletal system play important roles; (5) instrumentation for measurement and control of jolt/impact/vibration environments has become much less expensive; and (6) vibration controls can be implemented via engineering (e.g. coupler modification to reduce or absorb slack-action in trains and trucks, suspended vehicle cabs, and suspensions in seats that can provide isolation in one, two or three directions simultaneously) or by the use of personal protective equipment (e.g., anti-vibration gloves, anti-vibration tools, shock-absorbing shoe soles, shock-absorbing floor mats, and lumbar supports).

OCCUPATIONAL SETTING

Vibration is the periodic motion of a body in alternately opposite directions from a position of rest. Vibration is present in most work settings where mechanical equipment is used. There are two major types of vibration that have human health concerns. **Whole-body vibration** affects the entire body, and is usually transmitted in a sitting or standing position from a vibrating seat or platform. **Hand–arm vibration** (HAV) affects one or both upper extremities, and is usually transmitted to the hand and arm only from a motorized hand tool. WBV is generated by motor vehicle operation, including trucks, buses, and construction and agricultural (tractors, threshers or combines) equipment and heavy manufacturing equipment such as looms, large machine tools and presses. HAV is generated by any powered hand tool, including chippers, jackhammers, chain saws, trimmers and blowers, nut-tightening guns, polishers, grinders, and rivet guns. These tools have widespread use in industry, and may be electric, pneumatic or combustion engine powered. Workers using any powered hand tools have potential exposure.

OCCUPATIONAL VIBRATION MEASUREMENTS[6–8]

Vibratory motion is by definition a mathematical "vector quantity," which simply means that it is described by both a direction and a magnitude. At each measurement point, the total motion is described by six possible vector directions; three so-called linear directions with their magnitudes, and three rotational directions (pitch, yaw, and roll) with their magnitudes. In most human vibration work, only the linear directions are measured, reported, and eventually compared to health/safety standards. Figure 4.1 shows the internationally accepted "biodynamic coordinate system" used for head-to-toe or whole-body vibration and similarly for segmental, or hand–

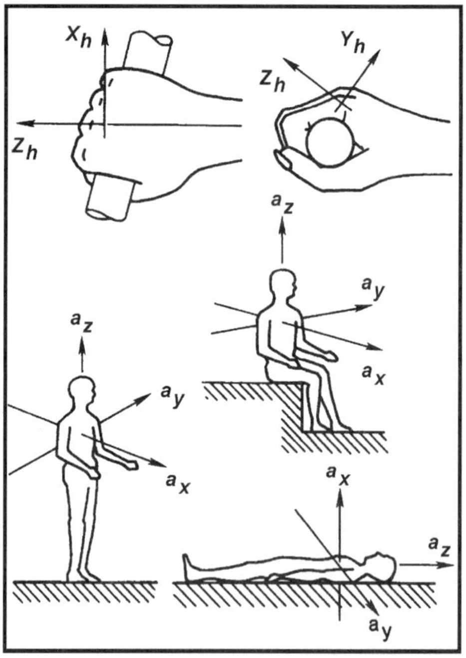

Figure 4.1 Hand–arm and whole-body biodynamc coordinate systems for human measurements.

arm vibration. For WBV the sternum is the reference point of the measurements. The vibration intensity or magnitude quantity of choice is "acceleration", where by definition the z direction is head-to-toe, the y direction is side-to-side (shoulder-to-shoulder), and the x direction is front-to-rear. Each of these three linear directions and corresponding acceleration magnitudes is separately measured and reported.

For HAV, the third metacarpal is the reference measurement point. The defined motion in the z direction is along the long bones of the forearm; the y direction is motion across the knuckles; and the x direction is motion through the palm. In many cases it is not practical to place measurement accelerometers on the third metacarpal, so measurements are obtained directly from the vibrating tool handle as the operator works and grips the handle (Figure 4.2). This tool handle frame of reference is called the "basicentric coordinate system". Finally, the reason for using these types of coordinate systems is to establish uniform measurement methods worldwide and a means of easily comparing measurement results to health and safety standards.

In practice, three-axis (triaxial) vibration measurements are simultaneously, but individually, obtained and recorded on separate channels (on a computer-based data acquisition system) or tracks (using a specialized digital audio tape recorder (DAT)) in accordance with the applicable standards. These recorded vibration axes (x, y, z) are then separately analyzed into their respective vibration "spectra". Each of these individual axis spectra is then compared to the appropriate health and safety standard(s) to determine if and by how much said standard(s) have been exceeded. If these results indicate that one or more vibration axes has indeed exceeded the standard(s), then the entire standard has been exceeded and vibration controls are most likely needed.

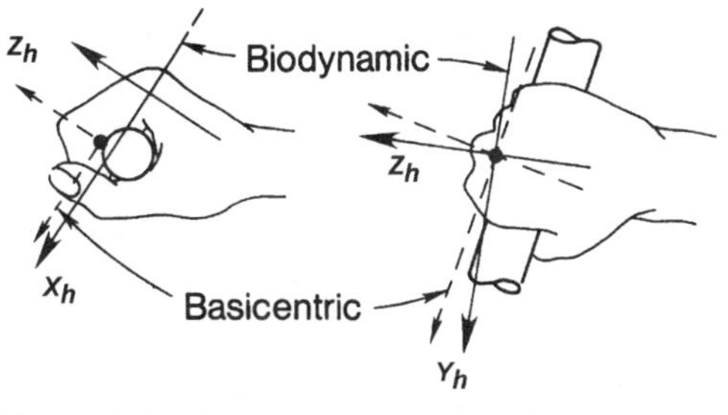

Figure 4.2 Biodynamic and basicentric coordinate systems for hand–arm vibration and tool measurements.

OCCUPATIONAL VIBRATION GUIDELINES USED IN THE USA

Occupational standards per se are actually guidelines and represent the state of knowledge available when the guideline was promulgated. Thus these standards/guidelines are periodically reviewed and updated as new information becomes available.

Currently, two WBV guidelines are used in the USA, ACGIH for WBV[9] and the ANSI S3.18.[10] For HAV there are three standards used: ACGIH for HAV,[11] ANSI S3.34[12] and NIOSH (National Institute for Occupational Safety and Health) Publication #89-106 criteria document for a standard for HAV.[13] Because of the complexity of these standards, the reader is advised and cautioned to obtain, read and thoroughly understand the actual standard(s) before beginning measurements and data analysis. For the reader's information, the 1997 ISO 2631[14] WBV guideline has been modified from its previous version, and has not yet been accepted for use in the USA. Its use has also been described as confusing by Griffin.[15] What follows only comprises the basic elements of the US standards.

Whole-body vibration standards

ANSI S3.18 and ACGIH for WBV are used extensively in the USA. Figures 4.3 and 4.4 show the basic elements of these standards. ANSI S3.18 calls these curves the FDP or "Fatigue, Decreased Proficiency" curves and they are the same as those used in the ACGIH for WBV. Figure 4.3 is used to evaluate WBV

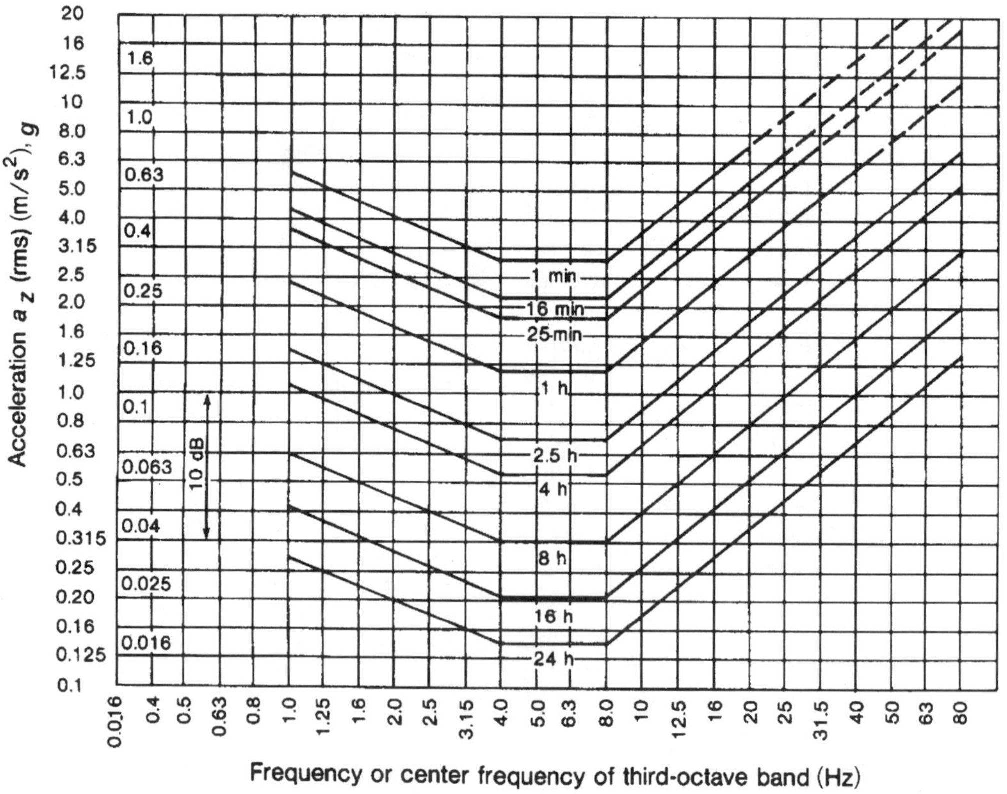

Figure 4.3 Whole body vibration weighted z (vertical axis) daily exposure curves used by ACGIH and ANSI S3.18 Fatigue, Decreased Proficiency Standards. (Adapted from American Conference of Governmental Industrial Hygienists and American National Standards Institute[9,10]).

spectral acceleration data in the vertical or z direction. Figure 4.4 is used separately, twice, first to evaluate spectral acceleration data in the side-to-side y direction and again to evaluate spectral acceleration data in the front-to-rear x direction.

Referring to Figure 4.3, the horizontal axis shows vibration frequency in so-called third-octave bands, extending from 1 to 80 Hz. The vertical axis of Figure 4.3 gives a measure of the vibration "intensity" magnitude in root-mean-squared (rms) acceleration in either g or meters per second per second (m/s^2), where $1\,g = 9.81\ m/s^2$. A "family" or set of parallel U-shaped daily exposure time curves is also shown in the graph. What Figure 4.3 really shows are vibration intensity (acceleration) limits as a function of vibration frequency and daily worker exposure time for the z axis. Figure 4.4 shows the same type of information for the x and y axes. If one or more of the axis accelerations exceeds the standard, then the entire standard has been exceeded and appropriate vibration control action needs to be taken.

Limitations Jolt/impact is emerging as an important problem. In the past, the standards/guidelines have tried to cope with jolt/impact by comparing its frequency-weighted peaks to the frequency-weighted root-mean-square of the signal of which it is a part over a period of time. This ratio is known as the crest factor.[10] The current standards/guidelines emphasize that if the crest factor exceeds 6 or the crest factor exceeds 9, then the exposed individual's risk is underestimated.[14] Jolt/impact can result from a combination of sinu-

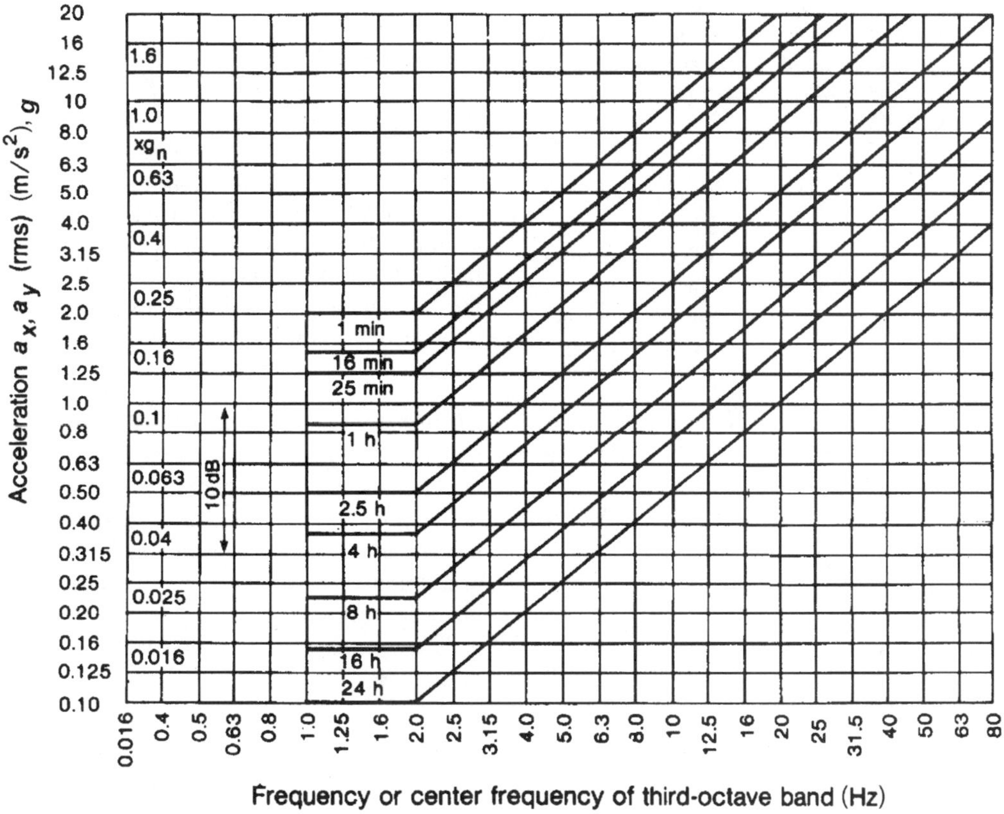

Figure 4.4 Whole body vibration weighted *x* and *y* (transverse axes) daily exposure curves used by ACGIH and ANSI S3.18 Fatigue, Decreased Proficiency Standards. (Adapted from American Conference of Governmental Industrial Hygienists and American National Standards Institute[9,10]).

soidal signals and is not fully addressed by the current standards. When Cohen et al[16] studied human responses to single versus combined sinusoidal WBV signals, it was clear that, for the same root-mean-square acceleration levels, people (healthy firefighters) were more sensitive to the combined vibration. The above standards are based on the assumption that workers in many jobs are exposed to sinusoidal vibration, but the scientific community knows that workers are exposed to non-sinusoidal vibration. For these jobs, the above standards do not fully address and likely underestimate the risk. To address these issues, research is ongoing,[17–20] and the ISO and ANSI are developing standards for exposure to non-sinusoidal vibration and repetitive mechanical jolt/impact. Exposure to seated, whole-body, non-sinusoidal vibration and jolt/impact conditions should be considered in the following, including, but not limited to, the marine, agricultural, rail-guided, rail-constrained, forklift, mining/quarrying, over-the-highway trucking, and heavy equipment vehicle environments.

Hand–arm vibration standards

Most HAV standards are aimed at reducing the probability of an attack of vibration white finger (i.e. beginning of the blanching process) occurring. The first HAV standard in the USA was introduced by the ACGIH in 1984. Table 4.1 shows the ACGIH limits for HAV. Weighted triaxial acceleration measurements are obtained over a third-octave band vibration

Table 4.1 ACGIH limit for exposure of the hand to vibration in either X_h, Y_h and Z_h directions. a_K and a_{Keq} are weighted accelerations calculated as defined by ACGIH for exposure of the hand and arm to vibration.

Total daily exposure duration[a]	Values of the dominant[b] frequency-weighted, rms, component acceleratioin which shall not be exceeded a_K, (a_{Keq})	
	m/s^2	g
4 hours and less than 8 hours	4	0.40
2 hours and less than 4 hours	6	0.61
1 hour and less than 2 hours	8	0.81
Less than 1 hour	12	1.22

[a]The total time for which vibration enters the hand per day, whether continuously or intermittently.
[b]Usually, if one axis of vibration axis exceeds the total daily exposure, then the limit has been exceeded.

frequency range of 5.6–1250 Hz. The total rms acceleration is determined separately for each of the three linear axes. If one or more of these values exceed those in Table 4.1, then the standard has been exceeded. For example, if the highest value measured on a tool in any one of the three axes is 20 m/s^2, then that tool could not be used during the work day, since the maximum Table 4.1 value is 12 m/s^2. If, however, a highest value of 7 m/s^2 were obtained, then this value is higher than the 6 m/s^2 permitted for 2–4 hour/day exposure, but is lower than the 8 m/s^2 permitted for less than 2 hour/day exposure. Thus, the worker could operate the tool for less than 2 hours daily.

Figure 4.5 shows the ANSI document S3.34, promulgated in 1986. This figure is similar in format to Figures 4.3 and 4.4. The horizontal axis is vibration frequency in third-octave bands, extending from 5.6 to 1250 Hz. Vibration acceleration intensity is given in m/s^2 rms on the vertical axis. A series of parallel exposure time-dependent "elbow-shaped weighted curves" is given in the graph, forming daily exposure zones. This same graph is used separately three times, once with the vibration spectra of the x axis, then with the y axis spectra, and again for the z axis spectra. If one or more exposure zones are exceeded by spectral peaks in any of the three axes, then the standard has been exceeded. The highest-occurring spectral peak(s) intersecting the least number of hours/day exposure zone curves is usually the limiting factor for permitted daily exposures.

Limitations The NIOSH HAV Criteria for a Standard #89-106[13] is different from ACGIH or ANSI documents. This is an interim, and rather more limited, standard, since the NIOSH has not as yet chosen a maximum permissible acceleration value for HAV. It relies instead on medical monitoring and engineering controls and an extended high-frequency range cutoff of 5000 Hz.[21–23]

WHOLE-BODY VIBRATION AND LOW BACK PAIN

Low back pain (LBP) is a significant disabling health problem. The medical, industrial and socio-economic consequences of LBP syndromes are staggering. In the USA, LBP is the leading cause of industrial disability payments and the second most common medical cause of work loss in industry, resulting in eight million workers being affected. Two million are chronically disabled by these processes. In the USA, one estimate of the total cost of LBP comes to nearly 80 billion dollars per year,[24] a cost more appropriate to natural disasters.

Epidemiologic studies have shown strong associations between back trouble and WBV.[25–29] Bovenzi and Zadini[26] followed a group of bus drivers. Bovenzi and Betta[25] followed a group of agricultural tractor drivers. Both studies correlated low back diagnoses with vibration exposure levels and concluded that the ISO 2631/1 vibration guidelines were not conservative enough corroborating the

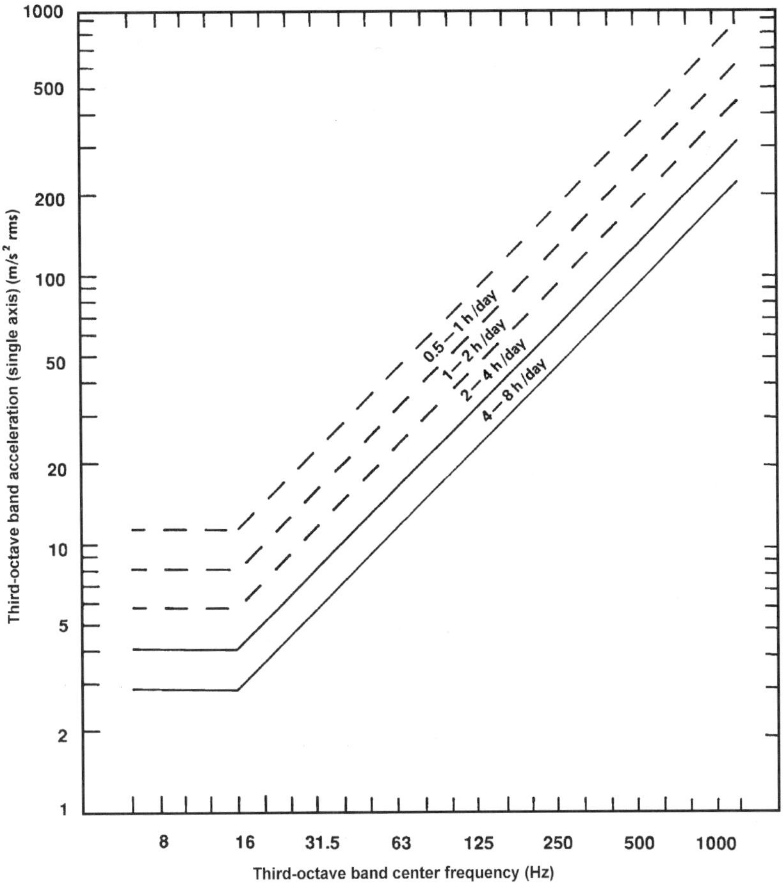

Figure 4.5 Hand-arm vibration weighted daily exposure zones used in ANSI S3.34 standard. (Adapted from American National Standards Institute[12]).

findings of Bongers and Boshuizen[30] in their work on LBP related to WBV exposure at work. In the more recent study,[25] Bovenzi and Betta recommended following the Commission of European Communities' 1993 A(8) action level of 0.5 m/s^2 WBV.[31] Magnusson et al[28] studied bus and truck drivers both in Goteborg, Sweden and in Vermont, USA, and found a significant correlation between LBP and history of WBV exposure. These important studies followed the guidelines set out by Hulshof and van Zanten,[32] and represent the beginning of the development of firm guidelines for reducing WBV in the workplace. These studies have begun to clarify the exposure–response relationship needed for setting exposure guidelines.[33] Germany has taken the step of designating diseases of the lumbar spine related to long-term, WBV exposure as occupational disease number 2110.[34]

With data collected for the 1988 Occupational Health Supplement to the National Health Interview Survey,[35,36] Behrens et al[37] reported that the highest prevalence of injury-related back pain and the overall prevalence of back pain due to repeated activities at work were associated with WBV work conditions.

More recently, Hoogendoorn et al[38] and Bovenzi and Hulshof[39] reviewed the literature associating mechanical factors with back trouble. They found there were sufficient, properly executed studies to conclude that low back trouble is associated with WBV exposure.

Etilogic factors

Bovenzi and Hulshof[39] indicate that Wilder[40] and Wilder and Pope[41] provide a reasonable hypothesis for the mechanism for low back problems in the WBV environment. The seated posture can be a mechanically extreme orientation for the lumbar intervertebral disc, increasing its internal pressure, and increasing its anteroposterior shear flexibility, while decreasing its resistance to buckling instability and stressing the posterior region of the disc. Vibration and impact are additional mechanical stressors that can lead to large stresses and strains in the disc and subsequent mechanical fatigue of the disc material.[40,41]

Mechanical studies have been performed to evaluate the effect of WBV in seated, standing and supine postures, in both single and multiple directions. The dynamic behavior of the human subject can be determined by two means: acceleration transmissibility and impedance. Using the former method, one compares the output acceleration resulting from the input or driving acceleration. At resonance, the ratio of output to input exceeds unity. For the impedance method, one computes the ratio of the force to move the body to its resulting velocity. This ratio, as a function of frequency, defines the mechanical response.

The degree to which an operator moves in a vibration environment is demonstrated by the magnitude of the acceleration transmissibility at the frequency of interest. Acceleration transmissibility is greatest at the resonant frequency, and many studies have shown transmissibilities greater than 1.0 for the first resonant frequency of the seated subject. Resonant frequencies were reported to occur between 4 and 6 Hz, which was usually attributed to the upper torso vibrating vertically with respect to the pelvis, and between 10 and 14 Hz, representing a bending vibration of the upper torso with respect to the lumbar spine.

It was also found that vibration response could be altered by posture.[42] Postures which are quite common in the occupational workplace (lateral bend and axial rotation) lead to greater transmission of vibrations. Typical examples would include the twisted posture of a tractor or forklift driver. We can gain considerable insight into the biomechanics and biodynamics of the spine if we look at the way in which it is assembled. The spine consists of a stack of bony elements that are each able to resist compression and bending. Between each of these relatively rigid elements is a softer, intervertebral disc, an element that allows relative motion between adjacent vertebrae. A network of ligaments and muscles, tensile elements, stabilizes the stack overall and regionally. According to Levin,[43] the similarity of the spine to a structure that consists of discontinuous compressive elements and continuous tensile elements is striking, and reminds us of the original insight and sculptures of Snelson[44] and the subsequent structures of Buckminster Fuller. In those sculptures and structures, a locally applied load affected the entire structure. Hence, disturbance or change in any of the structure's individual tensile or compressive elements would have an effect on the rest of the structure. The additional complexity in the spine relates to the need for the coordinated action of agonist and antagonist muscles. Not only do the muscles and ligaments play a stabilizing role in static and quasi-static conditions, but they must also provide stability in dynamic environments. When muscles are needed as active stabilizers, we must also be aware that muscles may become less effective when the stimulus rate exceeds the muscles' ability to respond. Therefore, the posture held in the workplace can influence the spine, its network of support, the load path in the trunk and the possibility of injury.

Using accelerometers and pins implanted in the lumbar region, Dupuis,[45] Panjabi et al[46] and Pope et al[47] showed that the resonant frequency of the seated operator was 4.5 Hz. However, Panjabi found little or no relative motion between L1 and L3. In contrast, by vibrating primates in seated postures, Quandieu and Pellieux[48] and Slonim[49,50] showed relative motion between lumbar levels and between the upper and lower spine via accelerometry. Coermann[51] also found relative motion between the pelvis and neck in humans by using mercury strain gauges between aluminum shells surrounding and tightly fitting the pelvic and chest regions. Zagorski et al,[52] using accelerometers taped to the backs of human subjects, found greater acceleration at L3 than at the sacrum, in the 2–5 Hz frequency range. Wilder et al[53] found relative motion on the surface of the lumbar region by means of filming seated subjects vibrating at their natural frequency as a moiré fringe pattern was projected onto their backs. Pope et al[47] found relative motion at the first natural frequency, using transducers rigidly fixed to the lumbar spinous processes and fixed to the skin. Under local anesthesia, a threaded K-wire was threaded transcutaneously into the spinous process at L3. The greatest transmissibility was reported at 4 Hz, and substantial differences were noted between the vertical displacement of the pelvis and the adjacent LED (light-emitting diode) marker and L3 and its adjacent LED marker. In a later study, using an intervertebral relative orientation sensing device, Pope et al[54] found that greater rotations and translations occurred at 5 Hz compared to 8 Hz, again confirming the effect of vibration at the natural frequency.

Muscle response

The electromyographic (EMG) signals of the erector spinae and external obliques were measured by Wilder et al[53] at each of three vibrational frequencies, first in the neutral position and then in varying body postures, as well as during the Valsalva maneuver. Wide variations were observed in the EMG activity with respect to body posture. Increased activity of external obliques was found in rotation and lateral bend, and during the Valsalva maneuver. A significantly increased myoelectric activity of the erector spinae was observed in women at the third resonant frequency. At this frequency, marked acceleration of the female breast mass could explain the increase in myoelectric activity.

The phasic activity of the erector spinae muscles was measured by Seroussi et al[56] in male subjects free of LBP. The ensemble-averaged EMG signals were converted to predicted torque using an in-vivo EMG-torque calibration technique. From these data, the phase relationship between the input signal to the platform and the resulting torque was established. Output data were the average, maximum and minimum torque as a function of frequency. Higher average EMG levels, or muscle torques, were found for the vibration condition. The time lag between the input displacement and the peak torque varied from 30 ms to 100 ms at 3 Hz and 70 ms to 100 ms at 10 Hz. At 10 Hz, the muscle contraction tended to coincide with the input signal, or to be 360° out of phase. At all other frequencies, it was out of phase. Seidel and Heide[57] have also monitored the timing of the back muscle response to a vibration stimulus and found that the muscles are not able to protect the spine from adverse loads.

At the vertical natural frequency, Pope et al[58] found significantly greater erector spinae muscle activity without any foot support than with foot support. Pelvic rocking, reduced with the aid of a foot support, was shown to be an important factor in the reduction of the first natural frequency response of the seated individual.

Magnusson et al[59] investigated the fatigue of the dorsal muscles under 5-Hz sinusoidal vibration. To increase the response, the subjects wore pouches placed anteriorly over the ribs with 10-kg weights placed inside. The median frequency of the myoelectric power spectrum was used to establish the fatiguing effect of vibration as measured over the 30-minute time interval. Among seated, non-vibrated subjects, no change was observed over a 30-minute observation period. In contrast, among subjects vibrated over the 30-

minute interval, a shift in the median frequency of erector spinae muscles was recorded in response to the vibratory input, suggesting muscle fatigue. In the industrial environment especially for those with awkward postures, vibration will lead to muscular fatigue.

Solomonow et al[60–62] and Gedalia et al[63] found that cyclic loading compromised the neuromuscular control system's reflexive ability to stabilize the lumbar spine. Wilder et al[64] also found that when subjects deliberately adopted an awkward posture in a well-configured, vertically vibrating seat, there was a significant increase in back muscle electrical activity.

Mechanical fatigue due to vibration loading

Although this chapter discusses soft tissue responses to impact/vibration, the review by Brinckmann et al[65] does include bony responses in the spine.

Just as static postures produce intradiscal pressures unique to them, vibration also has an effect on the intradiscal pressure. Hansson et al[66] vibrated pigs longitudinally while they simultaneously obtained measurements of intradiscal pressure. The vibration frequency used varied from 1 to 12 Hz. Intradiscal pressure was sensitive to frequency. Disc pressure peaked at 5 Hz and was 2.5 times that at 3 Hz, thus indicating a natural frequency similar to that of the seated human. If disc pressure in the human is similarly sensitive to vibration frequency, then vibrating at the natural frequency would introduce time-varying disc pressure as a fatigue factor.

Adams and Hutton[67] simulated a day of heavy flexion and torsion labor at a rate of 40 times/min with loads based on the person's body weight. Forty-one cadaveric lumbar motion segments, 12–57 years old, showed plainly visible distortion as a precursor to a disc herniation tracking tear. In further work, Adams and Hutton[68] produced disc prolapse in 6 of 29 specimens subjected to flexion and cyclic compression, which was increased at regular intervals. Five of the prolapses occurred at the posterolateral "corner" of the disc while the other occurred centrally.

Brown et al[69] produced a tear throughout the annulus (parallel to the endplate and to within 3.2 mm of the disc periphery) of a non-degenerated segment, with ligaments and posterior elements removed, as a result of 1000 cycles of 63.6-N compression load and 5° of forward flexion at a frequency of 1100 cycles/min.

Ten lumbar segments tested by Liu et al[70] experienced bony facet or vertebral body failures, or disc annular or facet capsular ligament tears, as a result of 0.5-Hz cyclic torque with a 445-N axial preload and testing until failure or elapsing of 10 000 cycles. Torques applied were ±11.3, 22.6 or 33.9 N-m. Generally, failures occurred in segments subjected to more than 1.5° axial rotation. Other responses to this testing included discharge of synovial fluid from the articular facet joint capsule and joint "looseness" at the end of the test.

Wilder et al[71] observed mechanical energy absorption by the spinal motion segment during cyclic loading of the spine.

Wilder et al[72] also produced disc herniations as a result of combined vibration loading. The herniations occurred in young calf discs subjected to a 9.5-Hz, combined flexion and lateral bend cyclic loading with constant superimposed axial rotation torque. Specimens were loaded from 6000 to 37 740 times at a frequency of 9.5 Hz. A motion segment from a 68-year-old human male was also tested in a similar combined loading mode. After 2778 cycles, the specimen failed suddenly through the disc as a result of a tear beginning in the posterolateral portion of the disc, a point through which clinically observed herniations occur.

Forty lumbar motion segments tested either in simulated vibrating or non-vibrating, sitting environments showed mechanical sensitivity to load exposure history.[73] When they were subjected to 5-Hz vibration loads corresponding to accelerations occurring at physiologic levels (the 8-Hour Fatigue, Decreased Proficiency Limit of ISO 2631[74]), significant mechanical changes were produced in the motion segments. In addition, the vibrated segments exhibited rapid, short-column buckling.

Vertebral buckling instability

A long, slender, flexible column has the potential to buckle or give way suddenly. Buckling can lead to a rapidly occurring mechanical failure. When a buckling event occurs, the column's mode of resisting a vertical load, applied coaxially changes, so that it must resist that same load with the column in a bent shape. When the transition from straight to bent occurs, the bent column is less stiff than the straight column and the point at which the load is applied accelerates rapidly. The reader can observe this using a screen door spring or a bamboo kebab skewer. Buckling can also occur in short columns. It is most easily understood in the case where the load vector becomes directed outside the base of support. This is why a catamaran-type sailboat will continue to capsize or tip over once its center of gravity is outside either of its hulls, its base of support. The intervertebral motion segment can be considered as a flexible, short column, susceptible to buckling, especially if its disc has been compromised by injury, fatigue, disease, or degenerative processes, thereby decreasing the size of its effective base of support.

Wilder[55] and Wilder et al[73] described experimental observations of short column buckling in individual lumbar motion segments, in-vitro, in response to simulated exposure to a seated vertical vibration environment. Often, those segments buckled in a combination of flexion and lateral bend, placing the posterolateral aspect of the intervertebral disc at risk of experiencing a tensile impact load. Since then, 82 segments have been tested with the additional condition of maintaining a simulated laterally bent posture.[40] Of the normal segments, 79% buckled due to vibration exposure in a simulated awkward posture, while only 10% buckled due to the same awkward posture maintained in a static environment. Typically, the buckling of the motion segments occurred in less than 0.1 s. This raises important challenges for the neuromuscular control system in terms of its ability to sense and prevent or control a buckling event in the lumbar spine.

Many researchers have pointed out the importance of synchronized lumbar and trunk muscle activity in the active stabilization of the lumbar spine, most appearing since the first edition of this chapter.[75–83] Cholewicki et al[75] concluded that the lumbar spine is susceptible to injury during a buckling event occurring during movements associated with actions as apparently simple as picking up a pencil. McGill[82] described an apparent buckling event and associated it with muscle activation behavior.

Impact as a sudden and unexpected load

If we consider that an impact event can be considered as a suddenly, and in many cases an unexpectedly, applied load, then we can apply another area of the literature to the understanding of the body's response to impact. Impact loading can come from many sources, such as: load shifting, slips, trips, stepping off an unexpected curb, mechanical slop in seats, sloshing of liquid in a tank trailer, and slack-action in train or truck couplings (Figures 4.6–4.8), to name a few. The trunk musculature around the lumbar region responds differently to sudden loads, depending on whether or not the load is expected. Whether the subject is standing or sitting, the pelvis acts as a foundation for the spine. The orientation of the pelvis is affected not only by trunk muscles located above the pelvis but also by leg muscles located below the pelvis. Sudden events such as slips, trips and falls affect actions of the leg and trunk muscles. In 1981, Manning and Shannon,[84] and in 1984, Manning et al[85] showed that slip events are considered to be first events that lead to back injuries and expressed concern that this was a neglected research area in back

Figure 4.6 Typical trailer hook in a coupler ring on a dolly to allow a tractor–trailer set to pull a second or third trailer. The constraining latch is raised and the compression pad is retracted for visualization purposes. Without the compression pad in place, the space between the pin and tongue ring would be a potential source of slack-action impact.

Figure 4.7 Typical train car couplers. The space provides a source for potential slack-action impact. Couplers attached to their cars using a properly-maintained drawbar cushioning apparatus tend to modify the impact from slack action. (Personal communication, Robert Hitson, 18 Dec 2000).

Figure 4.8 Side view of a style of train-car coupler that has a design that tends to modify slack-action impact. Couplers attached to their cars using a properly-maintained drawbar cushioning apparatus tend to modify the impact from slack action. (Personal communication, Robert Hitson, 18 Dec 2000).

injury etiology. When, in 1987, Marras et al[86] showed that sudden, unexpected loads applied at the hands lead to large overcompensations in the back muscles, he showed that the hazard is not necessarily the load applied. The hazard is the body's excessive reaction or overcompensation to the applied load. Mannion et al[87] continued along these lines and predicted disc overloading as a result of sudden, unexpected loading. These papers show that sudden loads and sudden movements of the body can place the back at risk of injury, as they each describe conditions that require the back muscles to respond rapidly to an imposed load or movement.

Responses to sudden loads applied at the hands are not necessarily symmetric either. Pelvic orientation,[88] load application location[89] and fatigue and hand dominance[90] can all affect the symmetry of the trunk muscle response.

Triggering buckling of an unstable system

The work of Fethke et al,[17] Morrison et al[18] and Robinson[19] shows that the responses of the trunk musculature of a seated subject to lateral impacts are asymmetric in response time, duration, and amplitude. This condition raises the serious possibility that an asymmetric muscle response could trigger a buckling event in the spine. This also corroborates the impression of a long-time operator of rail-constrained vehicles who felt that the toughest part of the job was maintaining awkward postures while also trying to remain prepared for an unexpected horizontal impact (Figure 4.7).

VIBRATION CONTROL WITH SOME EXAMPLES

There exists an international WBV (i.e. head–toe, fore–aft, side–side vibration exposure) guideline.[14] Also, there are two domestic WBV guidelines[9,10] in the USA. Workplace exposure to WBV continues to be indicated as an important issue in musculoskeletal disorders for the lower back.[38,39,41] Griffin, from the UK, claims that the most recent international guideline for human exposure to WBV, ISO 2631-1, is confusing.[15] We in the USA agree and therefore continue to adhere to the ACGIH and the ANSI S3.18 for WBV. Consistent with safe work practices, efforts should be made to isolate mechanically the individual from WBV exposure, in accordance with the ACGIH and ANSI S3.18.

From the work by Wilder and Pope,[41] it is clear that the characteristics of the seated posture held in the vibration environment are very important. Efforts should be made to eliminate awkward, asymmetric postures and provide lumbar support, an adjustable seat pan and seat back, arm rests and any other ergonomic modifications (including recommending an elastic industrial back support) that would tend to decrease intradiscal pressure, back muscle activity, and stress and strain on the posterior and posterolateral regions of the lumbar intervertebral discs.

Responses to impact environments are less well understood, and would benefit not only from current research efforts, but also from the use of participatory ergonomics. This implies working with the people doing the job, and asking for their insights and suggestions for improvements.[91] Once suggestions are implemented, follow-up evaluations should be conducted to ensure that the improvements are working. One company for which this author worked had a policy that required employees to stop using and report a worksite if it was perceived to be a problem.[92] Upon analysis of a forklift that led to back problems, it was discovered that the development of the lateral, repetitive impact condition was due to the introduction of a newer, lighter forklift, that had tires that wore out unevenly. This was a great relief to the company, as it was concerned that it would have to dig up and recast the entire concrete floor.

Experience with a container-shipping terminal operation in the eastern USA demonstrated that ergonomists can have a powerful and direct influence on the design of equipment to accommodate human shortcomings. This is done by working on specifications with groups about to buy equipment. The immediate impact is to guide design of equipment that is safer for their operators. The long-term impact is to optimize products for human use, one industry at a time. Once a company (or its compensation insurer) sees that the new specifications create a much better product, the forces of marketing, competition and concern about being behind the "state of the art" can take over and drive the improvement, distribution and market penetration process.

Until more is known about the combined effects of posture, jolt/impact and vibration, employers and equipment manufacturers should work with end-users to assess the success of a particular isolation solution. The process of product improvement can use the various domestic and international guidelines as a starting point. However, the improvement process should follow up and integrate the concerns and suggestions of the end-users, as they know the most about the environment. This is the basic concept behind epidemiologic surveillance. To equipment manufacturer's benefit, end-users' opinions and observations comprise a significant source of new or improved product ideas.

Recent data regarding lumbar supports are encouraging. Because the overcompensating response to sudden load can injure the back, the ability to reduce the magnitude of the back's response to sudden load is an important approach. In particular, refer to the work supporting the identification of the response to sudden symmetric or asymmetric loading.[84–88,93–95] There is empirical evidence to show that lumbar supports can be used as protective equipment, as they act as a barrier and limit the degree of back muscle overcompensation during the application of sudden symmetric or asymmetric loads applied at the hands.[94]

In addition to the successful use of lumbar supports in crane work,[96] the following appears to be an "ergonomic success story" regarding the use of lumbar supports associated with a seated, vibration environment.

Although this was not conducted as a formal study and can only be reported as a sanitized anecdote, it does have important ramifications. A trucking firm, employing approximately 3000 over-the-highway truck drivers, with facilities in 300 locations in the USA, faced challenging low back problems among the drivers. In 1992, the company tried out the use of lumbar supports by their drivers. In 1994, the use of lumbar supports by the drivers was made mandatory. The firm's risk manager estimated that approximately 2000 of the drivers complied. Over the next 5 years, the company saw a significant decrease in the frequency and severity of subsequent back problems in the drivers. Although encouraging, this would be a particularly important case to study, as no indication was given that confounding factors were involved. There are questions that should be asked of this situation. Were the seats and/or cabs improved? Was the work organization modified? Were material-handling devices added or improved? Was there an effect due to the extra attention paid to the drivers?

If none of the above reasons inspire employers or product manufacturers to improve this environment, then consider the issue from the business perspective of a cost–benefit ratio. If the cost of an air-ride suspension seat is $400 and the average benefit of having avoided a lost-time back injury is $23 716,[97] then the cost–benefit ratio is $400/$23 716 or 0.017 or 1.7%. This is another example of good ergonomics being good economics.

Once vibration measurements have been made and the data have been evaluated with regard to the appropriate standard and it is determined that vibration control is necessary, usually a series of multiple control steps are taken, depending on the problem.[7,98]

1. Reduce vibration exposure by placing the worker away from vibrating surfaces by using remote controls, closed circuit TV monitors, etc.
2. Reduce exposure time by modifying work organization, job sharing, etc.
3. If possible, mechanically isolate the vibrating surface, machine, etc.
4. Maintain mechanisms and replace worn-out mechanisms that contribute to production of jolt/impact and vibration.
5. In vehicles, use vibration-isolating "suspended or air-ride" seats and cabs and replace vehicle suspension systems as necessary.
6. Once changes have been made, repeat vibration measurements and compare data to WBV standards as well as previous data.
7. Ask equipment users to comment on the effectiveness of the solutions.
8. Institute a surveillance program.

For WBV applications, the above might necessarily be supplemented with additional engineering redesign methods using vibration damping techniques (i.e. converting vibration into a small amount of heat due to the deformation of a viscoelastic material) and/or vibration isolation (i.e. intentionally mismatching the vibration pathway between the vibrating source and the worker receiving this vibration).

PREVENTION

LBP can have many causes, and it is difficult to eliminate some of them. However, many industrial risk factors can be modified to reduce the rate of back disorders. Epidemiologic studies suggest that vibration is an important risk factor for LBP and that many vehicles subject the worker to levels of vibration greater than that recommended by the standards. Workers in occupations where vibration is combined with lifting, pulling or pushing may be especially prone to back disorders—for example, truck drivers who also load and unload trucks.

The epidemiologic data are supported by laboratory studies of spine changes that might produce back pain, e.g. lumbar disc flattening, disc fiber strain and height increase, and intradiscal pressures. It is apparent from WBV data that the human spinal system has a characteristic response to vibration in a seated posture. Resonances occur at uniform frequencies for all of the subjects. The first vertical resonance occurs within a band of 4.5–5.5 Hz.

These studies indicate that maximum strain or stretching occurs in the seated operator's lumbar region at the first natural frequency. In addition, back muscles are not able to protect the spine from adverse loads. At many frequencies, the muscles' responses are so far out of phase that their forces are added to those of the stimulus. The fatigue that was found in muscles after vehicular vibration is indicative of the loads in the muscles. Thus, it would be advisable to walk around for a few minutes before bending over or lifting after vibration exposure (e.g. unloading a truck). One fuel company had their tank-truck drivers take care of paperwork before handling hoses at a fuel delivery stop.[99] It would also be advisable for those exposed to prolonged vibration (e.g. long-distance driving) to take frequent breaks, including walking around for a few minutes.

The field of mechanics provides an encouraging note for attempts to improve the WBV environment. If one considers that the damage to the spine due to vibration occurs from the work performed on the body by the kinetic energy from the vibration, two things become very clear. In the simplest form, the work performed on the body is equal to the kinetic energy applied to the body. The equation for that kinetic energy is $\frac{1}{2}mv^2$ (work $=$ kinetic energy $= \frac{1}{2}\ mv^2$), where m is mass and v is velocity. Because velocity $=$ acceleration $\times$ time ($v = at$), solving the energy equation in terms of acceleration and time yields an equation where kinetic energy equals $\frac{1}{2}m(a^2t^2)$. Using this formulation, it is then apparent that the work performed on the body from the kinetic energy of vibration depends on the square of the vibration acceleration and the square of the vibration exposure duration (time). This is important, because it means that a small reduction in acceleration and/or exposure time can lead to relatively large reductions in energy absorbed by the worker. For example, a 10% reduction in either vibration acceleration or time of exposure to the vibration can result in a 19% reduction in the energy applied to the body. Reducing by 10% both the vibration acceleration and time of exposure to the vibration can result in a 34% reduction in the energy or work applied to the body. Inexpensive accommodations can therefore have a big effect.

There is a clear relationship between vibration environments and low back disorders. Jolt/impact is emerging as an important factor and is a subset of sudden/unexpected load conditions. More case-controlled epidemiologic studies in which relevant occupational exposures are quantified are needed. The relationship between intrinsically and extrinsically applied mechanical stresses, and the accompanying hard and soft tissue deformations, both acute and chronic, still needs greater definition. It is particularly important that seated vibration exposure not be used as a "work-hardening" treatment modality in trying to return someone with low back trouble to work.[100] It is also very important to realize that because the vibration standards/guidelines have limitations, it is critical to monitor the health and proficiency of people in the whole-body impact/vibration environment.

HAND–ARM VIBRATION

Adverse health effects from exposure to HAV have been recognized since 1911, when Loriga[101] reported "dead fingers" among the Italian miners who used pneumatic tools. Such tools had been introduced into the French mines in 1839 and were being extensively used by 1890. In the USA pneumatic tools were first introduced into the limestone quarries of Bedford, Indiana in about 1886. In

1918, Dr. Alice Hamilton and her colleagues subsequently investigated the health hazard from their use.[102] Since then there have been many reports of health hazards arising from the use of hand-held vibratory tools in the literature from all over the world.[6]

It is now evident that adverse health effects can result from almost any vibrating source if the vibration is sufficiently intense over the frequency range 6–5000 Hz for a significantly long period of time. The most important sources of HAV are mechanically driven tools (pneumatic, hydraulic, electric motor and gasoline-engine powered), e.g. grinders, drills, fettling tools, jack hammers, riveting guns, and chain saws. Users of brush saws and hedge cutters, and speedway (dirt-track) motorbike riders, are also at risk.

The predominant health effect is known as the hand–arm vibration syndrome (HAVS),[103,104] a disease entity with the following separate peripheral components:

- Circulatory disturbances: cold-induced vasospasm with local finger blanching—"white finger"
- sensory and motor disturbances: numbness, loss of finger coordination and dexterity, clumsiness and inability to perform intricate tasks
- Musculoskeletal disturbances: muscle, bone and joint disorders.

The vasospasm, also known as Raynaud's phenomenon, is precipitated by exposure to cold and/or damp conditions, and sometimes vibration exposure itself. The time period between first exposure to HAV and the onset of fingertip blanching is termed the latent interval. It may range from 1 month to 30 years, depending on the intensity of vibration entering the hand and the susceptibility of the worker. The blanching is restricted initially to the tips of one or more fingers, but progresses to the base of the fingers as the vibration exposure time increases. The thumbs are usually the last to be affected.

The blanching is accompanied by numbness, and as the circulation to the digits returns there is usually tingling and pain. Tingling and paresthesia may precede the onset of blanching in many subjects. These sensory symptoms and signs may be the predominant complaint in some patients, and their recognition as a distinct entity led to the revision of the Taylor–Pelmear classification for assessment of HAVS devised in 1968.[105] It was replaced in 1985 by the Stockholm classification,[106,107] based on the subjective history supported by the extensive results of a battery of clinical tests, to stage the severity of disease (Tables 4.2 and 4.3). The vascular and sensorineural symptoms and signs are evaluated separately, and for both hands individually.

In advanced cases, the peripheral circulation becomes very sluggish, giving a cyanotic tinge to the skin of the digits, while in the very

Table 4.2 The Stockholm Workshop Scale for the classification of cold-induced Raynaud's phenomenon in the hand–arm vibration syndrome.

Stage	Grade	Description
0		No attacks
1	Mild	Occasional attacks affecting the tips of one or more fingers
2	Moderate	Occasional attacks affecting the distal and middle fingers (rarely also proximal) phalanges of one or more fingers
3	Severe	Frequent attacks affecting all phalanges of most fingers
4	Very severe	As in stage 3, with trophic skin changes in the fingertips

The staging is made separately for each hand. In the evaluation of the subject, the grade of the disorder is indicated by the stages of both hands and the number of affected fingers on each hand, e.g. "2L(2)/1R(1)", "–/3R(4)".

Table 4.3 The Stockholm Workshop Scale for the classification of sensorineural effects of the hand–arm vibration syndrome.

Stages	Symptoms
OSN	Exposed to vibration but no symptoms
1SN	Intermittent numbness, with or without tingling
2SN	Intermittent or persistent numbness, reduced sensory perception
3SN	Intermittent or persistent numbness, reduced tactile discrimination and/or manipulative dexterity

The sensorineural stage is to be established for each hand.

severe cases trophic skin changes (gangrene) will appear at the fingertips. The toes may be affected if directly subjected to vibration from a local source, e.g. vibrating platforms, or they may be affected by reflex spasm in subjects with severe hand symptoms. Reflex sympathetic vasoconstriction may also account for the increased severity of noise-induced hearing loss in HAVS subjects.[108,109]

In addition to tactile, vibrotactile and thermal threshold impairment, which may vary from subject to subject, impairment of grip strength is a common symptom in longer-exposed workers.[110,111] Discomfort and pain in the upper limbs is also a common complaint. Bone cysts and vacuoles, although often reported, are more likely to be caused by biodynamic and ergonomic factors.[112,113]

Carpal tunnel syndrome (CTS) an entrapment neuropathy affecting the median nerve at the wrist, is often associated with HAVS.[114–116] Usually it is due to ergonomic stress factors, including the constant repetitive nature of the work, grip force, and mechanical stresses, e.g. torque and posture. The pathophysiology, etiology, clinical picture and treatment of CTS have been well reviewed by Carragee and Hentz.[117] When vibration is the primary cause of the median nerve neuropathy, the edematous reaction in the adjacent tissues and the nerve sheath compress the central axon.[118] The median nerve is affected together with the ulnar in two-thirds of the cases. Rarely, the ulnar nerve may be affected alone.[119]

Whether smoking accelerates the onset of HAVS has not yet been demonstrated conclusively, but this aggravating factor has been shown to increase the risk in several studies.[120,121]

DIAGNOSIS

The diagnosis of HAVS is based on a history of HAV exposure and the exclusion of other causes of Raynaud's phenomenon, i.e. primary Raynaud's phenomenon (Raynaud's disease or constitutional white finger), local trauma to the digital vessels, thoracic outlet syndrome, drugs and peripheral vascular and collagen diseases, including scleroderma. The diagnosis of HAVS is confirmed and the severity assessed by stage from the results of a battery of laboratory tests.[119,122–124]

Vascular tests should include some or all of the following: Doppler studies, plethysmography, finger systolic pressure measurement, and cold water provocation tests to verify that vasospasm occurs on cold exposure. Subjective sensorineural tests should include, for example, depth sense and two-point discrimination, fingertip vibration threshold measurement, thermal hot/cold perception, and current perception threshold tests. Objective nerve conduction tests should be undertaken to confirm the presence and severity of the neuropathy, which in HAVS normally affects both median and ulnar nerves. When the median nerve myelinated fibers are involved at the wrist level, there can be confusion with CTS nerve entrapment, because the symptoms and signs are similar.

PATHOPHYSIOLOGY

The pathophysiology of HAVS has been well reviewed by Gemne.[125] The basic mechanism is not yet fully understood. Owing to the mechanical stimulus, specific anatomic changes occur in the digital vessels, i.e. vessel wall hypertrophy and endothelial cell damage. In the initial stages there is extrusion of fluid into the tissues. This edema, together with the subsequent spasmodic ischaemia from the cold induces vasospasm, damages the mechanoreceptor nerve endings and non-medullated fibers. Subsequently, a demyelinating neuropathy of the peripheral nerve trunks develops.

The vasular response to cold is complex because, in addition to the diversity of receptor systems (adrenergic, cholinergic, purinergic, and serotonergic), there are several subtypes of specific receptors. The differential distribution and functional significance of the various receptor types is largely unknown. It is probable that the cold-induced pathologic closure of the digital arteries and end vessels is mainly mediated by alpha-2 adrenoreceptors in the wall of the arterioles and veins. It has been demonstrated that the alpha-2 receptors are more receptive to the cold stimulus. In HAVS, it is postulated that there is selective damage of alpha-1 receptors, so the cold stimulus is more effective. While arterial spasm is necessary to stop the bloodflow, vasospasm in the skin arterioles is essential to produce the blanching.

Cold, as well as vessel wall injury, causes platelet aggregation. The subsequent release of serotonin (5-hydroxytryptamine, 5-HT) promotes further release of 5-HT from the platelets, and the increased concentration stimulates smooth muscle to contract. Besides promoting contraction, serotonin may also contribute to vasodilation by inducing the release of endothelium-derived relaxing factor (ERDF) and prostacyclin from the endothelial cells. Acetylcholine and its agonist methacholine, acting through the muscarine receptors, also release ERDF, while nitric oxide and its agonists, nitroprusside and nitroglycerine, release prostacyclin. The prostacyclin and ERDF so released, besides inhibiting platelet aggregation, stimulate the production of cyclic adenosine monophosphate (cAMP) and cyclic guanosine monophosphate (cGMP) in the smooth muscle cell. The latter substances inhibit calcium utilization by the smooth muscle cells so they do not contract. A delicate balance between smooth muscle contraction and relaxation is produced by these mechanisms interacting simultaneously in the normal person.

TREATMENT AND MANAGEMENT

To reduce the frequency of blanching attacks, the central body temperature must be maintained and cold exposure must be avoided. Mittens rather than gloves should be worn if dexterity is not an issue. Discontinuation of smoking is an essential requirement, because of the adverse effect of nicotine and carbon monoxide on the digital arterial system.

To reverse the pathology and achieve recovery, further vibration exposure must be avoided. In the worker aged less than 40 years, this alone may be sufficient. If avoidance of HAV is not possible, a modified work routine should be practiced to reduce vibration exposure and slow the progression of this condition.

Recent advances in drug therapy have focused on three areas: (1) use of calcium channel antagonists to produce peripheral vasodilation, (2) use of drugs to reduce platelet aggregation in combination with the above, and (3) drugs to reduce blood viscosity and emboli formation. The preferred calcium channel antagonist is a slow-release product. In the more severe cases, additional medications are often prescribed for platelet deaggregation.

The necessity for drug therapy increases with the severity of the symptoms and the age of the subject. The results are encouraging for resolution of the vascular symptoms, particularly if vibration exposure is avoided. Unfortunately, drug intolerance to the earlier calcium channel antagonists caused some young patients to abandon their use prematurely. Where this has happened, the newer products should be tried. Recovery from the sensorineural effects has not been reported. It remains to be seen whether an improvement in the peripheral circulation will result in reversal of the sensorineural symptomatology. For the most part, HAVS is irreversible; therefore, prevention measures are the watchword in order to minimize the effects of HAV exposure.

When patients are thought to be suffering from HAVS, their employers should be advised to assess the work situation and introduce preventive procedures. All workers should be advised of the potential vibration hazard and receive training on the need to service their tools regularly, to grip the tools as lightly as possible within the bounds of safety, to use the protective clothing and equipment provided, to attend for periodic medical surveillance, and to report all signs and symptoms of HAVS as soon as they develop.

HAND–ARM VIBRATION CONTROL

As with WBV control, once vibration measurements have been made and the data have been evaluated with regard to the appropriate standard and it is determined that vibration control is necessary, usually a series of multiple control steps are taken, depending on the problem.[6,7,13,126]

1. Use only ergonomically correct anti-vibration (A/V) tools wherever possible.
2. If possible, do not use materials that wrap around tool handles and claim to significantly reduce HAV exposure. Usually, they increase the tool handle diameters, causing the worker to grip the tool handle more forcefully, thereby compressing the materials and reducing their limited usefulness. It is far better to use a well-designed A/V tool.
3. Use good-fitting "full finger protected" A/V gloves; do not use exposed finger gloves that only protect the palm. HAVS nearly always begins at the fingertips, advancing towards the root; only full-finger protected gloves offer the desired protection.
4. In factory situations, use suspended "tool balancers" to remove the weight of the tool from the operator.
5. Workers are advised to do the following as good work practices:
   (a) Let the tool do the work, gripping it as lightly as possible consistent with safe tool handling.
   (b) Operate the tool only when necessary and at reduced speeds if possible.
   (c) Properly maintain hand tools and replace as necessary.
   (d) Do not smoke, since nicotine, vibration and cold all constrict the blood vessels.
   (e) Keep your hands and body warm and dry.
   (f) Consult a physician if signs of digit tingling, numbness, or blanching occur.
   (g) Medical prescreening of workers is advised to minimize the risk to idiopathic Primary Raynaud's disease sufferers and others from operating hand tools that could exacerbate such a pre-existing condition.
   (h) Workers and others need to be made aware of this problem and its signs and symptoms.
6. Recently, some vibrating tool manufacturers have begun placing warning labels on their tools and in their instruction books.
7. Use HAV standards and limits as critical adjuncts to control measures.
8. Institute a surveillance program.

Finally, for HAV exposures, these controls may need additional engineering redesign methods using vibration damping techniques (i.e. converting vibration into a small amount of heat due to the deformation of a viscoelastic material) and/or vibration isolation (i.e. intentionally mismatching the vibration pathway between the vibrating source and the worker receiving this vibration) to be effective.

REFERENCES

1. Goldman DE, von Gierke HE. The effects of shock and vibration on man. In: Harris CM, Crede CE, eds. *Shock and vibration handbook* 2nd edn. New York: McGraw-Hill Book Company, 1976: 44-1–44-57.
2. Griffin MJ. *Handbook of human vibration.* London: Academic Press Limited, Harcourt Brace Jovanovich, 1990.
3. Abrams R, ed. *Seminars in perinatology*, Vol. 14, Special Supplement. Philadelphia: Saunders, 1990.
4. Abrams R, Wasserman DE. Occupational vibration during pregnancy. *Am, J. Obstet Gynecol* 1991; 164:1152.
5. Seidel H. Selected health risks caused by long-term whole body vibration. *Am. J. Indust Med* 1993; 23:589–604.
6. Pelmear P, Wasserman D, eds. *Hand–arm vibration; a comprehensive guide for occupational health professionals*, 2nd edn. Beverly Farms, Mass: OEM Publishers, 1998.
7. Wasserman D. *Human aspects of occupational vibration.* Amsterdam: Elsevier, 1987.
8. Wasserman DE, Wilder DG. Vibrometry. In: *The occupational ergonomics handbook.* Karwowski W, Marras WS, eds. Boca Raton: CRC Press, 1999:1693–705.
9. American Conference of Governmental Industrial Hygienists. Physical agents: Threshold limit values: ergonomic: whole-body vibration. *2001 TLVs and BEIs threshold limit values for chemical substances and physical agents and biological exposure indices.* Cincinnati, Ohio: American Conference of Governmental Industrial Hygienists, 2001: 122–9.
10. American National Standards Institute. ANSI: S3.18. *Guide for the evaluation of human exposure to whole-body vibration.* New York: ANSI, 1979.
11. American Conference of Governmental Industrial Hygienists. Physical agents: threshold limit values: ergonomic: hand–arm (segmental) vibration. In *2001 TLVs and BEIs threshold limit values for chemical substances and physical agents and biological exposure indices.* Cincinnati, Ohio: American Conference of Governmental Industrial Hygienists, 2001:118–21.
12. American National Standards Institute. ANSI: S3.34-1986. *Guide for the measurement and evaluation of human exposure to vibration transmitted to the hand.* New York: ANSI, 1986.
13. National Institute for Occupational Safety and Health: Criteria for a Recommended Standard. *Occupational exposure to hand–arm vibration.* DHHS (NIOSH) Publication No. 89-106. Cincinnati: NIOSH, 1989.
14. International Organization for Standardization. *Mechanical vibration and shock—evaluation of human exposure to whole body vibration.* Part 1. *General requirements.* International Standard 2631-1:1997. Geneva: ISO, 1997.
15. Griffin MJ. A comparison of standardized methods for predicting the hazards of whole-body and repeated shocks. *J. Sound Vibration* 1998; 215(4):883–914.
16. Cohen HH, Wasserman DE, Hornung RW. Human performance and transmissibility under sinusoidal and mixed vertical vibration. *Ergonomics* 1977; 20(3):207–16.
17. Fethke N, Wilder DG, Spratt K. Seated trunk-muscle response to impact. Adelaide, Australia: International Society for the Study of the Lumbar Spine, 9–13 April 2000; Presentation #51. *Society administration office*, Toronto.
18. Morrison J, Robinson D, Roddan G, et al. *Development of a standard for the health hazard assessment of mechanical shock and repeated impact in army vehicles: Phase 5. Report CR-96-1. Fort Rucker AL: US Army Aeromedical Research Laboratory, 1997.*
19. Robinson DG. *The dynamic response of the seated human to mechanical shock.* PhD Dissertation, School of Kinesiology, Simon Fraser University, Burnaby, British Columbia, 1999.
20. Sandover JI. High acceleration events: an introduction and review of expert opinion. *J. Sound Vibration* 1998; 215(4):927–45.
21. Pelmear P, Leong D, Taylor W, et al. Measure-

ment of vibration of hand-tools: weighted or unweighted? *J Occup Med* 1989; 31:903.

22. Starck J, Pekkarinen J, Pyykko I. Physical characteristics of vibration in relation to vibration-induced white finger. *Am Indust Hygiene Assoc J* 1990; 51:179.
23. Wasserman D. The control aspects of occupational hand–arm vibration. *Appl Indust. Hygiene* 1989; 4:22.
24. Cats-Baril WL, Frymoyer JW. The economics of spinal Disorders. In: Frymoyer JW, Ducker TB, Hadler NM, et al, eds. *The adult spine*. New York: Raven Press, 1991: 85–105.
25. Bovenzi M, Betta A. Low-back disorders in agricultural tractor drivers exposed to whole-body vibration and postural stress. *Appl Ergonomics* 1994; 25(4):231–41.
26. Bovenzi M, Zadini A. Self-reported low back symptoms in urban bus drivers exposed to whole-body vibration. *Spine* 1992; 17(9):1048–59.
27. Dupuis H, Zerlett G. *The Effects of whole-body vibration*. Berlin: Springer-Verlag, 1986.
28. Magnusson M, Wilder DG, Pope MH, Hansson T. Investigation of the long-term exposure to whole-body vibration: a 2-country study. Winner of the Vienna Award for Physical Medicine. *Eur J Physical Med Rehabilitation* 1993; 3(1):28–34.
29. Sandover J. Dynamic loading as a possible source of low-back disorders. *Spine* 1983; 8:652–58.
30. Bongers PM., Boshuizen HC. *Back disorders and whole-body vibration at work*. Dissertation, University of Amsterdam, 1990.
31. Commission of the European Communities (1993). Proposal for a Council Directive on the minimum health and safety requirements regarding the exposure of workers to the risks arising from physical agents. *Official J Eur Communities* 1993; No. C 77:12–29.
32. Hulshof C, van Zanten BV. Whole-body vibration and low back pain: a review of epidemiologic studies. *Int Arch Occup Environ Health* 1987; 50:205–20.
33. Wickström B-O, Kjellberg A, Landström U. Health effects of long-term occupational exposure to whole-body vibration: a review. *Int J Indust Ergonomics* 1994; 14:273–92.
34. Dupuis H. Medical and occupational preconditions for vibration-induced spinal disorders: occupational disease no. 2110 in Germany. *Int Arch Occup Environ Health* 1994; 66(5):303–8.
35. National Center for Health Statistics. *Current estimates from the National Health Interview Survey, 1988*. Series 10, no. 173. DHHS publication PHS 89-1501. Hyattsville, MD: National Center for Health Statistics, 1989.
36. National Center for Health Statistics. *Health conditions among the currently employed: United States. 1988*. Series 10, no. 186, DHHS publication PHS 93-1514. Hyattsville, MD: National Center for Health Statistics, 1993.
37. Behrens V, Seligman P, Cameron L, Mathias T, Fine L. The prevalence of back pain, hand discomfort, and dermatitis in the US working population. *Am J Public Health* 1994; 84(11):1780–85.
38. Hoogendoorn WE, van Poppel MNM, Bongers PM, Koes BW, Bouter LM. Physical load during work and leisure time as risk factors for back pain. *Scand J Work, Environ Health* 1999; 25(5)387–403.
39. Bovenzi M, Hulshof CTJ. An updated review of epidemiologic studies on the relationship between exposure to whole-body vibration and low back pain. *J Sound Vibration* 1998; 215(4):595–611.
40. Wilder DG. The biomechanics of vibration and low back pain. *Am J Indust Med* 1993; 23(4):577–88.
41. Wilder DG, Pope MH. Epidemiological and etiological aspects of low back pain in vibration environments—an update. *Clin Biomechanics* 1996; 11(2):61–73.
42. Wilder DG, Woodworth BB, Frymoyer JW, Pope MH. Vibration and the human spine. *Spine* 1982; 7(3):243–54.
43. Levin S. The icosahedron as a biological support system. Toronto: International Society for the Study of the Lumbar Spine, 6–10 June 1982. Society administration office, Toronto.
44. Snelson K. *Continuous tension, discontinuous compression structures*. United States patent #3,169,611 of 16 February 1965, filed 14 March 1960.
45. Dupuis H. Belastung durch mechanische

Schwingungen und moegliche Gesundheitsschäedigungen im Bereich der Wirbelsaule. *Fortschritte Med* 1974; 92(14):618-20.

46. Panjabi MM, Andersson GBJ, Jorneus L, et al. *In vivo* measurement of spinal column vibrations. *J Bone Joint Surg* 1986; 68A(5):695–703.
47. Pope MH, Svensson M, Broman H, et al, Mounting of the transducer in measurements of segmental motion of the spine. *J Biomech* 1986; 19(8):675–77.
48. Quandieu P, Pellieux L. Study *in situ et in vivo* of the acceleration of lumbar vertebrae of a primate exposed to vibration in the Z-axis. *J Biomech* 1982; 15:985–1006.
49. Slonim AR. *Some vibration data on primates implanted with accelerometers on the upper and lumbar spine: methodology and results in rhesus monkeys*. Technical Report TR-81-153. Wright-Patterson Air Force Base, Dayton: Air Force Aerospace Medical Research Laboratory, 1983.
50. Slonim AR. Some vibration data on primates implanted with accelerometers on the upper thoracic and lower lumbar spine: results in baboons. Technical Report TR-83-091. Wright-Patterson Air Force Base, Dayton: Air Force Aerospace Medical Research Laboratory, 1984.
51. Coermann RR. *Mechanical vibrations*. In: Occupational safety and health series. Geneva, Switzerland. The International Labour Office, 1970; 21:17–41.
52. Zagorski J, Jakubowski R, Solecki L, et al. Studies on the transmissions of vibrations in human organism exposed to low-frequency whole-body vibration. *Acta Physiol Polonica* 1976; 27:347–54.
53. Wilder DG, Frymoyer JW, Pope MH. The effect of vibration on the spine of the seated individual. *Automedica* 1985; 6:5–35.
54. Pope MH, Kaigle AM, Magnusson M, et al. Intervertebral motion during vibration. *J Eng Med, Proc Inst Mech Eng* 1991; 205:39–44.
55. Wilder DG. *On loading of the human lumbar invertebral motion segment*. Dissertation for degree of Doctor of Philosophy, Mechanical Engineering, Civil and Mechanical Engineering Department, University of Vermont, 1985. Abstract: *Dissertation Abstracts International* 1986; 46(12): 4328-B. Manuscript # DA8529728. Ann Arbor, University Microfilms International 1986.
56. Seroussi RE, Wilder DG, Pope MH. Trunk muscle electromyography and whole body vibration. *J Biomech* 1989; 22(3):219–29.
57. Seidel H, Heide R. Long-term effects of whole-body vibration: a critical survey of the literature. *Int Arch Occup Environ Health* 1986; 58:1–26.
58. Pope M, Wilder D, Seroussi R. Trunk muscle response to foot support and corset wearing during seated, whole-body vibration. *Trans Orthop Res Soc* 1988; 13:374.
59. Magnusson ML, Aleksiev A, Wilder DG, et al. European Spine Society—the AcroMed Prize for Spinal Research 1995. Unexpected load and asymmetric posture as etiologic factors in low back pain. *Eur Spine J* 1996; 5(1):23–35.
60. Solomonow M, Zhou B-H, Harris M, Lu Y, Baratta RV. The ligamento-muscular stabilizing system of the spine. *Spine* 1998; 23(23):2552-62.
61. Solomonow M, Zhou B-H, Baratta RV, Lu Y, Harris M. Biomechanics of increased exposure to lumbar injury caused by cyclic loading. Part 1. Loss of reflexive muscular stabilization. 1999 Volvo Award Winner in Biomechanical Studies. *Spine* 24(23):2426–34.
62. Solomonow M, Zhou B-H, Baratta RV, Lu Y, Zhu M, Harris M. Biexponential recovery model of lumbar viscoelastic laxity and reflexive muscular activity after prolonged cyclic loading. *Clin Biomechanics* 2000; 15:167–75.
63. Gedalia U, Solomonow M, Zhou B-H, Baratta RV, Lu Y, Harris M. Biomechanics of increased exposure to lumbar injury caused by cyclic loading. Part 2. Recovery of reflexive muscular stability with rest. *Spine* 1999; 24(23):2461–7.
64. Wilder DG, Tranowski JP, Novotny JE, et al. *Vehicle seat optimization for the lower back* Marseilles: International Society for the Study of the Lumbar Spine, 15–19 June 1993. Society administration office: Toronto.
65. Brinckmann P, Wilder DG, Pope MH. Effects of repeated loads and vibrations. In: Weisel SW, Weinstein JN, Herkowitz H, Dvorak J,

Bell G, eds. *The lumbar spine*, 2nd edn. Philadelphia: International Society for the Study of the Lumbar Spine, WB Saunders Co. 1996:181–202.

66. Hansson TH, Keller TS, Holm S. The load on the porcine lumbar spine during seated whole body vibrations. *Orthop Trans* 1988; 12(1):85.
67. Adams MA, Hutton WC. The effect of fatigue on the lumbar intervertebral disc. *Orthop Trans* 1983; 7(3):46l.
68. Adams MA, Hutton WC. Gradual disc prolapse. *Spine* 1985; 10(6):524–31.
69. Brown T, Hansen RJ, Yorra AJ. Some mechanical tests on the lumbosacral spine with particular reference to the intervertebral discs: a preliminary report. *J Bone Joint Surg* 1957; 39A:1135–65.
70. Liu YK, Goel VK, DeJong A et al. Torsional fatigue of the lumbar intervertebral joints. *Orthop Trans* 1983; 7(3):461.
71. Wilder DC, Woodworth BB, Frymoyer JW, et al. Energy absorption in the human spine. In: Paul I. ed. *Proceedings of The Eighth Northeast (New England), 1980 Bioengineering Conference*, 1980. Cambridge: MIT; 1980:443–5.
72. Wilder DG, Pope, MH, Frymoyer JW. Cyclic loading of the intervertebral motion segment. In Hansen EW, ed. *Proceedings of the Tenth Northeast Bioengineering Conference* 1982. New York: IEEE, 1982:9–11.
73. Wilder DG, Pope MH, Frymoyer JW. The biomechanics of lumbar disc herniation and the effect of overload and instability. American Back Society Research Award. *J Spinal Disord* 1988: 1(1):16–32.
74. International Standards Organization. *Evaluation of human exposure to whole-body vibration*. ISO 2631/1-1985. Geneva: ISO 1985.
75. Cholewicki J, McGill SM. Mechanical stability of the *in vivo* lumbar spine: implications for injury and chronic low back pain. New Concepts and Hypotheses. *Clin Biomechanics* 1996; 11(1):1–15.
76. Cholewicki J, Panjabi MM, Khachatryan A. Stabilizing function of trunk flexor-extensor muscles around a neutral spine posture. *Spine* 1997; 22(19):2207-12.
77. Cholewicki J, Juluru K, McGill SM. Intra-abdominal pressure mechanism for stabilizing the lumbar spine. *J Biomechanics* 1999; 32:13–17.
78. Crisco JJ, Panjabi MM. The intersegmental and multisegmental muscles of the lumbar spine. A biomechanical model comparing lateral stabilizing potential. *Spine* 1991;16(7):793–9.
79. Crisco JJ, Panjabi MM, Yamamoto I, Oxland TR. Euler stability of the human ligamentous lumbar spine. Part II: Experiment. *Clin Biomechanics* 1992; 7:27–32.
80. Gardner-Morse MQ, Stokes IAF, Laible JP. Role of muscles in lumbar spine stability in maximum extension efforts. *J Orthop Res* 1995; 13:802–8.
81. Gardner-Morse MG, Stokes IAF. The effects of abdominal muscle coactivation on lumbar spine stability. *Spine* 1998; 23(1):86–92.
82. McGill SM. The biomechanics of low back injury: implications on current practice in industry and the clinic. ISB Keynote Lecture. *J Biomechanics* 1997; 30(5):465-75.
83. Quint U, Wilke H-J, Shirazi-Adl A, Parnianpour M, Löer F, Claes LE. Importance of the intersegmental trunk muscles for the stability of the lumbar spine. A biomechanical study *in vitro. Spine* 1998; 23(18):1937–45.
84. Manning DP, Shannon HS. Slipping accidents causing low-back pain in a gearbox factory. *Spine* 1981; 6(1):70–2.
85. Manning DP, Mitchell RG, Blanchfield LP. Body movements and events contributing to accidental and non accidental back injuries. *Spine* 1984; 9(7):734–9.
86. Marras WS, Rangarajulu SL, Lavender SA. Trunk loading and expectation. *Ergonomics* 1987; 30:551–62.
87. Mannion AF, Adams MA, Dolan P. Sudden and unexpected loading generates high forces on the lumbar spine. *Spine* 2000: 25(7):842–52.
88. Aleksiev A, Pope MH, Hooper D, et al. Pelvic unlevelness in chronic low back pain patients—Biomechanics and EMG time–frequency analyses. Recipient of the 1995 Vienna Award in Physical Medicine and Rehabilitation. *Eur J Physical Med Rehabilitation* 1996; 6(1):3–16.
89. Schumacher C, Wilder DG, Goel VK, Spratt

K. Back muscle response to sudden load with a modified lumbar support. Adelaide, Australia: International Society for the Study of the Lumbar Spine, 9–13 April 2000; Presentation #277. Society administration office, Toronto.

90. Wilder DG, Aleksiev A, Magnusson M, Pope MH, Spratt K, Goel V. Muscular response to sudden load: a tool to evaluate fatigue and rehabilitation. *Spine* 1996; 21(22):2628–39.
91. Wilson JR. Participation—a framework and a foundation for ergonomics? *J Occup Psychol* 1991; 64:67–80.
92. Wilder DG. Validation of forklift ride reports. Adelaide, Australia: International Society for the Study of the Lumbar Spine, 9–13 April 2000; Presentation #184. Society administration office, Toronto.
93. Magnusson M, Hansson T, Broman H. Back muscle fatigue and whole body vibrations. *Ortho Trans* 1988; 12(3):598.
94. Schumacher C, Wilder DG, Spratt K. Erector spinae response to central and offset impact moments. Adelaide, Australia: International Society for the Study of the Lumbar Spine, 9–13 April 2000; Presentation #276. Society administration office, Toronto.
95. Thomas JS, Lavender SA, Corcos DM, Andersson GB. Effect of lifting belts on trunk muscle activation during a suddenly applied load. *Human Factors* 1999; 41(4):670–6.
96. Udo H, Yoshinaga F, Tanida H, et al. The effect of a preventive belt on the incidence of low-back pain (Part III): investigation in crane work. *J Sci Labour* 1993; 69(1):10–21.
97. National Safety Council. *Accident facts 1993 Edition*. Ithaca, Illinois: NSC, 1993: 38, 42, 49.
98. National Institute for Occupational Safety & Health. *Vibration syndrome*, Current Intelligence Bulletin #38. Cincinnati: DHHS/NIOSH Pub. #83-110, 1983.
99. Wald P. Personal communication, 1995.
100. Wasserman D, Wilder D, Pope M, Magnusson M, Aleksiev A, Wasserman J. Whole-body vibration exposure and occupational work hardening. *J Occup Environ Med* 1997; 39(5):403–07.
101. Loriga G. Il Lavoro Con i Martellie Pneumatici. *Boll Inspett Lovoro* 1911; 2:35–60.
102. Hamilton A. *A study of spastic anaemia in the hands of stonecutters*. Ind Accident Hyg Services, Bulletin 236, No 19, US Department of Labor, Bureau of Labor Statistics, 1918:53–66.
103. Gemne G, Taylor W, eds. Hand-arm vibration and the central nervous system. *J Low Freq Noise Vib* 1983; XI.
104. Brammer AJ, Taylor W, eds. *Vibration effects on the hand and arm in industry*. New York: John Wiley & Sons, 1982.
105. Taylor W, Pelmear PL, eds. *Vibration white finger in industry*. London: Academic Press, 1975:XVII–XXII.
106. Brammer AJ, Taylor W, Lundborg G. Sensorineural stages of the hand–arm vibration syndrome. *Scand J Work Environ Health* 1987; 13:279–83.
107. Gemne G, Pyykkö I, Taylor W, et al. The Stockholm Workshop scale for the classification of cold-induced Raynaud's phenomenon in the hand–arm vibration syndrome (revision of the Taylor–Pelmear scale). *Scand J Work Environ Health* 1987; 13:279–283.
108. Iki M, Kurumantani N, Satoh M, et al. Hearing of forest workers with vibration induced white finger: a five year follow-up. *Int Arch Occup Environ Health* 1989; 61:437–42.
109. Pyykkö I, Starck J, Färkkilä M, et al. Hand–arm vibration in the etiology of hearing loss in lumberjacks. *Br J Indust Med* 1981; 38:281–9.
110. Färkkilä M. Grip force in vibration disease. *Scand J Work Environ Health* 1978; 4:159–66.
111. Färkkilä M, Aatola S, Stark J, et al. Hand-grip force in lumberjacks: two year follow-up. 1986; *Int Arch Occup Environ Health* 1986; 58:203–8.
112. Gemne G, Saraste H. Bone and joint pathology in workers using hand-held vibratory tools—an overview. *Scand J Work Environ Health* 1987; 13:290–300.
113. James PB, Yates JR, Pearson JCG. An investigation of the prevalence of bone cysts in hands exposed to vibration. In: Taylor W, Pelmear PL eds. *Vibration white finger in industry*. New York: Academic Press, 1975: 53–51.
114. Färkkilä M, Koskimies K, Pyykkö I, et al. Carpal tunnel syndrome among forest workers. In: Okada A, Taylor W, Dupuis H, eds.

Hand–arm vibration. Kanazawa: Kyoei Press, 1990: 263–65.

115. Koskimies K, Färkkilä M, Pyykkö I, et al. Carpal tunnel syndrome in vibration disease. *Br J Indust Med* 1990; 47:411–16.
116. Wieslander G, Norback D. Gothe CJ, et al. Carpal tunnel syndrome (CTS) and exposure to vibration, repetitive wrist movements, and heavy manual work. A case-referent study. *Br J Indust Med* 1989; 46:43–7.
117. Carragee EJ, Hentz VR. Repetitive trauma and nerve compression. *Orthop Clin North Am* 1988; 19(1):157–64.
118. Lundborg G, Dahlin LB, Danielsen N, et al. Intraneural edema following exposure to vibration. *Scand J Work Environ Health* 1987; 13(4 Special lssue):326–9.
119. Pelmear PL, Taylor W. Clinical evaluation. In: Pelmear PL, Taylor W, Wasserman DE, eds. *Hand–arm vibration: a comprehensive guide.* New York: Van Nostrand Reinhold, 1992: 77–91.
120. Ekenvall L, Lindblad LE. Effect of tobacco use on vibration white finger disease. *J Occup Med* 1989; 31(1):13–16.
121. Virokannas H, Anttonen H, Pramila S. Combined effect of hand–arm vibration and smoking on white finger in different age groups. *Arch Complex Environ Studies* 1991; 3(1–2):7–12.
122. McGeoch KL, Taylor W, Gilmour WH. The use of objective tests as an aid to the assessment of hand–arm vibration syndrome by the Stockholm classification. In: Dupuis H, Christ E, Sandover J, Taylor W, Okada A, eds. *Proceedings of the 6th International Conference on Hand–Arm Vibration.* Bonn 1992:783–92.
123. Pelmear PL, Wong L, Dembek B. Laboratory tests for the evaluation of hand–arm vibration syndrome. In: Dupuis H, Christ E, Sandover J, Taylor W, Okada A, eds. *Proceedings of the 6th International Conference on Hand–Arm Vibration*. Bonn, 1992.
124. Pelmear PL, Taylor W. Hand–arm vibration syndrome—clinical evaluation. *J Occup Med* 1991; 33(11):1144–9.
125. Gemne G. Pathophysiology and pathogenesis of disorders in workers using hand-held vibratory tools. In: Pelmear PL, Taylor W, Wasserman DE, eds. *Hand–arm vibration: a comprehensive guide.* New York: Van Nostrand Reinhold, 1992; 41–76.
126. Wasserman D. To weight or not to weight...that is the question. *J Occup Med* 1989; 31:909.

5

MECHANICAL ENERGY

James Kubalik, M.S., C.S.P.*

OCCUPATIONAL SETTING

Mechanical energy impacting on the human body is the most frequent cause of direct physical injuries (Table 5.1). Our workplaces and individual workspaces are dynamic and have both direct and indirect exposure potential. The management of these interactions will determine whether the outcome is efficient production and productivity or a failure resulting in an employee injury and escalating costs to a company.

Our daily work activities are a series of interactions and physical contacts with equipment, work surfaces, chemicals and materials. Viewing the workplace and an individual's workspace as a 360° environment that moves, bends, twists and interacts with sources of mechanical energy provides a "graphic display" of the potential and magnitude of the exposure. With each direct physical contact and the subsequent transfer of sufficient mechanical energy, this potential can be manifest as a cut, bruise, strain, fracture, amputation or other physical injury or illness.

Perceptions that simply eliminating manual tasks will also eliminate all exposures to mechanical energy are often in error. In many cases we are substituting one source for another. Replacing tasks with tools (i.e. pallet jacks, forklifts, punch presses, computers) has also exposed workers to equipment or mechanical energy-related hazards. Often, the impacts of these changes are not proactively recognized and the trailing consequences of these changes are injuries and illnesses.

An aging workforce further compounds the results of uncontrolled exposures. Workers are less able to endure the consequences of physical and repetitive tasks (i.e. lifting, twisting, repetitive motions) and contact with equipment.

There are many sources of occupational exposure to mechanical energy. Uncontrolled and unmanaged interactions with mechanical energy cause many serious injuries and are the most common reason for a worker to seek

*Peter Wald and A.B. Barnes contributed to the previous edition of this chapter.

Physical and Biological Hazards of the Workplace, Second Edition, Edited by Peter H. Wald and Gregg M. Stave
ISBN 0-471-38647-2

Table 5.1 Types of direct injuries and their causes.

Injury type	Causes/Locations	Safeguards
Traffic accidents	Roadways Rail spurs	Bridges over crossings Signals Seat belts
Falls from heights	Platforms Walkways Aerial baskets Open structures Ladders	Railings Enclosures Fall protection systems
Slips, trips and falls	Slippery surfaces Cluttered work site	Abrasive surfaces Good housekeeping
Major crush injuries	(1) Forklifts (2) Cranes	Restraints Rollover protection Overhead guards Audible travel alarms Inspection Crane director Audible travel alarms
Explosions	(1) Flammables/combustibles (2) High-pressure steam/air/product systems	Fire protection program and systems Grounding No smoking Hydrotesting
Burns	Steam Hot surfaces Cryogens Open flame Electrical	Insulate process Insulating clothing Energy isolation (during maintenance) Electrical safety program (see also Explosions above)
Electrocutions	Any electrical processes Power tools Electric moving equipment Extension cords	Electrical safety program See chapter 17
Abrasions, lacerations and contusions	(1) Moving equipment—routine operation (2) Moving equipment—maintenance operations (3) Power and hand tools	(1) Guard openings (2) Two-handed "trip" operation (3) Enclosures (4) Interlocks (5) Automatic feeds (6) "Presence" sensors (7) Lockout/tagout
Musculoskeletal strain	Manual material handling	(1) Maintain and repair (2) Appropriate tool choice (3) Hand and eye protection (4) Solid work surface See Chapter 3

medical attention. These injuries and their causes are summarized in Table 5.1.

The magnitude and cost of these exposures cannot be underestimated. Based on the 1999 National Safety Council Injury Facts for 1998 injuries and illnesses, there were 1 833 400 days lost due to non-fatal occupational injuries. The total estimated cost of the 1998 injuries and illnesses was 125.1 billion dollars. Twenty-seven percent of the total injuries were the result of worker contact with objects and equipment. Mechanical energy was most likely a factor in these incidents.[1]

MEASUREMENT ISSUES

Mechanical hazards are measured in terms of the forces of kinematics and mechanics, which were reviewed in Chapter 1. In general, measurements are made during an accident investigation or while investigating a cluster of similar accidents. Examples of measurements include velocity, distance, acceleration, force, weight, temperature, pressure, and friction. These are all relatively simple measurements that do not require complex instruments.

There are no established "action levels" for workplace injuries. There are Occupational Safety and Health Administration (OSHA) regulatory standards that require control of hazards, many of which include exposure to mechanical energy. There are also general industry standards (i.e. ANSI) that address specific conditions and procedures, and a number of trailing indicators based on past occupational injury and illness experience such as the OSHA Log or worker's compensation injury and illness/claim experience.

Workplace safety observation techniques can be useful in identifying and proactively correcting unsafe conditions and acts. This is a growing field, and proactive safety observations are based on observing employee work habits, specifically safe and unsafe work behavior. The ratio of unsafe acts and conditions observed to the total number of "safety-related" observations can be used as a predictive safety indicator, and to develop an "action level". Once a baseline is established, any increase in the ratio is usually associated with an increase in workplace accidents. Likewise, a decreasing ratio often predicts a decrease in accidents. These observational techniques are a tool that can monitor real time safety conditions, and proactively target intervention".

EXPOSURE GUIDELINES

There are regulatory standards that indirectly cover injuries from mechanical energy. However, as stated above, there is no "action level" for workplace injuries. Injury statistics are used to calculate workers compensation insurance rates, and companies with higher injury rates will pay higher workers compensation premiums, so there is a direct financial incentive to keep these rates as low as possible. Increasingly, regulatory agencies will also target companies with high injury and illness rates.

Regulatory agency standards and recommended industry practices

There are a number of regulations that pertain to controlling worker access to moving parts of production equipment and the management of exposure to mechanical energy. The Code of Federal Regulations (CFR) contains a number of minimum safety and health standards and best practices that pertain to mechanical energy and overall employee safety and prevention of injuries. Many are contained in 29 CFR 1900–1910.999 (OSHA Act), which are summarized in Table 5.2. It is important to note that there are state OSHA programs, and their regulations may be more stringent than the Federal OSHA regulations. Readers are advised to refer to the pertinent state OSHA web sites for further information.

Designing processes to eliminate hazardous conditions and production methods are some of the best means to prevent accidents. When these conditions cannot be eliminated by design or engineering or substitution, then machine guarding and other controls should be considered. Regulations are intended to

Table 5.2 Table of Contents for 29 CFR Part 1910 Occupational Safety and Health Standards (http://www.osha=slc.gov/OshStd_data/1910.html).

1910—Table of Contents
1910 Subpart A—General (1910.1 to 1910.8)
1910 Subpart B—Adoption and Extension of Established Federal Standards (1910.11 to 1910.19)
1910 Subpart C—Adoption and Extension of Established Federal Standards (1910 Subpart C)
1910 Subpart D—Walking–Working Surfaces (1910.21 to 1910.30)
1910 Subpart E—Means of Egress (1910.35 to 1910.38)
1910 Subpart F—Powered Platforms, Manlifts, and Vehicle-Mounted Work Platforms (1910.66 to 1910.68)
1910 Subpart G—Occupational Health and Environmental Control (1910.94 to 1910.98)
1910 Subpart H—Hazardous Materials (1910.101 to 1910.126)
1910 Subpart I—Personal Protective Equipment (1910.132 to 1910.139)
1910 Subpart J—General Environmental Controls (1910.141 to 1910.147 App A)
1910 Subpart K—Medical and First Aid (1910.151 to 1910.152)
1910 Subpart L—Fire Protection (1910.155 to 1910.165)
1910 Subpart M—Compressed Gas and Compressed Air Equipment (1910.166 to 1910.169)
1910 Subpart N—Materials Handling and Storage (1910.176 to 1910.184)
1910 Subpart O—Machinery and Machine Guarding (1910.211 to 1910.219)
1910 Subpart P—Hand and Portable Powered Tools and Other Hand-Held Equipment (1910.241 to 1910.244)
1910 Subpart Q—Welding, Cutting, and Brazing (1910.251 to 1910.255)
1910 Subpart R—Special Industries (1910.261 to 1910.272 App C)
1910 Subpart S—Electrical (1910.301 to 1910.399)
1910 Subpart T—Commercial Diving Operations (1910.401 to 1910.441)
1910 Subpart Z—Toxic and Hazardous Substances (1910.1000 to 1910.1450 App B)

address proper controls should the design, engineering and substitution options not be feasible or possible.

Two standards of particular importance are the OSHA machine guarding standard (29 CFR 1910.211—Subpart O) and the OSHA lockout/tagout standard (29 CFR 1910.147—The Control of Hazardous Energy Lockout/Tagout).

These two standards outline the requirements for (1) guarding or controlling access to moving machine parts, and (2) the elimination or management of energy that can power machinery or exist within the equipment when primary sources of power are disconnected or interrupted. These regulations are excellent tools for managing mechanical energy and will be reviewed in some detail in the section on prevention later in this chapter.

PATHOPHYSIOLOGY OF INJURY

Machines are designed to apply large amounts of force to wood, metal, or other materials. When applied to bone and tissue, that same force can produce disastrous results. Direct injuries can result from dramatically different types of accidents, which were reviewed in Table 5.1.

Slips, trips and falls are common in commerce and industry, and result from the same primary and secondary causes as those in and around the home. In industry, however, there is always the potential for further trauma as a result of additional hazards in the environment, such as unguarded mechanisms, hot surfaces, or unprotected chemical processes. Electrical injuries and muscular strains due to material handling are covered in their own chapters in the physical hazards section.

TREATMENT

Medical personnel should be prepared to treat traumatic injury. If the injuries tend to recur in the same facility location, or the same type of injuries occur in different parts of the facility, targeted injury prevention programs should be implemented. This book will not attempt to deal with the treatment issues associated with direct injuries. Instead, the reader is advised to consult one of the many available texts on emergency treatment.

SURVEILLANCE PROGRAMS

Injury surveillance programs

Recognition and assessment of mechanical energy hazards are the first steps in successful management. Unfortunately, the first recognition of hazards is often realized after an injury, interruption of production or an alarming "near miss". Reactive assessments and incident investigations are an important part of identifying, measuring and improving. However, using these assessments or investigations as the primary methods to identify mechanical energy hazards is an ineffective and costly practice.

A basic review of injury and illness experience is the first step. Three readily available sources of information are: (1) the OSHA 200 Log, (2) the OSHA 101 or equivalent/accident/injury investigation reports, and (3) workers' compensation claim data. (Effective January 1, 2002 the OSHA 200 and 101 log form will be replaced with the OSHA 300 and 301 forms respectively, and new injury and illness reporting requirements will go into effect.)

These data will highlight your past experience, some of the causes associated with your experience and costs and your actions to prevent injuries and illnesses. After a thorough review of your loss experience, a series of inspections of production areas is highly recommended.

OSHA regulations require an employer with more than 10 employees (at any time during the calendar year immediately preceding the current calendar year) to keep and actively manage an OSHA 200 Log. The current year log is a real-time record of occupational injuries and illnesses (Figure 5.1).

Each OSHA recordable injury and illness must be recorded on the log within 6 working days of notification of the incident. A good-faith effort is required to maintain and update pertinent 200 Log information. Injury and illness recording requirements are contained in OSHA 29CFR Part 1904—Recording and Reporting Occupational Injuries, and additional guidance is available in the "Blue Book"—Record keeping Guidelines for Occupational Injuries and Illnesses.[2] (Note: New OSHA record-keeping regulations are due to go into effect on January 1, 2002. However, the Department will propose that the criteria for recording work-related hearing loss not be implemented for one year pending further investigation into the level of hearing loss that should be recorded as a "significant" health condition. In addition, the Department will propose a delay for one year the record keeping rule's definition of "muscloskeletal disorder" (MSD) and the requirement that employers check the MSD column on the OSHA Log. At the time of printing, a new standard had not been issued.) States with OSHA programs may also have requirements and must be included as applicable.

The 200 Log requirements and methods are outlined in the "Blue Book" and ensure a minimum of consistency in record-keeping. In addition, the Blue Book has a wealth of information on frequently asked record-keeping questions and answers. You may also contact your local OSHA consultation office for additional information and can contact Federal OSHA to request copies of any letters of interpretation on questions submitted by industry, or formally request them under the Freedom of Information Act.

Five years of updated OSHA logs must be readily available, so these documents are a good historical record of your health and safety performance. Keeping the 200 Log information up to date, especially after the change of a calendar year, is often a challenge but is required to meet regulatory require-

Log and Summary of Occupational Injuries and illnesses					
NOTE:	This form is required by Public Law 91-596 and must be kept in the establishment for 5 years. Failure to maintain and post can result in issuance of citations and assessment of penalties. (*See posting requirements on the other side of form*)			**RECORDABLE CASES:** You are required to record information about every occupational death; every nonfatal occupational illness; and those nonfatal occupational injuries which involve one or more of the following: loss of conciousness, restriction of work or motion, transfer to another job, or medical treatment (other than first aid) (*See definitions on the other side of form*)	
Case or File Number	Date of Injury or Onset of Illness	Employee's Name	Occupation	Department	Description of Injury or Illness
Enter a nonduplicating number which will facilitate comparisons with supplementary records.	Enter Mo/Day	Enter first name or initial, middle initial, last name	Enter regular job title, not activity employee was performing when injury occurred or at onset of illness. In the absence of a formal title, enter a brief description of the employee's duties.	Enter department in which the employee is regularly employed or a description of normal workplace to which employee is assigned, even though temporarily working in another department at the time of injury or illness.	Enter a brief description of the injury or illness and indicate the part or parts of the body affected. Typical entries for this column might be: Amputation of 1st joint right forefinger; Strain of lower back; Contact dermatitis on both hands; Electrocution - body.
(A)	(B)	(C)	(D)	(E)	(F)
					PREVIOUS PAGE TOTALS =>
					TOTALS (Instructions on other side of form) =>
OSHA No. 200					

Figure 5.1 The OHSA 200 Log and summary for recording work-related injuries and illnesses (US Department of Labor-Occupational Safety and Health Administration).

U.S. Department of Labor

For Calendar Year ____________ Page: _____ of _____

Company Name	Form Approved
Establishment Name	O.M.B. No. 1218-0176
Establishment Address	See OMB Disclosure
	Statement on reverse.

Extent of and Outcome of Injury						Type, Extent of, and Outcome of Illness												
Fatalities	Nontatal Injuries					Type of Illness							Fatalities	Nonfatal Illnesses				
Injury Related	Injuries with Lost Workdays				Injuries Without Lost Workdays	CHECK Only One Column for Each Illness (See other side of form for terminations or permanent transfers)							Illness Related	Illnesses with Lost Workdays				Illnesses without Lost Workdays
Enter Date of death. mm/dd/yy	Enter a Check if injury involves DAYS away from work or restricted work activity or both.	Enter a Check if injury involves DAYS away from work.	Enter number of DAYS away from work	Enter number of DAYS of restricted work activity	Enter a Check if no entry was made in column 1 or 2 but the injury is recordable as defined above.	Occupational Skin Disorder or Disease	Dust Disease of the lungs	Respiratory Conditions due to toxic agents	Poisoning (systemic effects of toxic materials)	Disorders due to physical agents	Disorders associated with repeated trauma	All other occupational illnesses	Enter DATE of death, mm/dd/yy	Enter a CHECK if Illness involves DAYS away from work, or DAYS of restricted work activity or both.	Enter a CHECK if Illness involves DAYS away from work.	Enter number of DAYS away from work.	Enter number of DAYS of restricted work activity	Enter a CHECK if no entry was made in columns 8 or 9
						(7)												
(1)	(2)	(3)	(4)	(5)	(6)	(a)	(b)	(c)	(d)	(e)	(f)	(g)	(8)	(9)	(10)	(11)	(12)	(13)

Certification of Annual Summary Totals by: ______________ Title: ______________ Date: ______________

OSHA 200 POST ONLY THIS PORTION OF THE LAST PAGE NO LATER THAN FEBRUARY 1

Figure 5.1 (***continued***).

OMB DISCLOSURE STATEMENT

Public reporting burden for this collection of information is estimated to vary from 4 to 30 (time in minutes) per response with an average of 15 (time in minutes) per response, including the time for reviewing instructions, searching existing data sources, gathering and maintaining the data needed, and completing and reviewing the collection of information. Persons are not required to respond to the collection of information unless it displays a currently valid OMB control number. If you have any comments regarding this estimate or any other aspect of this information collection, including suggestions for reducing this burden, please send them to the OSHA Office of Statistics, Room N-3644, 200 Constitution Avenue, N.W. Washington, D.C. 20210

Instructions for OSHA No. 200

I. Log and Summary of Occupational Injuries and Illnesses

Each employer who is subject to the recordkeeping requirements of the Occupational Safety and Health Act of 1970 must maintain for each establishment, a log of all recordable occupational injuries and illnesses. This form (OSHA No. 200) may be used for that purpose. A substitute for the OSHA No. 200 is acceptable if it is as detailed, easily readable, and understandable as the OSHA No. 200.

Enter each recordable case on the log within six (6) workdays after learning of its occurrence. Although other records must be maintained at the establishment to which they refer, it is possible to prepare and maintain the log at another location, using data processing equipment if desired. If the log is prepared elsewhere, a copy updated to within 45 calendar days must be present at all times in the establishment.

Logs must be maintained and retained for five (5) years following the end of the calendar year to which they relate. Logs must be available (normally at the establishment) for inspection and copying by representatives of the Department of Labor, or the Department of Health and Human Services, or States accorded jurisdiction under the Act. Access to the log is also provided to employees, former employees and their representatives.

II. Changes in Extent of or Outcome of Injury or Illness

If, during the 5-year period the log must be retained, there is a change in an extent and outcome of an injury or illness which affects entries in columns 1, 2, 6, 8, 9, or 13, the first entry should be lined out and a new entry made. For example, if an injured employee at first required only medical treatment but later lost workdays away from work, the check in column 6 should be lined out and checks entered in columns 2 and 3 and the number of lost workdays entered in column 4.

In another example, if an employee with an occupational illness lost wordays, returned to work, and then died of the illness, any entries in columns 9 through 12 would be lined out and the date of death entered in column 8.

The entire entry for an injury or illness should be lined out if later found to be nonrecordable. For example, an injury which is later determined not to be work related, or which was initially thought to involve medical treatement but later was determined to have involved only first aid.

III. Posting Requirements

A copy of the totals and information following the total line of the last page for the year, must be posted at each establishment in the place or places where notices to employees are customarily posted. This copy must be posted no later than February 1 and must remain in place until March 1. Even though there were no injuries or illnessed during the year, zeros must be entered on the totals line, and the form posted.

The person responsible for the annual summary totals shall certify that the totals are true and complete by signing at the bottom of the form.

IV. Instructions for Completing Log and Summary of Occupational injuries and illnesses

Column A - CASE OR FILE NUMBER. Self Expanatory

Column B - DATE OF INJURY OR ONSET OF ILLNESS

For occupational injuries, enter the date of the work accident which resulted in the injury. For occupational illnesses, enter the date of initial diagnosis of illness, or, if absence from work occurred before diagnosis, enter the first day of the absence attributable to the illness which was later diagnosed or recognized.

Columns C through F - Self Explanatory

Columns 1 and 8 - INJURY OR ILLNESS-RELATED DEATHS - Self Explanatory

Columns 2 and 9 - INJURIES OR ILLNESSES WITH LOST WORKDAYS - Self Explanatory

Any injury which involves days away from work, or days of restricted work activitiy, or both, must be recorded since it always involves one or more of the criteria for recordability.

Figure 5.1 ***(continued)***.

Columns 3 and 10 - INJURIES OR ILLNESSES INVOLVING DAYS AWAY FROM WORK - Self Explanatory

Columns 4 and 11 - LOST WORKDAYS -- DAYS AWAY FROM WORK.
Enter the number of workdays (consecutive or not) on which the employee would have worked but could not because of occupational injury or illness. The number of lost workdays should not include the day of injury or onset of illness or any days on which the employee would not have worked even though able to work. NOTE: For employees not having a regularly scheduled shift, such as certain truck drivers, construction workers, farm labor, casual labor, part-time employees, etc., it may be necessary to estimate the number of lost workdays. Estimates of lost workdays shall be based on prior work history of the employee AND days worked by employees, not ill or injured, working in the department and/or occupation of the ill or injured employee.

Columns 5 and 12 - LOST WORKDAYS -- DAYS OF RESTRICTED WORK ACTIVITY.
Enter the number of workdays (consecutive or not) on which because of injury or illness:
(1) the employee was assigned to another job on a temporary basis, or
(2) the employee worked at a permanent job less than full time, or
(3) the employee worked at a permanently assigned job but could not perform all duties normally connected with it.

The number of lost workdays should not include the day of injury or onset of illness or any days on which the employee would not have worked even though able to work.

Columns 6 and 13 - INJURIES OR ILLNESSES WITHOUT LOST WORKDAYS - Self Explanatory

Columns 7a through 7g - TYPE OF ILLNESS. Enter a check in only *one* column for each illness.
TERMINATION OR PERMANENT TRANSFER - Place an asterisk to the right of the entry in columns 7a through 7g (type of illness) which represented a termination of employment or permanent transfer.

V. Totals
Add number of entries in columns 1 and 8.
Add number of checks in columns 2, 3, 6, 7, 9, 10 and 13.
Add number of days in columns 4, 5, 11 and 12.
Yearly totals for each column (1-13) are required for posting. Running or page totals may be generated at the discretion of the employer.

In an employee's loss of workdays is continuing at the time the totals are summarized, estimate the number of future workdays the employee will lose and add that estimate to the workdays already lost and include this figure in the annual totals. No further entries are to be made with respect to such cases in the next year's log.

VI. Definitions
OCCUPATIONAL INJURY is any injury such as a cut, fracture, sprain, amputation, etc. which results from a work accident or from an exposure involving a single incident in the work environment. NOTE: Conditions resulting from animal bites, such as insect or snake bites or from one-time exposure to chemicals, are considered to be injuries.

OCCUPATIONAL ILLNESS of an amployee is any abnormal condition or disorder, other than one resulting from an occupational injury, caused by exposure to environmental factors associated with employment. It includes acute and chronic illnesses or diseases which may be caused by inhalation, absorption, ingestion, or direct contact.

The following listing gives the categories of occupational illnesses and disorders that will be utilized for the purpose of classifying recordable illnesses. For porposes of information, examples of each category are given. These are typical examples, however, and are not to be considered the complete listing of the types of illnesses and disorders that are to be counted under each category.

7a. Occupational Skin Diseases or Disorders. Examples: Contact dermatitis, eczema, or rash caused by primary irritants and sensitizers or poisonous plants; oil acne; chrome ulcers; chemical burns or inflamation, etc.

7b. Dust Diseases of the Lungs (Pneumaconioses). Examples: Silicosis, asbestosis and other asbestos-related diseases, coal worker's pneumaconioses, byssinosis, siderosis, and other pneumaconioses.

7c. Respiratory Conditions Due to Toxic Agents. Examples: Pneumonitis, pharyngitis, rhinitis or acute congestion due to chemicals, dusts, gases, or fumes; farmer's lung; etc.

7d. Poisoning (Systemic Effects of Toxic Materials). Examples: Poisoning by lead, mercury, cadmium, arsenic, or other metals; poisoning by

Figure 5.1 (***continued***).

carbon monoxide, hydrogen sulfide, or other gases; poisoning by benzol, carbon tetrachloride, or other organic solvents; poisoning by insecticide sprays such as parathion, lead arsenate; poisoning by other chemicals such as formaldehyde, plastics, and resins; etc.

7e. Disorders Due to Physical Agents (Other than Toxic Materials). Examples: Heatstroke, sunstroke, heat exhaustion, and other effects of environmental heat, freezing, frostbite, and effects of exposure to low temperatures; caisson disease; effects of ionizing radiation (isotopes, X-rays, radium); effects of nonionizing radiation (welding flash, ultraviolet rays, microwaves, sunburn); etc.

7f. Disorders Associated with Repeated Trauma. Examples: Noise-induced hearing loss; synovitis, tenosynovitis, and bursitis. Raynaud's phenomena; and other conditions due to repeated motion, vibration, or pressure.

7g. All Other Occupational Illnesses. Examples: Anthrax, brucellosis, infectious hepatitis, malignant and benign tumors, food poisoning, histoplasmosis, coccidioidomycosis, etc.

MEDICAL TREATMENT includes treatment (other than first aid) administered by a physician or by registered professional personnel under the standing orders of a physician. Medical treatment does NOT include first aid treatment (one-time treatment and subsequent observation of minor scratches, cuts, burns, splinters, and so forth, which do not ordinarily require medical care) even though provided by a physician or registered professional personnel.

ESTABLISHMENT: A single physical location where business is conducted or where services or industrial operations are performed (for example: a factory, mill, store, hotel, resturant, movie theater, farm, ranch, bank, sales office, warehouse, or central administrative office). Where distinctly separate activities are performed at a single physicial location, such as construction activities operated from the same physical locations as a lumber yard, each activity shall be treated as a separate establishment.

For firms engaged in activities which may be physically dispersed, such as agriculture; construction; transportation; communications and electric, gas, and sanitary services, records may be maintained at a place to which employees report each day.

Records for personnel who do not primarily report or work at a single establishment, such as traveling salesmen, technicians, engineers, etc., shall be maintained at the location from which they are paid or the base from which personnel operate to carry out their activities.

WORK ENVIRONMENT is comprised of the physical location, equipment, materials processed or used, and the kinds of operations performed in the course of an employee's work, wether on or off the employer's premisis.

Figure 5.1 (***continued***).

ments. A listing of commonly used safety statistics based on the OSHA 200 Log is included in Table 5.3.

The second source of information comprises the 101 equivalent documents/ accident and injury investigation reports. These reports will give you valuable information, including data on the number and types of injuries resulting in employees being unable to work (lost time), restricted in their regular work activity (unable to perform all their normal duties), and injuries requiring medical treatment by a qualified provider. Severity information on injuries and illness based on the days lost/restricted activity will also be included.

Your accident investigation reports of employee injuries, and non-injury accidents including property damage, equipment damage and, if available, near-miss incidents will help identify high-hazard work areas, tasks, equipment and overall accident/injury trends. Once trends have been identified, plans can be developed to address improvements.

An OSHA 101 document (Figure 5.2) or its equivalent (your accident/injury investigation report) is also required for each entry on the log. You can design your company incident investigation form to meet the OSHA 101 requirements. (Refer to the Bureau of Labor Statistic (BLS) "Blue Book" and State requirements if applicable.)

Your insurance carrier/broker can provide you with the third recommended source of information, your workers' compensation (WC) loss/claim data. These data will provide general information on types of incidents, causes, body part injured and costs incurred due to each reported WC claim. (Note: A number of state regulations may have a significant impact on the recording and reporting of WC claims. In addition, the recording

Table 5.3 OSHA Log-based safety performance metrics.

Lost workday case
an employee work-related injury or illness that is so severe that a worker is unable to come to work.

Restricted activity case
an employee injury or illness that is severe; a worker, although able to come to work, is unable to perform all his/her regular job duties.

Medical treatment cases
employee injuries and illnesses that result in a worker being able to work; however, due to his/her work-related injury, treatment by a medical professional is required (i.e. stitches, prescriptions). The worker is able to return to full duties after treatment.

Total employee hours worked
represents the total hours worked (exposure to workplace hazards) by an employee population represented in the calculation. note: 200 000 hours represents 100 employees working 2000 hours over a 1 year period. The rates can be reported as lost workday/restricted activity days per 100 employees or per hours worked.

OSHA lost workday case rate

$$\frac{\text{No. of lost workdays and restricted activity cases} \times 200\,000}{\text{Total employee hours worked}}$$

OSHA recordable rate

$$\frac{\text{No. of lost workdays} + \text{restricted activity} + \text{medical treatment cases} \times 200\,000}{\text{Total employee hours worked}}$$

and reporting requirements for OSHA Log maintenance and WC claims are different. Some injuries may be recorded and reported in one system but not the other (i.e. OSHA record-keeping requirements will capture restricted data activity cases, while workers' compensation will not.))

WC costs have a "long tail", meaning that it often takes 5–8 years before they are fully realized. Thus costs reported at the close of a fiscal year will significantly increase over a 5-year period. As a general rule, your compensation costs at the close of a year will easily double within this 5-year period. Your insurance carrier may have loss control services, and their professionals can assist you in the prevention, management and reporting of injuries and illnesses.

You may also use your WC experience modification rate (EX Mod) as an overall and general performance indicator. The EX Mod is based on a standard formula and, with industry data, it considers but is not limited to your payroll, industry performance and company WC performance for 3 previous and complete years (prior to your last year's experience). A modifier can be compared between companies. An Ex Mod of 1 means that your costs and experience are average for your industry/competitors. A rate below 1 means that your performance is better than your competitors'/industry average and your costs are lower. This would be a competitive advantage. If your rate is greater than 1, your costs are higher and you are paying more than your competitors. With 3 years of experience considered, 1 poor year will impact your costs for up to 3 years.

After a thorough review of your claim data, a walk/inspection of the active production areas, offices and especially the areas with serious or frequent incidents is recommended. These real-time observations of employee work practices, production processes and equipment operation will complement your experience review. Conducting this walk with a knowledgeable and experienced safety professional and production manager or foreman is highly recommended. Their observations and answers to your questions on production processes and safe work methods will be invaluable.

This form is required by Public Law 91-596 and must be kept in the establishment for 5 years. Failure to maintain can result in the issuance of citations and assessment of penalties.	Case or File No.	Form Approved O.M.B. No. 1218-0176

Employer

1. Name

See OMB Disclosure Statement on reverse.

2. Mail address (No. and street, city or town, State, and zip code)

3. Location, if different from mail address

Injured or Ill Employee

4 Name (First, middle, and last) | Social Security No.

5. Home address (No. and street, city or town, State, and zip code)

6. Age | 7. Sex (Check one) Male ☐ Female ☐

8. Occupation (Enter regular job title, not the specific activity he was performing at the time of injury.)

9. Department (Enter name of department or division in which the injured person is regularly employed, even though he may have been temporarily working in another department at the time of injury.)

The Accident or Exposure to Occupational Illness

If accident or exposure occurred on employer's premises, give address of plant or establishment in which it occurred. Do not indicated department or division within the plant or establishment. If accident occurred outside employer's premises at an identifiable address, give that address. If it occurred on a public highway or at any other place which cannot be identified by number and street, please provide place references locating the place of injury as accurately as possible.

10. Place of accident or exposure (No. and street, city or town, State, and zip code)

11. Was place of accident or exposure on employer's premises? Yes ☐ No ☐

12. What was the employee doing when injured? (Be specific. If he was using tools or equipment or handling material, name them and tell what he was doing with them.)

13. How did the accident occur? (Describe fully the events which resulted in the injury or occupational illness. Tell what happened and how it happened. Name any objects or substances involved and tell how they were involved. Give full details on all factors which led or contributed to the accident. Use separate sheet for additional space.)

Occupational Injury or Occupational Illness

14. Describe the injury or illness in detail and indicate the part of body affected. (E.g., amputation of right index finger at second joint; fracture of ribs; lead poisoning; dermatitis of left hand, etc.)

15. Name the object or substance which directly injured the employee. (For example, the machine or thing he struck against or which struck him; the vapor or poison he inhaled or swallowed; the chemical or radiation which irriatated his skin; or in cases of strains, hernias, etc., the thing he was lifting, pulling, etc.)

16. Date of injury or initial diagnosis of occupational illness | 17. Did employee die? (Check one) Yes ☐ No ☐

Other

18. Name and address of physician

19. If hospitalized, name and address of hospital

Date of report	Prepared by	Official position

OSHA No. 101 (Feb. 1981)

(See Next Page/Reverse)

Figure 5.2 The OHSA 101 form is used to complete the OHSA 200 Log. It must be kept on file with the OHSA 200 records illnesses (US Department of Labor-Occupational Safety and Health Administration).

Observations should be focused on the immediate work of employees and include the workers' 360° environment. For instance, observe the activity and interaction of the employees in their workspace, with their equipment, materials and equipment around them, to each side, overhead, underfoot and behind.

Count the work actions involved in lifting, using tools, moving materials and operating equipment. This will help you identify some

SUPPLEMENTARY RECORD OF OCCUPATIONAL INJURIES AND ILLNESSES

To supplement the Log and Summary of Occupational Injuries and Illneses (OSHA No. 200), each establishment must maintain a record of each recordable occupational injury or illness. Worker's compensation, insurance, or other reports are acceptable as records if they contain all facts listed below or are supplemented to do so. If no suitable report is made for other purposes, this form (OSHA No. 101) may be used or the necessary facts can be listed on a separate plain sheet of paper. These records must also be available in the establishment without delay and at reasonable times for examination by representatives of the Department of Labor and the Department of Health and Human Services, and States accorded jurisdiction under the Act. The records must be maintained for a period of not less than five years following the end of the calendar year to which they relate.

Such records must contain at least the following facts:

1) About the employer - name, mail address, and location if different from mail address.

2) About the injured or ill employee - name, social security number, home address, age, sex, occupation, and department.

3) About the accident or exposure to occupational illness - place of accident or exposure, whether it was on employer's premises, what the employee was doing when injured, and how the accident occurred.

4) About the occupational injury or illness - description of the injury or illness, including part of the body affected, name of the object or substance which directly injured the employee; and date of injury or diagnosis of illness.

5) Other - name and address of physician; if hospitalized, name and address of hospital, date of report; and name and position of person preparing the report.

SEE *DEFINITIONS* ON THE BACK OF OSHA FORM 200.

OMB DISCLOSURE STATMENT

Public reporting burden for this collection of information is estimated to average 20 minutes per response, including the time for reviewing instructions, searching existing data sources, gathering and maintaining the data needed, and completing and reviewing the collection of information. Persons are not required to respond to the collection of information unless it displays a currently valid OMB control number. If you have any comments regarding this estimate or any other aspect of this information collection, including suggestions for reducing this burden, please send them to the OSHA Office of Statistics, Room N3644, 200 Constitution Avenue, NW, Washington, DC 20210

DO NOT SEND THE COMPLETED FORM TO THE OFFICE SHOWN ABOVE

Figure 5.2 (***continued***).

sources of mechanical energy and the exposure potential. Your partner on the walk can assist you in identifying the work practices that are not consistent with safety requirements. Also count the pieces of powered equipment used in the areas you review. The more equipment and materials, the higher the potential for injuries.

These simple methods will help highlight the dynamics of your workplace, identify some of the sources of mechanical energy and assist you in identifying a need for further assessment. Positive results from any of these basic injury surveillance program components suggest the need for further investigation, and are summarized in Table 5.4.

Based on this initial assessment, a determination can be made of the need for further evaluation. Poor accident experience

Table 5.4 Indicators of a need for further assessment.

A high number of injuries and illnesses recorded on your OSHA 200 Logs (especially if the incidents are in the areas where you have identified powered equipment and work habits that are not consistent with company work practices)
A high percentage of workers using powered equipment, and working near such equipment
A high number of workers' compensation claims and costs. (An experience modification rate can be used as an indicator, i.e. an experience modification rate of 1.25 or higher)
A frequent number of employee actions requiring interaction with or touching equipment or materials (i.e. lifting, cutting, activating powered equipment, feeding raw materials into powered equipment) per unit of time (half-hour increments recommended)
Difficult environmental conditions, such as slippery or steep surfaces

combined with an active production area, frequent employee equipment and materials interactions and high numbers of items of powered equipment indicate a greater exposure to mechanical energy and the potential for injury.

Safety surveillance

An effort to identify the need for an energy management system entails a comprehensive review, which is best achieved with a team approach. Experienced and knowledgeable personnel are required to identify equipment and sources of mechanical energy. The team should consist of members from production management/foreman, experienced workers, safety and industrial hygiene professionals, maintenance and plant engineers, and, where appropriate, equipment manufacturers.

Select a team leader (preferably an operations manager/director), keep the number of members to a minimum and assign individuals specific project tasks. Using project management techniques, milestones and timelines will keep the process on schedule. If there are other plants with the same equipment, divide the tasks among plants.

Table 5.5 contains the key elements for conducting a more detailed assessment. This list should be used as a reference and sections used only as appropriate. It is not all-encompassing or complete; however, by researching your workplace needs and using parts of this list and adding others, you will customize the criteria for assessment. The assessments will identify the needs. Choosing corrective action and a long-term plan to implement and maintain a system will require the same team effort, commitment and participation needed for the assessment.

PREVENTION

The key to reducing injuries is preventing accidents. An effective prevention program presumes a surveillance component to identify high-risk potential hazard exposures and processes, and then targets interventions at these identified items. The surveillance program is a "leading indicator", because it predicts likely accidents before they happen, and allows us to prevent them.[3,4]

Johnson has defined an accident as an unwanted transfer of energy because of lack of barriers or controls, which produces injury to persons, property, or processes, and which is preceded by sequences of planning and operation errors which: (1) failed to adjust to changes in physical or human factors, and (2) produced unsafe conditions and/or unsafe acts.[5]

If there is a hazard present, the results of this sequence may be an injury. In the work environment, and to a certain extent any envir-

Table 5.5 Identifying pertinent equipment, operations and procedures.

1. Conduct a physical plant equipment and energy survey/inventory.
   1. Identify all powered equipment.
   2. Identify all sources of equipment and building power supplies. Identify the common sources of equipment energy, including but not limited to:
      1. Electrical
      2. Hydraulic
      3. Mechanical
      4. Pneumatic
      5. Others
   3. Evaluate individual equipment to identify all power sources. Potential sources of power, both external and within the equipment (including the above sources):
      1. Compressed (i.e. spring)
      2. Gravity actuated
      3. Partially cycled equipment
      4. Energized capacitors, or
      5. Any other source that might cause unexpected movement
   4. Identify type and points of potential contact between the worker and equipment, materials, power sources, and moving parts. (Ref: Table 5.8 and Figure 5.1).
   5. Evaluate equipment operating software and procedures. Many equipment operations are directed by software. The potential for unplanned equipment activation/action during maintenance, power interruptions and surges and changes due to software upgrades is increasing the possibility of injuries).
2. Inventory work areas to identify high to low employee work activity and describe the types of work conducted:
   1. Materials and movement due to production processes (i.e. lifting, raw materials)
   2. Interaction with equipment
   3. Climbing and walking
   4. Repetitive actions
   5. Manual processes
   6. Automated processes
   7. Work surfaces and walkways
3. Work procedures and processes
   1. Manuals, manufacturers and company recommended procedures
   2. Modifications of equipment and procedures
   3. Foreign-manufactured equipment and operating/maintenance procedures
   4. Equipment de-energizing and re-energizing procedures
4. Observe and evaluate employees' work activities:
   1. Work areas to identify sources of exposure to mechanical energy
   2. Equipment and work procedures
   3. Maintenance of equipment
   4. Pertinent employee training
   5. Job safety/hazard analysis (JSA/JHA) by trained professionals
   6. Controls
5. Detailed assessment of injury, illness and incident experience.
   1. OSHA 200 Logs and Metrics (see Tables 5.3 and 5.4)
   2. Workers' compensation experience and specific claims
   3. Incident reports. (i.e. injuries, illnesses, near misses, serious potential incidents, production interruptions)
   4. Trend analysis of incidents (refer to # 3 above)
   5. Regulatory citations and notices
6. Worker evaluations of work activity and work place hazards

Table 5.5 (*continued*)

7. Identification and "evaluation" of work place hazards
    1. Control of hazards
    2. Evaluation of work procedures
    3. Worker training
    4. Changes in process and equipment
    5. Unsafe acts and conditions
8. Regulations and industry best practices
    1. OSHA regulations on managing workplace safety and mechanical energy. (i.e. 29 CFR 1910 and 29CFR 1910.147 – Control of Hazardous Energy Lockout/blockout)
    2. Industry and Best Practices (i.e. ANSI[a], CMA[b], API[c], NFPA[d], ISO[e])

[a]American National Standards Institute
[b]Chemical Manufactures of America
[c]American Petroleum Institute
[d]National Fire Protection Association
[e]International Standards Organization

onment with which people interact, there are basically three categories of hazard:

1. **Inherent properties** of the mechanism, process, or other environmental variables, such as walking or working surface, electrical hazards, and the hazards of mechanical motion, weight, or radiation which are most likely to affect the end result of the sequence.
2. **Failures** of materials, equipment, safeguards or the human element that may be the proximate cause of an accident.
3. **Environmental stresses**, which are contributory to the sequence of events such as natural, ergonomic, thermal, physical energy or electromagnetic stresses.

Accident prevention depends on correcting both **unsafe conditions** and **unsafe acts.** Unsafe conditions can be corrected with process substitution, engineering design changes and "guarding". Guarding in the safety profession refers to: (1) machine guarding (which is really an engineering control) and (2) personnel guarding, or the use of personal protective equipment. Good guard design shields the hazard and anticipates how workers might disregard or inactivate the guarding.

Unsafe acts can be divided into: (1) lack of knowledge, which can be prevented by training; (2) inattention; and (3) deliberate acts. Some workers do not recognize hazards, some will defeat controls and guards to save time, and most people are not, and cannot be, 100% alert at all times. Training workers in the consequences of unsafe acts is the main method to reduce accidents from these acts. Deliberate acts such as bypassing guards and using "shortcuts" are common because of the human tendency to find the easiest or quickest (but not necessarily the safest) way to do a job.

The systems approach

Prevention of injuries and accidents utilizes various levels of safeguards to shield workers from hazards. The systems approach is a modern safety technique. In brief terms, it looks at the total system—people, materials, facilities, equipment, procedures, and process—to determine what could go wrong and in what way, what could cause it to go wrong, and what to do to keep it from going wrong. The approach should be and has been applied proactively from the design phase through the life of the system.

Systems are comprehensive and interrelated activities which, when acted upon in a consistent sequence, serve to complement each other to achieve continuous improvement. There are a number of approaches to developing a safety system, and the hallmarks include, but are not limited to:

1. Leadership—Management plays an active and visible role in establishing a team approach to developing, implementing and maintaining continuous measurable improvement.
2. Self-assessment and performance management—Comprehensive self-assessments identify risks and liabilities and opportunities to manage these risks and limit potential losses. Metrics are established to accurately measure and report performance with the goal of continuous measurable improvement.
3. Personnel development—Organizational needs are identified and personnel selected to match both immediate and future needs. Personnel are developed to grow and change with the needs of the organization and business climate, and as opportunities develop.
4. Design and operational integrity—Production and operational standards are identified, and immediate and long-range design, implementation and operational standards are implemented.
5. Planning and change management—Strategic plans are developed to meet organizational and operational needs and ensure high-caliber performance in a competitive business environment. Change is an anticipated and integral part of the planning process. Contingencies are designed and readily implemented to maximize opportunities and ensure best practices.
6. Incident investigation—System feedback mechanisms are established to promptly assess success and failures. Lessons learned are promptly communicated and leveraged throughout the organization.
7. Audits—Self-evaluations are periodically conducted to compare performance to standards and implementation of system best practices. Results are reported and improvements monitored until fully implemented.

Two specific examples of systems of prevention applied to mechanical energy are machine guarding and energy control.

Machine guarding

Mechanical energy and machine guarding are part of the greater system of engineering controls. The system consists of a single unit of equipment and the associated production process, including raw materials, manufacturing process to modify materials, employee interaction in the process and the general work environment. All these factors are interdependent. The work area must be assessed as a 360° environment, with special emphasis on exposure to mechanical energy to help identify mechanical exposure and guarding/controls. Failures in the system can result in contact with mechanical energy and ultimately injuries. A list of commonly encountered machines is included in Table 5.6, and definitions used in machine guarding are listed in Table 5.7.

One or more methods of machine guarding are required to protect employees.[6] Machine guarding regulatory requirements have been published and apply to broad classes of equipment. Common hazards have been recognized through industrial experience, and are roughly classified as point of operation, ingoing nip points, rotating parts, flying chips and sparks.

Point of operation presents the most common and generally obvious hazard potential. These hazards occur where materials are modified, and the employee is usually in close proximity to moving equipment and raw materials.[7] Table 5.8 lists regulatory requirements for specific controls designed to eliminate access to the points of operation, specifically the power press design criteria for the distance of guards from a "danger zone" or point

Table 5.6 Examples of machines that usually require point of operation guarding.

Guillotine cutters
Shears
Alligator shears
Power presses
Milling machines
Power saws
Jointers
Portable power tools
Forming rolls and calendars

Obtained from Federal OSHA 29 CFR.

where the employee can be in contact with moving machine parts (i.e. point of operation).

The point of operation of machines whose operation exposes an employee to injury must be guarded. The guarding device must meet regulatory design criteria or, in the absence of applicable specific standards, shall be so designed and constructed as to prevent the operator from having any part of his body in the "danger zone" during the operating cycle.

One or more types of controls, machine guarding and protective methods must be employed to properly protect employees and meet regulatory requirements. The most common mechanical hazards are present at the point of operation (see definitions), during power transmission and at a number of locations as materials are processed with equipment. Aside from traditional physical machine guards, there are a number of presence-sensing devices, hand removal restraint devices and two hand trip devices that help prevent contact with moving machine parts and tools. The most common devices are:

Table 5.7 Definitions for machine guarding.

Point of operation—the point at which cutting, shaping, boring, or forming is accomplished upon production materials. This is where exposure to mechanical energy is greatest and production material (or a body part) is actually positioned and work is performed.

Pinch point—any point other than the point of operation in which it is possible for a part of the body to be caught between the moving parts of a press or auxiliary equipment, or between moving and stationary parts of a press or auxiliary equipment or between the material and moving part or parts of the press or auxiliary equipment.

Safety system—the integrated system, including the pertinent elements of equipment, the controls, the safeguarding and any required supplemental safeguarding, and their interfaces with the operator, and the environment, designed, constructed and arranged to operate together as a unit, in such a way that one error will not cause injury to personnel due to point of operation hazards.

Nip-point belt and pulley guard—devices that enclose pulleys and are provided with rounded or rolled edge slots through which the belt passes.

Authorized person—an individual who has the authority and responsibility to perform a specific assignment and has been given authority by the employer.

Fixed barrier guard—a barrier guard fixed to a press frame which is "unmovable" and restricts the operator's access to moving machine parts such as the point of operation.

Interlocked press barrier guards—barrier guards attached to the press frame and interlocked so that the press stroke cannot be started under most conditions unless the guard itself, or its hinged or movable sections, are in place and enclose the point of operation or other hazardous machine parts.

Adjustable barrier guard—a barrier requiring adjustment for each job or die setup.

Adapted from Federal OSHA 29 CFR 1910.

Table 5.8 OSHA required openings in inches to guard a power punch press.

Distance of opening from point of operation hazard	Maximum width of guard/device opening
$\frac{1}{2}$–$1\frac{1}{2}$	$\frac{1}{4}$
$1\frac{1}{2}$–$2\frac{1}{2}$	$\frac{3}{8}$
$2\frac{1}{2}$–$3\frac{1}{2}$	$\frac{1}{2}$
$3\frac{1}{2}$–$5\frac{1}{2}$	$\frac{5}{8}$
$5\frac{1}{2}$–$6\frac{1}{2}$	$\frac{3}{4}$
$6\frac{1}{2}$–$7\frac{1}{2}$	$\frac{7}{8}$
$7\frac{1}{2}$–$12\frac{1}{2}$	$1\frac{1}{4}$
$12\frac{1}{2}$–$15\frac{1}{2}$	$\frac{1}{2}$
$15\frac{1}{2}$–$17\frac{1}{2}$	$1\frac{7}{8}$
$17\frac{1}{2}$–$31\frac{1}{2}$	$2\frac{1}{8}$

1. Presence-sensing devices—prevent normal press operation if the operator's hands are inadvertently within the point of operation, or prevent the initiation of a stroke, or stop a stroke in progress, if a body part, material or device passes into an electronic sensing field. Presence-sensing devices like machine guards have specific regulatory requirements (see 29 CFR—1910 Appendix A to 1910.217). The employer is ultimately responsible for ensuring (in this case) that the regulatory certification/validation requirements are met.
2. Hand removal devices—automatically withdraw the operator's hands if the operator's hands are inadvertently within the point of operation.
3. Two hand trip/control devices—require two hands to trip/operate the equipment. For instance, the operator must simultaneously depress two buttons located on each side of the operator. Thus the operator must have his hands out of the danger zone to activate the equipment. Design criteria for the use of two hand control devices and calculation of the distance from the sensing field to the point of operation are essential elements in a safety design.

Other common devices for preventing operator contact with mechanical energy include the following:

1. Special hand tools for placing and removing material into equipment points of operation. These tools are designed to permit easy handling of material without the operator placing a hand in the danger zone. Such tools shall not be in lieu of other guarding required by this section, but can only be used to supplement protection provided.
2. Gates or movable barrier devices are barriers arranged to enclose the point of operation. The gates must be in place before the press/equipment can start a part of or a new production/machine cycle.
3. Holdout or restraint devices are mechanisms that prevent an operator's hands from entering the point of operation. The operator's hands are secured with cables (i.e. cord, metal cable) and the cables are adjusted and secured to ensure that the operator cannot reach into a point of operation.
4. Pullout devices are mechanisms attached to the operator's hands and to a movable equipment part. These must be adjusted to each worker's unique physical characteristics; they pull the operator's hands away from the point of operation or danger zone. They operate inversely to the movement of the equipment. As a point of operation is reaching its most hazardous point (i.e. mechanical energy is transferred as work is performed) this device pulls the work-

er's hands back to a point beyond the danger zone.

5. Sweep devices are arms or bars that move the operator's hands to a safe position as the equipment cycles. They will literally sweep the operator's hands away from the danger zone as the machine cycles.

Managing hazardous energy

Lockout/tagout programs are one of the most common methods to control mechanical energy during equipment maintenance, adjustment or repair. This standard 29 CFR 1910.147—The Control of Hazardous Energy Lockout/Tagout applies during maintenance, repair or equipment adjustment and does not apply to normal production operations. Important definitions of the standard are listed in Table 5.9. The primary objective of the standard is to eliminate or manage the unexpected "energization" or start-up of the machines or equipment, or release of stored energy that could cause injury to employees.

Managing hazardous energy requires employers to establish programs and utilize procedures to control or eliminate uncontrolled hazardous energy. This program must include procedures, practices and training to ensure that contact with energized equipment or equipment components is managed and includes appropriate lockout and tagout devices that isolate and or disable machines or equipment to prevent unexpected energization, start-up or release of stored energy in order to prevent injury to employees.[8]

Employee protection and energy elimination and isolation are the goals of lockout/

Table 5.9 Definitions for lockout/tagout.[8]

Affected employee—an employee whose job requires him/her to operate or use a machine or equipment on which servicing or maintenance is being performed under lockout or tagout, or whose job requires him/her to work in an area in which such servicing or maintenance is being performed.

Authorized employee—a person who locks out or tags out machines or equipment in order to perform servicing or maintenance on that machine or equipment. An affected employee becomes an authorized employee when that employee's duties include performing servicing or maintenance covered under this section.

Energized—connected to an energy source or containing residual or stored energy. The source of energy can be electrical, mechanical, hydraulic, pneumatic, chemical, thermal, or other energy.

Energy-isolating devices are devices that physically prevent the transmission or release of energy, including but not limited to the following: a manually operated electrical circuit breaker; a disconnect switch; a manually operated switch by which the conductors of a circuit can be disconnected from all ungrounded supply conductors, and, in addition, no pole can be operated independently; a line valve; a block; and any similar device used to block or isolate energy. Push-buttons, selector switches and other control circuit-type devices are not energy-isolating devices.

Lockout/tagout devices are devices placed in accordance with an established procedure, ensuring that the energy-isolating devices and tags and the equipment being controlled cannot be operated until the lockout device is removed.

Energy control programs are employer-implemented programs consisting of energy control procedures, training, periodic inspections and lockout/tagout devices to ensure that before any employee performs any servicing or maintenance on a machine or equipment where the unexpected energizing, start-up or release of stored energy could occur and cause injury, the machine or equipment shall be isolated from the energy source and rendered inoperative.

tagout. This requires that the employer demonstrate that the tagout program will provide a level of safety necessary to ensure no contact occurs with energized equipment or machinery. The employer shall demonstrate full compliance with all tagout-related provisions of the standard. Additional means of protection must be considered including periodic inspection of the lockout tagout program.

Procedures need to be developed, documented and utilized for the control of potentially hazardous energy when employees are engaged in lockout/tagout activities.

The procedures need to clearly and specifically outline the scope, purpose, authorization, rules and techniques to be utilized for the control of hazardous energy, and the means to enforce compliance including, but not limited to, the following:

1. A specific statement of the intended use of the procedure.
2. Shutting down, isolating, blocking and securing machines or equipment.
3. Placement, removal and transfer of lockout/tagout devices.
   (a) Lockout tagout devices are locks, tags, chains, wedges, key blocks, adapter pins, self-locking fasteners or other hardware for isolating, securing or blocking of machines or equipment from energy sources, and must be provided. These tags must be standardized, easily identifiable and only used for controlling energy. Lockout devices shall be substantial enough to prevent removal without the use of excessive force or unusual techniques, such as with the use of bolt cutters or other metal-cutting tools.
   (b) Tagout devices, including their means of attachment, shall be substantial enough to prevent inadvertent or accidental removal.
4. Responsibility for ensuring that lockout/tagout devices are utilized and who will ensure that this is done.
5. Testing a machine or equipment to determine and verify the effectiveness of lockout devices, tagout devices, blocks and other energy control equipment.
5. Procedures for re-energizing equipment and removing lock-out/target devices

Training and communication

The employers should provide training to ensure that the purpose and function of all energy control programs are understood by employees and that the knowledge and skills required for the safe application, usage and removal of the energy controls are acquired by employees. Examples of the training recommendations for the lockout/tagout regulations are included in Table 5.10; there are generalizable to training for all control technologies.

Periodic inspection

The final component of the systems approach is to inspect existing controls to ensure continued optimum operation. The employer shall conduct a periodic inspection of the energy control procedure at least annually to ensure that all equipment lockout/tagout procedures and the requirements of this standard are fully implemented.

The periodic inspection shall be performed by an authorized employee other than the ones(s) utilizing the energy control procedure being inspected and shall be conducted to correct any deviations and or areas requiring improvement.

Where tagout is used for energy control, the periodic inspection shall include a review, between the inspector and each authorized and affected employee, of that employee's responsibilities under the energy control procedure being inspected.

The employer shall certify that the periodic inspections have been performed. The certification shall identify the machine or equipment on which the energy control procedure was

Table 5.10 Training and communication for lockout/tagout as a paradigm for hazards control training.

Each authorized employee shall receive training in the recognition of applicable hazardous energy sources, the type and magnitude of the energy available in the workplace, and the methods and means necessary for energy isolation and control.
Each affected employee shall be instructed in the purpose and use of the energy control procedure.
All other employees whose work operations are or may be in an area where energy control procedures may be utilized shall be instructed about the procedure, and about the prohibition relating to attempts to restart or re-energize machines or equipment which are locked out or tagged out.
Retraining shall be provided for all authorized and affected employees whenever there is a change in their job assignments, a change in machines, equipment or processes that present a new hazard, or a change in the energy control procedures.
Affected employees shall be notified by the employer or authorized employee of the application and removal of lockout devices or tagout devices. Notification shall be given before the controls are applied, and after they are removed from the machine or equipment.

being utilized, the date of the inspection, the employees included in the inspection, and the person performing the inspection.

Utilization of as many of these principles of prevention as possible will maximize production and safety, and minimize injuries from mechanical hazards in the workplace.

SUMMARY

Mechanical energy is an ever-present exposure in our industrial environment. Our changing workplaces and work activities present numerous challenges for effectively identifying and managing exposure to this energy. OSHA regulations provide guidance; however, this is not a substitute for an effective systems approach to identifying and actively managing exposures. This chapter provides general guidance and basic tools needed to begin the process of systematically managing mechanical energy, starting with recognition and quantification, and ultimately leading to elimination and control.

REFERENCES

1. National Safety Council. *Injury Facts—1999 Edition.* Chicago: NSC, 1999:51–3.
2. US Department of Labor, Bureau of Labor Statistics. *Recordkeeping Guidelines for Occupational Injuries and Illnesses (Blue Book).* OMB No. 1220-0029, September 1986, Washington, D.C.
3. Krieger GR, Montgomery JF. *Accident prevention manual for business and industry: engineering and technology,* 11th edn. Chicago: National Safety Council Press, 1997.
4. CoVan J. *Safety engineering*. New York: John Wiley & Sons, 1995.
5. Johnson WG. *MORT safety assurance systems.* New York: Marcel Dekker, 1980.
6. US. Department of Labor. *The principles and techniques of mechanical safeguarding,* Bulletin No. 197. Washington DC, US Department of Labor, 1971.
7. Wadden RA, Scheff PA. *Engineering design for the control of workplace hazards.* New York: McGraw-Hill, 1987.
8. American National Standards Institute. *Personnel protection—lockout/tagout of energy sources. Minimum safety requirements.* Z 244.1–1982. New York: ANSI, 1982.

Section II

The Physical Work Environment

6

HOT ENVIRONMENTS

Gail M. Gullickson, M.D., M.P.H.

Exposure to high ambient temperatures while working in hot indoor environments or while working outdoors in hot weather is a common and potentially fatal occupational hazard. Normally, the human body functions within a very narrow range of core body temperature. In the occupational setting, heat stress, from the combined effects of environmental heat, metabolic heat, and often the use of impervious clothing, can strain the body's ability to maintain heat balance, and the body's core temperature may begin to rise. Workplace heat exposure, in addition to causing heat-related illness, has been found to decrease productivity and to increase job-related accidents.[1]

The opinions and assertions contained herein are the private ones of the author and not to be construed as official or reflecting the views of the Department of Defense, the Uniformed Services University of the Health Sciences or the US Navy.

OCCUPATIONAL SETTING

No recent estimate of the number of workers exposed to hot environments has been published. In 1986, the National Institute for Occupational Safety and Health (NIOSH) estimated that 5–10 million Americans worked in jobs where heat stress was an occupational health hazard, and this range is probably still a reasonable estimate.[2] Table 6.1 lists some common work sites where workers are exposed to heat or to hot, humid environments.[3] Heat-related deaths are common in the USA, and the Centers for Disease Control and Prevention (CDC) reported an average of 371 deaths annually attributable to "excessive heat exposure".[4] Most of the heat-related deaths reported by the CDC were in non-working populations. Occupational- or exertional-related heat death, while known to occur, is infrequently reported in the medical literature. On the other hand, occupational heat-related illness is relatively common. A recent case series of heat-related casualties in the mining

Physical and Biological Hazards of the Workplace, Second Edition, Edited by Peter H. Wald and Gregg M. Stave
ISBN 0-471-38647-2

Table 6.1 Work sites with heat exposure.[3]

Iron and steel foundries
Non-ferrous foundries
Brick-firing and ceramics plants
Glass products facilities
Rubber products factories
Electrical utilities
Bakeries
Confectioneries
Commercial kitchens
Laundries
Food canneries
Chemical plants
Mining sites
Smelters
Steam tunnels
Fires (firefighting)
Outdoor operations
Surface mines
Agriculture sites
Construction sites
Merchant marine ships
Hazardous waste sites
Military training sites
Athletic competitions

industry reported the incidence of heat exhaustion in the summer months as 43 cases/million man-hours worked.[5] Military personnel, who are often required to achieve very high levels of work output in thermally extreme environments, have a long history of high risk for heat stress.[6] Reports from the military continue to provide information on the occurrence of heat injury in healthy, young individuals.[7,8]

In any heat hazard environment, whether indoors or outdoors, high humidity or heavy manual labor increases the workers' risk for heat strain. Specific occupational groups, such as firefighters are at exceptionally high risk for heat stress. Not only are they exposed to extremely high temperatures while fighting fires, but they also must perform demanding physical tasks while wearing fire-resistant protective clothing that diminishes the elimination of body heat. The hazardous waste clean-up industry is also a field where heat is a potentially significant occupational hazard. Workers are required to wear respiratory protection and full protective clothing. These requirements limit the use of conventional heat stress monitoring guidelines and place an additional strain on the hazardous waste clean-up worker.

MEASUREMENT ISSUES

Environmental heat

Ambient or environmental heat affecting the worker and the worker's ability to transfer body-generated heat to the environment is determined by four environmental factors: (1) air temperature, (2) air humidity, (3) air movement or velocity, and (4) radiant heat (solar and infrared). Microwave radiation may also be a source of environmental heat in some work situations (Chapter 15). Various measures of ambient heat load, reflecting the factors influencing heat transfer, are available for use in the industrial setting. The most commonly used measures of external heat are dry bulb temperature, wet bulb temperature, and wet globe temperature–are described in Table 6.2.

Metabolic heat

Heat production in the body is the by-product of normal basal metabolism, and averages approximately 1.5 kilocalorie (kcal) per min.[2] Under resting conditions, most heat production occurs in the liver, brain, heart, and skeletal muscles. During exercise or work, skeletal muscle activity greatly increases the production of metabolic heat. A measurement of metabolic heat production or, at least, an estimate of metabolic heat is essential in determining a worker's total heat stress and in calculating workplace heat exposure limits. Direct or indirect measurement of each worker's level of work and the kilocalories of heat produced in the occupational setting is not practical. Therefore, tables of the workload or energy cost of various tasks have been developed and are used to estimate metabolic heat (Table 6.3).

Table 6.2 Measures of external heat.[2]

Measure	Device	Comments
Dry bulb	Liquid-in-glass thermometer; thermocouple; resistance thermometer	Measures ambient air temperature and is temperature useful in determining comfort zone for lightly clothed sedentary workers. Does not measure effect of humidity, radiant heat or air movement on temperature
Wet bulb	Thermometer bulb or sensor covered by a wet cotton wick that is exposed to air movement	Measures effect of humidity on evaporation temperature and effect of air movement on ambient temperature. Natural wet bulb temperature is the term used if the wet bulb is exposed to prevailing natural air movement; may be a useful guide in preventing heat stress in hot, humid, still environments where radiant heat does not contribute to heat load, such as underground mines
Globe temperature	Black globe—temperature sensor in the center of a 15-cm hollow copper sphere painted flat black Wet globe—temperature sensor in the center of a 3-in copper sphere covered by a wet black cloth	Measures effect of radiant heat. Wet globe thermometer (Botsball) also reflects the effect of humidity and supposedly exchanges heat with the environment similarly to a nude man with totally wet skin

From American Conference of Governmental Industrial Hygienists (ACGIH®), 2000 Threshold Limit Values (TLVs®) for Chemical Substances and Physical Agents and Biological Exposure Indices (BEIs®). Reprinted with permission.

Table 6.3 Examples of activities with metabolic rate categories.

Categories	Example activities
Resting	Sitting quietly Sitting with moderate arm movements
Light	Sitting with moderate arm and leg movements Standing with light work at machine or bench while using mostly arms Using a table saw Standing with light or moderate work at machine or bench and some walking about
Moderate	Scrubbing in a standing position Walking about with moderate lifting or pushing Walking on level at 6 km/hour while carrying 3-kg weight load
Heavy	Carpenter sawing by hand Shoveling dry sand Heavy assembly work on a non-continuous basis Intermittent heavy lifting with pushing and pulling (e.g. pick-and-shovel work)
Very heavy	Shoveling wet sand

From American Conference of Governmental Industrial Hygienists (ACGIH®), 2000 Threshold Limit Values (TLVs®) for Chemical Substances and Physical Agents and Biological Exposure Indices (BEIs®). Reprinted with permission.

EXPOSURE GUIDELINES

Heat stress indexes

The guidelines currently used for worker exposure to heat stress are based on indexes developed through subjective and objective testing of workers or from combinations of external heat measurements. Today, the most commonly used index is the Wet Globe Bulb Temperature Index. An historically widely used heat stress index is the Effective Temperature (ET) Index.[2] Developed in the early 1920s, it was derived from studies on the subjective impressions of comfort and heat effects in sedentary subjects exposed to differing combinations of air temperature, humidity, and air motion. A third index, the Heat Stress Index (HSI), was developed by Belding and Hatch in the 1950s. Derived from the heat balance equation, it combines the total heat gain from radiation, convection and metabolism in terms of the required sweat evaporation necessary to maintain heat balance. The required component measurements make this index difficult to use, although these same measurements may be valuable in identifying the sources of the heat stress.[9]

Wet Globe Bulb Temperature Index

The Wet Globe Bulb Temperature (WGBT) Index is a commonly used heat stress index in the USA. It is recommended by the Occupational Safety and Health Agency (OSHA), NIOSH, and the American Conference of Governmental Industrial Hygienists (ACGIH), and required by the US armed forces.[10,11]

The WGBT Index is calculated from measurements of the natural wet bulb (NWB), the black globe (GT), and the dry bulb (TA) temperatures.

For outdoor environments with a solar heat source, the WBGT formula is:

$$\text{WBGT} = 0.7\text{NWB} + 0.2\text{GT} + 0.1\text{TA}$$

For indoor use or for outdoor settings without a solar load, the formula is:

$$\text{WBGT} = 0.7\text{NWB} + 0.3\text{GT}$$

The necessary measurements require relatively simple instrumentation and can be easily obtained in an industrial environment. Heat stress monitors that measure all three temperatures and calculate the WBGT Index temperature are also available. An example of an automated WBGT monitor is shown in Figure 6.1. The WBGT Index was developed for men exercising outdoors in military fatigues. Therefore, if different types of clothing are worn, correction factors are required. This index is not applicable in settings where sweat-impermeable clothing is required and may not be as effective as other indices in preventing heat casualties in extreme heat stress conditions. Much of the original work on the WBGT index was performed by the military in an effort to decrease heat-related casualties during training and war-time operations.[6,12]

Heat strain indicators

In addition to the heat stress indexes based upon external heat measurements, measures of heat strain, or the physiologic responses to environmental heat stress, have also been used to evaluate worker tolerance to heat exposure. The four main heat strain indicators are heart rate, body temperature, skin temperature, and hydration status. Historically, the cumbersome nature of monitoring equipment and its lack of durability have limited our ability to make real-time measurements of a workers' core temperature or heart rate. As technology improves and durable, convenient monitors and sensors become available, the physiologic measures of heat strain may take on increasing importance, especially for workers in heavy protective clothing.

Of all the physiologic measures, heart rate is the most easily measured. It is a reliable indicator of overall heat strain, rising with both increasing workload and increasing core temperature. Utilizing this physiologic response to heat, a method of measuring oral temperature, post-work heart rate, and recovery heart rate to monitor for heat strain has been developed. Heat stress is assumed not to cause progressive deterioration of the pulse if:

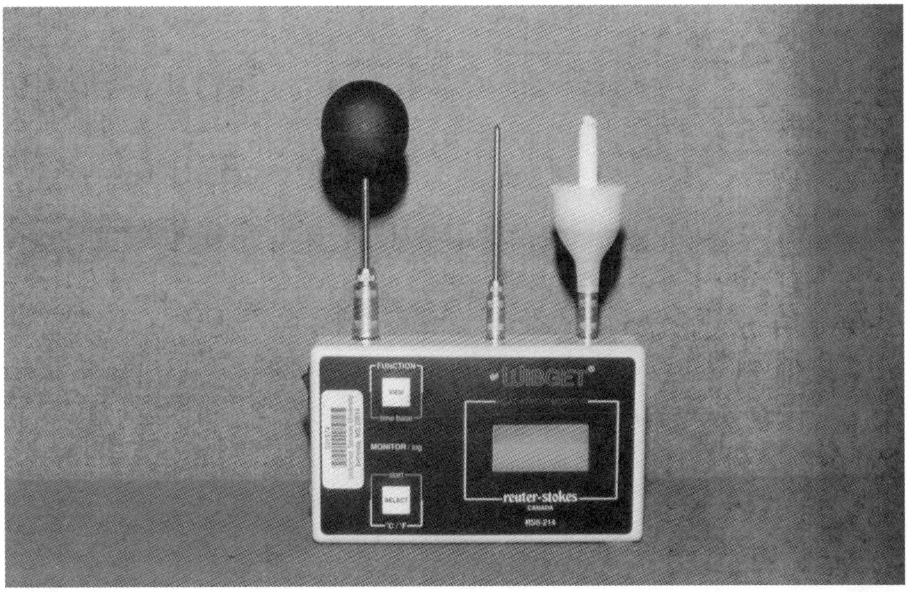

Figure 6.1 Automated WBGT monitor. From American Conference of Governmental Industrial Hygienists (ACGIH®), 2000 Threshold Limit Values (TLVs®) for Chemical Substances and Physical Agents and Biological Exposure Indices (BEIs®). Reprinted with permission.

(1) the oral temperature does not exceed 37.5°C (99.5°F) and if the heart rate at 30–60 s into rest is <110, or (2) the heart rate difference from the initial 30–60 s into rest and the heart rate 2.5–3 minutes into rest is >10 beats per min.[2] The availability of inexpensive electronic pulse monitors and timers has made this method easier, and it can be a valuable indicator of heat strain in certain settings.

Of all the measures of heat strain, core body temperature appears to be the most reliable indicator. The World Health Organization recommends that deep body temperature should not, under circumstances of prolonged daily work and heat, be permitted to exceed 38°C (100.4°F) rectally or 37.5°C (99.5°F) orally.[2] Though this heat strain index may seem ideal, monitoring internal or core body temperature with rectal or esophageal probes is not acceptable to many workers. Oral temperatures, while easy to obtain, may be inaccurate measures of core temperatures because of mouth breathing, drinking hot or cold liquids immediately before using a thermometer, or improper use or storage of the oral thermometer. An ingestible capsule containing a temperature sensor and a device producing a telemetry signal is available to monitor real-time internal temperatures as the capsule passes through the gastrointestinal tract. This type of temperature monitoring in the work setting may be an important advance; however, one study of experimental subjects found a difference between the sensor-measured internal temperature and commonly measured rectal temperature.[13] Therefore, additional research is needed before telemetry internal temperature monitoring can be advocated for heat-exposed workers. Tympanic membrane temperature monitors also provide an acceptable measurement of core temperature, but use of these monitors is uncommon because of ear discomfort and the need for a good seal in the ear canal.

Estimation of fluid loss or hydration status by regular weight measurements prior to work and throughout the day has also been proposed as an indicator of heat strain. Based upon weight measurements (assuming that the worker was fully hydrated before beginning work), the heat-exposed worker can be encouraged to drink liquids to maintain hydration and constant body weight throughout the day. Urine-specific gravity measurements before, during and after work have also been used to assess hydration status in workers.

EXPOSURE GUIDELINES

In the USA, the ACGIH guidelines are frequently used by industry to determine acceptable heat exposure for employees. These guidelines or threshold limit values (TLVs) permit working conditions "that nearly all adequately hydrated, unmedicated, healthy workers may be repeatedly exposed without adverse health effects."[10]

A recent revision of the ACGIH heat stress guidelines includes a decision tree for the

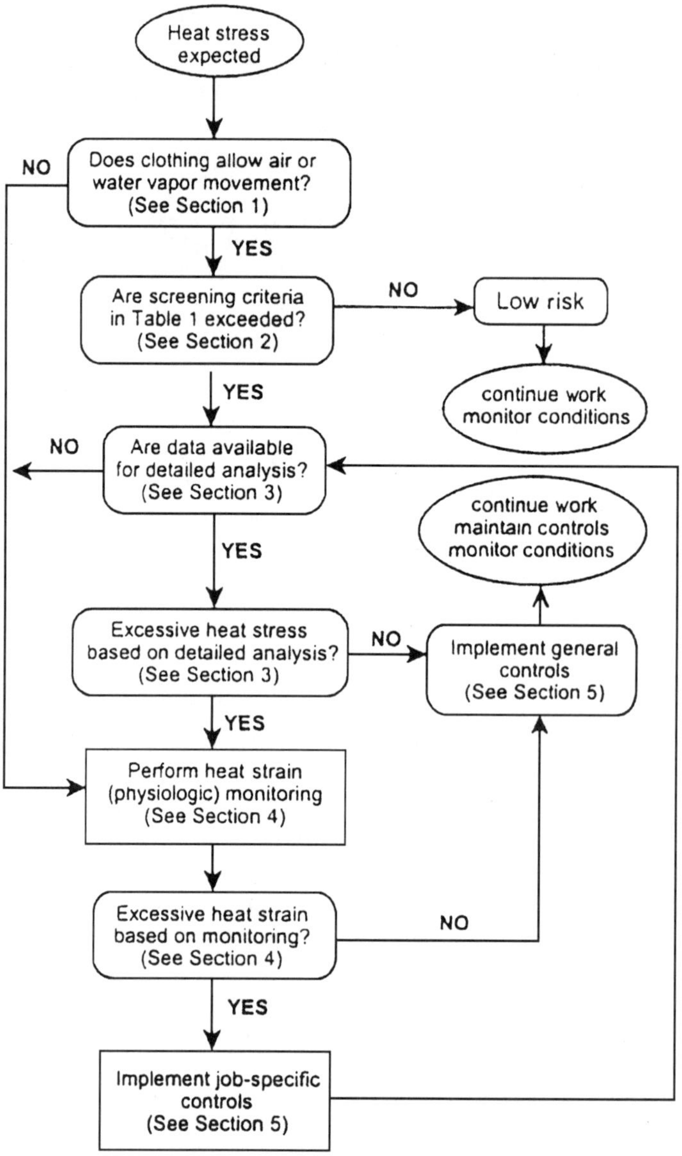

Figure 6.2 Heat stress decision tree. From American Conference of Governmental Industrial Hygienists (ACGIH®), 2000 Threshold Limit Values (TLVs®) for Chemical Substances and Physical Agents and Biological Exposure Indices (BEIs®). Reprinted with permission.

assessment of a worker's risk of heat stress (Figure 6.2). The previous ACGIH guidelines were similar to the NIOSH guidelines and recommended that workers in hot environments rest for a portion of each hour, with the amount of rest based upon WBGT Index, level of activity, and acclimation status of the worker. The current ACGIH recommendations still use a "work–rest regimen" but expand the guidelines for work situations where heat stress conditions (e.g. use of encapsulating clothing) exceed the typical work–rest cycle recommendations.

In general, the ACGIH decision tree incorporates an assessment of the type of clothing worn by the worker, the level of environmental heat or WBGT Index, an estimate of the worker's level of physical activity, an analysis of the work and work site, and if necessary, physiologic monitoring of the worker. After the initial assessment of clothing requirements, Table 6.4 provides work–rest cycles based upon worksite WBGT Index, work demand level and worker acclimation status. If the temperature index or work exceeds that outlined in Table 6.4, a detailed analysis of the heat stress potential of the work site is required. If the detailed analysis reveals excessive heat stress, physiologic monitoring of the workers is needed. The reader should consult an experienced health professional and review the specific guideline documentation before attempting to implement these guidelines in the workplace.[10]

NIOSH also has heat exposure guidelines published in the 1986 "Criteria for a Recommended Standard Occupational Exposure to Hot Environments."[2] In this document, the agency proposed a system of work–rest cycles, similar to the ACGIH guidelines in Table 6.4, to prevent heat-related illnesses. Work–rest time curves based upon WBGT Index measurements and metabolic heat estimates were developed for heat-unacclimatized and heat-acclimatized workers; they are called Recommended Alert Limits (RALs) and

Table 6.4 Examples of permissible heat exposure threshold limit values: Screening Criteria for Heat Stress Exposure (WBGT values in °C).

	Acclimatized				Unacclimatized			
Work demands	Light	Moderate	Heavy	Very heavy	Light	Moderate	Heavy	Very heavy
100% work	29.5	27.5	26		27.5	25	22.5	
75% work 25% rest	30.5	28.5	27.5		29	26.5	24.5	
50% work 50% rest	31.5	29.5	28.5	27.5	30	28	26.5	25
25% work 75% work	32.5	31	30	29.5	31	29	28	26.5

See Table 6.3 and the Documentation for work demand categories.

WBGT values are expressed in °C and represent thresholds near the upper limit of the metabolic rate category.

If work and rest environments are different, hourly time-weighted averages (TWA) should be calculated and used. TWAs for work rates should also be used when the work demands vary within the hour.

Values in the table are applied by reference to the "Work–Rest Regimen" section of the *Documentation* and assume 8-hour workdays in a 5-day workweek with conventional breaks, as discussed in the *Documentation*. When workdays are extended, consult the "Application of the TLV" section of the *Documentation*.

Because of the physiologic strain associated with heavy work among less fit workers regardless of WBGT, criteria values are not provided for continuous work and for up to 25% rest in an hour. The screening criteria are not recommended, and a detailed analysis and/or physiologic monitoring should be used.

From American Conference of Governmental Industrial Hygienists, (ACGIH®), 2000 Threshold Limit Values (TLVs®) for Chemical Substances and Physical Agents and Biological Exposure Indices (BEIs®). Reprinted with permission.

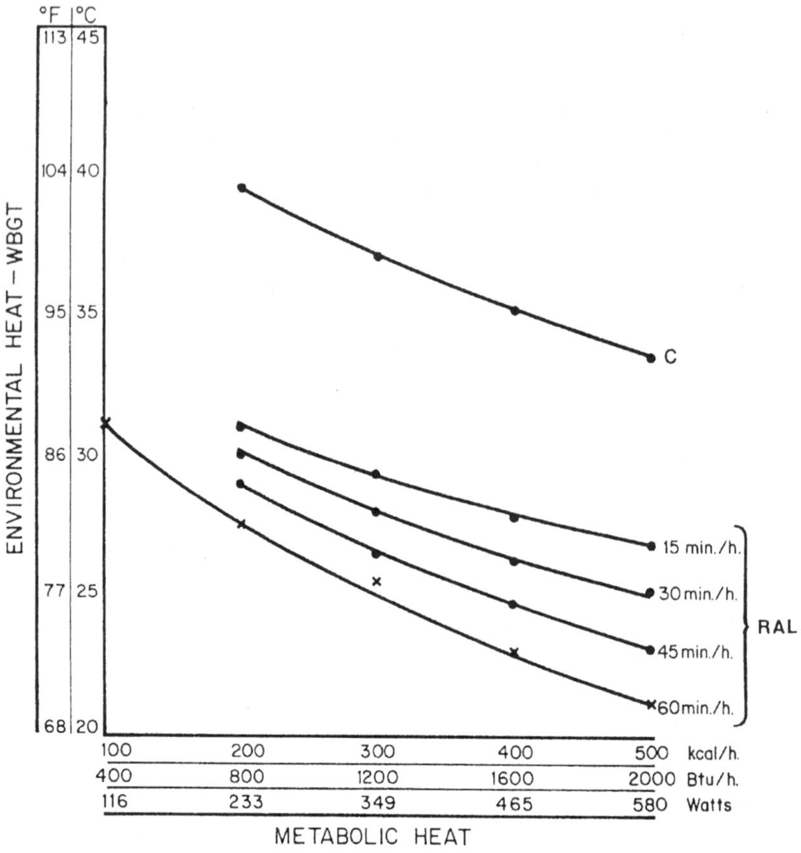

Figure 6.3 Recommended heat stress alert limits for heat-unacclimatized workers. C = ceiling limit; RAL = Recommended Alert Limit. Calculations are for a standard worker of 70 kg (154 lb) body weight and 1.8 m^2 (19.4 ft^2) body surface. Reproduced from National Institute for Occupational Safety and Health. Criteria for a recommended standard. Occupational exposure to hot environments. Revised criteria. DHHS (NIOSH) publication no. 86-113. Washington, DC: US Government Printing Office, 1986.

Recommended Exposure Limits (RELs), respectively (Figures 6.3 and 6.4). The NIOSH RALs and RELs were developed for healthy workers who are physically and medically fit for their level of activity and who are wearing customary work clothes (i.e. a long-sleeved shirt and trousers).

NORMAL PHYSIOLOGY

The normal human body maintains a narrow core body temperature range of 36.7–37°C (98–98.6°F) by oral measurement and 37.3–37.6°C (99–99.6°F) by rectal measurement. This narrow range of temperature control is maintained through the production and conservation of metabolic heat in cold ambient conditions or the transference of metabolic heat to the environment in hot ambient conditions. Under heat strain conditions, a series of physiologic mechanisms is initiated to bring internal heat to the body surface and to cool the body surface through the evaporation of sweat. In humans, the hypothalamus serves as the primary temperature regulator. Neurons in the preoptic area of the anterior hypothalamus, along with deep body temperature sensors, have the ability to detect small changes in blood temperature. As the blood temperature rises, sensing neurons in the hypothalamus

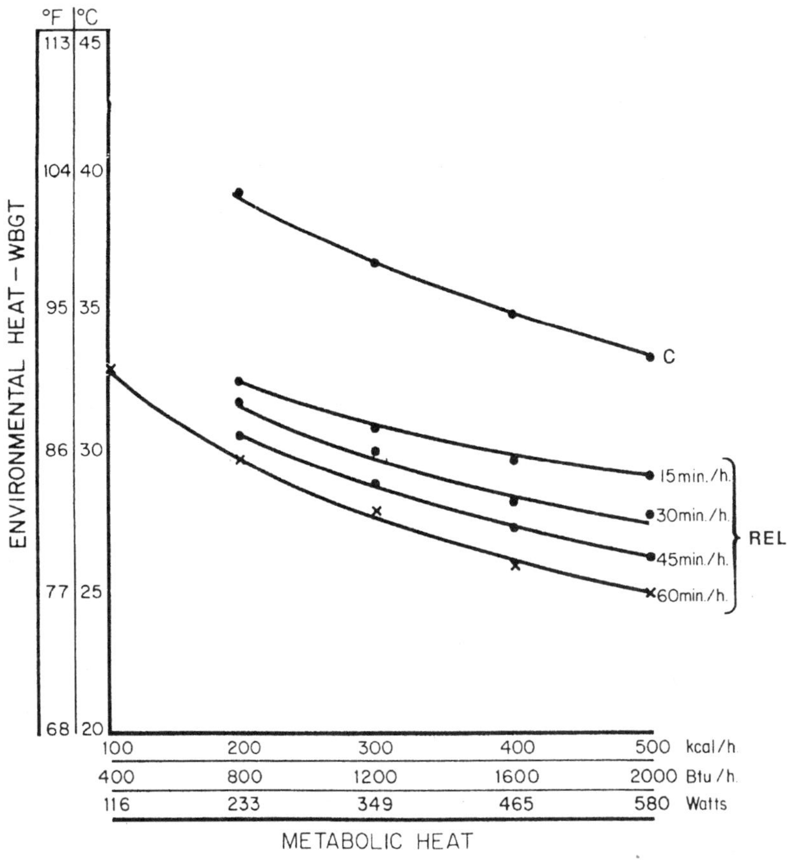

Figure 6.4. Recommended heat stress exposure limit for heat-acclimatized workers. C = ceiling limit; REL = Recommended Exposure Limit. Calculations are for a standard worker of 70 kg (154 lb) body weight and 1.8 m^2 (19.4 ft^2) body surface. Reproduced from National Institute for Occupational Safety and Health. Criteria for a recommended standard. Occupational exposure to hot environments. Revised criteria. DHHS (NIOSH) publication no. 86-113. Washington, DC: US Government Printing Office, 1986.

activate the autonomic nervous system, which in turn triggers changes in the circulatory, endocrine, and eccrine, or sweat gland, systems to rid the body of excess heat.

Heat stress places demands on many body systems. Proper functioning of the cardiac, pulmonary, renal, endocrine and autonomic nervous systems, the sweat glands, and the central nervous system thermoregulatory mechanism, are all necessary for toleration of heat exposure. The cardiovascular system plays a prominent role in heat dissipation. As internal body temperature rises, neuro-regulatory mechanisms increase cardiac output, through increases in stroke volume and heart rate, and dilate cutaneous blood vessels. Bloodflow is shunted from internal organs to the dilated blood vessels in the skin, where the heated blood can lose heat to the surroundings. To maintain blood pressure and bloodflow to exercising muscles and vital organs while blood is being shunted to the skin, adequate hydration is essential. The functioning of the sweat gland system is also essential in the dissipation of heat. Sweat evaporation is the body's primary method of heat loss. For each gram of water or sweat that evaporates from the skin, 0.58 kcal of heat is lost.[14] Under heat strain conditions, the sweat glands can be stimulated to produce up to 2 liters per hour (l/hour) in an acclimatized individual, and sweat rates of 1 l/hour would not be unusual in industrial workers.[2] Sweat rates this high cause significant losses of body water. Sweat also contains sodium, chloride and potassium and can account for measurable losses of these electrolytes from the body.[15]

After repeated exposures to heat, the human body gradually develops physiologic mechanisms that improve heat dissipation and tolerance. This adaptation to heat stress, or heat acclimatization, occurs over 5–14 days with daily exercise in hot ambient conditions and of sufficient duration and level of exertion to raise body temperature.[14] Once acclimatized, a worker exposed to the same levels of heat will have a lower core temperature, a lower heart rate, and an increased volume of sweat that contains a lower concentration of sodium. The reduced sweat sodium concentration is the result of aldosterone-mediated reabsorption of sodium and chloride ions in the sweat glands. With acclimatization, sweat rates can increase from 0.6 l/hour to 2 l/hour, and sodium chloride loss can decrease from 15–30 grams per day (g/day) to 3–5 g/day.[14] Overall, the sodium loss is reduced to <5 g/day after acclimatization. The average American diet provides 8–14 g of salt each day, which is adequate for the acclimatized worker. Before acclimatization, salt deficits may occur in heavily sweating workers.[2] However, except for increasing dietary salt during the period of acclimatization, salt supplementation is rarely required.

Generally, physically fit workers are better able to tolerate heat, and they achieve acclimatization more rapidly than non-physically fit workers. Physical training not only makes the muscles more efficient, so that they produce less heat per unit work, but also enhances the body's ability to sweat and dissipate heat. The adaptations of heat acclimatization are rapidly lost when the worker leaves the heat stress environment. Workers who return to hot environments after more than a few days away from the job—especially those who return after an illness—should be allowed to reacclimatize to the hot environment before resuming full-time work.

Heat exchange

Heat exchange between the body and the environment is influenced by air temperature and humidity, skin temperature, air velocity, evaporation of sweat, radiant temperature, and the clothing worn.[2] The heat balance equation incorporates the major modes of heat exchange or loss by the body (Table 6.5).

An additional mode of heat exchange, conductive heat exchange, where there is direct transfer of heat to air or objects in contact with the body, is rarely an important source of heat gain or loss in clothed workers in most occupational settings. Conductive heat exchange may be an important mode of heat transfer for workers, such as divers, working in hot or cold water. Since the human body requires evaporative and convective heat exchange to dissipate most excess internal heat, factors that diminish these modes of heat transfer will cause the body temperature to rise, resulting in hyperthermia. High ambient humidity and clothing are the primary factors in diminishing heat transfer for workers. Humidity decreases sweat evaporation, and

Table 6.5 Heat balance equation.

$$S = (M - W) + C + R - E$$

where:

S = amount of heat gained or lost by the body

$(M - W)$ = total metabolism − external work performed

C = convective heat exchange or heat transferred to or from the skin by the ambient air

R = heat gain from radiative heat exchange or heat transfer by electromagnetic radiation, such as heat from the sun or furnace, and heat lost by radiation from the body

E = evaporative heat loss as sweat evaporates from the surface of the skin

From American Conference of Governmental Industrial Hygienists (ACGIH®), 2000 Threshold Limit Values (TLVs®) for Chemical Substances and Physical Agents and Biological Exposure Indices (BEIs®). Reprinted with permission.

clothing insulates the body and skin from the surrounding air, impedes convective heat loss and interferes with the evaporation of sweat from the skin. Protective clothing, especially sweat-impermeable clothing, which effectively eliminates any body cooling from sweat evaporation, places workers at significant risk for heat strain and heat-related illnesses.

Numerous acute and chronic medical conditions, medications and individual characteristics may diminish the body's ability to cope with heat stress and dissipate internal heat (Table 6.6). Dehydration from any cause, whether due to increased sweating or associated with underlying illness, fever, vomiting or diarrhea, increases the risk of hyperthermia. Many medications, especially anti-hypertensive or cardiac medications that affect the cardiovascular or renovascular systems or medications with anticholinergic effects on sweat glands, may alter a worker's physiologic responses to heat and therefore increase the risk of heat strain. Previous heat-related illness, especially heat stroke, has historically been considered an indicator of both heat intolerance and a possible underlying defect in the individual's thermoregulatory system. One investigation of 10 individuals with previous exertional heat stroke found that nine of them readily acclimatized to heat while one displayed evidence of persistent heat intolerance for almost 1 year following heat stroke. The investigators concluded that a small percentage of individuals with previous heat stroke may be heat-intolerant.[16]

Table 6.6 Risk factors predisposing to heat disorders.

Individual factors
- Increased age
- Obesity
- Lack of acclimatization
- Use of water-impermeable/heavy clothing
- Fatigue/sleep deprivation
- Underlying medical conditions/states
- Dehydration
- Infection/fever/recent immunization
- Overuse of ethanol

Diseases
- Cardiovascular disease
- Renal disease
- Hyperthyroidism
- Diabetes mellitus
- Parkinson's disease
- Skin conditions limiting sweating, including sunburn
- Previous heat disorder
- Pesticide poisoning

Drugs
- Medications with anticholinergic effects
- Antispasmodics
- Tricyclic antidepressants
- Psychotropics
- Antihistamines
- Anti-hypertensive medications
- Diuretics
- Stimulants (decongestants, amphetamine, cocaine)

PATHOPHYSIOLOGY OF ILLNESS AND TREATMENT

The spectrum of heat-related disorders ranges from relatively harmless pruritus due to heat rash or miliaria to fatal heat stroke. In between these extremes are several other conditions caused by exposure to heat.

Heat-related skin conditions

Heat rash, or miliaria, is caused by sweat duct obstruction and resultant sweat retention within the sweat gland. Obstruction of the sweat duct leads to duct rupture within the skin and an inflammatory reaction surrounding the duct. Because the rupture of the sweat ducts may occur within different layers of the skin, three forms of miliaria—crystalline, rubra, and profunda—are described.[17]

Miliaria crystallina is a mild, asymptomatic skin condition consisting of small, clear vesicles resulting from sweat duct rupture within the surface layers of skin. Small erythematous macules resulting from sweat duct rupture within middle layers of the skin and associated with burning and itching is known as miliaria rubra, or "prickly heat". Miliaria rubra commonly affects the skin of the trunk and intertriginous areas of the body. Extensive cases of miliaria rubra involving large numbers of sweat glands can impede body heat dissipation and contribute to more severe heat-related illness. Miliaria profunda results from sweat duct rupture deep within the skin. The lesions, which usually appear only after prolonged periods of miliaria rubra, are small, white to flesh-colored papules and occur most commonly on the trunk. Sunburned skin and occlusive clothing that precludes the free evaporation of sweat increase the risk of all forms of miliaria. Treatment involves reducing sweating in the affected individual and keeping the skin cool and dry. In workers whose jobs require sweat-impermeable protective clothing, the treatment of miliaria may include a temporary transfer to a job not requiring protective clothing or into an air-conditioned workspace. Miliaria rubra usually resolves within a week. Complete resolution of miliaria profunda may take several weeks in a cool environment.

Heat edema

Heat edema is often not considered a true heat-related disorder. It is a rather common condition in which the extremities swell during the first 7–10 days of exposure to higher temperatures. Typically it is found in unacclimatized individuals, often women, who stand or sit for long periods during hot weather and is not associated with cardiac or renal impairment. The etiology of heat edema is uncertain, but it may involve local vasomotor changes or be associated with changes in aldosterone activity.[15] It is sometimes prominent in pregnancy, when heat can aggravate the underlying condition of pregnancy-associated edema. Heat edema usually resolves spontaneously within a few days as the individual acclimatizes. Diuretic therapy is thought not to provide significant relief and is not indicated. Symptomatic treatment—elevation of the legs, compression stockings, and gradual exposure to heat—is generally all that is required.

Heat cramps

Heat cramps are painful muscle spasms that occur during or following intense physical exercise in hot environments. The affected individual is usually acclimatized to heat and gives a history of heavy exertion in the heat, profuse sweating, drinking large quantities of water, and minimal salt or electrolyte replacement. Inadequate electrolyte replacement is probably associated with the underlying mechanism responsible for the muscle spasm.[15]

The muscles involved in the spasm are usually the same muscles used during the preceding exercise, such as the abdominal muscles or the large muscles of the thigh. Heat cramps may be heralded by fasciculations, and while multiple muscle spasms may occur simultaneously, usually only a small section of the muscle is involved. Individual heat-induced muscle spasms last less than a minute; but if untreated, attacks of intermittent heat cramps may last for 4–8 hours.[18]

Heat cramps respond to rest in a cool place and ingestion of 0.1% saline solution (1 teaspoon of salt in a quart of water) or fluids containing electrolytes.[19] Salt tablets should not be given. If nausea and vomiting preclude oral solutions, intravenous electrolyte solutions may be necessary. Heat cramps can be prevented by ensuring that workers, especially acclimatized workers, maintain adequate dietary salt intake in addition to adequate fluid replacement.

Heat syncope

Heat syncope occurs in individuals who stand for prolonged periods, who make sudden postural changes, or who exercise strenuously in the heat. The underlying mechanism of heat syncope is similar to that of orthostatic syncope. Because venous return to the heart is reduced by pooling of blood in dependent extremities or in dilated peripheral vessels, cardiac output is inadequate to maintain cerebral circulation and consciousness. Heat syncope is not associated with elevated body temperature, and the syncope victim may remember a typical prodrome of nausea, sweating and dimming of vision before loss of consciousness. Following syncope and falling to a recumbent position, consciousness returns rapidly. Heat syncope is generally a benign "faint". The primary health concern is the potential for falling and injury, especially for workers on roofs and scaffolding.

Following an episode of heat syncope, the worker should be allowed to recover in a cool area. To ensure that he or she has not been injured by the fall, a medical examination should be performed. The employer should make sure that the victim is hydrated and acclimatized before returning to a job that requires heavy exertion or standing in a hot environment.

Heat exhaustion

Heat exhaustion is a complex of symptoms of fatigue, nausea, headache and giddiness associated with findings of moist, clammy skin, rapid heart rate, low to normal recumbent blood pressures that may fall upon standing, and usually normal to slightly elevated rectal temperature (37.5–38.5°C or 99.5–101.3°F). Since the symptoms of heat exhaustion are similar to those of early heat stroke, all heat exhaustion victims must be evaluated to eliminate the diagnosis of heat stroke. Therefore, careful assessment of core or rectal temperature is essential, and the thermometer used must be capable of accurately recording temperatures above 40°C (104°F). Heat exhaustion occurs more commonly in workers who are unacclimatized to heat and who are without adequate water and salt replacement.[19] Although most types of heat exhaustion are of mixed etiology, two types of heat exhaustion—water-depletion heat exhaustion and salt-depletion heat exhaustion—are described. Water-depletion heat exhaustion occurs when water replacement is inadequate to compensate for the water lost in sweat. Dehydration is present and thirst may be a prominent symptom. Hypernatremia may be present and urine volume will be small. Salt-depletion heat exhaustion occurs when sweat fluid losses are replaced with water but salt intake is inadequate to replace sodium and chloride losses. Dehydration is usually not present and urine volume is normal.

Treatment of heat exhaustion must be individualized based upon the severity of the presenting symptoms and the underlying cause. Mild heat exhaustion may be treated by having the victim rest in a cool area and providing oral fluid and salt replacement. Severe cases of heat exhaustion—especially severe water-depletion type requiring intravenous fluids—need to be referred to an emergency room for careful replacement of body water. Timing of return to work after heat exhaustion has not been fully studied, but it seems prudent to allow at least 24–72 hours for full rehydration and correction of electrolyte abnormalities before the individual returns to work.

Heat Stroke

Heat stroke is a life-threatening medical condition. Symptoms include altered mental status and rectal temperatures >40°C (104°F) and often >42°C(107.6°F).[20] When the body's mechanisms to dissipate heat are overwhelmed by internal or external heat load, body temperature begins to rise. Continued increase in body temperature is associated with partial or complete collapse of the body's thermoregulatory mechanism, and heat stroke occurs.

Heat stroke is a medical emergency requiring immediate cooling of the affected individual. Classic heat stroke occurs during summer heat waves, predominantly in infants or elderly individuals. The condition is most common in poor, elderly individuals who take medications for underlying medical conditions and live in poorly ventilated housing.[21] After days of hot, humid environmental conditions, their ability to maintain body heat balance fails and heat stroke occurs. A second form of heat stroke, exertional heat stroke, occurs in workers, athletes or military recruits who perform vigorous exercise in hot, humid conditions. The symptoms and physical findings of both forms of heat stroke are similar, except that an exertional heat stroke victim may be able to sweat. Mental status changes are the predominant initial presenting symptom. Heat stroke victims may present with any form of mental status change, ranging from irrational behavior, poor judgment and confusion to delirium, seizures, and coma. The skin is usually flushed and hot. Sweating, characteristically absent in classic heat stroke, may be present in exertional heat stroke. The rectal temperature is >40°C (104°F) and may be much higher; the pulse is elevated; the blood pressure is normal or low; and hyperventilation is common. Nausea, vomiting and diarrhea may be present.

Initial laboratory findings often include proteinuria with red blood cells and granular casts also present in the urine, an elevated white blood cell count, and decreased platelet count. Serum electrolyte levels will vary with level of hydration, acid–base status, and underlying tissue damage. Serum enzymes—lactate dehydrogenase, creatinine phosphokinase, aspartate aminotransferase, and alanine aminotransferase—released from damaged muscle and liver cells are characteristically elevated. With hyperventilation, respiratory alkalosis may be present; but in exertional heat stroke, lactic acidosis may be the presenting acid–base abnormality. Abnormalities of blood clotting consistent with disseminated intravascular coagulation—including decreased fibrinogen, prolonged prothrombin time and partial thromboplastin time, and elevated levels of fibrin split products—may be present on initial evaluation.[15]

The treatment for all heat stroke victims is immediate initiation of cooling and appropriate resuscitation. The method of cooling used in military settings is immersion of the victim in ice water and massage of the skin of the extremities. Even though this type of cooling will cause cutaneous vasoconstriction and shivering, thereby decreasing conductive heat loss and increasing metabolic heat formation, it is still probably the most effective method of cooling outside of a healthcare facility.[22] Other rapid cooling methods used for heat stroke include packing the individual in ice or applying ice packs, and wrapping the victim in wet sheets or spraying with cold water and then vigorously fanning with cool, dry air. Specialized body-cooling units that use water spray to enhance evaporative heat loss are also effective in the treatment of heat stroke.[20] Once cooling has been initiated, the victim should be transferred immediately to a hospital for continuous monitoring of core temperature and definitive care. To decrease the possibly of over-cooling, cooling should be discontinued once the rectal temperature decreases to 39°C (102.2°F).[23] Endotracheal intubation and invasive cardiovascular monitoring are often needed during resuscitation and follow-up care.

The severity of sequelae and mortality in heat stroke is determined by the degree and duration of the elevated temperature. After initial resuscitation of an individual with heat stroke, failure or disruption of the function of multiple organ systems is common and should be expected. Frequent sequelae to heat stroke include liver function abnormalities, disseminated intravascular coagulation, rhabdomyolysis, and acute renal failure.[15] Other reported complications of heat stroke include pancreatitis, pulmonary edema, myocardial infarction, and central and peripheral nervous system damage.

MEDICAL SURVEILLANCE

Medical surveillance of heat-exposed workers is one aspect of the overall prevention of heat-related illness. Preplacement and periodic medical examinations of heat-exposed workers should ensure that they can meet the total demands and stresses of the hot job environment without putting their safety and health/or that of fellow workers in jeopardy.[2] The components of the preplacement and periodic medical examinations recommended by the NIOSH are provided in Table 6.7. There are no strict guidelines to identify which heat-exposed workers require medical surveillance. An initial medical history may be taken on all heat-exposed workers to identify those at risk for heat stress. The initial history should elicit information about underlying medical conditions, use of prescription and over-the-counter medications, and previous episodes of heat-related illness. Workers may not require regular medical surveillance if they (1) do not have medical conditions that increase their risk of heat injury, (2) are not exposed to heat above recommended guidelines, and (3) are properly trained in the prevention of heat-related illness. Special categories of workers likely to be exposed to extreme heat stress conditions, such as hazardous waste site workers who wear heavy protective clothing and respirators, should be screened carefully before placement and should receive periodic medical surveillance. One method for predicting heat tolerance has been tested in nuclear power workers.[24] Subjects pedal a bicycle ergometer against a fixed load (or bench-step at the same work rate) in a vapor barrier suit; then their heart rate is compared to a standard. Though this test is readily available, it has not found much use, possibly because of the time, equipment and supervision required to conduct the test.

HEAT EXPOSURE AND REPRODUCTION

Physiologic and hormonal changes during early pregnancy are associated with a slight increase in maternal resting core temperature; however, there is no evidence that the pregnant woman's ability to eliminate excess heat is diminished. In fact, physiologic adaptations during pregnancy—increase in blood volume, increase in cardiac output and resting pulse rate, and increase in cutaneous bloodflow—appear to offset the increased metabolic heat load associated with pregnancy.[25,26] In one study, physically fit, pregnant women appeared to maintain thermoregulation during exercise throughout their pregnancy at least as well as non-pregnant women.[27] During the late 3rd trimester of pregnancy, decreased venous return, due to the size of the uterus, may compromise cardiac output in the pregnant worker and impair heat tolerance.

In experimental animal studies on a variety of species, hyperthermia has been found to be teratogenic.[2] Early in the gestation, excessive heat exposure in animals is associated with structural defects, predominantly of the central nervous system and skeleton, and embryo death. Heat stress in animals later in gestation has been associated with retarded fetal growth and postnatal neurobehavioral defects. Studies of the effect of hyperthermia on the developing human fetus have been primarily concerned with the effect of illness-related fever during pregnancy. The results of these studies have

Table 6.7 NIOSH recommended medical surveillance for heat-exposed workers.

Preplacement medical evaluation		Periodic medical evaluation	
Component	Special emphasis on	Component	Special emphasis on
History		History	
Occupational	Previous heat exposure jobs Use of personal protective equipment	Occupational	Changes in job or personal protective equipment
Medical	Diseases of the following systems: cardiovascular, respiratory, endocrine, gastrointestinal, dermatologic, renal, neurologic, hematologic, reproductive	Medical	Change in health status Symptoms of heat strain
Personal habits	Alcohol and drug use	Personal habits	Update
Medications	Prescription and over the counter	Medications	Update
Characteristics	Height, weight gender, age	Characteristics	Update
Direct evaluation		Direct evaluation	
Physical examination	Cardiovascular, respiratory, nervous and musculoskeletal systems Skin	Physical examination	Systems emphasized in preplacement examination
Blood pressure		Blood pressure	
Clinical chemistry tests	Fasting blood glucose Blood urea Serum creatinine Serum electrolytes Hemoglobin Urinary sugar and protein	Clinical chemistry tests	Fasting blood glucose Blood urea Serum creatinine Serum electrolytes Hemoglobin Urinary sugar and protein
Mental status	Assessment of worker's ability to understand heat, communicate and respond to emergencies	Mental status	Reassessment of ability to understand heat
Detailed medical evaluation	Cardiovascular disease Pulmonary disease Medication use which might interfere with heat tolerance or acclimatization Hypertension Need to use respiratory protection History of skin disease that may impair sweating Obesity Women with childbearing potential	Detailed medical evaluation	Based upon changes in health

Table 6.8 Heat strain/heat-related illness prevention.[2]

Engineering controls
Decrease convection heat gain by worker
Cool air temperature to below mean skin temperature
Increase air movement (if ambient temperature is below skin temperature)
Decrease radiant heat gain by worker
Insulate hot surfaces
Use shielding between worker and heat source
Increase evaporative heat loss by worker
Eliminate humidity sources (steam leaks, standing water)
Decrease air humidity (ambient water vapor pressure)
Administrative and work practices controls
Limit workers' exposure to hot working environment
Use appropriate environmental monitoring
Work during cool parts of day or in the shade
Schedule hot work for cool seasons
Provide cool rest areas
Increase the number of workers for a given job
Use recommended work/rest regimens
Decrease the metabolic heat load
Mechanize heavy work when possible
Rotate heavy work over entire workforce or increase workforce
Decrease shift time; allow liberal work breaks; restrict overtime
Enhance tolerance to heat
Encourage physical fitness in workers
Require minimum level of fitness in certain jobs
Use heat acclimatization program for new workers or workers returning from vacations, layoffs, or illness
Encourage regular fluid and salt replacement
Health and safety training for supervisors and workers
Recognize signs and symptoms of heat intolerance
Emphasize acclimatization, fluid and salt replacement
Avoid conditions increasing risk of heat strain
Use control methods to prevent heat strain
Use protective clothing
Use buddy system, if applicable
Medical screening of workers with heat intolerance
Establish heat alert program
Establish heat alert committee
Reverse plant winterization measures
Ensure water sources, fans and air conditioners are working
Ensure that medical department is prepared to treat heat casualties
Establish criteria for heat alerts
Take all appropriate preventive measures during heat alerts
Post signs identifying heat hazard areas
Protective clothing and auxiliary body cooling
Water-cooled garments
Air-cooled garments
Ice-packet vests
Wetted overgarments
Aluminized overgarments

been inconsistent. Some investigators have reported an association between maternal fevers of 38.9°C (102°F) and abnormal fetal development, whereas others found no association between maternal hyperthermia and adverse pregnancy outcomes.[28] Heat exposure, in addition to its possible effects on female reproduction, is associated with decreases in sperm count and motility in male workers.[29] Considering the potential for reproductive effects due to heat in both females and males, preventive measures to limit excessive heat exposures are essential for all workers.

PREVENTION

The prevention of heat-related illnesses and conditions of unacceptable heat stress can be categorized into four basic areas of control: (1) engineering controls, (2) administrative controls, (3) work practices controls, and (4) protective clothing and devices.[2] Table 6.8 summarizes these prevention measures. Engineering controls, such as shielding and increased air movement, are the most desirable preventive measures. However, they are not effective in many outdoor work sites or work sites requiring full protective clothing. In these cases, administrative and work practices controls—such as reducing heat exposure, reducing work rates, enhancing fitness and heat tolerance, and offering special heat safety training—should be utilized to the fullest extent possible. Adequate fluid replacement in the heat-exposed worker is critical. The sensation of thirst has been proven to be inadequate to prevent hypohydration in heavily sweating individuals.[14] Therefore, workers need to be encouraged to drink adequate amounts of liquid to replace sweat losses. Any water or beverage provided for workers should be cool (10–15°C or 50–59°F), and it should be consumed in small volumes. The use of ice-cold water is not recommended. In the acclimatized worker, sweat rates of 1 l/hour are possible, and fluid replacement with approximately 5–7 oz every 15–20 minutes approximates fluid losses. Maximum gastric emptying in exercising individuals is 1–1.5 l/hour; therefore, larger volumes of fluid are not effective.[30] Overhydration and resultant hyponatremia, while rare, have been reported in marathon runners.[31] Numerous carbohydrate and electrolyte solutions are marketed for fluid replacement in athletes and workers. Studies have shown that the addition of carbohydrates and electrolytes to fluids may be beneficial for athletes who exercise strenuously for long periods.[30] For acclimatized workers whose diet includes sufficient calories and salt, water alone should provide adequate fluid replacement. Because of their diuretic effect, caffeinated beverages should be discouraged as a primary source of fluid replacement.

Personal protective equipment can be very effective in prolonging intervals of heat exposure. Ice-packet vests are the least cumbersome items to wear because they require no umbilical to provide cooling air or water. At work sites where radiant heat sources are the primary source of exposure, aluminized suits are a good choice.

REFERENCES

1. Dukes-Dobos FN. Hazards of heat exposure: a review. *Scand Work Environ Health* 1981; 7:73–83.
2. National Institute for Occupational Safety and Health. *Criteria for a recommended standard ... occupational exposure to hot environments.* Revised criteria. DHHS (NIOSH) publication no. 86-113. Washington, DC: US Government Printing Office, 1986.
3. Department of Labor, OSHA. *OSHA Technical Manual.* TED 1-0.15A. Washington, DC: OSHA, 1999.
4. Centers for Disease Control. Heat-related illness, deaths, and risk factors–Cincinnati and Dayton, Ohio, 1999 and United States, 1979–1997. *MMWR* 2000; 49:470–3.
5. Donoghue AM, Sinclair MJ, Bates GP. Heat exhaustion in a deep underground metalliferous mine. *Occup Environ Med*, 2000; 57:165–74.

6. Minard D, Belding HS, Kigston JR. Prevention of heat casualties. *JAMA* I957; 165:1813–1818.
7. Kark JA, Burr PQ, Wenger CB, et al. Exertional heat illness in Marine Corps recruit training. *Aviat Space Environ Med* 1996; 67: 354–60.
8. Bricknell MCM. Heat illness in the army in Cyprus. *Cccup Med* 1996; 46:304–12.
9. Ramsey JD, Bernard TE. Heat stress. In: Harris RL, ed. *Patty's industrial hygiene*, Vol. 2, 5th edn. New York: John Wiley & Sons, 2000:925–84.
10. American Conference of Governmental Industrial Hygienists. *2000 threshold limit values for chemical substances and physical agents and biological exposure indices*. Cincinnati: American Conference of Governmental Industrial Hygienists, 2000:180–8.
11. Department of the Army. *Prevention, treatment and control of heat injury.* Technical Bulletin MED 507 (TB MED 507). Washington, DC: Department of the Army, 1980. (Also identified as NAVMED P-5052-5 and AFO 160-1.)
12. Yaglou CP, Minard D. Control of heat casualties at military training centers. *AMA Arch Ind Health* 1957; 16:302–16.
13. Sparling PB, Snow TK, Millard-Stafford ML. Monitoring core temperature during exercise ingestible sensor vs. rectal thermistor. *Aviat Space Environ Med* 1993; 64:760–3.
14. Guyton AC. *Textbook of medical physiology.* 8th edn. Philadelphia: WE Saunders, 1991:797–803.
15. Knochel JP. Heat stroke and related heat stress disorders. *Dis Mon* 1989; 35:301–78.
16. Armstrong LE, DeLuca JP, Hubbard RW. Timecourse of recovery and heat acclimation ability of prior exertional heatstroke patients. *Med Sci Sports Exerc* 1990; 22:36–48.
17. Kanerva L. Physical causes and radiation effects. In: Adams RM, ed. *Occupational skin diseases,* 3rd edn. Philadelphia: WB Saunders, 1999:47–8.
18. Leithead CS, Lind AR. *Heat stress and heat disorders*. Philadelphia: FA Davis, 1964.
19. Callaham ML. Hyperthermia. In: Kravis TC, Warner CC, Benson DR, et al, eds. *Emergency medicine*. Rockville, MD: Aspen, 1987:629–37.
20. Khogali M, Hales JRS. *Heat stroke and temperature regulation*. New York: Academic Press, 1983.
21. Semenza JC, Rubin CH, Falter KH, et al. Risk factors for heat-related mortality during the July 1995 heat wave in Chicago. *N Engl J Med* 1996; 35:84–90.
22. Costrini AM. Emergency treatment of exertional heat stroke and comparison of wholebody cooling techniques. *Med Sci Sports Exerc* 1990; 22:15–18.
23. Yarbrough B, Bradham A. Heat illness. In: Rosen P, ed. *Emergency medicine: concepts and clinical practice,* 4th edn. St Louis: Mosby-Year Book, 1998:986–1000.
24. Kenny WL, Lewis DA, Anderson RK, Kamon E. A simple test for the prediction of relative heat tolerance. *Am Ind Hyg Assoc* 1986; 47:203–6.
25. Cunningham FG, MacDonald PC, Gant NE, et al. *Williams's obstetrics*, 20th edn. Stamford, CT: Appleton & Lange, 1997:191–226.
26. Vaha-Eskeli K, Errkola R, Seppanen A. Is the heat dissipating ability enhanced during pregnancy? *Eur J Obstet Gynecol Reproduc Biol* 1991; 39:169–74.
27. Jones RL, Botti JJ, Anderson WM, et al. Thermoregulation during aerobic exercise in pregnancy. *Obstet Gynecol* 1985; 65:340–54.
28. Paul ME. Physical agents in the workplace. *Semin Perinatol* 1993; 17:5–17.
29. Thonneau P, Bujan L, Multigner L, Mieusset R. Occupational heat exposure and male fertility: a review. *Hum Reprod* 1998; 13(8):2122–5.
30. Gisolfi CY, Duchman SM. Guidelines for optimal replacement beverages for different athletic events. *Med Sci Sports Exerc* 1992; 24:679–87.
31. Frizzell RT, Lang GH, Lowance DC, et al. Hyponatremia and ultramarathon running. *JAMA* 1986; 255:772–4.

7

COLD ENVIRONMENTS

Eric T. Evenson, M.D., M.P.H.

OCCUPATIONAL SETTING

Cold is a physical hazard that may affect workers, both indoors and outdoors, virtually anywhere in the world. Workers at risk include construction workers, farmers, fishermen, utility workers, lumberjacks, soldiers, petroleum workers, police, firefighters, postal workers, butchers, and cold storage workers.

Cold injuries may be either freezing (frostbite) or non-freezing (trench/immersion foot and hypothermia) and localized or systemic. While they occur sporadically in the civilian population, in either occupational or recreational settings, cold injuries have been a significant problem in military campaigns. The armies of Xenophon (400 BC), Hannibal (218 BC), Napoleon (1812–13) and Hitler (1941–42) all experienced significant numbers of cold injuries.[1]

MEASUREMENT ISSUES

Two important concepts for evaluating cold exposures are the core body temperature and the wind chill index. Core body temperature is measured by a low-reading rectal thermometer. In the hospital, an esophageal temperature probe is the preferred instrument to monitor core temperature. Ambient temperature is measured by thermometers capable of measuring temperatures down to at least −40°C (−40°F). The wind chill index is used to determine the risk of cold injury. It estimates the relative cooling ability of a combination of air temperature and wind velocity. The wind chill index is measured as the equivalent chill temperature (Table 7.1). In outdoor work situations, wind speed should be measured and recorded, together with air temperature, whenever the air temperature is below −1°C (30.2°F). The equivalent chill temperature should be recorded with these data whenever the equivalent chill temperature is below −7°C (19.4°F).

Physical and Biological Hazards of the Workplace, Second Edition, Edited by Peter H. Wald and Gregg M. Stave
ISBN 0-471-38647-2

Table 7.1 Cooling power of the wind on exposed flesh expressed as equivalent temperature (under calm conditions).

Estimated wind speed (in miles/hour)	Actual temperature reading (°F)											
	50	40	30	20	10	0	−10	−20	−30	−40	−50	−60
	Equivalent chill temperature (°F)											
Calm	50	40	30	20	10	0	−10	−20	−30	−40	−50	−60
5	48	37	27	16	6	−5	−15	−26	−36	−47	−57	−68
10	40	28	16	4	−9	−24	−33	−46	−58	−70	−83	−95
15	36	22	9	−5	−18	−32	−45	−58	−72	−85	−99	−112
20	32	18	4	−10	−25	−39	−53	−67	−82	−96	−110	−121
25	30	16	0	−15	−29	−44	−59	−74	−88	−104	−118	−133
30	28	13	−2	−18	−33	−48	−63	−79	−94	−109	−125	−140
35	27	11	−4	−20	−35	−51	−67	−82	−98	−113	−129	−145
40	26	10	−6	−21	−37	−55	−69	−83	−100	−116	−132	−148
(Wind speeds 40 miles/hour have little additional effect)	**Little danger** In <1 hour with dry skin maximum danger of false sense of security				**Increasing danger** Danger from freezing of exposed flesh within 1 min			**Great danger** Flesh may freeze within 30 sec				
Trenchfoot and immersion foot may occur at any time.												

Threshold limit values (TLVs) for cold stress are based on the wind chill index, and they require workplace temperature monitoring.[2] Suitable thermometry should be available at any workplace where the environmental temperature is below 16°C (60.8°F). Whenever the air temperature at a workplace falls below −1°C (30.2°F), the dry bulb temperature should be measured and recorded at least every 4 h.

EXPOSURE GUIDELINES

Cold stress TLVs are intended to protect workers from the severest effects of cold stress (hypothermia) and cold injury (frostbite) and to describe exposures to cold working conditions under which it is believed that nearly all workers can be repeatedly exposed without adverse health effects.[2] The objectives of TLVs are to prevent the core body temperature from falling below 36°C (96.8°F) and to prevent cold injury of body extremities. For a single, occasional exposure to a cold environment, a drop in core temperature to no lower than 35°C (95°F) should be permitted. However, some clinical signs and symptoms of cold injury would be expected at that temperature.

Whole-body protection, in the form of adequate, insulated, dry clothing, should be provided if work is performed in air temperatures below 4°C (39.2°F). For unprotected skin, continuous exposure should not be permitted when the equivalent chill temperature is less than −32°C (−26.5°F).

Special protection of the hands is required to maintain manual dexterity. If fine work is to be performed with bare hands for more than 10–20 minutes in an environment below 16°C (60.8°F), special provisions should be established for keeping the workers' hands warm. Metal handles should be covered by thermal insulating material at temperatures below −1°C (30.2°F).

Workers should wear anti-contact gloves to prevent contact frostbite. At temperatures below −7°C (19.4°F), a warning should be given at least daily to each worker by the supervisor to prevent inadvertent contact of bare skin with cold surfaces. If the air temperature is −17.5°C (0°F) or less, the hands should be protected by mittens. Workers handling evaporative liquids (gasoline, alcohol, or cleaning fluids) at air temperatures below 4°C (39.2°F) should take special precautions. Workers handling liquified gases (liquid natural gas, liquid oxygen, and liquid nitrogen) must also take special precautions, particularly in the event of a spill.

Provisions for additional total body protection, such as shielding the work area and wearing wind-resistant and water-repellent outer clothing, are required if work is performed at or below 4°C (39.2°F). If work is performed continuously in the cold at an equivalent chill temperature below −7°C (19.4°F), heated warming shelters (e.g. tents, cabins) should be available nearby and used at regular intervals (Table 7.2). The onset of heavy shivering, frostnip, a distant gaze, a feeling of excessive fatigue, drowsiness, irritability or euphoria are indications for immediate return to the shelter.

If available clothing does not provide adequate protection against the cold, work should be modified or suspended until adequate clothing is available or until weather conditions improve. If clothing is wet, workers should empirically change into dry clothing.

Workers should be instructed in safety and health procedures such as proper rewarming methods, appropriate first aid treatment, good clothing practices, eating and drinking requirements, recognition of impending frostbite and hypothermia, and safe work practices. New employees should be allowed to become accustomed to working conditions in the cold and use of the required protective clothing. The weight and bulkiness of clothing should be included in estimating the required work performance. For work at or below −12°C (10.4°F) equivalent chill temperature, there should be constant protective observation ("buddy" system or direct supervision). The work rate should not be so high as to cause heavy sweating that will result in wet clothing, and there should be frequent rest periods.

Eye protection should be provided to workers employed out of doors to protect against blowing snow and ice crystals, ultraviolet light, and glare. Additional workplace requirements exist for refrigerator rooms, working with toxic substances, and exposure to vibration.

NORMAL PHYSIOLOGY

Body temperature is the sum of heat produced internally, plus heat gain and loss from the environment. The body loses heat through radiation, conduction, convection, and evaporation. The most significant heat loss in the cold occurs with cold water immersion or with exposure to low air temperature and strong winds while in wet clothing.

The hypothalamus of the brain controls core body temperature in response to both heat and cold. The hypothalamus is responsible for initiating the body's two main defenses against cold: **peripheral vasoconstriction** and **shivering**. The body can maintain its core temperature by decreasing heat loss (peripheral vasoconstriction) and increasing heat production (shivering). Increasing physical activity also increases heat production. Peripheral vasoconstriction is the initial response to reduced skin temperature. Vasoconstriction directs blood away from the surface of the body to the core, where heat is more easily conserved. Shivering is produced by involuntary muscle contraction and results in increased metabolic heat production, which replaces heat being lost. There is an associated increase in the respiratory rate and heart rate. Shivering may increase the metabolic rate 2–5 fold.[3] However, if the core body temperature is reduced with continued cooling, the metabolic, respiratory and heart rates decrease.

With continued cold exposure, cold-induced vasodilation (CIVD) alternates with peripheral vasoconstriction to conserve core heat and, at the same time, intermittently save function of the extremities.[4] Shunting of blood from the skin to the body core results in cold

Table 7.2 Threshold limit values for work/warm-up schedule for 4-h shifts.

Air temperature—sunny		No wind		5 mile/hour wind		10 mile/hour wind		15 mile/hour wind		20 mile/hour wind	
°F	°C	Maximum work period	Number of breaks	Maximum work period	Number of breaks	Maximum work period	Number of breaks	Maximum work period	Number of breaks	Maximum work period	Number of breaks
−15 to −19	−26 to −28	Normal shifts	1	Normal shifts	1	75 min	2	55 min	3	40 min	4
−20 to −24	−29 to −31	Normal shifts	1	75 min	2	55 min	3	40 min	4	30 min	5
−25 to −29	−32 to −34	75 min	2	55 min	3	40 min	4	30 min	5		
−30 to −34	−35 to −37	55 min	3	40 min	4	30 min	5				
−35 to −39	−38 to −39	40 min	4	30 min	5						
−40 to −44	40 to −42	30 min	5								
≤ −45	≤ −43					Emergency work only					

Adapted from Saskatchewan Department of Labor, Occupational Safety and Health Regulation, Section 70(3) 1996. (www.labor.gov.sk.cq/safety/thermal/cold/index.htm)

diuresis and decreased fluid volume. Osbourne waves (J-waves) are positive deflections of the RT segment of the electrocardiogram that can be associated with either peripheral vasoconstriction or hyperkalemia. Nerves in the skin detect the sensation of cold. However, when the skin freezes, the sensation of cold is lost.

There is some evidence that humans are capable of minor physiologic adaptation and acclimatization or habituation to the cold.[5] However, the ability of humans to acclimatize to heat is much greater than their ability to acclimatize to the cold. Workers in a cold environment require 10–15% more calories.[6] Additional heat is expended working in heavy, protective clothing, and the body must use more calories to keep itself warm.

PATHOPHYSIOLOGY OF INJURY

Injuries from cold exposure are divided into two major types: non-freezing and freezing injuries. Non-freezing cold injuries include hypothermia, chilblains, pernio and trench/immersion foot. Freezing cold injuries include frostnip and frostbite. Other cold-associated problems include accidental injury, sunburn, snow blindness, carbon monoxide poisoning, and cold urticaria.

Hypothermia is the lowering of core body temperature below 35°C (95°F). Hypothermia results when the body is unable to produce enough heat to replace the heat lost to the environment. It may occur at air temperatures up to 18.3°C (65°F) or at water temperatures up to 22.2°C (72°F).

While the incidence of hypothermia is unknown, approximately 700 deaths in the USA each year are attributed to hypothermia, with most occurring in persons aged 60 years or older.[7] Death due to cold exposure occurs more frequently in males than in females. Hypothermia is associated with prolonged environmental exposure to the cold, physical exertion, wind, and becoming wet. The insulating capability of wet clothing is reduced because the layers of trapped, dead-space air are lost. Since the thermal conductivity of water is 20 times that of air, hypothermia occurs rapidly in cold water[8] (Table 7.3).

Table 7.3 Survival times in cold water.

Water temperature (°F)	Survival time (hour)
Over 70	Indefinite (depending on Fatigue)
60–70	Less than 12
50–60	Less than 6
40–50	Less than 3
35–40	Less than $1\frac{1}{2}$
Less than 35	Less than $\frac{3}{4}$

Adapted from United States Coast Guard. National Search and Rescue Manual (FM 20–150; NWP 376; AFM 64-2; CG 308), Amendment No. 6, April 1970.

Trench/immersion foot is due to prolonged exposure to cold water. It occurs in dependent parts of the lower extremities in relatively immobile workers who are partially immersed in cold water. Minor trench/immersion foot injuries occur after 3–12 hours of exposure; significant tissue damage occurs after 12 hours to 3 days of exposure; and severe, amputation-type injuries occur beyond 3 days of exposure (Figure 7.1).

Frostnip is a mild, reversible, superficial freezing cold injury that leads to no loss of tissue. Frostbite is the localized, irreversible freezing of tissue, with formation of ice crystals and disruption of the cells. The capillary walls, particularly the endothelial cells, of the frostbitten area are damaged, increasing cell wall permeability. Fluid is released into the tissues and is accompanied by local inflammation. The most peripheral parts of the body, the toes, fingers, nose, ears, and cheeks, are the most common sites of freezing cold injury. The incidence of freezing cold injuries in the USA is unknown (Figures 7.2 and 7.3).

Risk factors for cold injury may be associated with the agent (cold), the host (human), or the environment (wind chill, humidity, duration of exposure, amount of activity, and protective clothing). A decrease in the equivalent chill temperature and working with cold

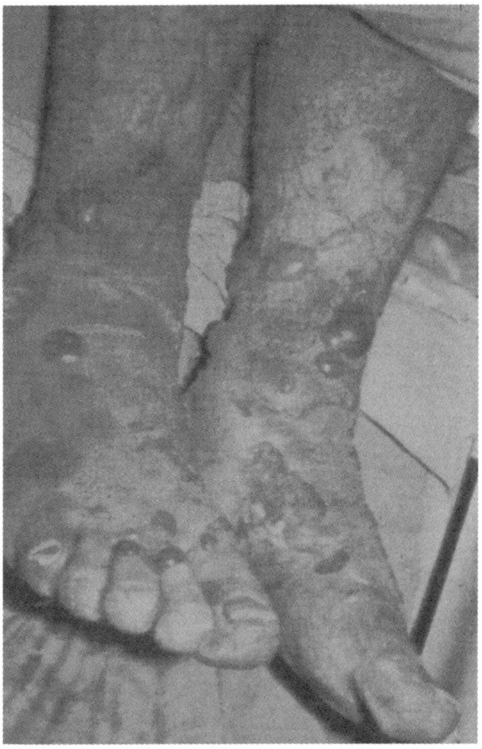

Figure 7.1. Superficial non-freezing cold injury (trench/immersion foot) in the distribution of a soldier's combat boot. Note the erythema, swelling and blister formation.

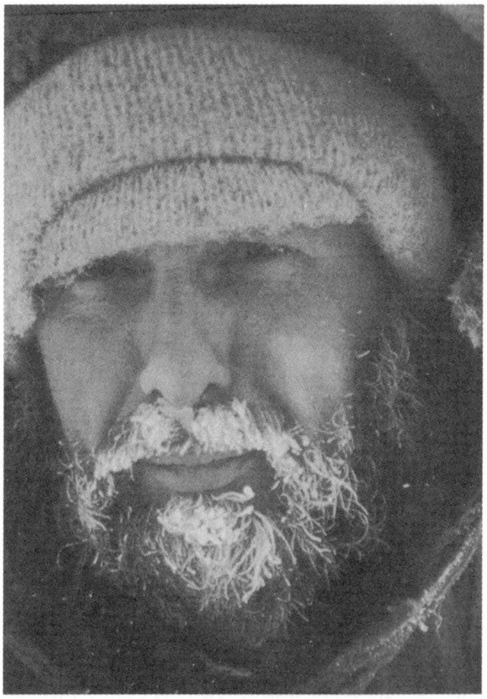

Figure 7.2. Superficial frostbite with a gray–white patch of skin on the right cheek on this mountain climber.

metal objects or supercooled volatile liquid fuels, which conduct heat away from the skin very rapidly, increase the risk of cold injury to unprotected skin. There is significant variation in individual susceptibility to cold injury. Several risk factors for cold injury in the host have been identified.[9] Poor physical condition, fatigue, age (the very young and the very old), inadequate caloric intake, injury, acute or chronic illness (e.g. angina or cardiovascular disease) and a previous cold injury are all risk factors associated with an increased risk of cold injury.

The body requires a large amount of fluids in cold weather. Because individual perception of thirst and the need to drink is suppressed in the cold, dehydration occurs when fluid intake is reduced. Dehydration results in decreased mental alertness, reduced work capacity, and decreased ability to support blood pressure as body temperature drops.

Alcohol, stimulants and prescription drugs also affect the body's cold adaption mechanisms. Alcohol impairs judgment and reduces awareness of the signs and symptoms of cold injury. It produces peripheral vasodilation, which interferes with peripheral vasoconstriction, increasing body heat loss. Alcohol also increases urine output, exacerbating dehydration. Caffeine may have similar effects on blood vessels and urine production. Nicotine increases the risk of a peripheral cold injury by inducing peripheral vasoconstriction, which increases the rate of skin cooling. The use of major tranquilizers (e.g. phenothiazines) also increases the risk of cold injury. Chlorpromazine suppresses peripheral shivering and produces vasodilation.[3]

Snow, ice and reduced visibility increase the incidence of accidents in the cold. Snow

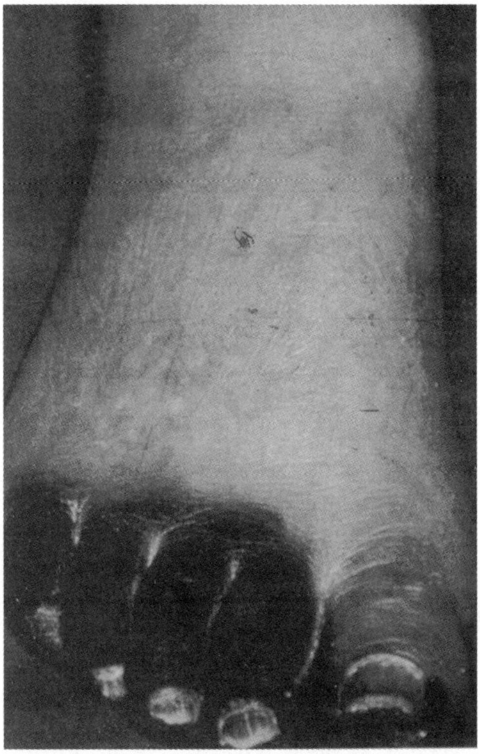

Figure 7.3. Deep frostbite. Black necrosis on the distal half of the great toe, and entire second through fifth toes.

blindness and sunburn may occur when the skin and eyes are unprotected from the ultraviolet rays of the sun and their reflection off the snow. The use of heaters and stoves increases the risk of fire and burns and also carbon monoxide poisoning, particularly in confined spaces. The wearing of heavy, protective clothing and gloves reduces mobility and manual dexterity.

DIAGNOSIS

Hypothermia is insidious in onset and may be difficult to identify. Hypothermia must be recognized to be treated. An appropriate clinical history, physical examination and rectal temperature must be performed to make the diagnosis.

The signs and symptoms of hypothermia may begin to appear at 36.1°C (97°F). Maximum shivering occurs at 35°C (95°F). Progressive decrease in core temperature results in confusion, unusual behavior, impaired coordination, slurred speech, drowsiness, weakness, lethargy, disorientation, and unconsciousness. There is slowing of the heart rate and respiratory rate. The pulse is weak, and blood pressure is decreased. Movements are slow, and deep tendon reflexes are reduced.

From 32.2°C (90°F) to 35°C (95°F) peripheral vasoconstriction and shivering occur. Between 25°C (77°F) and 32.2°C (90°F) shivering is diminished, and peripheral vasoconstriction is lost. Below 25°C (77°F), there is failure of all heat regulatory and heat conservation mechanisms.[10] There is loss of consciousness between 30°C (86°F) and 32°C (89.6°F). Loss of central nervous system function and coma occur below 28°C (82.4°F). Apnea occurs below 27°C (80.6°F) and asystole below 22°C (71.6°F).

Chilblains result from prolonged exposure to wet cold. Chilblains involve the dorsa of the hands and feet, and the skin is red, swollen and warm. There may be blister formation and ulceration. The skin may itch initially, and this may be followed by pain or numbness. Pernio is a severe form of chilblains, characterized by black eschars on the dorsa of the hands and feet, associated with significant pain. While there is usually no long-term impairment, chilblains may recur with subsequent cold exposure.

Trench/immersion foot has three stages. The ischemic stage lasts several days, and the affected area is swollen, cold, numb, and white or cyanotic. The hyperemic stage lasts 2–6 weeks, and there is pain and tingling in the affected area, with redness, swelling, blistering, and ulceration. The post-hyperemic stage may last for months and is characterized by paresthesias, pain, itching, numbness, and cold sensitivity, with gray–blue or black skin. A form of trench/immersion foot known as tropical immersion foot may occur at higher water temperatures.

Frostbite may be classified as superficial or deep.[11] Skin freezes at about −22°C (28°F). Superficial frostbite involves the skin and subcutaneous tissue, and there is no loss of tissue. The skin is gray–white, dry, and hard, with loss of sensation. With rewarming, there is pain, redness and swelling of the skin, and blisters containing clear fluid may form. Deep frostbite involves the skin and subcutaneous tissue, as well as deeper tissues, including muscle and bone. With deep frostbite, the affected area is pale, cold, and solid, and there is loss of tissue. Formation of deep hemorrhagic blisters, death of tissue (necrosis), and ulceration occur. Dry gangrene may develop, with auto-amputation of the dead tissue.

Trench/immersion foot and frostbite are primarily vascular injuries. The extent of a localized cold injury (freezing or non-freezing) cannot be determined until several days after rewarming the injured area. A line of demarcation separates dead from living tissue. The depth and extent of frozen tissue is crucial to the amount of tissue that can be salvaged. If blood vessels are frozen, necrosis will follow in that area.[12]

Among cold-associated problems, cold urticaria is characterized by local or systemic formation of wheals, with redness, swelling, and edema of the skin. The severity of cold urticaria is proportional to the rate of skin cooling and not to the absolute temperature.[10] Carbon monoxide (CO) poisoning is common in cold environments because of the use of unvented combustion heaters. CO poisoning should be considered in all cold, unresponsive patients. Symptoms include an initial headache, followed by confusion, dizziness, and somnolence.

TREATMENT

Hypothermia is a medical emergency. Only conscious patients with mild hypothermia (above 32.2°C (90°F)) should be rewarmed in the workplace. Their core temperature should be determined, and the patients should be prevented from losing additional body heat by insulating them, and then rewarming them with passive external rewarming. They may be placed in a sleeping bag, wrapped in blankets, or exposed to a radiant heat source. Shivering and voluntary physical activity, such as walking, should be encouraged to generate body heat. Warm, decaffeinated, non-alcoholic drinks should be provided to rewarm the body and replace lost fluids.

Severely hypothermic (below 32.2°C (90°F)) patients and unconscious victims of hypothermia are in a life-threatening situation. They should be handled carefully, insulated, provided with intravenous fluids (5% dextrose), and transported to definitive medical care for physiologic monitoring, controlled rewarming, and management of sequelae. Attempts to rewarm these hypothermia victims in the workplace should be avoided.[3]

In severely hypothermic individuals, a core body temperature of less than 25–26.1°C (77–79°F) is a poor prognostic sign. At this temperature, the myocardium is easily irritated and ventricular fibrillation or asystole are significant risks. Resuscitative measures such as active external rewarming, active core rewarming, cardiopulmonary resuscitation or defibrillation should not be performed unless cardiac monitoring capability is available.[3] Patients should be transported to a definitive care facility as quickly as possible. If cardiopulmonary resuscitation is started, it must be continued until the patient has been warmed to 36°C (96.8°F). Since metabolic processes are slowed with hypothermia, asystolic survival times are prolonged. Drowning may occur with sudden immersion in cold water, and resuscitation efforts should continue for the same reasons.

The degree of the cold and the duration of exposure are the two most important factors in determining the extent of a frostbite injury.[11] The keys to frostbite treatment are to protect the tissue from further injury and to increase bloodflow to the interface between injured and uninjured tissue.

Frostnip and mild frostbite may be rewarmed in the workplace by placing the injured part in the armpits or the groin. If there is absolutely no possibility of the tissue refreezing, the frozen tissue may be rewarmed outside the hospital. The rewarmed part should be insulated, and the patient transported to the hospital. Care must be taken not to apply excessive heat to rewarm frozen tissue, since this may produce a devastating secondary burn injury. Ideally, thawing of frozen tissue should occur in the hospital. Tissue should be rapidly rewarmed in a controlled-temperature water-bath (40–42.2°C (104–108°F)). Attention should be given to warming the whole body as well. Prevention of trench/immersion foot is most important, because treatment of this injury is relatively ineffective.

MEDICAL SURVEILLANCE

Selection of workers who will work in a cold environment requires: (1) determination of the physical and mental qualifications appropriate to the specific job, (2) medical evaluation of the individual's physical and psychological ability to work in the cold, and (3) identification of specific medical conditions which may be contraindications to working in the cold. Conditions which may preclude work in the cold include exertional angina, previous cold injury, asthma, peripheral vascular disease, coronary artery disease, alcohol abuse, use of tranquilizers, and thermoregulatory disorders.

Workers should be excluded from the workplace at −1°C (30.2°F) or below if they are suffering from any of these medical conditions, or if they are taking medication which either interferes with normal body temperature regulation or reduces tolerance to work in cold environments.[2] Workers who are routinely exposed to temperatures below −24°C (−11.2°F) with wind speeds less than 5 miles/hour, or air temperatures below −18°C (0°F) with wind speeds above 5 miles/hour, should be certified as medically cleared for such exposures.

PREVENTION

Cold injuries may be prevented by properly protecting workers from a cold environment through the use of appropriate protective clothing and shelter.

The proper clothing system for a cold environment is based on the principles of insulation, layering, and ventilation.[3] Insulation depends on clothing thickness, the properties of the material, and the amount of dead-space air trapped within the garment.[6] The inner layer of clothing, such as polypropylene, should wick moisture to the outer layers. The intermediate layers, such as wool or Thinsulate, provide insulation and may be increased or decreased for appropriate warmth. The outer layer, such as Gor-Tex, should be wind-resistant and water-repellent and allow water vapor and moisture, generated as perspiration, to pass through the layer. The outer layer should also be easily vented to release body heat and allow the evaporation of moisture.

A similar layering system should be used to protect the head, hands, and feet. Gloves should be worn. If fine manual dexterity is required, thin inner gloves may worn under heavier outer gloves or mittens. The outer gloves may be temporarily removed as needed. The head should be protected, since 30% of body heat is lost through the head. Workers at increased risk of cold injury may require additional clothing. All workers should be trained in the proper use of protective clothing.

When wearing cold-weather clothing, the mnemonic "COLD" should be used to guide appropriate clothing maintenance:

- Keep clothing **C**lean to ensure maximum insulation.
- Avoid **O**verheating by adding or removing insulating layers, as appropriate.
- Wear clothing **L**oose and in **L**ayers to allow free blood circulation, trap dead-space air, and adapt to changes in the workload or environment.

- Keep clothing **D**ry to ensure maximum insulation.

Shelter is used to reduce exposure to the cold, when periodic rewarming breaks must be taken. Workers must be encouraged to drink fluids at regularly scheduled times to avoid dehydration. Fluid intake should increase with decreasing temperature and increasing levels of exertion. At least 5–6 quarts of fluid should be drunk each day.[6] Warm, sweet drinks and soups provide calories and fluid volume. Coffee intake should be limited because of its diuretic effects, and alcohol is contraindicated. Caloric intake must be increased in the cold.

Protective eyewear should be used to protect against airborne particulates and ultraviolet radiation. Sunscreen will prevent sunburn, and moisturizers will reduce the effects of dry cold on the skin, lips, and nose. Self-aid and "buddy" aid, with frequent checks, should be used to identify early signs and symptoms of cold injury. Health and safety education on the diagnosis and treatment of cold injuries should be provided to workers. Training should be performed on how to wear and work in cold weather clothing. Most important of all, the worker must respect the cold and use common sense in dealing with it.

REFERENCES

1. Whayne TF, DeBakey ME. *Cold injury, ground type*. Washington, DC: US Army Medical Department, Superintendent of Documents, US Government Printing Office, 1958.
2. American Conference of Government Industrial Hygienists. *TLVs: Threshold limit values and biological exposure indices for 1999.* Cincinnati: ACGIH 1999.
3. Reed GR, Anderson RJ. Accidental hypothermia. In: Wolcott BW, Rund DA, eds. *Emergency medicine annual*: 1984. Norwalk, CT: Appleton, Crofts, 1984:93–124.
4. Hamlet MP. Human cold injuries. In: Pandolf KB, Swaka MN, Gonzalez RRI, eds. *Human performance physiology and environmental medicine at terrestrial extremes.* Indianapolis, IN: Benchmark Press, Inc., 1988:435–66.
5. Guyton AC, ed. *Text book of medical physiology.* Philadelphia: WB Saunders Co., 1971.
6. Young AJ, Roberts DE, Scott DP, Cook JE, et al. *Sustaining health and performance in the cold: a pocket guide to environmental medicine aspects of cold weather operations.* USARIEM Technical Note 93-2. Natick, MA: US Army Research Institute of Environmental Medicine, 1992.
7. Hypothermia associated deaths - United States, 1968–1980. *Morb Mortal Wkly Rep* 1985, Dec 20: 34(50): 753–4
8. United States Coast Guard. *National search and rescue manual* (FM 20-150; NWP 376; AFM 64-2; CG 308) Amendment No. 6. USCG, 1970.
9. Summer DS, Ciblez TL, Doolittle WH. Host factors in human frostbite. *Milit Med* 1974; 139:454–61.
10. Smith DJ, Robson MC, Heggers JP. Frostbite and other cold induced injuries. In: Auerbach PS, Geehr EC, Lewis E eds. *Wilderness medicine: management of wilderness and environmental emergencies,* 3rd edn. New York: Macmillan Publishing Co., 1995:101–18.
11. Washburn B. Frostbite: what is it—how to prevent it—emergency treatment. *N Engl J Med* 1962; 266:974–89.
12. Burtan RC. Work under low temperatures and reactions to cold stress. In: Zenz C, ed. *Occupational medicine: principles and practical applications,* 3rd edn. Chicago: Year Book Medical Publishers Inc., 1994:334–42.

FURTHER READING

Baker-Bloker A. Winter weather and cardiovascular mortality in Minneapolis, St Paul. *Am J Public Health* 1982; 72:3.

Bangs C, Hamlet M. Out in the cold—management of hypothermia, immersion, frostbite. *Top Emerg Med* 1980; 2:19–37.

Belding HS. Physiologic principles for protection of man living in the cold. In: Fisher FR, ed. *Man living in the arctic.* Washington, DC: National Academy of Sciences, National Research Council, 1961.

Centre for Disease Control. Hypothermia—United States. *MMWR* 1983; 32:46–8.

Centre for Disease Control. *Extreme cold: a prevention guide to promote your personal health and safety.* Atlanta: US Department of Health and Human Services, CDC, 1996.

Collins KJ. *Hypothermia: the facts.* New York: Oxford University Press, 1983.

Danzl DF, Pozos RS. Accidental hypothermia. *N Engl J Med* 1994; 331:1756–60.

Danzl DF, Pozos RS, Hamlet MP. Accidental hypothermia. In: Auerbach PS, Geehr EC, Lewis E, eds. *Management of wilderness and environmental emergencies,* 2nd edn. New York: Macmillian Publishing Co., 1991:1756–60.

Eisma TL. Handling the cold with dexterity. *Occup Health Safety,* 1991; 60(12):16–19.

Fitzgerald JC. Accidental hypothermia: a report of 22 cases and review of the literature. *Year book of medicine.* Chicago: Year Book Medical Publishers, Inc., 1982:127–50.

Gonzalez RR. Working in the north: physical aspects. *Arct Med Res* 1986; 44:7–17.

Kilbourne EM. Cold environments. In: Naji EK, ed. *The public health consequences of disasters.* New York: Oxford University Press, 1997:270–86.

Killian H, ed. *Cold and frost injuries.* New York: Springer-Verlag, 1981.

Mills WJ. Frostbite and hypothermia—current concepts. *Alaska Med* 1973; 15:26–147.

Mills WJ. Out in the cold. *Emerg Med* 1976; 8:134–47.

National Safety Council. *Pocket guide to cold stress.* Chicago: NSC, 1985.

Sellers EA. Cold and its influence on the worker. *J Occup Med* 1960; 2:115–17.

Steinman AM, Hayward JS. Cold water immersion. In: Auerbach PS, Geehr EC, Lewis E, eds. *Management of wilderness and environmental emergencies,* 2nd edn. New York: Macmillan Publishing Co., 1991:77–100.

Vaughn PB. Local cold injury—menace to military operations: a review. *Milit Med* 1980; 145:305–11.

8

HIGH-PRESSURE ENVIRONMENTS

David J. Smith, M.D., M.S.

Humans function well only within a narrow range of barometric pressures. Outside this range, they are subject to major physiologic stresses that occasionally result in disease. On land, workers are exposed to hyperbaric environments (i.e. increased barometric pressure) during tunneling projects that require the use of compressed air or when caissons are used to work in ground saturated with water. In addition, hyperbaric chamber support staff members are routinely exposed when they treat patients in hyperbaric medical treatment facilities. In the water, occupational exposures are diverse. Examples of exposures include breath-hold divers, such as the Ama pearl divers of Japan, and compressed gas divers, ranging from instructors of recreational SCUBA students, who breathe compressed air, to saturation divers supporting offshore oil exploration, who dive in excess of 1000 feet of sea water (FSW) while breathing artificial gas mixtures. Divers can also be found inland. These divers are involved in such jobs as inspecting dams and reservoirs, cleaning filters, maintaining fish farms, placing underwater demolitions, and conducting police searches.

The first practical method developed for conducting useful work underwater was the diving bell, which is essentially an upside-down cone. Invented by Smeaton in 1778, the diving bell was the forerunner of the modern caisson (caisse, in French, means "box").[1] This method is still used today in the construction of tunnels and bridge footings. In 1819 in England, Augustus Siebe invented the first practical diving dress; it consisted of a copper helmet bolted to a leather coverall. This appa-

The views expressed in this Chapter are those of the author and do not necessarily reflect the official policy or position of the Department of the Navy, Department of Defense, or the US Government.

Physical and Biological Hazards of the Workplace, Second Edition, Edited by Peter H. Wald and Gregg M. Stave
ISBN 0-471-38647-2

ratus gave humans the ability to walk and function relatively unencumbered underwater without holding their breath. In 1837, Siebe introduced an improved version of this gear, which served as the basic diving dress for deep-sea diving until the early 1980s and is still used by some divers today.[2] Modern surface-supplied equipment has incorporated various advances in technology, such as helmets made of lightweight composites with increased fields of vision and improved gas delivery systems that reduce breathing resistance.

Along with these advances, which have enabled divers and caisson workers to work at greater pressures or depths for longer periods of time, have come associated medical problems. The first descriptions of decompression-related disorders were made by Triger in 1841 among pressurized tunnel workers[3] and by Alfonse Galin 1872 with Greek sponge divers.[4] By the early 1900s, it was recognized that decompression related symptoms were due to inert gas and could be relieved by returning the individual to pressure. However, there was still no method for controlling the exposure to prevent the disease. The Royal Navy enlisted the help of the eminent physiologist J. S. Haldane to develop such a method. Haldane published his first set of decompression tables in 1908, based on experiments using sheep.[3] The tables incorporated delays on ascent (called stops) to allow time for excess inert gas dissolved in body tissues to be eliminated (off-gassing). Haldanian principles form the basis for most decompression tables used today.

OCCUPATIONAL SETTING

Diving

Diving operations can be classified into three basic categories: air, mixed gas, and saturation. Air diving is restricted to relatively shallow depths, due to the increasing narcotic effect of the nitrogen component of air as depth increases (the Occupational Safety and Health Administration (OSHA) restricts exposures on air to $<$190 feet, but will permit surface supplied air dives to 220 feet for less than 30 minutes.). Mixed-gas diving uses a breathing mixture other than air; it is employed principally when deeper working depths are required. It generally specifies the use of helium as the inert gas constituent; however, nitrogen–oxygen mixtures (nitrox) are becoming more common for relatively shallow applications, and hydrogen has been tested for deep commercial applications. Although greater depths are permitted in standard mixed-gas diving, they are also limited by the amount of decompression required. For a constant bottom time, the required decompression time increases significantly with depth, prolonging exposure to the environment and limiting the useful work period. To overcome this problem, saturation diving methods have been developed.

Saturation diving, a specialized extension of standard decompression diving, is based on the principle that at a constant depth, tissues will on-gas until the tissue partial pressure reaches equilibrium with the ambient pressure (see Henry's law, below). Once tissue equilibrium is reached, no net uptake will occur (unless the pressure is increased), resulting in no further increase in decompression time. At this point, the bottom time may be increased without additional decompression penalty. The divers are housed at "depth" for up to a month in a dry, pressurized chamber on the surface and transported under pressure to and from the work site via a diver transport capsule (Figure 8.1). Saturation diving techniques represent a cost-effective alternative to standard diving when deep or prolonged bottom times are required, since the need for repetitive, long, in-water decompressions is avoided.

Equipment

There are two basic classes of equipment employed by divers today: self-contained underwater breathing apparatuses (SCUBA) and surface-supplied equipment. Choice of equipment depends primarily on the job requirements and the employer's capabilities.

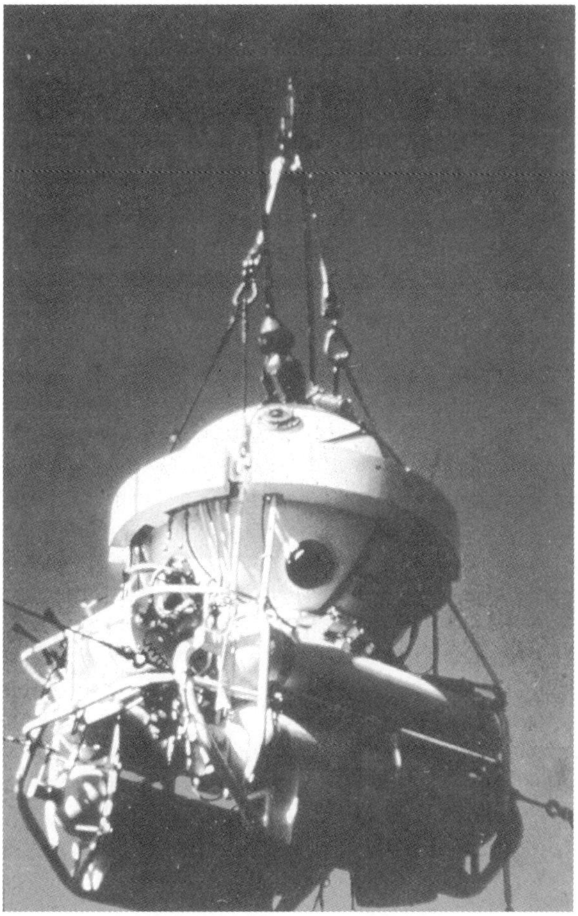

Figure 8.1 A saturation system diver transport capsule. (Courtesy of D. R. Chandler.)

Table 8.1 High-pressure and diving environments.

Class	Equipment type	Type of air supplied
SCUBA	Open-circuit Closed-circuit Semi-closed-circuit	Air
Surface-supplied	Lightweight Deep-sea Saturation	Air Mixed gas Mixed gas
Caisson	Shirt-sleeve	Pressurized air

Equipment generally used in high-pressure and diving environments is summarized in Table 8.1.

Modern SCUBA has evolved from equipment originally developed during World War II by Cousteau and Gagnon.[2] The advent of SCUBA ushered in a new era of underwater mobility. SCUBA equipment is designed either to be open-circuit, closed-circuit, or semi-closed (a hybrid between open-circuit and closed-circuit). Modern open-circuit SCUBA (Figure 8.2) is the principal equipment used in recreational diving; it is also used for many commercial applications. Closed-circuit SCUBA (Figure 8.3) removes the exhaled carbon dioxide from the breathing gas prior to returning the "scrubbed" gas to the diver. Oxygen is added as needed. Some closed-circuit rigs maintain a constant partial pressure of oxygen in the breathing gas, increasing depth capabilities. Closed-circuit equipment is generally more complex and expensive, but it has the advantage of less gas consumption without the generation of bubbles. The absence of bubbles is useful in marine research, marine photography, and various military applications.[5]

Many advances have been made in surface-supplied diving since the 1960s, including specialized materials for helmets and suits, hot water heating for suits, and advanced communication systems. Surface-supplied diving equipment can generally be divided into two types: lightweight and deep-sea. Deep-sea equipment can be further subdivided into air, mixed-gas, and saturation-capable. Lightweight equipment, such as a "band mask", is customarily used for work where surface communications are required, but significant diver protection is not necessary. Deep-sea diving equipment (Figure 8.4) affords increased protection, generally consisting of a "hard hat" offering maximum head/neck protection along with surface-to-diver communication, protective gloves and boots, weights, an emergency gas supply, and varying levels of thermal protection, as complex as suits that circulate hot water supplied from the surface.

As in most occupational settings, the choice of equipment used in diving generally depends on the characteristics of the environment and the work to be completed, along with the personal preferences and the capabilities of

Figure 8.2 SCUBA equipment. (Courtesy of U.S. Navy)

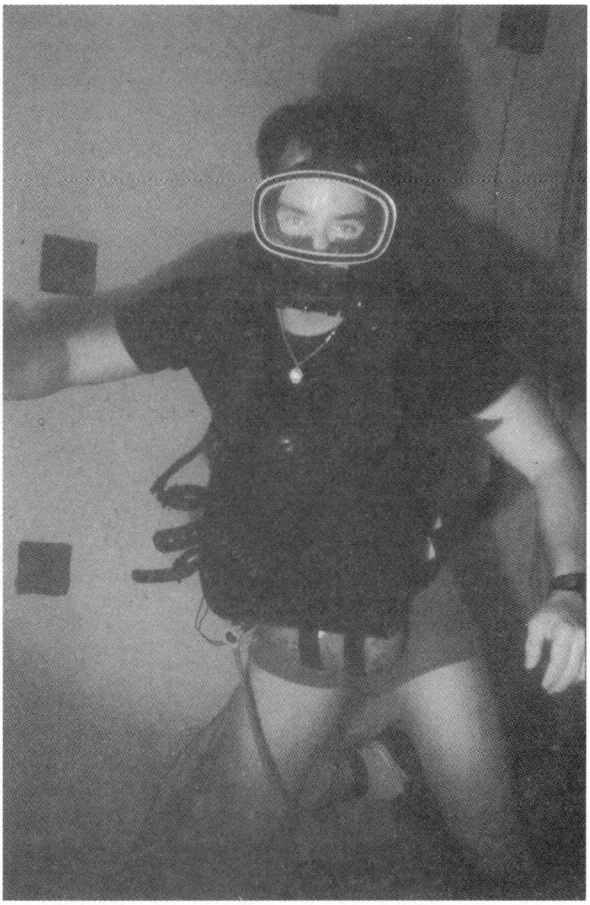

Figure 8.3 Diver using Drager LAR V closed-circuit SCUBA equipment. (Courtesy of T.J. Doubt.)

the individual diver. The principal advantages of SCUBA diving are its ease of use, transportability, and freedom of movement. Its disadvantages include limited gas supply, depth restrictions, minimal head protection, lack of communication, and inability to function safely in strong currents. Surface-supplied diving overcomes many of the disadvantages of SCUBA by providing: increased head protection, thus reducing the chance of drowning if the diver becomes unconscious; "unlimited" gas supply; better buoyancy control, thus enabling the use of heavy construction techniques; hard-wired communications; and greater depth capabilities (Figures 8.5 and 8.6). However, surface-supplied diving is more complex and costly, because it requires significantly more surface support in both equipment and personnel. In addition, it decreases the diver's mobility in the water and comfort on the surface (due to weight, etc.), and it can present significant noise hazards.

Caissons

A caisson is used for engineering projects where water or waterlogged soil precludes standard construction techniques. It can be envisioned as an inverted cone or diving bell that rests on the bottom. It is pressurized to exclude water and to allow a relatively dry working environment. Similar principles are also applied in the construction of tunnels

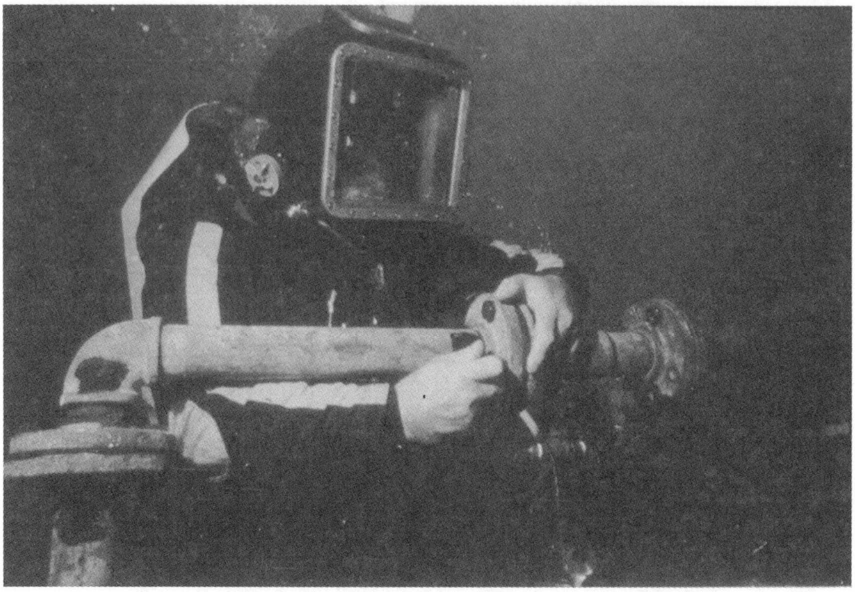

Figure 8.4 Diver using surface-supplied "hard" helmet rig with emergency gas supply. (Courtesy of D. R. Chandler.)

through waterlogged or unstable, "mucky" earth. Workers enter and exit the workplace through a pressurized lock system where decompression may be carried out on exit. At depth, the workers may labor for up to 8 h at pressure. Decompression is dependent on the exposure time and the pressure. The length of the decompression is determined from tables developed specifically for this purpose. Fire safety is of concern in pressurized environments, due to the higher partial pressures of oxygen; however, modern systems are constructed from steel, minimizing flammable materials.[6] The OSHA provides guidelines for safety in caisson work (29 CFR 1926.801).

The need for caisson techniques has been gradually declining as other engineering methods, such as pressure-balanced shields and unmanned excavating systems, have been developed to avoid the cost and complexity of caisson work. Currently, there is limited caisson work being done in the USA. Non-caisson construction is less costly since it eliminates both high-pressure equipment and the extra labor costs related to high-pressure work. However, with automated equipment, men occasionally need to enter a high-pressure compressed air environment to repair or maintain the equipment.[7]

The majority of hyperbaric exposures today occur in the undersea environment. Since caisson work is very limited, the remainder of this chapter will focus on undersea hyperbaric exposures. The pathophysiology, diagnosis, and treatment of hyperbaric exposures and their sequelae are similar in caisson and undersea workers.

MEASUREMENT ISSUES AND PHYSICS OF PRESSURE

Pressure is a measurement of force per unit area. Air pressure is the force exerted by the column of atmosphere above a particular point. The height of the column determines the pressure. Water pressure, measured by a gauge, is defined as the force exerted by the column of water above the submerged object. Pressure increases linearly with depth. Absolute pressure measures the force per unit area of the combined air and water column. The

Figure 8.5 Surface-supplied divers using hand tools. (Courtesy of U.S. Navy.)

absolute pressure at 33 ft of sea water is 2 atm, since 1 atm of pressure is contributed by both the air column and the water column. Table 8.2 defines some useful measurements of pressure in diving.

An understanding of the physiologic effects of pressure requires knowledge of basic physics. The relationship between pressure, volume and temperature is defined by the ideal gas law, which states:

$$PV = nRT$$

where P is absolute pressure, V is volume, n is the number of moles of the gas, R is the universal gas constant, and T is the absolute temperature.

Boyle's law defines the inverse relationship of pressure and volume where the temperature is held constant and the number of moles of gas is constant—i.e. no gas enters or exits the system ($P_1V_1 = P_2V_2$). For example, if a balloon is filled with 2.0 l of compressed gas at a depth of 99 ft (4.0 atm abs) and brought to 33 ft (2.0 atm abs), the gas volume will have expanded to 4.0 l ($V_2 = P_1V_1/P_2 = 4 \times 2/2 = 4$). If taken to the surface (1.0 atm abs), the volume will have expanded to 8.0 l ($V_S = P_1V_1/P_S = 4 \times 2/1 = 8$). It is important to note that the proportional change in volume per depth change increases toward the surface.

The law of partial pressures (**Dalton's law**) states that, in a mixture of gases, the total gas pressure is the same as the sum of the partial

Figure 8.6 Surface-supplied divers using air-powered tools. (Courtesy of U.S. Navy.)

Table 8.2 Units of pressure used in hyperbaric environments.

10.00 m of sea water (msw)[a] = 32.646 ft of sea water (fsw)[b]
10.00 msw = 1 bar = 100 kilopascals (kPa)[c]
1 atmosphere (atm) = 1.013 bar
1 atm = 760 torr (mm of mercury)
= 1033 cm of water
= 14.69 lb per in^2
= 33.08 fsw

[a] The definition of fsw (feet sea water) assumes a density (weight per unit volume) for sea water of 1.025 at 40°C.

[b] The definition of msw (meters sea water) assumes a density for sea water of 1.020 at 40°C.

[c] The unit for the pascal is defined as a Newton per meter squared.

pressures of the individual gases in the mixture ($P = p_1 + p_2 \cdots p_n$). As a result, the partial pressure of the gas can be calculated by knowing the total pressure of the mixture and the percentage makeup of a particular component.

Henry's law governs the solubility of gases in tissues and states that the amount of a gas that will dissolve in a liquid at a given temperature is almost directly proportional to the partial pressure of that gas. Therefore, as the pressure increases (depth increases), the amount of gas dissolved in tissues will increase. Once at a constant depth, gas will continue to increase in the tissues until equilibrium or saturation is attained (i.e. when the partial pressure of the gas in tissue equals the ambient pressure). Gases such as oxygen may be metabolized, whereas inert gases will not be. Subsequently, when the ambient pressure is decreased, inert gases will come out of solution. This law explains why nitrogen bubbles may form when you surface from an air dive and why a can of carbonated beverage fizzes when it is opened.

Archimedes' principle, which describes buoyancy, states that an object immersed in a fluid is buoyed up by a force equal to the weight of the volume of fluid that the object displaces. The effect of this principle is seen throughout diving. For example, inhaling a large breath will increase the diver's buoyancy (due to greater displacement of fluid).

EXPOSURE GUIDELINES

Commercial diving in the USA is covered by various regulations, including OSHA standard Title 29 Code of Federal Regulations 1910 Subpart T—Commercial Diving Operations, and Title 46 Code of Federal Regulations, Sub-chapter V—Marine Occupational Safety and Health Standards, Subpart B—Commercial Diving Operations. Diving contractors and operators generally have developed local standard operating procedures that define procedures and safety guidelines in more detail. Caisson work, and other high-pressure construction work is covered by OSHA Standard Title 29 CFR 1926 Subpart S—Underground Construction, Caisson, Cofferdams, and Compressed Air.

SPECIAL UNDERWATER STRESSORS

The worker in the undersea environment is exposed to a number of unique environmental stressors that may be enhanced by pressure. These environmental factors include sensory input modifications (sound, visual, proprioceptive), thermal challenges, and gas effects. Professionals generally do not have the luxury of waiting for ideal environmental conditions. As a result, most operations are conducted when at least one element is not ideal (Figure 8.7). Medical planning for diving operations should take this into consideration.

Sensory changes

Divers must tolerate seriously reduced sensory input while working, forcing increased alertness and vigilance.

Sound Sound travels approximately four times faster in water than in air; as a result, it travels further. Because of this increased speed, the normal delay cues required to place sound in three dimensions are in practice lost, making the localization of sound extremely difficult. Moreover, the increased density of air and increased middle-ear impedance inhibits air conduction, causing a $\leq$40 dB loss.[8] Sound does not transmit across an air–water interface efficiently. As a result, vocal communications through the water are effectively ruled out, thus adding to the sensory deprivation already present secondary to other effects.[8] In addition, many surface-supplied helmets are noisy secondary to their design, which compounds hearing difficulties. Divers

Figure 8.7 Surface-supplied USN MK-12 diver in typical operating conditions for professional divers. (Courtesy of D. R. Chandler.)

may sustain aural trauma from underwater blasts or activation of sonar devices while submerged.

Vision The underwater environment seriously affects vision in a number of ways. First, water transmits light poorly. This is caused principally by turbidity, i.e. suspended particles that obstruct light. At 10 m in clear water, ~60% of the light is filtered out, leaving only 40% visible light.[9] However, in highly turbid waters, like those found in many ports and inland waters, light may be reduced substantially more than this.

In addition, color vision is lost. As depth increases, long wavelengths (red end of the spectrum) are filtered initially, followed by the blues. In the absence of artificial light, perception of red is generally eliminated by 10 m, followed by yellow by ~20 m.

As the diver descends, the eyes must adapt from day to night light levels. However, the working diver generally descends faster than the retina can compensate, further compounding the visual adjustment.[10] Objects appear 25% larger and closer underwater because the refractive index of light is increased ~1.3 times over that in air. This causes problems with determination of distances and eye–hand coordination that improve somewhat, although not completely, with experience. Finally, if the cornea comes into direct contact with water—

e.g. in the absence of the air interface provided by a mask or helmet—a substantial hyperoptic refractive error of 40–50 diopters (D) results.[11]

Proprioception Proprioceptive inputs for orientation to the diver are reduced by buoyancy and the diver's protective clothing. As indicated above, the diver additionally receives fewer visual cues. On land, workers rely on visual, proprioceptive and vestibular inputs to maintain balance and orientation in space. This capacity is significantly impaired by the underwater environment.[9]

Thermal transfer Maintaining a constant body temperature while submerged is a challenge, even in relatively warm water. Heat is lost during water immersion, primarily through convection and conductance from the skin and through the respiratory tract. Water has a high coefficient of thermal conductivity (~25 times greater than air), and submersion effectively eliminates the normal protective air insulation layer. As a result, an unprotected diver in 27°C (80°F) water loses heat at the same rate as an unprotected subject in 6°C (42°F) air.[12] A water temperature of ~35°C (95°F) is required to keep a resting, unprotected, immersed diver thermoneutral.[13] To protect from this loss, passive insulation, such as neoprene, is used for mild to moderate cold shallow applications. Wet suits are generally made of neoprene, a material that is compressible and as a result loses insulation capacity with depth. For deeper, colder dives, dry suits (passive) and hot water suits (active) are used. Dry suits consist of an impermeable outer barrier and an insulating inner garment made of a variety of different materials. Dry suits can be inflated with pressurized gas as depth increases to maintain the insulation barrier. Active heating garments, such as hot water suits, which bathe the diver in warm water, are employed in very cold and deep diving, but they require more technical support.

At greater depths (>100 m), heat loss from the respiratory tract becomes significant secondary to the high heat capacity of the dense gas. This respiratory heat loss may not be sensed and therefore can cause asymptomatic hypothermia. Breathing gas is heated to prevent this phenomenon. Occasionally, in unique circumstances, the diver may be exposed to increased water temperatures. Such dives require special planning and measures to protect against hyperthermia, since the diver will be unable to self-regulate body heat by the evaporation of sweat. Hypothermia and cold exposure are discussed in detail in Chapter 7.

Gas effects

Table 8.3 summarizes the toxic effects of gases on divers.

Inert gas narcosis Inert gas narcosis is defined as the progressive development of symptoms of intoxication/anesthesia with increasing partial pressures of the gas. Inert gases vary substantially in their ability to induce narcosis at a given partial pressure of exposure. Xenon is anesthetic at 1 atm, whereas helium has little narcotic effect even at very great depths.[14] The narcotic effect of a specific inert gas is related to its lipid solubility; however, the precise pathophysiologic mechanism of narcosis is not well understood.

As a practical point, the depth or pressure to which a diver may descend while breathing air is restricted principally by nitrogen narcosis. Significant individual variability in response to the effect exists; but on average, at 100–200 ft (30–60 m), an individual feels lightheaded and euphoric (similar to alcohol intoxication) and experiences decreased reasoning capability, reaction time, and manual dexterity.[15] At 200–300 ft, reflexes slow, paresthesias may develop, and decrements in judgment and reasoning may produce dangerous overconfidence.[9] At 300–400 ft, marked impairment of judgment,[14] anesthesia, progressive depression of the sensorium with auditory and visual hallucinations and amnesia

Table 8.3 Toxic effects of gases on divers.

Gas	Causes/sources	Symptoms
Inert gas narcosis	Increased partial pressure of inert gas	100–200 FSW Lightheaded Decreased complex reasoning Loss of fine discrimination Euphoria 200–300 FSW Poor judgment/reasoning Slowed reflexes Paresthesia Dangerous marine life 300–400 FSW Progressive depression Auditory/visual hallucinations Loss of memory >400 FSW Unconsciousness
Oxygen toxicity Central nervous system (acute)	>1.3 atm abs partial pressure of oxygen Risks increase with increase in partial pressure	Visual disturbance Tinnitus/decreased visual acuity Nausea/vomiting Twitching Irritability/restlessness Vertigo Convulsions (may occur with no prodromal symptoms from above; no long-term consequences)
Pulmonary (chronic)	Prolonged exposure to >0.5 atm abs oxygen	Chest pain with inspiration Inspiratory irritation Coughing Progressive shortness of breath
Hypoxia	Hyperventilation prior to breath-hold diving Improper gas mixture	Unconsciousness (without symptoms)
Carbon dioxide toxicity	Equipment design Increased gas density Increased partial pressure of oxygen blunts response to CO_2 CO_2 contamination	Increased breathing rate Shortness of breath Headache Unconsciousness
Carbon monoxide toxicity	Use of improper compressor lubricants (flashable) Compressor failure Improper intake placement (e.g. next to combustion engine)	Headache Nausea/vomiting Unconsciousness

precede syncope. Finally, at depths >400 ft on air, the diver becomes unconscious.[9]

Primary prevention involves limiting the depth of exposure and substituting a less narcotic gas, such as helium. Individuals do not acclimatize to narcosis; however, repeated exposures help a diver adjust to this effect.

Oxygen toxicity Because oxygen is a mandatory constituent of breathing gas, it would be logical to postulate that diving on 100% oxygen (O_2) would solve the problems associated with inert gas during hyperbaric exposures. Applications of 100% O_2 breathing are restricted, however, because oxygen in the hyperbaric environment becomes increasingly toxic to the central nervous system (CNS). Breathing oxygen with a partial pressure as low as ~1.3 atm abs (equivalent on the surface to 130%) may cause acute CNS manifestations. The risk of CNS oxygen toxicity increases as the oxygen partial pressure increases. The most serious CNS manifestation secondary to oxygen is a tonic–clonic seizure, which may occur without prodromal symptoms. If a seizure occurs underwater, it may cause drowning. Prodromal manifestations include:[9]

- V = Visual disturbances, such as tunnel vision
- E = Ear problems, including tinnitus or decreased acuity
- N = Nausea and vomiting
- T = Twitching
- I = Irritability and restlessness
- D = Dizziness and vertigo

Manifestations of CNS O_2 toxicity are treated by decreasing the partial pressure of oxygen. Divers try to prevent toxicity by limiting their exposure time and the partial pressure of oxygen breathed. Factors that are thought to predispose to oxygen convulsions include exercise, carbon dioxide retention, and water immersion itself. No long-term health consequences from CNS oxygen toxicity have been demonstrated. Oxygen seizures per se are of less clinical importance than other seizures because the recipient is well oxygenated prior to their occurrence.

Prolonged exposures to partial pressures of oxygen >0.5 atm abs may cause clinically apparent pulmonary toxicity. Usually, only saturation diving scenarios and hyperbaric oxygen therapy, e.g. a prolonged therapy associated with decompression illness, are sufficiently prolonged to cause clinically apparent pulmonary toxicity. However, in rare cases, pulmonary oxygen toxicity may become a limiting factor in repetitive deep bounce dives, which require prolonged decompressions. Manifestations of pulmonary oxygen toxicity are the same as those of patients on ventilators exposed for prolonged periods to increased partial pressures of oxygen. These include: (1) substernal chest pain, which begins as irritation on inspiration and progresses with continued exposure to severe, burning chest pain during both inspiration and expiration; (2) coughing, which gradually increases in frequency and duration with exposure; and (3) progressive shortness of breath.[9] Physical examination is generally unremarkable and chest radiographs are clear except in severe cases. In practice, within a few hours after removal from exposure, symptoms begin to resolve, and they generally resolve completely within 24–48 hours (although if severe, the forced vital capacity may take substantially longer to recover). Limiting exposures is the only preventive measure available.

Hypoxia Hypoxia is particularly hazardous, because it can cause unconsciousness without warning. Although hypoxia rarely occurs in air diving, it may occur in mixed-gas diving secondary to procedural errors, such as improper gas mix or mechanical failures. Individuals are also prone to hypoxia during breath-hold diving, which is preceded by hyperventilation. This is a result of decreased pre-dive carbon dioxide (CO_2) levels secondary to hyperventilation. Since elevated CO_2 levels drive the need to breathe, low initial CO_2 levels permit divers to comfortably overstay their permissible dive time at depth (secondary to elevated partial pressure of oxygen at depth). On ascent, the individual becomes hypoxic as the partial pressure of oxygen decreases below safe levels.

Carbon dioxide toxicity Carbon dioxide build-up or retention is common in the hyperbaric environment. Causes include:

- Increased work of breathing caused by increased gas density (Boyle's law), increased breathing resistance and dead space as a result of the use of breathing equipment. Added dead space and breathing resistance is dependent on the design of the breathing equipment.
- Increased partial pressure of oxygen, which blunts the body's response to elevated partial pressures of CO_2.
- CO_2 in the breathing gas secondary to failure of CO_2 absorbent, poor gas analysis, or contaminated breathing gas.
- Deliberate reduction of ventilation. With some breathing equipment designs, ventilation to the helmet causes a great deal of noise, which interferes with communications. Divers sometimes reduce helmet ventilation to reduce noise levels and facilitate communications, thus predisposing to CO_2 build-up.

The early manifestations of CO_2 toxicity are shortness of breath, anxiety, and increased heart rate. These symptoms may go unnoticed when a diver is performing hard work. As the exposure increases, the worker may experience a headache (which presents as a mild to moderate throbbing during exposure, but may increase in severity postexposure). With higher levels, the diver becomes progressively confused, eventually losing consciousness. Eliminating the CO_2 build-up through ventilation or surfacing from the dive (i.e. removal from the work site) is the treatment of choice. Symptoms resolve rapidly after removal of exposure, with the possible exception of the headache.

CO_2 retention reduces the divers' exercise tolerance, increases the manifestations of N_2 narcosis,[16] and may increase the risk of oxygen toxicity and decompression illness (since cerebral vasodilation facilitates on-gassing).

Carbon monoxide toxicity The principle source of carbon monoxide (CO) in diving is contaminated breathing gas. For example, CO can be drawn into the compressor intake (e.g. from an idling truck in the vicinity of an intake) or via improper compressor maintenance (e.g. use of flashable lubricants or a failure of compressor rings). The OSHA requires that breathing gas be tested every 6 months and as needed to prevent unrecognized elevations of CO, oil mist, and CO_2.[17] The effects of CO may be masked in the diver while breathing contaminated gas because of the high partial pressure of oxygen in the breathing media. Unless the CO concentration is high, the symptoms of CO exposure will be delayed until the partial pressure of oxygen decreases—i.e. during ascent/decompression or after surfacing. However, if there is not a high degree of suspicion, manifestations may be confused with other diving-related disorders.

OTHER HAZARDS IN THE DIVING ENVIRONMENT

The elements of the professional diving environment are varied and diverse, but its hazards are similar to the occupational hazards commonly encountered on land. These hazards are summarized in Table 8.4, and most of them are covered elsewhere in the book.

In addition, it should be remembered that many divers do not dive full-time. They often have primary jobs that expose them to more traditional hazards, such as welding, degreas-

Table 8.4 General industrial hazards that may be encountered in the underwater work environment.

- Manual materials handling. While underwater, the diver is frequently involved in moving heavy loads. The normal complexity of safely handling materials is compounded, even when buoyancy aids and hoists are used. Lifting characteristics of materials are affected by their buoyancy. In addition, the diver's protective equipment is designed to be negatively buoyant in the water and is extremely heavy when carried out of the water by the diver. As a result, undue stress is often placed on the diver's shoulders and back by poor ergonomic design.
- **Welding, cutting** and **brazing** are frequently performed underwater, requiring both standard and environment-specific precautions. Underwater explosions secondary to build-up of explosive gases during these operations may occur.
- **Electrical hazards** may be compounded by the aquatic environment.
- **Ionizing radiation.** Divers should be issued personal dosimetric devices when working around ionizing radiation sources.
- **Underwater blasts.** Blast injuries may be compounded by the effects of pressure wave propagation.
- **Vibration.** Dives are frequently conducted in cold water, potentially compounding the effects of vibration.
- **Marine animal hazards** are frequently encountered in the work area.
- **Chemical exposures**, for example to hydrogen sulfides in sunken ships, may occur in enclosed spaces underwater. Many foams used in salvage contain isocyanates.
- **Pollution.** Many divers routinely work in contaminated waters. Therefore, meticulous ear and skin care are important. Preventive measures, such as vaccinations, may also be recommended in some circumstances.

ing, paint removal, or other chemical, physical, or biological hazards.

PATHOPHYSIOLOGY OF DIRECT PRESSURE INJURY

The health effects of direct pressure injury can be divided into two major groups: barotrauma and decompression illness. Barotrauma results from the expansion or contraction of gases in anatomic spaces, causing trauma. Decompression illness results when bubbles of inert gas form in body tissues. Body fluids become supersaturated with inert gases at high pressures, and this gas comes out of solution when the ambient pressure is decreased. A summary of direct pressure effects can be found in Table 8.5.

Barotrauma

Barotrauma may occur in any gas-filled space in the body. To sustain barotrauma, commonly referred to as a "squeeze", a pressure change must occur in an enclosed, gas-filled space that has (or has developed) rigid walls. The clinical consequences of this pressure change will depend on the location of the space, the magnitude of the pressure change, and the physiologic response mounted. Common locations for barotrauma include the ear (middle most commonly, outer and inner depending on circumstances), sinuses, teeth, gastrointestinal tract, and the lung. In addition, skin may be traumatized by pressure changes in gas pockets found in a suit or under a face mask.

Table 8.5 Human health effects of environmental pressure change.

Barotrauma site
Ear
Sinus
Lung
Skin
Teeth
Gastrointestinal tract
Manifestations of decompression illness
Pain
Neurologic effects
Pulmonary effects
Cutaneous effects
Lymphatic effects
Constitutional effects

Middle-ear squeeze is the most common form of barotrauma seen in diving. It occurs when the middle-ear space is not vented properly by the Eustachian tube. When an individual descends in a pressure column, whether to a dive site or in an airplane, the gas within the middle ear compresses in accordance with Boyle's law. To counterbalance this effect, the individual must introduce air into this space via the Eustachian tube, which joins the middle ear to the pharynx. This process, known as clearing the ears, may be accomplished by various methods such as yawning, swallowing, moving the jaw around, or a Valsalva maneuver. The ears will not clear if the Eustachian tube is blocked.

If descent continues without clearing, the diver will initially experience a sense of fullness and pressure, followed by sharp pain.[18] If descent is not stopped, the ear drum may rupture, allowing immediate equalization of pressure difference, or the middle-ear space may fill with blood, a non-compressible fluid to equalize the pressure. Middle-ear barotrauma occurs most frequently within the first 10–20 ft of descent, which is the period of greatest proportional pressure change. Following the dive, residual symptoms may include pain, a sensation of fullness in the ear, or a temporary, mild, conductive hearing loss across all frequencies, generally <20 dB (or greater if an ossicular rupture has occurred).[9] In a small proportion of cases, blood may be visible in the mouth or nose. Treatment depends on the amount of damage sustained. It may range from a mild squeeze, requiring no diving for 48–72 hours, to severe barotrauma, requiring diving restrictions up to 6 weeks or longer. To prevent this form of barotrauma, an individual should not dive when the ears do not clear properly, such as during periods of significant upper respiratory congestion. The absence of pre-dive symptoms does not guarantee adequate Eustachian tube function. In addition, descent should be stopped at the first sign of difficulty in equalizing (clearing ears).

Sinus barotrauma may occur during descent when the openings that vent the sinuses into the nasal cavity are obstructed. It presents as increasing pain over the effected sinus(es). On descent, if the sinus does not properly equalize, pressure in the sinuses decreases relative to ambient pressure. As a result, edema and hemorrhage of the mucosal lining of the sinus may occur. If a sinus opening subsequently becomes blocked (secondary to edema/hemorrhage) during a dive, sinus barotrauma may present during the ascent phase of a dive. In such cases, the sinus pain is caused by a relative increase in pressure within the sinus. This pain may continue for a number of hours after the dive. Relief may be accompanied by a discharge and often a high-pitched sound as gas leaves the sinus.

Pulmonary barotrauma is a very serious form of this disorder. Gas present in the lung expands during ascent. If the lung is allowed to overpressurize by as little as 90–110 cm H_2O (1 m of sea water (msw), or 3 ft), the lung may rupture.[19] This displaced gas causes a variety of sequelae that present individually or in combination. These include mediastinal/subcutaneous emphysema, pneumothorax, and arterial gas embolism.

After rupture, gas may migrate along the bronchial tree to the mediastinum. The result is mediastinal emphysema, which may remain asymptomatic or present as substernal burning chest pain. Mediastinal emphysema is thought to be the most common manifestation of pulmonary barotrauma. From the mediastinum, extrapulmonary gas may track into the neck, presenting as subcutaneous emphysema or occasionally as a voice change secondary to pressure directly on the larynx or the recurrent laryngeal nerve. Alternatively, the gas may be driven into the retroperitoneal region. The rupture may expel gas into the intrapleural space, causing a pneumothorax.[20] If this occurs at depth, the damage will be further exacerbated by ascent. Finally, the overpressurization may force gas into the pulmonary veins, causing an arterial gas embolism that presents as neurologic sequelae. Any type of cerebral neuro-logic manifestation is possible secondary to arterial embolization; symptoms range from subtle neurologic findings to hemiplegia, convulsions, coma, and death.

Any or all of these sequelae may occur simultaneously. Therefore, it is critical to perform a complete neurologic examination to rule out arterial embolization whenever pulmonary overinflation is suspected or detected. Individuals are at an increased risk of lung rupture: during: (1) diver training courses, particularly during underwater removal and donning of gear ("ditch and don"); (2) buoyant ascent training (diver may not exhale completely due to loss of buoyancy); (3) diving with predisposing lung pathology that impedes gas flow; and (4) panic/emergency/uncontrolled ascents. It is important to remember that lung rupture can occur after surfacing from a compressed gas dive from as little as 3–4 ft of water (1 msw), even when normal procedures are followed. This should not happen in breath-hold diving, since the volume of gas present on ascent will not exceed the original breath taken on the surface.

Barotrauma may also occur in a tooth, causing implosion or expulsion of amalgam/dental material. Gas can also expand in the gastrointestinal tract; however, this rarely causes more than mild discomfort.

Decompression illness

In accordance with Henry's law, hyperbaric workers will take up gas while breathing compressed gas at elevated pressures. Subsequently, when the diver ascends in the water column or the caisson worker leaves the work site, inert gas already present in the body expands and must be off-gassed. Decompression tables have been developed that provide rules for both ascent rates and stops in the pressure column, thus allowing time to asymptomatically off-gas the inert components on ascent. Rigorous adherence to these tables is a critical component to prevent decompression illness. The tables and diving computers used today are based principally on perfusion-limited theories (Haldanian principles), which have been modified as needed to reflect human experience. However, even when the tables are rigorously adhered to, decompression illness (DCI) can still occur. Disease is obviously more common if the tables are disregarded. Examples of commonly used tables in the USA today include the US Navy standard air tables[21] and the DCIEM air tables.[22] These tables are widely available in the sport and commercial diving community. There is an average predicted range of incidence of DCI on commonly used tables of <1–5%, but specific dive profiles within each table vary greatly.[23,24]

The overall incidence of DCI is virtually impossible to measure,[25] since records of the total numbers of divers (the denominator) are not routinely maintained. It has been estimated that the operational incidence in US Navy divers is <5 cases per 10 000 dives (0.05%), and the incidence within the US sports diving community has been estimated at <1 per 10 000.[26–28] US Navy analysis of dives not requiring decompression completed between 21 and 55 FSW from 1991–1994 showed an incidence of 2.9 cases/10 000 dives (0.029%), with the incidence increasing with depth.[29] Shields et al have documented an overall incidence of decompression sickness in the commercial UK sector of the North Sea over the period 1982–86 of 0.31%, and an incidence of 0.10% after the implementation of depth time restrictions.[30] Luby reported an incidence of approximately 4.0 cases/10 000 dives (0.04%) shallower than 100 ft and 10 cases/10 000 dives (0.1%) deeper than 100 ft in commercial diving in the Middle East.[31]

During and after ascent, bubbles may form in tissues. If bubbles form, they may either remain asymptomatic or cause clinically apparent damage. It is generally believed that bubbles form in the tissues, causing damage locally (authocthonous bubbles)[32], or that they are distributed widely by the venous system, principally filtered at the lungs. However, this filter may leak, particularly with an increased load. Alternatively, venous bubbles may traverse a patent foramen ovale. Once in the arterial system, the bubbles probably distribute commensurate with the organ's bloodflow, as with the bubbles associated with pulmonary barotrauma.[33]

Generally, any new neurologic manifestation presenting shortly after a hyperbaric exposure must be considered DCI until ruled out. The range of manifestations of bubble disease is infinite, but the manifestations may be grouped into the following categories: pain, neurologic, pulmonary, cutaneous, lymphatic, and constitutional.[34] Any combination of manifestations from these categories may be present. Limb pain is believed to be the most common manifestation.[35–37]

DCI pain generally presents as a deep, toothache-like, periarticular pain that is not affected by movement. Girdle pain, a distinct DCI pain syndrome, is characterized by a poorly localized, constricting sensation radiating from the back and often heralds the onset of severe neurologic manifestations. A wide variety of neurologic manifestations may be seen, including alterations in consciousness, higher-function abnormalities, derangements in sensory modalities, strength deficits, problems with special senses (audiovestibular particularly, including vertigo, tinnitus, and hearing loss), and loss of sphincter control (particularly bladder function).[34] Cutaneous presentations generally start with intense itching, most commonly on the torso, which progresses to an erythematous rash and may continue on to cyanotic marbling (mottling). Lymphatic disease presents as a painful swelling of an individual lymph node or group of lymph nodes, which on rare occasions may be accompanied by swelling/edema, presumably due to obstruction. Pulmonary manifestations include shortness of breath, cough, chest pain, or cyanosis. Most of these pulmonary manifestations are quite rare unless substantial decompression has been omitted. Constitutional symptoms, including fatigue, nausea, and anorexia, may accompany any of these manifestations.

A descriptive system of nomenclature for DCI is presented in Table 8.6. The evolutionary and clinical manifestation terms are used to form the label—e.g. acute progressive neurologic DCI, or acute static neurologic and limb pain DCI. "Acute" is used to discriminate the case from potential long-term health effects related to decompression.[34]

Traditionally, manifestations of arterial gas embolism have been distinguished from those of DCI, despite probable overlapping pathophysiology. As a result of experience in caring for the caisson workers digging the Dartford tunnel below the Thames River in London, Golding et al[38] further divided DCI into two categories based on presumed severity and disease location. Pain, cutaneous effects, and lymphatic manifestations were designated *as* type I decompression sickness, whereas neurologic and pulmonary manifestations alone or with any other combination of manifestations were called type II decompression sickness. Type II was believed to represent serious disease. This classification quickly became the standard, based on 35 cases of type II disease. It is still in frequent use today. However, this system has been demonstrated to give inconsistent diagnoses.[39–41] Furthermore, as can be seen from their clinical description and our present concepts of the pathophysiology, cerebral arterial gas emboli (CAGE) and DCI probably cannot be distinguished except in a few isolated circumstances.[34]

DCI can begin on ascent or after surfacing. When Francis et al reviewed 1070 well-documented cases of neurologic DCI, excluding all cases with histories thought to predispose to arterial gas embolism, they found that 50% presented within 10 min of surfacing, >85% of cases presented within 1 h of surfacing, and >95% presented within 6 h of surfacing.[42] However, symptoms attributable to DCI may present at 24–48 hours or more after a dive.[43] The classical arterial gas embolic mechanism secondary to barotrauma may occur as stated above when compressed gas is inhaled deeper than 3 ft and generally presents on surfacing or quickly thereafter. The inert gas mechanism, on the other hand, requires the diver to stay for

Table 8.6 A matrix for describing decompression illness.

Acute decompression illness

The five following terms are used to describe a case of DCI adequately:

1. Evolution. Used to summarize the development of a case from onset to the present moment, prior to recompression. Terms used to describe the evolution include:
    - Progressive
    - Static
    - Spontaneously improving
    - Relapsing
2. Manifestation. Used to describe the organ systems or parts of the body that are affected. Unlike "presenting" symptoms, which describe the initial manifestation of a medical condition, these terms are used to describe the disease complex at the time of the report. Manifestation terms include:
    - Pain
        - Limb pain
        - Girdle pain
    - Cutaneous
    - Neurologic
        - Audiovestibular
    - Pulmonary
    - Lymphatic
    - Constitutional
3. Time of onset. For each manifestation, record the time that has elapsed between surfacing from the dive and the onset of each principal manifestation.
4. Gas burden. An estimate of residual inert gas load present after a pressure exposure; gives some indication of the "sensitivity" of the exposure until a standard is established. An accurate record of the dive profile is probably the most useful measure at present.
5. Evidence of barotrauma. Clinical or radiographic evidence of barotrauma should be documented. When barotrauma cannot be diagnosed definitively, DCI terminology should be used; otherwise, the barotraumata are diagnosed as before.

Format for decompression illness "label"
Since lengthy descriptions are unwieldy for communication purposes, a specific abbreviated "label" is needed for each case.

The general form of the proposed label is as follows:
Acute [Evolution term], [Manifestation term(s)], decompression illness (see text).

Adapted from Francis, Smith and Sykes.[18]

a minimum time at depth to acquire an adequate gas burden. With normal diving, depths >33 FSW are generally required.[43]

A number of proposed predisposing factors for decompression illness have been identified in addition to the dive profile. They include individual susceptibility[23] (which is most likely multifactorial), patent foramen ovale,[44–46] obesity,[47–49] exercise at depth and during decompression,[23,50] and low ambient air temperature/wind chill.[51] Data are lacking to translate these associations, such as obesity and patent foramen ovale, into specific individual recommendations. Patent foramen ovale, for example, has a high prevalence in normal populations, including divers, despite a low incidence of resulting DCI. Therefore, most experts do not recommend screening divers for patent foromen ovale unless they have experienced recurrent episodes of "unexplained" neurologic DCI.[52] In addition, age, poor physical fitness, recent tissue injury, previous DCI and dehydration have been suggested as predisposing factors; however, little epidemiologic evidence exists to support these suggestions, and the studies are conflicting.

Diving while pregnant is a controversial issue, due to the paucity of available data, which are also conflicting. However, most hyperbaric authorities agree that a woman should not dive while pregnant.[53–56] Safe depth–time profiles have not been established.[57] The fetus may be susceptible to intravascular bubble formation, and these bubbles may have a deleterious effect on the nervous system of the fetus with a patent foramen ovale and ductus arteriosus. Moreover, there is concern about the potential effects of hypoxia during an unanticipated emergency while diving. The military and most commercial diving operations prohibit women from diving while pregnant.

Unique problems of saturation diving

Saturation diving techniques represent a cost-effective alternative to standard diving when deep or prolonged bottom times are required, since the need for frequent and long in-water decompressions is avoided. The worker essentially completes one decompression after a prolonged exposure. However, at greater depths, unique medical problems arise.

Compression at deep depths (>150 m) is associated with high-pressure nervous syndrome (HPNS) and compression arthralgia. The symptoms of HPNS frequently include a 5–8-Hz tremor, dizziness, nausea and associated vomiting, decreased mental alertness, and microsleep.[58] The manifestations are related to depth and rate of compression. Compression arthralgia comprises pains or ill-defined discomforts in joints on moving during and immediately after compression. The knees, hips and wrists are most commonly involved.[9] To avoid both HPNS and compression arthralgia, the compression rates are slowed in comparison to conventional diving.

Maintenance of thermal balance is a significant problem at depth. At 300 m, the comfort range varies by less than 2°C. Additionally, while the diver is lying down, the exposed surfaces may become cold secondary to thermal conduction, and the surfaces in contact with the mattress may become too warm. While the diver is actually working in the water, thermal balance is maintained with hot water suits and the breathing media are heated.

Because the saturation environment is humid and warm, pathogens grow well. Meticulous housekeeping is required to keep all divers within this closed community healthy. Otitis externa has been a particular problem, forcing the early completion of some dives. A preventive regimen using 2% acetic acid in aluminum acetate eardrops has been developed; this regimen has effectively controlled the problem in most situations.[59]

Communication is extremely difficult in a helium environment. Electronic “unscramblers” are mandatory for adequate communications between the diver and support personnel. All materials are transferred to depth and back via “medical locks”, which are small pressure-transfer chambers with doors on the outside and inside of the chamber. These locks can be pressurized to allow transfer of food, medical supplies, mail, and any important personal objects. Greater depths degrade the taste of food, requiring increased flavoring and spices. The heat of compression must be anticipated, because the additional heat generated will further cook food on descent.

Once they enter the chamber, the physician no longer has ready access to the divers; therefore, medical screening for deep saturation diving must be rigorous. The lack of access is due to the slow compression times required and subsequent decompression obligations incurred by the medical attendant, making access unfeasible. Pre-dive physicals should pay particular attention to the ears, skin, and respiratory tract. To be better prepared for emergencies, divers should be trained in various first aid and life-support techniques.

LONG-TERM HEALTH EFFECTS

Dysbaric osteonecrosis, or aseptic bone necrosis, is a well-recognized, relatively uncommon, long-term occupational hazard associated with compressed air work. The bone necrosis lesions occur principally in the femur, tibia, and humerus.[60] As a rule, shaft lesions and lesions that occur away from articular joints do not produce clinical symptoms. Juxta-articular lesions, on the other hand, can progress and produce debilitating disease. Juxta-articular lesions are more commonly found in compressed air workers than in divers. Necrosis is associated with increasing depth and duration of exposure, increasing age (although age may only be a surrogate measure of exposure), and a history of DCI, though not related to the site of DCI.[60] The lesions of dysbaric osteonecrosis are indistinguishable from other causes of aseptic necrosis. However, as McCallum and Harrison[60] note, "in men with a history of work in compressed air or diving, the probability of bone necrosis being due to compressed air is very high". The pathogenesis of the dysbaric form is not well understood. An excellent review of dysbaric osteonecrosis was completed by McCallum and Harrison.[60]

Divers are at risk for sensorineural hearing loss secondary to barotrauma, DCI, and potentially long-term noise exposure. Although the hearing threshold is increased underwater, a number of studies have shown that measured sound intensity levels of some diving helmets when pressurized are as high as 90–120 dB(A), depending on the application.[61,62] In addition, hyperbaric chambers during compression and some chambers during ventilation have high noise levels. Some investigators have documented standard threshold shifts with "normal" dive profiles.[62–65] The principal cause of the high noise level is the volume of gas flow required to sufficiently ventilate helmets to prevent CO_2 build-up. The frequency range of noise is 800–3000 Hz, which is within normal communication ranges. Redesign has significantly improved the noise levels of modern helmets.[63] Some epidemiologic studies have shown little difference in hearing acuity between professional divers and matched controls when corrected for age,[64] whereas others[65–68] have noted significant differences in divers. One recent age-adjusted prospective study in SCUBA divers demonstrated increased hearing loss in low frequencies only, suggesting a compression/decompression effect.[68] Although some of the high-frequency hearing loss seen in divers is probably due to other exposures, the studies imply that some of the losses may be due to workplace noise exposure.[69] Therefore, engineering controls and hearing protection should be instituted wherever possible.

Sequelae from episodes of acute DCI, particularly neurologic ones, are well recognized. However, there may be less obvious but potentially serious long-term health effects from diving, based primarily on descriptive and anecdotal evidence. This possibility has generated a number of hypotheses, which currently lack good epidemiologic support. Subsequent studies have frequently not supported the hypothesized findings. These effects include neuropsychiatric deficits, such as short-term memory deficits and emotional lability,[70–73] electroencephalographic abnormalities[74,75] (principally slow waves and spikes), and retinoangiography aberrations, including pigment and capillary changes in the retina.[76] Pulmonary function changes have been documented in groups of divers; these include increased vital capacities, which may be adaptive.[77] The clinical relevance and validity of these findings are still being investigated. An excellent review of the relevant literature can be found in Elliott and Moon.[78] This review concludes: "in the absence of a history of acute decompression illness, the possibility of a clinical syndrome among divers or ex-divers remains unproven. If it exists, the prevalence is unknown, and probably low".

TREATMENT OF BUBBLE-RELATED DISEASE

Because symptoms and signs can progress, a diagnosis of DCI should be treated urgently. The patient should be placed on oxygen at as high a partial pressure as reasonably feasible. Additionally, fluids (generally oral, but as appropriate for the condition of the patient) are pushed, and the individual is placed in a recumbent position. A hyperbaric chamber should be found and the patient transferred for recompression therapy (Figure 8.8). An excellent source of emergency information and referral is the Diver's Alert Network, The Peter B. Bennett Center, 6 West Colony Place, Durham, NC 27705; Telephone: (919) 684-8111; http://www.diversalertnetwork. org.

Many recompression tables exist; however, the US Navy Treatment Table 6 (Figure 8.9 and Table 8.7) is the principal therapeutic table used for the treatment of decompression illness. The patient is given oxygen at depth (60 ft) initially, with intermittent air breaks to reduce the incidence of oxygen toxicity. Response to therapy is generally excellent if recompression therapy is instituted promptly. Depending on the response to therapy, table modifications and follow-on hyperbaric treatments may be required if manifestations do not resolve or reccur.

MEDICAL SURVEILLANCE

Healthcare personnel must understand that few other workers experience the magnitude of physiologic stresses imposed routinely on divers and caisson workers. Medical surveillance requires a thorough understanding of the physiologic aspects of diving, along with the specific workplace hazards that may be encountered by the worker. In general, different classes of workers are exposed to varying levels or types of hazards, requiring a tailored approach. Most professional divers, including

Figure 8.8 Hyperbaric recompression chamber. (Courtesy of U.S. Navy.)

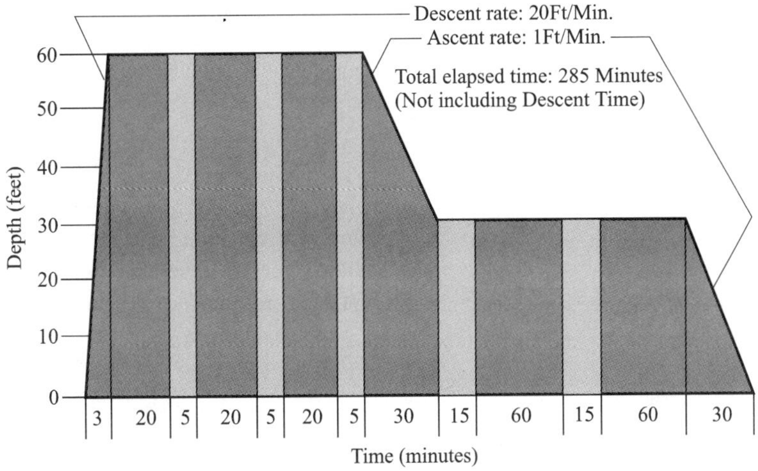

Figure 8.9 U.S. Navy Treatment Table 6. Source: U.S. Navy Diving Manual, Chapter 21. NAVSEA 0910-LP-0708-8000, Washington, D.C.: U.S. Government Printing Office, 1999.

military, commercial, and scientific divers, as well as caisson workers and hyperbaric chamber attendants, are provided with specific guidance and standards by their employer. For example, the Association of Diving Contractors periodically publishes medical requirements for their members. Title 29 CFR 1910 Subpart T, Appendix A provides examples of conditions that restrict or limit exposure to hyperbaric conditions. Other sources provide more detailed guidance.[79–83] Examination frequency varies depending on type of exposure and the regulations being followed. Re-examination must be completed after any significant illness or injury, particularly exposure-related injuries.

Training agencies for recreational divers provide medical guidelines for their members; however, there are no regulations in the USA covering medical standards for sports divers.

Healthcare providers who conduct medical surveillance on divers should have formal instruction in diving/hyperbaric medicine. This training enables them to correlate various medical conditions to the unique hyperbaric environment. One source for information on diving medicine courses, scientific meetings, general information on diving or hyperbaric medicine, or addresses of practitioners with an interest in diving or hyperbaric medicine is the Undersea and Hyperbaric Medical Society at 10531 Metropolitan Ave, Kensington, MD 20895, USA; Telephone: (301) 942-2980; http://www.uhms.org.

Medical history

The medical history is of primary importance in hyperbaric medical surveillance. It must be remembered that in the underwater environment, any condition that incapacitates an individual, even temporarily (e.g. seizure, fainting), may cause drowning in addition to the standard sequelae. Moreover, divers for the most part rely on the buddy system, which means that the divers must be able to help their buddies when they are in distress and must not endanger their buddies secondary to their own medical condition. In addition, a careful occupational history—including type of dives, number of dives, maximum depth obtained, and any untoward events—is uniquely important.

Physical examination

A complete physical examination should be completed, with an emphasis on ear, nose and throat, pulmonary, cardiovascular, skeletal and neurologic systems. Caisson and diving work-

Table 8.7 US Navy Treatment Table 6: Oxygen Treatment of Type II Decompression Sickness[a,b].

Depth (ft)[c]	Time (min)	Breathing media[d]	Total elapsed time (h:min)
60	20	O_2[e]	0:20[f]
60	5	Air	0:25
60	20	O_2	0:45
60	5	Air	0:50
60	20	O_2	1:10
60	5	Air	1:15
60 to 30	30	O_2	1:45
30	15	Air	2:00
30	60	O_2	3:00
30	15	Air	3:15
30	60	O_2	4:15
30 to 0	30	O_2	4:45

[a] Treatment of type II or type I DCI when symptoms are not relieved within 10 min at 60 ft.

[b] Extensions to Table 6: Table 6 can be lengthened up to two additional 25-min oxygen-breathing periods at 60 ft (20 min on oxygen and 5 min on air) or up to two additional 75 min oxygen-breathing periods at 30 ft (15 min on air and 60 min on oxygen) or both. If Table 6 is extended only once at either 60 or 30 ft, the tender breathes oxygen during the ascent from 30 ft to the surface. If more than one extension is done, the care-giver begins oxygen breathing for the last hour at 30 ft during ascent to the surface.

[c] Descent rate—25 ft/min. Ascent rate—1 ft/min. Do not compensate for slower ascent rates. Compensate for faster rates by halting the ascent.

[d] Care-giver breathes air throughout unless he has had a hyperbaric exposure within the past 12 hours, in which case he breathes oxygen at 30 ft.

[e] If oxygen must be interrupted because of adverse reaction, allow 15 min after the reaction has entirely subsided and resume schedule at point of interruption.

[f] Time at 60 f begins on arrival at 60 ft.

Adapted from the US Navy Diving Manual, Revision 4, chapter 21.

ers require a significant amount of cardiovascular reserve and aerobic work capacity, along with adequate dexterity and strength. During each examination, a neurologic examination should be completed to, at a minimum, document pre-existing deficits. A well-documented neurologic examination may prevent confusion during evaluation of symptoms and signs postdiving.

PREVENTION

The only certain method to prevent hyperbaric injuries is to simply avoid all high-pressure and diving work. Thanks to modern engineering techniques, the need for caisson work has been significantly decreased. However, preventing diving injuries by avoiding diving altogether is generally regarded as unfeasible. Therefore, the principal methods used to prevent diving injuries are: (1) extensive training, including both an academic understanding of principles and job-specific practical training; (2) dive planning, to include emergency procedures and appropriate use of tables; (3) maintenance of a high level of fitness; (4) meticulous care of equipment, along with ongoing improvement in equipment design; and (5) a healthy respect for the indigenous hazards of the profession. These same principles also apply to caisson work. Current research is directed toward the refinement of decompression models and tables, improving equipment design, enhancing treatment methods, and understanding the pathophysiology and associated risk factors of diving disorders.

REFERENCES

1. McCallum RI. Increased barometric pressure. In: Raffle PAB, et al, eds. *Hunter's diseases of the occupations*. Boston: Little, Brown, 1987:523–47.
2. Bachrach AJ. A short history of man in the sea. In: Bennett PB, Elliott DH, eds. *The physiology and medicine of diving*. 3rd edn. San Pedro, CA: Best Publishing, 1982:1–14.
3. Kindwall EP. A short history of diving and diving medicine. In: Bove AA, Davis JC, eds. *Diving medicine,* 3rd edn. Philadelphia: WB Saunders, 1997:3–11.
4. Gal A. Des dangers du travail dans l'air comprimée et des moyens de les prévenir. In: Bert P, ed. *Barometric pressure–researches in experimental physiology*. Columbus, OH: College Book Company, 1943:398.

5. Butler FK, Smith DJ. United States Navy diving techniques and equipment. In: Bove AA, Davis JC, eds. *Diving medicine*. 3rd edn. Philadelphia: WB Saunders, 1997:372–87.
6. Kindwall EP. Compressed air work. In: Bennett PB, Elliott DH, eds. *The physiology and medicine of diving*. 4th edn. Philadelphia: WB Saunders, 1993:1–19.
7. Kindwall EP. Compressed air tunneling and caisson work decompression procedures: development, problems, and solutions. *Undersea Hyper Med* 1997; 24(4):337–45.
8. Farmer JC. Vestibular and auditory function. In: Shilling CXV, Carlston CB, Mathias RA, eds. *The physician's guide to diving medicine*. New York: Plenum, 1984:192–8.
9. Flynn ET, Catron PW, Bayne CG. *Diving medical officer's student guide*. Memphis, TN: Naval Technical Training Command, 1981.
10. Somers LH. Diving physics. In: Bove AA, Davis JC, eds. *Diving medicine,* 3rd edn. Philadelphia: WB Saunders, 1997:15–25.
11. Kinney JAS. Physical factors in underwater seeing. In: Shilling CXV, Carlston CB, Mathias RA, eds. *The physician's guide to diving medicine*. New York: Plenum, 1984:199–205.
12. Mebane GY. Hypothermia. In: Bove AA, Davis JC, eds. *Diving medicine*. 3rd edn. Philadelphia: WB Saunders, 1997:207.
13. Craig AB, Dvorak M. Thermal regulation during water immersion. *J Appl Physiol* 1966;21:1577–85.
14. Bennett PB. Inert gas narcosis. In: Bennett PB, Elliott DH, eds. *The physiology and medicine of diving,* 4th edn. Philadelphia: WB Saunders, 1993:170–93.
15. Kiessling RJ, Maag CII. Performance impairment as a function of nitrogen narcosis. *J Appl Psychol* 1962; 46:91–5.
16. Hesser CM, Fagraeus L, Adolfson J. Roles of nitrogen, oxygen and carbon dioxide in compressed air narcosis. *Undersea Biomed Res* 1978; 5:391–400.
17. Code of Federal Regulations 29, part 1910, subpart T—*Commercial diving operations*. 1, Washington: US Government Printing Offices, 1992 July.
18. Francis TJR, Smith DJ, Sykes JJW. *The prevention and management of diving accidents*. INM Technical report R92004. Alverstoke, England: Institute of Naval Medicine, 1992.
19. Malhotra MC, Wright HC. The effects of a raised intrapulmonary pressure on the lungs of fresh unchilled cadavers. *J Pathol Bacteriol* 1961; 82:198–202.
20. Broome JR, Smith DJ. Pneumothorax as a complication of recompression therapy for cerebral arterial gas embolism. *Undersea Biomed Res* 1992; 19:447–55.
21. *US Navy Diving Manual*; rev. 4, Chapter 9. NAVSEA 0910-LP-708-8000, Washington, DC: Supervisor of Diving, US Navy 1999 (updates available at www.supsalv.org/divingpubs.html)
22. Lauckner GR, Nishi RY. *Decompression tables and procedures for compressed air diving based on the DCJEM 1983 decompression model*. No. 84-R-74. Toronto: DCIEM, 1984.
23. Vann RD, Thalmann ED. Decompression physiology and practice. In: Bennett PB, Elliott DH, eds. *The physiology and medicine of diving,* 4th edn. Philadelphia: WB Saunders, 1993:376–432.
24. Weathersby PK, Survanshi SS, Homer LD, Hart BL, Nishi RY, Flynn ET, Bradley ME. *Statistically based decompression tables. I. Analysis of standard air dives: 1950–1970*. NMRI Report 85-16. Bethesda, MD: Naval Medical Research Institute, 1985.
25. Sykes JJW Is the pattern of acute decompression sickness changing? *J R Navy Med Serv* 1989; 75:69–73.
26. Dembert ML. Individual factors affecting decompression sickness. In: Vann RD, ed. *The physiological basis of decompression*. Proceedings of the 38th Undersea and Hyperbaric Medical Society Workshop. Bethesda, MD: Undersea and Hyperbaric Medical Society, 1989:355–67.
27. Wilmshurst P, Allen C, Parish T. Incidence of decompression illness in amateur SCUBA divers. *Health Trends* 1994–5; 26(4):116–8.
28. Arness MK. Scuba decompression illness and diving fatalities in an overseas military community. *Aviat Space Environ Med* 1997; 68:325–33.
29. Flynn ET, Parker EC, Ball R. Risk of decompression sickness in shallow no-stop air diving: an analysis of US Navy experience 1990–94.

In: *Proceedings of the 14th meeting of United States–Japan Cooperative Program in Natural Resources (UJNR), Panel on Diving Physiology.* Panama City, FL: NOAA, 1997:23–38.

30. Shields TG, Duff PM, Wilcock SE, Giles R. Decompression sickness from commercial offshore air-diving operations on the UK continental shelf during 1982 to 1988. In: *Subtech '89*. Fitness for Purpose, Vol. 23. Amsterdam: Society for Underwater Technology, 1990:259–77.
31. Luby J. A study of decompression sickness after commercial air diving in the northern Arabian gulf: 1993–95. *Occup Med* 1999; 49(5):279–83.
32. Francis TJR, Dutka AJ, Flynn ET. Experimental determination of latency, severity, and outcome in CNS decompression sickness. *Undersea Biomed Res* 1988; 15:419–27.
33. Francis TJR. A current view of the pathogenesis of spinal cord decompression sickness in a historical perspective. In: Vann RD, ed. *The physiological basis of decompression*. Proceedings of the 38th Undersea and Hyperbaric Medical Society Workshop. Bethesda, MD: Undersea and Hyperbaric Medical Society, 1989:241–79.
34. Francis TJR, Smith DJ, eds. *Describing decompression illness*. Proceedings of the 42nd Undersea and Hyperbaric Medical Society Workshop. Bethesda, MD: Undersea and Hyperbaric Medical Society, 1991.
35. Rivera JC. Decompression sickness among divers: an analysis of 935 cases. *Mil Med* 1964; 129:314–34.
36. Kelleher PC, Francis TJR, Smith DJ, Hills RCP. INM diving accident database: analysis of cases reported in 1991 and 1992. *Undersea Biomed Res* 1993; 20(suppl):13 (abstract).
37. Denoble P, Vann RD, de L Dear G. Describing decompression illness in recreational diving. *Undersea Biomed Res* 1993; 20(suppl):14 (abstract).
38. Golding FC, Griffiths P, Hemplemen HV, Paton WDM, Walder DN. Decompression sickness during the construction of the Dartford tunnel. *Br J Indust Med* 1960; 17:167–80.
39. Kemper GB, Stegmann BJ, Pilmanis AA. Inconsistent classification and treatment of type I/type II decompression sickness. *Aviat Space Environ Med* 1992; 63:153 (abstract).
40. Smith DJ, Francis TJR, Pethybridge RJ, Wright JM, Sykes JJW. Concordance: a problem with the current classification of diving disorders. *Undersea Biomed Res* 1992; 19(suppl):47 (abstract).
41. Smith DJ, Francis TJR, Pethybridge RJ, Wright JM, Sykes JJW. An evaluation of the classification of decompression disorders. *Undersea Hyperbaric Med* 1993; 20(suppl):11 (abstract).
42. Francis TJR, Pearson RR, Robertson AG, Hodgson M, Dutka AJ, Flynn ET. Central nervous system decompression sickness: latency of 1070 human cases. *Undersea Biomed Res* 1988; 15:403–17.
43. Elliott DH, Moon RE. Manifestations of the decompression disorders. In: Bennett PB, Elliott DH, eds. *The physiology and medicine of diving,* 4th edn. Philadelphia: WB Saunders, 1993:492.
44. Moon RE, Camporesi EM, Kisslo JA. Patent foramen ovale and decompression sickness in divers. *Lancet* 1989; I:513–14.
45. Wilmshurst P, Byrne JC, Webb-Peploe MM. Relation between interarterial shunts and decompression sickness in divers. *Lancet* 1989; II:1302–6.
46. Gernompré P, Dendale P, Unger P, Balestra C. Patent foramen ovale and decompression sickness in sports divers. *J Appl Physiol* 1998; 84(5):1622–6.
47. Medical Research Council Decompression Central Registry, University of Newcastle-upon-Tyne. Decompression sickness and aseptic necrosis of bone. Investigations carried out during and after the construction of the Tyne Road tunnel (1962–66). *Br J Indust Med* 1971; 28:1–21.
48. Lam TH, Yau KP. Analysis of some individual risk factors for decompression sickness in Hong Kong. *Undersea Biomed Res* 1989; 16:283–92.
49. Dembert ML, Jekel JF, Mooney LW. Health risk factors for DCS. *Undersea Biomed Res* 1984; 11:395–406.
50. Van der Aue OE, Kellar RJ, Brinton ES. *The effect of exercise during decompression from increased barometric pressures on the incidence of decompression sickness on man.*

Report no. 8-49. Panama City, FL: United States Navy Experimental Diving Unit, 1949.

51. Broome JR. Climatic and environmental factors in the aetiology of DCI in divers. *Undersea Biomed Res* 1992; 19(suppl):17 (abstract).

52. Elliott DH. Medial evaluation for commercial diving. In: Bove AA, Davis JC, eds. *Diving medicine*, 3rd edn. Philadelphia: Saunders, 1997:364–5.

53. Fife WP, ed. *Effects of diving on pregnancy.* Proceedings of the 19th Undersea Medical Society Workshop. Bethesda, MD: Undersea Medical Society, 1978:15–19.

54. Fife WP, ed. *Women in diving*. Proceedings of the 35th Undersea and Hyperbaric Medical Society Workshop. Bethesda, MD: Undersea and Hyperbaric Medical Society, 1986:3–10.

55. Vorosmarti J, ed. *Fitness to dive*. Proceedings of the 34th Undersea and Hyperbaric Medical Society Workshop. Bethesda, MD: Undersea and Hyperbaric Medical Society, 1987:101–2.

56. Hill RK. Pregnancy and travel. *JAMA* 1989; 262:498.

57. Cresswell JE, St Leger-Dowse M. Women and scuba diving. *Br Med J* 1991; 302:1590–1.

58. Bennett PB, Rostain JC. The high pressure nervous syndrome. In: Bennett PB, Elliott DH, eds. *The physiology and medicine of diving*, 4th edn. Philadelphia: WE Saunders, 1993:194–237.

59. Thalmann ED. *A prophylactic program for the prevention of otitis externa in saturation divers*. Report no. 10-74. Washington, DC: Navy Experimental Diving Unit, 1974.

60. McCallum RI, Harrison JAB. Dysbaric osteonecrosis: aseptic necrosis of the bone. In: Bennett PB, Elliott DH, eds. *The physiology and medicine of diving*, 4th edn. Philadelphia: WB Saunders, 1993:561–84.

61. Summitt JK, Reimers SD. Noise: a hazard to divers and hyperbaric chamber personnel. *Aerospace Med* 1971; 42:1173–77.

62. Curley MD, Knafelc ME. Evaluation of noise within the MKl2 SSDS helmet and its effect on divers' hearing. *Undersea Biomed Res* 1987; 14:187–204.

63. Molvær OI, Gjestland T. Hearing damage to divers operating noisy tools under water. *Scand J Work Environ Health* 1981; 7:263–70.

64. Brady JI, Summitt JK, Berghage TE. An audiometric survey of navy divers. *Undersea Biomed Res* 1976; 3:41–7.

65. Edmonds C. Hearing loss with frequent diving (deaf divers). *Undersea Biomed Res* 1985; 12:315–19.

66. Molvær OI, Lehmann EH. Hearing acuity in professional divers. *Undersea Biomed Res* 1985; 12:333–49.

67. Molvær OI, Albrektsen G. Hearing deterioration in professional divers: an epidemiologic study. *Undersea Biomed Res* 1990; 17(23):1–46.

68. Haraguchi H, Ohgaki T, Okubo J, Noguchi Y, Sugimoto T, Komatsuzaki A. Progressive sensorineural hearing impairment in professional fishery divers. *Ann Otol Rhinol Laryngol* 1999; 108:1165–9.

69. Farmer JC. Otological and paranasal sinus problems in diving. In: Bennett PB, Elliott DH, eds. *The physiology and medicine of diving*, 4th edn. Philadelphia: WB Saunders, 1993:295.

70. Rózsahegyi I. Late consequences of the neurological forms of decompression sickness. *Br J Indust Med* 1959; 16:311–17.

71. Edmonds C, Boughton J. Intellectual deterioration with excessive diving. *Undersea Biomed Res* 1985; 12:321–6.

72. Edmonds C, Hayward C. Intellectual impairment in diving: a review. In: Bove AA, Bachrach AJ, Greenbaum U, eds. *Proceedings of the 9th International Symposium on Underwater and Hyperbaric Physiology.* Bethesda, MD: Undersea and Hyperbaric Medical Society, 1987:877–86.

73. Curley MD. US Navy saturation diving and diver neuropsychologic status. *Undersea Biomed Res* 1988; 15:39–50.

74. Todnem K, Nyland H, Skiedsvoll H, et al. Neurological long term consequences of deep diving. *Br J Indust Med* 1991; 48:258–66.

75. Murrison AW. The contribution of neurophysiologic techniques to the investigation of diving-related illness. *Undersea Biomed Res* 1993; 20(4):347–73.

76. Polkinghome PJ, Sebmi K, Cross MR, minassian D, Bird AC. Ocular hindus lesions in divers. *Lancet* 1988; 2:1381–3.

77. Crosbie WA, Reed JW, Clarke MC. Function characteristics of the large lungs found in commercial divers. *J Appl Physiol* 1979; 46:639–45.

78. Elliott DH, Moon RE. Long-term health effects of diving. In: Bennett PB, Elliott DH, eds. *The physiology and medicine of diving*, 4th edn. Philadelphia: WB Saunders, 1993:585–604.

79. Davis JC, ed. *Medical examination of sports scuba divers*, 2nd edn. San Antonio, TX: Medical Seminars, 1986.

80. Mebane GY, Mclver NM. Fitness to dive. In: Bennett PB, Elliott DH, eds. *The physiology and medicine of diving,* 4th edn. Philadelphia: WB Saunders, 1993:53–76.

81. Health and Safety Executive. *The medical examination of divers (MA 1)*, revised edn. London: Health and Safety Executive, 1987:1–9.

82. Elliott DH. Medical evaluation for commercial diving. In: Bove AA, Davis JC, eds. *Diving medicine*, 3rd edn. Philadelphia: WB Saunders, 1997:361–71.

83. Elliott DH, ed. *Medical assessment of fitness to dive.* Proceedings of an Internal Conference at the Edinburgh Conference Center, 1994. Kwell, England: Biomedical Seminars, 1995.

9

LOW-PRESSURE AND HIGH-ALTITUDE ENVIRONMENTS

Glenn Merchant, M.D., M.P.H., & T.M.*

The discussion of high-pressure environments in Chapter 8 noted that humans function well only within a narrow range of barometric pressures. Ascent to altitude places workers in an adverse environment and exposes them to multiple stressors—decreased barometric pressure, reduced oxygen levels (hypoxia), ionizing and non-ionizing radiation and low temperatures. Acclimated as most are to sea-level or near sea-level pressures, reduced barometric pressures and oxygen levels will produce a range of symptoms ranging from mild discomfort to severe disease, or even death.

OCCUPATIONAL SETTING

In 1986, it was estimated that more than 38 million people worldwide permanently resided above 2440 m (8000 ft). Tourism to mountainous regions of the western USA exposed an estimated 35 000 000 people to the hypobaric environment.[1] Occupationally, pilots and flight attendants have the greatest potential exposure, although actual incidents in commercial aviation are infrequent. Inside observers in hypobaric pressure chambers and scientists at research laboratories located at high altitudes routinely perform duties at decreased barometric pressures. Individuals who travel to mountainous regions as employees of the construction or travel industries are also at risk, depending on the altitude reached and the time taken to reach that altitude.

*Roy DeHart contributed to the previous edition of this chapter.

LOW-PRESSURE ENVIRONMENTS

Measurement issues

Barometric pressure is a measurement of the weight of the atmosphere at any given point. To ensure uniformity in the calibration of

Physical and Biological Hazards of the Workplace, Second Edition, Edited by Peter H. Wald and Gregg M. Stave
ISBN 0-471-38647-2

altimeters used in the aviation industry, the concept of standard atmosphere was universally accepted in the 1920s.[2] Standard atmosphere is defined as a sea-level pressure of 760 mmHg at a temperature of +15°C and a linear decrease in temperature as one ascends of 6.5°C per kilometer (Table 9.1). Equivalent units of standard atmosphere include 1 atmosphere (atm), 29.92 in Hg, 14.7 lb/in^2, 760 torr and 1013.2 mbar. Atmospheric pressure is denser at the lower altitudes, as the weight of the atmosphere above compresses the air below. On a standard day, 5486 m (18 000 ft) marks the mid-point of atmospheric pressure.

Exposure guidelines

There are no formal guidelines for hypobaric exposures. As the agency responsible for aviation safety in the USA, the Federal Aviation Administration (FAA) establishes requirements for supplemental oxygen use in both pressurized and unpressurized aircraft as detailed in Title 14 CFR Part 91.211.

Table 9.1 Altitude-pressure-temperature relationships.

Altitude	Pressure		Temperature
(m)	(mbar)	(torr)	(°C)
Sea level	1013	760	15.00
100	1001	751	14.35
200	989	742	13.70
300	977	733	13.05
400	966	724	12.40
500	954	716	11.75
1 000	898	674	8.50
2 000	795	596	2.00
3 000	701	525	−4.49
4 000	616	462	−10.98
5 000	540	405	−17.47
10 000	264	198	−49.90
15 000	121	90	−56.50
20 000	55	41	−56.50
25 000	25	19	−51.60
30 000	11	8	−46.64
40 000	2	2	−22.80
50 000	0.8	0.6	−2.50

Physiology and the physics of gases

The behavior and impact of gases in the body are largely the result of the three well-described gas laws previously discussed (Chapter 8): Boyle's law, Dalton's law and Henry's law. Boyle's law established the inverse relationship of the pressure and volume of a gas in a closed space at a constant temperature. Pressure reduction results in directly proportional expansion of a fixed mass of a gas. Dalton's law, or the Law of Partial Pressures, states that each gas in a mixture exerts pressure independent of the other gases present. Approximately 80% of the earth's atmosphere is composed of nitrogen. Oxygen fills the remaining 20%. Finally, Henry's law describes the behavior of gases dissolved in a liquid under pressure. If the pressure is increased, the amount of gas dissolved in a liquid will increase; conversely, if the pressure acting on the liquid's surface is reduced, gases will exit the solution.

Pathophysiology, diagnosis and treatment

Physiologically, hypoxia is the greatest threat to workers' survival in low-pressure environments. Hypoxia is covered on its own in the next section of this chapter. This section will focus on the deleterious behavior of gases contained in the body during exposure to reduced pressures.

Barotrauma (trapped gases) As an individual ascends to altitude, gases present in the body respond according to Boyle's law, i.e. gases expand inversely to the pressure acting on them. The middle ear, lungs, gastrointestinal tract and paranasal sinuses are gas-containing cavities and normally vent expanding volumes through physiologic openings such as the Eustachian tubes, mouth or rectum, or paranasal ostia. Pre-existing disease may interfere with the passage of expanding gases, creating a trapped gas syndrome. Trapped gases produce a range of symptoms, from mild discomfort to pain of such intensity as to interfere with job performance. Workers usually experience trapped gas symptoms on ascent. Exceptions include ear or sinus blocks, which occur on descent. Gas in the middle ear normally vents through the Eustachian tube on ascent. The same is true of air found in the paranasal sinuses, even in the presence of pre-existing disease, e.g. an upper respiratory infection. On descent, the volume in the middle-ear space or sinus is recompressed and unless one is able to equalize the space to ambient conditions, barotrauma may result from the negative pressure in the middle ear/sinus, with actual tearing of the mucosal lining.

Diagnosis Most trapped gas symptoms appearing during routine flight resolve on return to the surface, with the exception of ear or sinus blocks associated with descent. Patients presenting with flight-related symptoms should be carefully evaluated for pre-existing disease, especially those of the upper respiratory tract. In the case of descent-related ear block, direct visualization of the ear will reveal a retracted tympanic membrane with increased vascular marking in mild cases, and middle-ear effusion, hemotympanum or perforation in more severe cases. Patients experiencing sinus barotrauma will typically complain of sharp localized pain over the affected region, 80% of the time involving the frontal sinuses. Additional findings include epistaxis in 15% of patients.[3] A sinus series may reveal clouding or fluid levels in the affected sinus.

Treatment Symptomatic treatment of persistent symptoms includes temporary removal from flying duties, oral or nasal decongestants and analgesics. Otitic barotrauma symptoms usually resolve in a matter of days. Barosinusitis sequelae may persist for weeks. As it is impossible to directly measure the ability to equilibrate sinuses, an acceptable indirect measure of normal function is when the patient can comfortably equalize the ears through the Valsalva maneuver (forced expiration with the lips closed and the nostrils compressed). Until such time, workers should be removed from flying duties.

Aviation decompression illness The mechanisms of altitude- and diving-related decompression illness are identical—nitrogen present in the tissues leaves solution as the ambient pressure acting on the body is reduced on ascent. Decompression illness is more common in aviators operating unpressurized aircraft above 5500 m, although Voge reported a case occurring at 4268 m (14 000 ft).[4] Commercial aircraft protect occupants from decompression illness by maintaining cabin pressures at or below 2400 m (8000 ft). Risk factors for aviation-related decompression illness include altitude, duration and rate of exposure, physical exertion, low temperatures, age greater than 40 years, female gender and recent exposure to increased pressure, e.g. SCUBA diving.[5,6] A safe interval between diving and flying may be as short as 12 h, or as long as 48 h, depending on the number of dives and whether decompression was required. Divers should consult Navy Dive Table 9.5 for specific guidance (Chapter 8).

Rapid decompression Sudden decompression occurs when a pressurized aircraft suffers a structural or mechanical failure that results in loss of internal pressurization. Rapid changes in ambient pressure greatly increase the risk of decompression illnesses. The effect on passengers and crew is dependent on multiple factors, including the length of time of equilibration from the aircraft's internal pressure to the ambient external pressure and the differential in pressures. Extremely rapid decompression is termed an "explosive" decompression and can produce an instantaneous overexpansion of the lung with resultant pulmonary trauma, leading to a pneumothorax, subcutaneous emphysema, or air emboli entering the vascular system. Fortunately, such decompressions are limited to aircraft with small cockpit spaces flying at higher altitudes and are relatively rare.

Diagnosis and treatment Decompression illness symptoms might develop in flight, on descent, shortly after landing, or be delayed for several hours.[5] All workers experiencing a rapid decompression should be thoroughly evaluated for signs and symptoms of decompression illness. In a study of 447 cases of altitude-related decompression illness, Ryles observed that 83.2% had musculoskeletal involvement, 70% of the time involving the knees. Approximately 3% experienced pulmonary symptoms, 10.8% developed paresthesias and 0.5% had frank neurologic findings.[7] Any person with signs or symptoms suggesting post-flight decompression illness should be referred immediately to a hyperbaric chamber, as immediate recompression is the only appropriate therapy. Maintaining the lowest possible ambient pressure during transport is critical to avoid further sequelae. If aeromedical evacuation to a distant recompression chamber is necessary, the flight should be at the lowest possible altitude, preferably in an aircraft pressurized to sea level.

Prevention

Clearly, primary prevention is the method of choice for avoiding decompression sickness. Those exposed to a changing pressure environment should receive education and training. Workers must be aware of the risks associated with exposure to differing atmospheric pressures, as well as the signs and symptoms that are the first manifestations of decompression illness. All workers exposed to changing atmospheric pressure must follow the guidelines established for safe entry, work, egress and emergencies.

HYPOXIA

Most of the life-threatening effects of altitude are due to hypoxia. The response of humans to hypoxia is complex, and is heavily dependent on the severity and rate of exposure. The aviator who experiences sudden loss of cabin pressure at altitude has a different physiologic response from a traveler who has traveled for weeks on the ground to reach the same altitude. Broadly speaking, acute hypoxia occurs over seconds to an hour or two; chronic hypoxia occurs from many hours to days.

Pathophysiology of acute hypoxia

Ascent to altitude reduces both the ambient pressure and the oxygen content available for gas exchange (Table 9.2). At 2438 m, there is a 25% reduction in the partial pressure of oxygen entering the lungs; by 5500 m it is reduced by half. Commercial and military aviators are at increased risk of hypoxia, as jet aircraft routinely operate at altitudes exceeding 7315 m (24 000 ft). Loss of pressurization or failure of personal breathing equipment can result in loss of consciousness in a matter of minutes.

As the partial pressure of oxygen decreases, the body's ability to maintain adequate oxyhemoglobin saturation is impaired. Table 9.3 outlines the impact of hypoxia on oxygen availability to the arterial circulation on ascent to 6706 m (22 000 ft).

Physiological response to acute hypoxia

Acute hypoxia occurs in a series of stages, progressively affecting those tissues with the greatest requirement for oxygen, particularly the nervous system.

Indifferent Stage (surface to 3000 m) Early symptoms begin to appear, but are frequently unnoticed by the individual. Vision will be mildly impaired, especially night vision. Cognitive functioning is normal, with slight decrements in novel task performance. Respiratory rate and depth of inspiration and cardiac output begin to increase. Oxygen saturation is maintained at 90–98%.

Table 9.2 Atmospheric pressure and oxygen levels at altitude.

Altitude		Pressure		Ambient	
m	ft	PSIA	mmHg	PO_2 (mmHg)	PAO_2 (mmHg)
Sea level		14.69	759	159	103
610	2 000	13.66	706	148	93.8
1219	4 000	12.69	656	137	85.1
1829	6 000	11.77	609	127	76.8
2438	8 000	10.91	564	118	68.9
3048	10 000	10.10	522	109	61.2
3658	12 000	9.34	483	101	54.3
4267	14 000	8.63	446	93	47.9
4877	16 000	7.96	411	86	42.0
5486	18 000	7.34	379	79	37.8
6096	20 000	6.76	349	73	34.3
6706	22 000	6.21	321	67	32.8
7315	24 000	5.70	294	61	31.2

*PSIA = Pounds per square inch atmospheric.

Compensatory stage (3000–4500 m)
Oxygen saturation falls below 90%. Errors in skilled task performance appear along with euphoria and impaired judgment, although workers are frequently unaware of any deficiencies. Prolonged exposure produces a generalized headache.

Disturbance stage (4500–6100 m)
Oxygen saturation falls below 80%. Cerebral functions are severely impaired. Mental calculations become unreliable. Headaches increase in severity, and neuromuscular control is greatly diminished. Tunnel vision frequently occurs. Personality and emotional changes appear and range from elation or euphoria to belligerence. Increased respiratory drive leads to hyperventilation and hypocapnia. Paresthesias of the extremities and lips are followed in severe cases by tetany and carpopedal or facial spasms. The chances of recovery are poor, due to serious deficiencies in judgment and loss of muscular coordination.

Critical stage (above 6100 m). Oxygen saturation is below 70%. Comprehension and mental performance decline rapidly, and unconsciousness occurs within minutes, often without warning (Table 9.4).

Table 9.3 Oxyhemoglobin saturation at selected altitudes.

	Altitude			
Tissue level	Sea level	3048 m (10 000 ft)	5486 m (18 000 ft)	6706 m (22 000 ft)
Alveolus PO_2	100 mmHg	60 mmHg	38 mmHg	30 mmHg
Arterial PO_2	100 mmHg	60 mmHg	38 mmHg	30 mmHg
Venous PO_2	40 mmHg	31 mmHg	26 mmHg	22 mmHg
A–a gradient	60 mmHg	29 mmHg	12 mmHg	8 mmHg
Oxyhemoglobin saturation	98%	87%	72%	60%

Table 9.4 Duration of useful consciousness.

Altitude (m)	(ft)	Duration of useful consciousness
5486	18 000	20–30 min
6706	22 000	10 min
7620	25 000	3–5 min
9144	30 000	1–2 min
10 668	35 000	30–60 s
12 192	40 000	14–20 s
13 106	43 000	9–12 s

Treatment and prevention of acute hypoxia

Initial management of hypoxia is with the immediate use of 100% oxygen. Aircraft lacking an oxygen system or experiencing a depressurization should begin an immediate, emergency descent. Recovery usually occurs within seconds of supplemental oxygen use, although a transient worsening of symptoms may occur for 15–60 s.

Hypoxia prevention requires adequate oxygen during flight, through either individual oxygen systems or aircraft pressurization. In commercial aircraft, the high-flying passenger jet provides a pressurized cabin that rarely exceeds an altitude of 2400 m (8000 ft). In unpressurized aircraft, supplemental oxygen is required for the pilot at a cabin altitude of 14 000 ft or higher. If flight is maintained for >30 minutes at altitudes between 12 500 and 14 000 ft, the pilot must use oxygen. When flying above 15 000 ft, all occupants must have supplemental oxygen.[8] At an altitude of 10 363 m (34 000 ft) 100% oxygen is required in order to maintain adequate oxygenation, equivalent to sea level. Sustained flight at ambient altitudes of 13 700 m (45 000 ft) or higher requires use of pressure suits.

Medical surveillance and education

Commercial airline transport pilots are required to undergo FAA-approved physical examinations every 6 months to maintain their medical certification.[9] Annual flight physicals are mandatory for military aviators, who receive excellent medical surveillance, given the high ratio of flight surgeons to aircrew and robust prevention programs in place throughout the armed services. Pre-employment and periodic examinations pay great attention to otorhinolaryngeal conditions that might predispose aircrew to otitic or sinus barotrauma.

Primary prevention is the goal. Aircrews receive extensive training in the physiology of operating in hypobaric environments. The US Code of Federal Regulations (CFR), Title 14, Part 61.31 (g)(2)(i) indicates that "no person may act as pilot in command of a pressurized airplane that has a service ceiling or maximum operating altitude, whichever is lower, above 25,000 feet MSL unless that person has completed ground training that includes instruction on respiration; effects, symptoms, and causes of hypoxia and any other high altitude sicknesses; duration of consciousness without supplemental oxygen; effects of prolonged usage of supplemental oxygen; causes and effects of gas expansion and gas bubble formations; preventive measures for eliminating gas expansion, gas bubble formations, and high altitude sicknesses; physical phenomena and incidents of decompression; and any other physiological aspects of high altitude flight".

The FAA coordinates low-pressure chamber "flights" at US Air Force bases to allow civilian aviators to experience hypoxia in a controlled environment—information for such courses is available on the World Wide Web at http://www.cami.jccbi.gov/AAM-400/asemphys.html. The US armed services maintain a large network of hypobaric chambers, as military aviators are required to complete low-pressure training every 3–4 years. Aviators learn early in their career to refrain from flying when congested. Foods that cause excessive intestinal gas—beans, cabbage, cauliflower, carbonated beverages or peas—are also soon avoided.

The risk of decompression illness in flight can be minimized through several strategies.

Use of 100% oxygen for 30 minutes before flight will reduce the body's nitrogen load, as will use of oxygen throughout a flight. Limiting the altitude and duration of exposure will significantly reduce the incidence. SCUBA divers should allow a sufficient interval between diving and flight. A minimum of 12 hours should elapse following a no-compression dive with less than 2 hours total bottom time; 24–48 hours is recommended after more complex dive profiles.[5]

HIGH-ALTITUDE ACCLIMATIZATION AND ILLNESS

Acute exposure to high altitude is fatal to unprotected workers in a matter of minutes (Table 9.4), yet men have climbed the tallest peaks in the world with nothing more than thermal protection. The difference lies in man's ability to adapt to severely hypoxic conditions through progressive acclimatization over extended time periods (days to months).

Acute mountain sickness (AMS)

Acute mountain sickness is the most common altitude-related illness affecting travelers to high altitude. Chinese authors first described AMS in 32 BC. Jose de Acosta, a Jesuit priest living in Peru in the 16th century, provided a more complete description, based on his experiences in the Andes.[10] AMS consists of a group of symptoms occurring 6–48 hours after rapid ascent. At elevations over 3000 m, 25% of travelers will have mild symptoms, including headache, fatigue, nausea, malaise, loss of appetite and disturbed sleep.[11] Symptoms are so non-specific that patients often fail to recognize their condition. The most important reason to diagnosis AMS is that, unrecognized it may progress to HACE or HAPE (see below) if there is further increase in altitude without adequate time for acclimatization. A greater incidence has been reported in individuals in their early 20s.[12]

Pathophysiology of AMS No etiology has been firmly established for AMS. There is evidence of cerebral vasodilation with increased bloodflow, and leakage of proteins and fluid across the blood–brain barrier, but the exactmechanism remains elusive.[13]

Diagnosis and therapy Complete history and physical examination is usually adequate to rule out conditions with similar symptoms such as viral illnesses, exhaustion, dehydration or hangover. Mild AMS is limited to the symptoms described above and is best treated with arrest of ascent or a slight descent to allow time for acclimatization—usually 1–3 days is sufficient. Acetazolamide (Diamox) 125–250 mg twice a day has been effective in reducing symptoms.[14] Acetaminophen or non-steroidal anti-inflammatory drugs are indicated to manage headaches. Recent studies with theophylline have indicated reduced AMS symptoms in 14 subjects given 375 mg oral slow-release theophylline twice a day at simulated altitudes of 3454 m.[15]

Symptoms of moderate AMS include severe headache not relieved by medication, nausea and vomiting, progressive weakness and fatigue, shortness of breath and loss of coordination. Moderate AMS should be treated with descent. When descent is delayed, oxygen and acetazolamide should be considered.

High-altitude cerebral edema (HACE)

Pathophysiology HACE is a potentially fatal metabolic encephalopathy, believed to be of vasogenic etiology, with leakage of protein and water across the blood–brain barrier.[16]

Diagnosis HACE occurs in 2–3% of trekkers at altitudes of 5 500 m, although HACE can appear in individuals above 2 500 m.[14] Symptoms include severe headache, nausea, vomiting, ataxia, disorientation, hallucinations, seizures, stupor and coma. Mild AMS can progress to HACE in 12–72 hours.

Treatment HACE is life-threatening—definitive therapy is immediate descent with close supervision, as symptoms may worsen while descending. When descent is delayed due to the situation or patient's condition, oxygen and dexamethasone, 10 mg intravenously, then 4 mg intramuscularly every 6 hours is indicated.[11] Prognosis is poor once the patient becomes comatose.

High altitude pulmonary edema (HAPE)

Pathophysiology HAPE is the most frequent cause of death of the altitude illnesses. It frequently develops on the second night of exposure at altitudes above 2 500 m. In Colorado, 1 in 10 000 skiers will develop HAPE, with a higher incidence in younger men. HAPE victims develop substantial increases in pulmonary artery pressures, with increased vascular permeability. Fluid increases in the lungs, reducing oxygen exchange. Respiratory alkalosis and severe hypoxemia follow, with mean oxygen saturations of 56%.[14]

Diagnosis Early symptoms include dyspnea on exertion, fatigue, weakness and a dry cough. Signs typically are tachycardia, tachypnea, rales pink-tinged frothy sputum and cyanosis. Radiographs show patchy peripheral infiltrates, which may be unilateral or bilateral.[14]

Treatment Treatment comprises immediate descent to lower altitudes with close monitoring. Patients should be kept warm and minimize exertions. If descent is impossible due to conditions, nifedipine (10 mg every 4 hours)[17] and oxygen (4 to 6 l per minute) have been shown to improve patients. Descent remains the key to therapy.

Prevention of altitude illnesses

The rate of ascent and individual susceptibility are the primary determinants of altitude illnesses. The key to prevention is a gradual ascent. Workers traveling to altitudes above 3000 m should spend a night at 1 800 m, and then two or three nights at 2500–3000 m, before proceeding higher.[18] Travel above 3000 m should be on the ground. Once 3000 m is crossed further ascents should be limited to 300 m per day, with a rest day added for every 1000 m of elevation gained. Climbs exceeding 300 m/day can be accomplished by returning to lower altitudes at night to sleep. Sleep hypnotics and alcohol should be avoided, as they suppress breathing during sleep, worsening cerebral oxygenation. High-carbohydrate diets and avoidance of dehydration or overexertion have also been widely reported as helping to prevent high-altitude illnesses, although precise mechanisms are unknown.

OTHER ALTITUDE-RELATED CONDITIONS

Vision

Beck Weathers' experience on an ill-fated Mt Everest expedition in 1996 heightened the public's awareness of high-altitude effects on patients who have undergone eye surgery. Weathers, a Texas pathologist, suffered severe hyperopia at altitude following his radial keratotomy (RK), effectively disabling him.[19] Ng et al experimentally demonstrated reversible hyperopic changes in RK subjects exposed to altitudes of 4300 m.[20] Studies of post-photorefractive kerectomy (PRK) and laser in situ keratomileusis (LASIK) patients revealed no change in PRK patients, and a small, but, statistically significant, myopic change in LASIK patients.[21] The mechanism in all cases appears to be hypoxia-induced corneal hydration.[22]

Extreme altitudes have also been implicated in high-altitude retinopathy (HAR). Weidman and Tabor examined 40 climbers climbing Mt Everest. Fourteen of 19 climbers who ascended to altitudes between 4880 m and 7620 m developed HAR, and 19 of the 21 who exceeded 7620 m (25 000 ft) developed HAR.[23] Most patients were asymptomatic; descent was not required.

Pregnancy

Altitude has been implicated in a number of complications of pregnancy. Ali et al and Niermeyer have suggested in independent studies that there is an increased incidence in preterm labor among pregnant high-altitude travelers.[24,25] Palmer et al reported a 16% incidence of pre-eclampsia at 3100 m compared to a 3% rate at 1260 m at high and low altitudes in Colorado. Birth weight averaged 285 g less in those deliveries at 3100 m.[26]

Radiation exposure

It has long been known that high altitude exposes workers to elevated levels of cosmic radiation, especially in higher latitudes.[27] Aircrews on polar routes have significantly higher exposures than those flying equatorial routes, but the long-term impact on health is currently unknown. Gundestrup and Storm reported increased acute myeloid leukemia, malignant melanoma and skin cancer rates in Danish male jet cockpit crew members. The melanomas and skin cancers were attributed to sun exposure during vacations rather than occupational exposure at altitude.[28] Other studies have shown individual exposures to be well within current international recommended exposures,[29–32] although a pregnant flight attendant would have to change routes to remain under the exposure limits.[32]

REFERENCES

1. Hultgren HN. *High altitude medicine*. Stanford: Hultgren Publications, 1997:10–11.
2. Ward MP, Milledge JS, West JB. *High altitude medicine and physiology*. London: Chapman & Hall Medical, 1995:32–7.
3. O'Reilly BJ. Otorhinolaryngology. In: Ernsting J, Nicholson AN, Rainford DJ, eds. *Aviation medicine*. 3rd edn. Oxford: Butterworth-Heinemann, 1999:319–36.
4. Voge VM. Probable bends at 14,000 feet: a case report. *Aviat Space Environ Med* 1989; 60(11):1102–3.
5. Heimbach RD, Sheffield PJ. Decompression sickness and pulmonary overpressure accidents. In: DeHart RL, ed. *Fundamentals of aerospace medicine*. 2nd edn. Baltimore: Williams & Wilkins, 1996:131–61.
6. Weien RW, Baumgartner N. Altitude decompression sickness: hyperbaric results in 528 cases. *Aviat Space Environ Med* 1990; 61(9):833–6.
7. Ryles MT, Pilmanis AA. The initial signs and symptoms of altitude decompression sickness. *Aviat Space Environ Med* 1996; 67(10): 983–9.
8. Code of Federal Regulations 14, part 91, subpart C, section 91.211 *Supplemental oxygen*. Government Printing Office, Washington, DC, 25 April 2000.
9. Code of Federal Regulations 14, part 61, subpart A, section 61.23 *Medical certification and duration*. Government Printing Office, Washington, DC, 10 October 2000.
10. Hultgren HN. *High altitude medicine*. Stanford: Hultgren Publications, 1997:213–14.
11. Hultgren HN. *High altitude medicine*. Stanford: Hultgren Publications, 1997:212–48.
12. Hackett PH, Rennle, D. The incidence, importance, and prophylaxis of acute mountain sickness. *Lancet* 1976; 2:1149–55.
13. Hackett PH. The cerebral etiology of high-altitude cerebral edema and acute mountain sickness. *Wilderness Environ Med* 1999; 10(2):97–109.
14. Kloche DL, Decker WW, Stepanek J. Altitude-related illnesses. *Mayo Clin Proc* 1998; 73(10):988–93.
15. Fischer R, Lang SM, Steiner U, et al. Theophylline improves acute mountain sickness. *Eur Respir J* 2000; 15(1):123–7.
16. Hackett PH. High altitude cerebral edema and acute mountain sickness. A pathophysiology update. *Adv Exp Med Biol* 1999; 474:23–45.
17. Oelz O, Maggiorini M, Ritter M, et al. Nifedipine for high altitude pulmonary oedema. *Lancet* 1989 25; 2(8674):1241–4.

18. Arias-Stella J, Kryger H. Pathology of high altitude pulmonary edema. *Lancet* 1975; 2:758–61.
19. Krakauer J. *Into thin air*. New York: Anchor Books, 1997:246–9.
20. Ng JD, White LJ, Parmley VC, Hubickey W, Carter J, Mader TH. Effects of simulated high altitude on patients who have had radical keratotomy. *Ophthalmology* 1996; 103(3): 452–7.
21. White LJ, Mader TH. Refractive changes at high altitude after LASIK. *Ophthalmology* 2000; 107(12):2118.
22. Mader TH, Blanton CL, Gilbert BN, et al. Refractive changes during 72-hour exposure to high altitude after refractive surgery. *Ophthalmology* 1996; 103(8):1188–95.
23. Wiedman M, Tabin GC. High-altitude retinopathy and altitude illness. *Ophthalmology* 1999; 106(10):1924–6; discussion 1927.
24. Ali KZ, Ali ME, Khalid ME. High altitude and spontaneous preterm birth. *Int J Gynaecol Obstet* 1996; 54(1):11–5.
25. Niermeyer S. The pregnant altitude visitor. *Adv Exp Med Biol* 1999; 474:65–77.
26. Palmer SK, Moore LG, Young D, Cregger B, Berman JC, Zamudio S. Altered blood pressure during normal pregnancy and increased preeclampsia at high altitude (3100 meters) in Colorado. *Am J Obstet Gynecol* 1999; 180(5):1161–8.
27. Mohr G. The future perspective. In: DeHart RL, ed. *Fundamentals of aerospace medicine.* 2nd edn. Baltimore: Williams & Wilkins, 1996:37–55.
28. Gunderstrup M, Storm HH. Radiation-induced acute myeloid leukaemia and other cancers in commercial jet cockpit crew: a population-based cohort study. *Lancet* 1999; 354(9195):2029–31.
29. Bagshaw M, Irvine D, Davies DM. Exposure to cosmic radiation of British Airways flying crew on ultra-longhaul routes. *Occup Environ Med* 1996; 53(7):495–8.
30. Oksanen PJ. Estimated individual annual cosmic radiation doses for flight crews. *Aviat Space Environ Med* 1998; 69(7):621–5.
31. Tume P, Lewis BJ, Bennett LG, Pierre M, Cousins T, Hoffarth BE, Jones TA, Brisson JR. Assessment of cosmic radiation exposure on Canadian-based routes. *Health Physics* 2000; 79(5):568–75.
32. Waters M, Bloom TF, Grajewski B. The NIOSH/FAA Working Women's Health Study: evaluation of the cosmic-radiation exposures of flight attendants. *Health Physics* 2000; 79(5):553–9.

10

SHIFTWORK

Allene J. Scott, M.D., M.P.H., M.T. (ASCP)*

Shiftwork refers to hours of work occurring outside the regular daytime schedule, i.e. work schedules not falling between 6 a.m. and 5 p.m. According to the May 1997 *Current Population Survey by the US Bureau of Labor and Statistics*, considering only the primary job of full-time workers, 19% of male and 14% of female workers are shiftworkers. Of workers in service jobs, such as police, firefighters, and security guards, over half work shiftwork schedules. Among food service workers, 42% report non-daytime work schedules. Almost half, 47%, of employees of eating and drinking establishments are shiftworkers. Nearly one-third of laborers and operators in manufacturing types of industries, healthcare workers and transportation workers are shiftworkers. Less than 10% of professionals, managers and administrators and only 3% of construction workers can be categorized as shiftworkers.

*Joseph Ladou and Richard Coleman contributed to the previous edition of this chapter.

Alertness and performance of shiftworkers, particularly when working schedules involving the night (graveyard) shift, may be compromised. A substantial proportion of night workers and rotating shiftworkers do fall asleep on the job.[1] In addition to the related monetary cost to industry, estimated to be $70 billion per year,[2] public safety has been compromised by catastrophes involving chemical and nuclear power plant accidents and transportation accidents.[3]

Shiftwork also takes a toll on individuals and their families, due to conflicts between the work schedule and domestic responsibilities. Detrimental effects on general wellbeing, as well as increased risk for several physical and mental health conditions, have been associated with working shifts.[4]

OCCUPATIONAL SETTING

There are many different types of shiftwork schedules. The type of schedule is often deter-

Physical and Biological Hazards of the Workplace, Second Edition, Edited by Peter H. Wald and Gregg M. Stave
ISBN 0-471-38647-2

mined by traditional practices of the industry. For example, the Southern-Swing schedule, a weekly, backward rotation, has been commonly used for generations in the steel industry. Worker preferences, labor union interests and job-process needs are common factors influencing schedule design. More recently, shiftwork consultants have been called upon by larger businesses to design optimal shiftwork schedules based on chronobiological principles.

Common schedule designs include the following:

1. Fixed ("permanent")—only an individual employee works one shift. The number of days worked in a row and the number of days off vary between industries. "Permanent" night work is, in one sense, a misnomer, because almost all workers return to daytime activity on their days off, and thus still switch back and forth between night work and daytime activity.
2. Rotating—an individual employee regularly changes the shift worked. Rotating schedules vary in the direction and speed of rotation:

    - Direction of rotation: This may be clockwise (forward rotation, phase-delay) or counterclockwise (backward rotation, phase-advance).
    - Speed of rotation: This may be slow (greater than weekly), weekly, or rapid. Examples of rapidly rotating systems include the European Metropolitan (two identical, consecutive shifts are worked consecutively, followed by two of the next shift until all three shifts have been worked, 2-2-2) and the Continental (same as the Metropolitan, except three of one shift time are worked, 2-2-3) rotation schedules.

3. Split shifts—A hiatus of a few hours separates the work hours performed on the same day. For example, this may be done in the restaurant or transportation industries to cover main peak hours in business demand.
4. Alternative rotations—The increasingly popular 4-day workweek with 10–12 hours shifts may be used in a single-, two- or three-shift operation. The 8-day week with 4, 10-hour days followed by 4 days off is used primarily in firms operating for 10 hours per day, 7 days per week or operating 20 hours/day in two shifts. Twelve-hour shift schedules are common in hospital nursing schedules. Flexi-time, which originated in Europe, gives employees some choice in designing their own personal daily work hours to meet weekly requirements.

MEASUREMENT GUIDELINES

Several factors affect tolerance to nightwork and rotating shiftwork, including: (1) individual differences in susceptibility to physiologic disruption from altered sleep–wake cycles; (2) differences in social/family responsibilities and supportiveness; (3) effectiveness of off-the-job and on-the-job coping strategies for maximizing sleep and minimizing performance deficits; and (4) the schedule design.

In order to evaluate the appropriateness of a particular work schedule for a specific industrial site, information from several areas should be gathered. The general categories of information include: (1) demographics of the workforce; (2) frequency of cases of shiftwork intolerance from medical surveillance programs and/or medical claims; (3) rates of on-the-job fatigue-related accidents or performance deficits; and (4) information from confidential worker surveys concerning sleepiness on the job, occurrences of motor vehicle accidents or near misses driving home from each shift, general wellbeing, and worker satisfaction with the schedule.

Tepas et al[5] presents a survey method suggested for use in designing and evaluating shiftwork systems tailored to specific workers and plants. The work–sleep survey used has demonstrated that there are significant differences between industrial plants with respcct to demographics, worker habits, and worker preferences. "Before recommending a new work-system shift scheme to a plant, the complexity of shift-work issues must be recognized by assessing a wide range of personal, social, and health issues."[5] Education of workers with respect to the reasons for the selection and expectations of how the system will work is also important. Follow-up surveys, conducted after the system has been in operation for some time, are recommended to evaluate the effects on the workers' well-being and performance.[5] Kogi,[6] after reviewing an extensive survey of male and female shiftworkers in Japan working various rotating schedules in various jobs, made similar recommendations. He concluded that, besides the rotation type, other factors such as social and family life, commuting time and possibilities for anchor sleep (regularly scheduled sleep periods) during night shifts should be considered in determining ways to minimize the detrimental effects of shiftwork on the worker and his or her family.

Data necessary to measure exposure to shiftwork in addition to the number of years of shiftwork experience include schedule design variables such as the direction and frequency of shift rotation, the number of nights worked in a row, the number of days off after night shift assignment, the amount of overtime, and the length of shifts.

Direct comparisons of studies looking at performance or health outcomes of shiftworkers are often difficult, due to differences in schedule design. National data collection is limited. The United States Bureau of Labor Statistics does not include information in its accident statistics concerning the time of day of accidents or the shift system involved.[7] The Federal Occupational Safety and Health Administration's (OSHA) Form 2300 log of occupational injuries does not include the time of the accident, the number of hours worked preceding the accident, or the type of work schedule. However, the proposed OSHA Form 301 Incident Report (current Form 101) which employers are to maintain in their files, for 5 years does include the work start time and time of injury. Limited opportunities to look at national data concerning on-the-job accidents and scheduling factors may therefore exit.

Despite the inherent design problems in shiftwork research and the lack of national data collection, field and laboratory studies of 24-hour sleep and performance rhythms and of health outcomes in shiftworkers have provided information which has been used to make reasonable recommendations concerning shiftwork scheduling. These are presented in the following section.

EXPOSURE GUIDELINES (SCHEDULE DESIGN)

As is true of all Permissible Exposure Limits (PELs) and/or Threshold Limit Values (TLVs) for chemical exposures, general recommendations for scheduling design, which control exposure to nightwork, are applicable to most workers. TLVs are designed to protect nearly all healthy workers, with the recognition that a small percentage of workers may be susceptible to lower levels of exposure.[8] In the same way, shiftwork scheduling recommendations may not be adequate for workers with medical conditions, or physiologic traits which increase their individual susceptibility to shiftwork-related symptoms or aggravation of medical problems. In addition, with respect to performance on the job, the appropriateness of recommendations for the length of shift varies with the type of industry involved, e.g. healthcare work versus interstate driving.

Chronotoxicologic considerations

Exposure limits for chemical stressors in the workplace (TLVs and PELs) have been established for 8-hour workdays and under diurnal conditions. The American Conference of

Governmental Industrial Hygienists' guidance[8] for adjusting exposure limits for non-8-h workdays suggests that industrial hygienists follow the "Brief and Scala model", as described in Paustenbauch.[9] This model reduces the TLV proportionately for increased time of exposure and for decreased non-exposure time available for metabolism/elimination of the toxicant.

Longer shifts, with shorter non-exposure intervals in between, may result in higher biological levels being reached in the exposed workers. Therefore more frequent testing may be needed for workplace exposures requiring biological monitoring. In addition, the susceptibility to the adverse affects of the toxic materials may be greater when exposure occurs at night, just as response to medications varies with the time of administration. Smolensky and Reinberg[10] have prepared a detailed discussion of chronotoxicology as it relates to biological monitoring.

Musculoskeletal considerations

Work involving heavy physical labor and/or repetitive motion may need to be adjusted for production speed and number of breaks. Job assignments may need to be rotated to prevent repetitive musculoskeletal stress/injury. A model for special provisions for night workers who are also exposed to physical and/or toxic stressors has been provided by the Austrian "Night Shift/Heavy Work Law", which has been summarized by Koller.[11]

General scheduling considerations

Ideally, shiftwork schedules should be designed to minimize the potential negative effects of shiftwork on worker sleep, health, and performance. However, medical and performance/safety considerations are not the only driving force in scheduling design. Economic ramifications, production needs, labor union issues and worker preferences are important factors in influencing the final outcome. In order to help balance all of these considerations, shiftwork consultants are increasingly being hired to assist companies in redesigning shiftwork schedules.

Advantages and disadvantages of different schedules

Rotating schedules

Entrainment considerations Differences in recommendations concerning (1) whether permanent night shift schedules are preferable to rotation, (2) if rotation is chosen, and (3) what the speed of rotation should be reflect differing opinions about the effect of the schedule design on nocturnal adjustment by the night worker, and the desirability of re-entrainment to nocturnal work.[12] In order to maximize nightwork performance, rapid re-entrainment is desirable. However, minimizing circadian system disruption in order to avoid related deleterious effects on health and wellbeing conflicts with this goal.

The goal of entraining the night worker to the inverted day–night cycle is difficult to reach. It has been demonstrated that more than a week of consecutively worked nights is needed before complete adjustment of the circadian system begins to occur.[13,14] Most workers revert to day activity on their non-work days, and because considerable re-entrainment to diurnal activity occurs over only a couple of days off,[15] biological rhythm desynchronization can be expected to occur each time the night shift period is begun. Clearly, weekly and rapidly rotating schedules do not allow enough time for full adjustment to night orientation. When no more than two or three nights in a row are worked, little diurnal shifting of the circadian system will have occurred.[16] If the degree of adjustment to nightwork has been overestimated as suggested,[17] the rationale for recommending that "permanent" or slowly rotating shift systems be used in order to maximize adjustment of the circadian system is weakened.

Studies of long-term tolerance to shiftwork have suggested that shiftworkers with large amplitudes in circadian temperature rhythms (indicating resistance to adjustment) have better long-term tolerance to shiftwork.[18–20] Reinberg et al[18] have therefore suggested that, if tolerance to shiftwork over the long term is associated with large circadian amplitude of the temperature rhythm and therefore slow adjustment, rapid rotation is preferable to weekly rotation. Mills et al[21] suggested that if re-entrainment is accomplished, the worker will have trouble functioning in the diurnal world on days off.

Circadian and sleep debt considerations

Generally, the forward (clockwise, phase-delay) direction has been considered preferable, since the biological clock tends to be easier to set back than ahead. Jet lag and shiftwork laboratory studies suggest that it is easier to adjust to a phase-delay than to a phase-advance time-shift.[21–24] There are few shiftwork field studies reported which support the clockwise rotation over the counterclockwise.[21,25–27]

Knauth and Rutenfranz[28] have argued that a forward rotation allows less time for recovery from sleep debt accumulated during the workdays. Similarly, Folkard[29] has pointed out that when there is not a day off between shift changes, systems which rotate forward allow a break of 24 hours, while many backward rotations have a "quick return" after only an 8-h break.

With respect to the speed of rotation, Vidacek et al,[30] in their study of weekly rotating shiftworkers, found, for the night shift only, that a day-of-the-week effect was seen, with productivity increasing through the third day and then falling again over the last two days of the 5-day shift period, but not to the first-day low. They suggested that a weekly rotation might actually capitalize on circadian adjustment, while minimizing effects of sleep deprivation in comparison to rapidly rotating or more slowly rotating schedules. Wilkinson et al[31] found, for a slowly rotating system that performance on tests sensitive to the effects of sleep deprivation is significantly poorer on the seventh night, but not on the fourth night, compared to the first work night. Other studies have shown evidence of sleep deprivation after only two consecutive nights on the job,[32,33] and that, for rotating workers, sleep is most disturbed at the beginning of the night shift.[34] Czeisler et al[35] reported that most weekly rotators on a phase-advance schedule need 2–4 days for their sleep schedule to adjust to shift changes.

Chronic sleep deprivation has also been associated with several nights worked in succession in rotating systems.[36–38] Williamson and Sanderson[39] evaluated the effects of switching from a clockwise, slowly rotating (seven straight shifts) to a clockwise, rapidly rotating system with no more than 3 nights worked in a row. Workers reported improved sleep, and complaints of feeling tired at work and being irritable were eliminated after the switch. Smith et al[40] compared a slowly rotating, continuous 8-hour shift system involving seven shifts worked in succession, to rapidly rotating, continuous 8- and 12-hour shifts. After switching to the rapid rotations, day sleep was reported to be improved, fatigue decreased, and home and social life improved, and symptoms of circadian disruptions decreased.

Maasen et al[41] found that good sleep was obtained by workers on a weekly rotating system; however, the schedule provided 6 days off following the week of nightwork, and the morning shift did not start until 0800. Based on health, safety and productivity evaluations before and after the introduction of a new shift schedule, Moore-Ede's group has been successful with a 21-day, slowly rotating schedule which requires workers to maintain nocturnal orientation on their days off during the 3 weeks of nightwork.[42]

Finally, it should be noted that there are often other variables to be considered when contrasting rotating schedules with different speeds of rotation. Dahlgren[43] has concluded that "the speed of rotation in itself, without consideration of how the free days are organized is an insufficient criteria for judging the relative merits of different shift systems." Providing adequate time off after night shifts is necessary to prevent chronic sleep deprivation. Others have recommended that there should not be many night shifts in succession, and that at least 24 h should be given off after each night shift.[44,45] With respect to worker preference, several studies have found rapid rotations to be preferred by workers who have experienced them over weekly and permanent night shifts.[46–50,51,117]

Permanent shifts The total average duration of sleep has been found to be shorter for permanent night workers than for rotating ones, although sleep on the night shift was shorter for rotators.[52] Tepas et al[53] found measures of vigilance to be poorer for permanent night workers than for slowly rotating workers when on the night shift. Other evidence of chronic sleep deprivation was present for the permanent night workers but not for the rotators. Some research indicates that the sleep reduction associated with nightwork is greater for rotating workers; however, sleep deprivation performance deficits can be reversed during non-night shift rotation time. For the permanent night worker, sleep deprivation appears to be persistent.[54]

Length of shifts: 8 versus 12-h shifts
In addition to the nature of the shift rotation, the length of the shift may affect fatigue-related performance parameters. Several studies have found a negative impact of 12-hour shifts on performance and alertness parameters. Rosa et al,[55] in a laboratory, simulated work experiment, measured several performance parameters and evaluated worker reports of drowsiness and fatigue. The results were interpreted to indicate that 12-hour/4-day weeks were more fatiguing than 8-hour/6-day weeks. Using the same fatigue test battery, a worksite assessment was made evaluating the effects of a change from an 8-hour, three-rota or work shift schedule to a 12-hour, two-rota schedule.[56] An overall decline in performance and alertness was observed after the switch to the 12-hour shifts. Total sleep time was also noted to decrease across the 12-hour/4-day workweek.

There have been reports of fatigue effects on performance toward the end of 10- and 12-hour shifts.[57–60] A particular concern has been degree of fatigue and sleepiness sometimes experienced in the last hours of a long shift occurring at the same time as the circadian nadir of alertness. In a case–control study of large truck crashes on interstate highways, driving for over 8 hours was associated with an increased risk of crash involvement of almost two times that for drivers who had driven fewer hours.[61] Lisper et al[62] found, in a 12-hour observed driving test, that most episodes of falling asleep occurred after 8 hours of driving. Several recent studies, however, comparing 8-hour and 12-hour systems, have reported improvement or minimal deleterious effects on health, performance and alertness/sleep parameters with the introduction of a 12-hour shift.[63–68]

Compressed workweeks with 12-hour shifts are increasingly being used to cover 24-hour operations and may be recommended by shift-work-scheduling consultants.[69] Although compressed workweeks require working long shifts, they have become popular because they allow several days off in a row. Many workers consider the blocks of time off for family/social activity to be advantageous. In addition, almost a quarter of all night workers have been reported to hold a second job.[70] How much compressed work weeks encourage moonlighting and how much this adds to worker fatigue is yet to be determined.[58] Other advantages and disadvantages of 12-hour shifts have been described.[71] Inconsistencies in the literature on this topic have recently been reviewed.[60,72] Reasons for the sometimes conflicting findings may relate in part to differences in the perspective and priorities of employees in contrast to employers, differences in the demographic characteristics of workforces, and differences in production and safety issues for various types of industries. In order to avoid fatigue-related problems when using a 12-hour shift schedule system, Knauth[73] has suggested that shifts over 8 hours be used only if: (1) the nature of work and the workload are suitable;

(2) sufficient time for breaks are provided; (3) the design of the shift system minimizes the accumulation of fatigue; (4) coverage for absent workers is provided and overtime is not involved; (5) physical and toxic stressors are limited; and (6) a complete recovery is possible after the shift. Recent recommendations for shiftwork-scheduling design involving 12-hour shifts have also been made by Rosa and Colligan.[74] When 12-hour shifts are used, they advise that only two to three shifts be worked in a row, and for nightwork that two is probably best. A day or two off following the night shifts is also prescribed.

Other factors related to scheduling decisions

Early morning starting times Sleep deprivation and related fatigue have been associated with early starting times (before 7 a.m. to 8 a.m.) for the morning shift. In addition to being sleep deprived when working night shifts, rotating shiftworkers may have their sleep cut short when working first shifts requiring very early rising times, and permanent day workers on early starting 12-hour shifts have been reported to be more sleepy than night-working counterparts.[75,76] Day workers, who must begin work early in the morning, may have job/social bound sleep restrictions with significant consequences.[77,78] Knauth et al[79] have advised that morning shifts should not begin too early, to avoid an accumulation of sleep deficits.

Schedule predictability In order to maximize participation in family and social activities, work schedule predictability is necessary for making plans and keeping commitments. The Centers for Disease Control has recommended that rotating shift schedules be stable and predictable.[80] Reorganization of the on-call shifts may minimize the negative effects on social and family wellbeing.[81] Part of the reason for the popularity of 12-hour systems with workers may be that there is less variation in work times compared to 8-hour, three-shift rotations. In addition, some switches to 12-hour schedules come with less potential for required overtime and working back-to-back shifts.

Approaches to making schedule changes The actual shift system design chosen is not the only factor which will determine the success of a schedule change. The manner in which the decision was made may be equally important. A schedule system is more apt to be well accepted by the workers if their desires with respect to free time and family/social responsibilities are considered in its design.[82,83] The acceptance of an ergonomically "good" schedule may reflect whether workers or management initiated the introduction of the new schedule.[84] According to Kogi,[85] the trend in scheduling design is to use a participatory process. Recommended steps to follow when making shift schedule changes are to: (1) carry out a group study of operational needs, worker preferences, health and tolerance issues of the workforce, and potential options; (2) utilize joint planning to make plans for feasible options and specific measures; (3) provide for feedback and dialogue to build a consensus, and allow for adjustment and training; (4) implement jointly the new work organization, in a progressive fashion if appropriate; and (5) jointly evaluate the change, taking further action as needed.

NORMAL PHYSIOLOGY

Numerous psychological and physiologic variables have been found to have a demonstrable 24-hour, circadian (Latin: *circa* = about, and *dies* = a day)[86] rhythm, e.g. body temperature, the sleep–wake cycle, cardiovascular parameters, cognitive performance, endocrine and immunologic factors, therapeutic response to certain medications, and psychological variables.[87]

If these biological rhythms merely reflected responses to external time cues, they would be of little consequence for the shiftworkers. However, circadian rhythms have an endogenous component. The existence of a biological clock in humans has been repeatedly demonstrated in temporal isolation studies, in which subjects are separated from all environmental and social time cues.[88] Under normal nychthermal conditions (daytime activity and night-time sleep), the circadian system is synchronized with the 24-hour solar day by external triggers to which the biological clock is responsive. The normal phase relationships of the multitude of biological rhythms are affected by an orchestrated response to the internal pacemaker.

The time cues, which are capable of entraining the biological clock to an external periodicity, have been termed "zeitgebers" (German: time giver). Zeitgebers allow the biological clock, which runs slightly slower than the 24-hour day, to be reset and entrained to the 24-hour day.[89] Various agents have been shown to act as zeitgebers, including light, social factors, and behavioral patterns such as eating schedules, and sleep–wake schedules.[90, 91] Social cues, the sleep–wake schedule and the rest–activity cycle are relatively weak zeitgebers in comparison to sunlight (or electrical lighting of at least 7000–13 000 lux).[92] The phase-shifting effect of light on the circadian timing system is secondary to its suppressing action on the secretion of melatonin by the pineal gland (Figure 10.1.) Melatonin induces sleep and depresses the core body temperature.[93]

PATHOBIOLOGY (CIRCADIAN RHYTHMS AND SHIFTWORK)

The major function of the circadian system is the internal sequencing of physiologic events and metabolism. Biological processes are thus coordinated for optional functioning of the organism. Restorative functions of sleep are maximized by the normal phase relationship of biological rhythms during the night-time hours.

For the night shift worker, however, the activity at night will be out of phase with the circadian body temperature and other coupled rhythms. In addition, because individual biological rhythms re-entrain to a time-shift at different rates, each time the work schedule rotates

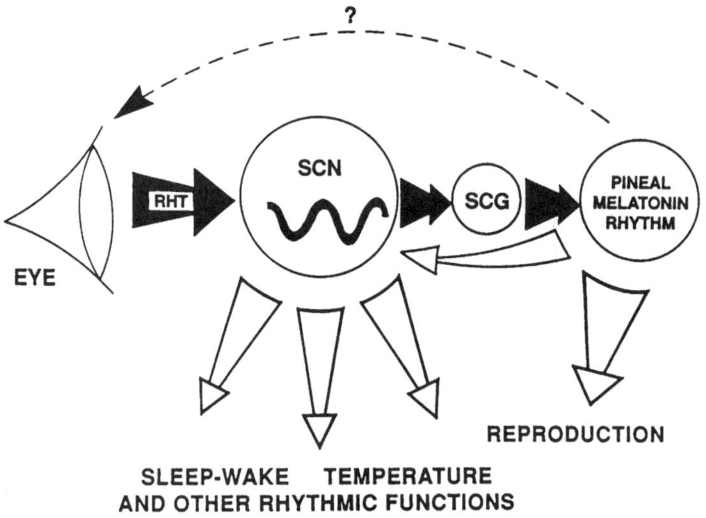

Figure 10.1 Generalized diagram of the circadian system rhythm-generating system in the suprachiasmatic nucleus (SCN). Melatonin feeds back at the level of the SCN to shift the circadian phase and also to influence rhythmicity at the retinal level. RHT = retinohypothalamic tract, SCG = superior cervical ganglion. Used with permission from Arendt and Deacon, Treatment of circadian rhythm disorders—Melatonin, *Chronobiology Int* 1997;14(2):185–204.

after the time-shift the circadian system will be in a desynchronized state for a period of time. For example, sodium and potassium excretion are closely linked in a stable rhythmic environment, but have significant differences in their rate of re-entrainment to phase shifts.[94] The sleep–wake rhythm adjusts faster than the body temperature rhythm, and activity re-entrains faster than many physiologic functions.[95]

Circadian rhythms are more easily re-entrained, after a time-shift, if all the important zeitgebers, including the light–dark cycle, are synchronously shifted, such as occurs with transmeridian flights. For shiftworkers, zeitgebers are shifted in a non-synchronized manner. Knauth and Rutenfranz[96] failed to find complete inversion of the body temperature rhythm in shiftworkers even after 21 consecutively worked night shifts, and concluded that the circadian system never fully adapts to nightwork. Other field studies of shiftworkers have also found adaptation to nightwork to be incomplete.[97–99]

Nightwork and sleep deprivation

Regular nightwork is associated with chronic sleep deprivation.[100,101] The sleep length of night workers is 15–20% that of day and afternoon workers, averaging 4–6 hours compared to 7–9 hours respectively.[102,103] In addition to being shorter than night-time sleep, day sleep is of a poor quality due to frequent awakenings and disruptions of the normal REM/non-REM sleep stage pattern.[104–106]

The etiology of the sleep problem of night workers is multifactorial. A major determinant of sleep duration and quality is the endogenous circadian system.[107–109] Job schedule requirements, domestic responsibilities and environmental conditions may also significantly contribute to the sleep problems of night workers.[110–112]

Chronic partial sleep deprivation can have significant negative effects on job performance and social functioning. Sleep deprivation is associated with increased irritability and generalized fatigue that can compromise social and domestic interactions.[113–115] In situations where falling asleep or decreased alertness threatens individual or public safety, shiftwork-related sleep deprivation is a public health concern.

DIAGNOSIS

Jet lag versus shift lag

The signs and symptoms of jet lag are an example of **desynchronosis** due to desynchronization of the normal phase relationships between biological rhythms within the circadian system, and to the external desynchronization between the circadian system and the 24-hour solar day–night cycle to which the bio-logical clock is normally synchronized (entrained).

Symptoms of jet lag include (in order of frequency as typically reported by frequent jet travelers): daytime sleepiness and fatigue, difficult in sleeping at night, poor concentration, slow physical reflexes, irritability, digestive system complaints, and feelings of depression. Not surprisingly, studies of shiftworkers have demonstrated that shiftworkers experience very similar symptoms,[116,117] that is, symptoms of "shift lag"[118] are essentially the same as those of jet lag. However, the symptoms of shiftwork-related desynchronosis have more health significance because: (1) the complaints of shiftworkers may not be as readily attributed to the work schedule; (2) jet-lag symptoms are limited to a few days following travel, while shiftwork-related desynchronosis is usually chronic; and (3) the conflict between the shiftworker's work schedule and the predominantly day-oriented social/business schedule is in opposition to achieving re-entrainment to nightwork.

Shiftwork intolerance

Fortunately for most shiftworkers, shift-lag symptoms are not debilitating, but for a significant minority, shiftwork-related symptoms are significant. Surveys of former shiftworkers indicate that, for some, health complaints increase with continued shiftwork and become severe enough to cause the worker to give up a job, often following medical advice.[119–121]

Up to 20% of shiftworkers may have a disproportionate amount of symptoms of illness when assigned to chronobiologically poorly designed schedules involving nightwork.[87,122] Clinical intolerance to shiftwork has been defined by the presence and intensity of the following set of medical complaints: (1) sleep alterations; (2) persistent fatigue (not disappearing after time off to rest); (3) changes in behavior; (4) digestive system problems; and (5) the regular use of sleeping pills (considered almost pathognomonic of shiftwork intolerance).[123,124] Askenazi et al[124] consider the presence of the symptoms in categories (1), (2) and (5) to be essential to classify a worker as shiftwork intolerant.

The term shiftwork maladaption syndrome (SMS) refers to the typical constellation of signs and symptoms seen in shiftwork-intolerant workers. In SMS, the symptoms are pronounced and worsen with continued exposure to shiftwork.[87,122] The longer the worker stays on shiftwork, the worse the symptoms become, and eventually the worker may be fired, quit his job, or be involved in an accident. A worker with SMS is likely to get into a vicious cycle due to the lack of recognition and understanding of the problem by himself/herself and/or by others not familiar with the condition. Inability to adjust family/social life to the work schedule and poor schedule design may significantly contribute to the degree of intolerance.[125] Diagnosis of SMS should be made only after a thorough occupational medical evaluation has been carried out and other relevant medical disorders ruled out.

Individual factors that predispose to shiftwork intolerance are not fully understood. In general, age over 40–50 years, extreme morningness and rigid sleep requirements are characteristics that have been associated with decreased tolerance for nightwork.[126–130] A recent review of research assessing circadian factors and shiftwork tolerance[17] suggests that: (1) individuals with small amplitudes of certain circadian rhythms, e.g. body temperature, may be more prone to desynchronization of rhythms when subjected to time-shifts; (2) some individuals are more likely to experience desynchronization of biological circadian rhythms unrelated to zeitgeber manipulation; and (3) certain individuals are particularly sensitive to rhythm desynchronization manifesting clinically significant symptomatology.

Factors affecting shiftwork tolerance, in addition to individual susceptibility, include social and family situations, working conditions, and shiftwork schedule arrangements.[131] Shiftwork tolerance must be assessed in the complex framework of interrelationships of these factors (Figure 10.2).

Shiftwork and specific medical disorders

Gastrointestinal (GI) disorders GI dysfunction is common in shiftworkers.[132–134] Gastritis or other digestive disorders have

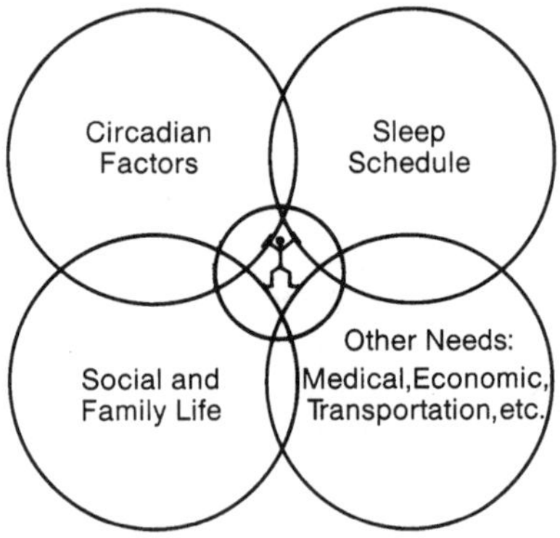

Figure 10.2 Factors influencing tolerance to shiftwork are shown in this diagram. These factors have a complex interrelationship mediated by the individual vulnerability of the shiftworker. Used with permission from Scott and Ladou. Shiftwork: effects on sleep and health. *Occup Med—State of the Art Reviews* 1990; 5(2):273–2.

been an explanation frequently given by shiftworkers for absenteeism and for switching to day work for health reasons.[135] While some studies have not found an increased incidence of peptic ulcer disease (PUD) in shiftworkers, the majority of studies addressing this outcome have found shiftworkers to be at greater risk of developing PUD than day workers.[136]

The etiology of GI disorders in shiftworkers is probably multifactorial, involving dietary factors, psychosocial stress, sleep loss, as well as circadian disruption. Clearly, the night workers' mealtimes are in conflict with the circadian rhythms of gastric acidity and gastric emptying.[137,138] Shiftworkers may alter their diet due to lack of eating facilities available during the night shift.[139–141]

Cardiovascular morbidity Shiftwork schedules have been reported to have a deleterious effect on the biomarkers of cardiovascular disease (CVD). Several studies suggest that shiftwork may increase low-density lipoproteins.[142,143] Although there are studies which have not found changes in lipid profiles related to shiftwork, the better studies indicate that shiftworkers are at higher risk of having somewhat higher levels of cholesterol as well as triglycerides.[144] Shiftwork has also been associated with an increased risk of developing hypertension.[145,146] and electrocardiographic changes (QT interval prolongation).[147]

Although early studies concluded that shiftwork was not associated with an increase in CVD,[148,149] most of the more recent, better epidemiologically designed studies have found an increased risk of CVD associated with shiftwork of, on average, around 40%.[150]

Analysis of data from the Helsinki Heart Study[151] demonstrated an interaction of shiftwork exposure with lifestyle factors known to increase the risk of coronary heart disease (CHD). For shiftworkers, the relative risk of CHD rose gradually with increasing numbers of adverse lifestyle factors, but for day workers, no clear dose–response pattern was found. The findings were interpreted to indicate that shiftwork may have a triggering effect on other lifestyle factors that increase the risk of CHD, and that active prevention is especially important for shiftworkers. Recent findings[152] suggest that stressful psychosocial work factors may act as a mediator of shiftwork-related effects on hypertension and lipoprotein levels.

Reproductive health

Preterm births and low birth weight McDonald et al[153] found rotating shiftwork to be associated with low birth weight, but not preterm birth. Later analysis[154] of data from this study, allowing for gestational age, suggested that shiftwork might slow fetal growth and increase the risk of preterm delivery. Nurimen[155] found a small association between shiftwork and small-for-gestational-age infants.

Axelsson et al[156] compared night and evening work, working irregular schedules outside the hours of 0645-1745, and rotating shiftwork to permanent day work. Working irregular hours was associated with an increased risk of low birth weight. Chinese investigators[157] found higher proportions of preterm birth (OR 2.0) and low birth weight (OR 2.1) for rotating shiftworkers.

Data on permanent nightwork and reproductive outcomes are very limited. Two studies[158,159] did not find an association with increased risk for adverse pregnancy outcomes. However, in one of these studies, only 4% of the female employees interviewed were night workers, some working only occasional nights. In the other study, 91% of the night workers questioned were part-time. Fortier et al[160] did find a modestly increased OR of 1.45 for delivering preterm, for women working regular evening or night shifts after the 23rd week of pregnancy.

Spontaneous abortion McDonald et al[161] found an association between rotating shiftwork and an increased risk of spontaneous abortion. Axelsson et al[162] found a statistically significant increased risk of spontaneous abortion for nightwork and three-shift work done during the first trimester. Infante-Rivard et al[163] compared the work schedules of women who had experienced pregnancy loss to those of gestationally age-matched controls. For fixed evening schedules, the adjusted OR was 4.17, and for fixed nights, 2.68. Other studies have demonstrated a significantly increased risk of miscarriage for shiftworkers.[164–166] Two other studies[167,168] did not find an increased risk of spontaneous abortion; however, as pointed out by Nurinem,[169] in these two negative studies, rotating shifts were part of broad exposure categories and were not analyzed explicitly.

Shiftwork and subfecundity The few studies which have been published studying time to pregnancy raise the possibility that shiftwork and long hours of work may reduce fecundity, particularly of the female partner.[165,170,171] Other studies have not found an association between subfecundity and shiftwork.[172–174]

Aggravation of exacerbation of medical disorders

The effectiveness of many medications varies with the time of day at which the medication is taken, due to circadian rhythms in metabolic parameters as well as target organ responsiveness. Irregular work schedules may preclude the application of chronopharmacologic principles for the best timing of taking medication. Workers with irregular schedules lack consistent routines, making it difficult to remember to take medications.

Shiftwork may also aggravate certain medical disorders related to circadian rhythm disruption, psychosocial stress and/or sleep deprivation. Table 10.1 summarizes these conditions that are potential contraindications for shiftwork. Koller[175] and Costa[176] have made similar recommendations, and common conditions that may be exacerbated or aggravated have also recently been reviewed.[177]

TREATMENT (COUNTERMEASURES)

In order for individual coping strategies to be effective, families must be involved. In addition, the shiftworker must remember the toll that the shiftwork schedules may take on the

Table 10.1 Potential contraindications for working night or rotating shifts.

Condition	Discussion—Remember to consult with treating physician, get evaluation with specialist if indicated, follow in medical surveillance program, may need to restrict from shiftwork
Prescription medication	If time of dosing is critical for drug effectiveness
Diabetes mellitus	Irregularity of meals and/or of taking medication may lead to poor control
Epilepsy	Sleep deprivation associated with nightwork may increase risk of seizure
Cardiovascular disease (CVD)	Uncontrolled hypertension, unstable angina, high risk for developing CVD
Psychiatric conditions	Bipolar disorder—irregular schedule may trigger manic behavior. Shift work may be associated with depression
Sleep disorders	Untreated sleep apnea, narcolepsy, circadian phase-related insomnia, long or rigid sleep requirements
Reproductive conditions	Poor pregnancy progress; history of spontaneous abortion, low birth weight, or preterm delivery while working shiftwork schedule
Asthma	Poor control when working shiftwork

family as well. The provision of educational programs for both the worker and family is essential for employees to successfully cope with shiftwork schedules. Educational materials addressing shiftwork issues, including countermeasures published in laymen's terms, are available which will assist employers and employees in this endeavor.[178–180] Strategies for coping with jet lag are reviewed in depth by Comperatore and Krueger[181] and Lamberg.[182]

Maximizing sleep

The most significant factor interfering with sleep for night workers is daytime noise. Even if the worker is not aware of actually being awakened by noise, sleep quality may be compromised.[183] Actions should be taken to soundproof the bedroom as much as possible. In addition to using sound damping items such as ceiling tiles and perhaps carpeting, white noise from a fan or air conditioner may be helpful. Family and even neighbor cooperation should be sought to help control noisy activities near the night worker's sleeping quarters. The phone and doorbell should not be audible in the bedroom. Earplugs can also be used to attenuate noise as long as the fit is comfortable.

Light exposure should be limited to as close to night-time conditions as possible. Lined drapery and window blinds or dark room shades are suggested. Eye shades are another option for decreasing light exposure.

Applying "sleep hygiene", techniques initially developed to help patients with insomnia, is also a recommended coping strategy for shiftworkers. Sleep hygiene is a program applying regular procedures and following behavioral rules that enhance the ability to fall asleep and stay asleep. Sleep hygiene recommendations have been recently summarized.[184]

Although the regular use of sleeping pills is contraindicated, short-acting hypnotics such as triazolam have been shown to improve quality and duration of daytime sleep.[185] Occasional use for a day or two, under a physician's care, may be useful when beginning a run of night shifts or following a transmeridian flight. However, caution must be exercised, as impaired cognition has been demonstrated to linger 8 hours after administration of a 0.5-mg dose.[186]

Caffeine and other alertness-enhancing drugs

Caffeine belongs to the xanthene class of drugs, which have been shown to cause phase shifts of the temperature rhythm in animals.[187] Caffeine's role in shiftwork coping relates to its stimulant effect in counteracting sleepiness, and to its ability to delay sleep onset at night.[188,189] Its effect in increasing alertness is most apparent after a time of abstinence. With repeated doses, the effect may diminish.[190] Caffeine does have beneficial effects on alertness and performance, and decreases sleep tendency as measured by multiple sleep latency tests.[191,192]

If caffeine is consumed, it should not be ingested closer than around 5 hours before bedtime.[193] The dose of caffeine should be limited to around 300 mg/day. Shiftworkers should limit use to the first half of night or evening shifts. It is important to avoid caffeine during the last half of the evening shift or night shift, since the worker's bedtime will come soon after getting home. Fruit juice is a good alternative drink for the second half of the shift.

Illicit drugs such as amphetamines should never be used. Owing to deleterious side-effects, stimulant diet pills should not be used to counteract shiftwork sleepiness. Research is being conducted on some new alertness-enhancing drugs that may, in the future, have some usefulness for occasional alertness promotion. Findings to date have recently been reviewed by Akerstedt and Ficca.[193]

Bright light and melatonin

In addition to its sleep-inducing property, animal and human laboratory studies suggest that melatonin may effect phase shifts when administered with appropriate timing.[194,195] Several field studies have demonstrated melatonin to be useful for ameliorating jet-lag symptoms. These have recently been reviewed in detail.[181,196] Five milligrams daily, orally administered, is the frequently used pharmacologic dose used in research protocols, although lower doses may also be effective.[197] Specific instructions (and side-effect warnings) for taking the hormone for eastbound and westbound flights, as given to subjects participating in jet-lag studies, are included in the review by Arendt and Deacon.[196] However, routine use of the product by travelers is not recommended.

There is limited research available on the use of melatonin in shiftwork situations. Beneficial effects on sleep and alertness have been reported to be associated with bedtime administration, but some performance measures may be adversely affected.[198] Although there are some shiftworkers reporting benefits from using over-the-counter (OTC) preparations for shift-lag symptoms,[199] its use cannot be recommended at this time.[178,196] Because it is a naturally available dietary supplement, melatonin is available OTC. However, since it is an OTC preparation, its purity and strength are not regulated by the US Food and Drug Administration.[200] The effects of long-term usage are not known, and inappropriately timed administration may be dangerous due to its sedative effect.

Exposure to bright light has been demonstrated to result in phase shifts of the circadian timing system. The timing of the exposure determines the direction of the shifts, i.e. either a phase-advance or phase-delay response.[201–203] Although appropriately timed, bright light exposure can enhance adjustment to night shifts, practical application for shiftworkers is a different matter. Not only is the timing of the light exposure critical, but prevention of outdoor sunlight exposure at times is necessary (e.g. on the commute home after dawn). In addition, there is considerable individual variation in the degree of phase-shift response.

The recently combined use of melatonin and light has received attention as potentially the most effective way to encourage re-entrainment after time-shifts.[204] Again, practical application for shiftworkers may not be realistic.

Diet and exercise

Good dietary habits and regular exercise are recommended in general for preventive health reasons. It may, however, be even more important for shiftworkers to follow preventive health recommendations for reducing the risk of coronary heart disease than for workers in general.[151] There is no consistent evidence to date that dietary manipulations, such as the "jet-lag diet",[205] can speed up adjustment to time-shifts. Nevertheless, it may be prudent, when trying to stay awake, to avoid foods high in carbohydrates, which may promote sleepiness. Heavy, greasy or otherwise difficult-to-digest meals should be avoided before bedtime and when working the night shift.[181,206]

Physical fitness training has been demonstrated in shiftworkers to reduce general fatigue and sleepiness at work, modestly increase sleep duration, and decrease musculoskeletal symptoms. Recommendations for exercise for shiftworkers include the following: (1) moderate physical exercise is preferred over intensive training; (2) exercise should be done a few hours before the main sleep period; and (3) for morning or day shifts, the best exercise time is after the shift. After night shifts, the exercise should be done before an evening nap.[128,207]

MEDICAL SURVEILLANCE

Preplacement assessment

Recommendations for medical evaluations of workers before they begin nightwork assignments have been made by occupational medicine practitioners and researchers involved in studying the health effects of shiftwork.[87,131,175,208,209] Recognition of this need is reflected in the International Labor Organization (ILO) 1990 Convention[210] (No. 171) provision for a health assessment for workers before beginning nightwork. In addition, the European Directive No 93/104/EC, "Concerning certain aspects of the organization of working time",[211] also considers it a right of workers to have a free health assessment before beginning their first assignment to the night shift.

Identification of individual characteristics that are associated with poor tolerance of nightwork is recommended, not with the goal of disqualifying workers for nightwork, but with the recognition that, for some, nightwork may medically not be advisable. In most situations, the preplacement examination will provide an opportunity to make susceptible workers aware of their situation, and if nightwork is to be tried, to plan appropriate medical supervision and develop coping strategies.[87,175] As is true in other situations where ongoing medical problems are identified which may preclude a work assignment, consultation with an appropriate specialist and review of medical records may be necessary before making a fitness-for-duty determination.

Medical surveillance and screening

Ongoing medical surveillance programs, including periodic medical screening examinations, and appropriate laboratory testing, have been recommended for rotating and permanent night workers. In addition, follow-up evaluations of day workers who have left shiftwork for medical reasons have also been advised.[131,175] Internationally, the ILO Night Work Convention also provides for health assessments of night workers at regular intervals as well as for work-related problems that may be secondary to the work schedule.

The frequency of medical surveillance examinations is somewhat arbitrary. However, recommendations are consistent in advising evaluation during the first few months after beginning shiftwork and at regular, but less frequent intervals, depending on the work schedule and the age of the worker. A reasonable schedule has been outlined by Harma.[128] Components of medical surveillance evaluations have recently been reviewed.[177]

Medical surveillance programs for shiftworkers should include educational/counseling opportunities related to the assessment and optimization of shiftwork coping strategies.[128,175] Aggregate results of medical surveillance examinations for individual workers should be collected. This information should be used to assess the overall health impact of the shiftwork schedule on the workforce.

PREVENTION AND ADMINISTRATIVE CONTROLS

Scheduling decisions

Scheduling decisions should be made with the goal of minimizing the potential negative impact of shiftwork on worker sleep, health, and performance. Although scheduling designs understandably reflect business needs and employee preferences, sleep and health considerations should not be secondary concerns. Before making schedule changes, the demographics of the workforce, including lifestyles, sleep habits, common medical problems and shift-scheduling preferences, as well as the type of work and the environment in which it is performed, should be assessed.[80,212]

Recognizing that "tailor-made" shift systems need to involve compromises between conflicting interests of employees, employers and ergonomic considerations, Knauth[73] has recently provided detailed practical recommendations for achieving ergonomically sound shift schedule systems. Four general categories of factors important in the evaluation of the degree of shift system compliance with ergonomic recommendations are reviewed: (1) the sequence of shifts, including the speed and direction of rotation and special cases; (2) the duration and distribution of working time, including the number of consecutive working days, shift duration, and time off; (3) the position of the working time, including the start of the morning shift and the end of the evening and night shift, and number of free weekends; and (4) short-term deviations from the established shift schedule resulting from wishes of the employees or from requirements of the employer.

It is clear from the above discussions that preparing schedule designs for shift systems is a complex matter. In addition, there may be individual workers with medical restrictions that need to be worked into particular rosters. Fortunately, computer software programs have been developed to assist in the process.[213,214]

Occupational health programs

Workplace facilities and environmental conditions can impact on tolerance to shiftwork and shiftworker performance. In order to assist employees in dietary countermeasures, equivalent canteen/eating facilities should be provided for night workers as for day workers. At a minimum, a microwave, refrigerator and vending machines with low-fat, nutritious foods should be available, including dairy products and fruit juices.

Other environmental factors should be assessed which can increase alertness on the job and help prevent episodes of falling asleep. For instance, bright, uniform lighting will enhance alertness. Non-variable background noise which promotes boredom may be replaced with judiciously selected music and, if appropriate, social interactions between workers. Keeping room temperatures below 70°F, and providing opportunities for physical activity, are recommended to maximize alertness on the night shift.[179,215]

Educational programs should be provided for workers and their families that provide information for shiftwork coping strategies. In addition, workers should be advised of the increased risk of motor vehicle accidents on the drive home when working night shifts. Provision of sleeping facilities for workers who need to sleep before driving home should be considered. The degree of driving risk for the individual workforce should be assessed, and the aggressiveness of preventive measures taken based on the findings. Some proactive companies, having recognized the difficulty that night workers with families face in obtaining child care, have established 24-hour child care facilities for their workers.[216] Monk and Folkard[180] have recommended that employers develop a "Shift Work Awareness Program" for coordinating educational and social support programs.

REFERENCES

1. Coleman RM, Demant WC. Falling asleep at work: a problem for continuous operations. *Sleep Res* 1986;15:265.
2. Problem proves costly on the job as productivity and safety decline. Wall Street Journal, 7 July 1988.
3. Scott AJ. Chronobiological considerations in shiftworker sleep and performance and shiftwork scheduling. *Human Performance* 1994; 7(3):207–33.
4. Scott AJ, Ladou J. Health and safety in shiftwork. In: Zenz C, ed. *Fundamentals of occupational and environmental medicine,* 3rd edn: St Louis: Mosby-Yearbook, 1994, 960–86.
5. Tepas DI, Armstrong DR, Carlson ML, Duchon JC, Gersten A, Lezotte DV. "Changing industry to continuous operations: different strokes for different plants". *Behav Res Methods, Instruments Computers* 198; 17(6):670–6.
6. Kogi K. Comparison of resting conditions between various shift rotation systems for industrial workers. In: Reinberg A, Vieux N, Andlauer P, eds. *Night and shift work: biological and social aspects.* Oxford: Pergamon Press, 1981, 417–24.

7. Bureau of Labor Statistics. Personal E-mail communication, 10 April, 2000.
8. American Conference of Governmental Industrial Hygienists. *1998 TLVs and BEIs, Threshold limit values for chemical and physical agents biological exposure indices.* Cincinnati: ACGIH, 1998.
9. Paustenbauch DJ. Occupational exposure limit, pharmacokinetics, and unusual work schedules. In: Harris RL, Cralley LJ, Cralley LV, eds. *Patty's industrial hygiene and toxicology,* 3rd edn, Vol. 3A. New York: John Wiley & Sons, 1994, 222–348.
10. Smolensky M, Reinberg A. Clinical chronobiology: relevance and applications to the practice of occupational medicine. *Shiftwork: Occup Med—State of the Art Reviews* 1990; 5(2):239–72.
11. Koller M. Occupational health services for shift and night workers. *Appl Ergonomics,* 1996; 27(1):31–7.
12. Monk TH, Folkard S. Circadian rhythms and shift work. In: Hockey GRS, ed. *Stress and fatigue in human performance.* Chichester: John Wiley & Sons, Inc. 1983, 97–121.
13. Knauth P, Rutenfranz J. Experimental shift work studies of permanent night and rapidly rotating shift systems: 1, Circadian rhythm of body temperature and re-entrainment at shift change. *Int Arch Occup Environ Health* 1976; 37:125–37.
14. Knauth P, Emde E, Rutenfranz J, Kieswetter E, Smith P. Re-entrainment of body temperature in field studies of shiftwork. *Int Arch Occup Environ Health* 1981; 49:137–49.
15. Van Loon JH. Diurnal body temperature curves in shiftworkers. *Ergonomics.* 1963; 6: 267–73.
16. Costa G, Ghirlanda G, Tarondi G, Mionors D, Waterhouse J. Evaluation of a rapidly rotating shift system for tolerance of nurses to nightwork. *Int Arch Occup Environ Health* 1994; 65:305–11.
17. Folkard S. The pragmatic approach to masking. *Chronobiology Int* 1989; 6(1):55–64.
18. Reinberg A, Vieux N, Andlauer P, Guillet P, Nicolai A. Tolerance of shift work, amplitude of circadian rhythms, and aging. In: Reinberg A, Vieux N Andlauer P, eds. *Night and shift work: biological and social aspects.* Oxford: Pergamon Press, 1981, 341–54.
19. Leonard R. Night- and shift-work. In: Reinberg A, Vieux N, Andlauer P, eds. *Night and shift work: biological and social aspects.* Oxford: Pergamon Press, 1981, 323–9.
20. Reinberg A, Motohashi Y, Bourdeleau P, et al. Alteration of period and amplitude of circadian rhythms in shift workers. *Eur J Appl Physiol* 1988; 57:5–25.
21. Mills JN, Minors DS, Waterhouse JM. Exogenous and endogenous influences on rhythms after sudden time shift. *Ergonomics* 1978; 21:755–61.
22. Klein KE, Wegman HM, Hunt BI. Desynchronization as a function of body temperature and performance circadian rhythm as a result of outgoing and homecoming transmeridian flights. *Aerospace Med* 1972; 43:119–32.
23. Klein KE, Herrmann R, Kuklinski P, Wegman HM. Circadian performance rhythms: experimental studies in air operations. In: Mackie R, ed. *Vigilance: theory, operational performance, and physiological correlates.* New York: Plennum Press, 1977, 111–32.
24. Bodanowitz M. The change of circadian rhythms of psychomotor performance after transmeridian flights. Translation of DLR-FB 73-52, *Die veranderung tagesperiodischer schwankungen der psychomotorischen leistung nach transmeridian flugen.* Bonn, Bad Godesberg: Deutsche Forschungs- und Versuchsanstalt fur Luft- und Raumfahrt, Institut fur Flugmedizin, 1973.
25. Orth-Gomer K. Intervention on Coronary risk factors by changing working conditions of Swedish policemen. In: Harvath M, Frankth E. *Psychophysiologic risk factors of cardiovascular diseases*—International Symposium, Suppl. 3, Prague: Avicenum-Czechoslavak Medical Press, 1982.
26. Orth-Gomer K. Intervention on coronary risk factors by adapting a shiftwork schedule to biologic rhythmicity. *Psychosom Med* 1983; 45:407–15.
27. Lavie P, Tzischinsky O, Epstein R, Zomer J. Sleep–wake cycle in shift workers on a "clockwise" and "counter-clockwise" rotation system." *Isr J Med Sci* 1992; 28(8–9): 636–44.
28. Knauth P, Rutenfranz J. Development of criteria for the design of shift work systems. *J Hum Ergology* 1992; 11(S):337–67.

29. Folkard S. Shift work—a growing occupational hazard. *Occup Health* 1989; 182–6.
30. Vidacek S, Kaliterna L., Radosevic-Vidack B, Folkard S. Productivity on a weekly rotating shift system: circadian adjustment and sleep deprivation effects? *Ergonomics* 1986; 29(12):1583-90.
31. Wilkinson R, Allison S, Feeney M, Kaninska Z. Alertness of night nurses: two shift systems compared. *Ergonomics* 1989; 32(3):281–92.
32. Knauth P, Landau K, Droge C, et al. Duration of sleep depending on the type of shift work. *Int Arch Occup Environ Health* 1980; 46:167–77.
33. Wojtczak-Jaroszowa J. Circadian rhythm of biological functions and night work. In: Wojtczak-Jaroszowa J, ed. *Physiological and psychological aspects of night and shift work.* Cincinnati: National Institute for Occupational Safety and Health, 1977, 3–12.
34. Dahlgren K. Adjustment of circadian rhythms and EEG sleep functions to day and night sleep among permanent nightworkers and rotating shiftworkers. *Psychophysiology* 1981; 18(4):381–91.
35. Czeisler CA, Weitzman ED, Moore-Ede MC, Zimmerman CJ. Human sleep: its duration and organization depend on its circadian phase. *Science* 1980; 210:1254–67.
36. Walker J. Frequent alternation of shifts on continuous work. *Occup Psychol* 1966; 40: 215–25.
37. Kiesswetter E, Knauth P, Schwarzenau P. Daytime sleep adjustment of shiftworkers. In: Koella WP, Ruther E, Schulz H, eds. *Sleep '84.* New York: Gustav Fischer Verlag, 1985, 273–75.
38. Kogi K. Estimation of sleep deficit during a period of shift rotation as a basis for evaluating various shift systems. *Ergonomics* 1978; 21(10):861–74.
39. Williamson AM, Sanderson JW. Changing the speed of shift rotation: a field study. *Ergonomics* 1986; 29(9):1085–96.
40. Smith PA, Wright BM, Mackey RW, et al. Change from slowly rotating 8-hr shifts to rapidly rotating 8-hour and 12-hour shifts using participative shift roster design. *Scand J Work Environ Health* 1998; 24(S3):55–6.
41. Maasen A, Meers A, Verhagen P. Quantitative and qualitative aspects of sleep in four shift workers. *Ergonomics* 1978; 21(10):861–74 (abstract).
42. Circadian Technologies, Inc. *Improving human performance and health in round-the-clock operations.* Seminar, Pittsburgh, PA, November 1987 (unpublished).
43. Dahlgren K. Adjustment of circadian rhythms to rapidly rotating shift work—a field study of two shift systems. In: Reinberg A, Vieux N, Andlauer P, eds. *Night and shift work: biological and social aspects.* Oxford: Pergamon Press, 1981, 357–65.
44. Rutenfranz J, Knauth P. Hours of work and shiftwork. *Ergonomics* 1976; 19(3):331–40.
45. Knauth P, Landau K, Droge C, et al. Duration of sleep depending on the type of shift work. *Int Arch Occup Environ Health* 1980; 46:167–77.
46. Ghata J, Reinberg A, Vieux N, et al. Adjustment of the circadian rhythm of urinary 17-OHCS, 5-HIAA, catecholamines, and electrolytes in oil refinery operators to a rapidly rotating shift system. *Ergonomics* 1978; 21(10):861–74 (abstract).
47. Wedderbrun AAI. How important are the social effects of shiftwork? In: Johnson LC, Tepas DI, Colquhoun WP, Colligan MJ, eds. *Advances in sleep research,* Vol. 7, *Biological rhythms, sleep, and shift work.* New York: Spectrum Publications 1981, 257–69.
48. Walker J. Frequent alternation of shifts on continuous work. *Occup Psychol* 1966; 40:215–25.
49. Conroy RT, Mills JN. *Human circadian rhythms.* London: J.& A. Churchill, 1970.
50. Reinberg A. Clinical chronopharmacology. In: Reinberg A, Smolensky M, eds. *Biological rhythms and medicine: cellular, metabolic, physiopathologic, and pharmcologic aspects.* New York: Springer-Verlag, 1983, 211–57.
51. Akerstedt T, Torsvall L. Experimental changes in shift schedules—Their effects on well-being. *Ergonomics* 1978; 21(10): 849–56.
52. Tepas DI, Walsh JK, Armstrong DR. Comprehensive study of the sleep of shift workers. In: Johnson LC, Tepas DI, Colquhoun WP, Colligan MJ, eds. *Biological rhythms, sleep and shift work—Advances in sleep research,* Vol. 7. New York: Medical & Scientific Books, 1981, 347–55.

53. Tepas DI, Walsh JK, Moss PD, Armstrong D. Polysomnographic correlates of shiftworker performance in the laboratory. In: Reinberg A, Vieux N, Andlauer P, eds. *Night and shift work: biological and social aspects.* Oxford: Pergamon Press, 1981, 179–86.
54. Tepas DI, Monk TH. Work schedules. In: Salvendy G, ed. *Handbook of human factors*, New York: John Wiley & Sons, 1987, 819–43.
55. Rosa RR, Wheeler DD, Warm JS, et al. Extended workdays: effects on performance and ratings of fatigue and alertness. *Behavior Res Methods, Instruments, Computers* 1985; 17(1):6–15.
56. Rosa RR, Colligan MJ, Lewis P. Extended workdays: effects of 8-hour and 12-hour rotating shift schedules on performance, subjective alertness, sleep patterns, and psychosocial variables. *Work Stress* 1989; 3(1):21–32.
57. Mills DQ. Does organized labor want the 4-Day week?. In: Poor R, Samuelson PA, eds. *4 Days, 40 Hours*. Cambridge MA: Bursk & Poor, 1970, 61–9.
58. Steele JL, Poor R. Work and leisure: the reactions of people at 4-day firms Poor R, Samuelson PA, eds. *4 Days, 40 Hour.* Cambridge MA: Bursk & Poor, 1970, 105–22.
59. Brief RS, Scala RA. Occupational health aspects of unusual work schedules: a review of Exxon's experiences. *J Am Ind Hyg Assoc* 1986; 47(4):199–202.
60. Colligan MS, Tepas DI. The stress of hours of work. *Am Ind Hyg Assoc J* 1986; 47:686–95.
61. Jones IS, Stein HS. *Effect of driver hours of service on tractor-trailer crash involvement.* Arlington, VA: Insurance Institute for Highway Safety, 1987.
62. Lisper HO, Laurell H, van Loon J. Relation between time to falling asleep behind the wheel on a closed track and changes in subsidiary reaction time during prolonged driving on a motorway. *Ergonomics* 1986; 29(3):445–53.
63. Laundry BR, Lees RE. Industrial accident experience of one company on 8-and 12-hour shift systems. *J Occup Med* 1991; 33(8):903–6.
64. Tucker P, Smith L, MacDonald I, Folkard S. Shift length as a determinant of retrospective on-shift alertness. *J Work Environ Health* 1998; 24(S3):49–54.
65. Tucker P, Barton J, Folkard S. Comparison of eight and 12 hour shifts: impacts of health, well-being, and alertness during the shift. *Occup Environ Med* 1996; 53:767–72.
66. Smith PA, Wright BM, Mackey RW, et al. Change from slowly rotating 8-hr shifts to rapidly rotating 8-hour and 12-hour shifts using participative shift roster design. *Scand J Work Environ Health* 1998; 24(S3):55–61.
67. Lowden A, Kecklund G, Aselsson J, Akerstedt T. Change from an 8-hour shift to a 12-hour shift, attitudes, sleep, sleepiness, and performance. *Scan J Work Environ Health* 1998; 24(S3):69–75.
68. Williamson AM, Gower CG, Clarke BC. Changing the hours of shiftwork: a comparison of 8- and 12-hour shift rosters in a group of computer operators. *Ergonomics* 1994; 37(2):287–98.
69. Mardon S, ed. *ShiftWork Alert,* 2(7) and (8) *Circadian Information.* Cambridge MA: Shiftwork Newsletter, 1997.
70. Finn P. The effects of shift work on the lives of employees. *Monthly Labor Rev* 1981; 104(10):31–5.
71. Tepas DI. Flexitime, compressed workweeks and other alternative work schedules. In: Folkard, S. Monk T. eds. *Hours of work: temporal factors in scheduling*, Chichester: John Wiley, 1985, 147–64.
72. Smith L, Hammond T, Macdonald I, Folkard S. 12-hr shifts are popular but are they a solution? *Intern J Indust Ergonomics* 1998; 21:323–31.
73. Knauth P. Changing schedules: shiftwork,. *Chronobiol international* 1997; 14(2):159–71.
74. Rosa R, Colligan M. *Plain language about shiftwork,* U.S. Dept. of health and human services, national institute for occupational safety and health, NIOSH Pub. No. 97-145, Cincinnati, Ohio (1997).
75. Folkard S, Arendt J, Clark M. Sleep and mood on a "weekly" rotating (7-7-7) shift system: some preliminary results. In Costa G, Cesana G, Kogi K, Wedderburn A, eds. *Shiftwork: health, sleep, performance*, Peter Lang: Frankfurt, 1989, 484–89.
76. Gillberg M. Subjective alertness and sleep quality in connection with permanent 12-hour day and night shifts. *Scand J Work Environ Health* 1998; 24(S3):76–81.

77. Folkard S, Barton J. Does the 'forbidden zone' for sleep onset influence morning shift sleep duration? *Ergonomics* 1993; 36:85–9.
78. Akerstedt T. Work schedules and sleep. *Experientia* 1984; 40:417–22.
79. Knauth P, Landau K, Droge C, et al. Duration of sleep depending on the type of shift work. *Int Arch Occup Environ Health* 1980; 46:167–77.
80. Center for Disease Control (CDC). Leading work-related diseases and injuries. *MMWR* 1986; 35(39):613–14; 619–21.
81. Imbernon E, Warret G, Roitg C, et al. Effects on health and social well-being of on-call shifts: an epidemiologic study in the French national electricity and gas supply company *J Occupational Medicine* 1993; 35(11):1131–7.
82. Ernst G, Rutenfranz J. Flexibility in shiftwork—some suggestions. *Ergonomics*, 1978; 21(10), 861–74 (abstracts).
83. Smith P. A Study of Weekly and Rapidly Rotating Shift Workers. *Ergonomics* 1978, 21(10), 861–74 (abstracts).
84. Northrup HR, Wilson JT, Rose KM. The twelve-hour shift in the petroleum and chemical industries. *Industrial and labor relations review* 1979; 32(3):312–26.
85. Kogi K. Improving shift workers' health and tolerance to shiftwork: recent advances. *Appl Ergonomics* 1996; 27(1):5–8.
86. Halberg F, Halberg E, Barnum CP, et al. Circadian rhythm—coined. In: Withrow RB, ed. *Photoperiodism and related phenomenon in plants and animals* AAAS, Washington D.C., 1959, 803–78.
87. Scott AJ, Ladou J. Shiftwork: effects on sleep and health. *Shiftwork: occup med, State of the Art Reviews* 1990; 5(2):273–99.
88. Wever R. Man in temporal isolation: basic principles of the circadian system. In: Folkard S, Monk TH, eds. *Hours of work—temporal factors in work scheduling* Chichester: John Wiley & Sons, Inc. 1985, 15–28.
89. Aschoff J. Circadian rhythms in man. *Science* 1965; 148:1427–32.
90. Vernibos-Danelles J, Winget CN. The importance of light, postural, and social cues in the regulation of the plasma cortisol rhythm in man. In: Reinberg A, Halberg F, eds. *Chronopharmacology, Proceedings of the seventh intern Cong. Pharmacology, Paris 1978,* New York: Pergamon Press, 1979, 101–6.
91. Webb WB, Agnew HW Jr. The Effects of a chronic limitation of sleep length. *Psychophysiology* 1974; 11(5):265–74.
92. Duffy JF, Kronauer RE, Czeisler CA. Phase-shifting Human Circadian Rhythms: Influence of sleep timing, social contact and light exposure. *J Physiol* 1996; 95(pt 1):289–97.
93. Arendt J, Deacon S. Treatment of circadian rhythm disorders—Melatonin. *Chronobiol Int* 1997; 14(2):185–204.
94. Webb WB. Sleep and biological rhythms. In: Webb, WB. ed. *Biological rhythms, sleep, and performance.* New York: John Wiley & Sons, 1982, 87–141.
95. Wever R. Phase shifts of human circadian rhythms due to shifts of artificial zeitgebers. *Chronobiologia* 1980; 7:303–27.
96. Knauth P, Rutenfranz J. Experimental shift work studies of permanent night and rapidly rotating shift systems: 1, Circadian rhythm of body temperature and re-entrainment at shift change. *Int Arch Occup Environ Health* 1976; 37:125–37.
97. Folkard S, Monk TH, Lobban MC. Short and long-term adjustment of circadian rhythms in permanent night nurses. *Ergonomics*, 1978; 21:785–99.
98. Akerstedt T, Patkai P, Dahlgren K. Field studies of shift work: II. Patterns in psychophysiological activation in workers alternating between night and day work. *Ergonomics* 1977; 20:849–56.
99. Reinberg A, Andlauer P, DePrins J, et al. Desynchronization of the oral temperature circadian rhythm and intolerance to shiftwork. *Nature* 1984; 308:272–4.
100. Tepas DI. Work/sleep time schedules and performance. In: Webb, WB. ed *Biological rhythms, sleep and performance.* Chichester: John Wiley & Sons, Inc., 1982, 175–204.
101. Tepas DI, Maham RP. The many meanings of sleep. *Work stress,* 1989; 3, 93–102
102. Akerstedt T. Work schedules and sleep. *Experientia* 1984; 40:417–22.
103. Tepas DI, Stock CG, Maltese JW, Walsh JK. Reported sleep of shift workers: a preliminary

report. In: Chase M, Mitler, M, Walter P, eds. *Sleep research*, Vol. 7. University of California, Los Angeles: Brain Information Research Institute, 1978.

104. Smith MJ, Colligan MJ, Tasto DL. Health and safety consequences of shift work in the food processing industry. *Ergonomics* 1982; 25(2):133–44

105. Tilley AJ, Wilkinson RT, Drud M. Night and day shifts compared in terms of the quality and quantity of sleep recorded in the home and performance measured at work: a pilot study. In: Reinberg A, Vieux N, Andlauer P, eds. *Night and shift work: biological and social aspects.* Oxford: Pergamon Press, 1981, 187–96.

106. Weitzman ED, Godmacher D, Kripke D, et al. Reversal of sleep–waking cycle: effect on sleep stage pattern and certain neuro-endocrine rhythms. *Trans Am Neuro Assoc* 1968; 93:153–7.

107. Akerstedt T, Gillberg M. Sleep disturbances and shiftwork. In: Reinberg A, Vieux N, Andlauer P, eds. *Night and shift work: biological and social aspects.* Oxford: Pergamon Press, 1981, 127–37.

108. Gillberg M, Akerstedt T. Body temperature and sleep at different times of day. *Sleep* 1982; 5(4):378–88.

109. Walsh JK, Tepas DI, Moss PD. The EEG sleep of night and rotating shiftworkers. In: Johnson LC, Tepas DI, Colquhoun WP, Colligan MJ, eds. *The twenty-four hour workday.* Proceedings of a Symposium on Variations in Work–Sleep Schedules. US Government Print Office, Washington, DC. 1981, 81–127.

110. Tepas DI, Sullivan PJ. Does body temperature predict sleep length, sleepiness, and mood in a lab-bound population? *Sleep Res* 1982; II:42.

111. Knauth P, Rutenfranz J, Schulz H. Experimental shift work studies of permanent night and rapidly rotating shift systems, II. Behavior of various characteristics of sleep. *Int Arch Occup Environ Health* 1980; 46:111–25.

112. Gadbois C. Women on night shift: interdependence of sleep and off-the-job activities. In: Reinberg A, Vieux N, Andlauer P, eds. *Night and shift work: biological and social aspects.* Oxford: Pergamon Press, 1981, 223–7.

113. Johnson LC, MacLeod WL. Sleep and awake behavior during gradual sleep reduction. *Percept Mot Skills* 1973; 36:87–97.

114. Grandjean E. *Fitting the task to the man, an ergonomic approach.* London: Taylor & Francis, 1982.

115. Cameron C. A theory of fatigue. *Ergonomics,* 1973; 16(5):633–48.

116. Winget CM, DeRoshia CW, Markley CL, Holley DC. A review of human physiological and performance changes associated with desynchronosis of biological rhythms. *Aviat Space Environ Med* 1984; 55(12): 1085–93.

117. Wojtczak-Jaroszowa J. Circadian rhythm of biological functions and night work. In: Wojtczak-Jaroszowa J, ed. *Physiological and psychological aspects of night and shift work.* Cincinnati: National Institute for Occupational Safety and Health, 1977, 3–12.

118. Kogi K. Introduction to the problems of shiftwork. In: Folkard S, Monk T.H, eds. *Hours of work.* New York: John Wiley & Sons, 1985, 165–84.

119. Smith MJ, Colligan MJ, Hurrell JJ Jr. A review of the psychological stress research carried out by NIOSH: 1971–1976. *New developments in occupational stress.* Cincinnati, Ohio: US. Department of Health and Human Services, 1980, 1–9.

120. Frese M, Okonek K. Reasons to leave shiftwork and psychological and psychosomatic complaints of former shiftworkers. *J Appl Psychol* 1984; 69(3):509–14.

121. Verhaegan P, Maasen A, Meers A. Health problems in shiftworkers. In: Johnson L, Tepas D, Colquhoun WP, Colligan MJ, eds. *Advances in sleep research,* Vol 7, *Biological rhythms, sleep, and shift work.* New York: Spectrum Publications, 1981, 271–87.

122. Moore-Ede MC, Richardson GS. Medical implications for shift work. *Annu Rev Med* 1985; 36:607–17.

123. Reinberg A, Motohashi Y, Bourdeleau P, et al. Internal desynchronization of circadian rhythms and tolerance of shiftwork. *Chronobiologia* 1989; 16:21–34.

124. Askenazi IE, Reinberg AE, Motohashi Y. Interindividual differences in the flexibility of human temporal organization: pertinence

to jet lag and shiftwork. *Chronobiol Int* 1997; 14(2):99–113.

125. Coleman RM. Shiftwork scheduling for the 1990s. In: *Personnel,* New York: American Management Association, 1989.

126. Reinberg A, Vieux N, Andlauer P, et al. Tolerance of shift work, amplitude of circadian rhythms, and aging. In: Reinberg A, Vieux N, Andlauer P, eds. *Night and shift work: biological and social aspects*, Oxford: Pergamon Press, 1981, 341–54.

127. Graeber RC, Lauber JK, Connel LJ, Gander PH. International aircrew sleep and wakefulness after multiple time-zone flights: a cooperative study. *Aviat Space Environ Med* 1986; 57(12):B3–9.

128. Harma M. Aging, physical fitness and shiftwork tolerance. *Appl Ergonomics* 1996; 27(1):25–9.

129. Akerstedt T, Froberg J. Interindividual differences in circadian patterns of catecholamine excretion, body temperature, performance and subjective arousal. *Biol Psychol* 1976; 4:277–92.

130. Costa G, Lievore F, Casaletti G, Gafuri E, Folkard S. Circadian characteristics influencing interindividual differences in tolerance and adjustment to shiftwork. *Ergonomics* 1989; 32:373–85.

131. Costa G. Guidelines for the medical surveillance of shiftworkers. *Scand J Work Environ Health* 1998; 24(S3):151–5.

132. Angersbach D, Knauth P, Loskant H, et al. A retrospective cohort study comparing complaints and diseases in day and shift workers. *Int Arch Occup Environ Health* 1980; 45:127–40.

133. Minors DS, Scott AR, Waterhouse JM. Circadian arrhythmia: shiftwork, travel, and health. *J Soc Occup Med* 1986; 36(2):39–44.

134. Colligan MJ, Frock IJ, Tasto D. Shift work—the incidence of medication use and physical complaints as a function of shift. In: *Occupational and Health Symposia—1978.* NIOSH Publ. No. 80-105. Washington DC: US Department of Health, Education, and Welfare, 1980, 47–57.

135. Walker J, De la Mare G. Absence from work in relation to length and distribution of shift hours. *Br J Indust Med* 1971; 28:36.

136. Costa G.The impact of shift and night work on health. *Appl Ergonomics* 1994; 27(1):9–16.

137. Moore JC, Englert E. Circadian rhythms of gastric acid secretion in man. *Nature* 1970; 226:1261–2.

138. Goo RH, Moore JG, Greenberg E, Alazraki NP. Circadian variation in gastric emptying of meals in humans. *Gastroenterology* 1987; 93:515–8.

139. Reinberg A, Migraine A, Apfelbaum C. Circadian and ultradian rhythms in the eating behavior and nutrient intake of oil refinery operators (Study 2). *Chronobiologia* 1979; 6(suppl 1):89–102.

140. Stewart AJ, Wahlquist ML. Effect of shiftwork on canteen food purchase. *J Occup Med* 1985; 27(8):552–4.

141. Tepas DI. Do eating and drinking habits interact with work schedule variables? *Work Stress* 1990; 4(3):203–1.

142. Knutsson A, Anderson H, Berglund U. Serum lipoproteins in day and shift workers: a prospective study. *B J Indust Med* 1990; 47:132–4.

143. DeBacker G, Kornitzer M, Peters H, Dramaix M. Relation between work rhythm and coronary risk factors. *Eur Heart J* 1984; 5(suppl 1): 307 (abstract).

144. Boggild H, Knutsson A. Shiftwork, risk factors and cardiovascular disease. *Scand J Work Environ Health* 1999; 25(2):85–99.

145. Lavie P, Chillag N, Epstein R, et al. Sleep disturbances in shift-workers: marker for maladaptation syndrome, *Work Stress* 1989; 3(1): 33–40.

146. Morikawa Y, Nakagawa H, Miura K. Relationship between shiftwork and onset of hypertension in a cohort of manual workers. *Scand J Work Environ Health* 1999; 25(2):100–4.

147. Murata K, Yana E, Shinozaki T. Impact of shift work on cardiovascular functions in a 10-year follow-up study. *Scand J Work Environ Health* 1999; 25(3):272–7.

148. Aanonsen A. Medical problems of shift-work. *Ind Med Surg* 1959; 422–7.

149. Taylor P, Pocock S. Mortality of shift and day workers: 1956–68. *Br J Indust Med* 1972; 29:201–7.

150. Boggild H, Knutsson A. Shiftwork, risk factors and cardiovascular disease. *Scand J Work Environ Health* 1999; 25:85–99.

151. Tenkanen L, Sjoblom T, Harma M. Joint effect of shiftwork and adverse life-style factors on the risk of coronary heart disease. *Scand J Work Environ Health* 1998; 24(5):351–7.

152. Peter R, Alfredsson L, Knuttsson A, Siefrist J, Werserholm P. Does a stressful psychosocial work environment mediate the effects of shift work on cardiovascular risk factors? *Scand J Work Environ Health* 1999; 25(4):376–81.

153. McDonald A, McDonald J, Armstrong B, et al. Prematurity and work in pregnancy. *Br J Indust Med* 1988; 45:56–62.

154. Armstrong G, Nolin A, McDonald A. Work in pregnancy and birthweight for gestational age. *Br J Indust Med* 1989; 46:196–9.

155. Nurimen T. Shift work, fetal development and course of pregnancy. *Scand J Work Environ Health* 1989; 15:395–403.

156. Axelsson G, Rylander R, Molin I. Outcome of pregnancy in relation to irregular and inconvenient work schedules. *Br J Indust Med* 1989; 46:393–8.

157. Xu X, Ding M, Li B, Christiani DC. Association of rotating shiftwork with preterm births and low birth weight among never smoking women textile workers in china. *Occup Environ Med* 1994; 51(7):470–4.

158. Saurel-Cubizolles M, Kaminski M. Pregnant women's working conditions and their changes during pregnancy: a national study in France. *Br J Indust Med* 1987; 44:236–43.

159. Axelsson G, Rylander R, Molin I. Outcome of pregnancy in relation to irregular and convienient work schedules. *Br J Indust Med* 1989; 46:393–8.

160. Fortier I, Marcoux S, Brisson J. Maternal work during pregnancy and the risks of delivering a small-for-gestational-age or preterm infant. *Scand J Work Environ Health* 1995; 21(6):412–8.

161. McDonald A, McDonald J, Armstrong B, et al. Fetal death and work in pregnancy. *Br J Indust Med* 1988; 45:148–57.

162. Axelsson, G. Ahlborg G Jr, Bodin L. Shift work, nitrous oxide exposure, and spontaneous abortion among Swedish midwives. *Occup Environ Med* 1996; 53:374–8.

163. Infante-Rivard C, David M, Gauthier R, et al. Pregnancy loss and work schedule during pregnancy. *Epidemiology* 1993; 4(1):73–5.

164. Axelsson G, Lutz C, Rylander R. Exposure to solvents and outcome of pregnancy in university laboratory employees. *Br J Indust Med* 1994; 41:305–12.

165. Uehata T, Sasakawa N. The fatigue and maternity disturbances of night workwomen. *J Hum Ergol (Tokyo)* 1982; 11:465–74.

166. Hemminki K, Kyyronen P, Lindbohm ML. Spontaneous abortions and malformations in the offspring of nurses exposed to anesthetic gases, cytostatic drugs, and other potential hazards in hospitals, based on registered information of outcome. *J Epidemiol Community Health* 1985; 39:141–7.

167. Eskenazi B, Fenster L, Wright S, et al. Physical exertion as a risk factor for spontaneous abortion. *Epidemiology* 1994; 5(1):6–13.

168. Bryant E, Love EJ. Effect of employment and its correlates on spontaneous abortion risk. *Soc Sci Med* 1991; 33:795–800.

169. Nurminen T. Shift work and reproductive health. *Scand J Work Environ Health* 1998; 24(suppl 3):28–34.

170. Ahlborg G Jr, Axelsson G, Bodin L. Shift work, nitrous oxide exposure and subfertility among Swedish midwives. *Int J Epidemiol* 1996; 25(4):783–90.

171. Bisanti L, Olsen J, Basso O, et al. Shift work and subfecundity: a European multicenter study. *J Occup Environ Med* 1996; 38(4):352–8.

172. Spinelli A, Figa-Talamanca I, Osborn J, et al. Time to pregnancy and occupation in a group of Italian women. *Int J Epidemiol* 1997; 26(3):601–9.

173. Olsen J. Cigarette smoking, tea and coffee drinking, and subfecundity. *Am J Epidemiol*, 1991; 133(7):734–9.

174. Tuntiseranee P, Olsen J, Geater A, et al. Are long working hours and shiftwork risk factors for subfecundity? A study among couples from southern Thailand. *Occup Environ Med* 1998; 55(2):99–105.

175. Koller M. Occupational health services for shift and night workers. *Applied Ergonomics* 1996; 27(1):31–7.

176. Costa G. Guidelines for the medical surveillance of shiftworkers. *Scand J Work Environ Health* 1998; 24(S3): 151–5.

177. Scott AJ. Shiftwork and health. In: Ladou J, ed. *Primary care: clinics in office practice—occupational medicine.* Philadelphia: Saunders WB, 2000; 27(4)1057–8.

178. Rosa R, Colligan M. *Plain language about shiftwork.* NIOSH Pub. No. 97-145. Cinninati, Ohio: US Department of Health and Human Services, National Institute for Occupational Safety and Health, 1997.

179. Monk TH. *How to make shift work safe and productive.* Des Plains, IL: American Society of Safety Engineers, 1988.

180. Monk TH, Folkard S. Strategies for the employer. *Making shiftwork tolerable.* London: Taylor & Francis, 1992, 69–75.

181. Comperatore CA, Krueger GP. Circadian rhythm, desynchronosis, jet lag, shift lag, and coping strategies. *Shiftwork: occup med, State of the Art Reviews* 1990; 5(2): 323–41.

182. Lamberg L. Coping with jet lag. *Body rhythms, chronobiology and peak performance.* New York: William Morrow & Co., 1994, 159–79.

183. Vallet M, Mouret J. Sleep disturbance due to transportation noise: ear plugs vs. oral drugs. *Experientia* 1984; 40:429–36.

184. Doghramji K, Fredman S. *Clinical frontiers in the sleep/psychiatry interface.* Satellite Symposium of the 1999. American Psychiatric Association Annual Meeting. Golden CO: Medical Education Collaborative, 1999.

185. Walsh JK, Muehlbach MJ, Walker PK. Acute administration of triazolam for the daytime sleep of rotating shiftworkers. *Sleep* 1984; 7:223–9.

186. Penetar DM, Belenky G, Garrigan JJ, et al. Triazolam impairs learning and fails to improve sleep in a long-range aerial deployment. *Aviat Space Environ Med* 1989; 60:594–8.

187. Ehret CF, Potter VR, Dobia KW. Chronobiological action of theophylline and of pentobarbital as circadian zeitgebers in the rat. *Science* 1975; 188:1212–14.

188. Dew PB. Behavioral effects of caffeine. In: Dew PB ed. *Caffeine—perspectives from recent research.* New York: Springer-Verlag, 1984, 86–103.

189. Bonnet MH, Arand DL. The use of prophylactic naps and caffeine to maintain performance during a continuous operation. *Ergonomics* 1994; 37:1009–20.

190. Kelly T, Gomez S, Engelland S, Naitoh P. Repeated administration of caffeine during sleep deprivation does not affect cognitive performance. *Sleep Res* 1993; 22:336.

191. Muehlbach MH, Walsh JK. The effects of caffeine on simulated Night-shift work and subsequent daytime sleep. *Sleep* 1995; 18(1):22–9.

192. Smith AP, Brockman R, Flynn A, et al. Investigation of the effects of coffee on alertness and performance during the day and night. *Neuropsychology* 1993; 27: 217–23.

193. Akerstedt T, Ficca G. Alertness-enhancing drugs as a countermeasure to fatigue in irregular work hours. *Chronobiol Int* 1997; 14(2):145–58.

194. Redman J, Armstrong S, Ng KT. Free-running activity rhythms in the rat: entrainment by melatonin. *Science* 1983; 219:1081–9.

195. Arendt J, Aldhous M, English J, et al. Some effects of jet lag and their alleviation by melatonin. *Ergonomics* 1987; 30:1379–93.

196. Arendt J, Deacon S. Treatment of circadian rhythm disorders—melatonin. *Chronobiol Int* 1997; 14(2):185–204.

197. Suhner A, Schlagenhauf P, Johnson R, et al. Comparative study to determine the optimal melatonin dosage form for the alleviation of jet lag. *Chronobiol Int* 1998; 15(6): 655–66.

198. Folkard S, Arendt J, Clark M. Can melatonin improve shift workers' tolerance of the night shift? Some preliminary findings. *Chronobiol Int* 1993; 10(5):315–20.

199. Mardon S, ed. Melatonin shows promise for shiftworkers, but long-term health effects are unknown. *ShiftWork Alert* 1996; 1(2):5–7.

200. Anonymous. Melatonin. *Med Lett. Drugs Ther* 1995; 37:111–12.

201. Eastman CI. Circadian rhythms and bright light: recommendations for shift work. *Work Stress* 1990; 4:245–60.

202. Czeisler CA, Kronauer RE, Allan JS, et al. Bright light induction of strong (type 0) resetting of the human circadian pacemaker. *Science* 1989; 244:1328–33.

203. Czeisler CA, Johnson MP, Duffy JF, et al. Exposure to bright light and darkness to treat physiologic maladaptation to night work. *N Eng J Med* 1990; 322(18):1253–9.

204. Cagnacci A. Influences of melatonin on human circadian rhythms. *Chronobiol Int* 1997; 14(2):205–20.

205. Ehret CF, Groh KR, Meinert JC. Considerations of diet in alleviating jet lag. In: Scheving LE, Halberg F, eds. *Proceedings of the NATO Advanced Study Institute on Principles and Application of Chronobiology to Shifts in Schedules with Emphasis on Man.* The Netherlands: Sijthoff and Noordhoff, Intern. Publishers, 1980, 393–402.

206. Leathwood P. Circadian rhythms of plasma amino acids, brain neurotransmitters, and behavior. In: Arndt J, Minors DS, Waterhouse JM, eds. *Biological rhythms in clinical practice.* London: Butterworth, 1989, 136–59.

207. Harma MI, Ilmarinen J. Physical training intervention in female shift workers: I. the effects of intervention on fitness, fatigue, sleep, and psychosomatic symptoms. *Ergonomics* 1988; 31:(1):39–50.

208. Rutenfranz J, Haider M, Koller M. Occupational health measures for nightworkers and shiftworkers. In: Folkard S, Monk TH, eds. *Hours of work: temporal factors in work scheduling.* Chichester: John Wiley & Sons, Inc., 1985, 199–210.

209. Rutenfranz J, Knauth P, Angerback D. Shiftwork research issues. In: Johnson LC, ed. *Advances in Sleep Research.* New York: Medical & Scientific Books, 1985, 165–96.

210. International Labour Office. *Night Work Convention No. 171.* Geneva: ILO, 1990.

211. European Council. Concerning certain aspects of working time. *Off J Eur Community* 1993; L307:18–24 (Council Directive 93/104/EC).

212. Siwolop S, Therrien L, Oneal M, Ivey M. Helping workers stay awake at the switch. *Business Week* 1986; 8 December, 108.

213. Schwarzenau P, Knauth P, Keisswetter E, et al. Algorithms for the computerized construction of shift systems which meet ergonomic criteria. *Appl Ergonomics* 1986; 17:169–76.

214. Gartner J, Wahl S. Design tools for shift schedules: empowering assistance for skilled designers & groups. *Int J Indust Ergonomics* 1998; 21:221–32.

215. Schwarzenau P, Knauth P, Keisswetter E, et al. Algorithms for the computerized construction of shift systems which meet ergonomic criteria. *Appl Ergonomics* 1986; 17:169–76.

216. Circadian Technologies, Inc. *Control room operator alertness and health in nuclear power plants.* Prepared for Electric Power Research Institute, EPRI Rep. No. NP-6748—project 2184-7. Palo Alto, CA 1990.

217. Mardon S, ed. Some companies meet shiftworkers' family needs with 24-hour child care. Should yours? *ShiftWork Alert* 1996; 1(2):9–11.

Section III

Energy and Electromagnetic Radiation

11

IONIZING RADIATION

Bryce D. Breitenstein, Jr., M.D., M.P.H., and James P. Seward, M.D., M.P.P.*

Although most physicians and other healthcare professionals do not encounter individuals injured by ionizing radiation in their practices, they can anticipate questions about radiation exposure and potential injury. In the rare event that a patient does present following radiation exposure or contamination with radioactive materials, the following information will be of assistance in the individual's case management. Important information on resources for emergency information and expert advice are included at the end of this chapter.

BACKGROUND RADIATION

Although this chapter is generally concerned with unusual exposures to ionizing radiation, we recognize that low-level exposure to radiation is unavoidable. This so-called background radiation is the source of most people's exposure during their lifetime. The average annual effective dose to individuals in the USA is ~0.36 rem/3.6 mSv. Of the total dose, approximately four-fifths (0.3 rem/3.0 mSv) is due to natural radiation. Two-thirds of the average annual dose from natural sources is due to radon (0.2 rem/2.0 mSv), which is potentially present in both homes and at work sites. The rest comes from cosmic radiation (0.03 rem/0.3 mSv), terrestrial radiation from naturally occurring radioactive thorium and uranium (0.03 rem/0.3 mSv), and radioactive potassium and carbon, and other radionuclides within the body (0.04 rem/0.4 mSv). Artificial radiation may come from medical uses (0.05 rem/0.5 mSv), or consumer products (0.01 rem/0.1 mSv); a very small component (<0.001 rem/0.01 mSv) is the result of occupational exposure, nuclear power, or radioactive fallout.

OCCUPATIONAL EXPOSURES

Exposure to ionizing radiation can occur in a number of different industries and industrial settings. Table 11.1 provides a listing of examples of occupations where ionizing radiation exposure is present.

*John H. Spickard contributed to the previous edition of this chapter.

Physical and Biological Hazards of the Workplace, Second Edition, Edited by Peter H. Wald and Gregg M. Stave
ISBN 0-471-38647-2

Table 11.1 Examples of occupational exposure to ionizing radiation.

Airline pilots and crew
Food irradiation facilities
Manufacture of consumer products (e.g. luminous dials, gas mantles)
Nuclear fuel cycle operators
Nuclear medicine
Research involving ionizing radiation
Uranium mining and milling
Use of X-ray equipment
Dentists and dental technicians
Industrial radiographers
Radiologists and radiation technicians

MEDICAL EXPOSURES

The use of ionizing radiation for medical diagnosis and treatment began with the discovery of X-rays by Wihelm Von Roentgen in Wurzburg, Germany in 1895. There is a wide range of current medical procedures that utilize ionizing radiation. Table 11.2 illustrates examples of these procedures and provides an estimate of the average radiation dose associated with the procedure. The information obtained from diagnostic radiologic procedures or the positive therapeutic effect from ionizing radiation treatment normally provides a benefit that offsets the risk of using ionizing radiation.

MEASUREMENT ISSUES AND THE PHYSICS OF IONIZING RADIATION

The energy in ionizing radiation is carried in packets by either electromagnetic radiation (X-rays and gamma rays) or moving particles (alpha and beta particles, neutrons and protons). The physical characteristics of these types of radiation are reviewed in Table 11.3. Ionizing radiation can be detected by a number of devices, but personal dosimeters are the instrument of choice to record individual exposures. Film badges have been almost completely replaced by thermal-luminescent dosimeters (TLDs) (Figure 11.1). TLDs use materials (most commonly lithium fluoride) that glow on heating after exposure to ionizing radiation. TLDs are useful to detect beta and gamma radiation. Polycarbonate "foils" are often incorporated into TLDs to detect neutrons. Neutrons disrupt the structure of the polycarbonate, and these trails can be visually counted under a microscope after "etch" development. Alpha particles cannot be counted directly with a dosimeter, but they can be identified by their associated low-energy gamma emissions. "Pancake" counters can directly measure surface alpha contamination. Whole-body counters are very sensitive gamma cameras that identify the low-energy gamma co-emissions from any internally deposited alpha emitters.

Table 11.2 Examples of medical exposure to ionizing radiation.[20]

Chest X-ray (single view)	10 mrem/0.1 mSv
Skull X-ray (four images)	15 mrem/0.15 mSv
Barium enema	54 mrem/0.54 mSv
Bone scan	440 mrem/4.4 mSv
Thallium cardiac stress test	440 mrem/44 mSv

Table 11.4 shows the units of radiation and radioactivity. The basic unit of radiation exposure is the rad (given in traditional units) or the gray (given in SI (System Internationale) units). Since some types of ionizing radiation are more effective at causing ionization than others, a second measure, rem/sievert, is used to quantify the dose equivalent. The dose equivalent is obtained by multiplying the radiation dose by a quality factor. For X-rays, the quality factor equals 1, and 1 rad equals 1 rem. There is an ongoing effort on the part of the scientific community to convert all radiation terms to the SI units of becquerel, gray, and sievert.[1] As with conversion of measurements to metric terms, the longer-standing terminology is still commonly in use. Scientific publications commonly require that units be given in SI units. Both units will be used here to facilitate familiarity with them.

Table 11.3 Types of ionizing radiation important to radiologic health.

Source	Symbol	Character	Mass[a]	Charge	Example
X-ray	X	Electromagnetic energy	0	0	X-ray tube
Gamma	γ	Electromagnetic energy	0	0	^{60}Co ^{192}Ir
Alpha	α	Particulate (helium nucleus)	4	++	^{239}Pu ^{212}Po
Beta	β	Particulate (electron)	1/2000	–	^{90}Sr ^{3}H
Neutron	n	Nucleus particle	1	0	^{235}U fission
Proton	p	Nucleus particle	1	+	Proton beam

[a]The atomic mass (AMU) is chosen so that a neutral carbon-12 atom has a relative mass exactly equal to 12. This is equal to ~1 AMU for both a proton and a neutron.

EXPOSURE GUIDELINES

Dose limits for ionizing radiation have been established for occupationally exposed individuals and for the general public. These limits are based on recommendations of the National Council on Radiation Protection and Measurements (NCRP)[2] and the International Commission on Radiological Protection (ICRP)[3]. The United Nations Scientific Committee on the Effects of Atomic Radiation (UNSCEAR) and the National Academy of Sciences (NAS) Committee on the Biological Effects of Ionizing Radiation (BEIR)[4] have periodically gathered and reviewed scientific data for estimating the genetic and somatic effects of exposure to ionizing radiation. The dose units that have been established by the Department of Energy (DOE) reflect the findings and recommendations of these advisory groups. The resultant dose limit guidelines are outlined in Table 11.5.

The ICRP[3] has suggested lowering the current recommended standards on the basis of the re-evaluation of the atomic bomb survivor cohorts from Japan. Their new recommen-

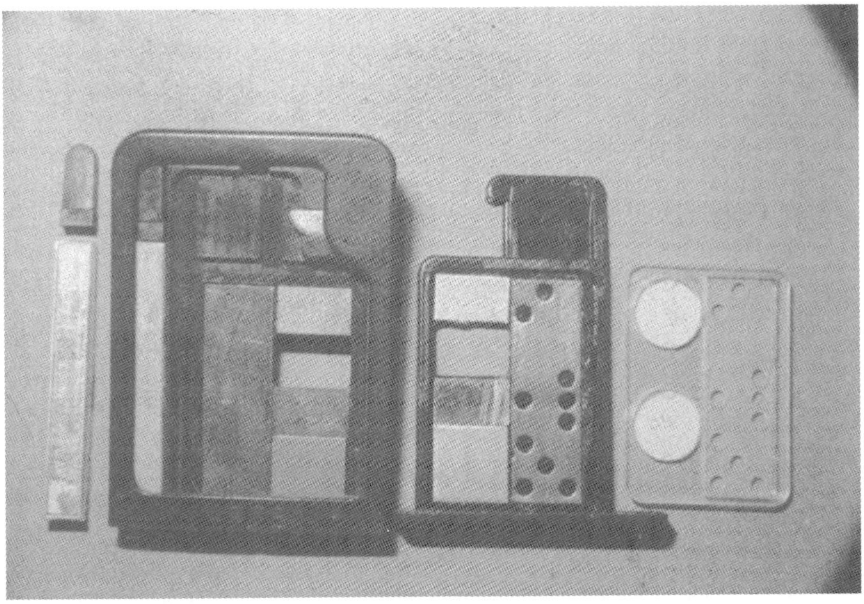

Figure 11.1 Interior of a thermal-luminescent dosimeter (TLD), showing areas for beta, gamma and neutron detection.

Table 11.4 Units of radiation and radioactivity.

Unit description	Unit name	Symbol	Definition
Activity	Curie	Ci	3.7×10^{10} disintegrations/s
	Becquerel	Bq	1 disintegration/s
Exposure	Roentgen	R	2.58×10^{-4} coulombs/kg
Absorbed dose	Rad[a]	rad	100 ergs/g of absorbing material
	Gray	Gy	100 rad
Dose equivalent	Rem[b]	rem	rad × Q (quality factor)
	Sievert	Sv	100 rem

[a]Radiation absorbed dose.
[b]Roentgen equivalent man.
Measurement terminology: 1000 millicuries in 1 curie; 1 000 000 microcuries in 1 curie; 1 000 000 000 nanocuries in 1 curie; 1 000 000 000 000 picocuries in 1 curie; 1 000 000 000 000 000 femtocuries in 1 curie.
1 Curie = Approximate disintegration rate of 1 g of ^{226}Ra or 3.7×10^{10} disintegrations per second.

dations call for a maximum of 10 rem exposure over 5 years, with a yearly average exposure of 2 rem. These new recommendations have not yet been accepted into national standards.

Other guidelines have been established for emergency occupational exposures. Only actions involving lifesaving response would justify exposures >10 rad (100 mGy), and even in these heroic efforts, dosage should not exceed 100 rad/1 Gy. In industry and in the private sector, every effort is made to limit exposure to ionizing radiation to a level that is as low as reasonably achievable (ALARA). This means that radiation exposures must be kept as low as possible while still allowing workers to get their jobs done. Similarly, radiation exposure to patients is minimized by limiting diagnostic radiologic procedures to those that are clearly necessary and by ongoing attempts to improve radiologic technology. For example, the current radiation dose received during mammography is significantly less than it was with the initial techniques.

It is standard industrial practice to remove a radiation worker from exposure during a specific period once he has reached a specified dose. For example, even though the radiation dose standard is 5 rem/50 mSv per year, workers are often reassigned from radiation work for the remainder of the year if their TLD

Table 11.5 Ionizing radiation exposure guidelines.

	Dose limit guidance (NRC, DOE, NCRP, ICRP)	
Category	Rems	Millisieverts (mSv)
Occupational exposure (annual)	5	50
Lens of eye	15	150
Other organs/tissues	50	500
Unborn child of worker	0.5	5
Members of the public (annual)	0.1	1
Cumulative dose (lifetime)	Age × 1	Age × 10

readings reach an exposure of 3 rem/30 mSv. Specific efforts are made to ensure that the occupational exposure of a pregnant worker does not exceed 0.5 rem/5 mSv during the entire gestation period.

PATHOPHYSIOLOGY AND HEALTH EFFECTS

Ionizing radiation consists of electromagnetic waves or moving particles that carry sufficient energy to produce ions in matter. Ionization occurs when enough radiation energy transfers to atoms in the material through which it is passing to displace an orbital electron, thus leaving these atoms as electrically charged ions. In tissue, the ionization of atoms within cells produces biochemical changes that may result in immediate or late biological effects. Except for extremely high radiation doses, the individual so exposed perceives no immediate effects.

Early and late health effects from ionizing radiation

The spectrum of health effects from ionizing radiation can be divided into the stochastic and non-stochastic effects. The non-stochastic effects are those that appear predictably as a function of the dose received. The stochastic effects, including cancer and certain birth defects, may or may not occur in an individual as a result of an exposure, but there is an increased risk in a population of similarly exposed individuals.

All organs may experience non-stochastic, dose-dependent effects from radiation exposure, and a few will be summarized here. The skin reacts to intense local exposures (upwards of 600 rad/6 Gy with local erythema; epilation may occur at even lower doses. Radiation causes death of hematopoietic cells in the bone marrow with doses as low as 300 rem/3 Sv causing a decline in white blood cell particularly lymphocyte and platelet counts. The testes (spermatogonia) and the ovaries (oocytes) are particularly sensitive to radiation; decreases in sperm counts can occur at low doses. Exposures in the range of 300–400 rem/3–4 Sv can result in permanent sterility in men and women. The gastrointestinal epithelium, with its high rate of turnover, is often affected by whole-body radiation, and a syndrome of diarrhea, hematochezia and malabsorption may result within several days of a significant dose. Very high doses of radiation can cause nausea and vomiting within hours of exposure. Cataracts of the lens of the eye can occur from acute exposures or significant smaller exposures over time. In addition, a dose-dependent relationship has been found between radiation exposure and reduced IQ in children of Japanese atomic bomb survivors who were irradiated between 8 and 15 weeks of gestation.

The stochastic health effects of ionizing radiation are related to mutagenic and carcinogenic events in the cell. These effects include malignancy (somatic cell mutation), hereditary effects (germ cell mutation), non-malignant changes, and developmental effects (i.e. on the fetus). Radiation exposure probably increases the risk of all, or almost all, solid tumors, although its documented relationship to certain tumors such as thyroid, breast, lung and bone has been more widely recognized. A variety of birth abnormalities are associated with prenatal irradiation, and these depend on the timing as well as the magnitude of the dose in relation to gestation. These stochastic health effects do not occur in all individuals who receive an ionizing radiation exposure. The probability of an adverse effect lessens as the exposure dose becomes smaller.

Determination of risk in terms of human health effects from low-level exposure is difficult. Since there are no human experimental studies of low-dose-rate radiation exposure, the dose–response curve for exposure at low levels (<0.1–0.2 Gy/10–20 rad) is incomplete.

The model for the complete dose–response curve relies on extrapolation from exposures at higher levels. The low end of the curve can be modeled as a linear curve fitted to high-dose-rate points, or as a non-linear curve. The model is also sensitive to the assumption of a threshold or no-threshold effect. Current extrapolations are based on populations of atomic bomb survivors, fluoroscoped tuberculosis patients, and patients receiving radiation therapy for ankylosing spondylitis, cervical cancer and tinea capitis of 0.5–2 Gy (50–200 rad). Relative risks have been calculated for leukemia, and cancer of the thyroid, breast, lung, and gastrointestinal organs (Committee on the Biological Effects of Ionizing Radiation (BEIR V) and UNSCEAR). The BEIR V panel used a linear model, but cautioned that:

> "... departure from linearity cannot be excluded at low doses below the range of observations. Moreover, epidemiologic data cannot rigorously exclude the existence of a threshold in the millisievert dose range. Thus the possibility that there may be no risks from exposures comparable to external natural background radiation cannot be ruled out."[4]

In addition, cancer from radiation has no special features differentiating it from malignancies from other causes. Since approximately 20% of deaths in the USA are due to cancer, the estimated increases in deaths due to radiation exposure would not be discernible because of the background "noise". BEIR V has made estimates for three exposure categories.[4] The estimates illustrate the difficulty of finding the small increases in cancer rates from ionizing radiation in a population with low exposures (Table 11.6).

There are special cases of long-term effects that may occur from internally deposited radionuclides, such as lung cancer in uranium miners and osteogenic sarcomas in radium dial painters. Underground miners of uranium and other minerals have an increased incidence of lung cancer as a result of chronic exposure to radiation from inhaled radon. Radon is a noble gas and a decay product of radium, which in turn is a decay product of uranium.[5–8] Based on the recognition of the health effects of exposure to radon and its products in miners, guidelines have been developed for environmental exposure to radon for the population at large.

Radon, as it decays, creates daughter products, which are metals. Because they are solid, they attach themselves to the nearest solid object, which is usually a dust particle. The average half-life of these decay products is approximately 30 min, which means that there is a 50% reduction in that particular isotope every 30 min. The primary hazardous radiation dose from the radon daughters is due to alpha particles. These particles can induce metaplasia and atypical cell growth in the tracheobronchial epithelium that may subsequently develop into bronchial carcinoma. Because cigarette smoking has a synergistic effect with this radiation exposure, smokers have an increased risk of cancer, along with a decreased latency period from time of original exposure to the expression of disease. The risk of cancer in the general population is dependent on the amount of cumulative radon daughter exposure, the age distribution of the population, the time since the start of exposure, and the extent of cigarette smoking.[1]

The term "working level" (wl) was developed to describe exposure of miners to short-lived radon daughters. It represents the quantity that releases 1.3×10^5 MeV of potential alpha energy in 1 liter of air. Exposure at this level for a working month of 170 hours has been defined as a working level month (wlm).

Radon exposure is a significant public health issue. Although direct epidemiologic confirmation has not been accomplished, radon is estimated to be an important cause of lung cancer, possibly second in importance after smoking.[9] Exposure in the home is probably the most important source of radon for most people. While there is substantial

Table 11.6 Risk models for low-level ionizing radiation exposures: excess cancer mortality and their uncertainty—lifetime risks per 100,000 exposed persons.[a]

	Male			Female		
Exposure Level	Total cancer mortality	Non-leukemia mortality[b]	Leukemia mortality	Total	Non-leukemia morality[b]	Leukemia mortality
Single exposure to 0.1 Sv (10 rem)	770	660	110	810	730	80
90% Confidence limits	540–124	420–1040	50–280	639–1160	550–1020	30–190
Expected cancer deaths without radiation	20 510	19 750	760	16 150	15 540	610
Radiation % of expected	3.7	3.3	15	5	4.7	14
Total years of life lost	12 000			14 500		
Average years of life lost per excess death	16			18		
Continuous lifetime exposure to 0.001 Sv/year (0.1 rem/year)	520	450	70	600	540	60
90% Confidence limits	410–980	320–830	20–260	500–930	430–800	20–200
Expected cancer deaths without radiation	20 560	19 760	790	17 520	16 850	600
Radiation % of expected	2.5	2.3	8.9	3.4	3.2	8.6
Total years of life lost	8100			10 500		
Average years of life lost per excess death	16			18		

(*continued*)

geographic variation in the presence of radon, geography alone is not a sufficient predictor, and home measurement is advisable, particularly in those areas known to be at elevated risk. Relatively inexpensive short-term monitoring devices are available to assess levels of radon in the air; these screening devices are usually placed in the basement or the area of greatest risk. The Environmental Protection Agency has recommended a level of 4 pico-

Table 11.6 (continued)

Exposure Level	Male: Total cancer mortality	Male: Non-leukemia mortality[b]	Male: Leukemia mortality	Female: Total	Female: Non-leukemia morality[b]	Female: Leukemia mortality
Continuous exposure to 0.01 Sv/year (1 rem/year) from age 18 until age 65	2880	2480	400	3070	2760	310
90% Confidence limits	2150–5460	1670–4560	130–1160	2510–4580	2120–4190	110–910
Expected cancer deaths without radiation	20 910	20 140	780	17 710	17 050	650
Radiation % of expected	14	12	52	17	16	48
Total years of life lost	42 200			51 600		
Average years of life lost per excess death	15			17		

[a]Based on an equal dose to all organs and the committee's preferred risk models—estimates rounded to nearest 10. Estimates for leukemia contain an implicit dose rate reduction factor, but a dose rate reduction factor has not been applied to the risk estimates for solid cancer.
[b]Sum of respiratory, breast, digestive and other cancers.
Source: Adapted from National Research Council, Committee on the Biological Effects of Ionizing Radiation (BEIR V). *Health Effects of Exposure to Low Levels of Ionizing Radiation.* National Academy Press, Washington, D.C., 1990, pp. 172–3.)

curies/liter of air as an action level to consider remedial action. Effective mitigation approaches exist to reduce residential radon levels. Smokers living in homes with elevated levels of radon should be strongly supported in efforts to quit.[10]

DIAGNOSIS AND TREATMENT

Difference between external ionizing radiation exposure and radionuclide contamination

In the initial patient evaluation, it is important to decide whether the individual has been exposed to an energy source (irradiated) or contaminated with radioactive materials that emit ionizing radiation.

Exposure to X-ray and gamma (photon) radiation is called external exposure, because this form of radiation lacks mass and is entirely composed of electromagnetic waves. Although it may produce ionization, it does not make matter or tissue radioactive. Therefore, rescue workers dealing with these victims do not need to be concerned about receiving radiation exposure once the victim has been removed from the field of ionizing radiation exposure.

Radionuclides, such as radioactive iodine (^{131}I) or plutonium (^{239}Pu), have a mass that is radioactive and can become distributed on the body surface (contamination) or internally deposited in the body by way of inhalation, ingestion, or through the skin by puncture, laceration, abrasion, or burn. The contaminating or internally deposited radionuclide continues to emit radiation in types (e.g. alpha, beta) and amounts that are characteristic of the specific radionuclide. The radionuclide can contaminate the surrounding area and other individuals in that area.

General principles of evaluation and treatment of radiation accidents

External ionizing radiation exposure

Rescue workers responding to a radiation incident should remove the victim quickly from the radiation field and limit medical care for injuries to lifesaving procedures until both the rescuers and the patient are removed from the radiation exposure.

It is important to gather information about the exposure, either from a health physicist or from any other person who is knowledgeable about the exposure circumstances. Radiation dosimetry, if available, will also help in the management of the case. If the patient has been irradiated, try to establish whether the exposure was limited to an extremity or a specific body area, or whether it could be described as whole-body exposure. Extremely high whole-body exposures can result in an acute radiation syndrome, including the development of signs and symptoms that indicate the severity of the exposure. Acute radiation syndrome findings are summarized in Figure 11.2.

Local injuries generally exhibit only local effects, and these are often delayed in nature. The higher the dose, the more quickly the physical changes are seen. For example, a dose of 2000 rad/20 Gy may produce reddening of the skin in 2–3 h, whereas a dose of

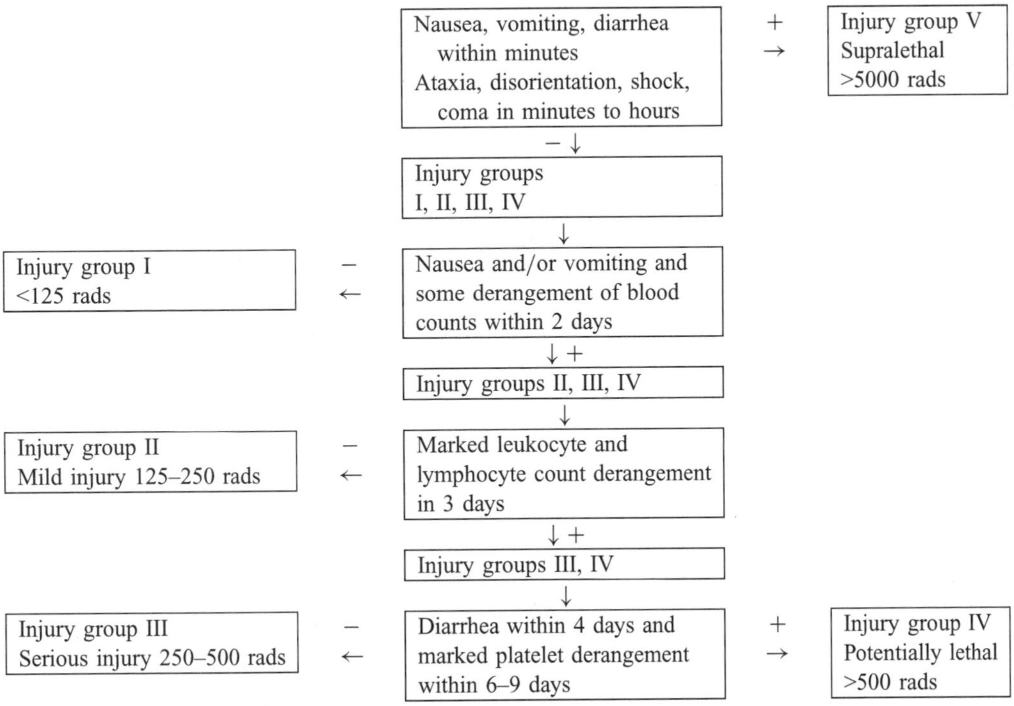

Figure 11.2 Acute radiation syndrome. Adapted from Thomas GE, Wald N. Diagnosis and management of accidental radiation injury. *J Occup Med* 1959; 1:421.

600–800 rad/6–8 Gy might result in a 2–3-day delay in the development of erythema. The extent of injury is dependent upon the energy of the radiation, the dose, and the dose rate. After an accident, it may not be possible to immediately reconstruct the exposure circumstances. Very often, local radiation injury may be observed concurrent with significant whole-body exposure. Careful clinical evaluation and observation for progressing signs and symptoms are helpful in managing the injury.

Usually, radiation exposure is not an acute medical emergency. Symptoms and effects are frequently delayed, so there is adequate time for evaluation, supportive treatment, and consultation. It is useful to collect a baseline blood specimen for complete blood count and differential diagnosis immediately after the patient arrives for evaluation, since serial leukocyte and lymphocyte counts can help in the assessment of the exposure (Table 11.7). Careful observation and documentation, particularly with the use of serial photographs, also helps in evaluating the progress of local injuries.

Chromosome studies can be a useful way to assess the dose received, particularly when the level of exposure is uncertain. A relatively new technique of chromosome painting called fluorescent in situ hybridization (FISH) has increased the sensitivity and ease of detecting aberrations. While analysis for dicentric and other chromosomal abnormalities has been used in the past, FISH uses chromosome-specific probes in peripheral blood lymphocytes to evaluate translocations, which were previously difficult to detect. Since chromosome translocations are relatively stable, their frequency can be used to estimate past radiation exposures.[11] FISH can be useful in the evaluation of relatively large radiation exposures, but it is not a particularly useful tool for low doses around the level of an annual occupational exposure limit.

High doses of radiation may result in the radiation hematopoietic syndrome, with failure of the blood-forming elements. With supportive therapy, some patients will survive the temporary decline in hematopoiesis until their endogenous marrow recovers; however, they are at great risk of infection and other complications. Donor bone marrow transplantation (BMT) was performed on a number of the Chernobyl incident victims; however, most of these individuals died of other radiation complications or of graft versus host disease. Two BMT survivors were later determined to have regenerated blood cells from their own marrow, although the donor cells may have helped them to survive for this to occur.

In the past 10 years there have been significant advances in hematology that offer promise for improved treatment of the hematopoietic syndrome. The science of autologous and allogenic stem cell transplantation has advanced; stem cells may now be harvested from peripheral blood and from cord blood.[12] The use of cytokines and colony-stimulating factors has contributed to the improvements in treatment. The number of radiation victims who ultimately benefit from BMT may remain small. Most victims with radiation exposure affecting the marrow can probably be supported with colony growth-stimulating factors until their own marrow recovers. In cases of severe exposure, individuals may die from other radiation-induced organ damage despite BMT.

Contamination with radionuclides

There are two ways in which a person may be exposed to radionuclides—by **external contamination** (i.e. contamination of the skin or exposed body parts only) or **internal deposition** (i.e. by inhalation, ingestion, wounds, or burns). In the worst case, both types of exposure occur together.

Contamination incidents present an additional level of concern over simple external radiation exposures, since they require prompt removal and containment of the radionuclides. Life- and limb-threatening conditions take precedence over contamination evaluation and control. However, early intervention not only permits the removal of external contamination, but may also help in reducing the amount of internally deposited radionuclides.

Most healthcare facilities have established emergency plans addressing chemical and radiologic hazards. These plans are usually required for accreditation by the Joint Commission on Accreditation of Health Care Organizations. Preplanning for management of patients contaminated with radioactive materials addresses the need for prompt decontamination and care of the patient as well as protection of medical personnel and facilities. Detailed guidance for this preplanning effort is available from a number of sources.[13–17] None of the required supplies are highly specialized, except for the instruments used for radiation detection. These are operated by a radiation safety officer (RSO), who should be formally designated at all healthcare facilities where radiotherapy devices are used or nuclear medicine is practiced. If no RSO has been named based on these activities, one should be designated in the facility's emergency plan.

The RSO should survey the patient as soon as other medical issues permit to assess the location, type and intensity of radiation contamination. The RSO or other trained individuals use specific radioactivity-detecting devices for this purpose. In most instances, radiation contamination does not pose an external radiation threat to the healthcare providers. However, they must be protected both from becoming contaminated and from internal deposition; this can usually be achieved with surgical gowns, gloves, caps and masks. Decontamination of patients can usually be performed readily using soap or detergents and water; in the case of wounds, copious irrigation and debridement are necessary. Collection of irrigation fluids and tissues and control of instruments, drapes and dressings also help to constrain the spread of contamination.

External radionuclide contamination alone (i.e. with no injury) requires identification of the involved skin surfaces to control and prevent spread of the contamination and to determine which areas to decontaminate. If the skin contamination is identified and removed promptly, the radionuclide is unlikely to cause damage to the skin or deeper structures. If potent chemicals are associated with the contaminating radionuclide—e.g. strong acids or bases—further damage to the skin or deeper structures may occur.

Internally deposited radionuclides require treatment based on the specific radionuclide. The advice of a physician trained in treating

Table 11.7 Dose–effect relationships following acute whole-body irradiation (X-ray or gamma ray).

Whole-body dose (rad/Gy)	Clinical and laboratory findings
<100/1	Often asymptomatic. Minor depressions of white blood and platelets detectable in a few persons, especially if baseline values established
100–200/1–1/2	Symptomatic course with nausea, vomiting and clear hematologic changes in a majority of exposed persons. Lymphocyte depression of ~50% within 48 hours
200–500/2–5	Serious disabling illness in most persons, with ~50% mortality if untreated. Lymphocyte depression equal to or greater than 75% within 48 hours
500+/5+	Accelerated version of acute radiation syndrome, with gastrointestinal complications within 2 weeks; bleeding and death in most exposed persons
5000+/50+	Fulminating course with cardiovascular, gastrointestinal and central nervous system complications resulting in death within 24–72 hours

Adapted from: Voeltz G. Ionizing radiation. In: Zenz, C. *Occupational Medicine*, 2nd ed, Chicago: Year Book Publishers, 1988:434.

this type of medical situation should be sought. For example, the chelator DTPA (diethylenetriaminepentaacetate) is used to remove internally deposited plutonium. This is similar to treating lead poisoning with DMSA (dimercaptosuccinic acid-succimer). Almost all accidental internal radionuclide depositions are accompanied by external contamination that will need to be identified, controlled, and removed. Chelation or other treatment for internal contamination must be administered as soon as possible to reduce internal deposition. Because this is a highly specialized activity, prompt consultation with experts in the management of internal contamination cases is critical.

Inadvertent radioactive iodine exposure is another situation in which medical intervention can reduce the internal deposition and limit the dose. Administration of iodine tablets or supersaturated potassium iodide solution (SSKI) is an important prophylactic measure which can reduce the absorption of the radioactive iodine into the thyroid gland. This measure, used after fission accidents, is likely to reduce thyroid cancer.

A comprehensive discussion of specific isotopes in interventions can be found in NCRP Report No. 65.[18] Some general guidelines are given in Table 11.8.

Anyone who cares for radionuclide-contaminated victims must be properly trained in radiation protection procedures in order to avoid becoming contaminated or incurring an internal radionuclide deposition.

Psychological aspects of radiation accidents Psychological considerations are important in all accidents, but they are even more critical for victims of radiation accidents. Taking great care to communicate clearly with the victim and offering a full explanation of the procedures should alleviate fear. The potential level of risk that might be anticipated from any radiation exposure should be discussed. If the victim is not fully informed, the adverse psychological stress can be the most serious consequence of the radia-

Table 11.8 Management of internal contamination.

Nuclides	Therapy	Comment
Transuranium elements: americium, californium, curium, plutonium	CaDTPA or ZnDTPA	CaEDTA may be used if DTPA is not immediately available, but it is less effective
Rare earths: cerium, lanthanum, promethium, scandium, yttrium	CaDTPA or ZnDTPA	CaEDTA may be used if DTPA is not immediately available. Consider stomach lavage and purgatives
Cesium	Prussian Blue	Consider stomach lavage and purgatives
Iodine	KI, NaI or SSKI	Success depends on early administration after exposure
Phosphorus	Aluminum hydroxide/phosphates	Severe overdose may be treated with parathyroid extract and oral phosphorus
Strontium	Aluminum phosphate/calcium	Barium sulfate and alginates are alternatives to block gastrointestinal uptake
Technetium	Potassium perchlorate	To reduce thyroid dose
Tritium	Forced water	Dilution

tion exposure, contamination, or deposition. Radiation incidents often generate substantial interest from the media, and it may be worthwhile preparing patients for this aspect of their exposure.

HEALTH PHYSICS AND MEDICAL SURVEILLANCE

Health physics

Any facility using radiation or radioactive materials is required to be licensed by the Nuclear Regulatory Commission (NRC). Licensing requires the establishment of a radiation safety committee and the appointment of an RSO. The RSO is responsible for (1) educating users about safety procedures, (2) monitoring environmental and personal exposure, and (3) ensuring that all recommended radiation safety policies, procedures and controls are followed. A dosimeter (TLD or film badge) is usually worn in the chest area of each employee who may be at risk. The dosimeters are collected and read at set intervals, i.e. weekly for higher-risk exposures, and monthly or quarterly for lower-risk exposures. Employees are informed of the results, and exposures greater than expected are investigated by an RSO or radiation protection specialist. Environmental monitoring for external hazards is usually performed with fixed equipment. Film badges, TLDs, or other monitoring instruments are placed outside the shielded area to ensure integrity of engineering controls. Equipment that generates X-rays must be kept in calibration; the health physics group usually performs this task. For radionuclide use, swipe samples can be taken at work areas to check for contamination.

Medical surveillance

It is standard practice for workers who regularly use radioactive materials or may be exposed to ionizing radiation to be medically evaluated regularly. Typically, this procedure includes a preplacement examination, periodic evaluations, and evaluations at employment termination or transfer. The medical surveillance is tailored to the type of exposure. For example, the NRC has published guidelines for the medical evaluation of nuclear power reactor operators.[19] As is evident from the example in Table 11.6, routine screening for cancer in workers with low levels of ionizing radiation exposure is not warranted.

For workers who work with radionuclides, periodic chest and whole-body scanning for detection of internal deposition of the radioactive material is standard practice. For those who work with radionuclides that can be excreted in the urine or feces, collection and radionuclide analysis of these excretions are done on a scheduled basis. The frequency of these examinations is dependent on the type and activity of the materials used; therefore, an expert in surveillance of these materials must be consulted to set up an appropriate schedule.

PREVENTION

Engineering controls such as lead shielding, high-density concrete walls, distance and safety interlocks are used to control external radiation exposure. Glove boxes and ventilated hoods are used to prevent the exposure of individuals who handle radionuclides.

Personal protective equipment designed to avoid radionuclide contamination or internal deposition includes special clothing and respirators that are worn only in the work area and then discarded or collected for cleaning when the worker leaves the area. The clothing includes a hood to shield any part of the head not covered by a respirator, coveralls, gloves, and booties. Masking tape is usually used to seal the areas between the hood and the respirator, the gloves and the sleeves, and the booties and the coverall trouser leg. In heavily contaminated areas, a supplied air respirator or Scott air pack may be used. Personal protective equipment for external exposures is widely used in industrial radiography and medical radiology. This equipment consists primarily of lead aprons, gloves, thyroid shields, and leaded glasses. Personal protective equipment for medical and research use of radionuclides includes gloves, laboratory coats, and eye/face protection.

Administrative controls must be built into standard safe radiation practices. Work practices should be used that treat radiation and radioactive materials as invisible hazards. These practices might include the following injunctions: (1) work only in designated areas; (2) use spill paper under radionuclide operations; (3) never mouth-pipette; and (4) handle all material to prevent secondary contamination of work areas, equipment, and personnel. The institution RSO and the radiation safety committee enforce operational policies and procedures required for licensing. As a final control, any worker who has reached a threshold exposure (action level) can be removed from potential exposure.

Ionizing radiation should be covered under an institutional reproductive health policy. All pregnant employees, or employees considering a pregnancy, should be evaluated on an individual basis. A reproductive health assessment should be obtained as soon as possible to determine the types and levels of potential exposure. This assessment will direct any job accommodations that might be necessary. In addition, some employers increase exposure monitoring during gestation to document ALARA exposures.

Fear of exposure to ionizing radiation during pregnancy is sometimes an issue. Employers should consider in advance whether they are willing to provide alternative duty for these workers (with a written physician's opinion), or if they will require the worker to use disability leave. In the absence of exposures above the NCRP recommended limit, and in the absence of other exposures, it is not necessary to restrict workers from their usual job duties.

EMERGENCY INFORMATION AND EXPERT ADVICE

Each state public health department has a radiation protection division that can provide guidance in setting up radiation protection programs. These professionals also help with the assessment of radiation accidents and in dealing with ionizing radiation problems in industry, healthcare settings, or research and development projects. In most states, the state radiation protection agency is involved with licensing, auditing, inspecting or accrediting groups that use ionizing equipment and materials.

The Oak Ridge Institute for Science and Education (ORISE) operates the Radiation Emergency Assistance Center/Training Site (REAC/TS). The US Department of Energy funds this facility. Expert advice and information on managing radiation accidents is available on a 24-h/7-day-per-week basis by calling 865-576-1005. The REAC/TS also has a web site at www.orau.gov/reacts/. In addition, the REAC/TS provides training courses in radiation protection, as well as courses that prepare hospital and emergency personnel to manage medical response in radiation accidents. Information about these courses can be obtained by writing to REAC/TS, ORISE, PO Box 117, MS 39, Oak Ridge, TN 37831-0117, USA, or by calling (865) 576-3131, or by e-mail: confinfo@orau.gov.

REFERENCES

1. National Council on Radiation Protection and Measurements. *SI units in radiation protection and measurements*. NCRP report no. 82. Washington, DC: National Council on Radiation Protection and Measurements, 1985.

2. National Council on Radiation Protection and Measurements. *Recommendations on limits for exposure to ionizing radiation*. NCRP report no. 91. Washington, DC: National Council on Radiation Protection and Measurements, 1987.
3. International Commission on Radiologic Protection. *1990 Recommendations of the International Commission on Radiologic Protection*. ICRP publication no. 60. New York: Pergamon, 1990.
4. National Research Council, Committee on the Biological Effects of Ionizing Radiation (BEIR V). *Health effects of exposure to low levels of ionizing radiation*. Washington, DC: National Academy Press, 1990.
5. Seve J, Kunze E, Placek V. Lung cancer in uranium miners and long-term exposure to radon daughter products. *Health Phys* 1976; 30:433–7.
6. Archer VE, Gilliam JD, Wagoner JK. Respiratory disease mortality among uranium miners. *Ann NY Acad Sci* 1976; 27(1):280–93.
7. Evans RD, Keane AT, Shanahan MM. Radiation effects in man of long-term skeletal alpha irradiation. In: Stover BJ, Jee WS, eds. *Radiobiology of plutonium*. Salt Lake City: Western Press, University of Utah, 1972, 431–86.
8. Martland HS, Conlon PO, Kreb JP. Some unrecognized dangers in the use and handling of radioactive substances. *JAMA* 1925; 85:1769–76.
9. Committee on Health Risks for Exposure to Radon, National Research Council. Health Effects of Exposure to Radon (BEIR VI), Washington, DC: National Academy Press, 1999.
10. American College of Occupational and Environmental Medicine. *Position Statement on Radon Exposure*. Chicago. ACOEM, 1992.
11. Natarajan A, Boei JJ, Darroudi F, et al. Current cytogenetic methods for detecting exposure and effects of mutagens and carcinogens. *Environ Health Perspect* 1996; 104(suppl 3):445–8.
12. Bishop MR. Potential use of hematopoietic stem cells after radiation injury. *Stem Cell* 1997; 15(suppl 2):305–10.
13. Mettler FA Jr, Kelsey CA, Ricks RC. Medical management of radiation accidents. Boca Raton, FL: CRC Press, 1990.
14. Ricks RC. *Hospital emergency department management of radiation accidents*. ORAU-224. Washington, DC: Federal Emergency Management Agency, 1984.
15. Bhattacharyya MH, et al. Guidebook for the treatment of accidental internal radionuclide contamination of workers. *Radiat Protect Dosimetry* 1992; 41:1–49.
16. American Medical Association. *A guide to the hospital management of injuries arising from exposure to or involving ionizing radiation*. Chicago: American Medical Association, 1984.
17. Voeltz G. Ionizing radiation. In: Zenz C, ed. *Occupational medicine*, 3rd edn; Chicago: Year Book, 1994, 426–62.
18. National Council on Radiation Protection and Measurements. *Management of persons accidentally contaminated with radionuclides*. NCRP report no. 65. Washington, DC: National Council on Radiation Protection and Measurements, 1980.
19. American National Standards Institute. *Medical certification and monitoring of personnel requiring operator licenses for nuclear power plants*. ANSI 3.4. New York: American National Standards Institute, 1983. (Adopted by the US Nuclear Regulatory Commission.)
20. Hall EJ. *Radiobiology for the radiologist*, 4th edn. J.B. Lippincott, 1994, 449.

12

ULTRAVIOLET RADIATION

James A. Hathaway, M.D., M.P.H., and David H. Sliney, Ph.D.

Ultraviolet (UV) radiation is that portion of the electromagnetic spectrum between visible light (about 400 nm) and the lower limit of ionizing radiation (about 100 nm). The energy of UV radiation photons increases as the wavelength decreases. The UV spectrum is divided into the following three bands: from 315 to 400 nm, it is called UV-A; from 280 to 315 nm, it is called UV-B; and from 100 to 280 nm, it is called UV-C.[1]

OCCUPATIONAL SETTING

Employees who work in the natural environment incur the greatest occupational exposure to UV radiation. Natural sunlight includes biologically significant amounts of energy in the UV-A and UV-B bands; the upper atmosphere filters out the UV-C radiation. Examples of such occupations include farmers and other agricultural and forestry workers, fishermen, outdoor construction workers, and lifeguards. Exposure is increased when UV radiation reflects off water, snow, or sand. The most common exposure to significant levels of non-natural UV radiation occurs among welders. Other workers may receive exposure from sources such as gas discharge lamps and carbon arcs.[2] Low-pressure mercury vapor lamps are used to control microorganisms in operating rooms, to control bacterial growth in meat, to prevent contamination in biological laboratories, to reduce airborne bacterial levels in air ducts, and to eliminate coliform bacteria in drinking water. High-pressure mercury vapor lamps are used for photochemical reactions and to identify minerals. High-pressure xenon arcs and carbon arcs have a broad spectrum of radiation, including visible and UV radiant energy. They are used as high-intensity light sources, such as searchlights, as well as in the printing industry.[3]

MEASUREMENT ISSUES

Measurement is not a significant issue when using many of these exposure sources, since it

Physical and Biological Hazards of the Workplace, Second Edition, Edited by Peter H. Wald and Gregg M. Stave
ISBN 0-471-38647-2

is already known that they emit harmful levels of UV radiation. In these cases, skin and eye protection is required. However, if direct measurement of UV radiation is needed, several UV meters are commercially available. Care should be taken to ensure that the UV meter is effective in the wavelength range of the source. UV detection devices include photodiodes, certain photomultipliers, and vacuum photodiodes. Selective filters are frequently used to isolate the part of the UV spectrum under study. Frequent calibration of meters is often necessary with heavy use. No single instrument perfectly matches the biological hazard action spectrum of UV radiation; therefore, it is necessary to calibrate meters to several wavelengths to evaluate broad-spectrum sources of UV.[4] Some recently developed detectors are remarkably well matched to the action spectrum.

EXPOSURE GUIDELINES

The American Conference of Governmental Industrial Hygienists (ACGIH) has published exposure limits called threshold limit values (TLVs) for UV radiation.[5] To protect against photokeratitis effects on unprotected eyes from UV radiation in the 320–400 nm range, total irradiance should not exceed 1.0 mW/cm^2 for periods >10 s (~16 mm) or >1.0 J/cm^2 for exposures less than ~10 s. Exposure limits for other wavelengths vary significantly by wavelength and duration of exposure. The tables in the ACGIH TLV guide should be used to determine exposure limits for a particular set of conditions. The International Commission on Non-ionizing Radiation Protection (ICNIRP) has also adopted these limits, with minor modifications in the UV-A region.[6]

NORMAL PHYSIOLOGY

Vitamin D is essential for the regulation of metabolism of bone minerals. UV irradiation of 7-dehydrocholesterol in the skin produces previtamin D_3, which is converted to vitamin D_3 (cholecalciferol). This vitamin and some vitamin D_3 from the diet are converted to circulating vitamin D.[7] Some level of ambient UV radiation below the TLV appears to be necessary to maintain good health.

Another normal physiologic reaction to UV radiation is the tanning of the skin. Tanning provides some protection against UV exposure. Exposure limits can be increased for tanned individuals.[5] However, tanning will not result from exposures below the TLV, and growing evidence suggests that a risk of skin cancer still exists even with careful tanning.[6]

PATHOPHYSIOLOGY OF INJURY

The penetration of UV radiation into human tissue is very limited. As a consequence, adverse health effects have historically been thought to be limited to acute and chronic skin and eye damage. The response of biological tissue to UV radiation is highly dependent on the depth of absorption, which in turn is wavelength-specific. The wavelengths of concern will be discussed along with each type of tissue damage.[8] Recent studies have demonstrated that UV radiation can suppress the immune response in the skin and may cause some systemic immunosuppression.[9,10] These effects may play a role in the development of skin cancer, and it has been speculated that they could alter host responses to infectious diseases.

Acute effects on the eye

The typical acute condition of the eye caused by UV exposure is photokeratitis of the cornea, commonly called welder's flash, arc eye, or flash burn. It is caused by UV radiation below 315 nm, usually from unprotected exposure to a welding arc. Between 2 and 24 hours after exposure, the worker experiences severe pain, redness, photophobia, and spasm of the eyelids if the exposure was severe. The condition usually clears up in 1–5 days, depending on the severity of the exposure. Healing is usually complete, and there is no residual injury.[11]

Acute effects on the skin

UV radiation, especially in the UV-C and UV-B bands, produces erythema of the skin (sunburn). If the exposure is more severe, edema and blistering will result. Although UV radiation above 315 nm is less efficient at producing erythema, there is sufficient energy in the UV-A band from tropical sunlight to produce a sunburn even if the UV-B is filtered out. The maximum effective wavelength for producing sunburn is 300–307 nm in sunlight and at shorter wavelengths in artificial light. The National Institute for Occupational Safety and Health (NIOSH) criteria document on UV radiation includes an extensive discussion of the histologic and cytologic changes in the skin induced by acute UV exposure.[12] These effects include an inflammatory response with edema, lymphocytic infiltrates, capillary leakage, and evidence of local dermal cell damage. The response to UV radiation is photochemical. A short-duration phase occurs in 1–2 hours; a later phase appears after 2–10 hours and may last for several days. The intensity and duration of the burn are proportional to the dose and wavelength of UV radiation; they are also related to the individual's skin pigmentation. The latent period to erythema becomes shorter with more intense exposures.[13]

An additional concern with acute exposure is the potential for photosensitivity reactions. Photosensitizing agents may have biological action spectra in the UV-A range, and they can act either systemically or locally. A number of medications (systemic photosensitizers) can predispose an individual to photosensitivity. Drugs such as sulfonamides, sulfonyureas, chlorothiazides, phenothiazines and tetracyclines are also well-known sensitizers. Occupational exposure to certain chemicals that may remain on the skin (local, or contact, sensitizers) can also act synergistically with UV radiation to produce erythema at much lower doses of UV radiation than would ordinarily be required. Coal tar products are well known for their photosensitizing properties. When they are combined with UV exposure, severe irritation and blistering can result.[14] Individuals with certain underlying diseases or a genetic predisposition may be unusually sensitive to the acute effects of UV radiation. Examples of conditions caused or aggravated by acute exposure to UV radiation include solar urticaria, polymorphous light eruption, the porphyrias, and systemic lupus erythematosus. Numerous other skin conditions can also be aggravated by the UV radiation from sunlight or artificial sources.

Chronic effects on the eye

UV radiation in the 295–400 nm range can cause photochemically induced opacities of the lens of the eye. The most effective wavelengths are 295–325 nm.[15] Radiation above 315 nm also causes cataracts in experimental animals. Some authorities have long theorized that ambient exposure to UV radiation is the primary cause of cataract development in older persons.[16,17] It has been recently reported that UV-B radiation may also be a risk factor for the development of pterygium.[18]

Chronic effects on the skin

Chronic exposure to UV radiation results in accelerated aging of the skin and an increased risk of skin cancer.[2,19] Prolonged exposure causes loss of elasticity, resulting in wrinkles. Actinic keratoses may form in the epidermis and are precursors of squamous cell carcinoma.

Three types of skin cancer have been associated with UV radiation—basal cell carcinoma, squamous cell carcinoma, and melanoma. Both basal cell and squamous cell carcinomas are related to the cumulative dose of UV radiation to the skin. They occur primarily on sun-exposed areas of the body, and it has been known for many years that persons in outdoor occupations are at a higher risk for these cancers.[20,21] Although UV radiation is important in the pathogenesis of melanoma, cumulative exposure to UV does not appear to be the primary factor, and nor has occupational exposure to UV radiation been correlated with the incidence of melanoma. Several studies have found that childhood exposures to sunlight that resulted in severe blistering sunburns were predictive of malignant melanoma risk.[22,23] The fact that melanoma is more common on the trunk than on commonly sun-exposed areas such as the face, back of neck, hands, etc. supports this theory. Melanoma is also rare on the buttocks and on women's breasts, which are typically covered during sunbathing.

Fair-skinned people have a higher risk of developing skin cancers than darker ones. Individuals of Celtic origin are also at greater risk. Genetic factors that affect DNA repair, such as xeroderma pigmentosa, can greatly increase the risk of skin cancer, and immunosuppression may also be an important factor. Kidney transplant patients have been identified as being at greater risk of developing skin cancer.[24]

Immunosuppression

The effects of UV radiation on the immune system have led to a new field of research called photoimmunology. In animal studies it was found that many UV-B-induced cancers were highly antigenic and were rejected when transplanted into normal syngenic animals. The UV radiation not only induced the tumors but also led to systemic T-lymphocyte-mediated immunosuppression, which reduced the host animal's ability to reject the tumor cells. This has raised the possibility that UV radiation-induced immunosuppression might also lead to a reduced response to infectious diseases.[16,17]

DIAGNOSIS AND TREATMENT

Detailed discussion of the treatment of chronic effects of UV radiation is beyond the scope of this book. The most significant acute effects are UV burns of the eye and skin (i.e. sunburn).

UV burns of the eye (photokeratitis) should be examined with a slit lamp using fluorescein stain. Diffuse punctate staining of the corneas will be seen within the palpebral fissure. Both eyes should be patched, and cycloplegic agents should be used. Local anesthetics should not be prescribed. Recovery is usually complete in 24 hours.[25]

Sunburn can range from mild to severe. Aspirin or other non-steroidal anti-inflammatory agents may be useful for fever and pain. Corticosteroids may be required for severe reactions.

MEDICAL SURVEILLANCE

Medical surveillance has generally not been recommended for workers who are exposed to UV radiation. It would not be useful for acute effects on the skin or eye and has not been demonstrated to be of particular value for chronic effects. Non-melanotic skin cancer and ocular cataracts both have relatively long latencies (time from exposure to the earliest manifestations of disease) in most people. Most of the disease attributable to occupational exposure would probably not be seen until after retirement; even if detected sooner, it would still be many years after the initial exposure. At best, medical surveillance might be expected to help detect some basal and squamous cell carcinomas when they are small and therefore more easily treated. It would also be possible to treat precursor lesions, such as actinic keratoses.

Periodic examinations of the skin have been suggested for high-risk groups such as coal tar workers with concurrent outdoor exposure. Skin cancers that are hard to see, such as those on the back of the neck or ears, could be detected earlier through this type of examination.

Early detection of cataracts for workers with UV exposure has not been specifically investigated. Studies on workers exposed to lasers or microwave radiation have attempted to assess early changes in the lens of the eye. These studies did not identify early changes that would be useful from a medical surveillance perspective.[26,27]

PREVENTION

Prevention of exposure to artificial sources of UV radiation is accomplished through a combination of engineering controls and personal protective equipment. Typical controls include the use of opaque shields or curtains when welding to eliminate exposure to coworkers. UV radiation used for germicidal purposes can usually be installed in ducts or recessed areas so that exposure to individuals in the same room is eliminated. Individuals performing operations such as welding use welding helmets and protective clothing.[28]

For outdoor work, simple measures such as long-sleeved shirts, broad-brimmed hats, canopies, and awnings provide significant protection. Sunscreens should also be used on exposed parts of the body. The wearing of tinted glasses has not been frequently recommended for protection from UV exposure but may be worth considering, especially for occupations near water, sand, or snow, where ambient exposures are amplified.[29] Eyeglasses with lenses made of glass normally provide substantial UV-B protection without special tinting, but tinted plastic lenses do not necessarily stop UV exposure. Plastic lenses usually state what degree of UV protection they provide, and this should be specifically checked. Tinted lenses should also be of a wrap-around design or have side-shields or large temples to block peripheral exposure of the eye.[29]

REFERENCES

1. Commission Internationale de l'Eclairage (International Commission on Illumination). *International lighting vocabulary*, 3rd edn. Publication no. 17. Paris: Commission Internationale de líEclairage, 1970.
2. Yost MG. Occupational health effects of non-ionizing radiation. In: Shusterman DJ, Blanc PD, eds. *Occupational medicine: state of the art reviews*. Philadelphia: Hanley & Belfus, 1992, 543–66.
3. Zenz C. Ultraviolet exposures. In: Zenz C, ed. *Occupational medicine*. 3rd edn. Chicago: Yearbook Medical Publishers, 1994, 463–7.
4. Wilkening GM. Nonionizing radiation. In: Clayton GD, Clayton FE, eds. *Patty's industrial hygiene and toxicology,* Vol. 1, Part B, 4th edn. New York: Wiley, 1991, 339–440.
5. American Conference of Governmental Industrial Hygienists. *Threshold limit values*. Cincinnati: American Conference of Governmental Industrial Hygienists, 2001, 155–8.
6. International Commission on Non-Ionizing Radiation Protection (ICNIRP). Guidelines on UV radiation exposure limits. *Health Physics* 1996; 71:978–82.
7. Federman DD. Parathyroid. In: Rubenstein E, Federman DD, eds. *Scientific American Medicine*; section 3, VI. New York: Scientific American, 1992, 1.
8. World Health World Health Organization. *Environmental health Criteria No. 160, Ultraviolet radiation.* Joint Publication of the United Nations Environmental Program, The International Radiation Protection Association and the World Health Organization. Geneva: WHO, 1994.
9. Kripke, ML. Ultraviolet radiation and immunology: something new under the sun-presidential address. *Cancer Res* 1994; 54:6102–5.
10. Beissert S, Scharz T. Mechanisms involved in ultraviolet light-induced immunosuppression. *J Invest Dermatol Symp Proc* 1999; 4:61–4.
11. Pitts DG, Tredici TJ. The effects of ultraviolet on the eye. *Am Ind Hyg Assoc* 1971; 32:235–46.
12. US Department of Health, Education and Welfare. *A recommended standard for occupational exposure to ultraviolet radiation*. HSM publication no. 73-11009. Rockville, MD: National Institute of Occupational Safety and Health, 1977.
13. CIE. *Erythemal reference action spectrum and standard erythemal dose*. CIE Standard S007-1998. Vienna: CIE, 1998; also available as ISO 17166, 1999.

14. Harber LC, Bickers DR. Drug induced photosensitivity. *Photosensitivity diseases; principles of diagnosis and treatment*. Philadelphia: WB Saunders, 1981, 121–53.

15. Pitts DG, Cullen AP. *Ocular effects from 295 nm to 335 nm in the rabbit eye*. DHEW (NIOSH) publication no. 177-30. Washington, DC: National Institute of Occupational Safety and Health, 1976.

16. Duke-Elder S. The pathological action of light upon the eye. Part II (continued)—action upon the lens: theory of the genesis of cataract. *Lancet* 1926; 1:1250–4.

17. Hanna C. Cataract of toxic etiology. In: Lerman S, ed. *Cataract and abnormalities of the lens*. New York: Grune & Stratton, 1975, 217–24.

18. Saw SM, Tan D. Ptergium: prevalence, demography and risk factors. *Ophthalmic Epidemiol* 1999; 6:219–28.

19. International Agency for Research on Cancer. *Solar and ultraviolet radiation. Monograph on the evaluation of carcinogenic risk to humans*. Vol. 55. Lyon: IARC, 1992.

20. Belisario JC. Effects of sunlight on the incidence of carcinomas and malignant melanoblastomas in the tropical and subtropical areas of Australia. *Dermatol Trop* 1962; 1:127–36.

21. Nicolan SG, Balus S. Chronic actinic chelitis and cancer of the lower lip. *Br J Dermatol* 1964; 76:278–84.

22. Shore RE. Nonionizing radiation. In: Rom WN, ed. *Environmental and occupational medicine*. Boston: Little, Brown, 1992, 1093–108.

23. Sober AJ, Lew RA, Kob HK, Barhill RL. Epidemiology of cutaneous melanoma. *Dermatol Clin* 1991; 9:617–29.

24. Walder BK, Robertson MR, Jeremy D. Skin cancer and immunosuppression. *Lancet* 1971; 2:1282–90.

25. Riordan-Eva P, Vaughan DG. Ultraviolet keratitis. In: Tierney LM, Mcphee SJ, Papadaki MA, eds. *Current medical diagnosis and treatment*. Norwalk, CT: Lange Medical Books, 1994, 171.

26. Friedman Al. The ophthalmic screening of laser workers. *Ann Occup Hygiene* 1978; 21: 277–9.

27. Hathaway JA, Stern N, Soles EM, Leighton E. Ocular medical surveillance on microwave and laser workers. *Occup Med* 1977; 19:683–8.

28. Tenkate TD. Optical radiation hazards of welding arcs. *Rev Environ Health* 1998; 13: 131–46.

29. Sliney DH. Eye protective techniques for bright light. *Ophthalmology* 1983; 90:937–44.

13

VISIBLE LIGHT AND INFRARED RADIATION

James A. Hathaway, M.D., M.P.H., and David H. Sliney, Ph.D.

Visible light is generally defined as that portion of the electromagnetic spectrum between approximately 380–400 nm and approximately 760 nm.[1] Some reference sources list the upper limit of the visible light band as 780 nm or 800 nm.[2,3] Infrared (IR) radiation is divided into the following three bands: IR-A is between 760 nm and 1400 nm, IR-B is between 1.4 μm (1400 nm) and 3 μm, and IR-C is between 3 μm and 1000 μm (1 mm). This ABC notation is sometimes referred to as near, middle and far IR.

OCCUPATIONAL SETTING

Visible light, along with the adjacent portions of the ultraviolet (UV) and IR bands of radiation, makes up much of the solar radiation reaching the surface of the earth. Outdoor occupations naturally have greater exposure to visible light and IR radiation. Visible light reflecting off sand and snow can create hazardous conditions that require eye protection. Ambient IR radiation can contribute to heat load, particularly in persons who work outdoors while wearing impervious clothing. Issues related to heat stress are covered in Chapter 6. Artificial sources of broad-spectrum intense visible light include arc welding or cutting, arc lamps, spotlights, gas and vapor discharge tubes, flash lamps, open flames and explosions.[4] Even though UV radiation is the main concern with many of these exposures, the potential for visible light-induced damage cannot be ignored.

IR radiation is emitted by many sources besides the sun. Artificial sources include heated metals, molten glass, home electrical appliances, incandescent bulbs, radiant heaters, furnaces, welding arcs, and plasma torches. Glassblowing and working in glass and steel plants are considered potentially hazardous due to excessive IR radiation.[5,6]

MEASUREMENT ISSUES

Among the adaptive responses to intense visible light are constriction of the pupil, light

Physical and Biological Hazards of the Workplace, Second Edition, Edited by Peter H. Wald and Gregg M. Stave
ISBN 0-471-38647-2

adaptation of the retina, squinting, and blinking. Intense light causes a natural aversion response, including shutting the eyes and turning away from the source of exposure. Measurement of continuous visible light emissions is usually not necessary to determine if the level of exposure is excessive or not, because the human eye itself provides adequate warning. Pulsed sources of visible light and sources that are turned on suddenly may present problems if the intensity of light is high enough to cause damage before an aversion response can take place. Usually such sources will be labeled with appropriate warnings; specific measurement of output levels will not be necessary.

Unfortunately, there are virtually no instruments designed as optical safety meters, so when measurements are necessary, a scientist experienced in radiometry may have to be consulted. A variety of instruments that use photodiodes or thermal detectors may be required for the measurement of visible light levels. These devices detect optical energy and convert the optical radiation to a measurable electrical signal. Similar instruments are also available to measure IR radiation. IR radiation is most frequently measured using thermal detectors such as thermopiles, or disc calorimeters.[7,8] These detectors measure heat from absorbed energy; they are suitable for the entire range of IR radiation, although the response time is slow. Lamp safety standards require the lamp manufacturer to perform detailed radiometric measurements of the optical radiation hazards and to place the lamp into one of four risk groups.

EXPOSURE GUIDELINES

Visible light and the near portion of the IR spectrum have threshold limit values (TLVs) developed by the American Conference of Governmental Industrial Hygienists (ACGIH).[9] These TLVs are for visible and near-IR radiation between 400 and 3000 nm. The TLVs apply to 8-hour exposures and require knowledge of the spectral radiance and total irradiance of the source as measured at the eyes of the worker. Moderately complex formulas and reference tables are required to calculate the TLV for each exposure situation; these calculations are beyond the scope of this book. TLVs can be calculated for three types of injury—retinal thermal injury from exposure to 400–3000-nm radiation, retinal photochemical injury from chronic blue-light (400–700-nm) exposure, and possible delayed effects leading to cataract formation from exposure to 770–1400-nm radiation. There are additional calculations for persons who have had a lens removed (cataract surgery) and not had a UV-absorbing intraocular lens surgically inserted. Such persons (although now rare) are at increased risk for photochemical retinal injury.

NORMAL PHYSIOLOGY

Life on earth would not be possible without visible light and IR radiation. IR radiation provides warmth, allowing a climate where life is possible; visible light provides the energy upon which life is based. Plant life uses chlorophyll to acquire energy from visible light. Using this energy, it converts carbon dioxide and water to carbohydrates in a process called photosynthesis. Plant-eating animals ingest stored carbohydrates in plants, and meat-eating animals acquire photosynthetically produced energy directly by feeding on plant-eating animals. The energy in all food consumed by humans is ultimately derived from visible light reaching the earth's surface. Most animal species have a sense of vision that is responsive in the near-UV, visible or near-IR portion of the electromagnetic spectrum. For humans, visual response defines the relatively narrow band of radiation called visible light (400–760 nm). Photons of light enter the eye and are focused by the cornea and lens onto the retina. In the retina, there are two types of photoreceptors—the cones, which are responsible for color vision and detailed visual acuity, and the rods, which allow peripheral vision and are responsive to lower light levels. The macula is an area on the retina that is densely packed with cones; it is the site of maximum visual acuity.

Photochemical reactions take place in both the cones and the rods. This stimulus results in a neurosensory transmission to the brain, where visual images are perceived. The retina and its photoreceptors are able to adapt to a wide range of light intensities. Adaptation to brighter levels of light typically occurs rapidly in a period of a few seconds. Adaptation to darkness requires many minutes, and in some cases more than an hour, to achieve maximum adaptation. In most circumstances, visible light is not hazardous. In addition to light adaptation, other normal protective mechanisms such as pupillary constriction, squinting, and blinking occur rapidly when bright light is encountered. When exposed suddenly to a highly intense visible light source, most people exhibit an aversion response that includes blinking and turning the head. This response typically occurs within 0.25 s; this time period is used to calculate exposure limits for radiation in the visible spectrum. The eyes are also naturally shaded from ambient sunlight by the eyebrows and the periorbital socket ridge.

Under some circumstances, visible light can be harmful—for example, when it is presented suddenly, as in a flash or explosion, or when equipment is first turned on. If the intensity is high enough to cause damage in <0.25 s, the natural protective mechanisms will be insufficient. It is also possible to create a hazardous situation by suppressing the aversion response and staring directly at a high-intensity light source such as the sun (solar maculopathy or eclipse photoretinitis) or a welding arc (welding-arc photoretinitis).

PATHOPHYSIOLOGY OF INJURY

Potential adverse health effects from overexposure to visible light or IR radiation occur primarily in either the eye or skin. Systemic effects of IR radiation from general body heating are considered in Chapter 6. Adverse effects can result from acute and chronic exposure. In the case of the eye, injury can result to different structures, depending on the wavelength of radiation.

Acute chorioretinal injury

Visible light and near-IR radiation from 400 to 1400 nm can be focused on the retina.

Sudden exposures to high-intensity sources of such radiation can cause adverse effects ranging from temporary flash blindness and afterimages to chorioretinal burns that produce scotomas (i.e. blind spots) in the field of vision. Retinal burns from gazing at the sun or observing a solar eclipse have been described throughout history. Artificial sources of luminance comparable to the sun have been developed in more recent decades. Even so, there have been fewer incidents of chorioretinal burns from artificial sources such as electric arcs, explosions and nuclear fireballs than from directly viewing the sun.[4]

Several factors are important in determining the exposure to the retina. These include (1) pupil size, (2) spectral transmission through the ocular media, (3) spectral absorption by the retina and choroid, and (4) the size and quality of the image. A dark-adapted pupil may be as large as 7 mm, as compared to a normal pupil size of 2–3 mm in outdoor sunlight. The area of a 7-mm pupil is about 12 times greater than that of a 2-mm pupil; thus, it allows that much more radiation to enter the eye. Although some radiation from 400 to 1400 nm can reach the retina, absorption in the ocular media (cornea, aqueous humor, lens, and vitreous humor) varies by wavelength. Optical transmission is greater from 500 to 900 nm, dropping about 50% to 1000 nm, rising again to 1100 nm, and dropping to low levels by 1200–1400 nm. Absorption of energy by the choroid and retina peaks around 500–700 nm, dropping gradually as the wavelength increases to 1000 nm, with a small rise peaking at about 1100 nm and with very little absorption past 1200 nm. The more energy that reaches and is absorbed by the retina and choroid, the greater the potential damage. For large, uniform images, the total absorbed dose per area on the retina is a good predictor of damage. Small images or images with "hot spots" blur as they are focused on the retina due to diffraction and therefore produce reduced peak retinal irradiance. Involuntary eye movements also spread the radiant energy over larger retinal areas.[10]

The mechanism of injury from accidental exposure to arc lamps or the sun was once thought to be primarily thermal, resulting in protein denaturation and enzyme inactivation. Today we know that most retinal injuries from staring at the sun or at a welding arc actually result from photochemical reactions, which dominate particularly with exposure to wavelengths of visible light between 400 and 500 nm.[11] However, thermal effects are still important. The threshold for thermal injury is dependent on light absorption, heat flow, and duration of exposure. Thermal injury is a rate-dependent process, so there is no single critical temperature that results in damage. In general, shorter exposures require higher temperatures to produce the same degree of damage.[4]

The degree of impairment caused by an acute chorioretinal injury depends on the size of the lesion in the retina and its location. If the source of exposure was directly viewed, as in gazing at a solar eclipse or looking at an explosion, the macula of the eye will be involved. Since fine visual acuity is dependent on intact macular function, damage in this area typically causes significant impairment of visual acuity. Injury to peripheral regions of the retina produces scotomas in the visual fields; but in many cases, peripheral lesions have minimal effects on overall visual function. Obviously, larger lesions cause more impairment than small lesions in the same location.

Chronic blue-light-induced retinal injury

Whereas thermal effects of visible and near-IR radiation on the retina are acute phenomena, the photochemical effects of blue-light photoretinitis are additive over time periods of seconds to hours and are probably partially additive even over many years. Exposure to light capable of causing thermal injury that does not cause actual injury is almost completely non-additive with subsequent exposures. In contrast, blue light—especially 400–500 nm radiation—can cause subclinical changes, which with repeated exposures can result in observable retinal damage. Subacute retinal injury due to photochemical mechanisms can occur at thresholds well below those of thermal injury. This threshold is only slightly higher than normal exposures to sunlight in outdoor work environments.[12] A number of mechanisms have been proposed to explain the effects, including photo-oxidative membrane damage, toxic chemical production in the outer retina, and metabolic disruption from extended overbleaching of retinal pigments. Ophthalmic examinations of experimental animals show both edema and pigmentary changes.

Some researchers believe that even typical, or "normal", outdoor exposures to sunlight can result in damage to the retina over a period of many years. They believe that macular degeneration, which is an important cause of blindness in older persons, is the result of lifelong exposure to the blue-light portion or possibly the entire visible spectrum of ambient sunlight. Many of these researchers regularly wear amber or red-tinted glasses to reduce blue-light exposure. Even though the link between macular degeneration and chronic blue-light or visible exposure must still be considered hypothetical, the results of subacute experiments provide support for the theory.[13]

Near-infrared exposure and cataracts

Near-IR radiation is capable of producing cataracts; such damage has been noted historically in glassblowers and furnace men. Radiation between 800 and 1200 nm is most likely responsible for temperature increases in the lens itself, because of its spectral-absorption characteristics. Visible wavelengths may also contribute to the problem, since heat absorbed by the iris could result in heat transfer to the lens.[14] Other structures of the eye, such as the cornea, absorb at longer wavelengths beyond 1200 nm and may also conduct thermal energy to the lens. Both mechanisms probably play a role, the relative importance of each being dependent on the wavelength characteristics of the exposure.[15]

Clinically, glassblowers' cataract has been described as a well-defined opacity in the outer layers of the axial posterior cortex of the lens, appearing as an irregular latticework with a cobweb appearance.[16] If exposure to IR between 700 and 800 nm or between 1200 and 1400 nm is a more significant factor, the cataracts are more likely to occur in the periphery of the lens.[17]

Acute skin, cornea, and iris injury

Both the skin and cornea of the eye are opaque to wavelengths >1400 nm. IR radiation in this region produces injury through thermal mechanisms, with absorbed radiation being converted to heat. Injury to the cornea is described as a gray appearance detectable by slit lamp that is caused by energy just above the threshold for injury.[18] Larger amounts of energy can produce extensive opacification of the cornea or even more severe injury. Focused sources of energy can create localized burns to the skin that resemble those caused by other sources of heat. There is some transmission of energy into the skin for radiation between 750 and 1300 nm, with maximum transmission at 1100 nm. At this wavelength, 20% of the energy will reach a depth of 5 mm. The nature of the injury will still be thermal. IR radiation below 3000 nm will penetrate into different depths of the cornea to varying degrees, depending on the specific wavelength. The iris of the eye can absorb energy and play a role only at wavelengths below approximately 1300 nm.

Solar urticaria and drug-induced photosensitivity

Although photosensitivity per se is primarily due to UV radiation, solar urticaria is often the result of visible light radiation, while drug-induced photosensitivity may be caused by visible light in the blue region, depending on the action spectrum of the specific drug. Solar urticaria is manifested by urticaria lesions on sun- or light-exposed areas of the body. Typically, the reaction begins as reddened skin; mild to moderate itching develops rapidly into urticaria lesions with edema.[19] The lesions resolve over several hours. Different parts of the body have variable degrees of susceptibility. Typically, chronically sun-exposed areas such as the face and arms are more tolerant to light exposure.

Some patients react only to UV radiation in the 320–400-nm region, whereas others have an action spectrum of 400–500 nm. Still other patients have a broad-action spectrum of 280–600 nm. The mechanism of action is believed to be immunologic; in some cases, sensitivity can be transferred by a patient's serum. An antigen may be formed in the skin of susceptible individuals following exposure to light, leading to an antigen–antibody reaction that produces the urticaria.[20] In some individuals, a non immunologic mechanism may be present where light causes the production of a substance that causes the urticaria directly.

Drug-induced photosensitivity may be caused by exposure to UV radiation or visible light in the blue region, depending on the action spectrum of the particular substance. For example, the action spectrum for pitch is 340–430 nm, and for dimethylchlorotetracycline it is 350–450 nm. Both of these are examples where visible blue light as well as near-UV can cause reactions. The clinical presentation may vary greatly. Lesions are usually on the light-exposed areas of the body, such as the face, the "V" of the neck, the back of the hands, and the extensor surfaces of the arms. Various degrees of redness, edema and vesicle formation may occur. In chronic cases, scaling and lichenification may occur.

Either phototoxicity (most common) or photoallergy may cause drug-induced photosensitivity. The former can be photodynamic, requiring oxygen, or it can be oxygen-independent. Phototoxicity is usually targeted at nuclear DNA or cell membranes. Photodynamic sensitizers interact with oxygen to form phototoxic compounds in the presence of UV or blue light. Oxygen-independent photosensitizers form toxic photoproducts even in the absence of oxygen. Photoallergy is the result of an immunologic response. The drug or chemical absorbs a photon of UV or blue light and is converted to a photoproduct that binds to a soluble or membrane protein to form an antigen.[21]

Porphyrias

While a number of conditions—such as systemic lupus erythematosus, atopic dermatitis, acne vulgaris, and herpes simplex—can be aggravated by exposure to UV radiation, the porphyrias are the result of blue-light interaction with porphyrins produced by aberrations in the enzymatic control of heme synthesis. The action spectrum is most predominant between 400 and 410 nm. Photons in this narrow wavelength band cause porphyrins to go to an "excited" state. Reactions with oxygen lead to peroxide formation, which in turn damages vital components of cell membranes, leading to cell death. The porphyrias may be due to either hereditary or acquired abnormalities in heme synthesis. They are classified into hepatic or erythropoietic categories, depending on the site of excess porphyrin. Most of the porphyrias are due to autosomal dominant defects in the enzymes responsible for heme synthesis. Porphyria cutanea tarda is the most common form and photosensitivity is the major finding. The disease, which usually manifests itself in middle age, may be triggered by exposure to certain medications such as barbiturates, phenytoin, and tolbutamide.

TREATMENT

Thermal burns to the skin from visible or IR radiation are treated like any thermally caused burn. If minor in nature, burns to the cornea are evaluated using fluorescein stain and slit lamp. Treatment focuses on the prevention of infection during healing. Typically it includes the use of cycloplegic agents and antibiotics in addition to patching of the eyes. Injuries to the deeper structures of the eye do not lend themselves well to specific treatment; they often result in permanent damage. Visual impairment depends on the extent and location of the injury.

Photosensitivity reactions can be treated with non-steroidal anti-inflammatory agents to control fever and pain. Corticosteroids may be needed for severe reactions and can be used both topically and systemically.

MEDICAL SURVEILLANCE

Medical surveillance has not been generally recommended for individuals exposed to intense levels of visible or IR radiation. Surveillance would not be appropriate for the acute effects of visible or IR radiation, and nor have specific subclinical effects been identified that would be useful for the surveillance of individuals with chronic exposures. Also, the magnitude of most occupational exposure is dwarfed by the contribution from ambient sunlight. Examination of the lens of the eye by slit lamp has shown a far greater prevalence of opacities of the lens in IR-exposed individuals than in controls.[15] However, it was not possible to demonstrate a dose–response relationship in these studies, and the changes noted were indistinguishable from naturally occurring cataracts. Although this type of examination has been worthwhile in epidemiologic studies, it is doubtful that it would be useful for individual medical surveillance.

PREVENTION

Exposure to artificial sources of visible and IR radiation can be prevented through engineering controls and protective equipment. Typical controls include barriers and reflectors or opaque shields to eliminate exposure to individuals. Viewing windows or ports can be equipped with glass or plastic with appropriate tinting materials to block the radiation. For visible light, neutral-density filters are commonly used. When eye exposure is the major concern, tinted glasses, goggles, or face shields can be used.[22] Reflective suits can help reduce thermal loading from exposures to the entire body and prevent burns.

To prevent photosensitivity reactions, exposure to sources of bright light, including sunlight, should be minimized. Simple measures include wearing long-sleeved shirts and broad-brimmed hats and using canopies or awnings. Sunscreen agents also offer some protection from blue-light photosensitivity reactions. Beta-carotene in doses of 60–80 mg/day can help to prevent photosensitivity reactions in persons with porphyria.

REFERENCES

1. Sliney DH, Moss E, Miller CG, Stephens JB. Semitransparent curtains for control of optical radiation hazards. *Appl Optics* 1981; 20:2352–66.
2. Yost MG. Occupational health effects of nonionizing radiation. In: Shusterman DJ, Blanc PD, eds. *Occupational medicine: state of the art reviews.* Philadelphia: Hanley & Belfus, 1992, 543–66.
3. Harber LC, Bickers DR, Kocherer I. Introduction to ultraviolet and visible radiation. *Photosensitivity diseases.* Philadelphia: WB Saunders, 1981, 13–23.
4. Sliney DH, Freasier BC. Evaluation of optical radiation hazards. *Appl Optics* 1973; 12:1–23.
5. Goldman H. The genesis of the cataract of the glass blower. *Am J Ophthalmol* 1935; 18:590–1.
6. Wallace J, Sweetnam PM, Warner CG, Graham PA, Cochrane AL. An epidemiologic study of lens opacities among steel workers. *Br J Ind Med* 1971; 28:265–71.
7. Wilkening GM. Nonionizing radiation. In: Clayton GD, Clayton FE, eds. *Patty's industrial hygiene and toxicology,* Vol. 1, Part B, 4th edn. New York: Wiley, 1991, 657–742.
8. Sliney DH, Wolbarsht ML. *Safety with lasers and other optical sources.* New York: Plenum, 1980.
9. American Conference of Governmental Industrial Hygienists. *Threshold limit values*. Cincinnati: American Conference of Governmental Industrial Hygienists, 2001, 151–54.
10. Ness JW, Zwick H, Stuck BE, et al. Retinal image motion during deliberate fixation: implications to laser safety for long duration viewing. *Health Physics* 2000; 78(2):131–42.
11. Ham WT, Mueller HA, Sliney DH. Retinal sensitivity to damage from short wavelength light. *Nature* 1976; 260:155–7.
12. Ham WT, Mueller HA, Williams RC, Gereraets WJ. Ocular hazards from viewing the sun unprotected through various windows and filters. *Appl Optics* 1973; 12:2122–9.
13. Mainster MA. Light and macular aging. *Lasers Light Ophthalmol* 1993; 5:117–9.
14. Goldman H. Genesis of the heat cataract. *Arch Ophthlmol* 1933; 9:314.
15. Lydahl E. Infrared radiation and cataract. *Acta Ophthalmol* 1984; 166(suppl):1–63.
16. Dunn KL. Cataracts from infrared rays (glass worker's cataracts). *Arch Ind Hyg Occup Med* 1950; 1:166–80.
17. Langley RK, Martimer CB, McCulloch C. The experimental production of cataracts by exposure to heat and light. *Arch Ophthalmol* 1960; 63:473–88.
18. Leibowitz HM, Peacock GR. Corneal injury—produced by carbon dioxide laser radiation. *Arch Ophthalmol* 1969; 81:713–21.
19. Botcherby PK, Gianelli F, Magnus I, et al. UV-A induced damage in skin cells from actinic reticuloid and normal individuals. In: Cronly-Dillon JR, Rosen DA, Marshall J, eds. *Hazards of light.* Oxford: Pergamon Press, 1973, 95–9.
20. Horio T, Minami K. Solar urticaria: photoallergen in a patient's serum. *Arch Dermatol* 1977; 113:157–60.

21. Harber LC, Bickers DR. Drug-induced photosensitivity. *Photosensitivity diseases*. Philadelphia: WB Saunders, 1981, 120–53.

22. American National Standards Institute. *Standard for occupational and educational eye and face protection*. ANSI Z 87.1. Washington, DC: American National Standards Institute, 1991.

14

LASER RADIATION

James A. Hathaway, M.D., M.P.H., and David H. Sliney, Ph.D

Lasers are devices that produce an intense, coherent, directional beam of light by stimulating electronic or molecular transitions to lower energy levels.[1] The beam of radiation emitted by lasers in common use may have a wavelength anywhere from the ultraviolet (UV) region of the electromagnetic (EM) spectrum to the far-infrared (IR) region. This includes numerous lasers operating in the visible light portion of the EM spectrum. Lasers vary widely in the intensity of their outputs; they may generate brief bursts or pulses of energy or operate continuously. The potential hazard of laser radiation depends on all of these factors.

OCCUPATIONAL SETTING

The use of lasers in industry, construction, research, medicine and the military is widespread and increasing. Lasers are used in alignment, welding, trimming, spectrophotometry, rangefinding, interferometry, flash photolysis, fiber optics communication systems, and surgical removal or repair procedures.[2,3] Low-power lasers are also widely used in commercial activities and consumer applications, including supermarket checkout counters, the detection of motor vehicle speed, as pointers for presentations, in CD-ROM drives for computers, and in CD, DVD, and laser disc players for home entertainment. Specific occupational titles may not be particularly helpful in identifying where lasers may be used. In industries using high-technology processes, various craftsmen, operators and service workers may be expected to use lasers.[4] Lasers are used for rangefinding in advanced weapon systems by military personnel. Maintenance personnel may actually be at higher risk of accidental exposure than the operators, because they may need to remove protective shielding and interlocks to repair the equipment. Similarly, the nature of laboratory research often precludes the use of engineering safeguards and may increase the risk of accidental exposure.[2] Medical uses usually require lasers with sufficient power to damage tissue. Accidental exposures have the potential to injure operating room personnel as well as patients.

Physical and Biological Hazards of the Workplace, Second Edition, Edited by Peter H. Wald and Gregg M. Stave
ISBN 0-471-38647-2

MEASUREMENT ISSUES AND CLASSIFICATION OF LASER POWER

In general, measurements of laser radiation are not necessary. The laser classification scheme described in the following paragraph was designed to minimize the need for measurements. It is the responsibility of laser manufacturers to perform measurements and classify their products. The classification system allows the user to determine potential risks and provide for the necessary safeguards, procedures, and personal protective equipment. Measurements are required only when information from a manufacturer is not available or when a laser system has been modified. Detailed information on measurement can be found in section 9 of the ANSI Z 136.1 standard[1] on the safe use of lasers. Appendix H4 of the same document provides a listing of catalogs on commercially available laser-measuring instruments.

The primary hazard from laser radiation is from exposure to the eye and, to a lesser extent, the skin. Therefore, the classification is based on the laser's capability of injuring the eye or the skin. Lasers manufactured in the USA are classified in accordance with the Federal Laser Product Performance Standard.[5,6] The actual process is somewhat complex, because numerous types of lasers have been developed that operate at different wavelengths. The threshold for biological injury varies with the wavelength of radiation. It is also dependent on the operating conditions of the laser—that is, on whether the radiation is continuous or pulsed. If it is pulsed, the duration and repetition rate of the pulse must also be considered. Details of the classification scheme are described in ANSI Z 136.1. The following outline provides a somewhat simplified view of the classification scheme:

- Class 1 laser—Will not produce injury even if the direct beam is looked at for the maximum possible duration inherent in the design of the laser. For many lasers, this essentially amounts to an unlimited viewing time.
- Class 2 laser—Will not produce injury if the direct beam is viewed for 0.25 s, the time period necessary for an aversion response. Class 2 lasers are limited to lasers emitting visible light on a continuous basis.
- Class 2A laser—Applies to lasers emitting visible light when the output is not intended to be viewed. The accessible radiation must not exceed that allowed for a class 1 laser for an exposure duration $\approx$1000 s.
- Class 3 laser—Can produce eye damage if the direct beam is viewed. Certain wavelengths may also damage the skin. This classification is subdivided into classes 3A and 3B. Class 3A, which is limited to the lower accessible outputs of this class, is believed to present less risk of actual injury from a practical standpoint. Class 3B represents those class 3 lasers with higher outputs where the risk of real ocular injury from even momentary viewing of the direct beam is high.
- Class 4 laser—Even the diffuse reflection of lasers with this level of power output can produce biological damage to the eye. The direct laser beam can injure the skin or pose a fire hazard.

Control measures apply primarily to lasers in classes 3B or 4. Limited precautions such as product labeling apply to class 2 and 3A lasers.

EXPOSURE GUIDELINES

Exposure guidelines have been developed by the American Conference of Governmental Industrial Hygienists (ACGIH) and by the American National Standards Institute (ANSI) Z 136 Committee on the Safe Use of Lasers.[1,7] Both of these organizations have issued guidelines for safe laser use. The Occupational Safety and Health Administration (OSHA) does not specifically regulate laser radiation, although the ANSI standard would be consulted in cases where the OSHA "general duty" clause is applied.

The ACGIH standards are called threshold limit values (TLVs). They vary, depending on the wavelength of the laser radiation and depending on whether the radiation is pulsed or continuous. Certain assumptions are also made regarding aversion time (0.25 s) and the size of the pupil under various exposure conditions. ACGIH tables 2 and 3 list the TLVs for either eye or skin exposure by wavelength and exposure time. The output of pulsed lasers is described in terms of energy (joules) and the output from continuous wave lasers is described in terms of power (watts). The TLVs are expressed as radiant exposure in joules per square centimeter (J/cm^2) or as irradiance in watts per square centimeter (W/cm^2). The ANSI Z 136 committee has labeled their standards maximum permissible exposures (MPEs). The MPEs for various conditions and types of lasers are listed according to wavelength in Tables 5.7 of the ANSI Z 136.1 standard.[1] The complexity of these tables and those of the ACGIH preclude them from being summarized here.

PATHOPHYSIOLOGY OF INJURY

Research on the biological effects of laser radiation has been directed toward determination of the thresholds for tissue damage. The threshold for identifying damage has typically been grossly apparent findings or findings observable using instruments such as microscopes, slit lamps, and ophthalmoscopes. The ANSI Z 136 committee used these data to determine exposure levels that produce damage 50% of the time. This would be analogous to an ED_{50}, or a dose that produces an adverse effect 50% of the time in experimental animals exposed to a chemical. A factor of 10 below the 50% damage level was then typically used to arrive at the MPE level, where the probability of damage was negligible. Actual regression lines were used to determine the slope of the dose–response curve where possible; when this slope was very steep, a factor <10 was used.[1] The principal biological hazards associated with laser radiation occur with acute short-term or intermittent exposures. Chronic effects are theoretically possible based on results of exposures to experimental animals or based on analogy to the chronic effects produced by ambient or artificial sources of UV, visible or IR radiation. Chronic exposure to laser radiation of sufficient power to be of concern is rare in occupational settings, because laser beams have very limited spatial extent. Therefore, we will focus on the acute biological effects of laser radiation.

Corneal damage from the infrared region (1400 nm to 1.0 mm)

Depending on the power level of the laser, tissue damage to the cornea from acute IR laser radiation can range from a minimal lesion involving only the epithelium, which appears as a small white area, to massive destruction of the cornea with severe burns to adjacent structures of the eye such as the conjunctiva and lids. Damage results from absorption of energy by tears and tissue water in the cornea. The heat is diffusely absorbed; a simple heat flow model is believed to explain the observed effects adequately.[1] Minor damage may heal completely within 48 hours; more severe damage will have permanent sequelae.

Corneal damage from the ultraviolet region (100–400 nm)

Biological damage from UV laser radiation is similar to that caused by other artificial or ambient sources of UV. Corneal effects following acute or subacute exposures include epithelial stippling, granules, haze, debris, exfoliation, and stromal haze and opacities.[8] Clinical symptoms and findings may include photophobia, tearing, conjunctival discharge, and redness. The damage caused by UV radiation is not due to heating effects but rather to photochemical denaturation of proteins and other macromolecules, such as DNA and RNA.

Retinal damage from the visible and near-infrared region (400–1400 nm)

The cornea, lens and ocular media are mostly transparent to visible light in the 400–700-nm wavelength range. Nearly all of the visible energy reaches the retina. Near IR radiation in the 700–1400-nm range also reaches the retina in significant amounts and produces damage similar to that caused by visible radiation. Clinically, the minimal lesion is a small white patch apparently caused by the coagulation of protein. It may be asymptomatic. It is visible within 24 hours of exposure. More significant exposures may produce immediate symptoms, such as loss of vision in the visual fields, spots (scotomata) in the field of vision, or persistent afterimages.[2] More severe exposure can cause substantial damage, including significant hemorrhage from the retina into the vitreous humor.

Laser radiation in the visible and near-IR can cause damage by a variety of mechanisms, depending on the type of laser. Damage has been attributed to thermal, thermoacoustic, and photochemical phenomena.[3] Lasers with short pulse durations of $<10^{-9}$ s may cause "blast" damage through non-linear mechanisms such as ultrasonic resonance and acoustic shock waves.[9] Most of the radiation is absorbed in the melanin granules of the retina in the retinal pigment epithelium and choroid. This structure underlies the cones and rods. When damage is caused by a thermal mechanism, it is due to protein denaturation. Damage caused by heat or photochemical mechanisms is similar to what would be expected from equally intense doses of noncoherent light.

Examination of an injured individual typically reveals a blind spot (scotoma) or spots in one or both eyes. Visual acuity may or may not be decreased, depending on the proximity of the injury to the macula. Obviously, injuries to or near the macula produce greater functional loss than injuries in the periphery of the retina. Funduscopic examination may show retinal or subretinal hemorrhages and hemorrhage into the vitreous. More minor injuries may not be immediately obvious on funduscopy, or they may present as minor retinal burns with edema. Healing takes place over a course of weeks. Some improvement in visual acuity may occur as the edema subsides. Generally, a blind spot remains in the visual field. The extent of functional loss depends on the size and location of the injury.

There are two reports on series of patients injured from exposure primarily to NdYag lasers operating at 1064 nm.[10,11] In one report, 8 of 12 patients had macular lesions. Visual loss ranged from minimal to severe. There was no improvement in vision over time, in spite of vasoprotective and corticosteroid treatments. In two cases, the extent of the injury was progressive. In one case, there was hemorrhage in the vitreous humor. In the other report, 25 of 31 eye injuries resulted in macular damage. Macular damage was progressive over a 1-week time period in seven cases. Ten eye injuries were followed for 4–10 years. The extent of injury remained stable over this time for nine of the eye injuries. Another report indicates that the presence of a hemorrhage, which may initially severely impair vision, does not preclude the possibility of a return to normal vision. When a short-pulse laser produces a microscopic retinal hole, recovery can be remarkable.[12]

Other ocular damage

Radiation in the near-UV zone and radiation in the zone between near-IR and IR has absorption characteristics such that significant levels of energy may be absorbed in structures of the eye between the cornea and the retina, including the lens and the iris. Acute damage to these structures would be expected from very high-energy lasers (e.g. at 1315 nm, iodine laser wavelength). Concurrent damage to the lens or the retina (depending on the wavelength) would also be expected. Chronic exposure to non-coherent sources of UV or IR radiation in these wavelength regions causes lenticular damage leading to cataracts. Theoretically, the same damage could be incurred from coherent laser radiation at similar wavelengths. In actual occupational settings, chronic exposure of unprotected workers is unlikely. Chronic effects on the lens have not been studied.

Skin damage

Laser radiation can cause injury to the skin. Higher levels of energy are required to produce skin damage than for eye injury. The focusing power of the cornea and lens of the eye increases the energy density reaching the retina, thus allowing lower levels of total energy to produce localized injuries. However, UV lasers can cause photochemical damage to the skin similar to acute sunburn. Visible and IR lasers can produce thermal burns from acute exposures. The power output of the laser determines whether accidental exposure to the skin produces a minor injury or a more severe one. Theoretically, chronic exposure to UV lasers would have the same risk of causing premature aging of the skin and increasing the risk of skin cancer; however, exposure conditions that would result in chronic exposures are unlikely, given the current uses of lasers.

TREATMENT

Individuals with suspected injuries to the retina should be referred to an ophthalmologist. In many cases, no treatment is required, but continued follow-up is important to evaluate functional loss (both visual acuity and blind spots in visual fields). Complications such as growth of new blood vessels in the vicinity of the injury may require treatment. More severe retinal injuries from very high-powered lasers require immediate evaluation by an ophthalmologist.

Minor UV injuries to the cornea can be treated in the same way as photokeratitis from other sources of UV radiation. Patching the eye and using cycloplegics are recommended. Anesthetic drops should not be used. Complete recovery takes about 48 hours. More severe corneal injuries from either UV or IR radiation require specialized treatment from an ophthalmologist.

The following precautions should be taken when dealing with eye injuries that require referral to an ophthalmologist:[13]

- Eye ointments should never be used, because they make clear visualization of the retina very difficult.
- Topical anesthetics should not be used to relieve pain from a UV injury.
- Prolonged use of these anesthetics can cause corneal breakdown and lead to blindness.
- Topical steroids should never be used unless prescribed by an ophthalmologist.
- If in doubt about the seriousness of an injury, err on the side of caution and refer the patient to an ophthalmologist.
- Keep in mind that some suspected laser-induced ocular injuries may not actually originate from laser exposure.[14]

Megadose intravenous methylprednisolone has been used in studies with cynomolgus monkeys to determine if it might improve healing of retinal laser burns caused by visible or near-IR laser radiation. An overall beneficial effect was noted. The authors indicated that the effect might be ascribed to the anti-inflammatory action, protection of microcirculation and anti-lipid peroxidation effects.[15]

Skin injuries from visible or IR radiation can be treated in the same way as localized thermal burns. Intramuscular vitamin E and/or use of vitamin E and an occlusive dressing has been reported to improve wound healing in miniature swine following exposure to IR lasers.[16] UV radiation can produce a localized injury equivalent to sunburn; it should be treated accordingly.

MEDICAL SURVEILLANCE

Medical surveillance requirements are included in the ANSI Z 136.1 standard.[1] They are required only for individuals working with class 3B or 4 lasers. No medical surveillance is required for use of class 1, 2A, 2, or 3A lasers. Examinations are required before work with lasers and after suspected injuries. No periodic examinations are required. For incidental personnel whose work makes it possible (but unlikely) that they will be exposed to laser radiation, only a check for visual acuity is required. For personnel who work routinely with lasers, initial screening is necessary, and additional screening may also be required. Initial screening consists of an ocular history, visual acuity test (far and near), Amsler grid test, and a test of color vision discrimination.

If visual acuity is 20/20 or better in both eyes, near vision is Jaeger 1 or better and the other two tests are normal, no further testing is required. Where necessary, additional screening should include examination of the eye with an ophthalmoscope to look for any abnormalities of the retina and optic disc. Deviations should be fully described. In addition, the ocular media should be examined for the presence of opacities. Dilation of the pupil is required.

The initial screening tests can be performed by a technician under the direction of a qualified physician, optometrist, or ophthalmologist. Evaluation following suspected laser injuries must be performed by an ophthalmologist.

The need for medical surveillance has been a controversial issue. Medical surveillance requirements have been under constant reevaluation. The requirements have been reduced as more information has become available. There are now many authorities who question the need for any medical surveillance.

Medical surveillance is ideally performed to detect reversible biological changes that may be an indication of overexposure to a specific agent. If this is not possible, medical surveillance can still be useful if early subclinical or mild conditions can be detected and interventions taken to eliminate further exposure. Medical surveillance is usually directed at detection of adverse effects that result from long-term chronic exposure. It is not effective when dealing with potential exposures that produce only acute effects. Although some wavelengths of laser radiation have the theoretical potential to produce chronic effects, the nature of their use makes this unlikely. A few studies have looked for possible chronic effects in laser workers, but no evidence of adverse chronic effects has been found.[17–20]

At this time, the preplacement evaluation of laser workers is required to establish a baseline against which the effects of accidental injury could be compared. Some investigators have argued that this evaluation is for the legal protection of employers and is not needed in terms of worker protection. It is hard to argue with this line of reasoning. A second reason for medical surveillance cited in the ANSI standard is the potential for some as yet unseen chronic effect from exposure to continuous wave lasers operating at selected wavelengths. The more time that passes without such effects being seen, the less tenable this reason becomes. Future revisions of the ANSI standard may drop the current limited requirements for medical surveillance in favor of an option that could be exercised at the discretion of the employer.

PREVENTION

Protecting the skin, and particularly the eye, from high-power laser radiation is critical because permanent damage—including blindness—can result. There are a number of excellent references on laser safety that can be consulted. Other references provide considerably more detail on preventive measures than is appropriate here.[1,21–26]

In their review of reported accidental exposures to laser radiation, the ANSI Z 136 committee specifically noted the following important causes of the incidents:[1]

- unanticipated eye exposure during alignment
- available eye protection not used
- equipment malfunction
- intentional exposure of unprotected persons
- operators unfamiliar with laser equipment
- improper restoration of equipment following service.

The ANSI committee also noted that several serious accidents were traceable to ancillary hazards such as electric shock, toxic gas exposure, and vaporized tissue exposure from medical procedures. These topics are covered in several reports.[1,22,25,27]

Obviously, the preferred method of prevention is to incorporate engineering control measures that limit access to laser radiation. Indeed, many laser systems arc designed to embed more powerful class 3B and 4 lasers within shields or enclosures. This safeguard eliminates the risk of accidental operator exposure, but it does not eliminate the risk to persons servicing the equipment. For many applications of lasers, however, it is not feasible to rely on enclosure; other methods of engineering control must be used, along with training, administrative procedures, personal protective equipment, and warning systems.

The ANSI Z 136.1 standard requires the appointment of a laser safety officer (LSO) to monitor and enforce the control of laser hazards. This task may involve training requirements, administrative procedures, standard operating procedures, and selection of engineering control measures.

Depending on how the laser is used, numerous engineering control measures may be necessary. Examples include: protective housing; interlocks on protective housings; interlocked service access panels; master switches that are disabled when the laser is out of use; interlocks, filters or attenuators for viewing portals and display screens and collecting optics; enclosed beam paths; remote interlock connectors; beam stops or attenuators; emission delay systems; and remote firing and monitoring.

Under certain circumstances, some of these engineering controls may not be feasible and alternative methods will be necessary. One important control measure is the establishment of what is called a laser-controlled area. Access to the area is limited to personnel who have been specially trained. These workers must have appropriate protective equipment, and they must follow all applicable administrative and procedural controls. Controlled areas need to be posted with warning signs. They must have limited access, be operated by qualified and authorized personnel, and be under the supervision of someone specially trained in laser safety. They should use beam stops of appropriate material, diffuse reflecting materials where feasible, and appropriate eye or skin protection. Furthermore, these areas limit the beam path to above or below eye level except as required for medical use; they eliminate the possibility of transmission of laser radiation through doors, windows, etc.; and they include a system that can disable the laser to prevent unauthorized use. Class 4 laser-controlled areas require safety controls to allow rapid egress, emergency alarms, non-defeatable area/entry controls where feasible, and other controls for particular operations. Inherent in the controlled-area concept is the need for rigorous compliance with training, administrative and procedural requirements. Protective equipment is mandatory whenever it is needed.

Industrial employers have generally complied well with the ANSI Z 136.1 consensus standard. This has not always been the case with research laboratories, particularly those in university settings. One recent article described a number of injuries in university research laboratories where persons using lasers had not received proper training and appropriate protective eyewear was not used.[2] The authors proposed a registration system for research lasers that would ensure that laser personnel receive proper training and that appropriate protective equipment is available.

REFERENCES

1. Laser Institute of America. *American national standard for the safe use of lasers*. ANSI Z 136.1. Orlando, FL: Laser Institute of America, 2000.
2. Barbanel CS, Ducatman AM, Garston MJ, Fuller T. Laser hazards in research laboratories. *J Occup Med* 1993; 35:369–74.
3. Wilkening GM. Nonionizing radiation. In: Clayton GD, Clayton FE, eds. *Patty's industrial hygiene and toxicology*, Vol. 1, Part B, 4th edn. New York: Wiley, 1991, 657–742.
4. US Department of Labor. *Guidelines for laser safety and hazard assessment*. OSHA instruction PUB 8-1.7. Washington, DC: US Government Printing Office, 1991.
5. Code of Federal Regulations, Title 21, Subchapter J, Part 1040. *Laser product performance standard*. Washington, DC: US Government Printing Office, 1999.
6. Code of Federal Regulations, Title 21, Parts 1000 and 1040. *Laser products: amendments to performance standard*. Washington, DC: US Government Printing Office, 1999.
7. American Conference of Governmental Industrial Hygienists. *Threshold limit values*, 2001. Cincinnati: American Conference of Governmental Industrial Hygienists, 2001, 132–9.
8. Pitts DG, Cullen AR. *Ocular ultraviolet effects from 295 nm to 335 nm in the rabbit eye*. DHEW (NIOSH) publication no. 177-30. Washington, DC: National Institute for Occupational Safety and Health, 1976.
9. Ham WT, Williams RC, Mueller HA, et al. Effects of laser radiation on the mammalian eye. *Trans NY Acad Sci* 1966; 28:517–26.
10. Pariselle J, Sastourne JC, Bidaux F, et al. Eye injuries caused by lasers in military and industrial environment. *J Fr Ophtalmol* 1998; 21:661–9.
11. Lui HF, Gao GH, Wu DC, et al. Ocular injuries from accidental laser injuries. *Health Phys* 1989; 56:711–6.
12. Hirsch DR, Booth DG, Schockett S, Sliney DH. Recovery from pulsed dye laser retinal injury. *Arch Ophthalmol* 1992; 110:6188.
13. Vinger PF, Sliney DH. Eye disorders. In: Levy BS, Wegman DH, eds. *Occupational health*, 2nd edn. Boston: Little, Brown, 1988, 387–97.
14. Mainster MA, Sliney DH, Marshall J, et al. But is it really light damage? *Ophthalmology* 1997; 104:179–80.
15. Takahashi K, Lam TT, Tso MO. The effect of high dose methylprednisolone on laser-induced retinal injury in primates: an electron microscopic study. *Graefes Arch Clin Exp Ophthalmol* 1997; 253:723–32.
16. Simon GA, Scmid P, Reifenrath WG, et al. Wound healing after laser injury to skin—the effect of occlusion and vitamin E. *J Pharm Sci* 1994; 83:1101–6.
17. Wolbarsht WL, Sliney DH. Historical development of the ANSI laser safety standard. *J Laser Appl* 1991; 3:5–11.
18. Hathaway JA, Stein N, Soles EM, Leighton E. Ocular medical surveillance on microwave and laser workers. *J Occup Med* 1977; 19:683–8.
19. Friedman AI. The ophthalmic screening of laser workers. *Ann Occup Hygiene* 1978; 21:277–9.
20. Hathaway JA. The needs for medical surveillance of laser and microwave workers. In: Tengroth B, ed. *Current concepts in ergophthalmology*. Sweden: Societas Ergophthalmologica Internationalis, 1978, 139–60.
21. Laser Institute of America. *Laser safety guide*. Cincinnati: Laser Institute of America, 1989.
22. Sliney DH, Wolbarsht ML. *Safety with lasers and other optical sources: a comprehensive handbook*. New York: Plenum, 1980.
23. Sliney DH, LeBodo H. Laser eye protectors. *J Laser Appl* 1990; 2:9–13.
24. Laser Institute of America. *American national standard for the safe use of optical fiber communication systems utilizing laser diode*

and LEP sources. ANSI Z 136.2. Cincinnati: Laser Institute of America, 1996.

25. Laser Institute of America. *American national standard for the safe use of lasers in health care facilities*. ANSI Z 136.3. Cincinnati: Laser Institute of America, 1996.
26. Thach AB. Laser injuries of the eye. *Int Ophthalmol Clin* 1999; 39:13–27.
27. Sliney DH, Clapham T. Safety of medical excimer laser with an emphasis on compressed gases. *Ophthalm Technol* 1991; 1423: 157–62.

15

MICROWAVE, RADIOFREQUENCY AND EXTREMELY LOW-FREQUENCY ENERGY

Richard Cohen, M.D., M.P.H., and Peter H. Wald, M.D., M.P.H.

Microwaves (MW) include that portion of the electromagnetic spectrum between 300 megahertz (MHz) and 300 gigahertz (GHz). Radiofrequency (RF) radiation comprises that portion of the electromagnetic energy spectrum in which wave frequency varies from 3 kilohertz (kHz) to 300 MHz. Extremely low-frequency (ELF) radiation includes frequencies <3 kHz; it commonly refers to radiation associated with electric power generation and transmission. This chapter will cover MW, RF and ELF radiation.

MICROWAVE AND RADIOFREQUENCY RADIATION

Occupational setting

The four types of devices that generate RF and MW energy are power grid tubes, linear beam tubes (klystrons), crossed-field devices, and solid-state devices. Sources of RF and MW energy can operate in three modes: continuous, intermittent, and pulsed. The continuous mode is used in some communication devices, the intermittent mode is used in heating devices, and the pulsed mode is used in radar and digital communication.

MW energy can be transmitted from the generating device through a wave guide or through a transmission line to an applicator or antenna. Microwaves are used to transmit signals in telecommunications, navigation, radar, and broadcasting (i.e. radio and television); they can also be used to produce heat in industrial and home microwave ovens and dielectric heaters (i.e. heaters used to heat electrically non-conductive materials by means of a rapidly alternating electromagnetic field).

Physical and Biological Hazards of the Workplace, Second Edition, Edited by Peter H. Wald and Gregg M. Stave
ISBN 0-471-38647-2

Cellular telephones operate at frequencies between 800 and 1000 MHz. Home microwave ovens use a microwave frequency of 2.45 GHz. Dielectric heaters are used in the manufacture of automobiles, furniture, glass fiber, paper products, rubber products, and textiles. RF dielectric heater applications include: sealing and molding plastics; drying glues after manufacturing; drying textiles, paper, plastic, and leather; and curing materials such as epoxy resins, polymers, and rubber. Video display terminals (VDT) can generate RF radiation (at ~10–30 kHz) because they have a cathode ray tube, which is a source of electrons (measured levels have been extremely low). Industrial welding also generates RF radiation, typically ~400 kHz. RF radiation is also used for diathermy (deep-tissue heating) applications in medical treatment.

Measurement issues

An electromagnetic wave results from the combination of electric and magnetic field vectors, each perpendicular to the other, producing a wave that travels or propagates perpendicular to the first two vectors (Figure 1.1). The power density or energy of the wave is derived from the measured intensity of the electric and magnetic field vectors. The total energy of the wave is expressed in milliwatts per square centimeter (mW/m^2). The individual field strengths can be measured; the electric field strength is measured in volts per meter (V/m), and the magnetic field strength is measured in amps per meter (A/m). Whole-body absorption is also measured/estimated as specific absorption rate (SAR) and expressed as watts per kilogram (W/kg).

Measurements are most often made using meters with frequency ranges of 2 kHz to 40 GHz. These meters yield point/spot measurements of the strength of either the electric or the magnetic fields. From there, the power density in milliwatts per centimeter squared is calculated. The instruments do not directly measure power density but have sensing probes that measure voltages or currents. These are usually displayed in volts per meter or amps per meter, which are then converted to power density as W/m^2 or mW/cm^2.

At frequencies above 300 MHz, the primary measurement is of the electric field. Below 300 MHz, separate electric and magnetic field measurements must be made and combined. The reasoning behind this methodology is complicated, but it is related to whether the measurement is taken in the "near" field or the "far" field. In the far field (greater than one wavelength from the emitter), the ratio between the magnetic and electric fields, is constant. When measured in the near field (less than one wavelength from the emitter), the ratio between the electric and magnetic field varies, and both need to be measured. At 300 MHz, the corresponding wavelength is 1 m. As the frequency decreases, the wavelength (and the length of the near field) increases. This increases the chance that measurements are made in the near field. Personal dosimeters are also available to measure exposures.

Exposure guidelines

The exposure standards for RF and MW radiation are based on the assumption that the primary way that energy is absorbed for these frequencies is by heat deposition. Power absorption (and heat deposition) is affected by the following factors:

- **frequency** of the radiation
- body **position** relative to wave direction
- **distance** between body and source (MW energy generally decreases with the inverse of the square of the distance from the energy source)
- Exposure **environment** (surrounding objects may reflect, resonate or modify incident waves)
- **Electrical properties** of the tissue (conductivity and dielectric constant).

The dielectric constant is a measure of the "permittivity" of the tissue; it measures the ratio of the amount of electric current that will flow in a specific medium versus the amount that will flow in a vacuum. Tissue electrical properties are a constant and depend on water content. Higher energy absorption occurs in tissues with higher water (high dielectric constant) contents, such as brain, muscles, and skin; and lower energy absorption occurs in tissues with lower water (lower dielectric constants) contents, such as bone and fat.

The boundaries between tissues can reflect the energy waves differently, resulting in "hot spots" that can cause localized injury. Changes in position can change the amount of energy absorbed, because the body acts as an antenna to receive the MW or RF energy. The effects of different body positions are addressed in the maximum power density exposure recommendations developed by the American Conference of Governmental Industrial Hygienists (ACGIH).[1]

OSHA regulations specify an exposure limit of 10 mW/cm^2 over a 6-min period (or longer) for frequencies between 10 MHz and 100 GHz.[2] There is also a consensus ACGIH standard (Table 15.1) that applies to a much broader frequency range. The ACGIH standards are derived on the basis of a SAR equivalent to 4 W/kg to which a safety factor of 10 is applied. This standard is based on data that show no effects from exposures at 4 W/kg over the 6-minute period. Accordingly, the resulting standard is based on an SAR of 0.4 W/kg. This relates to comparison energy values of 1 W/kg for an individual at rest and 5 W/kg when the person is exercising.

Because it is difficult to measure SAR directly, the standard is expressed in measurable quantities such as power density in mW/cm^2 or field strengths in V/m (or V^2/m^2) or A/m (or A^2/m^2). These power densities and field strengths represent allowable exposures that do not exceed the SAR. The standard, in effect, allows an equivalent power density of 1 mW/cm^2 for frequencies between 100 and 300 MHz. The exposures are averaged over 6 min; and the electric and magnetic fields must be measured separately below 300 MHz. Exposures at frequencies from 300 MHz to 3 GHz are calculated accord-

Table 15.1 Radiofrequency radiation standards.

Agency	Frequency	Power Density, S (mW/cm^2)	Electric field strength, E (V/m)	Magnetic field strength, H (A/m)	Averaging Time E^2, H^2 or S (minutes)
ACGIH	30 kHz–100 kHz		614	163	6
	100 kHz–3 MHz		614	16.3/f	6
	3 MHz–30 MHz		1842/f	16.3/f	6
	30 MHz–100 MHz		61.4	16.3/f	6
	100 MHz–300 MHz	1	61.4	0.163	6
	300 MHz–3 GHz	f/300			6
	3 GHz–15GHz	10			6
	15 GHz–300 GHz	10			616,000/f1.2
OSHA	10 MHz–100 GHz	10			

f = frequency in MHz
V/m = volt/meter
A/m = ampere/meter
cm^2 = milliwatts/square centimeter
ACGIH = American Conference of Governmental Industrial Hygienists.
OSHA = Occupational Safety and Health Administration.

ing to the formula $P = f/300$ (allowable power density in mW/cm^2 equals the frequency in MHz divided by 300).

The penetration of energy is a function of frequency with penetration depth of ~1.7 cm at 2.45 GHz in comparison to 2–4 cm of penetration at 0.915 GHz. Although there is greater penetration depth at lower frequencies, the resulting heating decreases with frequency. Below 10 MHz, the body is essentially transparent to RF radiation, and little heating takes place. The resonance frequency, or frequency that generates the greatest energy deposition, occurs at ~70 MHz.

Pathophysiology of injury

Biological tissues respond to MW and RF radiation exposure with the induction of their own electric and magnetic fields. Depending on the polarity of the biological molecules, rotation and agitation of molecules can occur, resulting in heat generation. Thermal injury can occur as a result of RF/MW exposure when exposures are in excess of $10\ mW/cm^2$. Because internal hot spots may result from internal resonance due to differences in dielectric properties or radiation reflection, there may be localized increases in energy absorption and heating. The net heating of the body is related to the amount of energy absorbed minus the amount lost through the usual heat-dissipating mechanisms (bloodflow, evaporation, radiation, convection, and conduction). The phenomenon of MW clicking ("hearing" MW radiation), originally thought to be a non-thermal response, appears to be caused by thermal elastic expansion and contraction of the cochlea.

The least likely tissues and organs to be affected are those with greater thermal regulatory ability, usually due to increased bloodflow and greater heat-dissipation potential. RF burns, which are the most frequently encountered industrial effect, usually involve the skin. Subcutaneous tissue heating usually occurs simultaneously with skin exposure. Where full-thickness skin burns occur and subcutaneous tissues are involved, healing may be prolonged due to the lack of base granulation tissue. Skin burns appear similar to a sunburn. The patient may initially present with a feeling of warmth, as if the skin or exposed portion were being heated. Within hours, redness and slight induration can occur. The course is usually characteristic of any thermal burn, but it may include vesiculation and ulceration. Similarly, thermogenic exposures have been associated with cataract formation in exposed workers, but cataracts have not occurred following exposures below recommended limits. There are a few case reports of massive exposures to MW and other RF sources that resulted not only in eye or skin burns, but also in symptoms of neurasthenia or post-traumatic stress-like disorders (recurrent headache, malaise, fatigue, depression). Hypertension and/or peripheral neuropathy have also been reported.[3,4]

The proliferation of cellular telephones and their potential MW emission while being held within a few centimeters of the brain has resulted in concern and investigation of possible health effects. Most of the laboratory and human data published to date have been negative.[5]

There have been anecdotal reports of non-thermal effects from MW and RF exposures. These include carcinogenic, reproductive, hematopoietic, immunologic, neurologic, neuroendocrine and psychological effects. These bioeffects have been reviewed extensively, but a specific non-thermal mechanism has not been identified. Because of inconsistent and conflicting animal and human data, genotoxicity, carcinogenicity, reproductive toxicity and other systemic/organ effects have not been clearly linked to non-thermal exposures.[6–9] Similarly, extensive research of reproductive effects in relation to VDT use has not found an association.[10]

Treatment

The medical response to MW/RF radiation exposure should involve: (1) removal from exposure, (2) determination of radiation frequency and exposure intensity, and (3) medical treatment for thermal injury to skin or subcutaneous tissues. High-intensity exposures can lead to deep-tissue injury. Localized subcutaneous hot spots and deeper penetration heating may make such thermal injury more difficult to evaluate. There have been reports of burning of the skin, with undamaged subcutaneous fat and burned muscle tissue below the fat layer. If a high-intensity exposure is suspected, tests for deeper tissue injury, such as CPK (creatine phosphokinase) to evaluate muscle injury, can be performed. Tests of specific organ function can also be ordered if injury is suspected. Routine burn management can be followed for superficial burns. Deeper or more serious burns should be referred to a burn specialist for specific medical treatment and follow-up.

Medical surveillance

Because no effects have been consistently demonstrated following long-term low-intensity exposures, periodic examination would not yield findings that would indicate a need for preventive actions. Given the inconsistencies and lack of scientific consensus regarding non-thermal effects, there is no basis for any periodic monitoring. Following an acute high-intensity exposure that results in thermal injury, appropriate follow-up should be instituted. Other than that predicted based on the thermal tissue effects, sequelae to that injury would not be expected.[11]

Prevention

Identification of RF/MW exposure in excess of recommended levels should be accomplished using available instrumentation that creates a plot of the potential fields and intensities. These measurements should occur at the time of initial equipment use and following any equipment changes thereafter.

Engineering controls include partial enclosure and elimination of leakage. Enclosing an area with wire mesh and sealing the seams with copper tape is a common engineering measure. Care must be taken to ensure that enclosures do not allow leakage. Where enclosures are not sufficient to reduce potential exposures, identification of the distance necessary for adequate energy dissipation can be effective. For example, hazard zones can be clearly marked surrounding the MW source. Although some personal protective equipment, such as eyewear and clothing, has been developed, its effectiveness is controversial, and it is not usually recommended.

There are no specific pregnancy-related recommendations for RF and MW exposures. The current recommendation of an SAR of 0.4 V/kg limits exposures to levels below those that cause significant thermal effects. Studies of the reproductive effects of MW and RF radiation have shown fetal loss and teratogenesis postexposure. However, most positive studies used exposures in the >100 V/kg range and were associated with internal temperature increases of 2–10°C.[12]

Heat has adverse effects on the testis and can also cause decreased spermatogenesis.[13] Exposures incapable of thermal effects are not considered reproductive hazards.

EXTREMELY LOW-FREQUENCY ELECTROMAGNETIC RADIATION—MAGNETIC FIELDS

Occupational setting

ELF energy is generated from electric power transmission and most household appliances. Workers with potential ELF exposure include: electrical and electronic engineers and technicians; electric power line, telephone and cable workers; electric arc welders; electricians; television and radio repair workers; power station operators; and motion picture projectionists. Magnetic fields and ELF electromagnetic radiation form the lowest end of the electromagnetic spectrum and include radiation at frequencies from 0 to 300 Hz. Concern regarding health effects in this portion of the electromagnetic spectrum centers around the power frequencies of 50 or 60 Hz (wavelength of 5000 km). Although it is composed of an electric and a magnetic field, the electric field does not have significant human tissue penetration.

Measurement issues

A magnetic field is formed whenever there is a flow of electric current. A static magnetic field occurs as a result of direct voltage or direct current (DC), while a time-varying magnetic field results from alternating current (AC). Magnetic field intensities are measured with a unit of magnetic flux called a Tesla or gauss (10 000 gauss = 1 Tesla, 10 gauss = 1 milliTesla). Magnetic fields freely penetrate many materials, including biological systems, and are very difficult to shield. Time-varying magnetic fields are of greatest interest because of their suggested association with molecular biological perturbation and health effects. Sources of time-varying magnetic fields include any device that utilizes an AC energy source, such as appliances, electrical equipment, and high-power transmission lines.

Two types of instruments are available for the measurement of magnetic fields; one is a multiturn loop used with a portable voltmeter, and the other is a gaussimeter. Otherwise, the measurement issues are similar to those discussed in the MW/RF section. Because cell membranes are relatively poor electrical conductors, the internal electric field is reduced between 10^6 and 10^8-fold compared to the external field. External fields of 1 million V/m would be needed to produce an internal field on the order of the existing transmembrane potential. Because of the tremendous attenuation of the electric field, it is not measured when evaluating ELF exposures. Personal dosimeters (EMDEX models A, B and C; Electric Field Measurements, West Stockton, MA, USA) are also available to measure exposures in the 35–300-Hz band of frequencies.

Exposure guidelines

The ACGIH has recommended threshold limit values for electric and magnetic fields in the 1–30-kHz range. At exposures between 1 and 300 Hz, the magnetic field limits are determined by the formula $B = 60/f$ where B is the limit in gauss and f is frequency; for frequencies between 300 Hz and 30 kHz, the limit is 0.2 mT(2 gauss). At 60-Hz, this would result in a 10-gauss (1-milliTesla) exposure.

The recommended standard for the electric field varies with frequency and is designed to both prevent induced internal currents and to eliminate spark discharges and other safety hazards that take place at field strengths >5–7 kV/m.

Natural background exposure from the static earth's magnetic field is ~450 mG, and it may change by as much as 0.5 mG per day due to changes in solar activity. The earth's electric field is ~120 V/m, which is comparable to the field found under a typical 12-kV urban power distribution line. Disturbances in the earth's local electric field are commonly found in the form of lightning, which needs at least 3 million V/m to ionize the air.

Electric and magnetic fields at the edges of a restricted right-of-way have been characterized by many public utilities. Typical electric and magnetic fields from the Bonneville Power Administration are reviewed in Table 15.2.[14] Table 15.3 summarizes some of the other typical electric and magnetic fields that might be encountered.

Pathophysiology and health effects

Many biological effects have been associated with ELF radiation (Table 15.4). Savitz and Calle first aroused interest in this area with their 1979 review of the incidence of leukemia in workers exposed to high electromagnetic fields.[15] Thereafter, this interest was heightened when a 2–4-fold increase in childhood

Table 15.2 Electrical and magnetic field strengths at ground level near high-tension transmission lines.[14]

	Distance from line				
	0	50 ft	65 ft	100 ft	200 ft
500-kV line					
Electric field (kV/m)	7		3	1	0.3
Magnetic field (mG)	70		25	12	3
230-kV line					
Electric field (kV/m)	2	1.5		0.3	0.05
Magnetic field (mG)	35	15	5	1	
115-kV line					
Electric field (kV/m)	1	0.5		0.07	0.01
Magnetic field (mG)	20	5		1	0.3

Table 15.3 Typical electric and magnetic fields.

Field source	Electric field (V/m)	Magnetic field (mG)
Home wiring	1–10	1–5
Electrical appliances	30–300	5–3000
Neighborhood distribution lines	10–60	1–10
Electrified railroad cars	–	10–200
High voltage transmission lines	1000–7000	25–100

(Adapted from Oak Ridge Associated Universities. *Health effects of low frequency electric and magnetic fields*. NTIS publication ORAU 92/F8, 1992. The highest appliance fields are recorded at the center of a spiral hot plate.

leukemia in the Denver area was attributed to ELF exposures in a report, also published in 1979, by Wertheimer and Leeper.[16] Their study linked high childhood cancer rates to ELF exposures by wiring code configurations (WCC), which were used as a surrogate for ELF exposure.

Good examples of childhood residential studies have been published by Savitz et al (Denver) in 1988, and London et al (Los

Table 15.4 Biological effects reported with ELF EMF fields.

Enhanced RNA synthesis in insect salivary gland culture
Decreased cell growth in slime molds
Enhanced DNA synthesis in mammalian cell culture in certain frequency windows
Lack of evidence for altering DNA structure
Promoting repair of non-union fractures
Epidemiologic evidence of cancer in humans
Changes in calcium ion flux from chick brain and embryo culture
Behavioral and EEG changes in mammals
Inhibition of melatonin secretion from the pineal gland

Adapted from Oak Ridge Associated Universities. *Health effects of low frequency electric and magnetic fields*. NTIS publication ORAU 92/F8, 1992.

Angeles) in 1991.[17,18] In the Denver study, spot magnetic field measurements were used in addition to WCC. In the Los Angeles study, spot measurements and 24-hour magnetic field measurements were recorded. Based on WCC data, the relative risk for leukemia in the high versus low current classification was 1.54 (95% confidence interval 0.9–2.63) in the Denver study, and 1.73 (95% confidence interval 0.82–3.66) in the Los Angeles study. Both these studies show a rise in the relative risk with increasing WCC. However, a significantly increased risk was not demonstrated in either study when the risk was assessed in relation to measured magnetic fields.

Since 1979, over 1000 articles have been published in the area of ELF effects. Research papers can generally be divided into the following areas: (1) human epidemiologic studies of cancer, focusing on childhood and adult residential exposures and on adult occupational exposures; (2) effects on growth control; (3) neurobehavioral effects; and (4) other physiologic effects.

Overall, research has yielded conflicting results about the presence (or absence) of an association between ELF and health effects. Current data do not allow a consensus.[19,20] For example, residential/non-occupational studies of adults have a number of epidemiologic shortcomings centered on the confounding factors of occupational versus home/other exposures, and they have not consistently shown an increased risk with measured ELF exposures. Non-occupational exposures include exposures from home wiring, electrical distribution lines and substations, transportation equipment (i.e. electric trains and buses), and home appliances (i.e. hot plates, refrigerators, hair dryers, electric blankets, and any other electrically powered device). Many of the studies negative for cancer were performed with field strengths well above what should be considered typical for an occupational or residential exposure. The two effects that have been reported at typical residential exposure levels (Table 15.4) are changes in calcium ion flux, and inhibition of melatonin secretion.[21,22]

The epidemiologic studies suggest a relatively weak and inconsistent association between ELF fields and leukemia and brain cancer. Kheifets et al reviewed occupational brain cancer and leukemia ELF research. They found small overall elevations in relative risks (<1.5) and inconsistent dose–response relationships.[23–25] Studies of ELF association with other cancers (e.g. breast) have been negative or inconclusive.[26,27] The effect of ELF on reproduction has also been studied. Huuskonen et al's review concluded that "the epidemiologic evidence does not, taken as a whole, suggest strong associations between exposure to ELF magnetic fields and adverse reproductive outcome. An effect at high levels of exposure cannot be excluded, however."[28] Other conditions, including multiple sclerosis and Alzheimer's disease have not been shown to be associated with ELF magnetic field exposure.[29,30] The biological plausibility of ELF effects rests on the ability of the magnetic field to interact with the body at the cellular level. Although the induced ELF fields are weak, the evidence of changes in calcium ion fluxes, increased rates of bone healing and changes in melatonin secretion suggest that magnetic fields are biologically active. Three additional recent large case–control studies also failed to find an association between cellular phone use and brain cancer.[31–33] Research is ongoing in all of these areas.

The 1999 report published by the National Institute of Environmental Health Sciences (NIEHS)/National Institutes of Health reflects the most recent consensus thinking on the biological effects of ELF magnetic fields.[34] The summary of the NIEHS report concludes:

> The scientific evidence that ELF-EMF exposures pose any health risks is weak. The strongest evidence for health effects comes from associations observed in human populations with two forms of cancer: childhood leukemia and chronic lymphocytic leukemia in occupationally exposed adults. While the support from individual studies is weak, the epidemiological studies demonstrate, for

some methods of measuring exposure, a fairly consistent pattern of a small, increased risk with increasing exposure that is somewhat weaker for chronic lymphocytic leukemia than for childhood leukemia. In contrast, the mechanistic studies and the animal toxicology literature fail to demonstrate any consistent pattern across studies although sporadic findings of biological effects (including increased cancers in animals) have been reported. No indication of increased leukemias in experimental animals has been observed.

The lack of connection between the human data and the experimental data (animal and mechanistic) severely complicates the interpretation of these results. The human data are in the "right" species, are tied to "real-life" exposures and show some consistency that is difficult to ignore. This assessment is tempered by the observation that given the weak magnitude of these increased risks, some other factor or common source of error could explain these findings. However, no consistent explanation other than exposure to ELF-EMF has been identified.

Epidemiological studies have serious limitations in their ability to demonstrate a cause and effect relationship whereas laboratory studies, by design, can clearly show that cause and effect are possible. Virtually all of the laboratory evidence in animals and humans and most of the mechanistic work done in cells fail to support a causal relationship between exposure to ELF-EMF at environmental levels and changes in biological function or disease status. The lack of consistent, positive findings in animal or mechanistic studies weakens the belief that this association is actually due to ELF-EMF, but it cannot completely discount the epidemiological findings.

The NIEHS concludes that ELF-EMF exposure cannot be recognized as entirely safe because of weak scientific evidence that exposure may pose a leukemia hazard. This finding is insufficient to warrant aggressive regulatory concern. However, because virtually everyone in the United States uses electricity and therefore is routinely exposed to ELF-EFM, passive regulatory action is warranted such as a continued emphasis on educating both the public and the regulated community on means aimed at reducing exposures. The NIEHS does not believe that other cancers or non-cancer health outcomes provide sufficient evidence of a risk to currently warrant concern.

ELF-EMF is complicated and will undoubtedly continue to be an area of public concern. The EMF-RAPID Program (http://www.niehs.nih.gov/emfrapid/home.htm) successfully contributed to the scientific knowledge on ELF-EMF through its support of high quality, hypothesis-based research. While some questions were answered, others remain. Building upon the knowledge base developed under the ELF-EMF through carefully designed, hypothesis-driven studies should continue for areas warranting fundamental study including leukemia. Recent research in two areas, neurodegenerative diseases and cardiac diseases associated with heart rate variability, have identified some interesting and novel findings for which further study is ongoing.

This is a very controversial area that deserves close monitoring of emerging research results.

Prevention

Even though there are no proven adverse health effects related to EMF ELF, a number of simple steps may be taken to reduce exposure without significant expense. This is known as the strategy of "prudent avoidance". Identifying high-voltage transmission equipment at the work site will help to focus on the kinds of monitoring that may be necessary. It would be prudent to reduce exposures to below the recommended levels. Distance is the best control since the field strength falls inversely with the square of the distance. Shielding is more difficult, because magnetic fields react differently to different metals and different metal configurations (e.g. screen, mesh, sheet metal). In some cases, rewiring to oppose adjoining field polarities may diminish exposure, but this procedure can often be prohibitively expensive.

REFERENCES

1. American Conference of Governmental Industrial Hygienists. *Documentation of threshold limit values for physical agents in the environment.* Cincinnati: American Conference of Governmental Industrial Hygienists, 2001.
2. 29CFR1910.97. Occupational Safety and Health Administration; http://www.osha-slc.gov/OshStd_data/1910_0097.html
3. Forman SA, Holmes CK, McMaramon TV. Psychological symptoms and intermittent hypertension following acute microwave exposure. *J Occup Med* 1982; 24:932–4.
4. Schilling CJ. Effects of exposure to very high frequency radiofrequency radiation on six antenna engineers in two separate incidents. *Occup Med* 2000; 50:49–56.
5. Masley ML, Habbick BF, Spitzer WO. Are wireless phones safe? A review of the issue. *Can J Public Health* 1999; 90(5):325–9.
6. Verschaeve L, Maes A. Genetic, carcinogenic and teratogenic effects of radiofrequency fields. *Mutation Res* 1998; 410:141–65.
7. Morgan RW, Kelsh MA, Zhao K, et al. Radiofrequency exposure and mortality from cancer of the brain and lymphatic/hematopoietic systems. *Epidemiology* 2000; 11(2):118–27.
8. Robert E. Intrauterine effects of electromagnetic fields (low frequency, mid-frequency RF, and microwave): review of epidemiologic studies. *Teratology* 1999; 59(4):292–8.
9. Repacholi MH. Low-level exposure to radiofrequency electromagnetic fields. *Bioelectromagnetics* 1998; 19:1–19.
10. COMAR Reports. Biological and health effects of electric and magnetic fields from video display terminals. A technical information statement. IEEE *Engng Med Biol* 1997; May/June:87–92.
11. Reeves GI. Review of extensive workups of 34 patients overexposed to radiofrequency radiation. *Aviat Space Environ Med* 2000; 71(3):206–15.
12. Polk C, Postow E. *Handbook of biological effects of electromagnetic fields*, 2nd edn. Bocca Raton, FL: CRC Press, 1996.
13. van Demark VVR, Free JR. Temperature effects. In: Johnson AD, Gomes WR, van Derman ML, eds. *The testis,* 3rd edn. New York: Academic Press, 1973.
14. Bonneville Power Administration. *Electric power lines,* 6th edn. DOE/BP-961. Portland, OR: Bonneville Power Administration, 1990, 4.
15. Savitz DA, Calle EE. Leukemia and occupational exposure to electromagnetic fields: a review of epidemiology studies. *JOM* 1979; 29:47–51.
16. Wertheimer N, Leeper E. Electrical wiring configuration and childhood cancer. *Am J Epidemiol* 1979; 109:273–84.
17. Savitz DA, Wachtel H, Barnes FA, et al. Case–control study of childhood cancer and exposure to 60-Hz magnetic fields. *Am J Epidemiol* 1988; 128:21–38.
18. London SJ, Thomas DC, Bowman JD, et al. Exposure to residential electric and magnetic fields and risk of childhood leukemias. *Am J Epidemiol* 1991; 134:923–37.
19. UK Childhood Cancer Study Investigators. Exposure to power-frequency magnetic fields and the risk of childhood cancer. *Lancet* 1999; 354(9194):1925–31.
20. Angelillo IF, Villari P. Residential exposure to electromagnetic fields and childhood leukaemia: a meta-analysis. *Bull WHO* 1999; 77(11):906–15.
21. Blackman CF, Benane SG, Elliott DJ, et al. Influence of electromagnetic fields on the efflux of calcium ions from brain tissue in vitro: a three-model analysis consistent with the frequency response up to 510 Hz. *Bioelectromagnetics* 1988; 9:215–27.
22. Burch JB, Reif JS, Noonan CW, et al. Melatonin metabolite levels in workers exposed to 60-Hz magnetic fields: work in substations and with 3-phase conductors. *J Occup Environ Med* 2000; 42(2):136–42.
23. Kheifets LI, Afifi AA, Buffler PA, et al. Occupational electric and magnetic field exposure and brain cancer: a meta-analysis. *JOEM* 1995; 37(12):1327–41.
24. Kheifets LI, Afifi AA, Buffler PA, et al. Occupational electric and magnetic field exposure and leukemia. *JOEM* 1997; 39(11):1074–90.

25. Kheifets LI, Gilbert ES, Sussman SS, et al. Comparative analyses of the studies of magnetic fields and cancer in electric utility workers: studies from France, Canada, and the United States. *Occup Environ Med* 1999; 56:567–74.
26. Sorahan T, Hamilton L, Gardiner K, et al. Maternal occupational exposure to electromagnetic fields before, during, and after pregnancy in relation to risks of childhood cancers: findings from the Oxford Survey of Childhood Cancers, 1953–1981 deaths. *Am J Ind Med* 1999; 35(4):348–57.
27. Caplan LS, Schoenfeld ER, O'Leary ES, et al. Breast cancer and electromagnetic fields–a review. *Ann Epidemiol* 2000; 10(1):31–44.
28. Huuskonen H, Linbohm ML, Juufilainen J. Teratogenic and reproductive effects of low-frequency magnetic fields. *Mutation Res* 1998; 410:167–83.
29. Graves AB, Rosner D, Echeverria D, et al. Occupational exposure to electromagnetic fields and Alzheimer disease. *Alzheimer Dis Assoc Disord* 1999; 13(3):165–70.
30. Johansen C, Koch-Henriksen N, Rasmussen S, et al. Multiple sclerosis among utility workers. *Neurology* 1999; 52(6):1279–82.
31. Muscat JE, Malkin MG, Thompson S, et al. Handheld cellular telephone use and risk of brain cancer. *JAMA* 2000; 284:3001–3007.
32. Inskip PD, Tarone RE, Hatch EE, et al. Cellular telephone use and brain cancer. *N Engl J Med* 2001; 344(1):79–86.
33. Harrington JM, Nichols L, Sorahan T, et al. Leukemia mortality in relation to magnetic field exposure: findings from a study of United Kingdom electricity generation and transmission workers, 1973–97. *Occup Environ Med* 2001; 58 (May):307–14.
34. National Institute of Environmental Health Sciences National Institutes of Health. *NIEHS Report on Health Effects from Exposure to Power-Line Frequency Electric and Magnetic Fields*. NIH Pub. No. 99-4493. NIEHS 1999. http://niehs-nik.gov/emfrapid/html/EMF_DIR_RPT/Report_18f.htm.

16

NOISE

Robert A. Dobie, M.D.

Approximately 5 million Americans have potentially hazardous noise exposures in the workplace.[1,2] Noise can be annoying and distracting. It interferes with spoken communication and masks the warning signals necessary for safety and productivity. As one of several generalized stressors, noise may contribute to cardiovascular disorders. However, the most important and best-characterized effect of excessive noise exposure is hearing loss. It is widely believed that reducing noise to a level low enough to prevent hearing loss will also prevent its other harmful effects (some important exceptions to this rule will be noted later). Therefore, this chapter will primarily deal with noise-induced hearing loss (NIHL), with special emphasis on risk assessment and prevention.

OCCUPATIONAL SETTING

Excessive noise is produced by an almost infinite variety of processes—anything that cuts, grinds, collides, explodes or just moves (itself, another object, or a gas or liquid) will make noise. The manufacturing industries listed in Table 16.1 are responsible for most hazardous occupational noise exposures in the USA. In some other industry groups such as fishing, forestry, construction, transportation, trade, and services, fewer than half of workers receive hazardous exposures. According to surveys, >50% of workers in such industries as textiles, lumber and wood, and mining receive hazardous exposures.[1]

Although these survey results are interesting, they are of limited use to the occupational physician. No industrial sector is completely free of hazardous noise exposures. Only assessment of individual workplaces—or, more appropriately, individual workers' exposures—can identify persons at risk for NIHL.

MEASUREMENT ISSUES

Fortunately, risk to hearing (unlike annoyance, sleep deprivation, and some other effects of noise) is relatively well predicted by three

Physical and Biological Hazards of the Workplace, Second Edition, Edited by Peter H. Wald and Gregg M. Stave
ISBN 0-471-38647-2

Table 16.1 Industries with substantial risks of hazardous noise exposure.[3]

Machinery
Food
Apparel
Textiles
Utilities
Chemicals
Paper
Tobacco
Leather
Transportation equipment
Electrical machinery
Fabricated metals
Primary metals
Printing and publishing
Lumber and wood
Rubber and plastics
Stone and glass
Furniture and textiles
Petroleum and coal

measurable physical properties of sound: **frequency**, **intensity**, and **time**. In general, more hazardous sounds are louder, longer, and more concentrated in the frequency range where we hear best.

Frequency

A vibrating object moves air molecules back and forth, creating a sound wave that propagates outward. The number of complete cycles or oscillations per second is the frequency of the sound, measured in hertz (Hz) or kilohertz (1 kHz = 1000 Hz). Normal young people can hear sounds ranging in frequency from about 20 Hz to 20 kHz, but our best hearing is in the 1–5 kHz region.

Only artificial objects like tuning forks and electronic oscillators put out pure tones, i.e. sounds having energy at only one frequency. Natural sounds like speech contain many frequencies simultaneously; indeed, it is the relative intensities of these different frequencies, or harmonics, that permit us to recognize speech sounds or to distinguish one musical instrument from another.

Intensity

Sound intensity (energy flow per unit area per unit time) is difficult to measure directly, but is directly proportional to the square of sound pressure, which is easily measured and thus much more commonly reported, in units of pascals (Pa) or micropascals ($1\,\text{Pa} = 10^6\,\mu\text{Pa}$). The range between the softest audible sound pressure and the loudest tolerable sound pressure is about 20 μPa to 20 Pa. To avoid having to either switch units or use too many zeroes, this million-to-one pressure range is compressed by using logarithms, just as is done when converting hydrogen ion concentration to pH.

Specifically, the decibel (dB) is defined as:

$$\text{dB} = 20\log_{10}\left(\frac{P}{P_\phi}\right)$$

For general-purpose sound level measurements, P_ϕ is set at 20 μPa, a barely audible level for the best-heard frequencies (inaudible for higher and lower frequencies). Decibels measured with $P_\phi = 20\,\mu\text{Pa}$ are identified as "sound pressure level" (SPL). Thus, a sound pressure (P) of 20 μPa would be 0 dB SPL.

$$\begin{aligned}\text{dB} &= 20\log_{10}\left(\frac{20}{20}\right) = 20\log_{10}(1)\\ &= 0\,\text{dB SPL}\end{aligned}$$

Sounds less intense than 20 μPa would represent negative values in dB SPL. In other words, ϕ dB does not represent the absence of sound but a sound whose pressure equals the reference pressure level. A very intense sound of 20 Pa (barely tolerable) would have a sound pressure level of 120 dB SPL:

$$\begin{aligned}\text{dB} &= 20\log_{10}\left(\frac{20\,000\,000}{20}\right) = 20(6)\\ &= 120\,\text{dB SPL}\end{aligned}$$

Since decibels are logarithmic, they combine in ways that may seem surprising. For example, if two sound sources which individually produce 80 dB SPL are turned on simultaneously, the result is only a 3-dB increase, to 83 dB SPL, rather than a doubling, as might be expected. Since doubling the distance from a sound source reduces the acoustic energy flow per unit area by a factor of 4 (area is proportional to distance^2), sound pressure level decreases by 6 dB. However, this only holds true outdoors; because of reverberation, sound levels change much less in most factories as one moves away from a sound source.

Sound pressure level is not enough to specify hazard to hearing: 120 dB SPL would be extremely loud (and hazardous) at a well-heard frequency such as 2 kHz, but it would be inaudible (and harmless) at 50 kHz. One could measure the sound pressure in each of several audible octave bands (an octave comprises a 2:1 frequency ratio, e.g. 500–1000 Hz), but this would be too cumbersome. Instead, hearing conservation professionals universally use decibels on the A-scale to combine frequency and intensity for a single-number measurement of potential hazard. A sound level meter operating in the A-scale mode uses electronic filters to cut out inaudible frequencies altogether, to partially remove poorly heard frequencies and to give full weight to the best-heard frequencies (1–5 kHz). Thus, two sounds with identical sound pressure levels could have different A-scale readings; the one with most energy in the 1–5 kHz range would be higher, more accurately reflecting the risk of NIHL (Table 16.2).

Sound level meters can be deceptively easy to use. However, microphone care, selection, calibration, and placement, timing of measurements, and proper interpretation of data, are just a few of the variables that can affect the reliability and validity of sound level measurements. Use of these instruments is normally best delegated to acoustic engineers, industrial hygienists, audiologists, or others who have been trained in their use.

Table 16.2 Effect of A-scale reading of two sounds with identical sound pressure levels.

Source	dB SPL	dBA
Jet	90	92
Diesel	90	85

Time

Even an A-scale reading is not enough; 5 minutes/day at 100 dBA is much less hazardous than 8 hours/day at 90 dBA. Recall that, using the logarithmic decibel scale, a 3-dB change is equivalent to a doubling (or halving) of sound energy per unit time. If the hazard were proportional to the total sound energy received by the ear, factoring in time would be simple: a twofold increase in exposure time would be equivalent to a 3-dB increase in sound pressure level (dBA). Indeed, many experts support the use of a 3-dB rule to relate time and level to overall hazard. However, there is considerable evidence that intermittent exposures are less hazardous than continuous exposures at the same level and total duration. Since high-level exposures are often interrupted, the Occupational Safety and Health Administration[2,3] (OSHA) has adopted a 5-dB rule in an attempt to incorporate the protective effects of intermittency. Under OSHA regulations, for example, a 90-dBA exposure for 8 h is considered as hazardous as a 95-dBA exposure for 4 h.

Time-weighted average (TWA) is a useful single number characterizing a day's exposure in terms of frequency, intensity, and time. This is the level (in dBA) which, if present for 8 h, would present a hazard equal to that of the exposure in question. Both of the previously mentioned exposures (90 dBA/8 h and 95 dBA/4 h) would be described as 90 dBA-TWA. Additional examples are shown in Table 16.3.

Calculating TWA for varying exposure times (e.g. 6 hours, 37 minutes) and combinations of exposures at different levels and durations is mathematically straightforward but cumbersome. Fortunately, most acoustical hazard assessment now uses noise dosimeters, wearable devices that use built-in microprocessors to automatically calculate TWA for a day's exposure.

Table 16.3 Time–intensity trading.[3]

Level (dBA)	Duration (hours)	TWA (dBA)
90	8	90
95	4	90
100	2	90
105	1	90
105	2	95
105	4	100
105	8	105

EXPOSURE GUIDELINES

The Occupational Safety and Health Act,[4] modified by the OSHA Hearing Conservation Amendment,[2,3] established extensive regulations for industries in the USA. The mining and railroad industries are subject to separate federal regulation by the Mine Safety and Health Administration and the Federal Railroad Administration, respectively.[5]

OSHA defines the permissible exposure level (PEL) as 90 dBA-TWA. Higher exposures must be reduced by engineering or administrative controls or by the use of hearing protection devices (HPDs). OSHA also recognizes a borderline or low-risk range of exposures at 85–90 dBA-TWA. Workers with daily exposures >85 dBA-TWA must be covered by hearing conservation programs (HCPs), which will be discussed later.

Exposures are sometimes described in terms of noise dose, where the PEL (90 dBA-TWA) is equivalent to a noise dose of 1.0 (or 100%). Exposures of 85 dBA-TWA and 95 dBA-TWA would be described as noise doses of 0.5 (50%) and 2.0 (200%), respectively. TWA and noise dose are interchangeable descriptions of noise exposure hazard, and most dosimeters will read out whichever the user prefers. OSHA requires that dosimeters incorporate all continuous, intermittent and impulsive sounds between 80 and 130 dBA into their calculations.

These guidelines are intended to protect almost all covered workers from substantial occupationally related NIHL. Note the "weasel words" in the preceding sentence. A few highly susceptible workers may incur mild NIHL at exposure levels below the PEL. Some workers are not covered by OSHA regulations; others may increase their risk by moonlighting at a second noisy job. Non-occupational exposures, especially hunting and target shooting, are often more hazardous to hearing than workplace exposures.

Observance of OSHA noise exposure guidelines will not always protect against safety hazards other than NIHL. For example, a brief period of intense noise (1 hour per day at 95 dBA) would neither exceed the PEL nor require an HCP. However, during that period, speech communication would be severely disrupted and warning signals could become inaudible.

Speech interference can be a useful clue to a potentially hazardous noise environment. People raise their voices to be heard over loud background noises. Most will need to shout to converse at arms length at sound levels >85 dBA; this is also the level at which OSHA requires hearing conservation programs for workers with 8-hour exposures.

NORMAL PHYSIOLOGY

The **outer ear** includes the pinna or auricle (the visible part) and the ear canal, a skin-lined tube leading to the ear drum; together, these structures provide resonances that enhance transmission of certain frequencies (around 3 kHz) and impair others. These effects are rather small; traumatic or surgical alterations of outer-ear structures produce only minimal hearing changes. Complete blockage of the ear canal with earwax or a foreign body will cause mild to moderate hearing loss.

The **middle ear** is separated from the ear canal by the ear drum, a thin membrane connected to a chain of three tiny bones, or ossicles: the malleus, incus, and stapes. The footplate of the stapes, which transmits the vibrations of the ear drum and ossicular chain into the inner ear, is much smaller in area than the ear drum. Thus, the pressure exerted on the inner-ear fluids is increased or amplified, much as the difference in cross-sections amplifies the force in a hydraulic system. Without this ingenious mechanical arrangement, most of the sound energy reaching the inner ear would be reflected back into the air. Two small muscles attach to the malleus and stapes and contract in response to loud sounds. This *acoustic reflex* stiffens the ossicular chain, impairing transmission of low-frequency sounds and slightly enhancing transmission of sounds above 2 kHz; the reflex offers some protection against NIHL, at least for low frequencies (1 kHz and lower). The middle ear is an air-containing space that receives a regular air supply via the Eustachian tube. Disruptions of the middle-ear mechanism (perforated ear drum, fixed or disconnected ossicles, fluid-filled middle-ear space) can cause mild to moderately severe hearing losses.

Hearing losses caused by outer-ear or middle-ear disorders are called conductive, because they interfere with normal air conduction of sound to the inner ear. Sounds presented by an oscillator held directly against the skull reach the inner ear by bone conduction and are heard normally by people with conductive hearing loss.

The **inner ear** contains organs of both hearing and balance; the latter will not be discussed here. The hearing organ, or cochlea, includes a spiral basilar membrane encased in a snail-shaped cavity in the temporal bone; on that membrane are spirally arranged hair cells, which change mechanical fluid vibrations into nerve impulses traveling to the brain. The hair cells of the base of the cochlear spiral respond best to high frequencies, whereas those at the apex respond only to low frequencies. At any given point along the spiral, there are three rows of outer hair cells, essential for hearing soft sounds, and one row of inner hair cells, which connect to almost all the nerves carrying sound to the brain.

Inner-ear hearing losses are usually lumped together as *sensorineural*, although the vast majority affect the hair cells (sensory) and very few directly affect the nerve cells (neural). The outer hair cells are more vulnerable to most diseases and injuries causing sensorineural hearing loss including age-related hearing loss, NIHL, and ototoxic drugs.

PATHOPHYSIOLOGY OF NOISE-INDUCED HEARING LOSS

Most hazardous noise exposures cause reversible inner-ear injury at first. Hair cells may lose their normal ability to respond to sound, accompanied by a temporary elevation of threshold (the softest sound that can be heard), or temporary threshold shift (TTS), lasting hours to days. Individuals often experience this as a muffling of sound, together with tinnitus (i.e. ringing in the ears) and fullness. After repeated TTS-inducing exposures, permanent hair cell loss and noise-induced permanent threshold shift (NIPTS) occurs. Some very brief but intense exposures, especially involving impulsive sounds such as gunfire, can cause immediate permanent threshold shift (PTS) without intervening TTS; this is called *acoustic trauma*, as distinct from ordinary NIHL.

We might expect NIHL to affect hearing for the same frequencies contained in the offending sound, and this is true, up to a point. However, most occupational and recreational noises contain a broad spectrum of frequencies; thus, the frequency pattern of NIHL is determined more by the sensitivity of the ear than by the frequency content of the noise. Other factors, such as the acoustic reflex, probably also play a role; the result is that for almost all cases of NIHL and acoustic trauma, the first and most severe effects are seen in the 3–6-kHz region.

Hearing loss is usually represented using an audiogram, a graph of hearing sensitivity (the softest sounds a patient can hear) as a function of frequency. The vertical axis plots thresholds in dB hearing level (HL). Recall that dB SPL implies a reference level of 20 μPa and that dBA represents a weighted average across the audible range of frequencies, suitable for noise hazard assessment. In contrast, dB HL implies a reference level of normal human (young adult) hearing; for each audiometric frequency, 0 dB HL is average normal hearing, whereas thresholds above 15 dB HL are abnormal for young adults (although not necessarily handicapping or even noticeable).

Figure 16.1 shows median audiograms for a group of retired jute-mill workers with noise exposure above 100 dB, and an age- and gender-matched control group.[6] The well-known 4-kHz dip is evident in the noise-exposed audiogram. NIPTS is simply the decibel difference between audiometric thresholds for noise-exposed and non-noise-exposed populations.

For repeated exposures to occupational noise, NIPTS grows gradually. After about 10 years, the growth decelerates markedly and NIPTS approaches a plateau. The International Organization for Standardization has published tables and formulae (ISO-1999[7]) that describe the growth of NIPTS over time for different frequencies and exposure levels. Some selected curves from ISO-1999 are reproduced in Figure 16.2. Note that all curves show the plateau effect, that the predicted NIPTS is much greater for 4 kHz than for 1 kHz, and that 40 years of exposure at 85 dBA produces only slight changes at 4 kHz (and none at 1 kHz). These curves represent median NIPTS; that is half of exposed persons would be expected to show

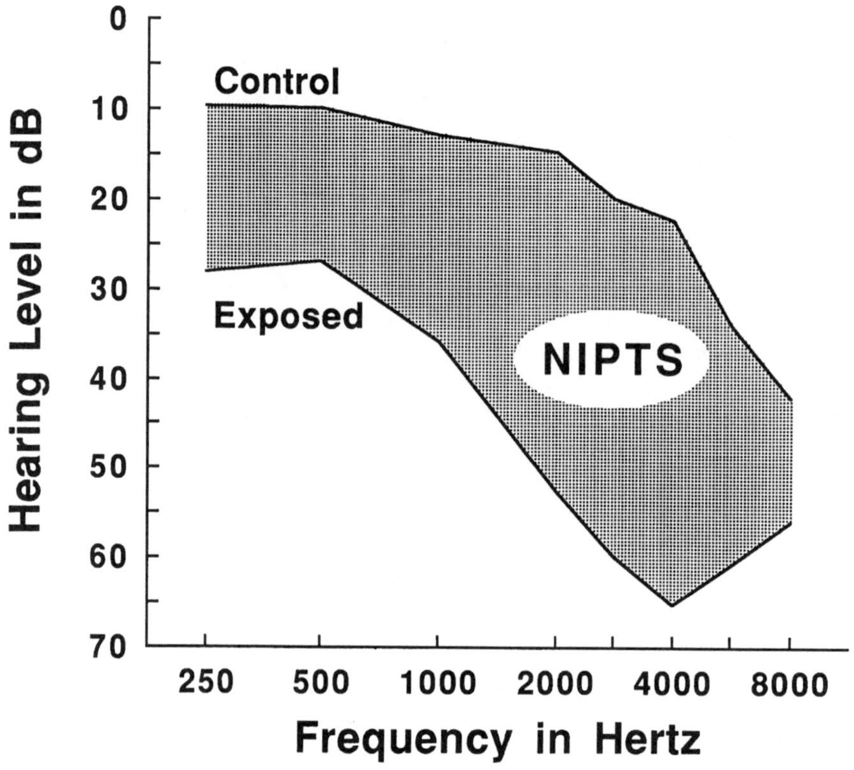

Figure 16.1 Median audiograms for noise exposed and control subjects.

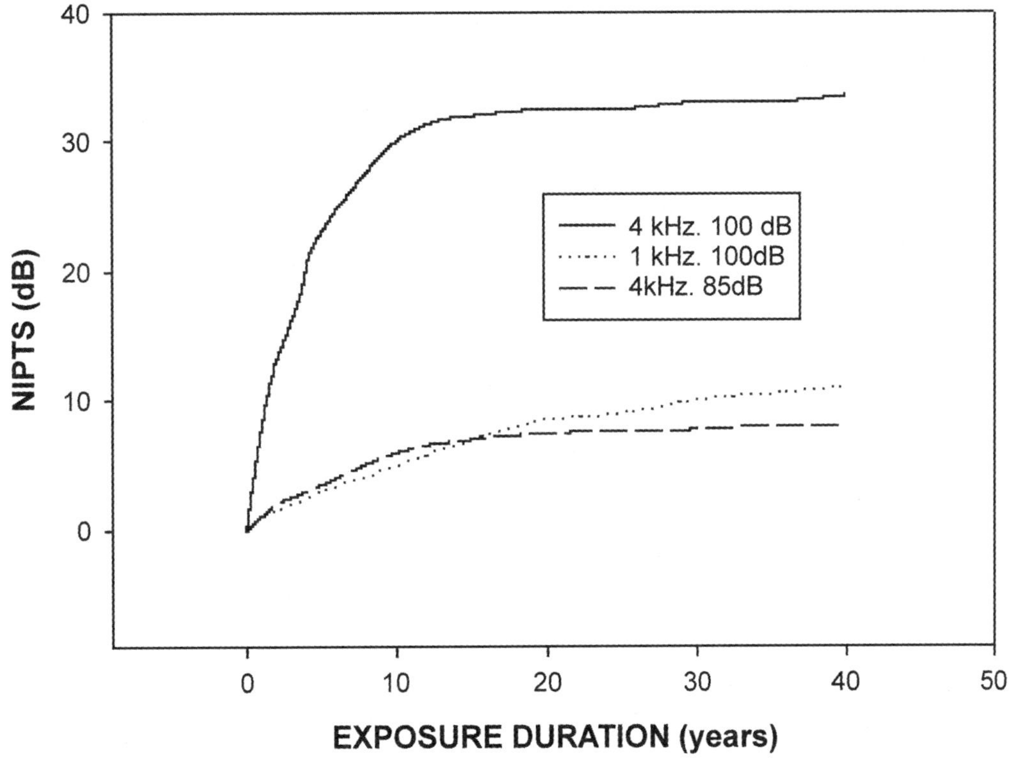

Figure 16.2 Noise induced permanent threshold shift as a function of time for different frequencies and exposure levels

greater losses and half would show lesser losses.

With prolonged exposure (or immediately in cases of acoustic trauma), frequencies below 3 kHz may demonstrate substantial losses. Most authorities consider the speech frequencies to be 0.5, 1, 2 and 3 kHz, because losses for these frequencies interfere with everyday speech communication, in quiet or noisy backgrounds. Most of the acoustic energy in normal speech is concentrated below 1 kHz, especially in vowel sounds. However, most of the information content is above 1 kHz, where the consonants have their peak energies. Thus, people with NIHL (or any other high-frequency sensorineural hearing loss) will often complain that they can hear speech but cannot understand it. This can lead to social isolation and depression.

NIPTS, as plotted in Figures 16.1 and 16.2, has not been directly measured in human epidemiologic studies. Rather, the hearing levels of a group of noise-exposed workers (known intensity and duration) are compared to hearing levels for non-noise-exposed workers with similar age, gender, and other characteristics. The differences between these two groups are reported as NIPTS. Implicit in this approach is the fact that people lose hearing as they age (high frequency > low frequency, men > women) and the assumption that age-related permanent threshold shift (ARPTS) and NIPTS are additive in decibels.

Figure 16.3 shows median ARPTS curves for both genders and two different databases included in ISO-1999, averaged across the speech frequencies. Note that ARPTS is an accelerating process. Contrast this with Figure

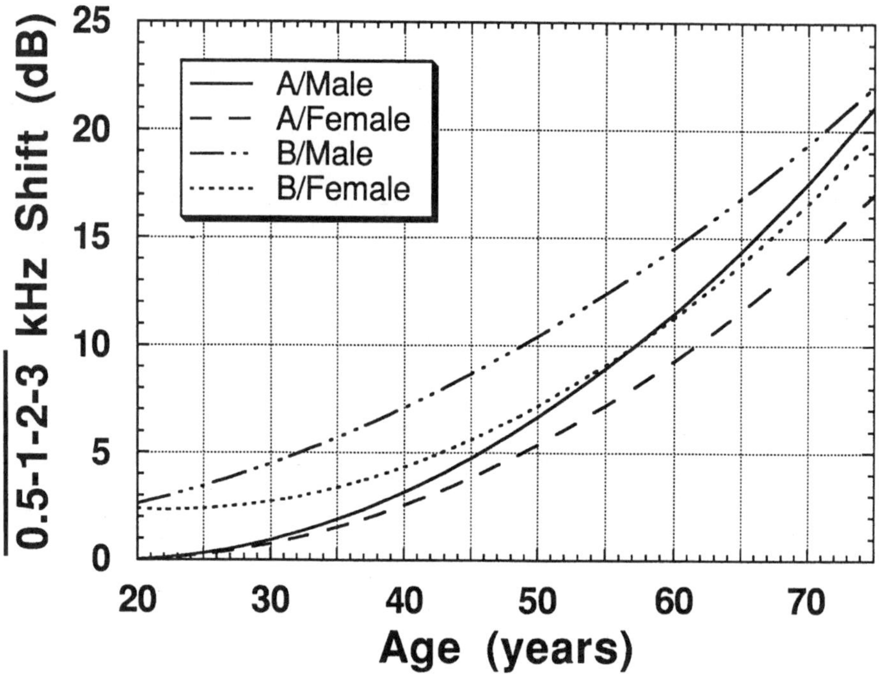

Figure 16.3 Speach frequency age-related permanent threshold shift as a function of age for two databases (A, B) and both genders.

16.4, which shows median speech-frequency NIPTS for exposures ranging from 85 to 100 dBA. NIPTS decelerates; approximately 60% of the 40-year total is present after 10 years.

DIAGNOSIS

The American College of Occupational Medicine[8] has enumerated a series of criteria for the diagnosis of occupational NIHL, which they have described as:

1. sensorineural
2. bilateral and symmetrical
3. not profound
4. not progressive after cessation of noise exposure
5. decelerating
6. greatest in 3–6-kHz range
7. stable after 10–15 years in high frequencies
8. less severe after interrupted exposures than after continuous daily exposures of the same level and duration.

Occupational NIHL is frequently diagnosed casually, without adequate attention to the patient's noise exposure history, both occupational and non-occupational. Employers and HCP managers can improve the quality of medical reports by providing examining physicians with noise exposure data, especially dosimetry, relevant to the individuals being evaluated.

Except in very young workers, noise and aging must be considered together. At retirement age, most workers exposed at levels below 100 dBA (the vast majority of noise exposed workers) will have more age-related

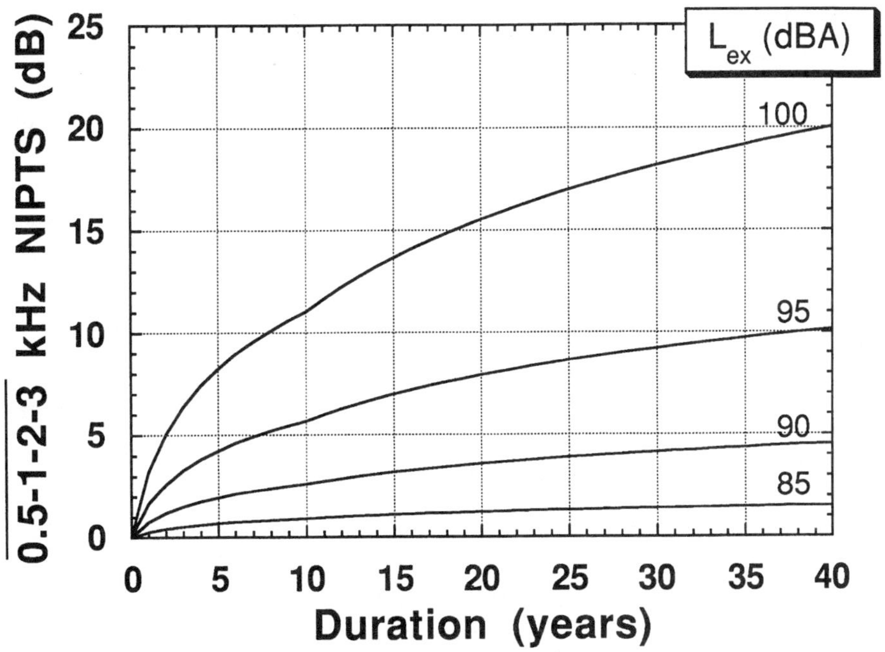

Figure 16.4 Speech frequency noise induced permanent threshold shift as a function of time for different exposure levels

loss than noise-induced loss. Statistical methods for estimating the relative contributions of aging, various periods of occupational exposure, non-occupational exposure and other otologic disorders have recently been developed.[9–11]

TREATMENT

No medical or surgical treatment has been shown to be helpful for NIHL, and none is usually offered. Some authorities have recommended treatments aimed at improving cochlear blood supply in cases of acoustic trauma, but NIHL remains a disorder without treatment of proven efficacy, for which most otolaryngologists offer only observation and serial evaluation.

NIHL can be palliated with hearing aids and assistive listening devices, such as amplified telephones. However, hearing aids rarely if ever compensate completely for sensorineural hearing loss; they can amplify inaudible sounds into the patient's audible range, but distortions usually persist, especially with loud sounds.

The lack of available treatment and the inadequacy of hearing aids combine to emphasize the role of prevention. A well-organized HCP can dramatically reduce the risk of NIHL and acoustic trauma for motivated workers.

MEDICAL SURVEILLANCE

The Hearing Conservation Amendment[3] requires HCPs for all workers with exposures exceeding 85 dBA-TWA. An HCP must include baseline and annual pure-tone air conduction audiometry (0.5, 1, 2, 3, 4, 6 kHz), with monitoring for standard threshold shifts (STS), defined as a 10-dB or greater change for the worse in either ear for the average of 2, 3 and 4 kHz. Audiometry within the HCP (usually on-site or in a mobile audiometric van) does not require bone conduction, speech or middle-ear tympanometric tests.

OSHA requires every HCP to be supervised by an otolaryngologist or other physician, or by an audiologist. If an STS occurs, the program supervisor may either elect to retest within 30 days or accept the STS without confirmation. An accepted or confirmed STS requires the initiation of earplugs or earmuff use for all workers in the 85–90 dBA-TWA zone who were not previously required to use HPDs, or the refitting of HPDs for those previously using them. Employee notification and counseling are required for all STSs. OSHA had announced that effective January 2002 that it would require all STSs to be recorded on Form 200 (proposed Form 300). The Department now proposes that the criteria for recording work-regulated hearing loss not be implemented for one year pending further investigation into the level of hearing loss that should be recorded as "significant" health condition.

The National Institute for Occupational Safety and Health (NIOSH) published revised Criteria for a Recommended Standard for occupational noise exposure in 1998,[12] reaffirming their support for an 85-dBA recommended exposure limit. With a 40-year lifetime exposure at 85-dBA NIOSH now estimates an excess risk of developing "material hearing impairment" of 8%—considerably lower than the 25% excess risk they calculate at the 90-dBA PEL currently enforced by OSHA. However, NIOSH changed their definition of material hearing impairment, to give more weight to high frequencies and less weight to low frequencies, prior to recalculating excess risk. When excess risk is calculated using the definition most widely used in the United States (that of the American Academy of Otolaryngology—Head and Neck Surgery), the corresponding figures are 3% for 85 dBA and 8% for 90 dBA.[5]

NIOSH now recommends an exchange rate of 3 dB for the calculation of TWA exposures to noise. The 5-dB exchange rate remains in the OSHA regulations, but the 3-dB exchange rate, while still controversial, has been increasingly supported by many groups national and international. NIOSH also recommends a new criterion for significant threshold shift: an increase of 15 dB in the hearing threshold level (HTL) at 500, 1000, 2000, 3000, 4000 or 6000 Hz in either ear, as determined by two consecutive audiometric tests. OHSA has not adopted this recommendation. In contrast with the 1972 criterion, the new NIOSH criterion no longer recommends age correction on individual audiograms. This practice is felt by some not to be scientifically valid; on the good side, it can decrease false-positives (STSs not due to noise), but on the bad side it can increase false negatives, losses in workers whose HTLs have increased because of occupational noise exposure. OSHA currently allows age correction as an option.

Referral is not required for STSs, but it is required when the program supervisor suspects a medical problem affecting the ears or believes that the audiometric data are questionable. The American Academy of Otolaryngology—Head and Neck Surgery[13] has published criteria for otologic referral from HCPs that have been adopted by many companies. These criteria recommend referral based on otologic symptoms, baseline audiometry (with special emphasis on substantial asymmetries, which often indicate serious otologic disease), and changes seen on periodic audiometry (including the lower frequencies, where communicatively significant or medically serious hearing losses are more likely to be seen).

PREVENTION

The OSHA Hearing Conservation Amendment requires more than periodic audiometry and medical or audiologic surveillance. After all, audiograms do not prevent hearing loss; they only measure it. Health professionals responsible for an HCP should carefully review the entire standard. The additional important elements of an HCP include:

1. Risk assessment using sound-level meters or dosimetry
2. Education and motivation of workers
3. Exposure reduction through engineering control, administrative controls, and use of HPDs
4. Record-keeping.

Education and motivation are important for two reasons. First, if the HCP includes the use of HPDs, there may be significant worker resistance and poor compliance unless they believe that hearing is really at risk, is worth saving, and can effectively be saved. Second, many—perhaps most—industrial workers have potentially hazardous non-occupational noise exposures; if they fail to reduce their recreational exposures, NIHL will continue to accrue despite reduction of on-the-job exposures to safe levels.

Engineering controls are obviously the most desirable way to reduce occupational noise exposures. If all workplaces can be brought below 85 dBA and no workers spend >8 hours per day on the job, the employer no longer needs an HCP at all. Lesser reductions are also valuable if they reduce the number of workers who must be enrolled in the HCP, the number who must use HPDs, or the HPD performance requirements. Cost is sometimes an issue, but a lot of newer equipment is specifically designed to reduce noise output.

Administrative controls involve reducing the individual's exposure to hazardous noise. Most experts in HCP design and administration have found that these changes are rarely feasible without unacceptable disruption of work routines.

In practice, then, most HCPs rely heavily on the use of HPDs to reduce exposures. Earplugs and earmuffs can both be effective if properly fitted and used. OSHA requires that employers offer their workers a variety of HPDs, recognizing that individuals and their jobs vary too much to permit specification of a single HPD for all. HPDs are rated by a *noise reduction rating* (NRR), which estimates the number of dB of attenuation obtained by proper use. However, most authorities believe that NRR numbers are usually too generous to apply to real-world situations. For conservative application, many recommend using half the NRR as a guide. For example, workers in a 100-dBA environment who use earplugs with a 30-dB NRR should be considered to be exposed to ~85 dBA (100 dB − 15 dB). Even this assumes that the plugs are actually being worn and have not been tampered with. Simple observation by supervisors is not enough to ensure proper use of HPDs. Earplugs are sometimes trimmed or perforated by uncooperative workers, and earmuffs can be stretched to reduce the spring force holding the muffs to the head, making them more comfortable but less protective. In either case, the worker would pass a cursory inspection.

Workers should be urged to use HPDs on and off the job whenever they are exposed to loud noises. For many workers, non-occupational exposures—especially shooting—are more hazardous than occupational exposures.

The speech interference and safety risks (difficulty in hearing warning signals) of loud noise have already been mentioned. Simply making the desired signal louder than the background noise; e.g. by shouting, can alleviate this problem. Once this has been done, the use of HPDs will not further degrade listening performance in normally hearing people. Both the signal and the noise will be reduced in intensity by the HPD, but as long as both are audible, the signal-to-noise ratio will be the same as without the HPD.

Persons with pre-existing high-frequency hearing loss may have poorer detection and discrimination of speech and warning signals when using HPDs. This occurs because the HPD may actually make high-frequency signals (including the high-frequency portions of speech) inaudible.

Intermittent noise poses special safety hazards because what is protective in noisy environments may disrupt communication in quiet ones. Workers who wear HPDs during quiet periods may have difficulty in hearing and understanding normal speech. HPDs can make the worker's own voice seem louder (try plugging your ears with your fingers while speaking at a constant level); thus, workers may speak less loudly than normal when wearing HPDs in quiet environments.

These risks may be managed by implementing special communication strategies, such as those often used in radio communication: use of a restricted message set to increase redundancy, spelling out words, and having important messages repeated back by the listener, to mention a few. Communication headsets are available that combine an earplug or earmuff, for attenuation of ambient noise, with a built-in radio receiver. In addition, active HPDs are available as earmuffs with built-in microphone–amplifier circuits. In quiet environments, these units amplify sound to overcome the attenuation of the earmuff, while in high-noise ones, the amplifier is automatically disabled. These units are particularly helpful for workers with hearing loss, who may need more amplification in quiet places, but still need protective attenuation in noisy ones. Provision of this type of equipment may constitute the sort of accommodation necessary to comply with the Americans with Disabilities Act (ADA). Earmuffs with active noise reduction circuits are also available. These headsets use noise cancellation technology, actively decrease low-frequency exposures, and are particularly helpful for pilots in general aviation (small planes) where cabin noise is dominated by low frequencies.

REFERENCES

1. Franks R. Number of workers exposed to occupational noise. *Seminars in Hearing* 1988; 9:287–97.
2. US Department of Labor—Occupational Safety and Health Administration. Occupational noise exposure: hearing conservation amendment. *Fed Register* 1981; 46:4078–179.
3. US Department of Labor—Occupational Safety and Health Administration. Occupational noise exposure: hearing conservation amendment final rule. *Fed Register* 1983; 48:9738–84.
4. US Department of Labor—Occupational Safety and Health Administration. Occupational safety and health standards, national consensus standards and established federal standards. *Fed Register* 1971; 36:10518.
5. Jayne R. Legal remedies for hearing loss. In: Dobie RA, ed. *Medical–legal evaluation of hearing loss*. 2nd edn. San Diego: Singular Thomson Learning, 2001.
6. Taylor W, Pearson J, Mair A, Burns W. Study of noise and hearing in jute weaving. *J Acoust Soc Am* 1965; 38:113–20.
7. International Organization for Standardization. *Acoustics: determination of occupational noise exposure and estimation of noise-induced hearing impairment*. ISO-1999. Geneva: International Organization for Standardization, 1990.
8. American College of Occupational Medicine—Noise and Hearing Conservation Committee. Occupational noise-induced hearing loss. *J Occup Med* 1989; 31:996.
9. Dobie RA. A method for allocation of hearing handicap. *Otolaryngol Head Neck Surg* 1990; 103(pt 1):733–9.
10. Dobie RA. The relative contributions of occupational noise and aging in individual cases of hearing loss. *Ear Hearing* 1992; 13:19–27.
11. Dobie RA. Diagnosis and allocation. In: Dobie RA, ed. *Medical–legal evaluation of hearing loss*. 2nd edn. San Diego, Singular Thomson Learning, 2001.
12. National Institute for Occupational Safety and Health. *Criteria for a recommended standard ... occupational noise exposure*. Revised criteria. DHHS (NIOSH) publication no. 98-126. Washington, DC: US Government Printing Office, 1998.
13. American Academy of Otolaryngology—Head and Neck Surgery—Medical Aspects of Noise Subcommittee. *Otologic referral criteria for occupational hearing conservation programs*. Washington, DC: American Academy of Otolaryngology—Head and Neck Surgery Foundation, 1983.

17

ELECTRICAL POWER AND ELECTRICAL INJURIES

Jeffrey R. Jones, M.P.H., M.S., C.I.H.*

Electrical hazards constitute a narrow but ubiquitous class of occupational physical hazards. Since the first report of death by electrocution in 1879, when a stage carpenter was killed after exposure to a 250-V AC generator, electric current has been responsible for a significant number of accidents that result in severe injury or death. Approximately 1500 cases of electrocution occur annually in the USA, including about 100 lightning-caused deaths.[1,2] Approximately 400 of these occur at work, and the vast majority could be prevented if currently mandated safeguards were used and proper procedures followed. Although we now know a great deal about how to prevent electrical injuries, electrical fatalities still account for a significant number of fatalities in the workplace. The majority of injuries and deaths occur while workers are performing duties they normally undertake in the course of their job, suggesting inadequate training or an underestimation of the risks inherent to working around electricity.[3] Because lightning injuries differ in significant ways from other electrical injuries, they will be discussed in a separate section at the conclusion of this chapter.

*Gary Pasternak contributed to the previous edition of this chapter.

ELECTROCUTION INJURIES

Occupational setting and risk factors

Electrical injury epidemiology The National Institute for Occupational Safety and Health (NIOSH) examined death certificates over a 13-year period to determine deaths at work from electrical causes (5348 total deaths, average of 411 deaths per year).[1] A subset of those deaths (224 cases) was examined in greater detail to determine actions

Physical and Biological Hazards of the Workplace, Second Edition, Edited by Peter H. Wald and Gregg M. Stave
ISBN 0-471-38647-2

and risk factors that underlay the electrocutions. The NIOSH found that electrocutions accounted for 7% of traumatic fatal injuries each year. Forty per cent of the deaths were in construction and 60% occurred in workers less than 35 years old. Ninety-nine per cent of the electrocutions were among men; 86% were Caucasian, probably reflecting the demographics of the workforce. Other industries with high rates included transportation/communication/public utilities 16%, manufacturing 12%, and agriculture/forestry/fishing 11%.

When the NIOSH examined some of the electrocutions in detail, several trends were noted. Of the total, 33% involved low voltages (less than 600 V) and 66% involved high voltage. Most of the high-voltage deaths involved voltages of 7200–13 800 (distribution voltages). Most of the low-voltage deaths involved 110–120 V. Utility linemen, who would be expected to have the most safety training, had the highest number of fatalities. In fifty-five per cent of the deaths evaluated, there had been a failure to use required personal protective equipment (gloves, sleeves, mats, blankets, etc.). The risk greatly increases when repairs are conducted under conditions of widespread damage to electrical transmission and distribution systems, as in the aftermath of a natural disaster such as a hurricane.[3] Laborers, who would be expected to have substantially less electrical safety training, had slightly fewer fatalities. In 35% of the incidents, there was no safety program or written safety procedures. When safety procedures did exist, there was a lack of enforcement or supervisory intervention. Supervisors were present at 53% of the incidents, and 17% of the victims were supervisors. Forty-one per cent of the victims had been on the job less than a year.

In the construction industry, electrocution is the second leading cause of death. Painters seem to be at particular risk, most likely from working near energized lines using ladders and other potentially conductive equipment.[4] The constantly changing construction work site exposes workers to temporary wiring, which may be substandard, and to harsh conditions that damage tool power cords and temporary power supplies. Portable arc-welding equipment is also responsible for numerous accidents, often as a result of improper grounding.

The greatest number of electrical injuries occurs in white men in the 25–34-year age group. Most injuries occur during the summer months, probably related to increased outdoor activity and the use of electrical equipment and machinery during this time of year. Increased sweating during the warmer months may also increase the severity of the electrical shock.[5] Alcohol and drugs have not been consistently found to be significant contributing factors in most work-related cases; however, widespread post-accident testing has not been routinely performed.[6,7]

Risk factors for electrocution The NIOSH[1] described five scenarios that accounted for the deaths they investigated:

1. Direct worker contact with an energized line (28%)
2. Direct worker contact with energized equipment (21%)
3. Boomed vehicle contact with an energized power line (18%)
4. Improperly installed or damaged equipment (17%), typically involving improper grounding
5. Conductive equipment, such as an aluminum ladder, contacting energized power lines (16%).

Risk factors for injury also include contact with moisture. When electric tools are handled with wet or perspiration-covered hands, resistance to electricity passing through the body is lowered. Risk is also heightened when workers wear clothing inadequate to protect them from electric arc and flash. Working on damp ground or where water is an integral part of a process, such as in irrigation, concrete work, and slaughtering, predisposes the victim to electric shock. Aerial power lines are a well-described hazard in the vicinity of irrigation pipes. In some agricultural regions, irrigation pipes have been the most common source of fatal human contact with electric lines.[8]

Exposure guidelines

The Occupational Safety and Health Administration (OSHA) has comprehensive standards covering electrical safety in both high and low voltages. The primary standards can be found in:

- Subpart S, 29 CFR 1910.302 through 1910.399, which applies to all electrical installations and related equipment regardless of when they were installed.
- Subpart K, 29 CFR 1926.402 through 1926.408 of the Construction Safety and Health Standards apply to installation of and work on electrical equipment, including equipment to supply power and light to job sites.

Training is an integral part of compliance with these regulations and is a critical element of any effective safety program.

Specific sections address, among other things: the need for lockout/tagout procedures (also covered in a specific lockout/tagout standard, CFR 1910.147—see Chapter 5); proper or assured grounding of equipment; use of protective equipment; work on overhead power lines; and use of portable extension cords and cables. Improper lockout/tagout is one of the most common citations issued by the OSHA. Protective equipment, such as rubber gloves and non-conductive sticks, must be adequate for the voltage levels encountered and must meet specific standards established by the American National Standards Institute (ANSI). Insulating gloves and other non-conductive personal protective equipment must be regularly inspected and periodically tested to ensure that they afford adequate protection. The OSHA requires the use of non-conductive ladders where the employee or the ladder could contact exposed electrical conductors; it further requires that all metal ladders be prominently marked with a warning label. Clothing must be fire-resistant when workers might be exposed to arc or flash.

Pathophysiology of injury

Basic concepts of electrical energy

In order to grasp the consequences of human exposure to electricity, it is important to understand certain basic concepts of electrical energy. The tissue damage produced by electricity is proportional to the intensity of the current that passes through the body, as stated in Ohm's law:

$$\text{resistance (ohms)} = \text{voltage}/\text{amperage}$$

Voltage (tension or potential) is the electromotive force of the system; amperage (intensity) is a measure of current flow per unit time; and an ohm is the unit used to measure the resistance to conduction of electricity.

Two types of current are encountered in the workplace. Voltage is supplied either as a continuous source (direct current, or DC), with electron flow moving in one direction, or as a cyclic source (alternating current, or AC), with a cyclic reversal of electron flow. Nearly all injuries are the result of contact with the much more common AC. The exception is lightning, which can be thought of as a high-voltage, DC shock.

AC is generated by large electromagnetic devices and generating stations, and it is transmitted many miles at very high voltages, typically greater than 100 000 V. Transformers reduce this voltage to 7620 V, the usual voltage feeding residential and industrial distribution lines. At the local level and at residences, the voltage is again decreased to 110–220 V for domestic use.[9] The frequency of most commercial AC current in the USA is 60 cycles per second or hertz (Hz); 50 Hz in much of the rest of the world. Sixty hertz means that the flow of electrons changes direction 60 times per second. The use of a 50–60-Hz frequency has evolved because it is optimal for the transmission and utilization of electricity and because it has advantages in terms of generation.

Pathophysiology of electrical injury

Whether or not death or injury occurs is directly related to the following related factors:

- type of current (AC versus DC)
- frequency of current
- voltage
- amperage
- duration of exposure
- exposure pathway through the body
- area of contact
- resistance at points of electrical contact and in tissues.

AC is significantly more dangerous than DC at similar voltages. AC can produce tetanic contractions that freeze the victim to the source of the current (an inability to "let go" of energized elements) and it can interfere with the respiratory and cardiovascular centers. Victims do not freeze to DC current as with AC, but they can be injured when hurled from the point of electrical contact by the shock or when burned.

The frequency of AC is of importance physiologically. The most critical frequencies are in the range 20–150 Hz. Human muscular tissue responds well to frequencies between 40 and 150 Hz. Sixty cycle household current lies directly within this range. If the shock spans the vulnerable period of the cardiac cycle, ventricular fibrillation may result. As the frequency increases beyond 150 Hz, human tissue response is decreased and the current is generally less dangerous.

Although the divisions are somewhat arbitrary, harmful effects from electrical current can be subdivided into the effects of high-voltage current, greater than 600 V, and the effects of low-voltage current, less than 600 V. Although most people recognize high voltage as dangerous, such as with overhead power lines, low-voltage current can prove equally deadly. Injury or death can occur with currents of very low voltage; instances of death have been reported with AC of 46 and 60 V. It has been claimed that any current greater than 25 V should be considered potentially lethal. Ventricular fibrillation is the most common cause of immediate death in high- and low-voltage injuries. Severe burns are more commonly associated with high-voltage currents.

The amount of current flow, expressed in amperes (A), is the single most important factor in injuries by electricity. Low-power alternating currents may be characterized by the thresholds of perception, "let-go", and ventricular fibrillation. The **threshold of perception** is the minimum current that causes any sensation in the human body. The **let-go threshold** is the maximum level of current flowing through an energized pathway at which a person taking hold of an element of the circuit is still able to release it. Currents greater than the let-go threshold are especially dangerous, because the body receiving the shock is unable to respond to it and break away. The International Electrotechnical Commission (IEC) has derived general values for the thresholds of perception and let-go. The threshold of perception is 0.5 milliamperes (mA) and the let-go threshold is 10 mA for signal frequencies in the range 15–100 Hz.

The threshold for an adverse effect is clearly related to the duration of exposure to the electrical energy. The IEC has derived values for the **threshold of ventricular fibrillation** by adapting results from animal experiments. For a 50/60-Hz signal, a current of 500 mA may cause fibrillation for a shock duration of 10 ms; likewise, 400 mA for 100 ms, 50 mA for 1 s, and 40 mA for greater than 3 s. The likelihood of ventricular fibrillation increases if the shock continues for a complete cardiac cycle.

The pathway of the current is critical to injury. Current passing through the head or chest is more likely to produce immediate death by affecting the central respiratory center, the respiratory muscles, or the heart. Arm-to-arm conduction can pass through the heart, and head-to-foot conduction can affect the central nervous system and the heart. Whereas a small current through the chest can be fatal, a large current passing through a single extremity may have little effect. The above values for the threshold of ventricular fibrillation are for current paths through the whole body, but fibrillation is only triggered by that level of current directly affecting the heart.[10] Non-fatal tissue injury is also determined by the current path. However, prediction of injuries from observation of the tissue path can be unreliable. Burn injuries have been shown to be more severe when the electrical energy traverses the long axis of the body.

The effect of electric current on the body depends on the resistance of the tissues involved. Intact dry skin has a relatively high resistance, but resistance falls dramatically where there is moisture, cuts or abrasions. A callused palm has a resistance of 1 000 000 ohms, dry skin may have 5000 ohms, and wet skin may have less than 1000 ohms. The resistance of individual tissues is thought to differ considerably and should play a role in differential injury to tissues. Resistance to current flow occurs in decreasing order through bone, fat, tendon, skin, muscle, blood vessels, and nerves. However, the theory that electric current always passes through the body along lines of least resistance may not be true; the path may largely depend on the voltage.

Controversy exists over the exact mechanism of tissue damage in electrical injury. The most widely held theory is that electrical energy is converted to thermal energy in tissues and causes damage by heating. The heat produced is a function of the current strength, duration of contact, and tissue resistance. However, it is possible that tissue damage is also partially a result of other undescribed electrical effects. Chronic neurologic changes seen in some patients are difficult to explain simply on the basis of thermal damage to tissue.[9]

Health effects Electric current can cause a spectrum of acute and chronic health effects and often results in multi-organ system manifestations.[9,11,12] Injury is most often the result of a direct effect on the heart, nervous system, skin, and deep tissues. Ventricular fibrillation, respiratory arrest and burns are the most common causes of immediate death. Burns may be minor, requiring minimal debridement, or may cause devastating destruction, requiring aggressive resuscitation and major limb amputation. Other health effects include vascular damage (such as thrombosis, rhabdomyolysis, and subsequent renal failure), fractures due to high temperature or tetanic contractions, cataracts, neuropsychological changes, degenerative neurologic syndromes, and associated trauma (such as from falling off a ladder).

Electrical injury resembles a crush injury more than a thermal burn. The extent of internal damage is often more severe than the cutaneous wound makes it appear. Small entry and exit wounds give little useful indication of the extent of underlying tissue damage and may not even be present in brief low-voltage injuries. In addition, there may be a direct electrical effect on the heart, nervous system, and skeletal muscles. Immediate death results from ventricular fibrillation, with or without asphyxia, from paralysis of the respiratory centers, or from prolonged tetany of respiratory muscles. Low-voltage injuries more commonly produce ventricular fibrillation rather than severe burns. Current passing through lower-resistance tissues causes necrosis of muscles, vessels, nerves, and subcutaneous tissues; it can also cause thrombosis and vascular insufficiency.

Cutaneous and deep-tissue effects Cutaneous injuries result from induced thermal burns, flame (so-called "flash") burns, and arc burns, a type of burn specific to electrical injuries and more common in high-voltage accidents. Arc injuries result from current coursing external to the body, jumping from its source to the victim, or arcing between different sites on an extremity. The flexor surface of the forearm is a frequent site of arcing injury. Arcing may cause significant flame burns from ignition of clothing or other materials. Frequently, an entrance and exit wound can be identified, but there may be severe deep-tissue damage with minimal cutaneous involvement. In severe burns, there may be extensive limb damage, requiring extensive debridement, fasciotomy, and limb amputation.

Cardiac effects The heart is particularly susceptible to electrical injury. A wide range of abnormalities may be seen, from arrhythmias to structural damage.[11] Ventricular fibrillation is a common cause of death in electric shocks. In non-fatal cases, arrhythmias are an important complication. The most common ECG abnormalities noted are sinus tachycardia and non-specific ST-T wave changes. Ventricular and atrial ectopy, atrial fibrillation, bundle branch blocks and ventricular tachycardia have all been reported. Generally, these abnormalities do not persist. Patients exposed to low voltages who are asymptomatic and initially have a normal ECG do not typically develop arrhythmias.[12]

High-voltage injury can cause myocardial necrosis, although the diagnosis can be difficult to make because of the absence of typical chest pain and ECG changes.[13] Evaluation of myocardial injury is based on measuring the cardiac fraction of the creatine kinase (CK) enzyme. Recent evidence suggests that a raised CK-MB (creative kinase MB isoenzyme) level, which is elevated in acute myocardial infarction, is not necessarily indicative of myocardial damage.[14] Myocardial infarction has been reported as a rare complication of electrical injury; it may be the result of electrically induced coronary vasospasm.[15] It has been suggested that extensive burns and entrance and exit wounds in the upper and lower parts of the body could predict patients at risk of myocardial involvement, and thus warrant intensive monitoring.[13]

Neurologic effects Acute central ner-vous system complications include respiratory center arrest or depression, seizures, mental status changes, coma, localized paresis, and amnesia. In mild cases, the patient may experience headaches, irritability, dizziness, and trouble in concentrating. These symptoms usually resolve in a few days. Peripheral nerve injuries are seen most often with extensive limb burns. Peripheral neuropathy may be seen following exposure to high current loads, even in the absence of extensive burns. Spinal cord damage is the most common permanent neurologic problem. It may cause progressive muscular atrophy or illness simulating amyotrophic lateral sclerosis or transverse myelitis. Symptoms characteristically appear after a latency period, with no (or minimal) neurologic symptoms in the acute stage. Some investigators have described a stereotyped generalized cerebral dysfunction leading to depression, divorce, and unemployment, as well as a high incidence of atypical seizures.[16] Complex regional pain syndrome types I and II (formerly RSD and causalgia, respectively) have been reported following electrical injury.[11]

Renal effects Electrical injuries produce a higher incidence of renal damage than other burns. Factors include shock, direct damage to the kidneys by high-voltage current, and the release of toxic products from the breakdown of damaged muscles. Myoglobinuria is commonly present and is proportional to the amount of muscle injury. Timely administration of fluids and diuretics is important in the prevention of pigment-induced acute renal failure.[17] The development of acute renal failure in electrical injury correlates poorly with the extent of surface burns, given that the volume of tissue destroyed is often much greater than observed in a surface burn.[18] The therapeutic implication is that the formula for estimating fluid replacement in surface burns may seriously underestimate the fluid required in patients with electrical burns.

Vascular effects Vascular complications have included immediate and delayed major vessel hemorrhage, arterial thrombosis, abdominal aortic aneurysms, and deep-vein thrombosis. This large- and small-tissue damage, typically leading to vascular insufficiency, may be responsible for the tissue damage in electrical injury that is not immediately apparent.[9]

Other organ systems Musculoskeletal injuries are often initially overlooked. They may include multiple fractures and dislocations. Shoulder and scapular fractures are frequently reported. Delayed diagnosis of femoral neck fractures has been reported. Injury may result from falls, being hurled from the electrical source, or simply abrupt muscle contraction. Cataracts, conjunctival burns and corneal burns can occur. Cataracts usually form 2–6 months after the shock, but they can appear immediately or many years later. Intra-abdominal injury should be expected in patients having burns of the abdominal wall. Stress ulcers of the duodenum (Curling's ulcer) can occur following severe burns.[11] Where arcing or burning has occurred, especially in vaults and other enclosed spaces, inhalation injury should be considered, since metals, dielectric fluids and a variety of other materials may have been vaporized.

Treatment

Treatment in the field involves rapidly and safely removing the victim from the source of the current, immediate and prolonged cardiopulmonary resuscitation (CPR), attention to other life-threatening injuries such as cervical spine injury, and initiation of advanced cardiac life support, if indicated. A rescuer can inadvertently become an electrocution victim. Do not touch victims who could still be in contact with the power source. Before touching the victim, the rescuer should attempt to turn off the electrical source. The victim can be separated from the power by using non-conductive materials such as rubber, a wooden tool handle, a mat, or heavy blankets. If it is not known whether the victim was thrown or fell, a cervical collar should be applied. Efforts should be made to revive victims who appear dead, since there is a good chance that some of them will respond to prolonged resuscitation attempts.[9]

Especially in high-voltage injuries, a thorough physical examination should be made to identify burns and other abnormalities, which will give clues to underlying trauma. Assume that a patient with burns is likely to have suffered physical trauma. In general, the greater the amount of energy that was absorbed, the greater the underlying tissue damage. A history of the events leading to and surrounding the injury can be useful in guiding the evaluation.

Ideally, fluid resuscitation should begin in the field, especially in high-voltage injuries, which can result in significant volume depletion secondary to exudation and sequestration of fluids in burned and damaged areas. Fluid requirements frequently are much greater than those recommended by formulas that predict fluid needs from the area of cutaneous burns. Because deep-tissue damage can occur with limited surface burns, fluid should be administered in sufficient volume to maintain a urine output of 50–100 ml/hour to prevent renal insufficiency from myoglobin deposition. Mannitol may help to ensure an adequate urine flow. Tetanus immunization should be administered if indicated.

There are no clear-cut criteria for the hospital admission of less severely injured patients. Inpatient observation is advisable for patients who show evidence of cardiac dysfunction, symptoms of neurologic impairment, presence of significant surface burns, suspicion of deep-tissue damage, or laboratory evidence of acidosis or myoglobinuria. The evolution of tissue injury and vascular necrosis is usually complete within 8–10 days post-exposure. A criterion for cardiac monitoring after electric injury has been suggested by Fish[9] following a review of the available literature:

1. Loss of consciousness
2. Cardiac dysrhythmias
3. Abnormal 12-lead electrocardiogram
4. Abnormality on mental status or physical examination
5. Burns or tissue damage that would be expected to cause hemodynamic instability or electrolyte imbalance.

Prevention

Occupational electrical injuries are generally preventable. Each electrical injury or fatality should be viewed as a sentinel health event. The incident should prompt an analysis of the work site with the intent of preventing any further injuries. Any injury suggests the need for a job-specific electrical safety analysis and provides an opportunity for preventive intervention in the workplace. The majority of these injuries occur either from lack of education concerning the specific hazards of the work or from failure to follow safe work practices.

Primary prevention of electrical injuries is the ultimate goal. Prevention can be accomplished through the use of engineering and administrative controls, personal protective equipment, and training. Lockout/tagout, the use of appropriate personal protective equipment, and an understanding of the hazards are key to injury prevention. Lockout/tagout and other administrative controls require commitment from management and workers and should be part of a written safety program. Their effectiveness depends on rigorous implementation and enforcement in the workplace. An example of an administrative control would be a scheduled preventive maintenance program for all power tools and electrical cords.

Training Persons exposed to electrical risk fall into two general categories: (1) those with training and experience in electrical work as a craft, such as electricians and linemen; and (2) those not engaged in "electrical work" who nonetheless use or work near equipment with potential electrical hazards. Injury prevention requires both education regarding hazards and attention to safe work practices. Many electrical injuries occur among non-utility/electrical workers, and these employees should be specifically targeted for primary prevention activities. Worksite electrical safety education should focus on the recognition of potential electrical hazards and how to avoid exposure to live electrical circuits.[17]

Proactive safety programs are needed in all occupations that have a high risk of electrical injury. The NIOSH and OSHA have produced numerous documents that contain elements of effective safety training, including electrical safety. Some state OSHA programs have stricter and/or more detailed requirements than the Federal program. In general, comprehensive safety programs should include written rules and safe work procedures for dealing with electrical hazards.[1,20]

Workers must be educated about the potential dangers of low voltage. They should be instructed to always ground hand tools properly, especially portable powered hand tools. Extension cords used to supply power to portable tools need particular attention, especially at temporary work sites, because grounding can fail. Using battery-powered or double-insulated power tools plugged into ground fault circuit interrupters (GFCIs) can help prevent electrical shock, even under adverse conditions.

Engineering controls Many options are available to reduce electrical hazards. The single most effective is to de-energize and lock out any active circuit or equipment that could be contacted during the work activity. Failing that, risk can be reduced through measures such as enhancement with visual markers, sleeving to insulate lines in high-risk areas, using ladders made of non-conducting materials, and using procedures to stabilize and prevent equipment from moving into power lines. Lookout workers can help guide aerial equipment. A minimum of 10 ft of clearance is required between operating aerial equipment and a power line.

Many accidents occur during the use of electrically powered machinery or portable powered hand tools. Repairing damaged power cords, maintaining proper grounding and using GFCIs can prevent the majority of these injuries. GFCI-protected circuits are required at construction sites, but can have much broader application, even at fixed locations. GFCIs represent a simple engineering control that could save dozens of lives each year.

LIGHTNING INJURIES

According to the National Weather Service, 100 000 thunderstorms annually produce approximately 30 million lightning strikes in the USA. These kill, on average, 82 people and injure 1000–2000 people per year. It is estimated that 25% of these cases are occupationally related.[21,22]

Occupational and geographic setting

The Centers for Disease Control summarized lightning-caused deaths from 1980 to 1995.[21] During this period, there were 1318 deaths (82 deaths per year, range 53–100 deaths). Occupationally, many of these accidents occurred in the agricultural and construction industries, but they also occurred in other jobs where people work outdoors, such as wild land firefighting, sailing, or other activities involving work on the open water. Most of the victims are injured during the summer months, when thunderstorms are most frequent. The states with the greatest number of deaths are Florida and Texas, but the highest death rates are in New Mexico, Arizona, Arkansas, and Mississippi.

Occupation may pose special risks to individuals. Working (and thus getting paid) may be an incentive to stay exposed. Workers may be forced to continue working when thunderstorms are near for fear of losing their jobs, thus prolonging their potential exposure to lightning. The hazard of lightning is often not realized by those exposed: lightning has struck 10 miles away from the rain of a thunderstorm.

Pathophysiology of injury

Lightning kills 30% of its victims, and 74% of survivors experience a permanent disability.[21] Sixty-three per cent of deaths occur within an hour of injury. The most common cause of death is immediate cardiopulmonary arrest. Lightning is dangerous due to high voltage, heat generation, and explosive force. Light-

Table 17.1 Lightning versus high-voltage electrical injury.[23]

Factor	Lightning injury	High-voltage injury
Energy level	Very high voltage and amperage	Lower
Duration of exposure	Brief instantaneous	Prolonged
Pathway	Flash over	Deep, internal
Burns	Superficial	Deep, major injury
Cardiac	Asystole more common	Ventricular fibrillation
Renal	Rare myoglobinuria	Myoglobinuric renal failure common
Fasciotomy and amputation	Very rare	Common, may be extensive
Blunt trauma	Explosive thunder effect	Falls, being thrown

Source: Adapted from Cooper M. Lightning injuries. In: Auerbach PS, Geehr EC, eds. *Management of wilderness and environmental emergencies.* 2nd ed. St. Louis: Mosby-Year Book, 1989: 171–93.

ning may also injure indirectly by starting forest and house fires, causing explosions, or by felling objects such as trees.

There are significant differences between injuries from electric current and injuries from lightning (Table 17.1).[23] The factor that seems to be most important in distinguishing lightning from electric current injuries is the duration of exposure to the current. Exposure to lightning current is nearly instantaneous, so that prolonged contact does not occur. The energy generally travels superficially over the surface of the body. Distinct entry and exit wounds are rare, and deep burns are infrequent. The explosive force of the lightning strike may cause significant blunt trauma if the victim is thrown by a direct strike or by the shock wave created by the flash. Clothing and shoes may be literally blown from the body. Injuries are classified from minor to severe. As with electrical injuries, several organ systems may be affected, and both acute and chronic effects are seen.

With minor injury, patients experience confusion and temporary amnesia. They rarely have significant burns but may complain of paresthesias and muscular pain. Patients usually recover completely. A ruptured tympanic membrane is a frequent finding, due to the explosive force of the lightning shock wave.

With moderate injury, patients may be disoriented, combative, or unconscious. Motor paralysis, more often of the lower extremities, may be seen, along with mottling of the skin and diminished or absent pulses due to arterial spasm and sympathetic instability. Hypotension should be ruled out and, if found, should prompt a search for inapparent blunt trauma and fractures. Victims may have experienced temporary cardiopulmonary arrest at the time of the strike. Respiratory arrest may be prolonged and lead to cardiac arrest from hypoxia. Seizures may also occur. Ruptured tympanic membranes are commonly found, and minor burns may become apparent after a few hours. Patients generally improve, although they may experience chronic symptoms such as sleep disturbances, weakness, paresthesias, and psychomotor abnormalities. As with electrical injuries, rare cases of spinal paralysis have been reported.

Severely injured patients may present in cardiac arrest, with either asystole or ventricular fibrillation. In fact, victims are unlikely to die unless cardiopulmonary arrest occurred at the time of the lightning strike.[24] Because

Table 17.2 Factors in the physical examination and work-up suggesting a high-risk electrical injury.

1. Factors suggesting significant effects on the patient
   A. Evidence of inhalation injury
   B. Dysrhythmias
   C. Confusion
   D. Abnormal physical examination (neurologic, orthopedic, vascular)
   E. Abnormal laboratory examination (UA, EKG, CK, CK-MB or troponins)
   F. Significant burn or other condition requiring treatment
2. Signs of deep-tissue injuries, especially in extremities
   A. Edema
   B. Ischemic changes
   C. Sensory or motor loss
   D. Full-thickness skin injury along possible path of current without associated flame burns
   E. Persistent flexion deformity

UA = Urinalysis; EKG = Electrocardiogram; CK = Creatine kinase; CK-MB = creatine kinase MB isoenzyme

Source: Adapted from Fish [12]

Table 17.3 Recommendations for preventing injuries from lightning strike.

1. During a storm, take shelter inside a home, large building, or vehicle
2. If outside and unable to reach shelter, do not stand under or near a tall tree or other structure in an open area. Go into a ravine or gully. Assume a squatting posture on the balls of your feet with your head down and your hands over your ears (the lightning crouch) to minimize the chance of a lightning strike. Do not lie flat
3. Get out of and away from open water
4. Get away from metal equipment or objects such as tractors, antennas, drainpipes, and metal stairs
5. Put down any objects that might conduct electricity (shovels, rakes, ladders, etc.)
6. In forested areas, seek shelter in a low-lying area under a thick growth of small trees
7. When indoors, avoid using electrical appliances or telephones

Source: Adapted from Center for Disease Control[21]

persons struck by lightning have a better chance of survival than persons suffering cardiopulmonary arrest from other causes, resuscitation should be started immediately.[12,21] Resuscitation may not be successful if there was a significant delay in initiating CPR. Direct brain damage may have occurred. Findings of blunt trauma suggest a direct strike. Long-term sequelae in survivors also include visual and hearing deficits, due most often to damaged tympanic membranes and cataracts.

Treatment

Initial treatment includes rapid attention to CPR and support, if necessary. In general, fluids should be restricted unless there is evidence of hypotension. Although minor burns may be present, vigorous fluid therapy and mannitol diuresis are not indicated unless myoglobinuria is found. Overhydration with resultant cerebral edema has probably killed more lightning victims than pigment-induced renal failure.[23] Fasciotomy is rarely needed in lightning injuries, since most burns are superficial. Fish[12] summarized factors suggesting a high-risk electrical injury (Table 17.2).

Prevention

Workers involved in activities in areas where lightning strikes are possible should be familiar with the preventive measures recommended by the Centers for Disease[21] (Table 17.3).

ACKNOWLEDGMENT

The author wishes to thank Gary Pasternak, the author of this chapter in the first edition, for providing a strong foundation on which to build.

REFERENCES

1. National Institute for Occupational Safety and Health. *Worker deaths by electrocution: a summary of NIOSH surveillance and investigative findings.* DHHS (NIOSH) Pub. No. 98–118. Cincinnati: NIOSH, 1998.
2. Mellen PE, Weed VW, Kao G. Electrocution: a review of 155 cases with emphasis on human factors. *J Forens Sci* 1992; 37:1016–22.
3. Harvey-Sutton PL, Driscoll TR, Frommer MS, Harrison JE. Work-related electrical fatalities in Australia, 1982–1984. *Scand J Work Environ Health* 1992; 18:293–7.
4. Centers for Disease Control. Up-date: work-related electrocutions associated with Hurricane Hugo—Puerto Rico. *MMWR* 1989; 38:718–20.
5. Suruda AJ. Work-related deaths in construction painting. *Scand J Work Environ Health* 1992; 18:30–3.

6. Baker SP, Samkoff JS, Fisher RS, Van Buren CB. Fatal occupational injuries. *JAMA* 1982; 248:692–7
7. Berkelman RL, Herndon JL, Callaway JL, Stivers R, et al. Fatal injuries and alchohol *Am J Prev Med* 1985;1:21–8.
8. Helgerson SD. Farm workers electrocuted when irrigation pipes contact power lines. *Public Health Rep* 1985; 100:325–8.
9. Fish R. Electric injury, Part I: Treatment priorities, subtle diagnostic factors, and burns. *J Emerg Med* 1999; 17(6):977–83
10. Robinson MN, Brooks CG, Renshaw GD. Electric shock devices and their effects on the human body. *Med Sci Law* 1990; 30:285–300.
11. Fish R. Electric injury Part II: Specific injuries. *J Emerg Med* 2000; 18(1):27–34.
12. Fish R. Electric injury Part III: Cardiac monitoring indications, the pregnant patient, and lightning. *J Emerg Med* 2000; 18(2):181–7.
13. Chandra NC, Siu CC, Munster AM. Clinical predictors of myocardial damage after high voltage electrical injury. *Crit Care Med* 1990; 18:293–7.
14. McBride JW, Labrosse KR, McCoy HG, et al. Is serum creatine kinase-MB in electrically injured patients predictive of myocardial injury? *JAMA* 1986; 255:764.
15. Xenopoulos N, Movahed A, Hudson P, Reeves WC. Myocardial injury in electrocution. *Am Heart J* 1991; 122:1481–4.
16. Hooshmand H, Radfar F, Beckner E. The neurophysiological aspects of electrical injuries. *Clin Electroencephalogr* 1989; 20: 111–20.
17. Gupta KL, Kumar R, Sekhar S, Sakhuja V, Chugh KS. Myoglobinuric acute renal failure following electrical injury. *Renal Failure* 1991; 13: 23–5.
18. DiVincenti FC, Moncrief JA, Pruitt BA. Electrical injuries: a review of 65 cases. *J Trauma* 1969; 9:497–507.
19. Jones JE, Armstrong CW, Woolard CD, Miller GB. Fatal occupational electrical injuries in Virginia. *J Occup Med* 1991; 33:57–63.
20. National Institute of Occupational Safety and Health. *Preventing falls and electrocutions during tree trimming*. DHHS (NIOSH) publication No. 92–106. Cincinnati: NIOSH, 1992.
21. Centers for Disease Control. Lightning-caused deaths 1980–1995. *MMWR* 1998; 47:391.
22. Duclos PJ, Sanderson LM, Klontz KC. Lightning-related mortality and morbidity in Florida. *Public Health Rep* 1990; 105:276.
23. Cooper MA. Lightning injuries. In: Auerbach PS, Geehr EC, eds. *Management of wilderness and environmental emergencies*. 2nd edn. St Louis: Mosby-Year Book, 1989:173–93.
24. Cooper MA. Lightning injuries: prognostic signs for death. *Ann Emerg Med* 1980; 9:134–8.

II

BIOLOGICAL HAZARDS

18

GENERAL PRINCIPLES OF MICROBIOLOGY AND INFECTIOUS DISEASE

Jerry J. Tulis, Ph.D., and Woodhall Stopford, M.D., M.S.P.H.

Common occupational and environmental biological hazards include microorganisms (viruses, rickettsia, chlamydiae, bacteria, fungi, and parasites), allergens of biological origin (e.g. the aeroallergenic fungi and animal dander), and the byproducts of microbial growth (e.g. the endotoxins and mycotoxins). Because of their invisible and frequently undetectable nature, biohazards are considered "silent hazards". Among the occupations associated with biohazards are the healthcare industry, agriculture, science and technology, livestock management, fish and shellfish processing, forestry, waste management, and recreation management.

Microorganisms are found everywhere in nature. They inhabit all environmental niches from the polar icecap to the tropics and deserts. Microorganisms are intimately associated with all living species. Many forms are present as the normal flora of the skin and body orifices, whereas others may cause disease. Most of the microorganisms found on earth, including most of the human and animal pathogens, belong to the mesophilic species, which survive best at ambient temperatures of 20–400°C. Microorganisms that require elevated temperatures for growth belong to the thermophilic species, those that thrive at lower temperatures belong to the psychrophilic species.

The human host is constantly exposed by a variety of routes to biological materials, including living microbes and their products. Humans are also the source of many microorganisms through the shedding process, whereby thousands of organisms are released continuously from the skin, mucous membranes, and body orifices. The great majority of these microorganisms are harmless

Physical and Biological Hazards of the Workplace, Second Edition, Edited by Peter H. Wald and Gregg M. Stave
ISBN 0-471-38647-2

to us and to our ecosystem. Most are saprophytic organisms that live on inanimate substrates and represent normal human and environmental flora. Nevertheless, pathogenic microorganisms have had a tremendous impact on humankind throughout history, with devastating pandemics of smallpox, yellow fever, influenza, AIDS, plague, tuberculosis, and malaria.

There is also a potential for new infectious diseases of occupational significance to emerge, such as hantaviruses. Such outbreaks could result from environmental changes, microbial adaptation, or population movements. Awareness of workplace biohazards is critical to the prevention of occupational disease; complacency can result in serious and life-threatening illness.

Microorganisms found in the workplace range from extremely small viruses and single-celled bacteria to multicellular fungi and parasites. To understand how these organisms produce disease, it is necessary to know how they function, how disease is transmitted, and what factors influence whether infection will develop.

ETIOLOGY OF DISEASE

When Robert Koch discovered the causative microbe of anthrax in 1876, he formulated criteria to establish that a specific microorganism was the cause of a clinically discernible disease. These criteria, known as Koch's postulates, stipulate that:

1. The specific organism must be found in diseased animals and not in healthy animals.
2. The specific organisms must be isolated from the diseased animal and grown in pure culture.
3. The identical disease must be produced upon inoculation of healthy susceptible animals with a pure culture of the originally isolated organism.
4. The identical organism must be isolated from the experimentally infected animal.

Microorganisms that can produce disease are classified into the categories of viruses, rickettsia, chlamydia, bacteria, fungi, and parasites.

Viruses

Viruses represent the smallest etiologic agent of human diseases (measuring 20–300 nm). They are responsible for the great majority of human infections, especially through inhalation. The classification of viruses depends on the following criteria: (1) morphology; (2) the presence of envelopes surrounding the viral capsid; (3) the type of genetic material (RNA or DNA); (4) organs and tissues preferentially infected; and (5) the nature of disease caused. All viruses are obligate intracellular parasites. Viruses cannot multiply outside the host cell. Their survival as naked particles in the environment is limited, ranging from several hours to a few weeks. Viruses can only infect cells where appropriate receptors are present. When a virus infects a host cell, it utilizes the metabolic machinery of that cell to replicate itself.

Viruses may contain either deoxyribonucleic acid (DNA) or ribonucleic acid (RNA) as their genetic material. If the virus contains DNA, this nucleic acid is a recognizable substrate for the host cell's DNA polymerase enzymes. These enzymes translate the DNA into messenger RNA (mRNA). The mRNA is then transcribed by the host cell's ribosomes into viral proteins. By contrast, the RNA in RNA viruses is not recognized by the host cell's ribosomes. These viruses, classified as retroviruses, contain a unique enzyme known as reverse transcriptase. This enzyme allows for transcription of infecting viral RNA to host cell DNA within infected cells. Once this DNA is created, the genetically programmed viral replication proceeds in a manner similar to that for DNA viruses. The replication process results in the production of many additional copies of the infecting virus. These are subsequently liberated into the surrounding milieu to infect other target cells. Some viruses, termed cytopathogenic viruses, cause the destruction of the host cell during the replication process.

Rickettsia

Rickettsiae are primarily intracellular parasites, although they are considerably more complex than the viruses. They are coccobacillary in morphology, contain both RNA and DNA, and resemble the Gram-negative bacteria. As intracellular parasites, the rickettsiae multiply through the process of binary fission. They are completely independent of host cell metabolic activity. Most rickettsial agents are transmitted to humans through arthropod vectors, e.g. the transmission of Rocky Mountain spotted fever by the *Dermacentor* tick. However, Q fever, caused by *Coxiella burnetii*, is readily transmitted through contaminated aerosols and has been responsible for numerous laboratory-acquired infections. Outdoor sites represent the greatest risk of rickettsial infection for workers, including those employed in agriculture, forestry, and construction.

Chlamydiae

Chlamydiae are usually classified as belonging to the domain of the bacteria, although they are separated by many authors for purposes of discussion. Like viruses, chlamydiae are obligate intracellular parasites; however, they differ from viruses by being susceptible to antibiotics. Among the vertebrate host range of the chlamydiae are birds, mammals, and humans. The leading sexually transmitted disease in the USA is trachoma, caused by *Chlamydia trachomatis*; worldwide, it is the primary cause of human blindness. The most important occupational disease caused by the chlamydiae is psittacosis, a zoonotic disease caused by *Chlamydia psittaci*. The primary reservoirs of *C. psittaci* are the psittacine birds (e.g. parrots) and domestic chickens and turkeys. Psittacosis is readily transmitted by aerosol, and the organisms remain stable in dried form for extended periods.

Bacteria

Bacteria are single-celled organisms. They have semirigid cell walls and a cell nucleus containing DNA that is not membrane-bound. They reproduce through the process of binary fission. Bacteria include thousands of species, encompassing numerous genera. In addition to genus and species classification, bacteria can be categorized in several ways.

Morphologically, bacteria are categorized as cocci, bacilli, and spirilla. The cocci, which are round, spheroidal or ovoid in shape, are found as single cells, doublets, tetrads, clusters, and chains. The bacilli, or rod-shaped bacteria, occur as coccobacilli, square-ended bacilli, round-ended bacilli, club-shaped bacilli, and fusiform bacilli. The spirilla include the corkscrew and comma-shaped organisms (e.g. the vibrios and spirochetes).

Bacteria may also be differentiated based on the results of a commonly used differential specimen stain, the Gram stain. Organisms detected microscopically with this technique are described as Gram-positive or Gram-negative. Approximately 67% of the cocci and 50% of the bacilli are Gram-positive. All spirilla are Gram-negative. The mycobacteria (organisms that cause tuberculosis and atypical tuberculosis) have a waxy envelope and do not stain readily with the Gram stain. To detect mycobacteria, a special acid-fast staining procedure is required.

Bacteria may also be differentiated based on biochemical characteristics. Organisms may exhibit characteristic patterns of sugar fermentation and metabolic product formation. They may also require specific substrates or nutrients for growth.

Other techniques used to classify bacteria include bacteriophage typing (classification of bacteria based on susceptibility to different strains or types of species-specific viruses) and bacterial chromosome analysis using restriction endonucleases.

All Gram-negative bacteria possess a lipopolysaccharide component of the cell wall that displays toxic properties. This is referred to as endotoxin. Although the potencies of endotoxins produced by different bacterial species vary, all endotoxins possess pyrogenic (fever-inducing) properties. In the occupational setting, exposure to aerosols contaminated with polluted water, animal feces or soil can result in human exposure to endotoxins.

Fungi

Fungi are composed of molds and yeasts; some species exhibit dimorphic properties, growing as either molds or yeasts depending on the substrate and temperature. Although thousands of fungal species are found in nature, <100 species are responsible for all human and animal diseases, and less than a dozen species are responsible for the majority of human mycotic infections. Fungal diseases are classified as mycoses, mycotoxicoses, and allergies.

The mycoses can be localized or systemic. The occupational mycoses transmitted by the respiratory route include blastomycosis, cryptococcosis, histoplasmosis, and coccidioidomycosis, which has been implicated as an etiologic agent in laboratory-acquired infections. All of these fungi are natural inhabitants of the soil and become aerosolized when the soil is disturbed, as occurs during construction, demolition, and other earth-moving activities. *Histoplasma capsulatum* and *Cryptococcus neoformans*, the causative agents of histoplasmosis and cryptococcosis, have a predilection for growth in soils contaminated with bird droppings; they are often found in the vicinity of poultry houses and bird-roosting areas.

Mycotoxicoses are intoxications resulting from exposure to fungal toxins (mycotoxins). Although numerous mycotoxins have been identified, the best studied are the aflatoxins, elaborated by species of *Aspergillus*. Besides possessing mutagenic, carcinogenic and teratogenic properties, the aflatoxins are acute toxins affecting various body organs. The substrates for these molds are extensive and include most agricultural products, e.g. corn and peanuts in the USA. Rigid standards have been imposed by the US Department of Agriculture to control aflatoxin levels of these commodities.

Fungal allergies are represented by clinical cases of allergic rhinitis, hypersensitivity pneumonitis, and asthma. The allergic manifestation is not an infection per se, and fungal viability is not required to induce allergic disease, since mycelial fragments, dead spores and other fungal debris can elicit a host response. Common aeroallergenic fungal genera include *Cladosporium*, *Penicillium*, and *Aspergillus*, among others. Clinical and laboratory findings indicate that building-related illness is sometimes due to the inhalation of mold-contaminated air, but so-called sick-building syndrome remains a poorly defined illness of unknown etiology.

Parasites

Parasites may be involved as etiologic agents of occupational infections as a result of travel to, or work in, endemic areas. Illnesses such as giardiasis and amebic infections can result from workplace exposure to contaminated water. Human parasites can be classified as the protozoa and the helminths. The protozoa are unicellular organisms composed of the amoebae, the ciliated protozoa, the flagellated protozoa, the malarial parasites, *Toxoplasma gondii* and *Pneumocystis carinii*. The helminths are multicellular parasitic worms.

Parasites may have complex life cycles that involve sexual and asexual reproductive states. Parasites may also infect both intermediate and definitive hosts. For different organisms, humans may serve the role as intermediate host, definitive host, or both.

TRANSMISSIBILITY OF DISEASE

The recognized routes of human exposure to etiologic agents of disease include (1) the respiratory route, (2) the oral route, (3) the contact route, (4) the parenteral route, and (5) transmission through arthropod vectors.

Etiologic agents of occupational disease are most frequently transmitted through the respiratory route. Aerosolized particles (bioaerosols) composed of infectious or airborne allergenic (aeroallergenic) agents are difficult to detect or control. Exposure may result in sporadic and multiperson exposure in indoor and outdoor environments. The great majority of documented laboratory-acquired infections have resulted from apparent respiratory exposure, including disease agents not normally transmitted through aerosols (e.g. Rocky Mountain spotted fever and rabies). Bioaerosol particles, which measure 0.5–5.0 microns in aerodynamic size, can readily penetrate deep into the respiratory tract, reaching the alveolar spaces.

Occupational exposure to infectious agents through the oral route occurs by the following mechanisms: sprays and splatters, ingestion while mouth pipetting, consumption of contaminated foods, or touching the nose or mouth with contaminated hands. Since mouth pipetting has been prohibited in most laboratories, infection by accidental ingestion has been reduced significantly. Occupational infection by enteric pathogens, hepatitis A virus (HAV), the listeriae and other agents continues to occur.

Contact exposure in the workplace has resulted in a variety of occupational infections, including tularemia, Newcastle disease, hepatitis B virus (HBV), human immunodeficiency virus (HIV), brucellosis, anthrax, glanders, erysipeloid, herpes, and leptospirosis. Transmission of disease organisms has occurred by contact with contaminated surfaces or fomites, exposure of mucous membranes and skin surfaces (including non-intact skin), and exposure to spatters and sprays of infectious agents. Routine handwashing practices and the use of gloves and other protective apparel can significantly reduce the spread of infectious organisms by contact.

Parenteral exposure to infectious organisms in the workplace results primarily from accidental needlestick or other penetrating trauma, such as skin puncture with sharp instruments or animal bites and scratches. Most workplace infections with HBV or HIV are the result of accidental needlestick or sharps injury.

Arthropods may serve as vectors in transmitting occupational infections. Examples of vector-borne diseases include the mosquito in malaria and the encephilitides, the flea in plague and tularemia, and the tick in Rocky Mountain spotted fever and Lyme disease. This route of infection is primarily associated with outdoor work, including forestry management and lumbering, agriculture, construction, and recreation management. Employees involved in outdoor activities and fieldwork need to be cognizant of vector-borne diseases, especially in endemic areas.

INFECTIVITY OF DISEASE

Following exposure to etiologic agents of disease, the infective process depends on a number of factors—namely, the resistance or susceptibility of the host, the exposure route and dose, and the virulence of the specific pathogen. Although host susceptibility is difficult to document, certain factors are recognized as being contributory, including age, race, gender, health status, underlying disease, pregnancy, vaccination status, and immunosuppression. The infectious dose varies significantly for different diseases, ranging from a single cell to millions of organisms. Moreover, the infectious dose differs by many orders of magnitude when exposure to the identical disease agent occurs through different routes of exposure. Exposure of non-human primates to *Francisella tularensis*, the causative agent of tularemia, results in disease with respiratory exposure to ~10 organisms. In excess of 100 000 are required to initiate disease when exposure occurs by ingestion. Additionally, although the infective dose of the bloodborne pathogens HBV and HIV are unknown, the significantly higher concentration of HBV in body fluids may be associated with the documented higher workplace infection rate of HBV. From the viewpoint of risk management, those pathogens that possess low infectious doses (e.g. tuberculosis, which has an infective dose of one tubercle bacillus) require a considerably higher level of infection control practices in the workplace.

Following exposure to disease-causing organisms, the infective process may lead to clinical, subclinical or asymptomatic disease, which occurs after an incubation period of several days to several months. Clinical disease, associated with hallmark signs and symptoms, often begins abruptly with elevated temperature and general malaise, whereas subclinical infections are generally milder, of shorter duration, and associated with fleeting symptoms. However, with most diseases, the majority of those infected experience asymptomatic disease. These persons are completely devoid of clinical symptoms and any outward appearance of illness. The diagnosis of clinical disease is aided by the presence of clinical findings, whereas asymptomatic disease is usually only recognized through specific serologic tests. The process of seroconversion and elevation in specific antibody titer represent important criteria in the screening of employees for occupational exposures to infectious organisms. Workplace monitoring of employees for asymptomatic disease has provided invaluable information on infections such as tuberculosis and the hemorrhagic fevers.

Although many infectious diseases are transmitted to humans by a primary route, some are transmitted by several routes. Thus, it is generally recognized that tuberculosis is transmitted via aerosol, HAV by ingestion, erysipeloid by contact, rabies by penetration and Lyme disease by a tick. However, in some occupational settings, especially in diagnostic, research and production facilities, employees may be exposed to pathogens by abnormal routes, thereby leading to infectious diseases with puzzling clinical symptoms. Moreover, because large concentrations of etiologic agents are grown and manipulated in these workplaces, the opportunity exists for doses far exceeding community exposures. Thus, in workplaces where large quantities of infectious agents are being used, vigilance must be exercised to prevent human exposures via abnormal routes or with an overwhelming exposure dose.

Opportunistic infections occur in individuals whose normal resistance to infection has been compromised, thereby making them susceptible to microorganisms that would not ordinarily cause disease. Those at higher risk include employees undergoing drug or steroid therapy resulting in transient immunosuppression, and those with underlying disease associated with a permanent state of immunosuppression. *Pneumocystis carinii* infection in HIV-infected individuals and *Aspergillus fumigatus* infection in bone marrow transplant recipients are examples of opportunistic infections.

Zoonotic infections result from human exposure to animal diseases. There are more than 200 recognized zoonoses. Data on laboratory-acquired infections have demonstrated that many were zoonotic in nature and represented all classes of infectious agents. The transmission of zoonotic agents can occur in numerous occupations, including veterinary practice, agriculture, animal husbandry and forest management. It also occurs in such workplaces as animal-holding areas, abattoirs, research laboratories, field operations, commercial fishing, and pet operations. It is imperative that specific infection control practices be instituted to protect workers from zoonotic infections, including the use of prophylactic vaccination, quarantine of feral animals, containment procedures, serologic screening, animal husbandry practices, vector management, and the use of personal protection equipment.

CLASSIFICATION OF MICROORGANISMS FOR LABORATORY WORK

The Centers for Disease Control and National Institutes of Health (CDC/NIH) classification of biosafety levels for infectious agents is based on a combination of pathogenicity and transmissibility. Combinations of engineering controls, work practices and personal protective equipment are recommended for each of the four biosafety levels (i.e. BSL1, BSL2, BSL3, and BSL4). The hierarchy of levels is based on the transmissibility of infectious agents by the aerosol route. For example, commonly used laboratory strains of *E. coli*, which are not virulent and not easily transmitted, are classified in BSL1. The bloodborne pathogens (e.g. HBV and HTV) are classified as BSL2 agents because they are not easily transmitted as aerosols but cause more serious and life-threatening disease. (For work with large quantities of these agents, such as viral cultures, they are classified as BSL3.) The aerosol-spread Venezuelan equine encephalomyelitis virus and the yellow fever virus, for example, are classified as BSL3 agents.

FURTHER READING

Joklick WK, Willet HP, Amos DB, Wilfert CM, eds. *Zinsser microbiology,* 20th edn. Norwalk, CT: Appleton & Lange, 1992.

Mandell GL, Douglas RG, Bennett JE, eds. *Principles and practice of infectious diseases,* 5th edn. New York: Churchill Livingstone, 2000.

Richmond JY, McKinney RW, eds. *Biosafety in microbiological and biomedical laboratories,* 4th edn. HHS publication no. (CDC) 93-8395. Washington, DC: US Government Printing Office, 1999.

19

CLINICAL RECOGNITION OF OCCUPATIONAL EXPOSURE AND HEALTH CONSEQUENCES

Gary N. Greenberg, M.D., M.P.H., and Gregg M. Stave, M.D., J.D., M.P.H.

Health effects associated with exposure to occupational and environmental biological hazards are mediated primarily by two distinct mechanisms. These mechanisms are infection by intact organisms and immunologic reaction to materials from biological sources. This chapter will describe these basic disease processes and their associated patterns of illness. Practical guidance will be offered as to when particular illnesses should be suspected and how they can be confirmed.

INFECTION

Infection versus colonization

Infection results when living microorganisms (usually viruses, bacteria, fungi, or parasites) establish an active and growing presence within the human host. This situation creates characteristic pictures of illness. Some disease elements are created by the behavior of, and damage caused directly by, the invading pathogens. Others result from the host's response to the organisms. The detection of disease requires knowledge of the microbiology of the attacking microorganism and an understanding of the human body's reactions. Although bacteria are the most commonly isolated source for infectious illness, only some interactions between mammalian organisms and bacteria produce disease. Infection, an event with important medical consequences, must be distinguished from colonization, a term used for the harmless presence of the microorganism in contact with human tissue.

The human host provides many microenvironments that act as sites for a complex microbiological ecology populated by numerous strains of bacteria. For example, several

Physical and Biological Hazards of the Workplace, Second Edition, Edited by Peter H. Wald and Gregg M. Stave
ISBN 0-471-38647-2

bacterial species, termed coliforms, are utilized by the colon to supplement the breakdown of food wastes. Coliforms represent harmless symbiotes as long as they remain in their usual habitat. However, these same bacteria can be the source of critical illness when opportunities place them at other biological locations, including the spaces outside the intestinal wall, the urinary system, or within the bloodstream.

Bacteriologic evaluation of patient specimens can be extremely useful in distinguishing between infection and colonization. Awareness of the source of the specimen is important for the selection of the appropriate test and interpretation of the results. Samples from many sites are commonly contaminated, including saliva, stool, vaginal secretions, and skin swabs. In these sites, the presence of bacteria (and often fungi) need not be interpreted as evidence of disease unless the organism is not part of the usual biology at that location. Clinicians must learn to recognize the species of normal flora specific to each body site and to distinguish them from the harmful organisms recognized as pathogens.

Systemic infection versus localized infection

When illness results from the presence of microorganisms within the host's tissues, infection is most easily diagnosed by the evaluation of the affected area. Some infections are localized and superficial, such as those involving the skin (e.g. cellulitis) or a mucous lining (e.g. streptococcal pharyngitis or strep throat). For other infections, the illness is diffuse, resulting from either a total body invasion of the organism (e.g. the spread of rickettsial organisms in Rocky Mountain spotted fever) or from the body's global reaction to infection.

Clinical manifestations of infections are based on both local and systemic mechanisms; they are mediated by the immune system and based on the effect of activated defensive cells. As the body recognizes the assault of foreign organisms, immediate reactions at the site are involved in a process known as inflammation. Redness and warmth arise from the stimulated local circulation, causing increased blood arrival through dilated capillaries. Swelling results from increased blood vessel permeability, permitting the escape of antibacterial proteins and plasma from the circulation into the surrounding area. Tenderness and limited local function are due to the presence of the offending pathogen and to the effect of the escaping local mediators from the host's activated protector cells.

In addition to the local effects of low-molecular-weight chemical signals, mediators circulate throughout the body and produce systemic signs of illness. Fever, which is mediated by the brain via its hypothalamus, shows that mediators of infection and inflammation have triggered a systemic response. Although it is unclear what advantage is gained by raising the body temperature, fever is one of the earliest and most common responses of infected mammals.

When stimulated by the arrival of infectious debris or cellular activating proteins, regional lymph nodes (collection and production sites for immune cells) enlarge. Another manifestation of infection is a more dynamic circulation, resulting in increased heart rate, reduced vascular resistance to bloodflow, and lowered blood pressure. Generalized muscular aches and stiffness occur in many areas not directly involved by the infection.

The next elements in the local infection process are much more common for bacterial organisms than for viruses or fungi. When a tissue fails to eradicate a local invasion, the process often results in a closed-space infection. Within this abscess are active and dead defensive cells and countless foreign organisms that are "walled off" from the nearby tissue. An abscess may either open spontaneously or require surgical drainage. When infections are especially severe, widespread bodily invasion via the circulatory system, or sepsis, can occur. This process, which allows seeding of distant tissues, is recognized as one of the most dangerous late stages of bacterial illness.

Sometimes, the cumulative volume of circulating infectious material and the massive release of immune mediators combine to produce the syndrome of septic shock. This is a dangerous picture of thready, weak pulse and poor circulatory perfusion, resulting in deteriorating function of vital organs, including the heart, brain, and kidneys. Septic shock requires prompt diagnosis followed by aggressive and intensive treatment, and it may often be fatal.

SPECIFIC CLINICAL DISEASES

Each component of the human host may be invaded by different bacteria, fungi, and viruses. For many body elements, the disease syndromes that result are clinically similar regardless of the attacking infections. The characteristic symptoms for these diagnoses are described here to provide a basis for the ensuing chapters, which will describe the consequences of infection due to specific organisms with such terms as meningitis, hepatitis, or pneumonia.

Upper respiratory infections

Although this most common of infections is often called a "cold" and is usually viral in origin, respiratory infections are not much different when caused by other invading organisms. Patients develop irritation of all respiratory surfaces, including the nose, throat, middle ear, and facial sinuses. These lining tissues become swollen and moist and may obstruct the passage of air. When lymphoid structures (tonsils, adenoids, and lymph nodes) are stimulated to respond protectively to the infection, they enlarge and cause additional symptoms of obstruction and pain. When a corridor for mucus drainage becomes persistently obstructed, immune mechanisms are rendered less effective, and bacteria that are not ordinarily pathogenic can become successful sources of infection and cause complications. This results in superinfections in areas where drainage is blocked, including facial sinuses and the middle ear.

Many of the symptoms of the respiratory infection are non-specific and result from circulating immune activators and foreign proteins. Fever, muscular aches, stiffness, chills, and headaches are common symptoms for any infection. They are associated with respiratory infection in most patients' minds only because colds and flu are such common forms of infection.

Bronchitis and pneumonia

Bronchitis and pneumonia are diagnoses that represent infection of lower respiratory structures. The clinical picture characteristically involves a productive cough, chest pain, shortness of breath, and fever. Bronchitis is associated with excess mucus (usually with infected material well represented) arising from within the chest. Wheezing (or asthma) represents an inappropriate muscular reflex of the contractile elements along the air passage, which narrows these passages and obstructs the exchange of air. Pneumonia represents infection of the surface of the lung where oxygen is transferred to the passing circulation. Many pneumonias involve "consolidation" or filling of the usually empty spongy lung tissue with the combined debris of the attacking organism and reactive immune cells. Pneumonia patients, with illness in one or more of the lungs' lobes, suffer obliteration of the respiratory surface in addition to the consequences of reduced airflow into the lungs, combining to cause respiratory insufficiency. This causes poor oxygenation (manifested by the desaturated blue color of circulating blood), weakness, shortness of breath, and risk of death.

Hepatitis

This infection is often subtle, only occasionally manifested by local tenderness over the liver itself (in the upper right corner of the abdomen). More universal among patients with liver disease is prominent and disabling fatigue, with loss of appetite and the onset of jaundice. This last sign represents the escape of bilirubin (a breakdown product of normal red blood cell turnover) from the liver's usual metabolic machinery. This displaced pigment accounts for the yellow skin and dark urine of hepatitis patients, and its absence from feces may be reported as "clay-colored" stools.

Liver failure is a rare but potentially fatal consequence of infection in this organ. Reduction in the metabolic and synthetic functions of the liver permit the development of clotting disorders, abdominal distension with extracellular fluid, deteriorating mental function, and unregulated blood sugar.

Dermatitis

Since skin is the outermost layer of the host, infectious rashes are the most easily noted of the body's reactions to invading organisms. Unfortunately, for many common patterns of illness, the skin produces a non-specific pattern of reaction, and non-infectious immune responses often confuse the patient and clinician regarding the origin of the illness. Furthermore, some rashes may be associated with infectious disease affecting body elements elsewhere, including systemic illnesses like measles, Lyme disease, and meningococcal meningitis.

Cellulitis is a primary infection of the skin appearing as a spreading area of redness. Often it originates where skin damage has occurred, usually from mechanical injury. Other obvious and direct skin infections may involve specific skin structures, such as hair follicles, sweat glands, and nail beds. Again, primary injury often triggers local infection at these sites. The organisms responsible are usually those already present at the skin's surface. Infection may also be caused by organisms introduced by the agent of mechanical injury (e.g. an animal tooth in a bite wound). Remote and systemic disease is also possible from skin infection. Rheumatic fever and toxic shock syndrome are both consequences of local infection at the skin. Tetanus organisms cause systemic illness unrelated to their direct skin effects. Illness results from the remote effects of tetanus toxin, released by the organism after successful infection of anaerobic spaces beneath the skin.

Central nervous system infections

The brain and spinal cord can be attacked by microorganisms in three distinct patterns of illness—meningitis, encephalitis, and abscess. The most common of these, meningitis, is an inflammation of the brain's lining and suspending fluids. Patients have severe headaches and are unwilling to stretch or fold these covering membranes by such activities as bending their necks or even turning their eyes. The brain itself shows normal function until late in the course of infection, when there is damage to the nervous tissue, affecting thought, movement, and behavior. The diagnosis requires finding evidence of infection in samples of spinal fluid.

Encephalitis involves a direct attack of a microorganism into the nervous tissue of the brain. It usually occurs in a diffuse pattern throughout the brain's substance. Some patients suffer damage in only one cerebral area, with symptoms relating to the specific brain structures involved. Since the brain is a fragile and shielded structure, diagnosis may require indirect testing, such as evaluation of cerebrospinal fluid or computerized brain imaging.

An abscess in the brain represents a circumscribed collection of infected material that not only causes local damage but, by expanding, causes compression of the brain as a whole, trapped in the skull's rigid compartment. This is a rare illness that requires specific treatment, possibly including drainage to empty the area where the infection is localized.

Gastroenteritis and dysentery

The presence of disruptive organisms in the upper gastrointestinal tract can result in painful abdominal distension and discomfort, caused by irritation of the gastric and esophageal lining. It can also cause reflex reactions from the brain, manifested as nausea, vomiting, and loss of appetite. In the lower intestinal tract, the presence of infection by pathogenic microorganisms can lead to different forms of diarrhea. If the organisms—or, more commonly, their secreted toxins—merely interfere with the intestine's ability to resorb liquid, then the patient suffers a watery diarrhea. Dehydration and electrolyte imbalance may result. This type of infection is often difficult to confirm, because the organism is not easily available for laboratory identification. If the infection actually attacks the intestine's wall, the patient loses more than the usual colonic contents. Dysentery is recognized by the presence of blood, inflammatory cells and mucus in the stool. In this form of illness, invasive microorganisms are more likely to be identified in laboratory evaluation.

"Flu-like illness"

Many illnesses with important consequences initially manifest themselves with widespread symptoms that are non-specific and undiagnosable. Even though the clinical picture may be different from the specific illness of influenza, whenever patients have prominent respiratory symptoms, the phrase flu-like illness will be applied to almost any fever-associated syndrome. The symptoms most often recognized to be "flu-like" are body aches, chills, stiffness, mild to moderate fever, headache, and fatigue. The term is rarely applied to illness with oniy respiratory symptoms, such as cough, nasal congestion, or sore throat.

IMMUNE MECHANISMS AND HYPERSENSITIVITY DISORDERS

The host immune system triggers a cascade of cellular, antibody and chemical activity in response to the presence of foreign materials. Though this system performs an essential protective function, its inappropriate activation by otherwise benign materials can sometimes lead to deleterious consequences. These responses may manifest themselves as allergy, asthma, arthritis, or other hypersensitivity disorders. Classical allergy, such as hay fever and laboratory animal allergy, results from antibodies of the immunoglobulin E (IgE) class. These antibodies are adherent to mast cells. When specific target proteins arrive at the cellular surface, IgE antibodies bind to their molecular targets and activate the mast cell's response. Mast cell products released include histamine, a small circulating compound that is responsible for many of the clinical manifestations of allergy. The propensity to develop an allergic response as a result of occupational exposure is not distributed uniformly in the population. A personal or family history of allergy, asthma, eczema or sinusitis (known collectively as atopy) is associated with an increased risk of developing allergy. However, history of atopy alone cannot predict whether an employee will develop symptoms from work exposures. Previously non-atopic individuals can also develop allergic illness. Not all immune reactions are mediated through IgE. In hypersensitivity pneumonitis, specific antibodies of the immunoglobulin G (IgG) class recognize foreign airborne antigens. The resultant antibody-triggered cascade of immune events can produce a devastating reaction in surrounding lung tissue.

Cellular-mediated immune mechanisms are slower and less well characterized than those initiated by circulating antibodies. The application of specific proteins to sensitized tissues causes activation of local immune cells, resulting in the activation of monocytes and the migration of macrophages to the area. These arriving cells cause changes in skin firmness.

SPECIFIC CLINICAL SYNDROMES

Upper respiratory allergy

Allergy to ragweed pollen ("hay fever") is a common environmental respiratory allergy.

Symptoms of this disorder are identical to those for allergy caused by other airborne proteins and result from direct contact of inhaled particles with the respiratory mucosa. Manifestations of exposure include increased production of tears and a continuous clear nasal discharge. The lining of the nasal passages may swell, resulting in obstruction. Patients commonly experience sneezing, along with itching of the eyes and throat.

Lower respiratory allergy

Asthma is an episodic illness. Symptoms result from constriction of muscle-lined air passages in the lungs. The ability to exhale air is limited. Patients experience shortness of breath, chest tightness, and possibly cough. Wheezing may be audible or appreciated only with a stethoscope. The inability to achieve adequate air exchange results in a diminished level of oxygen in the blood. In severe episodes, continued lack of adequate oxygen levels can have disastrous consequences. Regardless of the stimulating event, the symptoms of asthma are similar. However, there is great variability in the severity of symptoms, depending on exposure conditions and medical therapy.

Patients with acute hypersensitivity pneumonitis also commonly experience shortness of breath and cough. Unlike asthma, symptoms include chills, high fever, and muscle aches, which are signs of widespread reaction to active inflammation. Symptoms generally resolve after a brief illness lasting <8 hours, but some patients experience milder symptoms for several days. This illness may be confused with an infectious pneumonia caused by organisms actually present in the lung. Careful evaluation and a high degree of clinical suspicion are required to arrive at the correct diagnosis.

Skin reactions

Immunologic skin reactions are extremely varied and can often be confusing. However, the classic lesion associated with allergic contact dermatitis is the rash associated with poison ivy and poison oak. This red, itchy rash contains small, clear, fluid-filled raised blisters. For other allergic skin eruptions, the clinical findings are frequently much less specific, with evidence of only redness and possibly scaling.

The skin may also be active as part of a systemic response to ingestion or injection of antigens. One such manifestation may be widespread edema and diffuse hives accompanied by pale swelling and itching. In severe cases, this reaction may be associated with leakage of circulating blood plasma into peripheral tissues and consequent shock. Emergency medical treatment is required in these cases.

Another skin response is similar to that seen with the intradermal injection of purified protein derivative (PPD) in tuberculosis skin testing. The skin becomes firm and raised as a result of local cellular infiltration. The reaction may not be evident for 2–3 days following exposure.

Irritations

Non-immunologic individual variation in the response to other biological stimuli is also widely recognized. Wood and tobacco smoke and other irritants usually do not act as specific allergens and do not provoke the illness mechanisms described above. Nonetheless, variation among the doses tolerated by the human population is considerable. Even though irritation occurs without specific antibodies or immunologic mechanisms, there are sensitive individuals who react to many stimuli (including odors) with more severe symptoms than the population average, sometimes with easily recognized objective clinical findings.

LABORATORY CONFIRMATION OF INFECTIOUS AND HYPERSENSITIVITY DISEASES

Several technologies are available for the identification of microorganisms in human tissue. Each has specific advantages and they are commonly used in combination. Although microscopic evaluation of smears can be performed rapidly, it lacks sensitivity to small numbers of organisms and there is poor precision regarding the microbe's identity. Culturing of the organism on specific growth media permits recognition of even small numbers of pathogenic germs. It may be a slow process, dependent on the sometimes delayed growth of the microbes sought. Newer techniques, based on molecular biology concepts, have been introduced recently, but these methods are sometimes overly sensitive.

Microscopic visualization of the organism

The most rapid means for disease recognition is direct microbial identification in stained biological specimens. In pneumonia, for example, sputum samples are smeared onto a glass slide and allowed to dry. Specific stains and selective rinses are then applied before careful microscopic examination. This evaluation can reveal the nature and number of the organisms present and even indicate whether the microorganisms are pathogens or colonizers, based on whether they are being engulfed by the host's defensive cells. This technique not only provides a glimpse into the nature of the disease, but also discloses the possibility of mixed infection by several organisms and indicates the intensity of the battle between the microbes and the host cells.

For general bacterial evaluation, the Gram stain is used. This technique provides useful information regarding the organism's shape and the type of cell membrane of the microbes present. The Gram stain procedure is designed for a primary pigment to be selectively removed from the interior of certain Gram-negative bacteria, permitting the loss of the dark color and permitting staining only by the last stage, a light pink universal stain. The darkly stained bacteria are recognized as Gram-positive, whereas the paler organisms are Gram-negative. The organism's shape offers additional diagnostic clues. Rectangular and linear bacteria are characterized as rods; circles, coffee bean shapes, and clusters are cocci.

Special stains are required for certain organisms and situations. The acid-fast reaction is a special stain that provides the classic means to recognize mycobacteria (e.g. tuberculosis). Silver-based staining is used to identify microscopic protozoa, fungi, and spirochetal bacteria (e.g. syphilis and pneumocystis). Viruses, which are much smaller and necessarily intracellular, are more difficult to visualize by direct microscopic evaluation. Usually, the identity of the exact virus must be inferred from the clinical circumstances and the source tissue being evaluated. However, with the use of either special staining techniques or electron microscopy, the presence of viral organisms can sometimes be confirmed.

The most precise stains are based on the use of specific antibodies for individual organisms. Once these antibodies attach to their targets, they are linked to fluorescent compounds that allow detection in microscopic evaluations, even in fluids or tissue. This technique, called immunofluorescence, allows precise confirmation of actual microbial identity without the delay required for cultures to grow or the host's antibodies to emerge.

Growth and identification of microbial colonies

The organism's ability to multiply in the host tissue is the most common disease mechanism. Culture techniques use this same capability to identify organisms that would otherwise be missed by amplifying their numbers in specific artificial environments, called media. Once isolated and flourishing in the microbiological laboratory, pathogenic colonies can be tested for additional information regarding their precise speciation, antibiotic sensitivity, and biochemical activity. Molecular markers may be useful for epidemiologic evaluation.

The greatest weakness of culture as investigative tool is the delay before the organisms are sufficiently numerous to be detected. For bacteria, the lag is usually 24 hours. Most viruses fail to grow in laboratory settings, but even where it is possible, there is a similar delay. For fungi and mycobacteria, although a positive sample might be reported earlier, a sample cannot be considered "no growth" until it has been incubated for a full 6 weeks.

In addition, there are many occasions when a sample will fail to yield any growth, even in the presence of infection diagnosed by other means. Culture samples must be spared any risk of heat, cold, or drying. If the patient has taken antibiotics, if the sample is mishandled, or if non-pathogenic organisms are present that inhibit the pathogen's in vitro behavior, then the culture will not only be slow to yield the cause of the illness, but will also yield false-negative results.

After collecting a specimen with viable microbes, the next step in microbiological culture preparation is selection of the appropriate growth medium. This requires specific broths or agar gels with nutrients and cofactors designed to encourage the growth of even particularly fastidious organisms. In addition, when specimens are collected from a source known to be contaminated with selective non-pathogenic organisms (e.g. from the throat), antibacterial chemicals must be included to suppress their growth.

Finally, the environment for culture growth must be selected. When the organism sought is a tissue-invading bacterial pathogen, 37°C is used to simulate the human host. When a biological agent believed to be responsible for producing environmental allergies is cultivated, its disease-producing biology is different; the mechanism of disease involves the organism's growth at ambient temperature and the subsequent release of allergenic proteins into the environment. Environmental samples should be cultured at a temperature that accurately reflects the biology of the area where they were collected.

The results from cultures do not necessarily constitute a clinical diagnosis. For each biological sample, there are guidelines for culture isolates. For sputum, culture interpretation requires knowledge of whether the specimen included host cells that prove its origin from lung tissue (as opposed to mere saliva from the mouth). For urinary cultures, microscopic identification of cells from the vagina makes the interpretation of any culture results invalid. Because most infections are caused by only a single species, pure growth of even small numbers of organisms of the same species constitutes strong support for the presence of an important microbiological presence.

Even when a pure sample is achieved, the concentration of viable colonies can be an important interpretative fact. Rare stray organisms found in urine specimens do not constitute proof of disease. Useful culture reports must indicate, for example, that the culture showed 100 000 colonies of *E. coli* per ml. The report should name the organism, indicate the purity of its incubated growth, and describe the density of its presence in the specimen.

Clearly, there are many times when the mere identification of an organism in a sample constitutes a conclusive diagnosis of disease. Where a material is ordinarily sterile, the presence of a single colony is persuasive. Such samples include spinal fluid, liver biopsy material, and urine obtained via sterile catheter. In those cases, every microorganism must be considered a pathogen.

In other circumstances, the presence of any microbial colonies is sufficient proof for a firm diagnosis where the infecting agent is never part of the benign normal flora. The mere isolation of any of these organisms constitutes a firm diagnosis, regardless of concentration, coexisting organisms, and whether the specimen was otherwise contaminated by other body materials. These agents include *Neisseria gonorrhea* (the gonorrhea organism), *M geobacterium tuberculosis*, and the herpes simplex virus.

Diagnostic evaluation by immunity testing

For many infections, the responsible organisms cannot be identified by direct visualization or by culture because they are too fastidious or too few, or because they are located in unreachable anatomic sites. In these cases, including those caused by many viral organisms and several atypical bacteria, the best way to identify an illness is to monitor the patient's immune response to the microbe as indirect evidence of its presence.

A basic concept in biology is that each individual's immunity is developed as a consequence of its own exposure and infection experience. The immune system serves as a databank for the aggregate history of the host's exposures. Each foreign protein and macromolecule serves as a unique immune stimulus (called an antigen). Each exposure to a new antigen constitutes a new immunization and results in a uniquely responsive set of precise cellular and antibody reactions. These reactions create permanent changes in the way that the host reacts to the antigen on any future exposure. The sum of these learned reactions is retained in the organism's immunologic "memory". This represents a catalog of identifiable exposures, each of which resulted from a prior exposure and each of which can be tested and identified for proof of prior contact.

As an example, when a patient is effectively exposed to the mumps virus, a characteristic set of targeted proteins called antibodies are produced that react specifically to the viral proteins. Whether the exposure arrives as a vaccine or is the consequence of an actual infection, the result is the lifelong presence of identifiable circulating antibodies specifically reactive to this microbe.

Most immune testing evaluates the existence, quantity and type of circulating antibodies. The presence of each specific antibody proves prior exposure and infection. Proof that an antibody results from a new exposure or infection can be based on the fact that the production of immunoglobins of differing types occurs in predictable serial fashion. The classes of antibody are named with single-letter suffixes, e.g. immunoglobulin M (IgM). IgM is an antibody subtype with a transitory role, lasting only a few weeks, until Immunoglobulin G (IgG) is synthesized. Because of its transient presence, the identification of IgM against a microbe is itself proof of recent disease. IgG is the largest and longest-lasting of the antibody classes. Most antibody assays can be subdivided into IgG and IgM classes to provide information on the possibility of recent exposure. The presence of antibody is detected in laboratory agglutination reactions. Specific binding characteristics are used to clump the antibody proteins and their targets. The resulting aggregate effect proves the presence of antibody.

Evidence of recent immune activation may also be obtained by demonstrating a recent rise in antibody concentration. Antibody levels are usually expressed as titers, referring to serial dilutions of serum that demonstrate a positive reaction. For example, if four twofold dilutions of a sample produced a positive test result but the fifth did not, the result would be reported as positive at 1:16. Two samples collected 6 weeks apart (labeled acute and convalescent) constitute a matched set for analytic purposes. The customary threshold for a significant antibody elevation is a fourfold titer increase. Since the clinical consequences of exposure are likely to have run their course, the intervening delay is usually too long for clinical decisions concerning the affected patient. However, it may provide important documentation for workplace or public health use.

Immunoglobulins are also utilized in other diagnostic tests. Fluorescent compounds chemically linked to antibodies can be used to stain specific proteins for immunohistologic microscopic examination. This technique of linking a detectable moiety to an antibody is also used in enzyme-linked immunosorbent assay (ELISA). When an antigen from the biological substrate of interest binds to the

ELISA antibody, the linked enzyme generates a measurable product. This product may be a fluorescent compound or another easily measured chemical.

Although ELISA provides a rapid and relatively simple technique for antigen detection, it is not highly specific. A positive ELISA result may need to be confirmed by antibody electrophoresis testing (e.g. the Western blot assay). This more expensive and time-consuming assay increases specificity by providing a measurement of the size and the electric charge characteristic of the detected antibody.

Polymerase chain reaction testing

Biotechnology laboratory techniques have provided new opportunities to identify microorganisms. Based on the same means used for gene cloning and synthesis, methods now exist to identify segments of either DNA or RNA that represent specific identifiable microbial genes. Identification is accomplished by a process that detects and amplifies any genetic material present. This new capability can prove the presence of infecting organisms without relying on the immune system's response and without requiring successful growth of the organism outside the infected host. The process replaces whole-pathogen cultures with the detection of recognizable species-specific genes and gene segments. There are several advantages to this elegant laboratory tool. Confirming the presence of microbial DNA or RNA, even when the organism itself is too weak or too slow-growing to be cultured, enables the identification of organisms whose culture has never been possible. It can also provide results more rapidly than other diagnostic methods.

However, there is concern about the possibility of genetic contamination, where random bits of genetic material erroneously suggest that a particular microbe is present. More clinical experience is required before this technique achieves the reliability and standardization now available with either culture or antibody testing.

CLINICAL TESTING FOR HYPERSENSITIVITY

There are several mechanisms by which patients may become ill from biological sources without direct microbial infection. Each has a specific evaluative technology for recognition and diagnosis. None of these techniques to evaluate hypersensitivity is as well validated or standardized as the diagnostic tools for evaluation of infection. The testing techniques used for allergy and other forms of hypersensitivity require considerable judgment in their application. Consultation with an appropriate clinical specialist should be considered.

Allergy testing with "prick" skin tests

The mechanism by which traditional allergy occurs involves mast cell activation by the binding of a high-molecular-weight compound to the cell-attached IgE antibodies. The classical allergy test involves placing a drop of a dilute antigenic solution onto the skin, pricking the skin shallowly with a clean pin, and awaiting an immediate reaction. The response, when one occurs, shows the effects of the activated mast cell's release of histamine and other mediators, causing a localized reaction of swelling, reddening and notable itching.

When properly administered, this testing procedure is useful and reliable. Results are skewed by varying circumstances, including non-specific skin reactions, antihistamine use, and many complex issues involving the applied solution. Qualitative and quantitative problems may result in either positive or negative test errors. This test should be performed by an experienced clinician.

IgE evaluations with RAST testing

Because the actual mechanism of classical allergic symptoms involves IgE, diagnostic tools have focused on measuring circulating levels of this molecular class. Measurement of total IgE may be useful in some cases (e.g.

allergic bronchopulmonary aspergillosis). The concentration of IgE antibody directed against specific allergens can also be quantitated.

The technology most commonly used is called RAST: Radio (because the test involves radioactive methods in the laboratory) Allergo Sorbent Test. These tests are capable of accurately measuring very small concentrations of IgE antibodies in the serum of allergic patients. The resulting information has proven highly comparable with the results from skin testing, and both tests correlate with clinical diagnoses of allergy.

Patch testing and intradermal skin testing

Cellular-mediated immunity is also called delayed hypersensitivity, because it represents a slower mechanism of response. Although the recognition of the foreign agent still depends on the lymphocytes' molecular memory, the response utilizes an entirely different class of activated cells, and the disease is manifested by different mechanisms.

Because delayed hypersensitivity involves several cellular classes working together, testing currently requires measurement of the response by the intact host rather than any cellular extract or circulating antibody. Thus, the diagnostic tests require applying the potential offending allergen directly onto the patient and waiting 48 hours for cellular infiltration. The response considered to be diagnostic is the arrival of enough immune-activated cells to produce a circular area of irritation and hardening (induration).

Patch testing is performed to evaluate suspected cases of contact dermatitis. An extremely dilute antigen solution is applied to a gauze pad and held in place with a shallow aluminum protector during the test's incubation. To ascertain whether the reactions are in fact specific for the antigen in question, skin tests are always done simultaneously with control solutions. These include antigens known to produce positive results in the population at large as well as the saline preservative solution used for the allergen's dilution to determine non-specific responses. These skin test batteries thus often require an entire grid of applied patches, sometimes covering the patient's entire back for the 2-day waiting period.

When the tissue response in question involves organs other than the skin (including possible inapparent tuberculosis), stronger and more invasive dosing is performed, depositing 0.1 ml of the solution directly into the skin with a tiny needle (intradermal testing). In this case, the number of applied solutions is limited by the patient's discomfort, but control solutions may still be used. Many patients with suppressed immune response (including corticosteroid treatment, HIV infection, or even overwhelming systemic infection) will be unable to mount any cellular response, thus producing a false-negative result termed anergy. By including simultaneous doses of antigens with universal response (mumps, *Trichophyton*, *Candida*), the skin test battery can be self-validating.

For intradermal skin tests, the puncture site becomes the center of a spreading firm area. Reading the test simply involves measurement of the firm region's greatest diameter after a delay of 48–72 hours. The medical interpretation of skin testing requires additional consideration of the setting and the patient. Even the most common of intradermal skin tests, the PPD for tuberculosis, can be called positive at 5, 10 or 15 mm of induration, depending upon the clinical setting.

Exposure challenge testing

In cases where the clinician tries to evaluate environmental disease that may be explained by mechanisms of hypersensitivity, objective measures of dose may be totally misleading. Allergy responses by sensitized individuals are frequently many orders of magnitude more sensitive than the best industrial hygiene techniques, especially when others who share the exposure show no symptoms at all. Investigators of potentially allergenic environments must therefore cope with an obvious temptation—direct patient challenge.

There is a significant danger in sending potentially affected individuals into situations where they are suspected to be allergic to an airborne agent. Depending on the patient's prior reactions, such experiments must be done with ample opportunity for rescue and medical attention. They should only be done when the prior illness has been mild (e.g. skin rash) and when the symptoms are easily reversible. The exposure testing must progress in a stepwise fashion; the earliest exposures should be chosen to produce no response at all, even in a patient known to be allergic.

Exposure testing should be used only when no other means of evaluation is available and only with both medical guidance and the fully informed permission of the patient. It should be considered the choice of last resort in diagnostic techniques.

WHEN TO SUSPECT OCCUPATIONAL ILLNESS OF BIOLOGICAL ORIGIN

Most illness due to infection or allergy results from non-occupational exposures. However, certain settings or specific illnesses increase the likelihood that a medical problem has an occupational origin.

Unusual job activities

The health consequences of many workplaces can be predicted and prevented with planning and conscientious concern, but, despite such preventive measures, some employee populations remain at increased risk. Medical care obviously involves direct contact with patients whose illnesses may be transmissible. For organisms spread by airborne contagion (e.g. influenza or tuberculosis), the risk of contracting disease is greater for healthcare workers than for the population as a whole, because the disease is more concentrated among their clients. For other illnesses, healthcare workers are uniquely susceptible where exposure requires deposit or liberation of a pathogenic organism. Infection usually occurs as an untoward consequence of an invasive procedure (e.g. exposure to hepatitis B virus).

In these settings, clinicians and safety professionals must remain aware of the opportunity for illness to transform care providers into patients. Routine preventive measures, including vaccination against certain illnesses (e.g. influenza and hepatitis), universal precautions for blood and body fluids, and routine handwashing, are important.

Animal workers

Unfortunately, workers whose jobs require direct contact with other species are at risk for both allergy and infection. The proteins released from animal urine, skin and other tissues can easily become airborne, resulting in rashes, hives, allergic nasal and ocular symptoms, and even asthma. This form of hypersensitivity usually occurs immediately after exposure, facilitating the diagnosis. However, conditions such as asthma may be delayed. The lack of an immediate response does not automatically exclude an occupational association.

For infections, the parameters of risk attribution are reversed. Infection resulting from other species is rare and poorly recognized. Clinicians often fail to recognize the nature of illness and may not even know what microbial agents to suspect or what treatment is needed. When an agent is identified by culture or serologic means, it is not hard to determine that this unusual pathogen must have arisen as a consequence of work exposure. Populations at risk include workers at abattoirs, zoological parks, and veterinary clinics, and those involved in biology research. Pet owners also have large exposures to the possibility for allergy and for infection. Since many animal workers are also pet owners, both occupational and home environmental exposures need to be investigated when evaluating suspicious illnesses.

Workers handling waste and sewage

In the era before proper sanitation and sterilized water supplies, numerous infections associated with human waste posed community-wide dangers. The risk for diseases prevented by these techniques are now concentrated among the workers with potential exposure, usually in municipal water treatment centers. These agents include both viral organisms and bacteria. Recognition of disease in these workers is important to provide proper treatment and to minimize exposure for their coworkers.

Travelers

Geographic dislocation may result in environmental illnesses through a variety of mechanisms. Many areas of the world, including regions in the USA, contain unique pathogens in such high environmental quantities that the rate of pediatric infection is universal while the danger to adults exists only among new arrivals to the community. Usually fungal in nature, examples include histoplasmosis and coccidiomycosis, endemic in the Ohio River valley and in the American southwest, respectively. Even "traveler's diarrhea" can sometimes be explained by organisms that produce no symptoms among local inhabitants because they were naturally immunized years earlier.

Many illnesses are climate-specific and thus are seen in industrialized societies only among returning travelers, new immigrants, and visitors. Malaria and yellow fever represent important concerns for travelers to tropical areas, where insects act as potent vectors for disease. When these diagnoses are made, the link to foreign travel is essential. In addition to the area of the world, some consideration is required for the levels of accommodation that were present. Urban life, with its air-conditioned hotels and restaurant meals, represents a drastically reduced risk for tropical disease compared to traveling to remote villages and spending long, unprotected hours in the wilderness.

Sanitation and public health measures in other cultures are often less thorough than the norms in European and American societies. Water and food supplies are often the source of infection for adventurous travelers who sample local edibles contaminated with viable organisms that would not be present in their home food markets. Vibrio organisms and *Shigella* are much more common in settings where food and water regulations are lax for reasons of poverty, societal disruption, or crowding.

Travelers may also contract contagious illness from their new human contacts. The geographic migration of many illnesses (including measles, HIV, and resistant gonorrhea) has required migrating human hosts. Thus, an infection may result just as easily from contact with a newly arrived immigrant to the domestic environment as from visits to foreign regions.

Hypersensitivity is not usually related to travel. A period of several weeks is required for exposure to produce the necessary antibodies that create allergy, and travelers have by then become residents. Additionally, when travelers develop hypersensitivity-related illness, the best therapy is removal from exposure, which is easily accomplished when the visit is short-term by its nature.

Unusual clusters of disease events

Even among workers with unremarkable job activities, some evaluation is required in response to what appears to be an outbreak or cluster of disease. Even for illnesses that are common in our society, there is a poorly defined threshold when an investigation is required to explain the simultaneous development of numerous similar medical problems within a worker community. The rarity and nature of the illness, its prevalence within the at-risk workforce and the pattern of its occurrence and spread are important criteria in evaluating these situations.

In the office setting, occupational health professionals may be asked to evaluate health complaints that the occupants have ascribed to "sick-building syndrome". Symptoms reported commonly include headache, irritation of the eyes, nose, and throat, fatigue, and sensitivity to odors. Employees may report that symptoms occur only while they are in the building. Although controversy persists as to the most common etiology of this syndrome, it appears unlikely that such nonspecific symptoms are caused by a specific biological organism. Extensive searches for biological sources of illness are usually not warranted. A systematic review of employee complaints is an important first step in understanding the problem. If indicated, an evaluation of the ventilation system should be undertaken. Efforts should be directed toward providing adequate airflow and air exchanges, as well as appropriate regulation of temperature and humidity. Altering these physical aspects of the ambient environment may reduce occupant complaints. Psychosocial factors, including a variety of workplace stressors, may also contribute to health complaints. An evaluation of the role of psychosocial issues should be conducted contemporaneously with the rest of the evaluation.

By contrast, "building-related illness" describes a situation where building occupants have specific clinically diagnosable illnesses, such as hypersensitivity pneumonitis. Biological organisms may either cause or contribute to these illnesses. A thorough evaluation for potential sources of contamination should be pursued. Molds (usually fungal colonies and spores) are commonplace. Although they are common even in well-maintained office settings, their environmentally released dose is greatly magnified wherever imperfect ventilation and filtration are permitted. Clusters of symptomatic workers with "hay fever" symptoms (nasal congestion, tearing, sneezing, and coughing) should prompt an assessment of the air purity for potential contamination with invisible microbes and proteins. Evidence of either water condensation or prior flooding makes it even more likely that the symptoms can be explained by occupational exposures. These factors suggest a need for special testing for environmental flora that may be present either on surfaces or in the air.

Situations involving hypersensitivity pneumonitis (such as "farmer's lung") present a rare but more critical problem. Because of the delayed onset of their illness, the sensitive workers do not develop immediate symptoms; therefore, they may not associate their illness with any particular activity. Several episodes of illness, either in just a single worker or among a work team, may occur before a connection can be made to the work environment. In these cases, the evaluation of the environment and the patient should be coordinated, with open communication between the clinician and the environmental health professional.

There are occasions when work-related controls must be considered for an epidemic of infection as a result of a common source illness. Food-related illnesses can be introduced into the workplace by any common eating opportunity, including vending machines, in-house cafeterias, or popular neighborhood restaurants. Direct contagion must be considered for outbreaks of conjunctivitis ("pinkeye") among workers using shared optical devices, such as microscopes.

Rare or severe diseases

In some cases, the patient's diagnosis is sufficiently unusual on its own that the occupational environment must be considered to explain the source of disease. Just as with chemical exposures, where the development of peripheral neuropathy or bladder cancer requires a thoughtful assessment of the potential for environmentally triggered illness, there are certain diagnoses where occupational and environmental causes must be conscientiously sought, even without explicit hints or leads. Recurrent asthma and respiratory compromise, even in just one worker, represents such a commonly environmentally mediated danger that an investigation of workplace environment is a reasonable supplement to medical management. An inspection for potential organic contaminants or dusts is a prudent adjunct to the treatment of this hazardous and progressive condition. A diagnosis of Legionnaires' disease should prompt a consideration of where the pathogen was acquired. Tuberculosis in a worker should prompt an assessment of coworkers who might have provided or received the organism in the work setting. This public health response is similar for those with a shared home environment and will likely be performed by the same governmental prevention specialists. The need to identify those with recent exposure and early infection is very important, both to the individual and to the rest of the work community, since curative treatment abolishes the risk of further exposure in only a few days.

EVALUATION OF SUSPECTED OCCUPATIONAL ILLNESS

Clinical suspicion of a possible occupational illness should be heightened whenever a patient is a member of a group at increased risk for exposure to biological hazards, or falls into one of the other categories described above. Specific evaluation will vary greatly, depending upon the clinical presentation and differential diagnosis. When the illness is suspected to be occupational in origin, a detailed history should be taken to establish how the exposure occurred. In addition, a walkthrough evaluation of the work site should be considered. The work site visit should allow for a thorough understanding of job functions and work practices. Depending on circumstances and available resources, the walkthrough team may include occupational physicians, occupational health nurses, industrial hygienists, or biohazard scientists. The visit should result in a determination of the need for further action, possibly including additional diagnostic testing and a trial of early worksite remediation. Environmental sampling, commonly used for the evaluation of chemical exposure, should be used with caution in the evaluation of biological hazards. Although it may be useful in some situations, the ubiquity of microbes and the lack of "normal" values renders interpretation difficult.

Certain occupational diseases must be reported to the state health department as a matter of law. Many health departments are staffed by experts who can also assist with the evaluation of occupational and environmental illness. Additionally, they may coordinate relevant public health measures to protect other workers and the community. Other government resources include the Centers for Disease Control and Prevention (CDC) and the National Institute for Occupational Safety and Health (NIOSH). The CDC and the state public health department can be particularly helpful where there is an opportunity for prevention or research or when specialty evaluations would contribute to the resolution of the situation or crisis. The role of NIOSH includes the evaluation of how the job contributed to the illness. This agency is also interested in studying potential new disease mechanisms in order to develop health and safety standards that will prevent any recurrence of the illness.

FURTHER READING

Mandell GL, Douglas RG, Bennett JE, eds. *Principles and practice of infectious diseases*, 5th edn. New York: Churchill Livingstone, 2000.

Wilson JD, Braunwald E, Isselbacher KJ, et al, eds. *Harrison Principles of internal medicine*, 15th edn. New York: McGraw-Hill, 2000.

20

PREVENTION OF ILLNESS FROM BIOLOGICAL HAZARDS

Linda M. Frazier, M.D., M.P.H., Gregg M. Stave, M.D., J.D., M.P.H., and Jerry J. Tulis, Ph.D

Once we become aware of the potential biological hazards in a particular work setting, we can develop an effective plan to prevent occupational illness. Prevention of illness from biological hazards is accomplished by a combination of the three classic prevention strategies—primary, secondary and tertiary prevention.

Primary prevention aims to prevent illness before the disease process begins. Strategies include vaccination and measures to limit potentially hazardous exposure to biological agents and organisms. Exposures can be limited by using engineering controls (including ventilation and containment systems), proper work practices, and personal protective equipment (such as gloves, uniforms, laboratory coats, safety glasses, and respirators). Environmental monitoring may be useful to determine if controls are effective in reducing potential exposures.

Secondary prevention entails intervention when the physiologic changes that precede illness are recognized or when subclinical illness develops. Secondary prevention is most effective when a surveillance system detects these events systematically. Medical screening must therefore focus on both the results for individuals and those for the group (epidemiologic evaluation).

Tertiary prevention is directed at limiting the consequences of clinical illness once it has occurred. It may involve medical treatment, work restrictions, and/or removal of the worker from further potential exposure. Specific preventive practices vary, depending on the work setting and the level of hazard.

OCCUPATIONS WITH POTENTIAL BIOLOGICAL HAZARDS

Occupational biological hazards are those encountered when the workplace has greater risk of exposure than the surrounding community. Thus, the common cold is not usually

Physical and Biological Hazards of the Workplace, Second Edition, Edited by Peter H. Wald and Gregg M. Stave
ISBN 0-471-38647-2

considered to be an occupational biological hazard even though one employee can contract a cold from another, because cold viruses are ubiquitous in the community at large.

Biological hazards may be found in diverse work settings. In some settings, such as research laboratories conducting studies on specific biological agents or organisms, the hazards are clearly identified. On farms and at zoos, specific zoonoses (animal infections that may be transmitted to humans) may be a risk. However, in most settings, the hazard is an indirect consequence of the work or a risk that arises in the work environment. Before we develop a prevention program, we need to evaluate the setting for reasonably anticipated hazards.

THE OSHA BLOODBORNE PATHOGEN STANDARD

The Occupational Safety and Health Administration (OSHA) has issued one standard to date that addresses biological hazards. The Bloodborne Pathogen Standard (29 CFR 1910.1030) was issued in December 1991 and became effective in March 1992. The standard applies to all employers with one or more employees where employees may have exposure to blood-borne pathogens. Bloodborne pathogens are defined as pathogenic microorganisms that are present in human blood and can cause disease. These pathogens include, but are not limited to, hepatitis B virus (HBV) and human immunodeficiency virus (HIV). The standard applies not only to hospitals and doctors' offices but also to many other work settings, such as clinical and research laboratories, mortuaries, emergency response teams, lifeguarding, and medical equipment maintenance.

The first requirement of the Bloodborne Pathogen Standard is the performance of an Exposure Determination. Employers must evaluate the potential for employees to be exposed to blood-borne pathogens. Occupational exposure means reasonably anticipated skin, eye, mucous membrane or parenteral contact with potentially infectious materials on the job. Even employees who use personal protective equipment, such as gloves, are considered to be potentially exposed. If employees have the potential for occupational exposure to bloodborne pathogens, then the employer must develop a written Exposure Control Plan. Blood borne pathogen exposure can occur from handling substances other than blood. Also, body fluids from deceased individuals can be infectious.

An employee is at risk of exposure if he or she handles these substances. Exposure risk is negligible for personnel who work in healthcare settings but who do not handle body substances—e.g. telephone repair personnel. In contrast, many employers recruit worksite first aid teams that include employees whose usual work does not involve contact with human body fluids. If the emergency response duties of these volunteers may lead to contact with blood or body fluids of injured coworkers, the employer should provide appropriate training and personal protective equipment and offer the hepatitis B vaccine. (OSHA issued a ruling in June 1993 that permits employers to delay the vaccination of first aid providers in specific situations. Employers considering this option should carefully review the practical implications of this policy.)

Although the issue is not specifically addressed by the OSHA standard, tissues or cell lines derived from human or primate sources are also potentially infectious. HIV does not replicate in cells outside the host, so cell lines of relatively recent origin are potentially infectious, whereas later generations will not be infectious because of a dilution effect. HBV can persist in cell lines.

When workers can become exposed to bloodborne pathogens occupationally, the employer's written Exposure Control Plan must include certain critical elements that are described in detail in the standard. Issues to be addressed include engineering controls, personal protective equipment, and work practice controls that focus on universal precautions. Universal precautions means that all blood and body fluids should be regarded as potentially infectious; it is not sufficient to use precautions with some samples and not with others.

The employer must provide training (and annual retraining) in the use of personal protective equipment, safe storage and transport of body fluids, safe disposal of potentially infectious wastes, effective decontamination of contaminated work surfaces, and prohibition of storage or consumption of food and drink in areas where there is a reasonable likelihood of exposure. Hepatitis B vaccine must be provided promptly to employees at reasonable risk of exposure by the employer at no cost. Employees who refuse the vaccine should sign an OSHA-specified declination form. A procedure to evaluate employees who have had an exposure, to determine the potential infectivity of the source and to provide appropriate medical care for the exposed worker is also required. The employer must keep records documenting training, vaccination (or declination), and postexposure evaluation.

In 2000, several states and the US Congress considered laws and regulations to encourage or require the use of newer needlestick and sharps injury prevention technology. The federal Needlestick Safety and Prevention Act was passed in November 2000, and went into effect on April 18, 2001. Legal requirements include the use of needleless systems and other engineering approaches that effectively reduce the risk of an exposure incident. The standard requires that frontline employees who are using the equipment have the opportunity for input into purchasing decisions. The new needlestick log will help both employees and employers track all needlesticks to help identify problem areas or operations. As a result, injuries that do not meet the definitions of injuries under the OSHA record-keeping standard will still be collected. The updated standard also includes provisions designed to maintain the privacy of employees who have experienced needlesticks

OSHA GUIDELINES FOR TUBERCULOSIS

In late 1993, the OSHA issued mandatory guidelines (revised February 1996) for an enforcement policy intended to protect workers from tuberculosis (TB). These guidelines are based primarily on the Centers for Disease Control and Prevention 1990 Guidelines for Preventing the Transmission of Tuberculosis in Health-Care Settings with Special Focus on HIV-Related Issues.[1] The guidelines pertain to employers in settings where workers are at increased risk of exposure, such as healthcare facilities, correctional institutions, homeless shelters, drug treatment centers, and long-term care facilities.

OSHA guidelines require employee training and information on the signs and symptoms of tuberculosis, hazards of transmission, medical surveillance, and site-specific controls. Employers must institute a program of early identification of suspected cases. Medical surveillance should include preplacement evaluation, periodic Mantoux testing, and management of persons with positive test results. Persons who are infectious must be treated in respiratory isolation rooms under negative pressure. A formal OSHA compliance directive will be developed after the Centers for Disease Control completes the second edition of its guidelines. A draft of the guidelines includes a requirement to create a TB Infection Control Plan. Elements of the plan include a risk assessment, administrative controls, engineering controls, use of respiratory protection, employee education and training, and medical surveillance.

In 1997, OSHA proposed a comprehensive standard for preventing TB transmission among healthcare workers.[2] This standard is expected to be finalized soon. It differs from the CDC guidance in the areas of risk assessment, medical surveillance, and respiratory protection. It also contains medical removal protection for employees. Employers that have developed exposure control plans should review them in the context of the proposed OSHA regulations.

PREVENTION OF EXPOSURE TO BIOLOGICAL AGENTS

Hazardous exposures to biological agents occur mainly through inhalation and ingestion, although skin contact (or penetration) can cause illness with some agents. These routes of exposure can be eliminated through engineering, administrative or work practice controls, and the use of personal protective equipment.

Engineering controls

The preferred preventive measure for prolonged or highly hazardous potential exposures is the use of engineering controls. Workplace controls are intended to contain biohazards at their source, reduce their airborne concentration, and limit their movement through the work site. Heating, ventilation and air-conditioning (HVAC) systems must also be appropriately designed and maintained to prevent contamination by fungi and bacteria (including *Legionella pneumophila*). For indoor settings, such as medical or research facilities, room ventilation can be engineered to provide directional and single-pass airflow. In hospitals, air exhausted from high-risk infectious disease isolation rooms can be further decontaminated by filtration. Use of ultraviolet light to treat exhausted air is under study. In research and clinical laboratories, handling infectious agents in a biological safety cabinet (BSC) can prevent inhalation exposures. For bioaerosol control, the correct type of unit must be used.

Class I cabinets provide personnel protection but little or no product protection. Room air flows into this open cabinet and is ducted through a high-efficiency particulate air (HEPA) filter. HEPA filters clean air supplied to the work zone, providing product protection; and the HEPA filtration of exhaust air provides environmental protection. This filtration system traps all microorganisms, including viruses, with 99.97% efficiency at the 0.3 micron particle size and essentially 100% capture of particles larger than 0.3 microns. The class I cabinet is designed for work with low to moderate-risk biological agents. It can be used to house various aerosol-generating equipment, including blenders, centrifuges, and mixers. Since the cabinet work zone is not protected from external contamination by the inward flow of unfiltered laboratory air, the cabinet should not be used for work that requires aseptic conditions.

Class II laminar flow cabinets are the most commonly used laboratory containment devices. An air barrier at the front opening of the cabinet provides personnel protection. The air circulating in the workspace is HEPA-filtered, providing protection from contamination for the biological material inside the cabinet. The exhaust is also passed through a HEPA filter and either returned to the room or ducted outside. Class II cabinets are classified as A or B, based on design, airflow, and exhaust. The class II type A cabinet is used for work with biological agents in the absence of volatile or toxic chemicals and radioisotopes, since cabinet air is recirculated within the work zone. These cabinets may be exhausted to the room or externally via ductwork. Class II type B cabinets are ducted directly to the exhaust system; the plena remain under negative pressure.

Class III cabinets are totally enclosed gas-tight ventilated chambers. They are used in laboratories for work with organisms that are highly infectious through the airborne route.

Clean benches are not considered BSCs. They are designed only to protect the product from contamination by providing positive-pressure airflow. Using a clean bench to handle an infectious organism would cause the organism to be exhausted onto the user.

Other engineering controls include special containers for waste and sharps disposal, needleless systems, and devices such as self-resheathing needles.

Administrative controls

Administrative control focuses on maintaining good work habits to minimize exposures due to spills, accidental releases, or other causes. Hands should be washed frequently, work surfaces should be decontaminated properly, and under no circumstances should food, beverages or tobacco products be stored or consumed in the same work area as biohazardous agents. Access to biohazard work areas should be restricted to employees who have had appropriate safety training and who have the necessary personal protective equipment. In laboratories, mouth pipetting should be prohibited.

Personal protective equipment

The use of personal protective equipment (PPE) is indicated whenever the hazards cannot be eliminated through the use of facility design and other engineering controls. Gloves should always be worn when handling infectious agents or secretions from potentially infectious patients or animals. Protective clothing is desirable in many instances, including use of reinforced hand and arm wear (using leather or steel mesh) for certain animal-handling tasks where there is a risk of bite or laceration. Eye protection is important when working with certain airborne biological hazards. Instead of ordinary safety glasses, goggles or face shields should be employed when potentially infectious particulates may arise, such as when performing dental or surgical procedures on potentially infectious patients or animals.

Protection from inhalation exposure to biologicals can be accomplished by wearing a respirator. Because even the most lightweight respirators can be somewhat uncomfortable after prolonged periods of use, engineering controls are preferred except for short-term control. Surgical masks only protect the patient, animal or product from exposure to the worker's exhaled organisms. The worker breathes unfiltered air that enters the airway from around the sides of the mask. To protect the worker from biological hazards in the environment, one of many varieties of certified respirators must be used, such as a HEPA filter mask or dust/mist respirator. Respirator selection should be specific for the hazard and work situation. Employees using respirators are required by OSHA regulation to obtain medical clearance and to attend a training program. Training must include a fit test for the specific type of respirator being worn. These requirements apply to all respirators, including the simple dust mask.

Waste handling

The proper handling, decontamination or containment, and disposal of biological waste is an important infection control measure in all work settings. In medical facilities and laboratories, wastes that are potentially infectious must be initially segregated from other wastes and placed in identifiable biohazard storage bags, affixed with the international biohazard symbol. All sharps must be placed in hard-walled, leakproof and secure containers. Contaminated needles should not be cut or recapped prior to disposal.

Decontamination can be accomplished by means of sterilization, disinfection, sanitization, or antisepsis. **Sterilization** means the eradication of all living microorganisms and spores. **Disinfection** means elimination of most biological organisms, although hardier organisms and spores may survive. **Sanitization** is the lowest level of disinfection and removes most pathogenic organisms. **Antisepsis** means reducing bacterial counts by applying compounds to skin or other body tissues.

Disinfectants have been classified according to chemical composition and level of activity. Commercially available disinfectants often combine one or more agents. The high-level disinfectants possess a broad spectrum of antimicrobial properties and are recommended for use in the destruction of mycobacteria and the blood-borne pathogens; they are often referred to as **mycobactericidal** or **tuberculocidal**. The low-level disinfectants are recommended for use in sanitation and other public health applications. The resistance of microorganisms to chemical disinfection, from most resistant to most sensitive, is as follows: bacterial spores, tubercle bacilli, fungal spores, hydrophilic viruses, mycelial fungi, lipophilic viruses, Gram-negative vegetative bacteria, and Gram-positive vegetative

Table 20.1 Disinfectants and their uses.

Disinfectant	Antimicrobial Activity	Use
Chlorine-liberating halogens	Hypochlorites at 1–5% aqueous concentration possess wide spectrum of activity against microbials, including HIV and HBV	A 1:1000 dilution of household bleach recommended for use against blood-borne pathogens. Chlorination of potable water conducted at 0.2 ppm available chlorine
Formaldehyde	Bactericidal, tuberculocidal and virucidal; hours of exposure required for destruction of bacterial spores; aqueous formaldehyde (formalin) is 37% formaldehyde with 10–15% methanol in water	Usefulness limited by toxicity and odor. Formalin is used as a spray and surface disinfectant. Vapor phase formaldehyde is used to routinely disinfect biological safety cabinets and other enclosures
Alcohols (ethanol, propanol, and isopropanol)	Bactericidal, tuberculocidal and virucidal; devoid of sporicidal activity; most effective concentration is 70% in water	General disinfection of surfaces and equipment. Rapid destruction (in seconds) of vegetative bacteria, fungi and certain viruses. Leaves little to no residue
Glutaraldehyde	Broad spectrum of antimicrobial activity; hours of exposure required for destruction of bacterial spores; used as 2% alkaline glutaraldehyde	Excellent high-level disinfectant for inhalation therapy equipment and other devices. Residues need to be removed with sterile water wash
Iodophors	Not effective as a disinfectant	Primarily used as an antiseptic
Phenolics	Antimicrobial properties of phenolics vary considerably	Used primarily for housekeeping and sanitizing applications (e.g. Lysol)
Quaternary ammonium compounds, e.g. benzalkonium chloride (monoalkyldimethyl benzyl ammonium salt)	Possess detergent and surfactant properties; antimicrobial properties are questionable	Popular for sanitizing

bacteria. Common disinfectants and their uses are listed in Table 20.1.

Sterilization can be accomplished by several techniques. Steam autoclaving is a commonly used method to sterilize cultures and stocks of microorganisms, laboratory ware, and contaminated devices and instruments. Dry heat sterilization (i.e. 160°C for 1–2 hours) can be used to sterilize glassware and metallic instruments when corrosive effects of steam on sharps and cutting edges are undesirable. However, the penetrability and killing effects of dry heat are poorer than those of steam autoclaving or gaseous sterilization. Ethylene oxide, a commonly used gaseous sterilant with high penetrability, is found in most commercial and hospital sterile processing units. Excellent containment is required for these sterilization machines to avoid worker exposures. Ionizing radiation is gaining worldwide acceptance as a commercially feasible sterilization procedure. Attributes of radiation sterilization include penetrability, final package processing, and lack of toxic residues.

SURVEILLANCE

Surveillance for infectious organisms has become a common practice in hospital settings since Semmelweis and others began promoting handwashing and aseptic surgical technique in the 19th century. Hospital infection control programs were initiated in the 1950s in response to the first epidemics of antibiotic-resistant staphylococcal infection among hospitalized patients. An initial enthusiasm for routine environmental culturing has been replaced by monitoring programs that tabulate rates of nosocomial infection among patients.

Target infections are usually surgical wound infections, urinary tract infections, bacteremias, and pneumonias. Now that tuberculosis rates are rising again in the USA, some hospitals are also attempting to determine if nosocomial tuberculosis infection is occurring, especially among patients with the acquired immunodeficiency syndrome (AIDS). When rates are found to be elevated, patient care practices are reviewed to correct deficiencies in urinary catheter care, intravenous equipment care, respiratory therapy, surgical care, patient isolation, or handwashing. In agriculture, animal breeding, veterinary practices, and related settings, infected animals should be segregated and promptly diagnosed. To prevent zoonoses among workers, animals should be treated or killed and disposed of properly.

In industrial settings, occupational disease surveillance generally has a different target group. Workers themselves are monitored to detect disease caused by a work exposure, such as development of elevated blood lead levels among battery-manufacturing workers. Surveillance of only a few infectious diseases (e.g. tuberculosis) is conducted in this manner.

Tuberculosis monitoring through surveillance of workers by tuberculin skin testing has three goals. First, certain workers who convert from skin test negative to positive can be treated with antimicrobial therapy to prevent development of active TB in the worker. For workers whose previous skin test reactivity is unknown and who experience an acute exposure to tuberculosis occupationally, a skin test should be done immediately and then again in 6–12 weeks to check for tuberculin test conversion. Second, early identification of potentially infectious healthcare, food service or other workers with extensive contact with the public can prevent the infection of patients or others. Third, skin test conversion rates provide a measure of the quality of infection control procedures in healthcare workplaces.

HBV surveillance is generally limited to workers who have been exposed to a patient's blood or body fluids, such as from a needlestick. Surveillance of healthcare workers for antibody conversion could be used to assess the quality of infection control procedures. However, this is usually not done because of its expense, because of the sometimes difficult task of determining if a conversion is work-related, and because surveillance for occupational exposures such as needlesticks is more efficient.

When an exposure occurs, a targeted post-exposure follow-up and treatment protocol should be initiated for the employee. The protocol includes a baseline visit in which acute treatment is based on the source's HBV status and the employee's vaccination status. The employee and source should also be evaluated for HIV at baseline. During follow-up over 3–6 months, the employee is then assessed for seroconversion from either virus.

Mandatory screening of physicians, dentists and other healthcare workers for HIV has been hotly debated. Surveillance has not been required or recommended to date, because the risk for transmitting the virus to patients is very low. However, professionals who are infected should not perform exposure-prone invasive procedures and should refrain from patient contact when they have open skin lesions. There is no justification for HIV screening of workers who do not have patient contact.

In addition to healthcare workers, laboratory workers and agricultural workers can also be at risk of contracting infectious diseases occupationally. Infections that could occur from processing human tissue specimens include HBV, HIV tuberculosis, and the following bacterial and fungal pathogens: *Brucella* species, *Francisella tularensis*, *Shigella*, *Salmonella*, *Coccidioides immitis*, *Blastomyces dermatitidis*, and *Histoplasma capsulatum*.

Laboratory animals can potentially transmit to humans HBV, simian immunodeficiency virus, rabies, plague, tuberculosis, or other infections. Wild animals and farm animals or their products can potentially transmit anthrax, brucellosis, erysipeloid, leptospirosis, plague tularemia, candidiasis, coccidiodomycosis, dermatophytoses, histoplasmosis, hookworm, toxoplasmosis, ornithosis, Q fever, Rocky Mountain spotted fever, viral encephalitis, hantavirus, paramyxovirus, rabies, or parasitic infections. Safe handling procedures and vaccination are recommended for preventing these occupational infections rather than surveillance among workers. Universal precautions when handling macaque monkeys are essential, including physically and chemically restraining the monkeys and use of goggles and arm-length reinforced leather gloves. Relying on periodic serologic testing in the colony to determine which monkeys need to be handled cautiously is hazardous, because monkeys may seroconvert between testing but may appear clinically free from infection.

VACCINATION

Vaccinations are given in occupational settings for four common indications. First, employees may be vaccinated to protect them from an infectious organism such as tetanus or hepatitis B when they are at increased risk of being exposed in the workplace. Some vaccines are given as part of a postexposure protocol. Second, healthcare workers may be vaccinated against agents such as rubella or influenza to prevent them from inadvertently passing the infection to patients. Third, some company health units, as part of a corporate wellness program, may provide vaccinations against community-acquired infections such as tetanus to employees who are not at increased occupational risk of the infection. Fourth, vaccinations against agents such as yellow fever may be provided to prepare employees for international travel.

Vaccines commonly administered in occupational settings are listed in Table 20.2. For laboratory workers who handle unusual organisms, consult the chapters on specific organisms later in this book. The infectious disease sections of the Centers for Disease Control can provide helpful information. For information on dosage, administration, and contraindications, consult the package insert for each vaccine. For further information about commonly used vaccines, including those that generally do not need to be administered for occupational indications, consult the American College of Physicians Guide for Adult Immunization.[3] Vaccines used before international travel are listed separately in Table 20.3.

SPECIAL SITUATIONS

Immunocompromised workers

Many individuals remain in the workforce even after developing health problems that may lead to immune system compromise. The most highly publicized group comprises those with HIV infections, but other conditions can also confer some degree of immune dysfunction. Individuals with diabetes mellitus, splenic disorders, renal dysfunction, alcoholism or cirrhosis do not generally require special work restrictions or special work-related immunizations, but they should be followed closely by a personal physician,

Table 20.2 Vaccines and immunobiologicals commonly administered in occupational settings.

Vaccine or immuno biological	Type	Worker groups	Immuno compromised employees	Comments
Hepatitis A vaccine	Inactivated whole virus	Institutional workers (caring for developmentally challenged), child care workers, laboratory workers handling HAV, primate handlers working with animals that may harbor HAV	OK to give	
Hepatitis B vaccine	Recombinant DNA vaccine	Healthcare workers, other workers handling human blood or body fluids	OK to give	No risk of acquiring HIV from vaccine
Hepatitis B immune globulin	Pooled human antiserum	Postexposure, hepatitis B	OK to give	Also known as HBIG
Immune serum globulin	Pooled human antiserum	Postexposure, hepatitis A	OK to give	
Measles, mumps, rubella	Attenuated live viruses	Healthcare workers, day care workers	Do not give	Can test employee for rubella immunity in lieu of vaccination
Polio vaccines	Oral = attenuated live virus Parenteral = killed virus	Laboratory workers handling polio cultures	Do not give the oral or live vaccine	
Rabies vaccine	Inactivated virus	Workers handling animals which may have contracted rabies in the wild	Do not give	
Rabies immune serum globulin	Human antiserum	Postexposure	OK to give	
Tetanus, diptheria vaccine	Killed bacteria and toxoid	Animal handlers, postexposure, corporate wellness program	OK to give	Should have booster every 10 years
Tetanus toxoid	Human antiserum	Postexposure	OK to give	
Varicella vaccine	Attenuated live virus	Healthcare workers	Do not give	
Vaccinia vaccine	Attenuated live virus	Workers handling vaccinia cultures	Do not give	

HAV, hepatitis A virus.

Table 20.3 Vaccines and immunobiologicals commonly administered for international travel.

Vaccine or immuno biological	Type	Required versus recommended	Immuno compromised employees	Comments
Yellow fever	Live attenuated virus	May be required	Do not give	
Cholera	Killed virus	May be required; physician statement contraindicating use for specific patient may be accepted	OK to give	Low efficacy, high side-effects. No longer recommended by travel authorities but some countries still require
Typhoid				
Killed bacteria	Not required but highly recommended for certain high-risk locales	OK to give killed vaccine only	Live attenuated vaccine under development	
Polio	Live attenuated virus (oral) or killed (parenteral)	Recommended for certain high-risk locales	Do not give live attenuated	If not fully immunized previously, complete primary series
Tetanus and diptheria	Killed bacteria and toxoid	Recommended	OK to give	
Immune serum globulin	Pooled human antiserum	Recommended for certain high-risk locales	OK to give	Do not administer at same time as some live virus vaccines, may suppress immune response to them
Hepatitis A vaccine	Inactivated whole virus	Recommended for certain high-risk locales	OK to give	May be administered concomitantly with immune globulin if needed
Hepatitis B vaccine	Recombinant DNA vaccine	Recommended for certain high-risk locales	OK to give	6 months required for full series
Rabies vaccine	Inactivated virus	Recommended if high-risk animal contact is likely	Do not give	
Measles, mumps, rubella	Attenuated live viruses	Recommended for certain high-risk locales	Do not give	
Meningococcal vaccine	Mixed polysaccharides	Recommended for certain high-risk locales	OK to give	
Malaria	Chemoprophylaxis, not a vaccine	Recommended for certain high-risk locales	OK to give	

Table 20.4 Immunocompromising conditions with potential occupational significance.

Condition	Occupational significance
HIV 1 infection	Review work practices Do not administer live vaccines If employee is a healthcare provider, restrict from performing exposure-prone invasive procedures
Organ transplantation, receiving immunosuppressive drugs	Review work practices Consult transplant physician before administering vaccines to avoid non-specific immunologic response which may trigger allograft rejection
High-dose chronic corticosteroid therapy	Review work practices
Malignant disease, receiving immunosuppressive chemotherapy	Consult oncologist to determine if employee is immunosuppressed Review work practices Consult oncologist before administering live vaccines
Congenital immunodeficiency diseases	Review work practices Consult treating physician before giving vaccines Do not give immune globulin to persons with selective IgA deficiency

who may administer vaccines against pneumococcus, influenza, or other agents. Conditions in which special occupational considerations may be warranted are listed in Table 20.4.

Individuals with potentially immunosuppressive conditions should receive training in techniques to prevent exposure to infectious agents in the workplace, including the proper use of personal protective equipment. The need for special vaccines or surveillance should be reviewed, bearing in mind that certain vaccines—especially live virus vaccines—may be contraindicated. Consideration should be given to restricting employees from high-risk work exposures for which protective vaccinations are contraindicated. Although some employees may request to avoid certain other infectious agents and some employers may be able to accommodate these requests, standards generally do not exist for when immunocompromised employees absolutely must be restricted from working with specific infectious agents.

Pregnant workers

Some infectious diseases acquired during pregnancy can cause direct harm to the fetus or substantial maternal morbidity with indirect consequences for the developing fetus. Common agents of concern from the occupational standpoint are rubella, human parvovirus B19, cytomegalovirus, varicella zoster, hepatitis B, coccidiomycosis, and toxoplasmosis. Prenatal screening can determine if a pregnant woman is susceptible to contracting any of these agents during pregnancy. Prenatal infection with Lyme disease, malaria or viral encephalitis has also been associated with adverse fetal outcome, whereas perinatal transmission has not been demonstrated for polio, rabies, or influenza.

Standards are evolving for restricting occupational exposures among pregnant workers. Some authorities focus on educating workers about optimal work practices, because it has not been demonstrated that pregnant workers are any more likely than non-pregnant workers to contract the infections of concern, even though the health consequences of becoming infected may be serious. Other authorities recommend restricting susceptible pregnant employees from working with patients in acute aplastic crisis (human parvovirus B19), adult patients or children shedding cytomegalovirus, or persons with chickenpox or herpes zoster (varicella zoster virus).

The potential reproductive risks of uncommon infectious agents that could be encountered in occupational settings should be evaluated individually. The American College of Obstetricians and Gynecologists maintains a resource center in Washington, DC that can provide technical bulletins on perinatal care. Although female employees are often the focus of concern about occupational reproductive issues, men are also susceptible to reproductive hazards. The best-known infectious reproductive hazard for men is mumps; mumps can cause orchitis, which may lead to sterility.

Workers concerned about contracting disease from coworkers

Serious illness in an employee can generate substantial concern among coworkers. If an employee looks sick but the diagnosis is not known, coworkers may contact company health or safety personnel. Ethical issues about the confidentiality of the employee suspected of illness must then be addressed. HIV infection is an apt example. Even if coworkers suspect that an individual has the infection, there is no reason to invade his or her privacy to confirm or allay fears of infection through casual contact. The virus is not transmitted through handshakes, work surfaces, or telephones. While the virus has been isolated in very small quantities from saliva, there has never been a documented case of salivary transmission through food or eating utensils, or even through kissing on the lips. There is no evidence that the virus will replicate in insects, let alone be transmitted to humans from insects. There is no reason to restrict HIV-infected individuals from engaging in their normal work in order to allay unrealistic fears of contagion among coworkers. Coworker concerns should instead be addressed rapidly and decisively through education.

In general, employees with common viral respiratory infections are not restricted from work, because such viruses are endemic in the community. Employees with suggestive symptoms should be evaluated to rule out tuberculosis. If tuberculosis is confirmed, the individual should be restricted from work until his or her sputum becomes free of acid-fast bacilli. Coworkers should be skin-tested.

Occasionally, employees may become concerned that a coworker with a rash could have an infectious condition, such as scabies. Coworker anxiety may be accompanied by itching. Although scabies transmission is unlikely in the absence of personal contact, it may be helpful for medical personnel to determine the specific diagnosis in the "index" employee, to inform the work unit if treatment of coworkers is indicated, and to provide education on the myriad causes of rash that are not contagious.

International travel

International travel has become commonplace for business purposes. To prevent unnecessary illness or injury, a preventive health review before travel is strongly advised. Information on health and safety conditions for each country to be visited can be obtained from the Centers for Disease Control. The CDC maintains an international travel hotline and travel website. It issues a publication entitled *Health Information for International Travel* and has a database, accessible by computer, that is updated monthly. A printout can be obtained for each country that includes advice on infectious disease hazards, required and recommended immunizations, malaria prophylaxis, and food and water safety; it also details any recent incidents of civil unrest. This list underscores the importance of taking health precautions beyond vaccinations required for visa purposes. Vaccination requirements can also be obtained from embassies, the World Health Organization (Albany, NY), and sometimes from local health departments.

Travel can be categorized as high- or low-risk based on the country to be visited and on the person's itinerary during his or her stay. Travel to Westernized countries or to first-class hotels in some developing countries carries a lower risk than travel to rural areas of developing countries. Immunizations can be separated into those that may be required for visa purposes (e.g. yellow fever and cholera vaccines) and those that are recommended for personal protection (e.g. diphtheria-tetanus, immune serum globulin). The more commonly used vaccines are listed in Table 20.3. For information on contraindications, dosage, and timing of vaccine administration, see thc product information in the Physicians Desk Reference (PDR). The American College of Physicians Guide for Immunization provides a good overview of major aspects of preventive health during foreign travel.

For persons with chronic health problems, it is also advisable to investigate health insurance coverage for foreign travel, to ensure that medications are carried in their original prescription containers, and to obtain the location and telcphone numbers of the US embassies in each country to be visited.

Required reporting

Certain issues related to infectious diseases require reporting by health professionals, including infections that are diagnosed in occupational health units. Each state can provide lists of infections that must be reported to local health departments. Physicians and other healthcare providers must also maintain permanent records of immunization and report certain adverse effects of vaccination to the US Department of Health and Human Services.

Concern about bioterrorism

Concerns about biological warfare and bioterrorism have existed for several decades, and have been heightened by the horrific events of September 11, 2001, and their aftermath.

Several countries are known to have had biological warfare programs and stocks of agents are known to exist. At least 35 agents and organisms have been classified as possible bioterrorism concerns. The most widely discussed are anthrax, botulism, plague, tularemia, and smallpox (Table 20.5). An epidemic of anthrax occurred in the former Soviet Union following an accidental release from a military facility in Sverdlovsk in 1979. As of this writing, in October 2001, there have been exposures to anthrax sent through the mail in the US, resulting in several cases of cutaneous anthrax and three fatalities due to inhalation anthrax. These episodes have led to heightened fears about the possibility of further small and also large-scale bioterroism activities.

While publicized incidents of bioterrorism lead to significant fears, it should be kept in mind that the actual risk of being involved in an event is extremely small. The dissemination of large quantities of bioterroism agents and organisms fortunately has many significant technical challenges.

A prudent response to concerns involves different actions for individuals, professionals, and organizations. In general, people should be reassured that their personal risk is very low, and that they should take reasonable precautions in everyday life. This includes not handling suspicious looking mail and packages, and accessing their local emergency response system as needed. The practices of hoarding antibiotics or using antibiotics without a medical diagnosis should be discouraged.

For the medical and emergency response community, there is a need to learn to recognize the signs and symptoms of bioterrorism agents and organisms (Table 20.5.). This is of special concern because many of these diseases are otherwise uncommon, or, as in the case of smallpox, have not been seen for decades.

Significant improvements in the public health infrastructure are needed to enhance readiness for bioterrorism. Increased funding of these activities is likely in light of the events of 2001.

Table 20.5 Potential Biological Warfare Agents

Disease	Incubation	Symptoms	Signs	Diagnostic tests	Transmission and Precautions	Treatment (Adult dosage)	Prophylaxis
Inhaled Anthrax	2–6 days Range: 2 days to 8 weeks	Flu-like symptoms Respiratory distress	Widened mediastinum on chest X-ray (from adenopathy) Atypical pneumonia Flu-like illness followed by abrupt onset of respiratory failure	Gram stain ("boxcar" shape) Gram positive bacilli in blood culture ELISA for toxin antibodies to help confirm	Aerosol inhalation *No person-to-person transmission* Standard precautions	Mechanical ventilation Antibiotic therapy Ciprofloxacin 400 mg iv q 8–12 hours Doxycycline 200 mg iv initial, then 100 mg iv q 8-12 hours Penicillin 2 mil units iv q 2 hours – possibly add gentamicin	Ciprofloxacin 500 mg po bid or doxycycline 100 mg po bid for ~8 weeks (shorter with anthrax vaccine) FDA-approved vaccine: administer after exposure if available
Botulism	12–72 hours Range: 2 hours – 8 days	Difficulty swallowing or speaking (symmetrical cranial neuropathies) Symmetric descending weakness Respiratory dysfunction No sensory dysfunction No fever	Dilated or un-reactive pupils Drooping eyelids (ptosis) Double vision (diplopia) Slurred speech (dysarthria) Descending flaccid paralysis Intact mental state	Mouse bioassay in public health laboratories (5–7 days to conduct) ELISA for toxin	Aerosol inhalation Food ingestion *No person-to-person transmission* Standard precautions	Mechanical ventilation Parenteral nutrition Trivalent botulinum antitoxin available from State Health Departments and CDC	Experimental vaccine has been used in laboratory workers
Plague	1–3 days by inhalation	Sudden onset of fever, chills, headache, myalgia **Pneumonic:** cough, chest pain, hemoptysis **Bubonic**: painful lymph nodes	**Pneumonic:** Hemoptysis; radiographic pneumonia – patchy, cavities, confluent consolidation **Bubonic:** typically painful, enlarged lymph nodes in groin, axilla, and neck	Gram negative coccobacilli and bacilli in sputum, blood, CSF, or bubo aspirates (bipolar, closed "safety pin" shape on Wright, Wayson's stains) ELISA, DFA, PCR	*Person-to-person transmission in pneumonic forms* Droplet precautions until patient treated for at least three days	Streptomycin 30 mg/kg/day in two divided doses × 10 days Gentamicin 1–1.75 mg/kg iv/im q 8 hours Tetracycline 2–4 g per day	Asymptomatic contacts; or potentially exposed Doxycycline 100 mg po q 12 hours × 7 days Ciprofloxacin 500 mg po Tetracycline 250 mg po q 6 hours × 7 days Vaccine production discontinued

Tularemia "pneumonic"	2–5 days Range: 1–21 days	Fever, cough, chest tightness, pleuritic pain Hemoptysis rare	Community-acquired, atypical pneumonia Radiographic: bilateral patchy pneumonia with hilar adenopathy (pleural effusions like TB) Diffuse, varied skin rash May be rapidly fatal	Gram negative bacilli in blood culture on BYCE (Legionella) cysteine- or S-H-enhanced media Serologic testing to confirm: ELISA, microhemagglutination DFA for sputum or local discharge	Inhalation of agents *No person-to-person transmission but laboratory personnel at risk* Standard precautions	Streptomycin 30 mg/kg/day IM divided bid for 10–14 days Gentamicin 3-5 mg/kg/day iv in equal divided shoulders × 10–14 days Ciprofloxacin possibly effective 400 mg iv q 12 hours (change to po after clinical improvement) × 10–14 days	Ciprofloxacin 500 mg po q 12 hours × 2 weeks Doxycycline 100 mg po q 12 hours × 2 weeks Tetracycline 250 mg po q 6 hours Experimental live vaccine
Smallpox	12–14 days Range: 7–17 days	High fever and myalgia; itching; abdominal pain; delirium Rash on face, extremities, hands, feet; confused with chickenpox which has less uniform rash	Maculopapular then vesicular rash – first on extremities (face, arms, palms, soles, oral mucosa) Rash is synchronous on various segments of the body	Electron microscopy of pustule content PCR Public health lab for confirmation	*Person-to-person transmission* Airborne precautions Negative pressure Clothing and surface decontamination	Supportive care Vaccinate care givers	Vaccination (vaccine available from CDC)

Courtesy of Michael Hodgson, M.D., Office of Public Health and Environmental Hazards, Veterans Health Administration, Washington, D.C.

The animal care community can provide an early warning, since some of the organisms involved are animal pathogens. Veterinarians, farmers, and others who work with and care for animals need to recognize potential public health implications of certain problems seen in animals.

Organizations and companies that handle mail and packages need to develop prudent handling procedures to recognize and isolate suspicious items. Medical, maintenance, safety, and security staffs, and emergency response personnel should receive appropriate training.

Finally, concerns about bioterrorism should be kept in perspective in the context of the everyday risks. Most preventable morbidity and mortality is due to addictions (especially tobacco), modifiable lifestyle factors, and treatable diseases and risk factors. Individuals, healthcare personnel and health systems should maintain and increase the focus on these more mundane issues, as they will ultimately have the greatest impact on life and health.

REFERENCES

1. Centers for Disease Control. Guidelines for preventing the transmission of tuberculosis in health-care settings with special focus on HIV-related issues. MMWR 1990; 39:RR-17. Department of Labor, Occupational Safety and Health Administration. Occupational exposure to bloodborne pathogens; final rule. 29 CFR Part 1910.1030. Washington, DC: Department of Labor 1991.
2. Occupational Safety and Health Administration. Occupational Exposure to Tuberculosis; Proposed Rule. *Fed Register* 10/17/1997; 62:54159–308.
3. American College of Physicians Task Force on Adult Immunization and the Infectious Diseases Society of America. *Guide for adult immunization*, 2nd edn. Philadelphia: American College of Physicians, 1990.

FURTHER READING

American College of Obstetricians and Gynecologists. *Perinatal viral and parasitic infections*. ACOG technical bulletin no. 117. Washington, DC: American College of Obstetricians and Gynecologists, 1993.

Burge HA, Feeley JC. Indoor air pollution and infectious diseases. In: Samet JM, Spengler JD, eds. *Indoor air pollution: a health perspective*. Baltimore: The Johns Hopkins University Press, 1991:273–84.

Committee on Hazardous Biologic Substances in the Laboratory, National Research Council. *Biosafety in the laboratory: prudent practices for the handling and disposal of infectious materials*. Washington, DC: National Academy Press, 1989.

Richmond JY, McKinney RW, eds. *Biosafety in microbiological and biomedical laboratories*, 4th edn. Publication no. 017-040-00547-4. Washington, DC: US Government Printing Office, 1999.

Haley RW. The development of infection surveillance and control programs. In: Bennett JY, Brachman PS, eds. *Hospital infections*, 3rd edn. Boston: Little, Brown, 1992:63–77.

Livingston EG. Infectious agents and non-infectious biologic products. In Frazier LM, Hage ML, eds. *Reproductive hazards of the workplace*. New York, Wiley & Sons, 1998:463–505.

Miller BM, ed. *Laboratory safety: principles and practices*. Washington, DC: American Society of Microbiology, 1986.

North Carolina Department of Labor, Division of Occupational Safety and Health. *Farm safety*. NC-OSHA industry guide no. 10. Raleigh, NC: 1990.

Peter G, ed. *Report of the American Academy of Pediatrics on infectious diseases*, 22nd edn. Elk Grove Village, IL: American Academy of Pediatrics, 1991:348–51.

Sack RB, Barker LR. Immunization to prevent infectious disease. In: Barker LR, Burton JR, Zieve PD, eds. *Principles of ambulatory medicine*, 3rd edn. Baltimore: Williams & Wilkins, 1991:349–57.

Sears SD, Sack RB. Medical advice for the international traveler. In: Barker LR, Burton JR, Zieve PD, eds. *Principles of ambulatory medi-*

cine, 3rd edn. Baltimore: Williams & Wilkins, 1991:358–74.

Wilson ML, Reller LB. Clinical laboratory-acquired infections. In: Bennett JV, Brachman PS, eds. *Hospital infections*, 3rd edn. Boston: Little, Brown, 1992:59–74.

SUGGESTED WEB SITES

Centers for Disease Control and Prevention
http://www.cdc.gov

Occupational Safety and Health Administration
http://www.osha.gov

National Institutes of Health
http://www.nih.gov

Bioterrorism Web Sites

http://www.ama-assn.org/ama/pub/category/6206.html
http://www.bt.cdc.gov
http://biotech.law.umkc.edu/blaw/Bioterror.htm
http://biotech.law.umkc.edu/blaw/govdocs.htm
http://www.hopkins-biodefense.org
http://miemss.umaryland.edu/WMDSupplement.pdf
http://www.nbc-med.org
http://www.usamriid.army.mil/education/bluebook.html

21

VIRUSES

George W. Jackson, M.D.

ARBOVIRUSES

Common names for disease: Dengue fever, Yellow fever, Eastern and Western equine encephalitis, St Louis encephalitis, West Nile encephalitis, Japanese encephalitis

Classification: Family—Flaviviridae and Togaviridae

Occupational setting

Travelers to and workers in Africa, tropical North and South America, and Asia are at risk for Dengue and Yellow fever. Workers exposed to frequent mosquito bites are at risk for the encephalidities in various parts of the USA and around the world. Locations of viral outbreaks have changed and spread. For example, West Nile virus was first identified in the USA in 1999.

Exposure (route)

The arboviruses are transmitted by arthropod bites (various types of mosquitoes). Dengue and Yellow fever are associated with the domestic mosquito (*Aedes aegypti*).[1]

Pathobiology

The flaviviruses and togaviruses are enveloped, single-stranded RNA viruses that are sensitive to heat and detergents.[1]

The incubation period for yellow fever is 3-6 days, followed by the abrupt onset of fever, chills, headache, malaise, nausea, and vomiting. Yellow fever virus causes damage to the liver, kidneys, heart, and gastrointestinal tract. Degeneration of these organs results in the jaundice, hematemesis, and cardiovascular changes characteristic of yellow fever. Case fatality is approximately 5% overall and 20% in patients who develop jaundice. Yellow fever is endemic only in tropical South America and Africa.[2]

Physical and Biological Hazards of the Workplace, Second Edition, Edited by Peter H. Wald and Gregg M. Stave
ISBN 0-471-38647-2

Dengue fever is an illness with an incubation period of 5–8 days. The virus is endemic and epidemic in the tropical Americas, Africa, and Asia. Dengue is characterized by fever, chills, headache, prostration, nausea, vomiting, cutaneous hyperesthesia, and altered taste. The fever is commonly bimodal; a maculopapular rash accompanies the second episode. Hemorrhagic manifestations occur in some cases. A variant of classic dengue is dengue hemorrhagic fever, which is similar to the viral hemorrhagic fevers (e.g. Lassa, Ebola). Dengue is most commonly a self-limited illness requiring only limited supportive measures, but, in rare cases, it can result in severe hemorrhage and fatal shock.[3]

The encephalidities may present primarily as meningitis or encephalitis or both. They are characterized by fever and headache and then signs of meningeal irritation and/or altered levels of consciousness. Lethargy and drowsiness is followed by evidence of abnormal mental status and, in severe cases, focal and generalized seizures. Cerebrospinal fluid usually reveals a lymphocytic pleocytosis with a slight increase in protein and a normal glucose level. The initial step in evaluation is to exclude non-viral causes of meningitis/encephalitis. Herpes virus infection must also be excluded.

Diagnosis

Yellow fever must be differentiated from malaria, hepatitis, typhoid, and the viral hemorrhagic fevers. Specific diagnosis is based on viral isolation or detection of the viral antigen by a specific ELISA test. Dengue must be differentiated from malaria, typhus, influenza, and a variety of other viral syndromes. The hemorrhagic variant is similar to the other hemorrhagic fevers. The specific diagnosis can be obtained by viral isolation or a serologic test.

Treatment

Supportive care should be provided. No specific treatment is available.

Prevention

Protection from mosquito bites in endemic areas is a requisite for travelers and workers in those environments. A vaccine is available for yellow fever that provides long-term (10 years) immunity. In addition to travelers and workers in endemic areas, researchers working with this virus should also be vaccinated. Adverse reactions to the vaccine are infrequent, with <0.5% having to curtail activities post-vaccination. Immediate hypersensitivity has been seen primarily in persons with egg allergy. Individuals with altered immune states should not be vaccinated.[2,4]

REFERENCES

1. Tsai T. Flaviviruses (Yellow Fever, Dengue, Dengue Hemorrhagic Fever, Japanese Encephalitis and St Louis Encephalitis. In: Mandell GL, Bennett JE, Dolin R, eds. *Principles and practice of infectious disease.* 5th edn. New York: Churchill Livingstone, 2000, 1714–36.
2. Centers for Disease Control. Yellow fever vaccine. *MMWR* 1990; 39(RR-6):1–6.
3. Hayes EB, Bugler DJ. Dengue and dengue hemorrhagic fever. *Pediatr Infect Dis* 1992; 11:311–7.
4. Lange WR, Beall B, Deny SC. Dengue fever: a resurgent risk for the international traveler. *Am Fam Phys* 1992; 45:1161–8.

ARENAVIRUSES

Common names for disease: LCMV—lymphocytic choriomeningitis; Tacaribe viruses—South American hemorrhagic fevers; Lassa virus—viral hemorrhagic fever, Lassa fever

Classification: Family—Arenaviridae

Occupational setting

Healthcare workers may be at risk. These illnesses may be of importance to healthcare and research workers and travelers working in areas with endemic infection, including Europe, West Africa, and the Americas. Researchers working with arenaviruses have become infected.[1]

Exposure (route)

Small-particle aerosols are involved in person-to-person spread. Direct contact with rodents and rodent bites is the mode of transmission in endemic areas. Aerosolized virus from rodent excreta can result in the transmission from the rodent reservoir to humans.[2]

Pathobiology

The arenaviruses are enveloped single-stranded RNA structures with a natural reservoir in rodents. The various agents in this group exhibit specificity for rodent species. LCMV infects *Mus musculus*; Lassa infects *Mast omys natalensis*; and the American hemorrhagic fever viruses infect *Calomys masculinus* and *Calornys callosus*.[3]

LCMV infection is primarily seen in Europe and the Americas. A spotty distribution of infected mice has been seen when studies have been conducted in urban settings. Aerosol spread, direct rodent contact and rodent bites are the method of human infection. The incubation period is highly variable, but commonly ranges from 5 to 10 days.

Lymphocytic choriomeningitis begins as a low-grade fever with headache and myalgias. Lymphadenopathy and a maculopapular rash may develop during the initial 3–5-day period. A 2–4-day decrease in fever is followed by several days of recurrent high fever and more severe headache. A minority of patients develop clinical meningitis during this second phase. Rare complications include encephalitis, orchitis, pericarditis, and arthritis.[4] Congenital LCMV infection is proposed as a cause of central nervous system (CNS) disease in infants.[5]

The Lassa fever virus causes an illness characterized by an insidious onset with fever, sore throat, and malaise. Joint pain, headache, cough, and GI symptoms frequently follow. The initial aspect of the illness lasts for 1 week and then resolves in mild cases. Four to five per cent of cases develop the severe manifestations of mucosal bleeding, eighth-nerve deafness, and pleural/pericardial effusions. The presence of retrosternal pain and proteinuria in this syndrome is helpful diagnostically. The incubation period is approximately 1–3 weeks.[6]

The South American hemorrhagic fevers have a higher overall mortality rate than Lassa. Hemorrhage due to capillary leakage and/or a neurologic syndrome are seen in the more severe cases and are associated with the highest mortality.

Incubation of the virus is variable, depending on route of transmission, but ranges from 2 to 9 days.[7]

Diagnosis

Diagnosis of these illnesses is generally made by clinical and epidemiologic parameters. Specific diagnosis is by the isolation of virus from blood, urine, or Cerebrospinal fluid (CSF). The presence of IgM antibody to the specific virus and a titer rise in IgG antibody, acute versus convalescent, are diagnostic.[8]

Treatment

Recovery without treatment is the outcome of most infections with these viruses. Intensive supportive therapy reduces mortality in the severe cases. Ribavirin is reported to reduce morbidity and mortality with early administration in cases of viral hemorrhagic fever.[9]

Medical surveillance

Surveillance is indicated in hospital or research workers with possible exposure. Employees with a febrile illness should be fully evaluated. Workers exposed to foreign work environments in which these viruses are endemic should be evaluated thoroughly if a febrile illness occurs within the incubation period.

Prevention

No effective vaccines are available. Careful handwashing and barrier personal protective equipment are indicated for healthcare employees. Initial reports of high levels of person-to-person transmission of Lassa virus in hospitals have not recurred in more sophisticated healthcare settings. Patient isolation for blood and body fluids and respiratory transmission is indicated. All secretions and contaminated materials should be treated as biological hazards. High-level biosafety containment (BL3 or BL4) is indicated for laboratory research and viral isolation with these viruses, because aerosol infectivity is high. These viruses are readily inactivated by soaps and detergents, 10% bleach solutions, and other common disinfectants.[10]

REFERENCES

1. Barry M, Russi M, Armstrong L, et al. Treatment of a laboratory acquired Sabia virus infection. *N Engl J Med* 1995; 333: 294–6.
2. Howard CR, Simpson DIH. The biology of arenaviruses. *Gen Virol* 1980; 51:1.
3. Peters CJ. Lymphocytic choriomeningitis virus, Lassa virus and the South American Hemorraghic fevers. In: Mandell GL, Bennet JE, Dolin R, eds. *Principles and practice of infectious disease,* 5th edn. NewYork: Churchill Livingstone, 2000, 1855–62.
4. Lehmann-Grube E. *Lymphocytic choriomeningitis virus*. New York: Springer, 1971.
5. Barton LL, et al. Lymphocytic choriomeningitis virus: an unrecognized teratogenic pathogen. *Emerg Infect Dis* 1995; 1(4).
6. McCormick JB. Clinical, epidemiologic, and therapeutic aspects of Lassa fever. *Med Microbiol Immunol* 1986; 175:153.
7. Maiztegui JI. Clinical and epidemiologic patterns of Argentine hemorrhagic fever. *Bull WHO* 1975; 52:567.
8. Jahrling PB, Niklasson BS, McCormick JB. Early diagnosis of human Lassa fever. *Lancet* 1985; 1:250–2.
9. McCormick JB, King U, Webb PA, et al. Lassa fever. Effective therapy with ribavirin. *N Engl J Med* 1986; 314:206.
10. Centers for Disease Control. Management of patients with suspected viral hemorrhagic fever. *MMWR* 1988; 37:53:1014.

CYTOMEGALOVIRUS (CMV)

Common names for disease: CMV mononucleosis, heterophile-negative mononucleosis

Classification: Family—Herpesviridae

Occupational setting

Child day care workers are potentially at greater risk of exposure than other occupational groups because of the high prevalence of CMV infection in children. Seroconversion rates among susceptible workers range from 8% to 20% per year, as compared to 2% for women of the same age in the general population.[1] Although the potential for occupational exposure to CMV among healthcare workers has been a concern in the past, several recent studies have found that the risk of CMV acquisition among healthcare workers is not higher than that of the general population.[2–4]

Exposure (route)

Transmission requires intimate contact with secretions or body fluids of an infected individual. The virus has been isolated from blood, saliva, urine, tears, breast milk, cervical secretions, and semen, suggesting multiple possible modes of transmission. Congenital infection is well documented.[4]

Pathobiology

CMV is a member of the herpes virus family and is a double-stranded DNA virus. Infection is usually asymptomatic; only 10% of adults develop a mononucleosis syndrome. This may include fever, malaise, hepatitis, lymphadenopathy, or splenomegaly. Infrequent manifestations of CMV infection include GuillainBarre syndrome, meningoencephalitis, myocarditis, thrombocytopenia, hemolytic anemia, and granulomatous hepatitis. After primary infection, the virus becomes latent, but it can reactivate with production of infectious virions. CMV is the most common cause of congenital viral infection in the USA, affecting about 1% of all newborns. However, 90–95% of these infants are asymptomatic at birth. Symptomatic congenital infection can be severe, with mortality of 10–20% and major morbidity in 90% of survivors. Symptomatic infection usually results from primary CMV infection during pregnancy, but it occurs in only 2–4% of cases.[4]

Diagnosis

The diagnosis of CMV infection requires laboratory confirmation, but it should be suspected in cases of mononucleosis where tests for EpsteinBarr virus (EBV) are negative (e.g. monospot, heterophile). A rise in CMV-specific IgG titer over time or the presence of IgM antibody usually reflects active CMV infection. Viral culture remains the gold standard in the diagnosis of CMV but its practicality is limited because of the time and costs involved.[4]

Treatment

CMV infection is usually mild in the immunocompetent host and is treated with rest and supportive care. Specific treatment is usually not indicated. Several antivirals are used for severe complications of CMV infection, usually in immunocompromised patients. No treatment has been shown to protect against congenital infection.[4]

Medical surveillance

Screening programs to identify susceptible workers are not recommended by the CDC.[5] For pregnant healthcare workers whose antibody status is negative or unknown, there are no data to indicate that job reassignment or modification alters the risk of infection.[4] This issue remains controversial; some authors recommend counseling of child day care workers who could become pregnant about prevention and potential consequences of CMV infection during pregnancy.[6]

Prevention

The only known effective method of reducing the risk of CMV infection is to stress careful handwashing and strict adherence to universal infection precautions.[5]

REFERENCES

1. Reves R, Pickering L. Impact of child day care on infectious diseases in adults. *Infect Dis Clin North Am* 1992; 6:239–50.
2. Balcarek K, Bagley R, et al. CMV infection among employees of a children's hospital. *JAMA* 1990; 263:840–4.
3. Sepkowitz K. Occupationally acquired infections in health care workers. *Ann Intern Med* 1996; 125(11).
4. Pomeroy C, Englund J. CMV: Epidemiology and infection control. *Am J Infect Control* 1987; 15:107–18.
5. Williams W. CDC guidelines for the prevention and control of nosocomial infections. *Am Infect Control* 1984; 12:34–63.
6. Pass R, Hutto C, Lyon M, Cloud G. Increased rate of CMV infection among day care center workers. *Pediatr Infect Dis J* 1990; 9(7):465–70.

David P. Siebens

FILOVIRUSES

(Ebola and Marburg viruses)

Common name for disease: Viral hemorrhagic fever

Classification: Family—Filoviridae

Occupational setting

Researchers studying the viruses and primate handlers are at risk from exposure. Healthcare personnel treating patients with these infections may also be at risk.

Exposure (route)

The route of acquisition of natural infection is unclear, but person-to-person transmission occurs by direct contact with blood and other body fluids. The natural reservoir of the viruses is also unclear, but both are associated with cynomolgus or green monkeys, and these viruses have been imported in sick primates. In the laboratory, these viruses have been shown to be highly infectious via the respiratory route.

Pathobiology

These viruses are enveloped, bacillus-like rods with an RNA genome. Because of their highly hazardous nature, these viruses are investigated in high-level (BL4) microbiological containment facilities.[1] Both viruses are associated with East and South Africa but probably are present in Asia as well. After an incubation period of 4–16 days, both viruses cause a similar illness that is characterized by fever, headache, myalgia, and conjunctivitis. Nausea, vomiting and diarrhea appear in 2–3 days, followed by a maculopapular rash and hemorrhage from various body parts.[2] The case fatality rate is very high—approximately 20% for Marburg and approaching 90% for some strains of Ebola. (By contrast, one strain of Ebola, Ebola Reston, appears not to be pathogenic for humans.) Patients who develop hemorrhage related to the processes of disseminated intravascular coagulation (DIC) and/or platelet dysfunction generally do not recover.

Diagnosis

Because of the rarity of these illnesses, the clinical diagnosis is difficult and generally missed except in those cases where there is a high index of suspicion due to known exposure. A rapidly progressing, severe viral illness in a primate handler or exposed healthcare worker should be evaluated for possible hemorrhagic fever. Viral diagnostic studies such as IgM or rising IgG antibodies by indirect immunofluorescence or viral culture can provide specific diagnosis.[3]

Medical surveillance

Surveillance of workers who have had possible exposure to these viruses should be instituted for the development of febrile illness. Any illness should be thoroughly evaluated.

Treatment

There is no known effective treatment.

Prevention

Control of exposure is the only effective approach to prevention. All primates entering the USA are quarantined for at least 35 days to control this transmission route. Animal handlers must follow work practice guidelines established by the Centers for Disease Control, with personal protective equipment as indicated.[4] Healthcare workers must institute high-level enteric and respiratory isolation for patients with known infection. Universal precautions provide some protection in the unknown case. Disinfection of equipment and fomites is necessary. Research on these viruses is conducted only in a few BSL4 facilities using stringent precautions. These viruses are sensitive to soaps, detergents and 10% bleach solutions.[5]

REFERENCES

1. Sanchez A, Kiley MP. Identification and analysis of Ebola virus proteins. *Virology* 1987; 157:414.
2. Peters CJ. Marburg and Ebola virus hemorrhagic fevers. In: Mandell GL, Bennett JE, Dolin R, eds. *Principles and practice of infectious disease*. 5th edn. New York: Churchill-Livingstone, 2000, 1821–3.
3. Fisher-Hock SP, Platt GS, Neild GH, et al. Pathophysiology of shock and hemorrhage in a fulminating viral infection (Ebola). *J Infect Dis* 1985; 152:887–94.
4. Centers for Disease Control. Update: Ebola-related filovirus infection in nonhuman primates. *MMWR* 1990; 39:2:22–30.
5. Centers for Disease Control. Update: Management of patients with suspected viral hemorrhagic fever—United States. *MMWR* 1995; 44(25):475–9.

HANTAVIRUSES

Common names for disease: Hemorrhagic fever with renal syndrome, nephropathia epidemica, hantavirus pulmonary syndrome (HPS)

Classification: Family—Bunyaviridae

Occupational setting

Outdoor workers in endemic areas may be at risk from exposure. The risk may be greater for rodent control workers. Laboratory workers have become infected while working with infected wild and laboratory rodents.[1–3]

Exposure (route)

Unlike for other Bunyaviridae, there is no insect vector. Infection is spread from infected rodents to humans through direct contact with infected rodent secretions, urine, and feces, or by inhalation of airborne infected dust particles. These infectious aerosols may develop either by direct aerosolization of infected rodent urine (or other potentially infected body fluids), or by secondary aerosolization of dried rodent excreta. Rodent bites have resulted in disease transmission. Person-to-person transmission is rare.[1,4]

Pathobiology

The hantaviruses are spherical enveloped RNA viruses with a three-part genome. In general, the hantaviruses fall into two groups; old world and new world. There are greater than 25 subgroups. The old world group is primarily associated with hemorrhagic fever with renal syndrome (HFRS), while the new world group express as the hantavirus pulmonary syndrome (HPS). Various rodent species tend to harbor a specific viral subgroup; for example, the Sin Nombre group is found in deer mice. Most of the rodent reservoirs are not common in urban areas.[1]

The Hantaan and Seoul viruses cause a severe form of hemorrhagic fever with renal syndrome (HFRS). It is endemic in Scandinavia, eastern Europe, and the Far East. During the Korean War, thousands of military personnel were infected.

The incubation period for HFRS is 12–21 days, with a mild, non-specific prodrome followed by the febrile, hypotensive, oliguric, diuretic and convalescent phases. The febrile phase may include chills, high fever, lethargy, dizziness, headache, photophobia, myalgia, abdominal and back pain, anorexia, nausea, and vomiting. Most patients experience a slow, uneventful recovery. Others go on to develop the classical symptoms of hypotension, hemorrhage, and renal failure. Estimates of the case fatality rate range from 1% to 10%.[1,5,6]

The pathogenesis of the illness is complex. Vascular dysfunction with impaired vascular tone and increased vascular permeability produce hypotension and shock. Hemorrhage may result from disseminated intravascular coagulopathy. Renal dysfunction probably

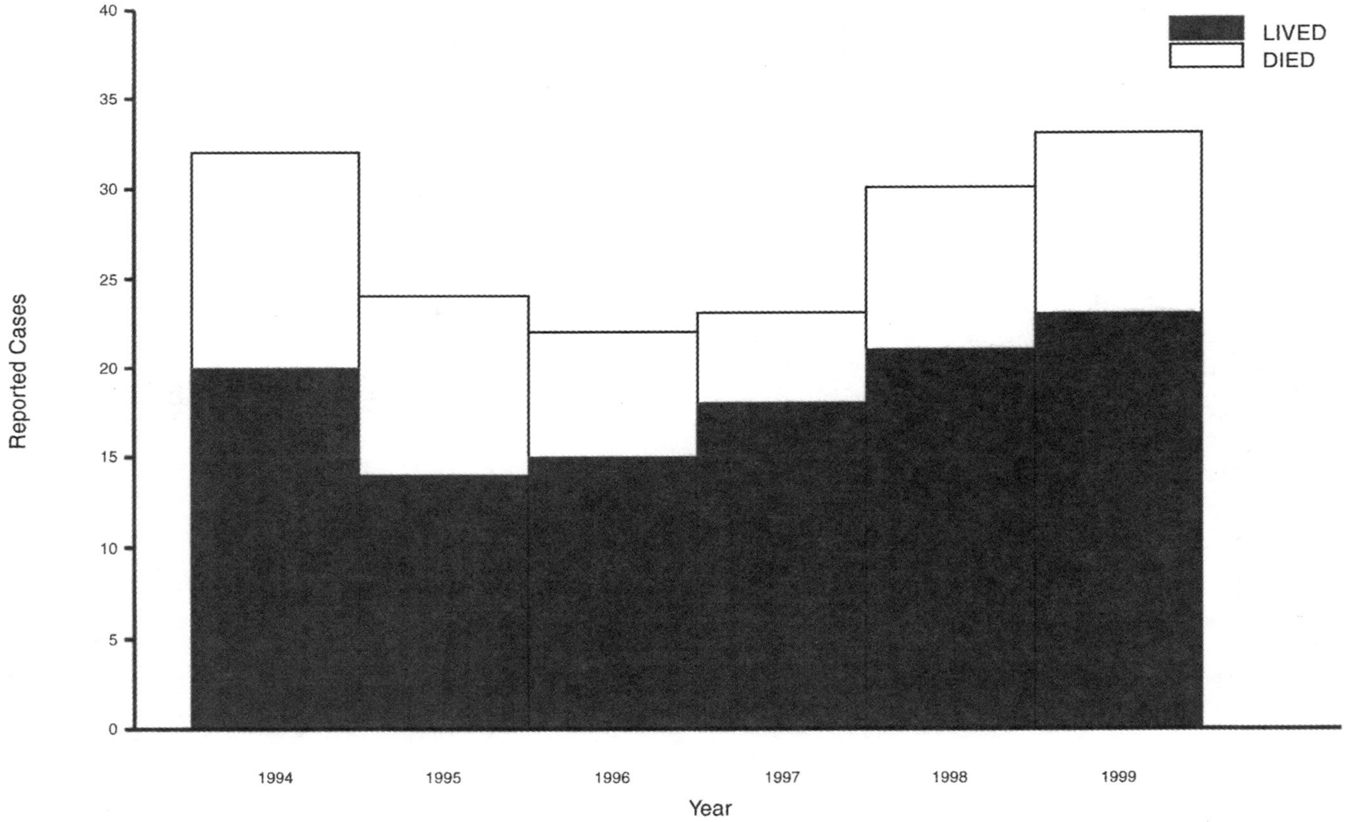

Figure 21.1 Hantavirus Pulmonary Syndrome—reported cases by survival status, by year, United States, 1994–1999. Source: MMWR 2001: 48(53):44.

results from antibody complex deposition, as opposed to a direct effect of the virus.[7]

A less severe illness, nephropathica epidemica, is caused by the Puumala virus in Europe. Patients experience the acute onset of fever, headache, oliguria, and back pain. Most patients develop thrombocytopenia, and one-third of them demonstrate evidence of bleeding. Most patients recover in 6 months. The case fatality rate is ~0.2%.[8]

The Prospect Hill virus is endemic in some rodent populations in the USA. It can produce human infection but has not produced disease.[4,9]

Before 1993, no cases of human disease due to hantaviruses had been reported in the USA. In 1993, an outbreak of an acute illness occurred initially in the Four Corners region of the southwest. It was characterized by the abrupt onset of fever, myalgias, headache, and cough, followed by the rapid development of respiratory failure. Hemorrhage and renal failure were not components of the syndrome.[4] The causative agent was identified as a previously unknown hantavirus carried by the deer mouse (*Peromyscus maniculatus*).[4] This syndrome, now known as hantavirus pulmonary Syndrome (HPS) has a 1–2 week incubation. The onset is non-specific with fever and myalgias. Progression to dsypnea, pulmonary edema, shock and death may occur in as many as 50% of cases (Figure 21.1). Hantavirus disease has now been identified in various sites in the Americas.

Diagnosis

During the prodrome, specification of hantavirus infection is difficult, but soon after presentation, shortness of breath and a cough may be present. The index of suspicion is raised by myalgias of large muscle groups plus gastrointestinal symptoms such as nausea, vomiting and abdominal pain. Interstitial edema is frequently present on chest X-ray. Bilateral alveolar edema develops later, and pleural effusions may be seen. Specific diagnosis is made with IgM testing of acute-phase serum. Sin Nombre virus antigen, a common subgroup, will cross-react with other hantaviruses which cause HPS in the Americas.[5]

Treatment

Treatment consists primarily of supportive care. This may include mechanical ventilation, renal dialysis, and transfusions, as appropriate. In the USA, ribavarin is available through an Investigational New Drug (IND) protocol,[4] although initial studies have not shown a dramatic response.[10]

Prevention

Control of rodent populations in endemic areas is helpful but may not be practicable. Rodent nests and dead rodents can be wetted with disinfectant prior to removal. Activities that may result in contact with rodents or aerosolization of rodent excreta should be avoided. For workers involved in animal control activities in endemic areas, work practices and personal protective equipment should protect from exposure by direct skin contact and inhalation. In the healthcare setting, universal precautions are considered adequate to prevent exposure. In the research setting, use BSL2/3 procedures and facilities.

REFERENCES

1. Hart CA, Bennett M. Hantavirus infections: epidemiology and pathogenesis. *Microbes Infect* 1999; 1: 1229–37.
2. Desmyter J, LeDuc JW, Johnson KM, et al. Laboratory rat associated outbreak of haemorrhagic fever with renal syndrome due to Hantaan-like virus in Belgium. *Lancet* 1983; 2:1445–8.
3. Lloyd G, Jones M. Infection of laboratory workers with hantavirus acquired from immunocytomas propagated in laboratory rats. *J Infect* 1986; 12:117–25.
4. Centers for Disease Control. Outbreak of acute

illness—Southwestern United States, 1993. *MMWR* 1993; 42:421–4.
5. Peters CJ, et al. Spectrum of hantavirus infection: hemorrhagic fever with renal syndrome and hantaviruses pulmonary syndrome. *Ann Rev Med* 1999; 50:531–45.
6. Niklasson BS. Haemorrhagic fever with renal syndrome, virological and epidemiological aspects. *Pediatr Nephrol* 1992; 6(20): 1–4.
7. Cosgriff TM. Mechanism of disease in hantavirus infection: pathophysiology of hemorrhagic fever with renal syndrome. *Rev Infect Dis* 1991; 13:97–107.
8. Doyle TJ, et al. Viral hemorrhagic fevers and hantavirus infections in the Americas. *Infect Dis Clin North Am* 1998; 12(1):95–110.
9. Yanagihara R. Hantavirus infection in the United States: epizootiology and epidemiology. *Rev Infect Dis* 1990; 12:449–57.
10. Huggins JW, Hsiang CM, Cosgriff TM, et al. Prospective, double-blind, concurrent, placebo-controlled clinical trial of intravenous ribavarin therapy for hemorrhagic fever with renal syndrome. *J Infect Dis* 1991; 164:119–27.

HEPATITIS A VIRUS (HAV)

Common name for disease: Infectious hepatitis

Classification: Family—Picornaviridae; genus—enterovirus

Occupational setting

Among six groups identified by the CDC as being at increased risk for hepatitis A, two have occupational implications: people from developed countries who travel to, or work in, developing countries that have high or intermediate endemicity, and those who work with non-human primates[1] Workers in food handling, day care centers, and healthcare institutions do not display increased prevalence rates, although these work settings may be subject to outbreaks or play a crucial role in transmission.[2] An increased risk for workers exposed to sewage has been suggested by some earlier studies, but not confirmed by more recent CDC data.[1]

Exposure (route)

Infection is transmitted almost exclusively by the fecal–oral route. Transmission by saliva or blood transfusion is possible but uncommon.

Pathobiology

This 27-nm, symmetrical RNA enterovirus is very stable against heat, cold, and non-ionic detergents. Pathogenesis appears to be mediated through a T-cell mechanism rather than by a direct cytopathic effect.

Hepatitis A occurs worldwide, with cyclical epidemics and late autumnal peaks. Under-reporting is a significant concern. For instance, the 30 021 US cases reported to the National Notifiable Diseases Surveillance System in 1997 are estimated by the CDC to represent 90 000 symptomatic cases and 180 000 persons infected with HAV that year. From 1987 to 1997, the reported US incidence averaged 10.8 cases per 100 000 population. Counties exceeding that average are concentrated in the western USA (Figure 21.2). The serologic prevalence of anti-HAV antibodies, suggesting prior HAV infection, varies directly with age, and is found in approximately one-third of the US population.

Geographically, anti-HAV antibody prevalence varies widely (from 10% in the USA to 90% in Israel and Yugoslavia among 18–19-year-olds). Fewer than 5% of infections are thought to be clinically recognized worldwide. Infection is typically asymptomatic in children. Adults usually have symptoms and/or jaundice, often not identified as signaling clinical hepatitis.[2] Fecal shedding has usually ceased when adults are diagnosed (may be detectable as late as 2 weeks after symptom onset), but it persists longer in preterm infants and, perhaps, in children.[3,6]

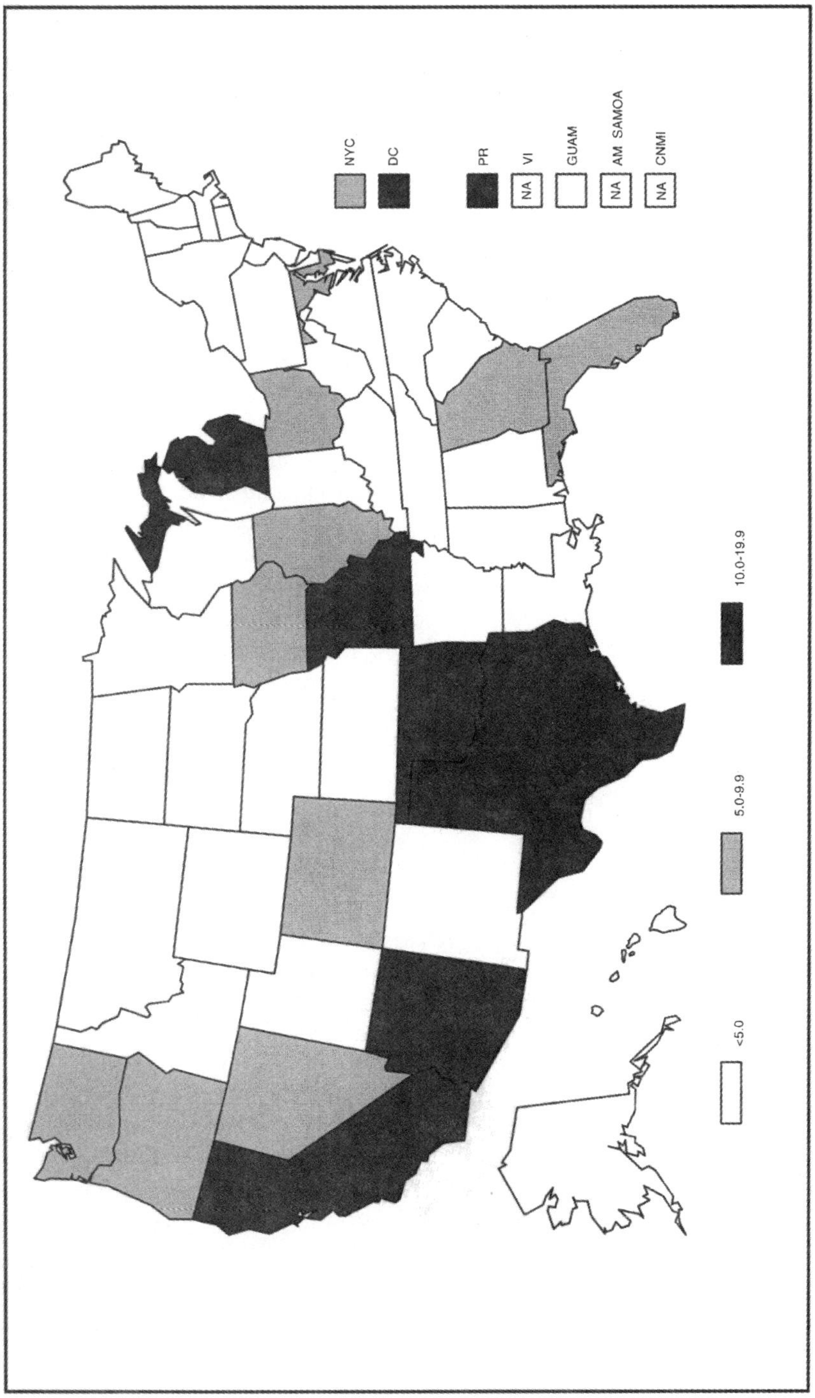

Figure 21.2 Hepatitis A—Reported cases per 100,000 population, United States, 1999. Source: MMWR 2001:48(53):47.

The incubation period lasts 15–45 days. A prodrome variably includes anorexia, nausea, fatigue, arthralgia, headache, upper respiratory symptoms, low-grade fever, and photophobia; jaundice follows the prodrome after 1–2 weeks, often with hepatomegaly that persists into the early recovery phase. Fulminant disease is rare, with an estimated case-fatality rate of 0.3% (1.8% for adults over 50).[1] There is occasional recurrence, but no recognized chronic or carrier state.

Diagnosis

Clinical features plus elevated alkaline aminotransferase (ALT), aspartate aminotransferase (AST) and bilirubin support the diagnosis. Elevated anti-HAV IgM antibody confirms acute infection and is reliably present at onset of symptoms.

Treatment

Supportive care includes fluid replacement, high-calory diet, and some degree of restricted physical activity.

Medical surveillance

Surveillance of exposed susceptible healthcare and institutional workers may be indicated during the incubation period.

Prevention

Two inactivated virus vaccines have been licensed in the USA since 1995. These vaccines are highly immunogenic, producing protective anti-body levels in 94–100% of calcium at 1 month after the first dose. A second dose is recommen-ded at 6–12 months for long-lasting immunity.

Comprehensive sanitation and hygiene measures remain of great importance, along with early recognition, investigation, and appropriate immunization efforts for outbreaks. The CDC recommends hepatitis A vaccine for persons at increased risk and others wishing to obtain immunity. However, the CDC makes formal vaccination recommendations based on prior occupational risk only for those who work with HAV-infected primates, or in an HAV research laboratory.

Enteric infection control procedures are crucial for occupations at risk, since transmission occurs primarily during the presymptomatic incubation phase. HAV retains its infectivity on environmental surfaces for at least several days, is effectively transmitted on human hands, and resists many common disinfectants.[4] Effective decontaminating agents for environmental surfaces include quaternary ammonium compounds with 23% HCl, 2% glutaraldehyde, or hypochlorite with >5000 ppm free chlorine.[5]

Travelers in developing countries or other areas where sanitation is of concern should scrutinize (if not avoid) local water (for drinking or food preparation), milk, shellfish, and uncooked fruits and vegetables. In all settings (and notably in day care of diapered children), thorough handwashing with running water and single-use towels are essential after potentially infectious contact and before handling food.[6]

Susceptible travelers or expatriates to locations of intermediate or high endemicity should receive a first dose of hepatitis A vaccine at least 4 weeks prior to departure, or passive immunization with 0.02 ml/kg immune globulin (IG) alone, or supplemental IG at a separate injection site if the interval between the first dose of hepatitis A vaccine and departure is too short to produce sufficient immunity. If IG is used alone for a prolonged trip (more than 3 months), the initial dose can be increased to 0.06 ml/kg, and immunization with IG repeated at 5-month intervals. IG preparation in the USA involves several procedures to assure it is not a conduit for other viral disease.[1] IG preparations in other countries have occasionally been the conduit for infectious agents.

For susceptible persons, postexposure prophylaxis with 0.02 ml/kg may be up to

85% effective in preventing clinical hepatitis A if administered up to 2 weeks after exposure. There is no apparent harm in administering hepatitis A vaccine or IG to persons with existing immunity (e.g. from prior undiagnosed infection). Therefore, any decisions about serologic testing before immunization should be based on projected cost reduction. The vaccine should not be used without IG for postexposure prophylaxis. Previously unvaccinated households and sexual contacts of persons with serologically confirmed hepatitis A are candidates for postexposure prophylaxis.

Some organizations have elected to vaccinate food service or child care workers because of the consequences of infection for the population served, even though these occupations are not necessarily at increased risk by themselves.

REFERENCES

1. Bell BP, Wasley A, Shapiro CN, Margolis HS. Prevention of hepatitis A through active or passive immunization: recommendations of the Advisory Committee on Immunization Practices. *MMWR* 1999; 48:1–37.
2. Forbes A, Williams R. Changing epidemiology and clinical aspects of hepatitis A. *Br Med Bull* 1990; 46:303–18.
3. Rosenblum LS, Villarino ME, Nainan OV, et al. Hepatitis A outbreak in a neonatal intensive care unit: risk factors for transmission and evidence of a prolonged viral excretion among preterm infants. *J Infect Dis* 1991; 164: 476–82.
4. Mbithi JN, Springthorpe, Boulet JR, et al. Survival of hepatitis A on human hands and its transfer on contact with animate and inanimate surfaces. *J Clin Microbiol* 1992; 30:757–63.
5. Mbithi JN, Springthorpe, Satter SA. Chemical disinfection of hepatitis A virus on environmental surfaces. *Appl Environ Microbiol* 1990; 56:3601–4.
6. Hadler SC, McFarland L. Hepatitis in day care centers: epidemiology and prevention. *Rev Infect Dis* 1986; 8:548–57.

Samuel D. Moon

HEPATITIS B VIRUS (HBV)

Common name for disease: Serum hepatitis

Classification: Family—Hepadnaviridae; genus—hepadnavirus type 1

Occupational setting

Healthcare workers are at increased risk of infection if they are exposed to blood products, especially surgeons, pathologists, dentists and dental hygienists, personnel in emergency departments, personnel in labor and delivery, laboratory technicians and researchers, and hemodialysis staff. Other occupations with increased risk include hospital custodial staff, morticians, and employees of institutions that care for the mentally disabled.

The prevalence of HBV surface antigen (HBsAg) in the serum of healthcare workers with no or infrequent blood contact (0.3%) is no higher than in the general population; that in healthcare workers with frequent blood contact—the group at highest risk of work-related HBV infection—is 1–2%.[1] Some non-occupational groups with high prevalence include immigrants and refugees from areas of high HBV endemicity (13%), patients in hemodialysis units (3–10%) and institutions for the mentally disabled (10–20%), intravenous drug users (7%), sexually active homosexual men (6%), household contacts of HBV carriers (3–6%), male prisoners (1–8%), and hemophiliacs.

In the USA the reported clinical cases of HBV infection have decreased from a high of 26 611 in 1985 to 10 637 in 1996. (Figure 21.3) Worldwide HBV infection continues to be a major problem with increasing numbers of HBV carriers.

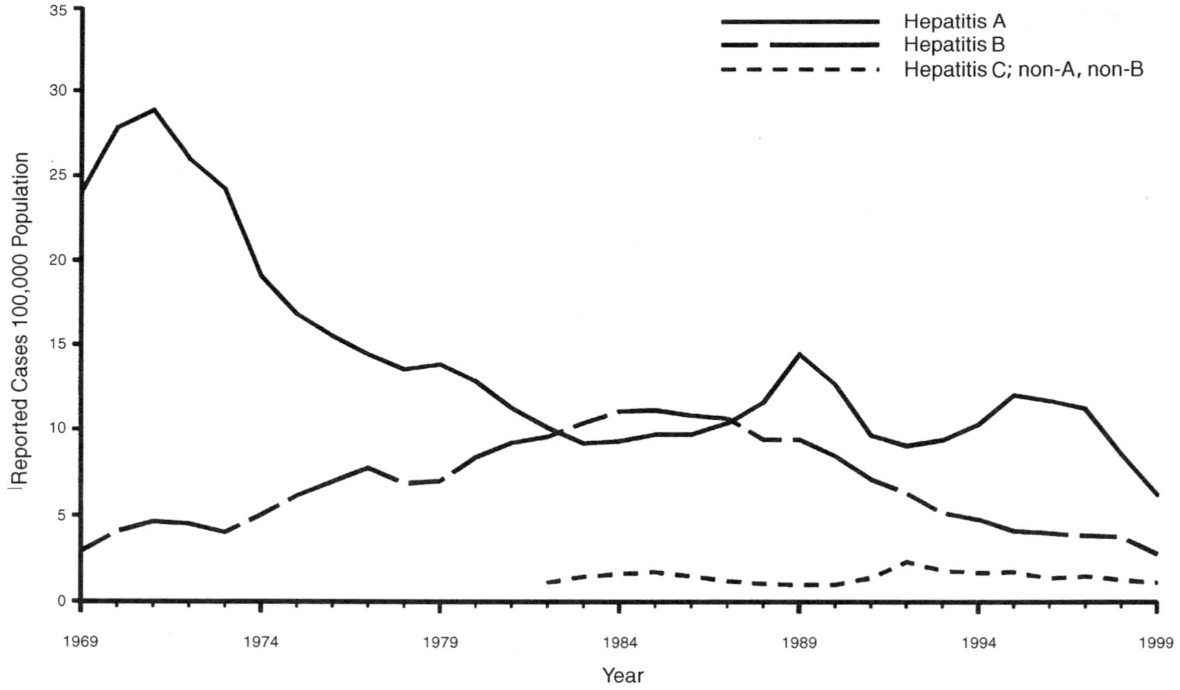

Figure 21.3 Hepatitis—reported cases per 100 000 population by year, United States, 1969–1999. Source: MMWR 2001:48(53):46.

Exposure (route)

Transmission is by parenteral, sexual and perinatal routes. Blood, serum, saliva, semen and vaginal fluid are known to contain HBV. The risk of HBV infection following percutaneous exposure to blood that is positive for hepatitis B e antigen (HbeAg) is 19–27%.[2] Contact of infectious material with skin breaks or mucous membranes, including the eye, can result in infection. There is no evidence for fecal-oral transmission, and transmission via oral secretions is rare (in one recorded case, disease was transmitted via a bite from an infected patient[3]). Indirect transmission of HBV infection can occur from blood-contaminated instruments and other objects.[4]

Transmission of HBV from healthcare workers (usually chronically infected ones) to patients appears to be a rare event.[4] Known and suspected cases have been associated with invasive procedures, either those with high risk of sharps injury or when failure to wear gloves on hands led to some compromise of skin barrier integrity (cuts, dermatoses, bleeding warts). Although transmission risk is low with proper infection control technique,[5] it has been documented in two cases where transmission from infected surgeons to patients occurred despite double-gloving and elimination of certain high-risk procedures.[6]

Healthcare workers who are hepatitis B antigen positive must be carefully evaluated for risk of transmission and may need to be restricted from high-risk, exposure-prone invasive procedures. Most healthcare centers have a confidential process for evaluating this situation.

Pathobiology

HBV is a 42-nm, double-shelled DNA virus. The outer shell, containing HBsAg, surrounds a 27-nm inner core containing the core antigen HBCAg. HBeAg is an internal component or degradation product of the core. All three antigens are immunologically distinct.[7]

The incubation period ranges from 4 to 12 weeks; acute infection lasts for 2–12 weeks; and early recovery or convalescence takes 2–16 weeks. Only 20–50% of cases of acute HBV infections show clinical jaundice (bilirubin 2.5 mg/dl). Cases without evident jaundice (non-icteric cases) are frequently missed because symptoms may be absent or mild. These cases are the most likely to progress to a chronic carrier state (chronic HBV infection). The characteristic hepatic pathology of acute infection includes panlobular, lymphocytic, hepatocellular necrosis, and cholestasis. The overall case fatality rate is low; 90% recover completely.

The clinical syndrome is quite variable and cannot be distinguished from that of hepatitis A in any one individual. Preicteric symptoms and signs, in order of frequency, include malaise, anorexia, nausea, and right upper quadrant abdominal pain. A "flu-like" syndrome of headache, myalgia, weakness, fever and chills is less frequent and is more suggestive of hepatitis A. The serum transaminases AST (SGOT) and ALT (SGPT) show variable elevations during the preicteric phase and peak during clinical jaundice. Five to ten per cent of patients with viral hepatitis (including all types) show a "serum sickness-like" syndrome as the first indication of illness, with the triad of fever, rash, and arthritis.

The icteric phase begins 3–10 days after the onset of other symptoms and may last 13 weeks. Scleral icterus, dark urine and light stools may be seen; the liver may become enlarged and tender; and mild weight loss is common. The serum bilirubin is typically in the 5–20 mg/dl range. Lymphopenia is followed by a relative lymphocytosis. A prolonged prothrombin time (PT), low serum albumin, hypoglycemia level and a very high serum bilirubin level may reflect extensive hepatocellular necrosis and a worse prognosis.

Complications include chronic infection with normal liver function tests, chronic persistent hepatitis (not progressive), chronic active hepatitis (may progress to severe disease and cirrhosis), and fulminant hepatitis. The carrier state increases the risk of hepatocellular carcinoma. Fulminant hepatitis is characterized by hepatic failure and encephalopathy with lethargy, somnolence, and personality changes, progressing in some cases to stupor and coma; mortality is high.

Delta hepatitis virus (HDV) may be seen as a coinfection or superinfection with HBV infection and generally increases the severity of disease. HDV is found only in the presence of HBsAg, which it requires for its replication. Prevalence of HDV is low in the US general population but higher in groups with multiple parenteral exposures, such as intravenous drug users and people with multiple blood transfusions.

Diagnosis

Diagnosis of HBV infection is by specific serology. While epidemiology, clinical presentation and laboratory features direct the differential diagnosis to include HBV infection, many cases are subclinical and, less commonly, may occur with no elevation of serum transaminase activity.[7] Figure 21.4 shows the pattern of serologic changes with time in a typical acute HBV infection with a normal recovery.

HBsAg is the first serologic marker in acute infection; its presence in the serum is diagnostic of active HBV infection. HBeAg appears shortly after HBsAg; its presence indicates high infectivity. The presence of IgM anti-HBc antibody is diagnostic of acute infection, whereas IgG anti-HBc in the presence of HBsAg and HBeAg is diagnostic of chronic infection, or the carrier state. A small but significant percentage of chronic carriers are HBsAg negative and can transmit HBV in donated blood screened for HBsAg.[7] Individuals recovering from acute HBV infection may be considered no longer infectious only after anti-HBs appears and HBsAg is not detectable. HBsAg should be tested every 1–2 months following an acute infection until it disappears. HBsAg persisting 6 months after the acute infection indicates that a carrier state has developed. Five to ten per cent of patients with acute infections become carriers; HbsAg and HBeAg become negative in carriers at rates of 2% and 10% per year, respectively.[7] What this means in terms of potential to transmit the virus is uncertain. Table 21.1 shows patterns that may be seen with a standard HBV serology screen.

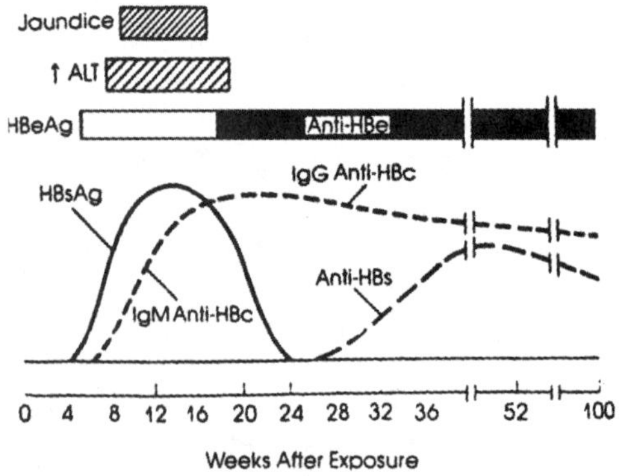

Figure 21.4 Scheme of typical clinical and laboratory features of acute viral hepatitis type B. Source: Reprinted, with permission, Harrison's principles of internal medicine, 12th ed.

Treatment

There is currently no specific treatment for acute HBV infection. General supportive care is indicated. Interferon-α_{2B} is a safe and effective treatment for chronic hepatitis B infection with active liver disease.[8]

Medical Surveillance

Based on current assessment of risk of HBV transmission and on the diversion of resources that would be necessary, mandatory HBV serology testing programs are not recommended by the CDC. However, the CDC does recommend that healthcare workers performing exposure-prone procedures know their HBV immunity status, and, if not immune, their HBsAg status, and, if that is positive, their HBeAg status. Characteristics of exposure-prone invasive procedures include digital palpation of a needletip in a body cavity or the simultaneous presence of the healthcare worker's fingers and a needle or other sharp instrument or object in a poorly visualized or highly confined anatomic site.[6]

Table 21.1 Hepatitis B serologic patterns.

HbsAg	Anti-HBs	HBeAg	Anti-Hbe	Anti-HBc	Interpretation
+	−	−	−	−	Late incubation or early acute HBV, infective
+	−	+	−	−	Early acute HBV, infective
+	−	+	−	IgM	Acute HBV, infective
−	−	−	−	IgM	Serologic window (late acute HBV), infective
−	−	−	+	+	Late acute HBV, low infectivity
−	+	−	+	IgG	Recovery from HBV, not infective, immune
−	+	−	−	IgG	Recovery from HBV not infective, immune
+	−	+	−	IgG	Chronic HBV (carrier), infective
+	−	−	+	IgG	Late acute HBV or chronic HBV, low infectivity
−	−	−	−	IgG	Low-level HbsAg carrier (possibly infective) or remote past infection (not infective, immune
−	+	−	−	−	Active immunization (vaccination) or passive immunization (postHB immunoglobulin infusion) or remote past infection

Prevention

The Occupational Safety and Health Administration (OSHA) mandates (29CFR 1910.1030) that universal blood and body fluid precautions be implemented as part of a written exposure control plan by any employer having employees with "any reasonably anticipated" occupational exposure. Occupational exposure is defined as "skin, eye, mucous membrane, or parenteral contact with blood or other potentially infectious materials that may result from the performance of an employee's duties". Other potentially infectious material includes semen, vaginal secretions, saliva where dental procedures are involved, unfixed tissues or organs, and cerebrospinal, synovial, pleural, pericardial, and amniotic fluids. When a fluid is contaminated with blood, or where it is difficult or impossible to differentiate between body fluids, the material should be considered potentially infectious.

The efficacy of universal precautions, where all human blood and certain body fluids are treated as if known to be infectious for HlV HBV and other blood-borne pathogens, is limited by compliance. Workers should understand the importance of, and be well trained in, the implementation of work practice controls, such as puncture precautions, handwashing and skin-washing, environmental controls, and standard infection control practices, and in the proper use of barrier precautions (personal protective equipment, or PPE). Gloves, masks, protective eyewear or face shields, barrier gowns or aprons, surgical caps/hoods, and shoe covers should be used appropriately.

All healthcare workers should take precautions to prevent injuries: caused by needles, scalpels, and other sharp instruments or devices used during procedures; when cleaning used instruments; during disposal of used needles; and when handling sharp instruments after procedures. To prevent needlestick injuries, needles should not be recapped, purposely bent or broken by hand, removed from disposable syringes, or otherwise manipulated by hand. After they are used, disposable syringes and needles, scalpel blades and other sharp items should be placed in puncture-resistant containers for disposal; the puncture-resistant containers should be located as close as practical to the use area. Large-bore reusable needles should be placed in a puncture-resistant container for transport to the reprocessing area.

Table 21.2 Recommended postexposure prophylaxis for exposure to hepatitis B virus.

Vaccination and antibody response status of exposed workers*	Treatment		
	Source HBsAg† positive	Source HBsAg† negative	Source unknown or not available for testing
Unvaccinated	HBIG§ × 1 and initiate HB vaccine series¶	Initiate HB vaccine series	Initiate HB vaccine series
Previously vaccinated			
Known responder**	No treatment	No treatment	No treatment
Known nonresponder††	HBIG × 1 and initiate revaccination or HBIG × 2§§	No treatment	If known high risk source, treat as if source were HBsAg positive
Antibody response unknown	Test exposed person for anti-HBs¶¶ 1. If adequate,** no treatment is necessary 2. If inadequate,†† administer HBIG × 1 and vaccine booster	No treatment	Test exposed person for anti-HBs 1. If adequate,¶ no treatment is necessary 2. If inadequate,¶ administer vaccine booster and recheck titer in 1–2 months

* Persons who have previously been infected with HBV are immune to reinfection and do not require postexposure prophylaxis.
† Hepatitis B surface antigen.
§ Hepatitis B immune globulin; dose is o.06 mL/kg intramuscularly.
¶ Hepatitis B vaccine.
** A responder is a person with adequate levels of serum antibody to HBsAg (i.e., anti-HBs $\geq$ 10 mlU/mL).
†† A nonresponder is a person with inadequate response to vaccination (i.e., serum anti-HBs $<$ 10 mlU/mL).
§§ The option of giving one doe of HBIG and reinitiating the vaccine series is preferred for nonresponders who have not completed a second 3-dose vaccine series. For persons who previously completed a second vaccine series but failed to respond, two doses of HBIG are preferred.
¶¶ Antibody to HBsAg.
Source: MMWR 2001: 50(RR-11):22.

Instruments and devices used in patient care that enter the patient's vascular system or other normally sterile areas of the body, or touch non-intact skin or mucous membranes, should be sterilized in accordance with normal sterilization procedures before being used for each patient. Those that at most touch intact mucous membranes should be sterilized when possible. They should undergo high-level disinfection if they cannot be sterilized before being used for each patient. Those that do not touch the patient or that only touch intact skin of the patient need only be cleaned with a detergent or as indicated by the manufacturer.[9]

Environmental surfaces contacted by blood or other potentially infectious materials should be cleaned of visible material and disinfected immediately or after the procedures. All surfaces that may have been so contaminated should be disinfected at the end of each shift. Medical and other equipment surfaces should undergo intermediate-level disinfection, as should any spill of blood or other patient material. A low-level disinfectant may be used for routine housekeeping surfaces (countertops, floors, sinks, etc.). Intermediate-level disinfection should be used for routine decontamination in serology and microbiology sections of clinical and research laboratories.[10]

HBV vaccination must, per 29CFR 1910.1030, be made available to at-risk employees within 10 days of employment unless they are already immune or the vaccine is medically contraindicated. Participation is voluntary, but it should be encouraged, especially for those at particularly high risk for exposure. CDC guidelines strongly recommend that employees be tested for adequate titer response after vaccination. Employees who are non-responders should be re-vaccinated and retested.

Prophylaxis must be considered when a significant exposure has occurred. This may be defined as blood or other body fluid contamination via the percutaneous route (e.g., needlestick or incised, bitten, broken, or otherwise compromised skin) or the permucosal route (splash into eye or mouth). The decision on whether to proceed with prophylaxis will depend on the availability of the source of the blood or body fluid, the HbsAg status of the source person, and the HBV vaccination and vaccine-response status of the exposed person. Table 21.2 summarizes recommendations for specific prophylaxis based on these considerations. A thorough review of all aspects of exposure to hepatitis B viruses is provided by the US Public Health Source guidelines for management of occupational exposures to HBV, HCV and HIV.[11] This publication is available on the web at www.CDC.gov/MMWR.

Concern about the small—but very real—risk of HBV transmission to patients mandates a prevention plan. The current available data do not justify automatic exclusion of HBV carriers from healthcare duties, because transmission requires direct parenteral or mucosal membrane exposure to HBV. Although HBV carriers who are HBeAg negative are considered less infectious than those who are HBeAg positive, there have been well-documented cases of transmission during surgery.[12] Thus all healthcare workers who are HbAg positive and potentially perform exposure-prone procedures must be evaluated regarding risk. Testing for viral load of those who are HbeAg negative can provide useful information.

REFERENCES

1. Centers for Disease Control. Protection against viral hepatitis; recommendations of the Immunization Practices Advisory Com-mittee (ACIP). *MMWR* 1990; 39(S-2):1–26.
2. Werner BG, Grady GF. Accidental hepatitis-B-surface-antigen-positive inoculations: use of e antigen to estimate infectivity. *Ann Intern Med* 1982; 97:367–9.
3. Centers for Disease Control. Hepatitis-B transmitted by human bite. *MMWR* 1974; 23:45.
4. Weber DJ, Hoffmann KK, Rutala WA. Management of health care worker infected with human immunodeficiency virus: lessons from nosocomial transmission of hepatitis B virus. *Infect Control Hosp Epidemiol* 1991; 12: 625–30.
5. LaBrecque DR, et al. The risk of hepatitis B transmission from health care workers to patients in a hospital setting—a prospective study. *Hepatology* 1986; 12:205–8.
6. Centers for Disease Control. Recommendations for preventing transmission of human immunodeficiency virus and hepatitis B virus to patients during exposure-prone invasive procedures. *MMWR* 1991; 40(RR-8):1–8.
7. Hoofnagle JH. Hepatitis. In: Mandell GL, Bennet JE, Dolin R, eds. *Principles and practice of infectious disease*. 5th edn. New York: Churchill Livingstone, 2000, 1652–85.
8. Niederan C, et al. Long-term follow-up of HbeAg-positive patients treated with interferon alfa for chronic hepatitis B. *N Engl J Med* 1996; 334:1422–7.
9. Centers for Disease Control. *Guidelines for infection control in hospital personnel.* GPO no. 544-436/24441. Atlanta, GA: Public Health Service, 1985:1–20.
10. Favero MS, Bond WW. Sterilization, disinfection and antisepsis in the hospital. In: *Manual of clinical microbiology.* Washington, DC: American Society for Microbiology, 1991, 192.
11. Centers for Disease Control and Prevention. Updated US Public Health Source Guidelines for the Management of Occupational Exposure to HBV, HCV and HIV and Recommendations for Postexposure Prophylaxis. *MMWR* 2001; 50(No. RR-11).
12. The Incident Investigation Teams. Transmis-

sion of hepatitis B from four infected surgeons without hepatitis B e antigen. *N Engl J Med* 1997; 336:178–83

HEPATITIS C VIRUS (HCV)

Common names for disease: Hepatitis C

Classification: Family—Flaviviridae

Occupational setting

Healthcare workers and others with exposure to blood are at risk.

Exposure (route)

Parenteral inoculation of blood (i.e. transfusion or intravenous drug abuse) is the primary route of infection. Other body fluids are potentially infective; however, one study did not demonstrate virus activity in semen or saliva.[1] Sexual transmission may occur but is infrequent.[2,3] Risk of transmission of HCV is generally considered to be greater than that for HIV and less than that for HBV. The average incidence of transmission is 2%, although one study reports an incidence of 10%.[4]

Pathobiology

HCV is a single-strand RNA virus that can exist in a single infected individual as more than one quasispecies. Six genotypes plus subtypes have been proposed. Specific testing methods defined HCV in the late 1980s as the cause of most previously labeled non-A non-B hepatitis cases. It is especially prevalent in Asia and Africa; the general population prevalence in the USA is 1–2%. High-risk populations such as injecting drug users have a higher prevalence.

After an incubation period of 1–8 weeks, the virus causes an illness that is not clinically different from other forms of viral hepatitis. Symptoms are non-specific and may include fatigue, nausea, anorexia, and low-grade fever. More specific findings, such as hepatomegaly and jaundice, are infrequent. Although the initial hepatitis is commonly mild, progression of the illness in a fluctuating, very low-grade manner to chronic liver disease is characteristic and common (70–75%). Older patients may be at greater risk for significant chronic liver disease. Studies have shown an association between hepatocellular carcinoma and HCV.[5]

Diagnosis

Hepatitis C presents as a mild illness, and clinical differentiation from other viral hepatitis is not possible. Serologic testing with ELISA methods results in significant false-positive results[6] and must be confirmed with a recombinant immunoblot assay (RIBA) and/or PCR-RNA methods. Although PCR-RNA methods will detect virus as early as 1–2 weeks postexposure, prior to the appearance of liver abnormalities, the methodology is not FDA approved and can prove to be erroneous.

Medical surveillance

The CDC does not recommend routine testing of patients or workers. Healthcare workers with HCV infection who perform exposure-prone surgical procedures should be evaluated by an expert panel to consider risk of transmission. Although the risk of transmission from healthcare worker to patient appears to be low, one report describes transmission during an exposure-prone surgical procedure to five patients.[7]

Healthcare workers or others exposed to blood or body fluid from an HCV-positive source should be followed for 6 months with periodic testing. Testing for liver enzyme abnormality and HCV antibody is recommended at least at 3 and 6 months. Some centers conduct more frequent testing or do PCR-RNA testing at 2–3 weeks following exposure (PCR-RNA methodologies are not yet FDA approved). There is no prophylaxis recommended. Early treatment prior to liver abnormality is suggested by some to reduce chronic disease;[8] however, evidence at this time suggests no difference in outcome with treatment during acute infection versus early chronic infection. There is no evidence that postexposure prophylaxis with interferon is of value. A thorough review of aspects of exposure to HCV is provided by the US Public Health Service Guidelines for Management of Occupational Exposure to HBV, HCV and HIV. This publication is available on the web at www.CDC.gov/MMWR.[10]

Treatment

Interferon-α and interferon-α plus ribavirin are current treatment interventions for infected patients. About 50% of patients will respond, and of those 50% will relapse within 6 months. Repeat treatment and, in selected cases, liver transplant are long-term treatment options.[9]

Prevention

The OSHA mandates (29CFR 1910.1030) that universal blood and body fluid precautions be implemented as part of a written exposure control plan by an employer having employees with "any reasonably anticipated" occupational exposure. Occupational exposure is defined as "skin, eye, mucous membrane, or parenteral contact with blood or other potentially infectious materials that may result from the performance of an employee's duties". Other potentially infectious materials include semen, vaginal secretions, saliva where dental procedures are involved, unfixed tissues or organs, and cerebrospinal, synovial, pleural, pericardial and amniotic fluids. When a fluid is contaminated with blood, or where it is difficult or impossible to differentiate between body fluids, the material should be considered potentially infectious.

The efficacy of universal precautions, where all human blood and certain body fluids are treated as if known to be infectious for HIV, HBV, HCV and other blood-borne pathogens, is limited by compliance. Workers should understand the importance of, and be well trained in, the implementation of work practice controls, such as puncture precautions, handwashing and skin-washing, environmental controls, and standard infection control practices, and in the proper use of barrier precautions and personal protective equipment (PPE). Gloves, masks, protective eyewear or face shields, barrier gowns or aprons, surgical caps/hoods and shoe covers should be used appropriately.

All healthcare workers should take precautions to prevent injuries caused: by needles, scalpels, and other sharp instruments or devices during procedures; when cleaning used instruments; during disposal of used needles; and when handling sharp instruments after procedures. To prevent needlestick injuries, needles should not be recapped, purposely bent or broken by hand, removed from disposable syringes, or otherwise manipulated by hand. After they are used, disposable syringes and needles, scalpel blades and other sharp items should be placed in puncture-resistant containers for disposal; the puncture-resistant containers should be located as close as practical to the use area. Large-bore reusable needles should be placed in a puncture-resistant container for transport to the reprocessing area.

Instruments and devices used in patient care that enter the patient's vascular system or other normally sterile areas of the body, or touch non-intact skin or mucous membranes, should be sterilized in accordance with normal sterilization procedures before being used for each patient. Those that at most touch intact mucous membranes should be sterilized when possible. They should undergo high-level disinfection if they cannot be sterilized before being used for each patient. Those that do not touch the patient or that only touch intact skin of the patient need only be cleaned with a detergent or as indicated by the manufacturer.

Environmental surfaces contacted by blood or other potentially infectious materials should be cleaned of visible material and disinfected immediately or after the procedures. All surfaces that may have been so contaminated should be disinfected at the end of each shift.

Medical and other equipment surfaces should undergo intermediate-level disinfection, as should any spill of blood or other patient material. A low-level disinfectant may be used for routine housekeeping surfaces (countertops, floors, sinks, etc.). Intermediate-level disinfection should be used for routine decontamination in serology and microbiology sections of clinical and research laboratories.

There is no currently accepted prophylaxis for exposure to hepatitis C.

REFERENCES

1. Fried MW, Shindo M, Tse-Ling F, et al. Absence of hepatitis C viral RNA from saliva and semen of patients with chronic hepatitis C. *Gastroenterology* 1992; 102:1306–8.
2. Scully U, Mitchell S, Gill P. Clinical and epidemiologic characteristics of hepatitis C. *Can Med Assoc* 1993; 148:1173–7.
3. Weinstock HS, Bolan G, Reingold AL. Hepatitis C virus infection among patients attending a clinic for sexually transmitted diseases. *JAMA* 1993; 269:392–4.
4. Bruix J, Barrera JM, Calvet X, et al. Prevalence of antibodies to hepatitis C virus in Spanish patients with hepatocellular carcinoma and hepatic cirrhosis. *Lancet* 1989; 2:1004–6.
5. Trepo C, Zoulim F, Alonso C, et al. Diagnostic markers of hepatitis B and C. *Gut* 1993; 34:S20–5.
6. Mitsui T, et al. Hepatitis C virus infection in medical personnel after needlestick accident. *Hepatology* 1992; 16:1109–14.
7. Esteban JL, et al. Transmission of hepatitis C virus by a cardiac surgeon. *N Engl J Med* 1996; 334:555–60.
8. Cammá C, et al. Interferon as treatment of acute hepatitis C. A meta-analysis. *Digestive Diseases and Sciences* 1996; 41:1248–55.
9. Centers for Disease Control. Prevention and control of hepatits C. *MMWR* 1998; 47(RR19): 1–39.
10. Centers for Disease Control and Prevention. Updated US Public Health Service Guidelines for the Management of Occupation and Exposure to HBV, HCV and HIV and Recommendations for Postexposure Prophylaxis. *MMWR* 2001; 50(No. RR-11).

HERPES B VIRUS

Common names for disease: Monkey B virus, B virus
Classification: Family—Herpesviridae; genus/species—Herpesvirus simiae

Occupational setting

Persons at risk include zoo attendants, nonhuman primate handlers, veterinarians, and researchers.

Exposure (route)

Transmission is believed to occur by exposure to contaminated monkey saliva via bites or scratches. Transmission has also occurred by needlestick, following exposure to contaminated cell cultures of simian origin, and (in one case) after cleaning a monkey skull. A case of human-to-human transmission has also been reported.

Pathobiology

Herpes simiae is a zoonotic alpha herpes virus that is enzootic in Asian monkeys of the genus Macaca.[1] Primary infection in monkeys frequently appears as buccal mucosal lesions. Subsequently, the virus remains latent in the host and may reactivate spontaneously or in times of stress. This may result in shedding of virus in saliva or genital secretions. The first case of infection in humans was reported in 1932 in a monkey handler who developed encephalitis. The disease is characterized by a variety of symptoms, most of which develop within 1 month of exposure.[2] Among the symptoms are vesicular skin lesions at or near the site of inoculation. Neurologic symptoms, such as dysesthesias and numbness, often develop at the site of inoculation. Subsequent symptoms often include myalgias, headaches, and dizziness. Symptoms may progress rapidly, leading to permanent neurologic impairment or death. Of the cases reported to date, over half have been fatal, and the majority of the survivors have had significant neurologic impairment.

In 1998, the first case of occupational B virus infection was reported following mucocutaneous exposure without injury. A 22-year-old primate worker sustained an ocular splash while moving an animal within a cage. She was first evaluated for eye symptoms 10 days after exposure and was hospitalized 5 days later. Despite institution of intravenous antiviral therapy, she developed acute demyelinating encephalomyelitis, which was fatal.[3]

Diagnosis

Wounds resulting from potential sources of B virus should be cultured. Lesions on mucous membranes of monkeys should also be cultured.[4] If vesicles are present, a Tzanck smear can be performed to look for multinucleated giant cells. A simple, rapid test based on restriction endonuclease analysis of labeled infected cell DNA is available. Serial serum samples should also be obtained from individuals inadvertently exposed to see if a rise in antibody titer occurs. Laboratory assistance may be obtained from the Southwest Foundation for Biomedical Research (San Antonio, TX).[4]

Treatment

Wounds and exposed mucosa should be thoroughly cleaned immediately with soap and water. The Guidelines for Prevention and Treatment of B-Virus Infections should be followed.[5] Until recently, no treatment was available. However, acyclovir has shown promise in a few cases.[5,6] Ganciclovir is a possible alternative. Contact the CDC for the latest recommendations on dosages. Postexposure prophylaxis with acyclovir should be strongly considered for a high-risk source and route of exposure.[5]

Medical surveillance

Preplacement and annual serum samples should be obtained and frozen for future analysis from all individuals working with macaques.

Prevention

Primate handlers should wear long-sleeved garments and leather gloves.[2,6] In addition, a face mask and safety glasses should be worn. Individuals should be trained in the proper use of mechanical and chemical restraints and squeeze cages. Research laboratories should evaluate the feasibility of acquiring and maintaining a B virus-free colony of monkeys. Routine screening of macaques for B virus is not recommended. In situations where laboratory studies may lead to immunosuppression, the investigator may want to determine the infection status of the animal to be used, since viral shedding may be enhanced in such situations. Animal cages should be thoroughly cleaned and free of sharp edges at corners.

Access to animal quarters should be restricted to individuals properly trained in procedures to avoid risk of infection. Workers must be educated to notify their supervisor immediately in the case of an animal bite, scratch, or mucous membrane exposure. There is no vaccine currently available to prevent B virus infection.

REFERENCES

1. Weigler BJ. Biology of B virus in macaque and human host: a review. *Clin Infect Dis* 1992; 14:555–67.

2. Centers for Disease Control. B virus infection in humans—Michigan. *MMWR* 1989; 38:453–4.
3. Centers for Disease Control. Fatal cercopithecine herpesvirus1 (B virus) infection following a mucocutaneous exposure and interim recommendations for worker protection. *MMWR* 1998; 47:1073–83.
4. Hilliard JK, Munoz RA/1, Lipper SL, Eberle R. Rapid identification of herpesvirus simiae (B virus) DNA from clinical isolates in nonhuman primate colonies. *J Virol Methods* 1986; 13:52–62.
5. Holmes GP, Chapman LE, Stewart JE, et al. Guidelines for the prevention and treatment of B-virus infections in exposed persons. *Clin Infect Dis* 1995; 20:421–39.
6. Centers for Disease Control. B virus infection in humans—Pensacola, Florida. *MMWR* 1987; 36:289–96.

Carol Epling
Rickey L. Langley

HERPES SIMPLEX VIRUS (HSV)

Common names for disease: Cold sores, fever blisters, genital herpes, herpetic whitlow

Classification: Family—Herpesviridae

Occupational setting

Healthcare workers, particularly dental, nursing and respiratory care personnel, appear to be at greatest risk of occupational exposure. Transmission of herpes simplex virus (HSV) from patients to healthcare workers is well documented and appears to occur more frequently than transmission from workers to susceptible patients.[1–3]

Exposure (route)

The principal mode of spread is by direct contact with infected secretions. The virus enters the body via muscosal surfaces or abraded skin.

Pathobiology

HSV is a member of the Herpesviridae family and is a double-stranded DNA virus. After primary infection, the virus establishes latency in sensory ganglia and can reactivate. There are two subtypes, HSV-l and HSV-2. HSV-1 is most commonly associated with oralfacial lesions and HSV-2 with genital lesions; however, this distinction is becoming increasingly blurred.

Primary infection with HSV-1 is frequently asymptomatic, but it may present as gingivostomatitis and pharyngitis, usually in children.[4] About 90% of adults have antibodies to HSV-1 by the 4th decade of life. Typical herpetic lesions are small vesicles on an erythematous base that subsequently ulcerate and then crust over. Recurrent oral lesions occur in 20–40% of the population, most commonly at the vermillion border of the lip.

Primary infection with HSV-2 is unusual before puberty, though perinatal transmission does occur. HSV-2 antibodies are detectable in 10–40% of the general adult population. Primary genital infection is manifested by painful, vesicular lesions that subsequently ulcerate and are frequently accompanied by fever and constitutional symptoms. Recurrent episodes, usually without constitutional symptoms, occur in 60–90% of those with primary infection.

Herpetic whitlow, HSV infection of the finger, is of particular occupational importance. It is a common manifestation of primary HSV-1 infection in dental and medical personnel, but it is typically associated with HSV-2 recurrence in the general population. Whitlow presents with pain, erythema, swelling and pustular lesions that are frequently difficult to distinguish from pyogenic bacterial infections. Accurate diagnosis is important, since incision and drainage, which may be appropriate for bacterial infection, may complicate HSV infection.

Diagnosis

Diagnosis is usually made clinically on the basis of the typical, vesicular lesions on an erythematous base. Rapid laboratory confirmation can be obtained by a Tzanck preparation, though this smear cannot distinguish HSV from other herpes virus infections. Definitive laboratory diagnosis, when clinically indicated, is best performed by viral culture.

Treatment

Several antivirals are effective in limiting the severity and duration of both primary and recurrent genital infections. The benefit is modest, however, and treatment is not recommended for mild episodes. Treatment of primary infection does not reduce the probability of recurrence. Chronic suppressive treatment may be helpful in patients with severe, frequent episodes. Treatment of recurrent labial HSV is not generally recommended. Antivirals may be beneficial in cases of herpetic whitlow.

Prevention

Since transmission occurs through direct contact, institution of universal infection precautions minimizes the risk of transmission. Adoption of universal precautions may be having an effect among denntists. In contrast to older studies, a recent study of seroprevalence for HSV-1 in dental personnel demonstrated no increased risk of infection relative to controls.[4] The importance of avoiding contact with all secretions and regular handwashing should be stressed. In one study, HSV could be cultured from 6 of 9 individuals with active oral lesions.[3] Thus, workers with active oral lesions should not care for high-risk patients until the lesions have crusted over. If this is impractical, all lesions should be covered by dressings or a mask. Vaccination may be possible in the future, but there is currently no effective vaccine available.

REFERENCES

1. Corey L. Herpes simplex virus. In: Mandell G, Bennett J, Dolin R. *Principles and practice of infectious disease*. 5th ed. Philadelphia: Churchill Livingstone, 2000:1564–80.
2. Perl T, Haugen T, et al. Transmission of herpes simplex virus type 1 infection in an intensive care unit. *Ann Intern Med* 1992; 117:584–6.
3. Adams G, Stover B, et al. Nosocomial herpetic infections in a pediatric intensive care unit. *Am J Epidemiol* 1981; 113:126–32.
4. Turner R, Shehab Z, et al. Shedding and survival of herpes simplex virus from fever blisters. *Pediatrics* 1982; 70:547–9.

David P. Siebens

HUMAN IMMUNODEFICIENCY VIRUS (HIV-1)

Common names for disease: Acquired immunodeficiency syndrome (AIDS), AIDS-related complex

Classification: Family—Retroviridae; Subfamily—Lentivirus

Occupational setting

Healthcare workers are at increased risk of infection if they are exposed to blood products, especially surgeons, pathologists, dentists and dental hygienists, personnel in emergency departments, personnel in labor and delivery, laboratory technicians and researchers, and hemodialysis staff. Other occupations with increased risk include hospital custodial staff, laboratory workers, morticians, and emergency response personnel.

Exposure (route)

The virus can be spread through blood, semen or other infected body fluids via percutaneous exposure, mucous membrane exposure, exposure to abraded or otherwise non-intact skin, breastfeeding, perinatal routes from an infected mother, or sexual activity. Non-bloody saliva is not an effective mode of transmission. In the healthcare setting, the risk of infection following a needlestick exposure has been estimated at approximately 1 : 300. The risk of infection varies with the type of exposure, for example, the estimated risk with percutaneous puncture with a hollow core bloody needle is 0.3%, while that for mucous membrane exposure is approximately 0.09%.[1]

By 2000, there had been 56 well-documented seroconversions among healthcare workers who had had a direct exposure to HIV-infected blood or body fluids. Three of these cases involved laboratory workers handling concentrated viral sources. There are another 138 healthcare workers who apparently have been infected as a result of occupational exposure but the direct exposure documentation is not available.[2]

Pathobiology

HIV is a retrovirus whose inner core contains two copies of a single strand of RNA. This group of viruses use reverse transcriptase to encode DNA, which is incorporated in the host cell. HIV is initially widely disseminated in the lymphoid organs. HIV primarily infects $CD4^+$ T-cells, which results in cell destruction and ultimately immunosuppression. The immunosuppressed patient is vulnerable to multiple opportunistic infections. HIV disease is classified, based on the $CD4^+$ T-cell count and the clinical status of the patient into nine categories. Acquired immunodeficiency syndrome (AIDS) is defined by a $CD4^+$ T-cell count of less than 200 and/or specific clinical conditions.

The primary infection with HIV ranges from asymptomatic seroconversion to a severe illness resembling infectious mononucleosis requiring hospitalization. AIDS or the immunocompromised phase of this illness presents typically as an opportunistic pneumonia but can initially present as other infectious diseases and malignancies.

The global epidemic of HIV infection involves an estimated 40 million individuals. AIDS is the fifth leading cause of death among Americans aged 25–44. It is estimated that HIV infection prevalence in the USA is 0.3–0.4% (Figure 21.5).

Diagnosis

The initial infection is frequently misdiagnosed, as it is commonly a mild illness. AIDS is diagnosed when opportunistic infections occur. HIV is usually diagnosed based on the presence of HIV antibody (measured by an enzyme immunoassay), which may not be detectable for 4–8 weeks after infection. HIV is detected earlier by the p24 antigen test and HIV-RNA polymerase chain reaction assay.

Treatment

There is no cure for HIV infection; however, treatment with nucleoside analog reverse transcriptase inhibitors and protease inhibitors has markedly altered the course of this illness. Many individuals with AIDS have been able to re-establish immune competency, although the virus is still present. Early aggressive treatment has slowed progression and prevented/delayed the AIDS stage of disease. New treatment for opportunistic infection in immunocompromised individuals has reduced morbidity and mortality.

Prophylactic treatment for workers exposed to HIV-infected blood or body fluids has been recommended by the CDC since the mid-1990s.[3] In an exposure event, the source of exposure should be tested for the presence of HIV antibody. Initial testing with a rapid methodology followed by the more time-consuming, standard ELISA method allows for prompt initiation of prophylaxis and information for the exposed worker. Initiation of prophylaxis within 2 h of exposure is recommended. The details of decision-making for prophylaxis are provided in Tables 21.3 and 21.4.

The exposed worker should be tested for evidence of seroconversion (HIV antibody) at three and six months. Some centers test at 12 months also. A thorough and detailed review of all aspects of exposure of HIV is provided

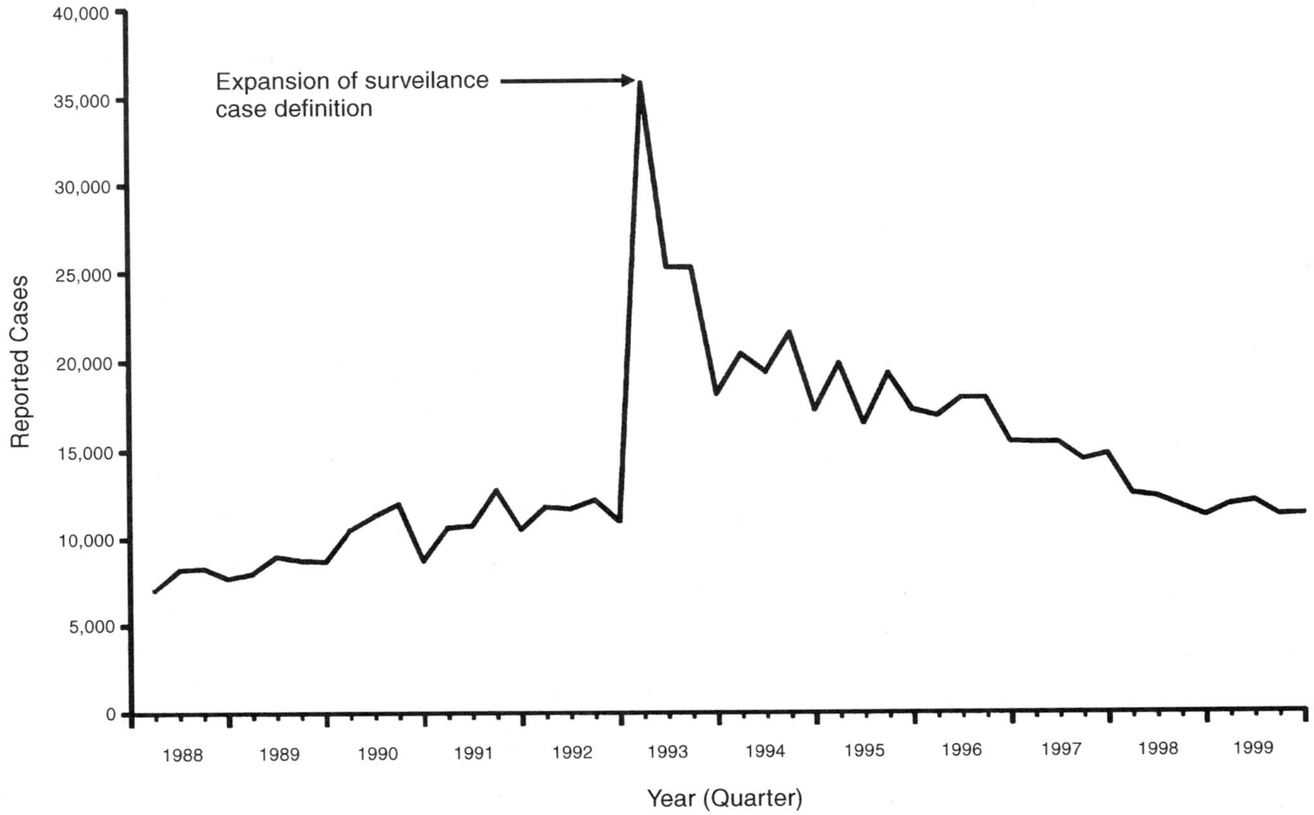

Figure 21.5. Acquired Immunodeficiency Syndrome (AIDS)—reported cases by quarter, United States, 1988–1999. Source: MMWR 2001:48(53):19.

TABLE 21.3 Recommended HIV postexposure prophylaxis for percutaneous injuries.

	Infection status of source				
Exposure type	HIV-positive Class 1*	HIV-positive Class 2*	Source of unknown HIV Status†	Unknown source§	HIV-negative
Less severe¶	Recommended basic 2-drug PEP	Recommend expanded 3-drug PEP	Generally, no PEP warranted; however, consider basic 2-drug PEP** for source with HIV risk factors††	Generally, no PEP warranted; however, consider basic 2-drug PEP** in settings where exposure to HIV-infected persons is likely	No PEP warranted
More severe§§	Recommend expanded 3-drug PEP	Recommend expanded 3-drug PEP	Generally, no PEP warranted; however, consider basic 2-drug PEP** for source with HIV risk factors††	Generally, no PEP warranted; however, consider basic 2-drug PEP** in settings where exposure to HIV-infected persons is likely	No PEP warranted

* HIV-Positive, Class—asymptomatic HIV infection or known low viral load (e.g., <1500 RNA copies/mL). HIV-Positive, Class 2—symptomatic HIV infection, AIDS, acute seroconversion, or known high viral load. If drug resistance is a concern, obtain expert consultation. Initiation of postexposure prophylaxis (PEP) should not be delayed pending expert consultation, and, because expert consultation alone cannot substitute for face-to-face counseling, resources should be available to provide immediate evaluation and follow-up care for all exposures.

† Source of unknown HIV status (e.g., deceased source person with no samples available for HIV testing).

§ Unknown source (e.g., a needle from a sharps disposal container).

¶ Less severe (e.g., solid needle and superficial injury).

** The designation "consider PEP" indicates that PEP is optional and should be based on an individualized decision between the exposed person and the treating clinician.

†† If PEP is offered and taken and the source is later determined to be HIV-negative, PEP should be discontinued.

§§ More severe (e.g., large-bore hollow needle, deep puncture, visible blood on device, or needle used in patient's artery or vein).

Source: MMWR 2001:48(53):24.

by the updated US PHS Guidelines for Management of Occupational Exposures to HBV, HCV, and HIV. This publication is available on the web at www.CDC.gov/MMWR.[4]

Medical surveillance

The CDC does not recommend routine screening for HIV infection at most work sites. Screening of workers who handle concentrated sources of HIV, such as those in research laboratories, may be appropriate. This is only useful if done periodically as well as at the time of hire. Healthcare workers who are identified as HIV infected must be evaluated if they do exposure-prone invasive procedures. An expert panel evaluates the risk to patients and considers any restrictions on the health care worker's activities. The risk of transmission is considered to be extremely low (less than that of other surgical risks). There have been several 'look-back' studies of thousands of patients of infected physicians and dentists, with no cases of transmission.

HIV-infected workers who are immunocompromised need to be evaluated in the

TABLE 21.4 Recommended HIV postexposure prophylaxis for mucous membrane exposures and nonintact skin* exposures.

	Infection status of source				
Exposure type	HIV-positive Class 1†	HIV-positive Class 2†	Source of unknown HIV status§	Unknown source¶	HIV-negative
Small volume**	Consider basic 2-drug PEP††	Recommend basic 2-drug PEP	Generally, no PEP warranted; however, consider basic 2-drug PEP†† for source with HIV risk factors§§	Generally, no PEP warranted; however, consider basic 2-drug PEP†† in settings where exposure to HIV-infected persons is likely	No PEP warranted
Large volume¶	Recommend basic 2-drug PEP	Recommend expanded 3-drug PEP	Generally, no PEP warranted; however, consider basic 2-drug PEP†† for source with HIV risk factors§§	Generally, no PEP warranted; however, consider basic 2-drug PEP†† in settings where exposure to HIV-infected persons is likely	No PEP warranted

* For skin exposures, follow-up is indicated only if there is evidence of compromised skin integrity (e.g., dermatitis, abrasion, or open wound).

† HIV-Positive, Class 1—asymptomatic HIV infection or known low viral load (e.g., <1500 RNA copies/mL). HIV-Positive, Class 2—symptomatic HIV infection, AIDS, acute seroconversion, or known high viral load. If drug resistance is a concern, obtain expert consultation. Initiation of postexposure prophylaxis (PEP) should not be delayed pending expert consultation, and, because expert consultation alone cannot substitute for face-to-face counseling, resources should be available to provide immediate evaluation and follow-up care for all exposures.

§ Source of unknown HIV status (e.g., deceased source person with no samples available for HIV testing).

¶ Unknown source (e.g., splash from inappropriately disposed blood).

** Small volume (i.e., a few drops).

†† The designation "consider PEP" indicates that PEP is optional and should be based on an individualized decision between the exposed person and the treating clinician.

§§ If PEP is offered and taken and the source is later determined to be HIV-negative, PEP should be discontinued.

¶¶ Large volume (i.e., major blood splash).

Source: MMWR 2001:48(53):25.

context of risk to their health when exposed to tuberculosis in patient care or animal care settings.

Prevention

The prevention of HIV and other blood-borne pathogen transmission is specified in OSHA regulation 29 CFR 1910.1030. In addition, the CDC has released guidelines regarding precautions to reduce the spread of HIV infection.

All workers should routinely use appropriate barrier precautions to prevent skin and mucous membrane exposure when contact is anticipated with blood or other body fluids of any patient (or other worker). Gloves should be worn when touching blood and body fluids, mucous membranes, or non-intact skin of all patients, when handling items or surfaces soiled with blood or body fluids, and when performing venipuncture and other vascular access procedures. Gloves should be changed after contact with each patient. Masks and protective eyewear or face shields should be worn during procedures that are likely to generate droplets of blood or other body fluids to prevent exposure of mucous membranes of the mouth, nose, and eyes. Gowns or aprons should be worn during procedures that are likely to generate splashes of blood or other body fluids.

Hands and other skin surfaces should be washed immediately and thoroughly if contaminated with blood or other body fluids. Hands should be washed immediately after gloves are removed.

All healthcare workers should take precautions to prevent injuries caused: by needles, scalpels, and other sharp instruments or devices during procedures; when cleaning used instruments; during disposal of used needles; and when handling sharp instruments after procedures. To prevent needlestick injuries, needles should not be recapped, purposely bent or broken by hand, removed from disposable syringes, or otherwise manipulated by hand. After they are used, disposable syringes and needles, scalpel blades, and other sharp items should be placed in puncture-resistant containers for disposal; the puncture-resistant containers should be located as close as practical to the use area. Large-bore reusable needles should be placed in a puncture-resistant container for transport to the reprocessing area.

Although saliva has not been implicated in HIV transmission, to minimize the need for mouth-to-mouth resuscitation, mouthpieces, resuscitation bags or other ventilation devices should be available for use in areas where the need for resuscitation is predictable.

Healthcare workers who have exudative lesions on their hands should refrain from all direct patient care and from handling patient care equipment until the condition resolves.

As with most safety issues, engineering controls are more effective than PPE. The introduction of various safety devices such as needleless intravention systems has reduced blood/body fluid exposures in many healthcare work sites.[5] Thorough training is necessary when introducing new devices to prevent a flare of injuries related to the new device.

Exposure-prone procedures should be identified by medical/surgical/dental organizations and institutions where the procedures are performed. Healthcare workers who perform exposure-prone procedures should know their HIV antibody status. Healthcare workers who are infected with HIV should not perform exposure-prone procedures unless they have sought counsel from an expert review panel and been advised under what circumstances, if any, they may continue to perform these procedures. Such circumstances could include notifying prospective patients of the healthcare workers' seropositivity before they undergo exposure-prone invasive procedures.[6]

There have been no documented cases of HIV transmission involving environmental surfaces. HIV is rapidly inactivated via contact with many common germicides at low concentrations. Household bleach (sodium hypochlorite) diluted with water at concentrations of one part bleach to 10 parts water can inactivate HIV.

Spills or patient-derived infected material can be cleaned by removing visible material before chemical decontamination. Spills of concentrated research laboratory solutions (which may contain HIV in concentrations thousands of times higher than clinically encountered) should be chemically decontaminated, then cleaned, then re-decontamined. Cleaning spills of infective material requires the use of personal protective devices.

REFERENCES

1. Chiarello LA, Gerbearding JL. Human immunodeficiency virus in health care settings. In: Mandell GL, Bennett JE, Dolin R, eds. *Principles and practice of infectious diseases,* 5th edn. New York: Churchill Livingstone, 2000, 3052–66.
2. Centers for Disease Control and Prevention. Public Health Service Guidelines for the

Management of Health Care Worker Exposure to HIV and Recommendations for Postexposure Prophylaxis. *MMWR* 1998; 47(RR-7).

3. Centers for Disease Control and Prevention. Case control study of HIV seroconversion in health care workers after percutaneous exposure to HIV-infected blood. *MMWR* 1995; 44:929–33.
4. Centers for Disease Control and Prevention. Updated U.S. Public Health Source Guidelines for the Management of Occupational Exposures to HBV, HCV and HIV and Recommedations for Postexposure Prophylaxis. *MMWR* 2001; 50 (RR-11):1–53.
5. Younger B, Hunt EH, Robinson C, McLemore C. Impact of a shielded safety syringe on needlestick injuries. *Infect Control Epidemiol* 1992; 13:349–53.
6. Centers for Disease Control. Recommendations for preventing transmission of human immunodeficiency virus and hepatitis B virus to patients during exposure-prone invasive procedures. *MMWR* 1991; 40:1–9.

HUMAN T-CELL LYMPHOTROPHIC VIRUS TYPE I (HTLV-I) AND TYPE II (HTLV-II)

Common names for disease: HTLV-1—adult T-cell leukemia, HTLV-1 myelopathy, tropical spastic paraparesis; HTLV-II—not clearly associated with any diseases

Classification: Family—Retroviriae; genus —oncovirus

Occupational setting

Healthcare workers exposed to blood and body fluids, laboratorary workers, researchers and other workers with blood/body fluid exposure potential are all at risk. One case of seroconversion after accidental inoculation has been reported. However, no seroconversions have occurred among 31 laboratory and healthcare workers exposed to HTLV-1 via puncture wounds.[1] Although there is no reported occupational transmission of these viruses, the potential exists for transmission by blood or body fluid inoculation. The potential for exposure is greater in areas where the virus infection is endemic, including southern Japan, the Caribbean islands, South America, and Africa.[2] Seropositivity in the USA is <0.1% in the general population and as high as 49% in a study of intravenous drug users. About half of US volunteer blood donors seropositive for HTLV-I/II are infected with HTLV-II.

Exposure (route)

The transmission of HTLV-I/II occurs through mechanisms similar to HIV. Infection may occur through transfusion of blood (including the sharing of contaminated needles), sexual transmission, and mother-to-child transmission during breastfeeding.

Pathobiology

HTLV-1 and HTLV-II are retroviruses containing a double RNA strand. They use reverse transcriptase to encode DNA that becomes incorporated in the host cell. HTLV-1 preferentially attacks the helper ($CD4^+$) T-lymphocytes and probably the neural dendritic cells. Studies in Japan suggest that 24% of infected individuals will eventually develop disease.[3] There is no clinical syndrome yet associated with initial infection. A pro-longed latent period prior to clinical disease is most common.[4] A number of clinical syndromes are associated with HTLV-I infection, including acute and chronic adult T-cell leukemia and HTLV-I myelopathy (tropical spastic paraparesis). The leukemias present either as a fulminant illness, progressing to death in a number of months, or as a slow, smoldering leukemia. The myelopathy presents initially with weakness and stiffness of the lower extremities, often with paresthesias and low back pain.[5] A number of other syndromes reflective of immunodeficiency may be associated with HTLV-I.

HTLV-II preferentially infects the suppressor ($CD8^+$) T-lymphocyte. This virus is not clearly associated with human disease.

Diagnosis

Initial screening with an ELISA will not differentiate between HTLV-I and HTLV-II. Confirmation with a radioimmune assay or Western blot must be carried out. Suspicion of the existence of HTLV-I infection should be high in any illness associated with immune system suppression and helper T-cell reduction.

Treatment

There is no treatment for these viral infections. Supportive therapy for the resulting clinical syndrome is indicated.[6]

Medical surveillance

Surveillance for seroconversion is indicated for individuals with a known parenteral or mucous membrane exposure to infected blood or body fluids. There is no basis for routine testing of workers. There is no current standard regarding healthcare workers who are known to be infected with HTLV-I/II.

Prevention

Control of exposure is the only effective approach. Because HTLV-I and HTLV-II are blood-borne pathogens, the requirements of OSHA regulation 29CFR 1910.1030 are applicable. However, these agents are not specifically discussed in the regulations. Approaches that are protective of HIV transmission will provide protection from these viruses. Transmission is estimated to be less frequent than with HIV exposure. Disinfection is accomplished with 10% sodium hypochlorite solution and other common germicides.

REFERENCES

1. Amin RM, et al. Risk of retroviral infection among retrovirology and health care workers. In: *American Society for Microbiology 92nd General Meeting*, New Orleans, 1992 (abstract T-20).
2. Madeline MM, Wiktor SZ, Goedert JJ, et al. HTLVI and HTLVII world-wide distribution. *Int J Cancer* 1993; 54:255–60.
3. Quinn TC, Zacarias RK, St John RK. HW and HTLV-I infections in the Americas. *Medicine* 1989; 68:200–1.
4. Dixon AC, Dixon PS, Nakamura JM. Infection with the human T-lymphocyte virus type I. *West J Med* 1989; 151:632–7.
5. Hollsberg P, Hafler DA. Pathogenesis of diseases induced by human lymphotropic virus type I infections. *N Engl J Med* 1993; 328:1173–82.
6. Centers for Disease Control and Prevention. Recommendations for counseling persons infected with human T-lymphotrophic virus, types I and II. *MMWR* 1993; 42(RR-9):1–13.

INFLUENZA VIRUS

Common names for disease: Influenza, flu, grippe

Classification: Family—Orthomyxoviridae; genera—influenza virus types A, B, and C

Occupational setting

The risk of infection is increased for employees in hospitals, nursing homes, and child care facilities, because these infections may have a high prevalence among the patients and children.

Exposure (route)

The virus is transmitted predominantly by airborne small-particle aerosols (mean diameter, $<10\,\mu m$). These aerosols are generated by the coughing, sneezing or talking of infected persons.

Pathobiology

Influenza type A and type B are the two clinically important influenza genera. Both are medium-size, encapsulated RNA viruses that cause an acute, febrile illness. The respiratory tract is the primary site of infection, producing clinical syndromes ranging from common cold symptoms to pharyngitis, croup, tracheobronchitis, and pneumonia. Reye's syndrome is a serious hepatic and Central nervous system (CNS) complication of influenza that predominantly affects children. There is strong epidemiologic evidence linking the development of Reye's syndrome with concurrent aspirin use. Influenza C is an infrequent, non-epidemic infectious agent that produces afebrile, common cold symptoms.[1]

Influenza (types A and B) is distinctive among viral pathogens for its epidemic nature and for the excess mortality it produces through pulmonary complications. For healthy adults under age 65, "flu" is usually a self-limited, albeit very unpleasant, illness. This is not the case for population subgroups at risk for influenza-related complications—each influenza epidemic causes tens of thousands of excess deaths in the USA. These high-risk groups include persons over age 65, residents of chronic care facilities, adults and children with chronic pulmonary or cardiovascular disease (including asthma), those under age 18 receiving long-term aspirin therapy, and persons with other chronic diseases such as renal dysfunction, diabetes, or immunosuppression.[1]

Frequent changes in the virus's antigenicity are an important feature of influenza (particularly type A) and help explain why the illness is still a major epidemic disease for humans. This nearly annual alteration in the virus's makeup introduces a "new" virus to which the population at large often has little resistance. These genetic variations are referred to as antigenic drift or antigenic shift, based on the magnitude of the change. Antigenic shifts have been associated with pandemics of influenza.

Diagnosis

Although viral cultures and serology can be useful tools in studying influenza, most diagnoses are made on clinical and epidemiologic grounds. Classic cases of "flu" begin with the abrupt onset of fever, chills, myalgias, headache, and malaise. Respiratory symptoms such as sore throat, rhinorrhea and hoarseness develop, as well as a non-productive cough that often persists during a convalescent phase of 2–3 weeks. When influenza has been confirmed in a community, as many as 85% of persons presenting with these symptoms will have the disease. The majority of people presenting with less severe upper respiratory symptoms during such periods are also likely to be infected.

Treatment

Amantadine and rimantadine are chemically related agents possessing antiviral activity against influenza A only. Both drugs have been shown to be effective at reducing symptom duration when begun within 48 hours of the onset of illness. When administered prophylactically throughout an epidemic, they are 70–90% effective in preventing clinical infections with influenza A.[3] The most common (5–10%) side-effects are mild, reversible CNS symptoms such as dizziness, insomnia, and nervousness. Two recent additional medications, which inhibit viral neuraminidase of both influenza A and B viruses, have few side-effects. An inhaled powder (zanamivir) and oral tablet (oseltamivir phosphate) must be started within 48 hours of onset of illness to shorten the overall length of influenza symptoms.

Medical surveilliance

Although an acute care hospital might consider a prospective influenza surveillance program that included patients and employees, there are few data to support the efficacy of this strategy in reducing nosocomial infection rates. Facilities with an aggressive policy against influenza generally direct resources toward achieving a high vaccination rate for their employees. Nosocomial infection rates have been reported as 0.3 per 100 hospital admissions during the influenza season.[2]

Prevention

The administration of inactivated vaccine for influenza A and B is the best option for reducing the impact of influenza. The recommendation to vaccinate healthcare workers (and others in close contact with high-risk individuals) is based primarily on reducing the risk of transmission of the virus from employees to patients.[2] Because of the small size of respirable aerosols, surgical masks do not significantly reduce the transmission rate between individuals.

Vaccination is a logical means of reducing employee absenteeism due to influenza; however, the only placebo-controlled clinical trial of immunization in hospital employees demonstrated a beneficial trend that did not reach statistical significance in the vaccinated group (six versus eight illnesses in 179 employees).[4] The authors postulated an antigenic shift of the prevalent strain away from the vaccine type as an explanation for this blunted benefit. They also pointed out that the immunization program was carried out at minimal cost and risk. For employees who provide essential community services, any reduction in illness and absences is important, and the cost/benefit ratio for the use of influenza vaccine favors immunization. Worksite provision of influenza vaccination to employees is considered to be a cost-effective means for corporations to reduce illness and absenteeism during the influenza season.[5] The economic benefit of vaccination of non-high-risk workers is not seen consistently, especially in years in which the frequency of influenza is low.

Vaccination itself can be associated with mild adverse reactions. Studies have shown that 0–5% of employees missed work time after their immunization. Use of the split virus preparation appears to decrease the occurrence of the most objectionable of these vaccine side-effects.[6] Amantadine or rimantidine may be useful for preventing influenza A in persons who cannot be immunized because of an anaphylactic hypersensitivity to eggs or other vaccine components.

REFERENCES

1. Treanor JJ. Influenza virus. In: Mandell GL, Bennett JE, Dolin R, eds. *Principles and practice of infectious disease.* 5th edn. New York: Churchill-Livingstone, 2000, 1823–49.
2. Prevention and Control of Influenza: Recommendations of the Advisory Committee on Immunization Practices. *MMWR* 1998; 47(RR-6): 1–26.
3. Dolin R, Reicliman RC, Madore HP, Maynard R, Linton PN, Webber-Jones J. A controlled trial of amantadine and rimantadine in the prophylaxis of influenza A infection. *N Engl J Med* 1982; 307:580–3.
4. Olsen GW et al. Absenteeism among employees who participated in a workplace influenza immunization program. *J Occup Environ Med* 1998; 40(4):31140(4):311–16
5. Bridges CB, et al. Effectiveness and cost benefit of influenza vaccination of healthy working adults. *JAMA* 2000; 284(13): 1655–63.
6. Al-Marzou A, Scheifele DW, Soong T, Bjornson. Comparison of adverse reactions to whole-virion and split-virion influenza vaccines in hospital personnel. *Can Med Assoc J* 1991; 145:213–18.

MEASLES VIRUS

Common names for disease: Measles, red measles, rubeola

Classification: Family—Paramyxoviridae; genus—morbillivirus

Occupational setting

Employees in the healthcare setting (particularly pediatrics personnel), child care workers and school employees are at increased risk of exposure.[1]

Exposure (route)

Measles is transmitted through the respiratory route as aerosol droplets. It is one of the most communicable of the infectious diseases (Figure 21.6).[2]

Pathobiology

The measles virus is a small pleomorphic sphere with a single strand of RNA in a coiled helix of protein. Its natural hosts are humans. Measles, or rubeola, is often a severe illness complicated by bronchopneumonia and otitis media. Encephalitis is seen in one of every 1000 cases. Death occurs in one of every 1000 reported cases due to respiratory or neurologic complications. Measles during pregnancy results in increased rates of premature labor, spontaneous abortion, and low birth weight, but it is not associated with congenital abnormalities. Risk of death is higher in infants and adults. The incubation period is 10–14 days, after which a prodromal phase of malaise, fever, anorexia, cough, and coryza begins. A few days of prodrome are followed by the development of Koplik's spots (small, red, irregular lesions with blue–white centers) in the mouth and an erythematous, maculopapular rash primarily on the face and trunk. The illness lasts 7–10 days in most cases. Although viral shedding is greatest during the late prodrome due to coughing, the contagious period extends from several days before the development of the rash to several days after.

Diagnosis

Measles is a clinical diagnosis based on the progression of a viral prodrome with cough, coryza and conjunctivitis to Koplik's spots (which are diagnostic) and the maculopapular rash beginning on the face. A marked leukopenia is frequently seen. Viral isolation is not a clinically practical approach. Acute and convalescent serologic titer change can also be used to establish the diagnosis.

Treatment

There is no specific treatment other than supportive therapy for fever and dehydration.

Medical surveillance

Surveillance of susceptible workers exposed to measles should be conducted during the incubation period.

Prevention

A live vaccine has been available for measles since 1963. The first dose is recommended at 12–15 months of age; a second dose is given at school age, 5–6 years. Documentation of two doses of measles vaccine should be required of all healthcare personnel who have patient contact.[3] Documentation of measles disease, laboratory evidence of measles immunity, or birth before 1957, is an alternative to vaccination.[4] Documentation of vaccination of children entering day care and school provides protection for personnel in these settings; however, they should be encouraged to ensure their own immunity for personal protection.

The vaccine produces an inapparent, mild, non-communicable infection with antibody titer development in 95% of individuals. After primary vaccination, 5–15% of individuals develop a fever and rash. These adverse reactions do not occur with the second vaccination unless the individual has not developed immunity with the first dose. Contraindications to vaccination include altered immunocompetence, recent administration of immune globulin, and severe febrile illness. Individuals with a history of anaphylactic reaction to eggs or neomycin should not be vaccinated. Pregnancy is a theoretical contraindication with the monovalent measles vaccine but a necessary contraindication if administered together with

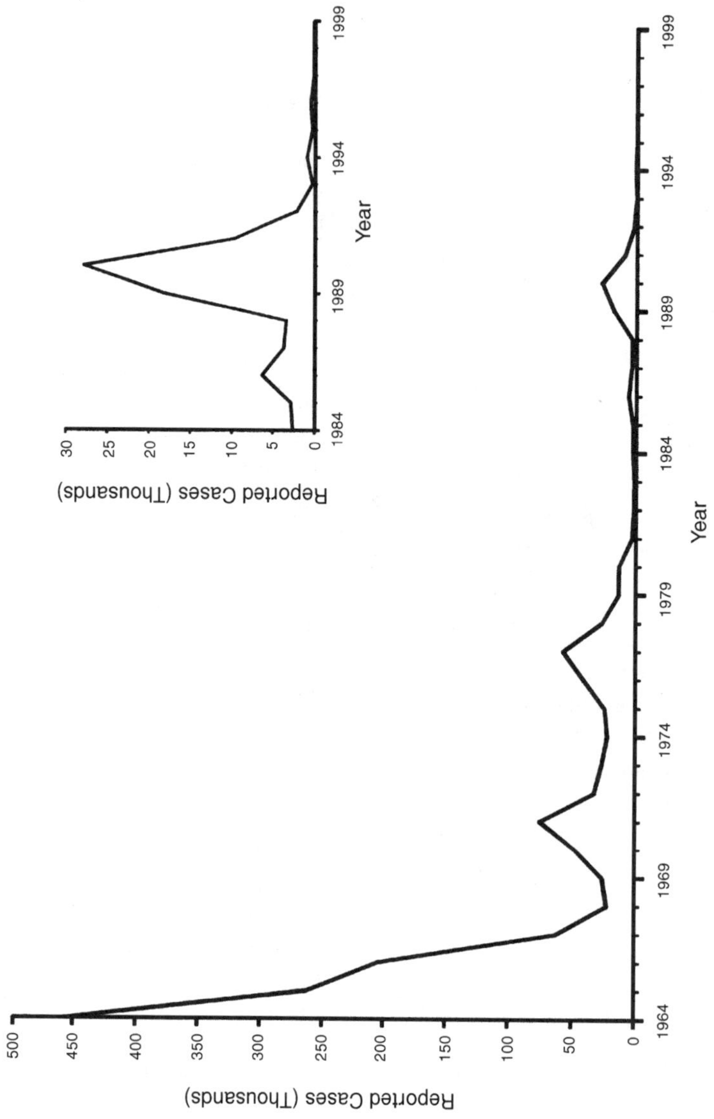

Figure 21.6 Measles—reported cases (thousand) by year, United States, 1964–1999. Source: MMWR 2001:48(53):51.

mumps and rubella vaccines. Women should be advised not to become pregnant for at least 30 days after vaccination.[3]

The infected worker must be excluded from the work setting until 7 days after the appearance of the rash. Non-immune personnel who are exposed to an active case of measles must be excluded from work from day 5 through day 21 after exposure.[4] If live measles vaccine is given within 72 hours of exposure, it may provide some protection.[3]

Concern regarding a link between measles vaccine and autism or inflammatory bowel disease has been raised. Epidemiologic investigation indicates that a causal association is unlikely.[5]

REFERENCES

1. Steingart KR, et al. Transmission of measles virus in health care settings during a community wide outbreak. *Infect Control Hosp Epidemiol* 1999; 20:1151–9.
2. Sienko DG, Friedman C, McGee HB, et al. A measles outbreak at university medical settings involving healthcare providers. *Am J Public Health* 1987; 77:1223.
3. Centers for Disease Control. Measles prevention: recommendations of the Immunization Practices Advisory Committee. *MMWR* 1989; 38:5–9.
4. Weber DJ, Rutala WA, Orenstein WA. Prevention of MMR among hospital personnel. *J Pediatr* 1991; 119:322–5.
5. Taylor B, et al. Autism and measles, mumps and rubella vaccine: no epidemiologic evidence for a causal association. *Lancet* 1999; 353:2026–9.

MUMPS VIRUS

Common name for disease: Mumps

Classification: Family—Paramyxoviridae; genus—paramyxovirus

Occupational setting

Persons working in child care, schools, health care, research and laboratories are at risk from exposure.

Exposure (route)

Mumps virus can be transmitted by direct contact with saliva or other recently contaminated materials. The virus is transmitted in droplet nuclei; entry is by nose or mouth. In contrast to measles or chickenpox, substantial contact is needed for transmission (Figure 21.7).

Pathobiology

The mumps virus is an enveloped, spherical RNA virus. Humans are the only natural host. A carrier state is not known. An average incubation of 16–18 days (range 24 weeks) is followed by a prodrome of low-grade fever, headache, malaise, and anorexia. Enlargement of the parotid gland with tenderness follows within 1–2 days. The illness resolves in ~1 week. Involvement of other glands occurs infrequently.[1] Epididymitis–orchitis occurs in about one of four cases in postpubertal men; oophoritis occurs infrequently (<5%) in postpubertal women.[2] Meningitis and encephalitis are infrequent complications. Mumps meningitis may occur without parotitis. Mumps and mumps meningitis resolve with no sequelae. Mumps encephalitis is a severe illness that rarely results in death. Mumps orchitis frequently causes testicular atrophy.

Diagnosis

The clinical findings with a history of exposure are the basis for diagnosis. Acute and convalescent serologic studies are definitive. Glandular involvement results in elevated amylase levels. In mumps meningitis, lumbar puncture reveals that the cerebrospinal fluid (CSF) typically contains a leukocyte count of 10–2000 white cells per mm^3 (predominantly lymphocytes), a normal to slightly increased

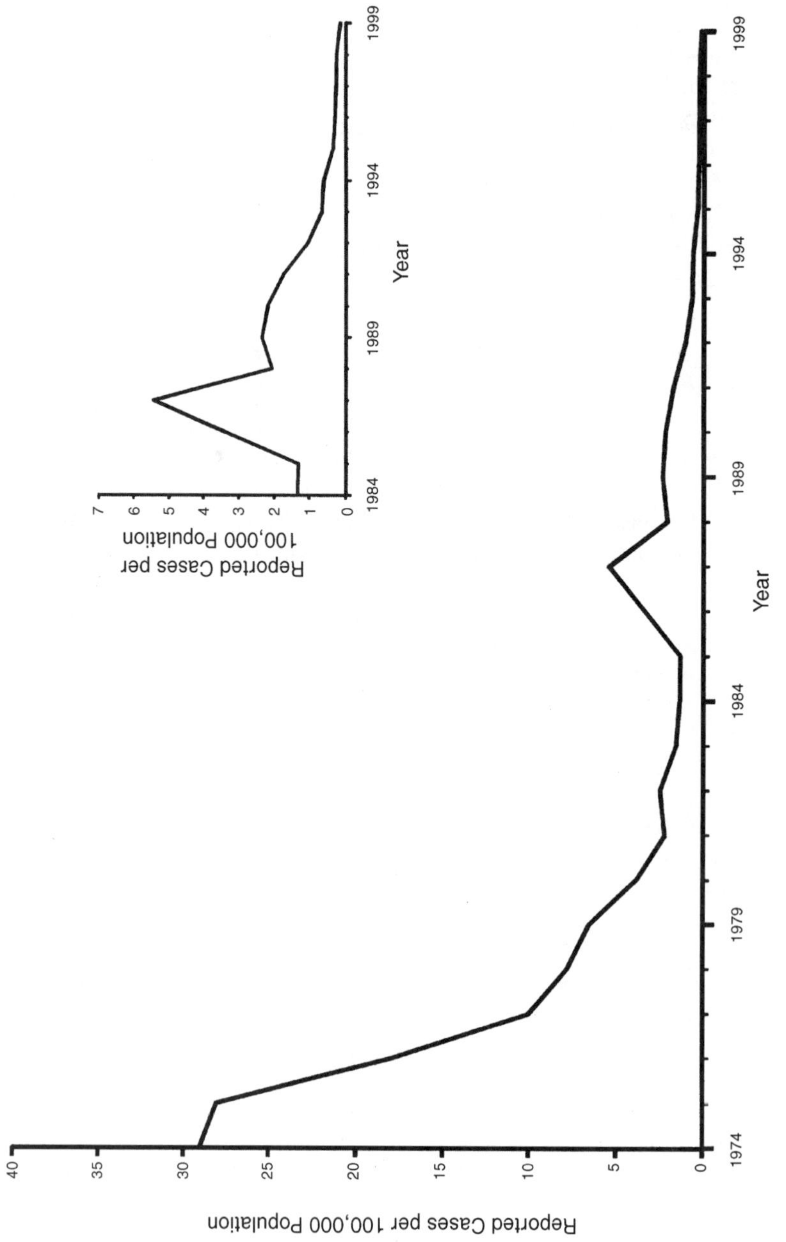

Figure 21.7 Mumps—reported cases per 100 000 population by year, United States, 1984–1999. Source: MMWR 2001:48(53):53.

protein level, and a depressed glucose level (<10 mg/100 ml).

Treatment

There is no therapy other than supportive measures, which are dependent on the type of glandular involvement.

Medical surveillance

Surveillance of workers exposed to mumps should be conducted during the incubation period. No other surveillance is indicated.

Prevention

Workers who are potentially exposed to individuals with mumps, such as healthcare workers and research/laboratory workers, should be immunized against mumps or asked to provide documentation of immunity. To protect workers, children entering day care centers or school should show evidence of immunization. Day care and school personnel may wish to ensure their immunity with revaccination. Documentation of mumps disease, or laboratory evidence of mumps immunity, is an alternative to vaccination. Among older adults, ~80% have no history of the illness or immunization.[3]

Vaccination in children is combined with measles and rubella at 12–15 months. The vaccine is a live, attenuated virus commercially produced since 1967. The vaccine produces an inapparent, mild, non-communicable infection with antibody titer development in 95% of individuals.[4] Adverse reactions are rare but include parotitis and low-grade fever. Contraindications to immunization include altered immunocompetence, recent administration of immune globulin, and severe febrile illness. Pregnancy is a contraindication to vaccination.[5]

The infected worker must be excluded from the work setting until 9 days after the onset of parotitis, although viral shedding begins 1–2 days or more before any evidence of illness.[6] Exposed workers in whom immunity cannot be established must be excluded during the lengthy incubation period, which averages 12–26 days.[7]

Outbreaks of illness in the work setting were documented during the hiatus between the introduction of the vaccine in 1967 and legislation requiring vaccination for school entry in the mid- to late 1970s. The reduction in community infection may have produced a group of unvaccinated and unexposed young adults who could provoke an outbreak situation in a work group. Employers should report to public health authorities and follow up on cases of suspected mumps among workers.[8]

REFERENCES

1. Baum SG, Litman N. Mumps virus. In: Mandell GL, Bennet JE, Dolin R, eds. *Principles and practice of infectious disease*, 5th edn. New York: Churchill Livingstone, 2000, 1776–81.
2. Beard CM, Benson RC, Kelalis PP, et al. The incidence and outcome of mumps orchitis in Rochester, Minnesota, 1935 to 1974. *Mayo Clin Proc* 1977; 52:3–7.
3. St Geme JW, Yamaguchi T, Eisenklaur EJ, et al. Immunologic significance of mumps virus skin test in infants, children and adults. *Am J Epidemiol* 1975; 101:253.
4. Buynak EB, Hilleman MR. Live attenuated mumps virus vaccine. *Proc Soc Exp Biol Med* 1966; 123:768.
5. Williams WW, Preblud SR, Reichelderfer PS, et al. Mumps virus. *Infect Dis Clin North Am* 1989; 3:701–21.
6. Henle G, Henle XV, Wendell KK, et al. Isolation of mumps virus from human beings with apparent or inapparent infections. *J Exp Med* 1948; 88:223.
7. Weber DJ, Rutala WA, Orenstein WA. Prevention of MMR among hospital personnel. *Pediatrics* 1991; 119:322–5.
8. Kaplan KM, Marder DC, Cochi SL, et al. Mumps in the workplace. *JAMA* 1988; 260:1434–8.

NORWALK VIRUS (AND OTHER ENTERIC VIRUSES)

Common name for disease: Acute viral gastroenteritis

Classification: Family—Calciviridae

Occupational setting

Persons in healthcare settings (particularly nursing and military) in areas with limited sanitary facilities are at risk from exposure. Norwalk and Norwalk-like viruses are a cause of food-borne illness outbreaks, which can occur in workplace cafeterias.

Exposure (route)

Fecal-oral contamination is the common route of exposure. Infection occurs through ingestion of contaminated food and water.[1]

Pathobiology

Norwalk virus is a non-enveloped RNA virus with a small, round structure. The virus has been shown to cause structural changes in the lining cells of the stomach and small intestine.

Following an incubation period of 1–2 days, an illness characterized by nausea, vomiting, and watery diarrhea lasts for 12–60 hours. Abdominal pain is accompanied by fever and an elevated white count. Depending on the duration and severity of diarrhea and vomiting, dehydration may become a problem. In the healthy adult, this illness rarely results in mortality; however, lost work time can be substantial.[2]

Diagnosis

Conclusive diagnosis is by electron microscopy of stool samples.[3] Reverse transcriptase–polymerase chain reaction (RT-PCR) has been used in a research setting for detection of the virus.[4] Pre- and post-convalescent serum may be tested for IgG antibody using ELISA to document the etiology of illness.[5]

Treatment

There is no treatment other than supportive care, which may include fluid and electrolyte replacement.

Medical surveillance

No routine surveillance is indicated. In outbreak situations, surveillance of staff for infection by symptom reporting may facilitate control of nosocomial spread. Suspected food-borne illness outbreaks in the workplace should be investigated.

Prevention

Although the possibility of vaccine development exists, it is remote. Excellent sanitation and personal hygiene is the principal means of control. Food service workers with symptoms of viral gastroenteritis should be evaluated by a physician. If Norwalk or other enteric viruses are suspected, the worker should remain out of work for at least 3 days after symptoms resolve to avoid spreading the infection. Precautions against secondary spread to family members should also be taken at home. In the institutional setting, the use of personal protective equipment (PPE) and enteric isolation is indicated for control. In addition to fecal–oral spread, there is evidence of spread in vomitus.[6] Because these viruses are small and non-lipid, disinfection is accomplished with high-level disinfectants.[7]

REFERENCES

1. Taterka JA, Cuff CE, Rubin DH. Viral gastrointestinal infections. *Gastroenterol Clin North Am* 1992; 21:307–10.
2. LeBaron CW, Furietan NP, Lew JF, et al. Viral agents of gastroenteritis. *MMWR* 1990; 39(RR-5):1–24.

3. Miller SE. Detection and identification of viruses by electron microscopy. *J Electron Microsc Tech* 1986; 4:265–301.
4. DeLeon R, Matsui SM, Bane RS, et al. Detection of Norwalk virus in stool specimens by reverse transcriptasepolymerase chain reaction. *J Clin Microbiol* 1992; 30:3151–7.
5. Monroe SS, Stine SE, Jiang XI, Estes MI, Glass RI. Detection of an antibody to recombinant Norwalk virus antigen in specimens from outbreaks of gastroenteritis. *J Clin Microbiol* 1993; 31:2866–72.
6. Ho MS, Glass RI, Monroe SS, et al. Viral gastroenteritis aboard a cruise ship. *Lancet* 1989; 2:961–5.
7. Favero MS, Bond WW. Sterilization, disinfection and antisepsis in the hospital. In: Balows A, ed. *Manual of clinical microbiology.* Washington DC: American Societyof Micro-biologists, 1991, 183–200.

PARVOVIRUS B19

Common names for disease: Fifth disease, erythema infectiosum

Classification: Family—Parvoviridae; genus—parvovirus

Occupational setting

Employees in health care, day care centers, primary and secondary schools and research laboratories may be at risk from exposure.

Exposure (route)

The presence of parvovirus B19 in respiratory secretions and blood, combined with observed disease patterns, suggests that transmission occurs in the following ways: (1) most often by direct contact with respiratory secretions; (2) with transfusion of contaminated blood products (especially clotting factor concentrates); (3) infrequently through direct contact with or aerosol transmission from concentrated B19 antigen (based on suspected cases in laboratory workers); and (4) to the fetus in ~20% of serologically confirmed maternal infections.

Pathobiology

This small, single-stranded, non-enveloped, heat-stable DNA virus is the only confirmed human pathogen in its genus.[1]

Erythema infectiosum occurs worldwide, year-round, and most commonly in the school age years. Parvovirus B19 IgG antibody prevalence is 15–60% at ages 5–19 and 30–60% in adults. During outbreaks, seroconversion rates are 10–60% in students, 20–30% in school staff, and up to 36% (from a highly contagious source) in hospital care-givers. In normal adults and children, 20% of infections are asymptomatic.[2] Incubation usually ranges from 6 to16 days (shorter for red blood cell aplasia, longer for immunologically mediated symptoms).

Parvovirus B19 causes erythema infectiosum, fetal hydrops and death, arthropathy, and (in susceptible hosts) transient aplastic crisis and chronic red blood cell aplasia. Evidence also suggests an association with other disorders ranging from neuropathy to acute myocarditis and vasculitis.

Primary identified disease mechanisms are erythroid aplasia and immune response. Parvovirus B19 infects erythroid progenitor cells (the precursors of red blood cells) and replicates, leading to cell lysis. Normally, the associated erythroid aplasia (sometimes with neutropenia and thrombocytopenia) is self-limited and clinically inapparent. Knowledge of immunopathologic mechanisms is sketchy. Appearance of IgM antibodies signals recovery from erythema infectiosum but coincides with the onset of arthropathic syndromes.

The most frequent manifestation of parvovirus B19 infection is erythema infectiosum. It is common in children, with a mild, non-specific prodrome and a "slapped-cheek" facial rash. A lacy trunk and extremity rash often fades and reappears. Pruritus is common and atypical rashes may occur.

Parvovirus B19 arthropathy is relatively common in adults. It is characterized by a symmetrical peripheral polyarticular distribution. There is no preceding rash in up to 50% of cases. Arthralgic symptoms typically resolve in 2 weeks to a few months, but in rare cases they may persist for years.

Certain conditions predispose the host to rather specific sequelae of erythema infectiosum. Chronic hemolytic disorders predispose the infected host to transient aplastic crisis due to pure red blood cell aplasia. Parvovirus B19 causes 80–93% of transient aplastic crises in known risk groups (e.g., persons with spherocytocytosis sickle cell disease). It presents with pallor, fever, malaise, headache, severe anemia, and low reticulocyte count. The crisis typically resolves with supportive treatment, without which it may be fatal.

Infection during pregnancy may lead to fetal hydrops and death. Fifty to sixty-five percent of pregnant women are immune to parvovirus B19, based on IgG seropositivity. Risk of maternal infection ranges from ~5% following casual exposure to 20% following intense, prolonged exposure to children with erythema infectiosum. The risk of fetal death following maternal infection appears to be <10% and is greater in the first 20 weeks of gestation.[3–5]

Immunodeficient hosts with parvovirus B19 infection are at risk of chronic bone marrow failure, secondary infection, and pancytopenia.

Diagnosis

For erythema infectiosum, diagnosis is made by clinical features in the context of an exposure history. IgM antibody confirms acute infection; the more sensitive DNA hybridization test, the most sensitive polymerase chain reaction tests or a microscopic examination of infected tissue may be useful for high-risk hosts or where immunocompromised state limits antibody response.[6] Test availability is limited but increasing. The CDC performs antibody testing through state health departments.

Treatment

Treatment is directed toward management of complications in special hosts. Immune globulin may speed recovery in some immunocompromised patients. With maternal infection during gestation, pregnancy termination is not typically recommended. Fetal status is usually monitored closely for signs of fetal hydrops, with cordocentesis and intrauterine transfusion being considered in some situations.[7]

Medical surveillance

In the absence of well-established guidelines, medical authorities recommend warning pregnant women in high risk categories (e.g., healthcare workers, elementary school teachers, child care workers, and mothers of school age children), and offering them serologic screening as needed.[7] Determination of immune status has been recommended for parvovirus B19 research workers.[8]

Prevention

There is active investigation of methodologies for screening blood donors and products for parvovirus B19, with particular interest in plasma pools and products. Vaccine development is also in the early stages.[9] Infectivity with erythema infectiosum generally precedes development of rash; thus, universal precautions alone will suffice in most clinically apparent cases. However, high levels or chronic viremia may occur in some hosts. With such patients (e.g., transient aplastic crisis or chronic parvovirus B19 infection), the CDC recommends private patient rooms, contact isolation, and respiratory protection.[10] In one recent study involving only two infected patients, where isolation was not instituted at the onset of hospitalization, there was no new seroconversion among 87 hospital workers caring for the patients with transient aplastic crisis.[11]

Heat (600°C for 20 min) reportedly inactivates the virus. EPA-registered hospital disinfectants are used for decontamination. Some parvovirus B19 research scientists employ and advise viral inactivation by gamma irradiation and use of safety cabinets.

The American Academy of Pediatrics (AAP) and the American College of Obstetrics and Gynecology (ACOG) have advised that pregnant personnel should not care for patients with transient aplastic crisis, but adherence to this advice is not universal. (The latest ACOG technical bulletin on this topic does not address this issue.[3]) No routine pregnancy-related restrictions are recommended for other work settings. Instead, emphasis is on risk education and fastidious infection control practices.[12,13]

REFERENCES

1. Torok TJ. Parvovirus B19 and human disease. *Adv Intern Med* 1992; 37(43):15–5.0
2. Plummer FA, Hammond GXV, Forward K, et al. An erythema infectiosum-like illness caused by human parvovirus infection. *N Engl J Med* 1985; 313:74–9.
3. American College of Obstetrics and Gynecology. *Perinatal viral and parasitic infections.* Technical bulletin no. 177. Washington DC: A909, 1993: 1–7.
4. Jensen IP, Thorsen P, Jeune B, Moller BR, Vestergaard BF. An epidemic of parvovirus B19 in a population of 3,596 pregnant women: a study of sociodemographic and medical risk factors. *Br J Obstet Gynecol* 2000: 107:637–43.
5. Makhseed M, Pacsa A, Ahmed MA, Essa SS. Patterns of parvovirus B19 infection during different trimesters of pregnancy in Kuwait. *Infent Dis Obstet Gynecol* 1999: 7:287–92.
6. Sevall JS, Ritenbaus J, Peter JB. Laboratory diagnosis of parvovirus B9 infection. *J Clin Lab Anal* 1992; 6:171–5.
7. Levy R, Weissman A, Blomberg G, Hagay ZJ. Infection by parvovirus B19 during pregnancy: a review. *Obstet Gynecol Survey* 1997: 52:254–9.
8. Cohen RJ, Brown KE. Laboratory infection with human parvovirus. *J Infect Dis* 1992; 24:113–14.
9. Prowse C, Ludlam CA, Yap PL. Human parvovirus B19 and blood products. *VoxSanguinis* 1997; 72:1–10.
10. Centers for Disease Control. Risks associated with parvovirus infection. *MMWR* 1989; 38:81–7.
11. Ray SM, et al. Nosocomial exposure to parvovirus B19: low risk of transmission to healthcare workers. *Infect Control Hosp Epidemiol* 1997; 18:109–14.
12. Peter G, Lepow ML, McCracken GIl, et al. eds. *Parvovirus.* Report of the Committee on Infectious Disease. 22nd edn. American Academy of Pediatrics, 1991, 348–50.
13. American Academy of Pediatrics and American College of Obstetrics and Gynecology. Perinatal infections. In: Freeman RK, Poland RL, eds. *Guidelines for perinatal care.* 3rd edn. Elk Grove Village, IL: American Academy of Pediatrics, 1992, 128.

Samuel D. Moon

RABIES VIRUS

Common name for disease: Rabies

Classification: Family—Rhabdoviridae

Occupational setting

Persons at risk include veterinarians, animal handlers, laboratory workers, foreign travelers to endemic areas, recreational hunters, animal control officers, outdoor workers, and recreational enthusiasts who are potentially in contact with rabid animals. (Figure 21.8) Physicians, nurses, therapists and laboratory workers are also at risk if they provide care for patients with rabies.

Exposure (route)

Rabies is almost always transmitted to humans from an animal bite inoculation with virus-

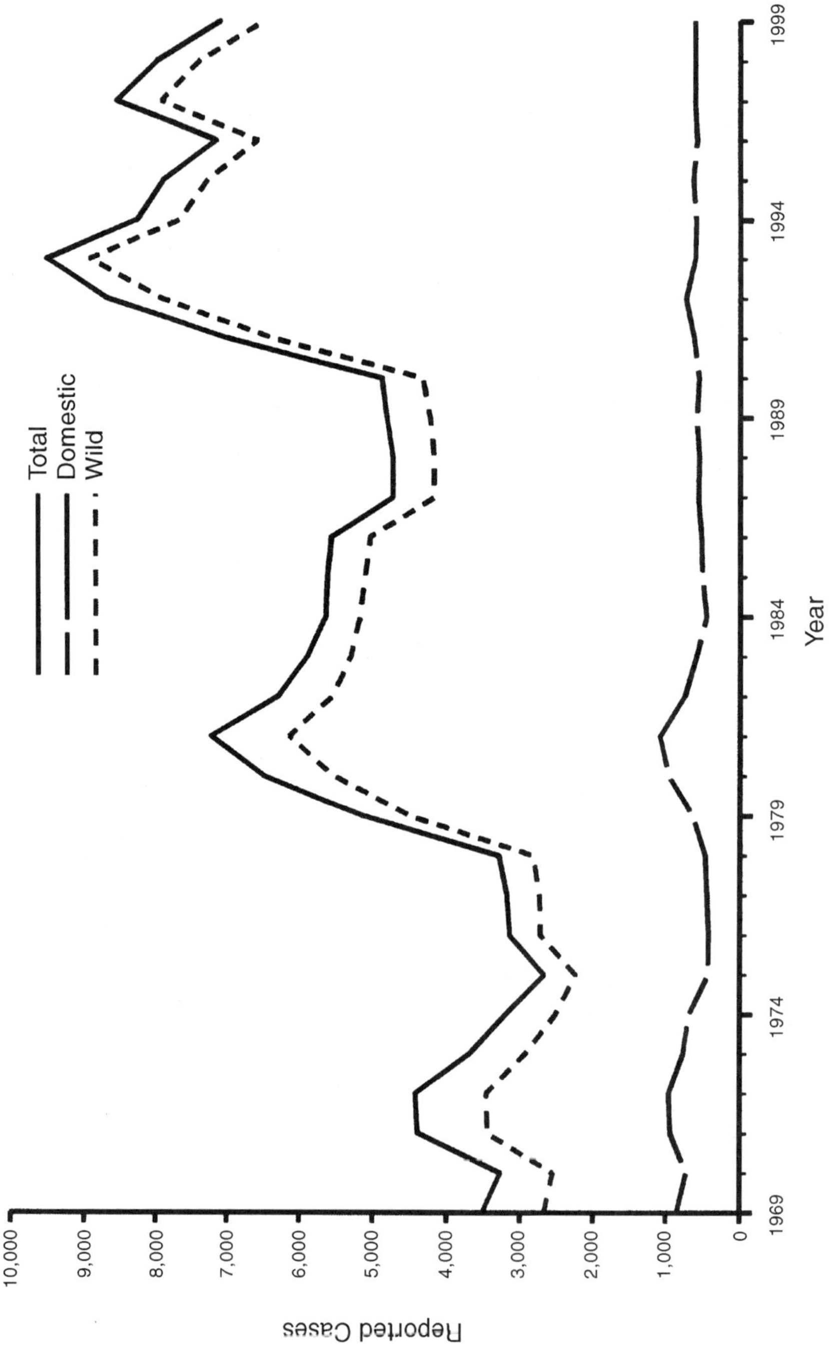

Figure 21.8 Rabies—reported wild and domestic animal cases by year. Source: MMWR 2001:48(53): 59.

laden saliva. Non-bite exposures to saliva or other potentially infectious materials, such as brain tissue from a rabid animal, may occur through abraded skin, open wounds, or mucous membrane inoculation. Although non-bite exposures constitute a sufficient reason to initiate postexposure prophylaxis in some circumstances, these exposures rarely cause rabies.[1] Rare cases have been attributed to aerosol transmission in rabies research laboratories.[2] Two cases of rabies have also been attributed to probable airborne exposures in a bat-infested cave.[3]

Rabies virus is present in a variety of human fluids and tissues during the first 5 weeks of illness, but there are only eight well-documented cases of human-to-human transmission, all in corneal transplant recipients.[4] Although it has never been documented, human-to-human transmission of rabies following saliva exposure remains a theoretical possibility. Thus, in these circumstances, adherence to contact isolation and use of personal protective equipment (PPE) standard precautions will minimize the risk of exposure.[5]

Pathobiology

Rabies is a DNA-containing virus. Virions measure 180×75 nm; they are cylindrical with one round or conical end and one flat end, giving each particle the shape of a bullet. Five proteins have been identified from purified rabies virus, including glycoprotein (G), nucleocapsid (N) protein, viral polymerase (L), and two smaller proteins (NS (P) and M).[6] Monoclonal antibodies directed against both nucleocapsid and glycoprotein antigens have been developed. The virus is rapidly inactivated by desiccation, ultraviolet and X-radiation, sunlight, trypsin, beta-propiolactone, ether, and detergents.

Following inoculation, the virus replicates locally and then spreads via peripheral nerves to the CNS, where it causes encephalitis. The incubation period is usually 20–90 days; the incubation period is shorter when the bite is on the head or face rather than on an extremity. During the incubation period, the infected individual is usually well except for symptoms related to local wound healing.

After a latent period lasting from days to several weeks, months, or (in some cases) years, the virus spreads via peripheral nerves to the spinal cord and CNS, especially the limbic system. The virus is normally present in the CNS in high titer before the development of systemic symptoms. When symptoms occur, the virus has already traveled down efferent nerves to nearly every organ and tissue, including, most important for the life cycle of the virus, the salivary glands.

Clinical rabies usually begins with generalized, non-specific constitutional symptoms, including malaise, fatigue, headache, anorexia, and fever. In 50% of patients, pain or paresthesia occurs at the site of the exposure and may be the first rabies-specific symptom.[2] Other non-specific symptoms, including sore throat, cough, abdominal pain, nausea and vomiting, diarrhea, or chills, have also been associated with the prodrome, which normally lasts 2–10 days. This prodrome is followed by the neurologic phase, which may include intermittent hyperactivity, hallucinations, disorientation, bizarre behavior, seizures, nuchal rigidity, or paralysis. The hyperactive episodes may occur spontaneously, or they may be precipitated by a variety of tactile, auditory, visual or other stimuli. Diaphoretic spasm often leads to the classical hydrophobia. Other symptoms seen during the acute neurologic phase include fever, muscle fasciculation, hyperventilation, hypersalivation, and convulsions.

The acute neurologic phase usually lasts 2–7 days, with longer duration in the paralytic forms. In patients in the USA who do not receive intensive supportive care, the average duration of illness until death is 7 days. For those who receive intensive support, the duration of illness averages 25 days.[8]

Three cases of recovery from presumed rabies have been reported. In each case, the patient had received some form of pre-exposure or postexposure prophylaxis and intensive life support care.[6] With these rare exceptions, rabies is considered a fatal disease. Therefore, emphasis is placed on prevention rather than treatment.

In the USA and other parts of the developed world, animal vaccination programs have greatly reduced the incidence of rabies among domestic dogs and cats. Wild animals now constitute the most important potential source of infection for both humans and domestic animals in the USA. Most of the animal rabies cases in the continental USA occur in skunks, foxes, raccoons, and bats. Only Hawaii remains consistently rabies-free. Bats are increasingly implicated in human rabies transmission. Recent epidemiologic data suggest that transmission of rabies virus can occur from minor or unrecognized bites from bats. From 1980 to 1997, 21 (58%) of the 36 human cases of rabies diagnosed in the USA were associated with bats.[4] The dog remains the major source of human exposure to rabies outside the USA and is a concern for foreign travelers. Rodents, such as squirrels, hamsters, guinea pigs, gerbils, chipmunks, rats, mice, and rabbits, are rarely infected with the rabies virus and have not been known to transmit rabies to humans. The marked decrease of rabies cases among domestic animals in the USA has drastically decreased human infections. However, outside the USA and Europe, dogs remain the most common source of infection for humans. In fact, 12 (33%) of the human rabies deaths reported to the CDC from 1980 to 1997 appear to have been related to rabid animals outside the USA.[9]

Diagnosis

No tests are currently available to diagnose rabies in humans before the onset of clinical disease. The virus is immunologically protected and does not usually stimulate antibody production until after invasion of the CNS. The rapid fluorescent focus inhibition test (RFFIT) is the standard test for measuring rabies-neutralizing antibody. The most reliable and reproducible of the direct immunofluorescent studies that can aid in patient diagnosis is the neck skin biopsy. In this procedure, a 68-mm full-thickness wedge or punch biopsy of skin is taken from the posterior aspect of the neck above the hairline. This can be temporarily stored at 70°C until shipped to a laboratory familiar with the immunofluorescent technique. Ante-mortem studies have shown that rabies virus can be isolated from human saliva, brain tissues, CSF, urine sediment, and tracheal secretions. The RT PCR test is emerging as the diagnostic procedure of choice in suspected rabies cases and can be performed on saliva, CSF or tissue.[6]

Treatment

There is no specific treatment once clinical rabies is established. Treatment consists of respiratory and cardiovascular support. Passive rabies immune globulin has been used in several cases with no clear benefit. To prevent secondary bacterial infection of the patient and to prevent exposure of hospital staff to rabies virus, the patient should be isolated. To avoid exposure, hospital personnel should wear face masks, gloves, and gowns. The administration of vaccine after the onset of clinical illness has not been successful. Interferon has been noted to have activity against rabies virus in tissue culture, but it has not been successful in the treatment of patients and has not altered the course of the disease.[6]

Prevention

Human rabies prevention consists of pre-exposure vaccination and postexposure therapy and prophylaxis. There are currently two types of rabies-immunizing products available in the USA. Human rabies immune globulin (HRIG) provides rapid short-term passive immunity (half-life, 21 days). Rabies vaccines include the human diploid cell vaccine (HDCV), the rabies vaccine absorbed (RVA) and the purified chick embryo cell vaccine (PCEC). These vaccines induce an active immune response with the production of neutralizing antibodies, usually within 7–10 days, which persists for at least 2 years. Recent studies suggest that antibody responses may persist for up to 8 years following vaccination.[9]

All types of rabies vaccines are considered equally efficacious and safe when used as indicated. HDCV, RVA and PCEC should be administered through the intramuscular route. Deltoid muscle administration is recommended. Gluteal administration is to be avoided because of concerns about efficacy. HDCV can also be given via the intradermal route in a lower-volume dose, but this route of administration has not been adequately studied for RVA and PCEC.

TABLE 21.5 Rabies preexposure prophylaxis guide—United States, 1999.

Risk category	Nature of risk	Typical populations	Preexposure recommendations
Continuous	Virus present continuously, often in high concentrations. Specific exposures likely to go unrecognized. Bite, nonbite, or aerosol exposure.	Rabies research laboratory workers;* rabies biologics production workers.	Primary course. Serologic testing every 6 months; booster vaccination if antibody titer is below acceptable level.†
Frequent	Exposure ussually episodic, with source recognized, but exposure also might be unrecognized. Bite, nonbite, or aerosol exposure.	Rabies diagnostic lab workers,* spelunkers, veterinarians and staff, and animal-control and wildlife workers in rabies-enzootic areas.	Primary course. Serologic testing every 2 years; booster vaccination if antibody titer is below acceptable level.†
Infrequent (greater than population at large)	Exposure nearly always episodic with source recognized. Bite or nonbite exposure.	Veterinarians and animal-control and wildlife workers in areas with low rabies rates. Veterinary students. Travelers visiting areas where rabies is enzootic and immediate access to appropriate medical care including biologics is limited.	Primary course. No serologic testing or booster vaccination.
Rare (population at large)	Exposure always episodic with source recognized. Bite or nonbite exposure.	US population at large, including persons in rabies-epizootic areas.	No vaccination necessary.

* Judgment of relative risk and extra monitoring of vaccination status of laboratory workers is the responsiblility of the laboratory supervisor.

† Minimum acceptable antibody level is complete virus neutralization at a 1 : 5 serum dilution by the rapid fluorescent focus inhibition test. A booster does should be administered if the titer falls below this level.

Source: Centers for Disease Control. Rabies prevention—United States. MMWR 1999; 48: No. RR-1, p. 6.

Adverse reactions to HDCV, RVA and PCEC are far less common and serious than reactions to previous vaccines. They include mild local reactions, such as pain, erythema, and swelling at the injection site. Systemic reactions such as headache, nausea, abdominal pain, muscle aches and dizziness have been reported in 5–40% of recipients. Three cases of neurologic illness resembling Guillian–Barre syndrome that resolved without sequelae have been reported.[10] An immune complex-like reaction has been reported in 6% of persons receiving booster injections of HDCV. Symptoms include generalized urticaria, sometimes accompanied by arthralgia, myalgia, vomiting, fever, and malaise, but in no cases were these reactions life-threatening.[11]

Corticosteroids, other immunosuppressive agents, antimalarials and immunosuppressive illness can interfere with the antibody response, so it is especially important that these individuals be tested for effective antibody titers. Pregnancy is not considered a contraindication to postexposure or pre-exposure prophylaxis in high-risk individuals. Pre-exposure vaccination should be offered to high-risk groups such as veterinarians, animal handlers, laboratory workers, and foreign travelers on extended visits to foreign countries where canine rabies is endemic. Pre-exposure prophylaxis may provide protection to persons with inapparent exposures to rabies; it can also provide protection when postexposure therapy is delayed. However, pre-exposure vaccination does not eliminate the need for additional therapy after a known rabies exposure. Table 21.5 shows the CDC guidelines for pre-exposure rabies prophylaxis, and Table 21.6 shows the CDC pre-exposure prophylaxis for primary and booster vaccination series. For laboratory personnel handling rabies virus, BSL3 containment is recommended.

Although rabies has the highest case fatality rate of any known human infection, it can almost always be prevented if exposures are recognized and postexposure prophylaxis is initiated. Each year, 20 000 people receive antirabies postexposure prophylaxis in the USA. Even though most of these individuals have not had significant rabies exposure, there have been no post-exposure vaccine failures in the USA since HDCV was licensed.[9] Postexposure prophylaxis begins with effective local wound cleansing and, in most cases, administration of both HRIG and rabies vaccine. Table 21.6 summarizes postexposure prophylaxis recommendations following animal bites and guidelines on the evaluation and disposition of the animal. Table 21.7 outlines the postexposure treatment regimen for both previously vaccinated individuals and those not previously vaccinated.

Table 21.6 Rabies preexposute prophylaxis schedule—United States, 1999.

Type of vaccination	Route	Regimen
Primary	Intramuscular	HDCV, PCEC or RVA; 1.0 mL (deltoid area), one each on days 0,* 7, and 21 or 28
	Intradermal	HDCV; 0.1 mL, one each on days 0,* 7, and 21 or 28
Booster	Intramuscular	HDCV, PCEC or RVA; 1.0 mL (deltoid area), day 0* only
	Intradermal	HDCV; 0.1 mL, day 0* only

HDCV = human diploid cell vaccine; PCEC = purified chick embryo cell vaccine; RVA = rabies vaccine adsorbed.

* Day 0 is the day the first dose of vaccine is administered.

Source: MMWR 1999: 48 (RR-1): 5.

Table 21.7 Rabies postexposure prophylaxis guide—United States, 1999.

Animal type	Evaluation and disposition of animal	Postexposure prophylaxis recommendations
Dogs, cats, and ferrets	Healthy and available for 10 days observation	Persons should not begin prophylaxis unless animal develops clinical signs of rabies.*
	Rabid or suspected rabid	Immediately vaccinate.
	Unknown (e.g., escaped)	Consult public health officials.
Skunks, raccoons, foxes and most other carnivores; bats	Regarded as rabid unless animal proven negative by laboratory tests†	Consider immediate vaccination.
Livestock, small rodents, lagomorphs (rabbits and hares), large rodents (woodchucks and beavers), and other mammals	Consider individually.	Consult public health officials. Bites of squirrels, hamsters, guinea pigs, gerbils, chipmunks, rats, mice, other small rodents, rabbits, and hares almost never require antirabies postexposure prophylaxis.

* During the 10-day observation period, begin postexposure prophylaxis at the first sign of rabies in a dog, cat, or ferret that has bitten someone. If the animal exhibits clinical signs of rabies, it should be euthanized immediately and tested.

† The animal should be euthanized and tested as soon as possible. Holding for observation is not recommeded. Discontinue vaccine if immunofluorescence test results of the animal are negative.

Source: MMWR 1999: 48 (RR-1): 7.

REFERENCES

1. Afshar A. A review of non-bite transmission of rabies virus infection. *Br Vet J* 1979; 135:142–8.
2. Winkler XVG, Fashinell TR, Leffingwell L, Howard P, Conomy P. Airborne rabies transmission in a laboratory worker. *JAMA* 1973; 226:1219–21.
3. Centers for Disease Control. Rabies prevention United States, 1991. *MMWR* 1991; 40:1–14.
4. Centers for Disease Control. Human rabies prevention—United States, 1999. *MMWR* 1999; 48(RR-1): 1–17.
5. Garner JS. The Hospital Infection Control Practices Advisory Committee. Guideline for isolation precautions in hospitals. *Infect Control Hosp Epidemiol* 1996; 17:54–80.
6. Bleck TP, Rupprecht CE. Rabies virus. In Mandell, Douglas, and Bennett, eds. *Principles and practice of infectious diseases*. 5th edn. Philadelphia. Churchill Livingstone, 2000, 1811–2004.
7. Murphy FA. The pathogenesis of rabies virus infection. In: Kaprowski H, Plotkin S, eds. *World's debt to Pasteur*. New York: Alan R Liss, 1985, 153–69.
8. Anderson U, Nicholson KG, Tauxe RV, et al. Human rabies in the United States, 1960–1979: epidemiology, diagnosis, and prevention. *Ann Intern Med* 1984; 100:728–35.
9. Roumiantzett AN, et al. Experience with pre-exposure rabies vaccine. *Rev Infect Dis* 1988;10(suppl 4):5751–6.
10. Bernard KW, Smith PW, et al. Neuroparalytic illness and human diploid cell rabies vaccine. *JAMA* 1982; 248:3136–8.
11. Centers for Disease Control. Systemic allergic reactions following immunization with human diploid cell rabies vaccine. *MMWR* 1984; 33:185–7.

Dennis J. Darcey

RESPIRATORY SYNCYTIAL VIRUS (RSV)

Common name for disease: Respiratory syncytial virus pneumonia

Classification: Family—Paramyxoviridae; genus—pneumovirus

Occupational setting

Hospital personnel and child care personnel may be at risk. In one report, nearly half of the nurses, residents and medical students on children's hospital wards developed RSV infections during an RSV outbreak.[1] Because repeated infection in the older population may result in very little titer rise, serologic titer change in adults provides a less reliable diagnosis. Adult illness associated with exposure to sick infants with RSV is the basis for a presumptive diagnosis.

Exposure (route)

Infection is transmitted by the respiratory route. Large-droplet inoculation is the primary mode. Small-particle aerosols may also transmit infection, but this means is less contagious. Transmission can also occur through contact with contaminated fomites and skin.[2]

Pathobiology

RSV is a single-strand RNA virus measuring 100–300 nm. On electron microscopy, the virus is pleomorphic with spherical and filamentous forms. RSV is the primary cause of lower respiratory tract disease in children. Most children contract this disease within the first few years of life; however, immunity is incomplete and thus reinfection is common. Repeat infections with RSV are common, and they affect all age groups. Inoculation occurs primarily through the eyes and nose, and less so via the mouth. Subsequent infection after the primary one, which usually occurs during early childhood, produces less severe illness. The incubation period is 3–5 days (average 4 days). Viral shedding in adults continues for an average of 3–6 days.

In children, the primary infection is manifested as pneumonia, bronchiolitis, tracheobronchitis, or upper respiratory tract illness with fever and otitis media. Infection in otherwise healthy adults usually presents as upper respiratory tract illness with nasal congestion and cough. The illness tends to be more prolonged than the common cold; symptomatology averages 9 days. Incapacitation is seen in about half of adults with RSV. In some adults, pulmonary function change and airway hyperactivity can be seen 8 weeks after the illness.[3]

Diagnosis

Diagnosis of RSV is clinical and epidemiologic. Community evidence of RSV disease in the infant population associated with lower respiratory findings provides a good presumptive diagnosis. Viral isolation of nasal washing may be done, or, more commonly, an immune fluorescent assay using monoclonal antibodies, which provides more definitive information.

Treatment

Most healthy adults require no treatment for this self-limited illness. Elderly or debilitated individuals may require supportive therapy. Infants are commonly treated with a small-particle aerosol of ribavarin. However, a number of studies suggest that ribavirin treatment does not improve outcome except in immunocompromised hosts.

Prevention

RSV is sensitive to temperature and pH changes. The virus is inactivated by a variety of detergents. The virus in patient secretions survives <30 hours on countertops and <1 hour on cloth and paper. Infectivity of the virus on hands is usually <1 hour.[4]

Vaccination with various vaccine preparations has not been effective in reducing disease; in fact, some studies showed that it may have enhanced pathogenesis.[5] Effective control of transmission in the hospital or other child care settings is dependent on the interruption of hand-carriage of virus from one individual to another and the interruption of self-inoculation of the eyes or nose. Excellent handwashing before and after contact is the most important measure. Using protective clothing and eye–nose goggles for close contact work may be helpful.[6] Because of the type of contact and undiagnosed status, child care workers outside of the hospital setting probably cannot effectively interrupt transmission. Standard laboratory practice should prevent transmission in the clinical laboratory.

REFERENCES

1. Hall GB, Geiman JM, Douglas RG. Control of respiratory syncytial viral infections. *Pediatrics* 1978; 62:730.
2. Hall CB. The nosocomial spread of respiratory syncytial viral infections. *Annu Rev Med* 1983; 34:311–19.
3. Hall WJ, Hall CB, Speers DM. Respiratory syncytial virus infections in adults. *Ann Intern Med* 1978; 88:203.
4. Hall GB, Geiman JM, Douglas RG. Possible transmission by fomites of respiratory syncytial virus. *Infect Dis* 1980; 141:98–102.
5. Kapikian AZ, Mitchell RH, Chanock RM, et al. An epidemiologic study of altered clinical reactivity to RSV infection in children previously vaccinated. *Am J Epidemiol* 1969; 89:405–21.
6. Graman PS, Hall CB. Epidemiology and control of nosocomial viral infections. *Infect Dis Clin North Am* 1989; 3:815–23.

Simoes E. Respiratory syncytial virus infection. *Lancet* 1999; 354:847–52.

RUBELLA VIRUS

Common names for disease: German measles, rubella, third disease

Classification: Family—Togaviridae; genus—rubivirus

Occupational setting

Healthcare workers, teachers and day care workers may have an increased risk of contact with unvaccinated populations.

Exposure (route)

Rubella is transmitted almost entirely by infected airborne droplets.

Pathobiology

Rubella virus is a 60–70-nm spheroidal, enveloped, single-stranded RNA virus.

The rubella virus causes a disease with usually trivial symptoms for the primary host, but it produces catastrophic consequences for secondary victims in utero. The congenital rubella syndrome produces fetal and neonatal damage, with recognizable cardiac, ocular and CNS malformations. When epidemics of the disease occurred (the last was in 1964, before the licensing in 1969 of an effective vaccination), many thousands of simultaneous birth defects were found among affected pregnancies. Prevention efforts are presently focused entirely on reducing the possibility of infection for pregnant women.

The primary illness of rubella is preceded by an incubation period of 14–21 days and is commonly asymptomatic.[1] For those with recognized illness, early symptoms of fever and malaise often precede lymph node enlargement and eventual rash. The rash is very similar to rubeola (measles), commonly beginning on the face and spreading downward during the few days of its duration. Complications are infrequent; they can include arthritis and delayed blood clotting (with resulting hemorrhage).

Ill effects to the fetus are determined by the gestational timing of the infection. Just as with chemical sources of early fetal danger, early gestational exposure leads most commonly to spontaneous abortion. Somewhat later infections cause recognizable structural damage. Heart problems occur in up to 30–35% of fetuses infected during the 3rd month of pregnancy. Late infection (after the 20th week) causes less structural damage but it carries a 10% risk of producing functional damage to the CNS, especially deafness.

Fetal injury does represent a true infection with the virus, since the virus can be isolated from affected tissues. Damage may continue to be discovered over ensuring years, with structural manifestations first noted at entrance to school. Juvenile diabetes mellitus is more common in these children, implying that there is damage to the pancreatic insulin-secreting cells.

The virus is spread by affected individuals as contaminated airborne droplets, through contact with nasopharyngeal secretions. The period of greatest risk for contagion is closest to the time that the rash is evident. Experimental studies have shown that the virus can be isolated from normal individuals several days before the occurrence of the rash and for up to 2 weeks after the onset of symptoms. Adults and children should be considered contagious until at least 4 days after the rash develops. Newborns with congenital infection may remain infectious for several months.

Diagnosis

Postnatal diagnosis is achieved by recognition of the rash and confirmed by documentation of an elevation in antibody titer. The virus can also be isolated from the pharynx and urine. In anticipation of the rare but critical situation of a potential rubella viral illness in pregnancy, obstetricians routinely obtain early pregnancy serologic data to assess antibody titers against this organism. Rising values following a suspicious illness or exposure (remembering that many cases are asymptomatic to the adult) are considered diagnostic.

Treatment

No antiviral therapy is available for postexposure treatment of exposed pregnant patients. For others, the illness is so mild that isolation from others who may be pregnant is the only necessary clinical management.

Medical surveillance

Documentation of adequate rubella vaccination is universally required by public health agencies and clinical credentialing bodies for all personnel in healthcare facilities.[2,3] Vaccinations should be routinely provided for workers with inadequate medical documentation and low serum antibody titers.

Surprisingly, even in successfully vaccinated individuals, elevations in antibody titer can be seen following viral exposure, sometimes with isolated live virus. This implies that antibody-mediated immunity is incomplete and that the virus can be reacquired by those with normal immune function and successful vaccination. Five cases of post-rubella syndrome have been noted among “immune” patients whose antibody rise followed viral exposure during pregnancy. Thus, obstetric monitoring of rubella antibody levels must be quantitative rather than qualitative. More broadly, the need to isolate infected individuals remains relevant even when the potentially exposed population has been shown to have antibody.

Prevention

The rubella vaccine is a live, attenuated virus that confers excellent resistance against subsequent clinical illness and greatly reduces the risk of incurring and transmitting the illness. Even though the target population for protection is women of childbearing years, all children should be vaccinated and all healthcare workers should be required to document immunity. The rationale for vaccinating men and boys is merely to reduce the availability of the virus to susceptible women.

Rubella has been cited as an illness where comprehensive eradication is a possibility.[4] This ideal is not yet a reality, but the results achieved by the present immunization efforts have been impressive. From the peak epidemic of rubella in 1964, when 30 000 infants were affected, rates of congenital rubella syndrome have fallen to less than a dozen annually. The

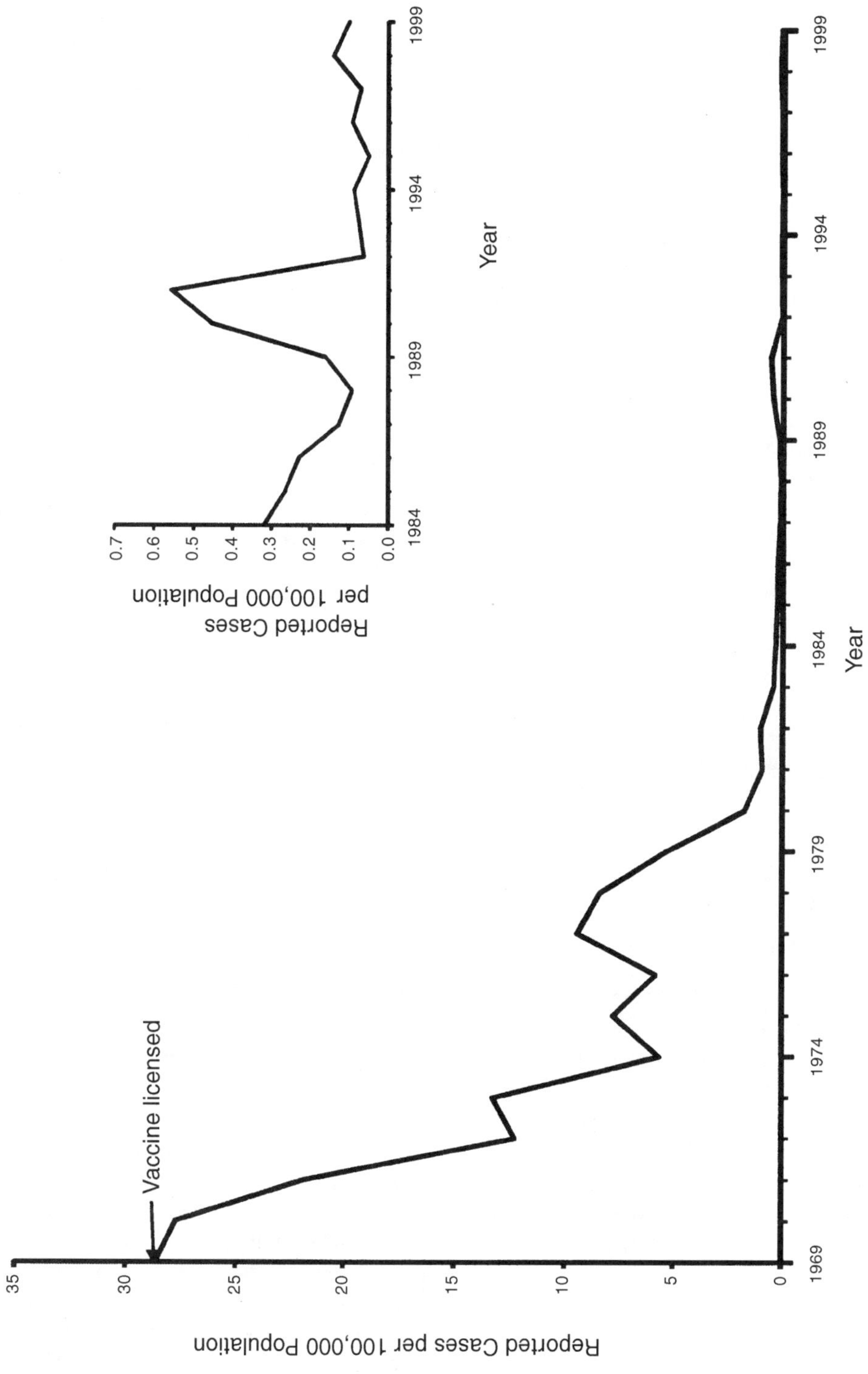

Figure 21.9 Rubella—reported cases per 100 000 population by year, United States, 1969–1999. Source: MMWR 2001: 48(53): 61.

diagnosis of rubella requires reporting to the CDC, which tallies overall rates. Although only 137 cases were reported in 1992, 1256 were found the year before—a climb from the previous low of 188 in 1988.[5]

Because the vaccine is a live virus, theoretical considerations require that it not be administered to individuals already or soon to be pregnant. Despite these protective concerns, inadvertent episodes of vaccination during pregnancy have occurred without subsequent congenital rubella syndrome.[6]

REFERENCES

1. Gershon AA. Viral diseases: togaviridae: rubella virus (German measles). In: Mandell GL, Douglas RG, Bennett JE, eds. *Principles and practice of infectious diseases*, 2nd edn. New York: Wiley, 1985, 926–31.
2. Centers for Disease Control. *Immunization recommendations for health-care workers*. Atlanta: Centers for Disease Control, Division of Immunization, Center for Prevention Services, 1989.
3. Centers for Disease Control. Recommendations of the Immunization Practices Advisory Committee: update on adult immunization. *MMWR* 1991; 40(RR-12):1.
4. Centers for Disease Control. Update: International Task Force for Disease Eradication, 1992. *MMWR* 1992; 41:691,697–8.
5. Centers for Disease Control. Update: changes in notifiable disease surveillance data—United States, 1992–1993. *MMWR* 1993; 42(42):824–6.
6. Mann JM, Preblud SR, Hoffman RE, et al. Assessing risks of rubella infection during pregnancy: a standardized approach. *JAMA* 1981; 245:1647.

Gary N. Greenberg

SIMIAN IMMUNODEFICIENCY VIRUS (SIV)

Common name for disease: Simian AIDS

Classification: Retroviridae

Occupational setting

Veterinarians, non-human primate handlers, animal care-takers and virus researchers are at risk from exposure.

Exposure (route)

Simian immunodeficiency virus (SIV) is transmitted via needlesticks, possibly bites and scratches, and mucous membrane contact with certain non-human primate body fluids. SIV culture material and any equipment that comes into contact with these materials must be considered infectious.

Pathobiology

SIV belongs to the family of Retroviridae and is closely related to human immunodeficiency virus types I and II.[1,2] Originally reported in 1985, it was initially called simian T-lymphotropic virus type III (STLV-III). All lentiviruses appear to replicate in and persist in cells of the monocyte/macrophage lineage. SIV is also tropic for CD4-positive leukocytes. Serologic surveys indicate that numerous species of wild and captive non-human primates are infected with SIV. Several species of macaques, African green monkeys, mangabeys, mandrills, guenons, talapoins and colobus monkeys have been infected with SIV. Infection of macaque monkeys can cause a chronic disease resembling human AIDS. Opportunistic infections, diarrhea, thymic atrophy, decrease in CD4 cell numbers, lymphomas and encephalitis occur in these monkeys.

There has been evidence of seroconversion in a laboratory worker who was exposed by a needlestick to blood from an SIV-infected macaque.[3] Despite immediate scrubbing of the wound, inflammation and swelling developed at the puncture site and persisted for several weeks. Seroactivity to SIV developed within 3 months, peaked between the 3rd and 5th months, and declined afterwards. Attempts to isolate the virus and to find SIV provirus by PCR were unsuccessful. Also, attempts to transmit SIV by inoculation of a macaque with the worker's blood were unsuccessful. It was concluded that the worker had not become permanently infected with SIV.

A subsequent survey of SIV researchers revealed that three of 472 had antibodies to SIV.[4] In 1994, there was a first report of an SIV laboratory worker who developed actual infection with SIV.[5] This researcher developed a severe dermatitis of the forearms and hands, and continued to handle clinical specimens from infected monkeys without wearing gloves. SIV was successfully isolated from the worker's peripheral blood mononuclear cells. However, two monkeys inoculated with the researcher's blood remained seronegative. The researcher's infection had not resulted in clinical illness at the time of the report.

Diagnosis

Serologic procedures to test for SIV antibody are used. Gene amplification may allow differentiation of specific virus gene sequences from specimens of exposed individuals.[3] There is serologic cross-reactivity between antibodies to SIV and HIV-2.

Treatment

There is no specific treatment for the virus. Opportunistic infections, if they occur, should be treated with appropriate antibiotics.

Medical surveillance

Serum should be collected and stored at 6-month intervals for individuals performing research with SIV.[6] Serum banking should also be undertaken on individuals whose work entails exposure to non-human primates. If a worker is inadvertently exposed to SIV-contaminated material, he or she should undergo a medical evaluation and serum examination for antibody against SIV. Seronegative workers should be re-tested at 6 weeks, 12 weeks, and 6 months. They should seek medical attention for any acute illness that develops within 12 weeks of exposure.

Prevention

During work with clinical specimens, laboratory coats, gowns or uniforms should be worn along with protective eyewear, face masks, and gloves.[4] If SIV is being propagated in research laboratories or procedures are performed that may generate aerosols, activity should be performed in BSL2 facilities, with additional practices and containment equipment recommended for BSL3. Activities involving large-volume production should be conducted in BSL3 facilities.

BSL2 standards are recommended when handling infected animals and when performing activities involving clinical specimens.[4] Work surfaces should be decontaminated and hands washed immediately after handling infectious material, even when gloves have been worn.

REFERENCES

1. Desrosiers RC, Danielle MD, Li Y. Minireview: HIN-related lentiviruses of non-human primates. *AIDS Res Hum Retroviruses* 1989; 5:465–72.
2. Desrosiers RC. Simian immunodeficiency viruses. *Annu Rev Microbiol* 1988; 42:607–25.
3. Khabbaz RF, Rowe T, Murphey Corb M, et al. Simian immunodeficiency virus needlestick accident in a laboratory worker. *Lancet* 1992; 340:271–3.
4. Centers for Disease Control. Anonymous survey for Simian immunodeficiency virus (SIV) seropositivity in SIV laboratory researchers—United States. *MMWR* 1992; 41:814–15.
5. Khabbaz RF, Heneine XV, George JR, et al. Brief report: infection of a laboratory worker with simian immunodeficiency virus. *N Engl J Med* 1994; 330:172–7.
6. Centers for Disease Control. Guidelines to prevent simian immunodeficiency virus infection

in laboratory workers and animal handlers. *MMWR* 1988; 37:693–703.

Ricky L. Langley
Carol A. Epling

VACCINIA

Common name for disease: None

Classification: Family—Poxviridae; genus—orthopoxvirus

Occupational setting

Vaccinia is related to, but not the same as, cowpox, a rare zoonosis from domesticated animals. Vaccinia is also a highly effective immunizing agent against the closely related and far more dangerous smallpox virus (variola). Since global eradication of naturally acquired smallpox occurred in 1977, there is no longer a routine requirement for population immunization. Vaccinia immunization is now quite rare, except for selected researchers and perhaps for special military troops.

Smallpox is still considered by some to be a potential biological warfare threat; therefore, military forces may stockpile vaccinia immunizations for emergency use or actually immunize certain high-risk troops.

Today, vaccinia is encountered commonly in certain research laboratories. Genetically engineered (recombinant) vaccinia will express foreign DNA materials without hindering the virus's ability to replicate. For this reason, vaccinia is popular in basic research and the preparation of experimental immunizations.

It is possible for researchers and vaccine developers to become infected with their cultures, whether recombinant or not. Occasionally, vaccinia exposures can have serious consequences for exposed workers. For this reason, the Centers for Disease Control recommend that laboratory research personnel who handle live vaccinia virus or monkeypox also be immunized with vaccinia virus. The vaccine may be obtained from the Drug Service of the Centers for Disease Control. This recommendation for vaccination is controversial, as will be discussed.

Exposure (route)

Most vaccinia exposure is by intentional immunization. Immunized individuals harbor live virus in pustules for several weeks. Live virus may then infect others by pustule-to-wound exposure or mucous membrane exposure. In addition, laboratory uses of vaccinia may require large quantities of virus, which can enter the human worker via mucous membrane or wounds, or perhaps by inhalation. Immunization is applied directly to the skin, which is punctured several times in a small area. This procedure is repeated every 3 years to ensure immunity. Vaccinia immunization is relatively contraindicated in patients with dermatologic problems, for reasons described below. Immunization is absolutely contraindicated in patients with any causefor immune suppression or previous complication of immunization, such as vaccinia encephalitis.

Pathobiology

Vaccinia is a DNA-containing orthopoxvirus. Healthy, non-immune individuals respond to primary vaccinia challenge with a mild fever and sometimes lymphadenopathy. A red papule forms at the site of inoculation 3–5 days after exposure. This vesiculates several days later and becomes pustular by the 9th or 11th day. The pustule dries after 2 weeks and drops off when encrusted, usually about 3 weeks after vaccination. The axillary lymphadenopathy often associated with the process may be persistent.[1,2] Generalized vaccinia is the widespread but usually mild dissemination of vaccinia lesions in patients without pre-existing skin disease. It occurs in ~3:100 000 primary vaccinations. Other complications of infection or immunization are rare.

They include systemic vaccinia (also called progressive vaccinia or vaccinia gangrenosum). Systemic vaccinia leads to underlying destruction of skin, subcutaneous tissues and even viscera of immunosuppressed individuals. Patients with agammaglobulinemia or T-cell deficiencies and those receiving immunosuppressive therapy are at greatest risk. The disease is usually fatal. Vaccinia infection of eczematous skin or other problem skin (eczema vaccinatum) is also a serious complication. Accidental autoinoculation of eyelids, vulva or other normal mucous membrane is another rare skin complication. Unintentional exposures via transmission from infected individuals or from laboratory exposures are rare but may be serious if they involve eyes or if the exposed individual is immunosuppressed.

Post-vaccinal encephalomyelitis is the most dreaded complication of immunization of patients with normal immune systems. Its incidence is still debated; it is said to occur in 1:10 000 to 1:25 000 vaccinations, or perhaps as infrequently as 1:300 000 vaccinations[3] with at least 25% mortality and an equal risk of other severe neurologic sequelae.

Diagnosis

Systemic vaccinia or other complications may follow vaccination, laboratory exposure, or secondary exposure to live vaccinia in a carrier. Diagnosis requires clinical suspicion and recognition of the typical lesion and its spread. Suspicion should be heightened for vaccinia research workers, recently immunized workers, and contacts of those recently immunized (or otherwise infected). History of exposure is the critical diagnostic clue. In addition, the virus can be isolated in scrapings and vesicule fluids of lesions, and recent exposure (infection) can be inferred from sequential serologic testing.

Treatment

Treatment is supportive. Vaccinia immune globulin can be used for serious complications.

Prevention

In the vaccinia research laboratory, prevention consists of engineering controls, good work practices, and sensible immunization policies. Work practices are standard for the BSL2 laboratory, appropriate to vaccinia's status as a potential human pathogen.[4]

The primary engineering control of the recombinant laboratory is the biosafety cabinet. Class I and II cabinets provide partial containment and should prevent most inoculation. If vaccinia incorporate recombinant material, which may somehow confer increased risk of infection or else increased virulence with infection, then a class III biosafety cabinet (also known as a glove box) should be used for full containment. Class III cabinets are also indicated if manipulations of highly concentrated cultures are required in the laboratory.

Administrative controls are also critical. No one who is pregnant, immunosuppressed, or has a serious skin problem should work with vaccinia. Protective immunization is also contraindicated for immunosuppressed workers.

The CDC recommends vaccinia vaccine for laboratory workers who directly handle vaccinia or recombinant vaccinia cultures or infected animals. Recipients are given controlled percutaneous doses—2.5×10^5 pock-forming units.[5] The following rationales for immunization (infection) with a live organism are suggested by the CDC:

1. Immunization may offer some protection to recipients against inadvertent inoculations with larger amountsfrom highly concentrated laboratory cultures.
2. Immunization may protect against inad-

vertent exposure via undesirable routes (such as the eye).
3. Immunization may protect against seroconversion to foreign antigen expressed by a recombinant vaccinia.

The CDC has made these recommendations, and they should certainly not be ignored. However, physicians and scientists should also be aware that strong counterarguments have been made in the name of public health:[6]

1. Immunization is certain infection, whereas inadvertent inoculation has been rare. The most feared side-effects relate to frequency and dose of infection, not the route of infection. Immunization represents maximum frequency of infection. It is unknown whether the iatrogenic risk of immunization outweighs the protective benefits for the immunized worker.
2. The population not immunized is increasingly "vaccinia-naive". Contacts by those immunized with immunosuppressed, vaccinia-naive individuals will be common. It is unknown whether the purported benefits for immunized workers will outweigh risks to secondary contacts. In the era when the military was still immunizing recruits, severe secondary infections of vaccinia-naive contacts did occur.
3. Protection against seroconversion to foreign antigen is now considered a potential benefit of planned immunization. This protection may not be beneficial if an important recombinant vaccinia-based immunization against a more important disease (such as HIV) is developed. Best evidence suggests the possibility that vaccinia-naive populations will mount a better response to vaccinia-based immunizations.[7] It would be ironic if the scientists who develop a vaccinia-based HIV immunization cannot fully respond and benefit because of previous vaccinia immunization.

Physicians responsible for the health care of vaccinia laboratory areas should offer the vaccine and explain the benefits as described by the CDC, as well as the risks and public health concerns.

REFERENCES

1. Benenson AS, ed. *Control of communicable disease in man*. 15th edn. Washington, DC: American Public Health Association, 1990.
2. Ray CG. Smallpox, vaccinia, and cowpox. In: Petersdorf R, et al, eds. *Harrison's principles of internal medicine*. 10th edn. New York: McGraw-Hill, 1983, 1118–21.
3. Friedman HM. Smallpox, vaccinia, and other pox viruses. In: Wilsoni D, et al, eds. *Harrison's principles of internal medicine*. 15th edn. New York: McGraw-Hill, 2000.
4. Liberman DE, Fink F. Containment considerations for the biotechnology industry. In: Ducatman AM, Liberman DE, eds. *The biotechnology industry*. Philadelphia: Hanley & Belfus, 1991, 271–83.
5. Centers for Disease Control. Vaccinia (smallpox) vaccine. Recommendations of the Immunization Practices Advisory Committee. *MMWR* 1991; 40:1–10.
6. Anonymous. Pros and cons of vaccinia immunization. *Occup Med* 1992; 34:757.
7. Cooney EL, Collier AC, Greenberg PD, et al. Safety of and immunological response to a recombinant vaccinia virus vaccine expressing HIV envelope glycoprotein. *Lancet* 1991; 337:567–72.

VARICELLA-ZOSTER VIRUS (VZV)

Common names for disease: Chickenpox (varicella), shingles (herpes zoster)

Classification: Family—Herpesviridae

Occupational setting

Infection with varicella-zoster virus (VZV) is so common that it is likely to be encountered in any occupational setting, usually from a community-acquired infection. Infection of susceptible health care workers is a known cause of nosocomial spread. Other high risk settings include teachers of young children, daycare employees, residents and staff members in institutional or correctional settings, and military personnel (Figure 21.10). Primary infection in pregnant women is of particular concern because of the risks of congenital and neonatal infection.

Exposure (route)

VZV is transmitted from person to person by direct contact, droplet or aerosol from vesicular fluid of skin lesions, or by secretions from the respiratory tract. The virus enters the host via the respiratory tract. The virus is labile, and transmission by inanimate objects is unlikely to occur.[1,2]

Pathobiology

VZV is a member of the Herpesviridae family. It is a double-stranded DNA virus. After primary infection (chickenpox), the virus establishes latency in the dorsal root ganglia. Primary infection confers lifelong immunity to varicella, but reactivation of the virus causes herpes zoster (shingles) in about 15% of the population.

Varicella is characterized by sudden onset of mild fever, constitutional symptoms, and a pruritic skin rash. The rash is typically maculo-papular for a number of hours, then vesicular for 3–5 days, with subsequent crusting. It begins on the trunk and face and extends centripitally. The hallmark of the rash is the presence of lesions in all stages of development at the same time.

Varicella is extremely contagious; secondary attack rates among susceptible household contacts range as high as 90%. The incubation period ranges between 10 and 21 days. Patients are infectious for 48 hours prior to developing a rash and until the vesicles crust over. It is typically a disease of childhood; 90–98% of adults have immunity. Although primary infection in adults is less common, the complication rate is higher. Adults account for only 2% of varicella cases but almost one-quarter of the mortality. The most frequent complications in immunocompetent adults are pneumonitis and encephalitis. Additional complications, including visceral organ involvement, tend to occur in neonates and immunocompromised patients.[3] Maternal varicella is associated with a higher rate of spontaneous abortion and, rarely, a congenital varicella syndrome characterized by skin and eye lesions, limb deformity, and CNS damage. This syndrome is estimated to occur in 2% of women infected in the 1st trimester. Neonatal varicella is associated with a high morbidity and mortality. It tends to occur when the onset of maternal varicella is 5 days before to 2 days after delivery. Zoster is characterized by a painful, unilateral, vesicular rash in a dermatomal distribution. It most frequently occurs on the trunk, with lesions starting on the back and extending forward. Pain usually precedes the rash by 2–3 days. The total duration of the episode in the normal adult is 10–15 days. Zoster can occur at any age, but the peak incidence is in the 6th decade.

Potentially serious complications can occur with involvement of the ophthalmic branch of the trigeminal nerve. Zoster ophthalmicus, which can be sight-threatening, is often heralded by lesions on the tip of the nose. Post-herpetic neuralgia is a frequent complication; 25% of patients over the age of 50 experience pain that persists for over a month.

Diagnosis

Diagnosis is generally made on the basis of history and physical examination. Specific laboratory testing is usually unnecessary, but viral culture can be useful when the diagnosis is in doubt.

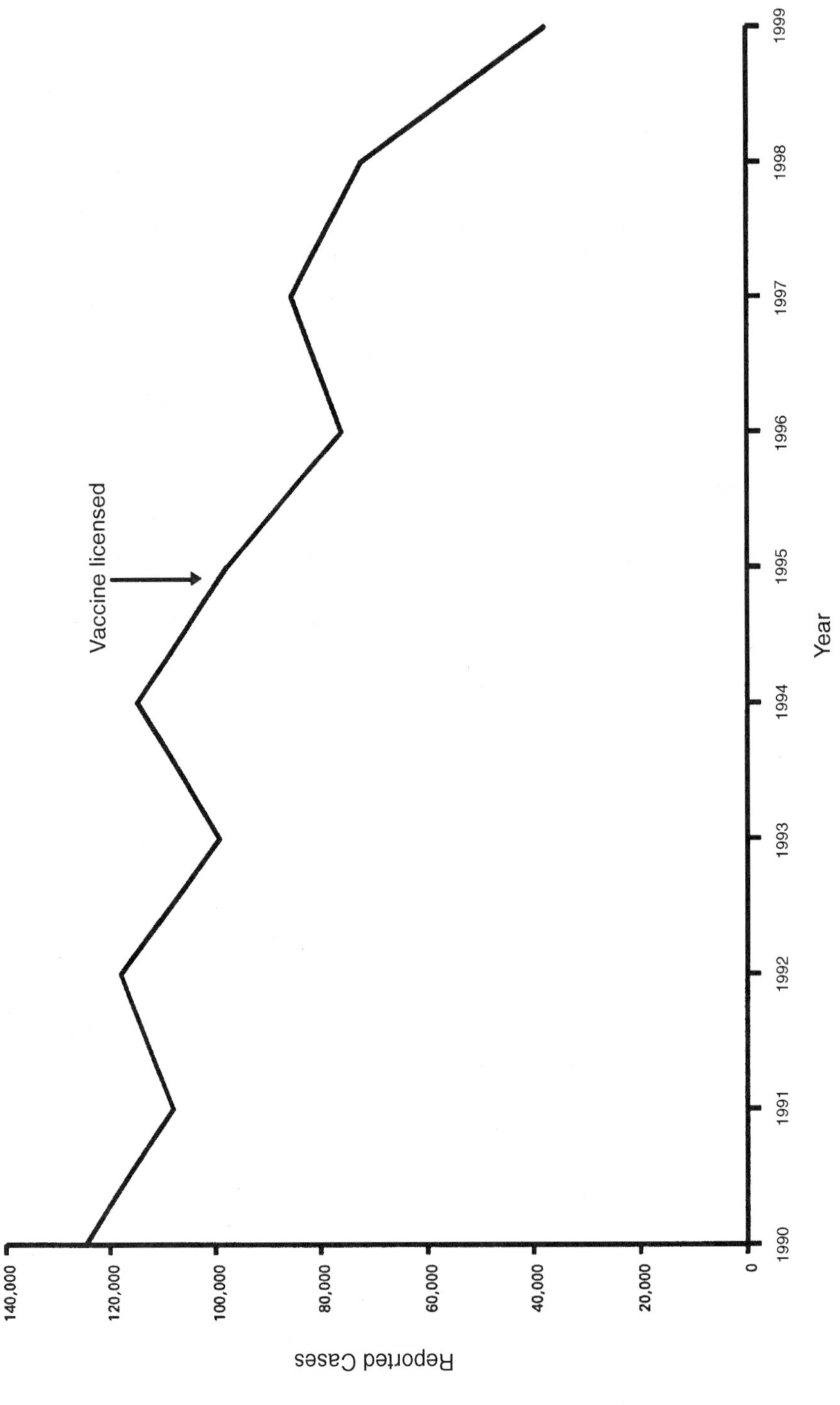

Figure 21.10 Varicella (Chicken pox)—reported cases from selected US states, (n = 7), 1990–1999. Source: MMWR 2001: 48(53): 80.

Treatment

Management of varicella in the immunocompetent patient involves supportive care and aggressive hygiene to avoid bacterial superinfection of skin lesions and to limit transmission. Pruritis can be treated with topical or systemic antipruritic drugs. Fever should be treated with acetaminophen; aspirin is contraindicated because of the risk of Reye's syndrome. Antiviral drugs are generally not used for varicella in immunocompetent patients. Several antivirals are licensed for the treatment of herpes zoster. Treatment is most effective when begun in the first few days of the illness. Antivirals have been shown to decrease the duration of the episode and reduce the likelihood of post-herpetic neuralgia. Postherpetic neuralgia is notoriously difficult to treat and may require judicious use of narcotic analgesics. Tricyclic antidepressants, phenothiazines, and antiepileptics may also be effective.

Medical surveillance

Surveillance for susceptibility to varicella infection is important among health care workers to prevent nosocomial transmission. The CDC recommends screening of all health care workers, with serologic testing of workers with a negative or equivocal history of varicella.[4,5] Other authors recommend serologic screening of all health care workers regardless of the history of varicella or prior vaccination, pointing out that rarely, workers with a positive history of varicella or varicella vaccination are susceptible to infection.[6,7] In a study of house officers at a teaching hospital, 4% lacked immunity to varicella.[7]

Prevention

The mainstay of prevention is surveillance and vaccination of susceptible workers in high risk occupations who have no contraindications to vaccination. Varicella vaccine was approved in the United States in 1995. It is an attenuated virus vaccine, which induces high antibody titers in 99% of adults after two doses. Breakthrough VZV infection can occur, but is substantially less severe among vaccinees. Vaccination provides 70–90% protection against infection and 95% protection against severe disease for 7–10 years after vaccination. It is not yet known whether longer term immunity wanes. There have been concerns about transmission of vaccine strain varicella and vaccine strain zoster, but generally the benefits of vaccination outweigh the risks. Vaccine strain transmission has occurred in only 3 of 15 million does administered, and cases of transmission have occurred in the absence of a vesicular rash. Vaccine associated zoster occurs at a rate of 2.6/100 000 vaccinations. Comparison with the rate of zoster from natural infection (215/100 000 person years) is difficult because of the shorter follow up period for vaccinees.[2]

CDC recommendations[3] for prevention and managing clusters of VZV infection in hospitals include:

- isolating patients who have varicella and other susceptible patients whoe are exposed to VZV
- controlling air flow
- using rapid serologic testing to determine susceptibility
- furloughing exposed susceptible personnel or screening thse persons daily for skin lesions, fever and systemic symptoms
- temporarily reassigning varicella-susceptible personnel to locations remote from patient areas.

Effective prevention strategies must account for the fact that most varicella cases, even those among health care workers, result from community-acquired infection.[8] Post exposure vaccination of susceptible, exposed individuals can be effective in preventing or decreasing the severity of varicella if given within 3, and possible up to 5 days following exposure, and should be considered.[2]

Varicella immune globulin has been shown to prevent or ameliorate varicella in high-risk, exposed individuals. It is usually used for newborns of mothers who develop varicella 5 days before and 2 days after delivery and for exposed, susceptibe, immunocompromised adults. Its use in normal adults must be evaluated on an individual basis. Its utility in the occupational setting is limited by the fact that it does not necessarily prevent varicella, and may prolong the incubation period by a week or more thus lengthening the period of furlough or reassignment.[3]

REFERENCES

1. Centers for Disease Control and Prevention. Prevention of Varicella; Recommendations of the Advisory Committee on Immunization Practices. *MMWR* 1996; 45(RR-11): 1–36.
2. Centers for Disease Control and Prevention. Prevention of Varicella: Update Recommendations of the Advisory Committee on Immunization Practices. *MMWR* 1999; 48(RR-6):1–12.
3. Whitley R. Varicella-zoster virus. In: Mandell GL. Bennett JE, Dolin R. *Principles and practice of infectious disease*. 5th ed. New York: Churchill-Livingstone, 2000: 1580–6.
4. Centers for Disease Control and Prevention. Immunization of Health Care Workers. Recommendations of the Advisory Committee on Immunization Practices and the Hospital Infection Practices Advisory Committee. *MMWR* 1997; 46(RR-18): 1–51.
5. Weber D, Rutala W, Hamilton H. Prevention and control of varicella-zoster in health care facilites. *Infect Control Hosp Epidemiol* 1996; 17:694–705.
6. Gallagher J, Quaid B, Cryan B. Susceptibility to varicella zoster virus in health care workers. *Occupational Med* 1996; 46(4):289–292.
7. Alagappan K, Fu L, Strater S, et al. Seroprevalence of varicella antibodies among new house officers. *Ann Emerg Med* 1999; 33(5):516–9.
8. Josephson A, Karanfil L, Gombert M. Strategies for the management of varicella-susceptible healthcare workers after a known exposure. *Infect Control Hosp Epidemiol* 1990; 11(6): 309–13.

David P. Seibens

22

BACTERIA

Christopher J. Martin, M.D., M.Sc., and John D. Meyer,* M.D., M.P.H.

ACINETOBACTER SPECIES

Common name for disease: None

Occupational setting

Acinetobacter species are ubiquitous in nature. They are commonly isolated from work settings with moist environments and micro-environments (e.g. swine confinement buildings,[1] wastewater treatment plants,[2] composting plants,[3] poultry-processing plants,[4] foundries,[5] cotton mills,[6] metal-working operations[7] and bakeries[8]).

Exposure (route)

Inhalation is the main route of exposure in the occupational setting.

*The authors are indebted to Brian Schwartz, Clifford Mitchell, Virginia Weaver and Marianne Cloeren for a previous version of this chapter

Pathobiology

Acinetobacter Iwoffii has been implicated in an outbreak of hypersensitivity pneumonitis in workers in an automobile parts manufacturing plant using metal-working fluids.[7] *Acinetobacter* species, particularly *Acinetobacter baumannii*, are common causes of nosocomial pneumonia and other infections, particularly in intensive care unit (ICU) settings.[9] They may rarely cause community acquired pneumonia in persons with impaired host defenses.[10]

Exposure to water aerosols from environments with polymicrobial contamination, including *Acinetobacter*, has been associated with a spectrum of respiratory diseases (asthma, hypersensitivity pneumonitis, bronchitis) in several settings.[7] Such workplaces also have the potential for exposure to other known sensitizers such as fungi, in addition to endotoxin and non-sensitizing irritants (oil mists, additives). Therefore, the precise role of *Acinetobacter* in such outbreaks is uncertain.

Physical and Biological Hazards of the Workplace, Second Edition, Edited by Peter H. Wald and Gregg M. Stave
ISBN 0-471-38647-2

Diagnosis

Infections with this organism are diagnosed using standard isolation and culture methods of appropriately selected clinical specimens.

Treatment

Since many *Acinetobacter* strains have developed multidrug resistance, therapy depends on the clinical setting (nosocomial versus community) as well as results of susceptibility testing. In mild infections involving susceptible strains, monotherapy may consist of trimethoprim–sulfamethoxazole or ciprofloxacillin. In serious infections, combination therapy with an effective β-lactam such as imipenem and an aminoglycoside may be indicated. *Acinetobacter* species with resistance to all routinely tested antibiotics have been described in nosocomial outbreaks.[11]

Medical surveillance

There are no recommended medical surveillance activities.

Prevention

Engineering controls and work practices should be aimed at reducing microbial contamination of water and other media. Aerosolized processes involving contaminated water are of particular concern. The use of air-purifying respirators may also be appropriate.

REFERENCES

1. Cormier Y, Tremblay G, Meriaux A, et al. Airborne microbial contents in two types of swine confinement buildings in Quebec. *Am Ind Hyg Assoc J* 1990; 51:304–9.
2. Laitinen S, Kangas J, Kotimaa M, et al. Workers' exposure to airborne bacteria and endotoxins at industrial wastewater treatment plants. *Am Ind Hyg Assoc J* 1994; 55(11):1055–60.
3. Lundholm M, Rylander R. Occupational symptoms among compost workers. *J Occup Med* 1980; 22:256–7.
4. Lenhart SW, Olenchock SA, Cole EC. Viable sampling for airborne bacteria in a poultry processing plant. *J Toxicol Environ Health* 1982; 10:613–19.
5. Cordes LG, Brink EW, Checko PJ, et al. A cluster of Acinetobacter pneumonia in foundry workers. *Ann Intern Med* 1981; 95:688–93.
6. Delucca AJ, Shaffer GP. Factors influencing endotoxin concentrations on cotton grown in hot, humid environments: a two year study. *Br J Ind Med* 1989; 46:88–91.
7. Zacharisen MC, Kadambi AR, Schlueter DP, et al. The spectrum of respiratory disease associated with exposure to metal working fluids. *J Occup Environ Med* 1998; 40(7):640–7.
8. Domanska A, Stroszejn-Mrowca G. Endotoxin in the occupational environment of bakers: method of detection. *Int J Occup Med Environ Health* 1994; 7(2):125–34.
9. McDonald A, Amyes SG, Paton R. The persistence and clonal spread of a single strain of Acinetobacter 13TU in a large Scottish teaching hospital. *J Chemother* 1999; 11(5):338–44.
10. Yang CH, Chen KJ, Wang CK. Community-acquired Acinetobacter pneumonia: a case report. *J Infect* 1997; 35(3):316–78.
11. Go ES, Urban C, Burns J, et al. Clinical and molecular epidemiology of acineto-bacter infections sensitive only to polymyxin B and sulbactam. *Lancet* 1994; 12:1329–32.

BACILLUS ANTHRACIS

Common names for disease: Anthrax, woolsorter's disease (inhalation anthrax)

Occupational setting

Anthrax is a zoonotic disease that is transmitted to humans via contact with animals (agricultural anthrax) or animal products (industrial anthrax). Important sources of exposure are raw wool, goat hair, animal bone products, and hides and skins imported from areas where anthrax is enzootic, especially Africa and Asia. Shepherds, farmers and workers in manufacturing plants using the above materials are at highest risk for occupational anthrax[1] and, in the past, textile mills that used these animal products presented a significant occupational hazard. Anthrax is most commonly associated with domestic herbivores, especially cattle, but also sheep, goats, horses, and mules.[2]

Exposure (route)

Transmission occurs via direct contact with contaminated animal products through abrasions in the skin, by indirect contact with a contaminated environment, or by contact with airborne *B. anthracis* during the processing of animal products.[2] Anthrax can also be acquired from direct contact with infected animals, primarily cattle, sheep, or goats, in the agricultural setting. There are no known cases of human-to-human transmission.

Pathobiology

B. anthracis is a Gram-positive, non-motile, spore-forming bacillus; spores may persist for several years in the industrial or agricultural environment and are resistant to most antibacterial measures, including drying, heat, ultraviolet light, gamma radiation, and many disinfectants.[1] *B. anthracis* produces an exotoxin with three components (termed edema factor, lethal factor, and protective antigen) that is an important contributor to the organism's virulence. Production of a capsular material is also a factor in virulence; attenuated strains appear to be those expressing poorly encapsulated biotypes.

Anthrax exists in three primary forms: cutaneous, inhalational, and gastrointestinal. Cutaneous anthrax, which accounts for more than 95% of anthrax cases, results from introduction of the organism into the skin, most commonly on the head and neck or upper extremity, via a wound or a penetrating animal fiber. The bacteria germinate and multiply in the subcutaneous tissue, with production of toxin and tissue necrosis. A slowly enlarging papule is first noticed, which then vesiculates, eventually bursting to form a black eschar around which smaller vesicles may appear. The lesion is generally painless and may be associated with impressive local edema, regional lymphadenopathy, and septicemia. Antibiotic therapy will eliminate the bacteria, although the eschar is toxin-mediated and therefore may not appear to respond to appropriate treatment. The case fatality ratio is 20% without treatment, and mortality is rare with appropriate therapy.[2] The diagnosis should be considered in any patient with a painless ulcer with vesicles and edema who has a history of exposure to animals or animal products.

Inhalation anthrax is a very rare disease that is almost always fatal. Less than one case per year has been reported during the past 20 years in the USA.[1,3] After inhalation of aerosolized contaminated particulate matter, *B. anthracis* is phagocytized in the terminal alveoli by macrophages and carried to mediastinal and hilar lymph nodes. There the bacteria germinate and multiply, producing large amounts of toxin. The result is a hemorrhagic, edematous mediastinitis, evidenced by widening of the mediastinum on chest X-ray. The initial phase of the illness (3–4 days) resembles an upper respiratory tract infection, followed within several days by severe respiratory distress, and death usually within 24 hours of the onset of pulmonary symptoms. Inhalation anthrax, similar to infection in other primary sites, can lead to hemorrhagic meningitis, after hematogenous spread of bacteria to the meninges.

Gastrointestinal anthrax, never reported in the USA but still reported in developing countries, occurs after ingestion of contaminated meat. Depending on the location where organisms are deposited in the gastrointestinal tract, oropharyngeal or abdominal forms of the disease may result. Symptoms and signs of the former include fever, anorexia, cervical or submandibular lymphadenopathy, and edema. Mesenteric lymphadenitis and intestinal ulceration occur in abdominal anthrax; the main clinical features include nausea, vomiting, anorexia, fever, abdominal pain, hematemesis, and bloody diarrhea.[2] Ascites, septicemia, intestinal perforation, shock, and death may ensue, with a case fatality ratio ranging from 25% to 75%.

Diagnosis

In cutaneous anthrax, the lesion has a characteristic appearance, and organisms can be obtained for microscopy and cultured from exudate. Inhalation anthrax is characterized by mediastinal widening on chest X-ray, but even early diagnosis rarely influences the uniformly fatal outcome. Gastrointestinal anthrax is non-specific in its signs or symptoms; organisms may be identified from ascitic fluid, vomitus, or feces. The organism can be identified in cerebrospinal fluid in anthrax meningitis. Elevated antibody levels to the major immunogenic components of *B. anthracis* (capsular antigens and exotoxin components) may be identified by enzyme-linked immunosorbent assay (ELISA). These tests are most useful for diagnosis of past infection or vaccination; during acute infection, antibodies to toxins may not appear until too late in the clinical course for serologic testing to be of diagnostic value.

Treatment

The treatment of choice is a quionotone or penicillin with supportive therapy. Doxycycline may be used in penicillin-allergic patients. Care should be taken in culturing or debriding eschars and other lesions, as excision of skin lesions has been reported to increase the severity of symptoms.

Medical surveillance

There are no recommended medical screening activities. Cases of anthrax should be immediately reported to the local health authorities in most of the USA and most other countries. Anthrax should be recognized as a Sentinel Health Event (Occupational) (SHE-O) in shepherds, farmers, butchers, handlers of imported hides or fibers, veterinarians, veterinary pathologists, and weavers.[4]

Prevention

Formaldehyde is the preferred method for decontamination of anthrax-contaminated materials, but gamma irradiation, steam under pressure, and ethylene oxide have also been used to sterilize contaminated raw hair and wool. Effective vaccines are available for humans and animals, directed against both *B. anthracis* and its toxins. Worker training and education, protective equipment use, such as gloves, and respirators if aerosols are produced, and work practices are all effective in the prevention of the disease.[2] Prophylactic antibiotics and hyperimmune serum have not been shown to be effective.

REFERENCES

1. Shafazand S, Doyle R, Ruoss S, et al. Inhalational anthrax: epidemiology, diagnosis and management. *Chest* 1999; 116:1369–76.
2. Dixon TC, Meselson M, Guillemin J, Hanna PC. Anthrax. *N Engl J Med* 1999; 341:815–26.
3. Weinberg N. Respiratory infections transmitted from animals. *Infect Dis Clin North Am* 1991; 5:649–61.
4. Rutstein DD, Mullan RJ, Frazier TM, et al. Sentinel Health Events (Occupational): a basis for physician recognition and public health surveillance. *Am J Public Health* 1983; 73:1054–62.

BORRELIA BURGDORFERI

Common names for disease: Lyme borreliosis, Lyme disease

Occupational setting

Borrelia burgdorferi, the causative agent of Lyme disease, is the most common vector-borne disease of humans in the USA. Outdoor work in endemic areas presents a hazard for transmission by tick bite. The disease is found worldwide, with important foci in several areas of Europe, Asia, and North America. The distribution of the disease parallels the distribution of the principle vectors and hosts.[1–3] In the USA, 92% of cases have been reported from two main geographic regions: the northeastern coastal states from Virginia to Maine (particularly New York, Connecticut and Rhode Island); and the upper Midwest, primarily Wisconsin and Minnesota (Figure 22.1). Areas of the west, including northern California and Oregon, are a third important location of the disease. However, Lyme disease has been reported from the majority of states; these cases may represent cases acquired outside the area of residence, sporadic cases in areas of low endemicity, or problematic diagnoses. In several states where cases have been reported, competent tick vectors have not been identified and the causative agent of the disease has not been recovered from ticks, animals, or humans.[3]

Outdoor workers have been reported to have a 4–6-fold elevation of risk for clinical Lyme disease or seropositivity for antibodies to *B. burgdorferi*.[4,5] Seroprevalence of antibody to *B. burgdorferi* measured by ELISA or indirect fluorescent antibody has ranged from 5.6% to 35% in populations with varying degrees of risk in the USA and Europe.[6,7] The annual or seasonal cumulative incidence of seroconversion in several longitudinal studies of *B. burgdorferi* infection has ranged from 0.4% to 10%.[6] Seroconversion was most often associated with asymptomatic infection, though the incidence of non-clinical disease (seroconversion only) ranged from 26% to 98%.

Exposure (route)

Lyme disease is transmitted by ticks of the *Ixodes ricinus* complex. These include *I. scapularis* (previously known as *I. dammini*) in the northeastern and upper midwestern USA, *I. pacificus* in the western USA, *I. ricinus* in Europe, and *I. persulcatus* in Asia.[2] Although *B. burgdorferi* has been found in other species of ticks, and in mosquitoes and deer flies, there are no convincing epidemiologic data that insects or arthropods other than *I. ricinus* ticks are important vectors of the disease.

The tick vectors of Lyme disease have a 2-year lifespan which takes place in three stages: larval, nymph, and adult. Larval and nymphal ticks feed only once during the summer season before overwintering and emerging into the next stage the following year. The sub-adult stages of *Ixodes* ticks are not species-specific in seeking hosts for their blood meals, and have been found on many species of mammals, birds, and reptiles. However, larvae and nymphs most frequently feed on *Peromyscus leucopus*, the white-footed mouse, an important natural reservoir for *B. burgdorferi*. Adult *I. dammini* feed principally on the white-tailed deer, *Odocoileus virginianus*.[7]

Ixodes ticks generally feed on their hosts for 3–5 days, increasing many-fold in size and weight during this time. Studies in rodents suggest that *B. burgdorferi* is not transmitted from tick to host until after 24–48 h of feeding.[8]

Pathobiology

Lyme disease is caused by *B. burgdorferi*, which, like *Leptospira* and *Treponema*, belongs to the eubacterial phylum of spirochetes.[1] *Borrelia* species are fastidious, microaerophilic bacteria that are difficult to culture

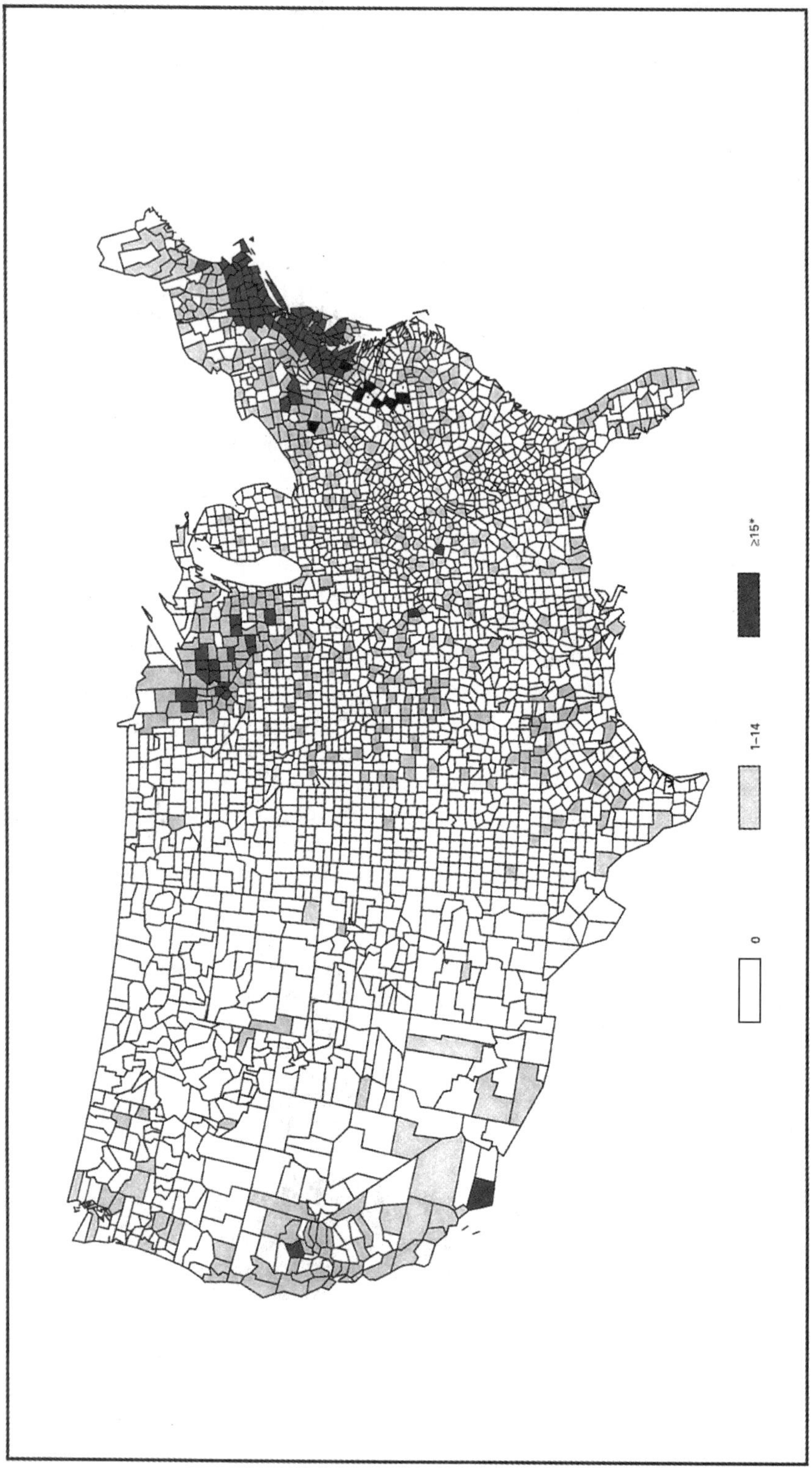

Figure 22.1 Lyme – reported cases by county, United Sates, 1999. Source: MMWR 2001:48(53):49.

from infected patients, and hence diagnosis is based most frequently on serologic criteria rather than isolation of the organism from cases.

Lyme disease generally occurs in stages: early localized infection (previously termed stage I); early disseminated infection (stage II); and late-stage infection (persistent infection or stage III).[1,2] The clinical manifestations are stage-specific. After a tick bite that results in the transmission of spirochetes (usually from an attachment that lasts 24–48 hours), a distinctive rash, erythema migrans (EM), develops in 60–90% of patients within 3–30 days. The EM rash is characterized by a red expanding border surrounding a clearing center with the appearance of a "bull's-eye" in an archery target. The rash can eventually reach 20 cm in diameter. This early, or localized, infection may be accompanied by fever and constitutional symptoms such as malaise and fatigue. In untreated patients, EM can resolve within 3–4 weeks. Diagnostic tests generally have poor sensitivity at this stage of the illness.

Within days or weeks, *B. burgdorferi* can disseminate throughout the body to distant sites, including myocardium, retina, bone, synovium, spleen, liver, meninges, skin, peripheral nerves, and brain. Secondary annular lesions, the most characteristic sign of early-disseminated infection, can appear in up to half of patients at this stage.[2] Patients at this stage are generally ill with severe malaise and fatigue. The manifestations of this disseminated stage can be protean, but several syndromes are of interest. Involvement of the heart can produce myopericarditis and conduction abnormalities (the most characteristic being atrioventricular nodal block) which can be asymptomatic. Migratory pain in the joints is often prominent at this stage as well, with occasional brief attacks of frank arthritis. The nervous system is also an important site of early disseminated infection, with meningitis, cranial neuritis (Bell's palsy), peripheral neuropathy, encephalitis, radiculoneuritis and myelitis all common syndromes. Without therapy, many of these second-stage manifestations can last weeks or months, recur, or become chronic.

Late or persistent infection is characterized by episodes of arthritis that begin months to years after the initial infection. The brief episodes of arthritis characteristic of early disseminated infection tend to last longer, up to several months, and chronic arthritis, defined as a year or more of continual joint inflammation, can occur. The severity of chronic arthritis or the frequency of episodic arthritis generally decreases after 1–2 years. *B. burgdorferi* may also cause late syndromes involving the central and peripheral nervous systems. Progressive encephalomyelitis, characterized by spastic paraparesis, bladder dysfunction, cranial nerve palsies, cognitive impairment (including dementia), and ataxia, is the best described.[9] More subtle chronic neurologic manifestations, including memory loss and behavioral changes, have been described, although evidence of deficits on neuropsychological testing is lacking.[10]

B. burgdorferi can be transmitted transplacentally and fetal infection can result, although it is probably not common. Reported outcomes have included congenital malformations and fetal demise. Population-based epidemiologic studies have not found associations between congenital malformations and serologic evidence of *B. burgdorferi* infection.

Diagnosis

Diagnostic difficulties associated with Lyme disease have been well reported in the lay and scientific literatures. As noted previously, *B. burgdorferi* can be difficult to culture from patients, so serologic tests for detection of antibody to *B. burgdorferi* are the only currently available diagnostic aids. Current recommendations are for initial testing with ELISA or indirect fluorescent antibody assay, followed by testing with the more specific Western immunoblot test to confirm positive results or further test equivocal ones.[11] Detectable antibody response is usually found by 6 weeks after infection, usually after the point at which an EM rash is noted, and therefore the diagnosis of early Lyme disease should be made primarily on clinical grounds. Antibody response may be blunted in those patients receiving early antibiotic treatment. Specific antigens diagnostic of *B. burgdorferi* infection will become apparent in early disseminated or later disease on Western blot testing. Antibodies will persist for months to years after either successful or untreated infection, and therefore seroreactivity cannot be used as an indicator of active disease. Routine use of antibody screening in high-risk populations is therefore problematic as a surveillance tool.

Treatment

Lyme disease is treatable at all stages. Response to therapy is generally better the earlier it is initiated, with more rapid and complete resolution of symptoms. It should be noted that many of the treatment recommendations have not been subject to controlled clinical trials to identify the best choice or the optimal duration of therapy. For early Lyme disease, localized or disseminated, oral doxycycline is generally the preferred therapy for adults. Amoxicillin and azithromycin are acceptable alternatives. For disseminated infection characterized by objective neurologic abnormalities, intravenous ceftriaxone is most often recommended, with the exception that isolated facial nerve palsy with normal cerebrospinal fluid can be treated with oral antibiotics. Lyme disease producing mild cardiac abnormalities, such as first-degree atrioventricular block, can be treated with oral antibiotics, but high-degree blocks are better treated with intravenous antibiotics. The treatment of late-stage arthritis can be more difficult, with slow resolution of symptoms regardless of the efficacy of treatment, and a significant proportion of patients not responding. Both long-term oral regimens (such as doxycycline for 30–60 days) and intravenous antibiotics have been used successfully to treat Lyme arthritis. Treatment of pregnant patients remains unresolved, with some clinicians advocating high-dose intravenous penicillin for all women with Lyme disease during pregnancy.

The use of prophylactic antibiotics for asymptomatic individuals with a history of tick bite in an endemic area remains controversial. The low risk of acquiring *B. burgdorferi* infection after a single tick bite (from 1% to 5%) indicates to many that prophylactic therapy is not warranted.[12] Cost-effectiveness analysis suggests that prophylactic therapy for tick bites would decrease overall costs and minimize morbidity from Lyme disease and antibiotic side-effects if the risk of infection after the tick bite exceeded 3.6%, a risk found only in highly endemic areas.[13] Early attention to symptoms and signs in persons who have sustained a tick bite or have noted removing a tick is recommended in place of tick-bite prophylaxis in most instances.

Medical surveillance

There are no currently recommended medical screening activities for Lyme disease. Routine use of anti-*B. burgdorferi* antibody screening is not recommended because of low sensitivity in early infection and problems in distinguishing past asymptomatic infection from current infection. Cases of Lyme disease must be reported to the local health authorities in the USA and some other countries.

Prevention

Although tick populations can be controlled for a short time with a number of commercially available pesticides, repeated applications appear to be necessary, which lessens the desirability of this strategy.[7] Personal preventive measures are thus preferred. It is important to emphasize that studies in animals suggest that the spirochete is not transmitted from tick to host until after 24–48 hours of feeding.[8] Thus, careful examination of the skin for attached ticks at the end of each workday should greatly reduce the risk of Lyme disease in outdoor workers. Other preventive behaviors—wearing light-colored clothing for easier visualization of ticks, tucking trousers into socks, and wearing long-legged trousers and long-sleeved shirts—are recommended for their ease and low cost.

Insect repellents containing high concentrations of DEET (*N,N*-diethyl-*m*-toluamide) are effective after application to the skin, but systemic toxicity has been reported.[7] Permethrin-containing products (e.g. Permanone), when applied to clothing, provide protection from questing ticks even after several launderings because of binding of the insecticide to the clothing fibers. Whereas DEET repels ticks, permethrin rapidly kills *I. dammini*, and studies have revealed a marked decrease in the number of viable ticks in field studies of permethrin-treated clothing.

Two Lyme disease vaccines which use recombinant *Borrelia* outer-surface protein antigen have been developed recently; the Food and Drug Administration has licensed one, marketed as LYMErix, for use in the USA. The vaccine is administered on a three-dose schedule at 0, 1 and 12 months; few side effects other than pain and tenderness at the injection site were noted. The vaccine efficacy in protecting against "definite" Lyme disease (presence of an EM rash, or objective signs of disseminated Lyme disease with a positive Western blot) in endemic areas was 49% after two doses and 83% 2 years after full immunization.[14] The vaccine is recommended, or should be considered, for persons aged 15–70 years who reside, work or relax in areas of high or moderate risk, specifically those who engage in recreational, property maintenance, occupational or leisure activities that result in frequent or prolonged exposure to tick-infested habitats.[14] The benefit of vaccination for those who have neither frequent nor prolonged exposure, over that provided by basic personal protective measures, is not known. Although specific recommendations for the occupational setting have not appeared, consideration should be given to vaccinating workers in endemic areas whose work places them in forest, woodland or littoral settings.

REFERENCES

1. Steere AC. Lyme disease. *N Engl J Med* 1989; 321:586–96.
2. Spach DH, Liles WC, Campbell GL, et al. Tick-borne diseases in the United States. *N Engl J Med* 1993; 329:936–45.
3. Orloski KA, Hayes EB, Campbell GL, Dennis DT. Surveillance for Lyme Disease—United States, 1992–1998. *MMWR* 2000; 49(SS-3): 1–11.
4. Schwartz BS, Goldstein MD. Lyme disease in outdoor workers: risk factors, preventive measures, and tick removal methods. *Am J Epidemiol* 1990; 131:877–85.
5. Smith PF, Benach JL, White DJ, et al. Occupational risk of Lyme disease in endemic areas of New York state. *Ann NY Acad Sci* 1988; 539:289–301.
6. Fahrer H, van der Linden SM, Sauvain MJ, et al. The prevalence and incidence of clinical and asymptomatic Lyme borreliosis in a population at risk. *J Infect Dis* 1991; 163:305–10.
7. Schwartz BS, Goldstein MD. Lyme disease: a review for the occupational physician. *J Occup Med* 1989; 31:735–42.
8. Piesman J, Mather TN, Sinsky RJ, Spielman A. Duration of tick attachment and *Borrelia burgdorferi* transmission. *J Clin Microbiol* 1987; 25:557–8.
9. Ackermann R, Rehse-Kupper B, Gollmer E, et al. Chronic neurologic manifestations of erythema migrans borreliosis. *Ann NY Acad Sci* 1988; 539:16–23.
10. Shadick NA, Phillips CB, Sangha O, et al. Musculoskeletal and neurologic outcomes in patients with previously treated Lyme disease. *Ann Intern Med* 1999; 131:919–26.
11. Centers for Disease Control and Prevention. Recommendations for test performance and interpretation from the Second National Conference on Serologic Diagnosis of Lyme Disease. *MMWR* 1995; 44:590–1.
12. Shapiro ED, Gerber MA, Holabird NB, et al. A controlled trial of antimicrobial prophylaxis for Lyme disease after deer-tick bites. *N Engl J Med* 1992; 327:1769–73.
13. Magid D, Schwartz BS, Craft J, Schwartz JS. Prevention of Lyme disease after tick bites: a cost-effectiveness analysis. *N Engl J Med* 1992; 327:534–41.

14. Centers for Disease Control and Prevention. Recommendations for the use of Lyme Disease vaccine. *MMWR* 1999; 48(RR-7):1–21.

BRUCELLA SPECIES

Common names for the disease: Brucellosis, Bang's disease, Mediterranean fever, undulant fever, Neapolitan fever, Malta fever, Gibraltar fever

Occupational setting

Brucellosis is a zoonotic infection that arises via transmission of several species of the genus *Brucella* from different animal reservoirs. *Brucella abortus* infects cattle, and occasionally hogs, *B. suis* infects hogs, and less frequently cattle, caribou, reindeer and hares, while *B. canis* infects dogs and foxes. The most serious infections have been caused by *B. melitensis*, which infects goats and sheep. Human brucellosis is rare in the USA, where the predominant species are *B. abortus* and *B. suis*. The epidemiology of the disease depends on regional patterns of domestic livestock farming, as well as the animal disease control methods in place. Brucellosis cases in the USA have declined from a peak of 6341 cases in 1947 to the current plateau of 100–200 cases per year,[1] primarily as a result of animal control methods, including vaccination, inspection, and prompt segregation of diseased animals (Figure 22.2).

Brucellosis can be transmitted by consumption of unpasteurized milk; there are still some cases in the USA that have been traced to eating soft cheeses, usually imported from Mexico or Italy.[1] This mode of transmission is still the most frequent source worldwide. However, currently the majority of cases of brucellosis in the USA are caused by occupational exposure to the secretions and excretions of infected animals. Activities with particularly high risk of exposure include animal slaughter, milking, handling bull semen, and handling aborted animal fetal tissue and placentas.[2] Jobs with the highest risk of exposure are livestock handling and slaughterhouse work; other occupations with significant risk are veterinarians, government meat inspectors, farmers, dairy workers, and microbiology laboratory personnel.[3–5] Cases acquired during work at a pharmaceutical company producing *Brucella* vaccines for animals have been reported.[3] Among slaughterhouse workers, those on the kill floor have the highest risk, probably due in part to aerosolization of infected fluids.

Exposure (route)

Occupational infection can be acquired by transmission through the skin, particularly in slaughterhouse, and abattoir workers, and veterinarians, who may have cuts, abrasions, or other skin breaks. The bacteria can also enter the body by penetrating the mucosa of the mouth or throat, or through the conjunctiva, if infected material is splashed or sprayed. Infection may also occur through inhalation of contaminated aerosols. Ingestion is less likely in the occupational setting, but remains an important route of exposure in cases transmitted by infected dairy products.

Pathobiology

Brucella organisms are small, non-motile, Gram-negative rods. From the portal of entry, *Brucella* organisms invade the lymphatic system, and overcome the lymph barriers to spread hematogenously. The most commonly infected organs are the spleen, liver, lymph nodes and bone marrow. Bone involvement may lead to osteomyelitis. Other organs that are less commonly involved are the heart (including valves), joints, testes, ovaries, prostate, lungs and kidneys. Pathologic examination of the infected tissue reveals non-specific granulomata.

The clinical course is variable, and because few infected persons can recall an exposure

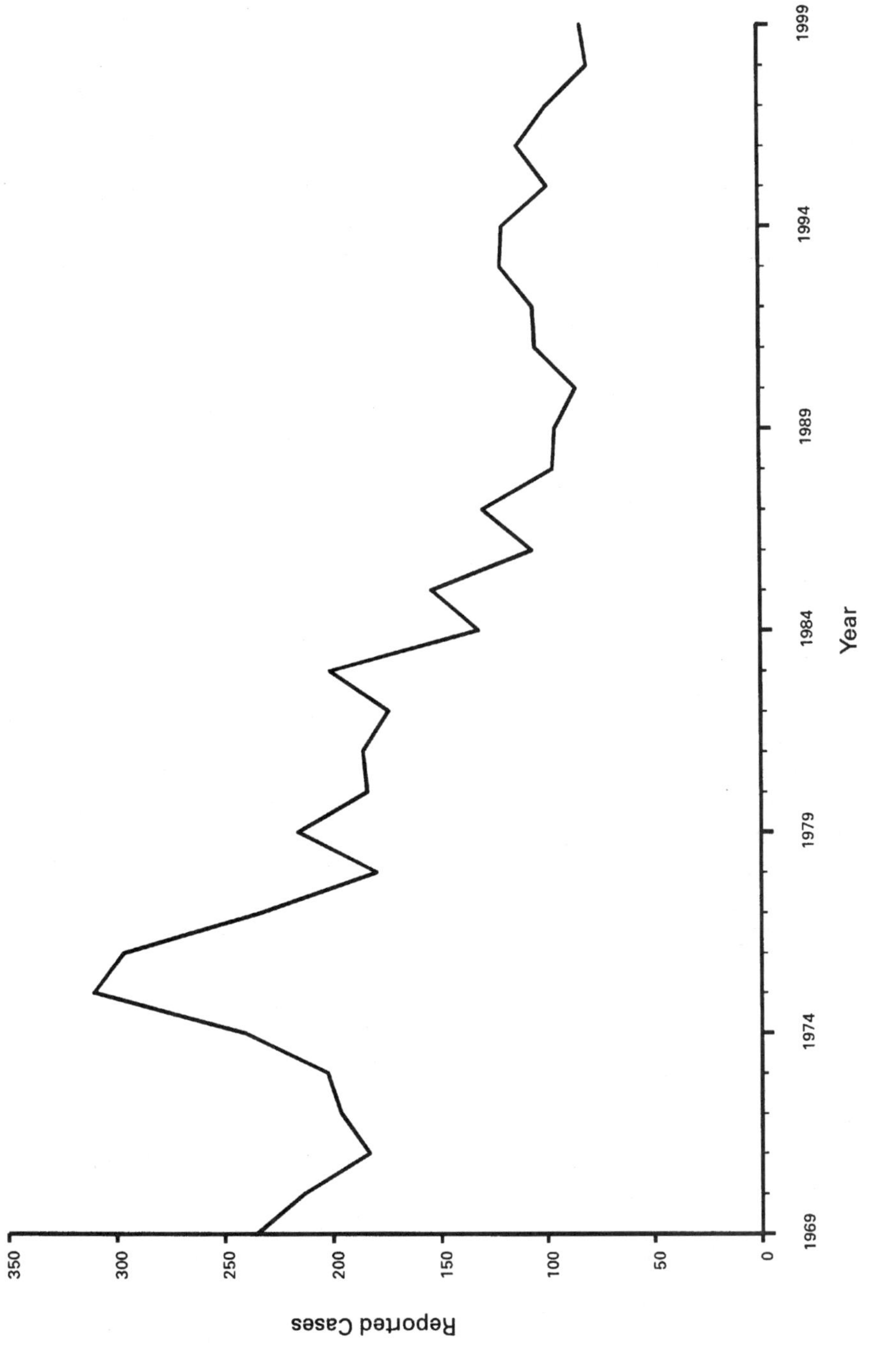

Figure 22.2 Brucellosis – reported cases by year, United States, 1969–1999. Source: MMWR 2001:48(53):29.

incident, it is difficult to determine an incubation period, which may range from 5 days to several months. The usual symptoms of acute brucellosis may be non-specific and include intermittent fever, chills, weakness, malaise, myalgias, headache, anorexia, and weight loss; cases may have gastrointestinal, neurologic, respiratory, or dermal symptoms and signs.[6] In unusual cases there are localizing symptoms or signs such as bone pain, pain in a nerve root distribution, arthritis, splenomegaly, or heart murmur. Miscarriage may occur in pregnant women who acquire the infection. Infection usually results in immunity, including cross-immunity to other *Brucella* species, but relapse and even reinfection have occurred. Subclinical infection can occur, as evidenced by serologic studies of slaughterhouse workers, many of whom demonstrated an antibody response but no history of disease.[7]

Diagnosis

Infection with *Brucella* species can be determined by standard laboratory methods. Definitive diagnosis is made by isolating *Brucella* organisms in cultures of blood or other infected tissue, which is often difficult, given the slow growth and fastidious nature of these bacteria.[35] Agglutination tests for *Brucella* antigen can detect infections arising from *B. abortus*, *B. melitensis* and *B. suis*, although findings may be non-specific, since titers also rise after infection with *Yersinia*, *Salmonella*, and *Francisella tularensis*, after *Brucella* antigen skin testing, and after vaccination against cholera. ELISA testing for *Brucella* antigens is widely used for serologic diagnosis; polymerase chain reaction (PCR) tests are being developed but require greater standardization before their widespread use. Skin testing is generally not helpful, since a positive test may reflect past infection, and a negative test may occur as a consequence of immune suppression during acute *Brucella* infection.

Treatment

The recommended treatment for brucellosis is a combination of rifampin and tetracycline for 6 weeks.[6] Severe infections with meningitis, endocarditis, osteomyelitis or arthritis should be treated for 8 weeks with a multiple drug regimen which includes rifampin, a tetracycline, and an aminoglycoside. Treatment for meningitis should be continued for 8 months.

Medical surveillance

Skin testing may have some utility as a surveillance tool to document a negative response so that a positive response during a later illness may aid in diagnosis. A positive skin test result does not imply immunity, as reinfection can occur. Reporting of brucellosis cases is required in most of the USA and in most other countries. Brucellosis is a Sentinel Health Event (Occupational) (SHE-O) for shepherds, veterinarians, laboratory workers, and slaughterhouse workers.[8]

Prevention

The most important preventive measures, which have resulted in a dramatic decline in human brucellosis in the USA, are the vaccination and careful inspection of animals at risk, along with immunologic testing of cows' milk and blood for evidence of *Brucella* infection. Diseased animals are segregated or slaughtered. Despite these eradication measures, work practice measures are essential for protection against remaining diseased animals. Kill floors should be isolated from other areas of the slaughterhouse, and be under negative-pressure ventilation with entry restricted to essential personnel. All workers handling animal products, including milk, and especially placenta, uterine discharges, and blood, should wear heavy gloves, aprons, and goggles.[2] The use of high-top boots should be considered as well. Areas where exposure is likely should be posted with information about brucellosis, including routes of exposure, disease symptoms, and preventive activities. Work sites should have accessible handwashing facilities, first aid kits for prompt treatment of wounds, and separate areas, isolated from animal work, for eating, drinking and smoking. These latter activities should also be prohibited in the work area. Microbiology laboratory personnel should be instructed in the proper handling of *Brucella* specimens, which should include working in a biosafety hood, and wearing gloves and goggles.

REFERENCES

1. Fox MD, Kaufmann AF. Brucellosis in the United States, 1965–1974. *J Infect Dis* 1977; 136:312–6.
2. Kligman EW, Peate WF, Cordes DH. Occupational infections in farm workers. *Occup Med State Art Rev* 1991; 6:429–46.
3. Olle-Goig JE, Canela-Soler J. An outbreak of *Brucella melitensis* infection by airborne transmission among laboratory workers. *Am J Public Health* 1987; 77:335–8.
4. Huddleson IF, Munger M. A study of an epidemic of brucellosis due to *Brucella melitensis*. *Am J Public Health* 1940; 3:944–54.
5. Buchanan TM, Hendricks SL, Patton CM, et al. Brucellosis in the United States, 1960–1972: an abattoir-associated disease. *Medicine* 1974; 53:427–39.
6. Corbel MJ. Brucellosis: an overview. *Emerg Infect Dis* 1997; 3:213–21.
7. Henderson RJ, Hill DM. Subclinical *Brucella* infection in man. *Br Med J* 1972; 3:154–6.
8. Rutstein DD, Mullan RJ, Frazier TM, et al. Sentinel Health Events (Occupational): a basis for physician recognition and public health surveillance. *Am J Public Health* 1983; 73:1054–62.

CAMPYLOBACTER SPECIES

Common name for disease: Campylobacteriosis

Occupational setting

Although *Campylobacter* is one of the most common causes of infectious diarrhea, only a very small number of cases are linked to outbreaks.[1] Available studies have failed to identify occupational risk factors for sporadic cases.[2] Point-source outbreaks in the workplace have been related to food[3] or water[4] contamination. Direct, occupationally acquired infections have been described in farm workers handling turkeys.[5] Outbreaks have also been documented in child day care centers[6] and prisons,[7] raising the possibility of illness being acquired by workers in these sectors. Industrial hygiene surveys of slaughterhouses have found widespread contamination of equipment and air with *Campylobacter*.[8] As with other enteric pathogens, international travel is a well-known risk factor for illness from *Campylobacter*.[9]

Exposure (route)

Spread occurs through fecal–oral transmission. Poultry and to a lesser extent swine are the most common animal reservoirs, although almost any mammal is a potential source, including household pets. Disease in humans is most frequently caused when meat products are consumed following improper preparation and storage. Contaminated water and improperly pasteurized milk are other possible modes of infection.[1]

Pathobiology

Campylobacteria are motile, Gram-negative, curved rods. *C. jejenui* is the species responsible for the vast majority of *Campylobacter* illness. The major reservoir for *C. jejuni* is poultry. A closely related species more commonly found in swine, *C. coli*, produces an illness clinically indistinguishable from that produced by *C. jejuni*. *C. upsaliensis* has also been implicated in one outbreak of gastroenteritis,[43] while *C. fetus* may additionally cause bacteremia and infections, such as endocarditis, abscesses, septic arthritis and abortions, at other sites. Recently, a new *Campylobacter*-like organism, with a proposed name of *Campylobacter lanienae*, has been isolated from the feces of healthy abattoir workers.[10]

Acute enterocolitis is generally caused by *C. jejuni*, or, in about 10% of cases, *C. coli*. Following an incubation period of approximately 3 days, the most common symptoms are diarrhea (with or without blood), abdominal pain, and fever lasting about 7 days.[11] However, this represents the midpoint on a continuum of presentations, which may range from an asymptomatic carrier state to severe prolonged diarrhea. Abdominal pain may be quite prominent, mimicking a surgical abdomen or inflammatory bowel disease.[12] Fecal excretion of bacteria usually ceases within 2–3 weeks.

It is now well recognized that *C. jejuni* gastroenteritis may be followed by the development of Guillain–Barre syndrome usually within 12 weeks of the infection.[13] This association may also exist for *C. coli*.[14] The mechanism is thought to result from the development of antibodies to *Campylobacter* lipopolysaccharides which cross-react with similar antigens in peripheral nerve tissue.[15] Reactive arthritis or Reiter's syndrome, most frequently involving the knee, may also follow *Campylobacter* infection in HLA B27-positive individuals.[16]

Diagnosis

The presence of the bacteria on microscopic examination of stool specimens supports the diagnosis, and isolation of the organism in culture confirms it. At least two stool samples should be taken. PCR has recently been shown to be reliable and sensitive when compared to culture and may be used increasingly in the future.[17] Serologic testing is not recommended for routine clinical use.

Treatment

Supportive therapy consisting of fluid and electrolyte replacement is the primary consideration in the treatment of *Campylobacter* enteritis. Antibiotic therapy is recommended in patients with high fever, or bloody or profuse diarrhea, or in those whose symptoms are worsening at the time of presentation, or not improving within 1 week. Erythromycin is the treatment of choice. Ciprofloxacin is also effective, although increasing resistance rates have been noted. Antibiotic treatment eliminates excretion of the bacteria.

Medical surveillance

Since transmission by asymptomatic workers is considered uncommon, routine surveillance through stool cultures is not recommended.[18] Cases should be reported to the local health authorities.

Prevention

Proper preparation and storage of food at the workplace will prevent this and other causes of infectious gastroenteritis. Work practices should emphasize good hygiene with strict handwashing after contact with potentially infected material. Eating and smoking should be avoided in the workplace, and gloves should be worn. In slaughterhouses, work practices and engineering controls should be directed towards minimizing fecal contamination.

Employees in jobs with a risk of transmission (e.g. food handlers and personnel caring for immunocompromised patients) may be required to have a negative stool culture before returning to work after *Campylobacter* enteritis.

Travelers should take appropriate preventive precautions common to all enteric pathogens, with particular attention to avoiding undercooked poultry.

REFERENCES

1. Pebody RG, Ryan MJ, Wall PG. Outbreaks of campylobacter infection: rare events for a common pathogen. *Commun Dis Rep CDR Rev* 1997; 7(3):R33–7.

2. Brieseman MA. A further study of the epidemiology of Campylobacter jejuni infections. *NZ Med J* 1990; 103(889):207–9.
3. Murphy O, Gray J, Gordon S, Bint AJ. An outbreak of campylobacter food poisoning in a health care setting. *J Hosp Infect* 1995; 30(3):225–8.
4. Rautelin H, Koota K, von Essen R, Jahkola M, Siitonen A, Kosunen TU. Waterborne Campylobacter jejuni epidemic in a Finnish hospital for rheumatic diseases. *Scand J Infect Dis* 1990; 22(3):321–6.
5. Ellis A, Irwin R, Hockin J, Borczyk A, Woodward D, Johnson W. Outbreak of Campylobacter infection among farm workers: an occupational hazard. *Can Commun Dis Rep* 1995; 21(17):153–6.
6. Goosens H, Giesendorf BA, Vandamme P, et al. Investigation of an outbreak of Campylobacter upsaliensis in day care centers in Brussels: analysis of relationships among isolates by phenotypic and genotypic typing methods. *J Infect Dis* 1995; 172(5):1298–305.
7. Fernandez-Martin JI, Dronda F, Chaves F, Alonso-Sanz M, Catalan S, Gonzalez-Lopez A. Campylobacter jejuni infections in a prison population coinfected with the human immunodeficiency virus. *Rev Clin Esp* 1996; 196(1):16–20.
8. Berndtson E, Danielsson-Tham ML, Engvall A. Campylobacter incidence on a chicken farm and the spread of Campylobacter during the slaughter process. *Int J Food Microbiol* 1996; 32(1–2):35–47.
9. Gallardo F, Gascon J, Ruiz J, et al. Campylobacter jejuni as a cause of traveler's diarrhea: clinical features and antimicrobial susceptibility. *J Travel Med* 1998; 5(1):23–6.
10. Logan JM, Burnens A, Linton D, Lawson AJ, Stanley J. Campylobacter lanienae sp. nov., a new species isolated from workers in an abattoir. *Int J Syst Evol Microbiol* 2000; 50 (Pt 2):865–72.
11. Melby K, Dahl OP, Crisp L, Penner JL. Clinical and serological manifestations in patients during a waterborne epidemic due to Campylobacter jejuni. *J Infect* 1990; 21(3):309–16.
12. Cooper R, Murphy S, Midlick D. Campylobacter jejuni enteritis mistaken for ulcerative colitis. *J Fam Pract* 1992; 34(3):357, 361–2.
13. Rees JH, Soudain SE, Gregson NA, Hughes RA. Campylobacter jejuni infection and Guillain–Barre syndrome. *N Engl J Med* 1995; 333(21):1374–9.
14. Bersudsky M, Rosenberg P, Rudensky B, Wirguin I. Lipopolysaccharides of a Cam-pylobacter coli isolate from a patient with Guillain–Barre syndrome display ganglio-side mimicry. *Neuromusc Disord* 2000; 10(3):182–6.
15. Yuki N. Molecular mimicry between gangliosides and lipopolysaccharides of Campylobacter jejuni isolated from patients with Guillain–Barre syndrome and Miller Fisher syndrome. *J Infect Dis* 1997; 176(suppl 2):S150–3.
16. Peterson MC. Rheumatic manifestations of Campylobacter jejuni and C. fetus infections in adults. *Scand J Rheumatol* 1994; 23(4):167–70.
17. Vanniasinkam T, Lanser JA, Barton MD. PCR for the detection of Campylobacter spp. in clinical specimens. *Lett Appl Microbiol* 1999; 28(1):52–6.
18. Skirrow MB. Foodborne illness—Campylobacter. *Lancet* 1990; 336:921–3.

CLOSTRIDIUM BOTULINUM

Common names for diseases: Botulism, infant botulism, wound botulism

Occupational setting

A toxin formed by the bacterium *Clostridium botulinum* causes botulism. It is primarily a foodborne illness, but cases of wound botulism, may arise in wounds contaminated with soil or gravel.[1] In the past ten years, cases of wound botulism have increased markedly, and it is now the leading cause of botulism among adults in the United States (Figure 22.3). The organisms are ubiquitous in most soils, and have been found in agricultural products, in the intestinal tracts of animals, including fish, and in marine sediment.[2,3] Type A botulinum toxin is manufactured for cosmetic and therapeutic use, including treatment of conditions involving involuntary muscle spasm,

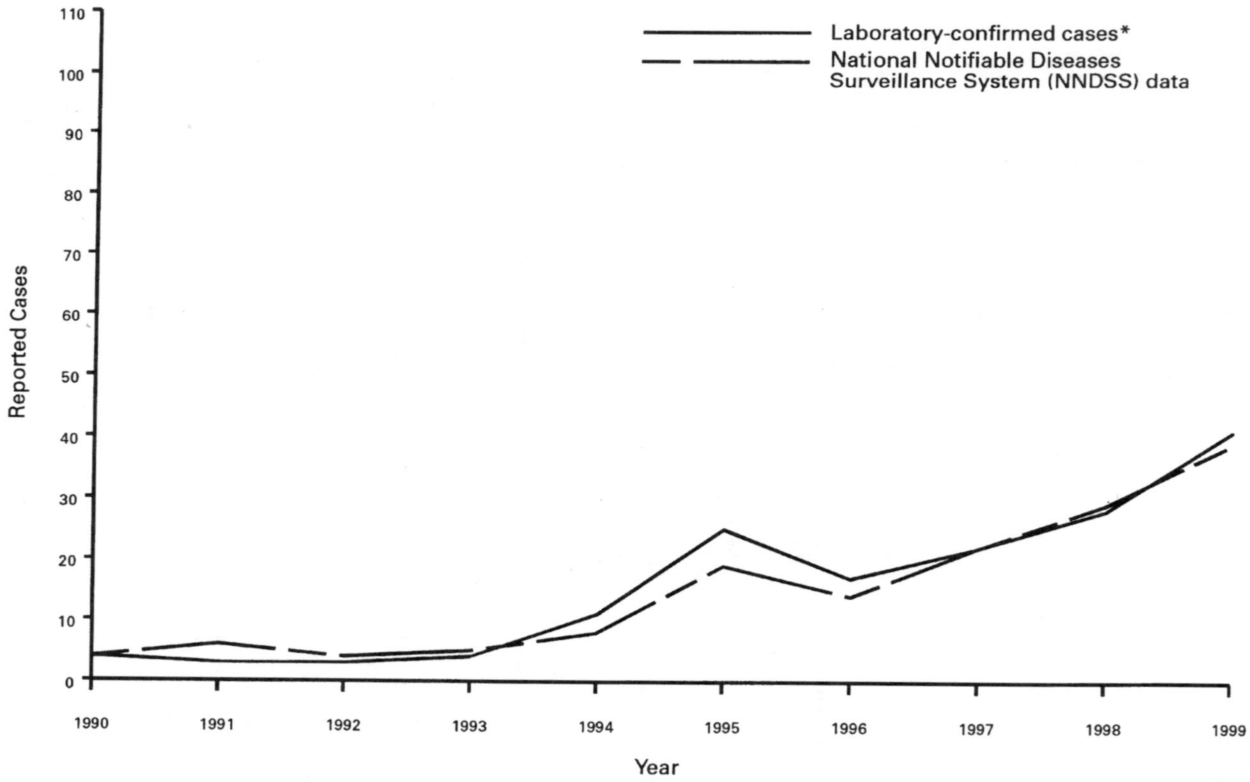

Figure 22.3 Botulism – cases from wounds and unspecified causes (excludes infant and foodborne cases) United States, 1990–1999. Source: MMWR 2001: 48(53):28.

including torticollis and dysphonia. Personnel involved in this manufacturing process are potentially at risk of intoxication, but there have been no reported cases to date.[4]

Exposure (route)

Intoxication usually occurs after ingestion of food containing preformed toxin elaborated by *C. botulinum*. Foods most commonly associated with botulism are home-canned foods, but cases have also occurred after ingestion of salt-cured and fresh uneviscerated fish products. The toxin may be formed by ingested bacteria, which colonize the immature (infant) or abnormal intestinal tract, or by bacteria introduced into a wound.

Pathobiology

C. botulinum is a spore-forming, obligate anaerobic bacillus. There are seven types of toxin, designated A–G, which may be elaborated by the bacillus, but most human cases are caused by types A, B and E, with rare cases due to types F and G. In a conducive environment, such as the hypoxic atmosphere produced by canning or in the reduced oxygen tension of a deep or gangrenous wound, clostridial spores germinate and multiply. The growing bacterial colonies release a potent neurotoxin that acts at peripheral cholinergic synapses to block the release of acetylcholine, causing impaired vision, and diffuse muscular weakness. Death is usually due to respiratory failure from paralysis of the diaphragm and respiratory muscles.

Diagnosis

Clinical suspicion, based on the symptoms and signs of botulism intoxication, is supported by demonstrating the specific toxin in serum, stool or the suspected food source. *C. botulinum* can sometimes be cultured from the stool in cases in infants or those arising in patients with intestinal anomalies. The presence of *C. botulinum* spores in the implicated food is less helpful than finding toxin, as the spores are ubiquitous and are not themselves harmful. In cases of suspected wound botulism, serum should be tested for toxin, and the wound cultured for the organism.

Treatment

After collection of serum for specific toxin identification, all suspected cases of botulism should be treated as soon as possible with intravenous or intramuscular trivalent botulinum antitoxin (types A, B and E), available from local health departments. This treatment should not be withheld while laboratory confirmation is pending. Cases of wound botulism should be treated with antitoxin as well as wound debridement or drainage, and antibiotics (penicillin is the first choice for clostridial infection). Because of the risk of respiratory failure, patients with known or suspected botulism should be admitted to an ICU.

Medical surveillance

There are no recommended medical screening activities for botulism. All cases of botulism, confirmed or suspected, must be reported immediately to the local public health authorities in most of the USA and in most other countries.

Prevention

Appropriate food handling can prevent the majority of cases. Public notification and recall of tainted products are essential after identification of commercial food sources of poisoning. Tracing of others who may have consumed contaminated food is important when botulism has been identified in commercially prepared or distributed foods. The public should be educated about the risk of botulism being present in bulging containers, such as cans, but they should be aware that this sign of contamination is often absent. Those involved in home canning should be educated about the proper time, temperature, and pressure needed to destroy spores. Uneviscerated fish products should be avoided because of the risk of contamination. Prompt cleaning of wounds and careful attention (including irrigation or debridement) to wounds that are not healing may prevent wound botulism. Botulism associated with toxin manufacture and use is, to date, only a theoretical problem, and should be preventable by maintaining strict containment procedures in handling or production of the toxin.

REFERENCES

1. Benenson AS, ed. *Control of communicable diseases manual*, 16th edn. Washington, DC: American Public Health Association, 1995.
2. Centers for Disease Control. Fish botulism—Hawaii 1990. *MMWR* 1991; 40(24):412–4.
3. Centers for Disease Control. Outbreak of type E botulism associated with an uneviscerated, salt-cured fish product—New Jersey 1992. *MMWR* 1992; 41(29):521–2.
4. Hambleton P. *Clostridium botulinum* toxins: a general review of involvement in disease, structure, mode of action and preparation for clinical use. *J Neurol* 1992; 239:16–20.

CLOSTRIDIUM DIFFICILE

Common names for disease: C. difficile colitis, pseudomembranous colitis, antibiotic-associated colitis

Occupational setting

Although generally associated with individual antibiotic use, and therefore not usually transmissible, nosocomial spread of diarrhea arising from *Clostridium difficile* has been documented. Hospital personnel caring for patients with *C. difficile* colitis often develop skin colonization, and have on occasion developed diarrhea from infection.[1,2] Workers in day care centers and chronic care facilities, where hygiene is difficult and *C. difficile* infection can spread quickly, are also at risk.

Exposure (route)

Cases usually arise from clostridial overgrowth in the gut of severely ill or debilitated individuals or in those taking oral antibiotic therapy. In occupational and nosocomial cases of *C. difficile* enteritis, the route of infection is fecal–oral.

Pathobiology

C. difficile is an aerobic, spore-forming, Gram-positive bacillus, which forms part of the normal colonic flora in many healthy adults. Although the organism has been isolated from the intestinal flora of animals, they are not considered an important reservoir in human disease.

Colitis due to *C. difficile* is more common in elderly and debilitated patients, in women, in patients with cancer or burns, and in people who are taking antibiotics. Antibiotic-associated colitis occurs when alteration of the normal intestinal flora allows overgrowth of *C. difficile* and elaboration of toxins into the intestinal lumen. Clinical effects arise from the toxins, as the bacteria themselves are rarely invasive.

There is a wide range of symptoms arising from *C. difficile* infection, with mild cases of diarrhea being the most common. In antibiotic-associated cases, symptoms usually begin 4–9 days after starting antibiotics. A typical patient with *C. difficile* colitis presents with profuse, foul-smelling diarrhea, which may be watery or green and mucoid. There is usually crampy abdominal pain, with fever and abdominal tenderness on examination. Acute arthritis may accompany *C. difficile* colitis.

Diagnosis

Diagnosis is confirmed by isolation of *C. difficile* from the stool by special culture techniques, or more often by demonstration of the presence of toxin in the stool. The diarrheal stool may have gross or occult blood, and fecal leukocytes are a common finding. Gram stain of the stool is usually not helpful. Other laboratory findings that may be helpful are leukocytosis and hypo-albuminemia. On sigmoidoscopic examination, there are often whitish plaques resembling membranes on the walls of the distal descending colon, sigmoid colon, and rectum, which give rise to the name pseudomembranous colitis for *C. difficile* enteritis. More proximal lesions in the colon occur much less commonly. Not all cases of antibiotic-associated colitis have pseudomembranes.

Treatment

Mild to moderate cases of antibiotic-associated colitis may be treated with discontinuation of antibiotics, and supportive therapy. Cholestyramine binds *C. difficile* toxin and may be useful in mild cases, but should not be used with antibiotics because it may bind them and result in their extended presence in the colon. Improvement can be expected within 48 hours, with resolution of symptoms in about a week.

If the inducing antibiotic cannot be discontinued, or if improvement does not occur after discontinuing antibiotics, specific antibiotic therapy against *C. difficile* should be started. Patients presenting with high fever, leukocytosis and severe abdominal pain and tenderness should also receive antibiotics. The drug of choice in severe cases of *C. difficile* colitis in adults is oral vancomycin for 5–7 days. An alternative in mild to moderately ill non-pregnant adults is oral metronidazole. Antidiarrheal agents should not be used, since they may cause toxin retention. Recurrence from the persistence of spores in the gut is not uncommon, but usually responds to retreatment.

Medical surveillance

There are no recommended routine screening or surveillance activities for workers at risk, but surveillance of stools of patients is sometimes used to identify carriers so that enteric precautions among staff of institutions may be taken.

Prevention

The carrier state cannot be readily eradicated with antibiotics, so the major tool in prevention of occupational and nosocomial transmission is handwashing and environmental disinfection.

REFERENCES

1. Strimling MO, Sacho H, Berkowitz I. *Clostridium difficile* infection in health-care workers. *Lancet* 1989; 2(8667):866–7.
2. Delmee M. *Clostridium difficile* infection in health-care workers. *Lancet* 1989; 2(8671): 1095.

CLOSTRIDIUM PERFRINGENS

Common names for diseases: Gas gangrene, enteritis necroticans (Darmbrand, Pig-bel), food poisoning

Occupational setting

There are three major categories of disease caused by *Clostridium perfringens*, two of which are foodborne. The common foodborne illness, a mild self-limited intestinal disorder, occurs in institutional food-preparation areas, so occupations at risk may include workers who regularly eat at institutional cafeterias.[1] The less common foodborne illness, known as enteritis necroticans, is not known to be associated with work. The most serious disease caused by *C. perfringens* is gas gangrene, which usually occurs after traumatic injury or surgery.[2] Occupations at risk are those with high risk of traumatic injury, such as agriculture and the armed forces. As is the case with other *Clostridium* species, the bacteria are ubiquitous, but are found primarily in the soil and in vertebrate (including human) intestines.

Exposure (route)

In cases of foodborne illness, the route of infection is ingestion. In gas gangrene, *C. perfringens* is introduced into a wound from an external source such as soil, or the wound may be seeded from the patient's own colonic flora.

Pathobiology

C. perfringens is an anaerobic, spore-forming, Gram-positive bacillus. It is a common intestinal inhabitant in normal humans. It can be found in blood cultures of patients without evidence of clostridial infection and thus it is considered to cause a transient, clinically insignificant bacteremia. It is also commonly found as part of the mixed flora found in many anaerobic infections.

Diseases caused by *C. perfringens* result from the effects of toxins produced after germination and replication of the spores in a hospitable anaerobic environment. There are several types of *C. perfringens*, which are categorized by the toxins produced and their clinical effects.[1]

The most common, but least serious, syndrome is food poisoning, caused by *C. perfringens* type A. When spore-contaminated food (usually meat) is inadequately heated, or inadequately reheated after slow cooling, spores germinate and produce toxin, which is then ingested with the contaminated food. The circumstances under which the toxin is produced are common in institutional kitchens preparing food in large batches. After an average incubation period of 7–15 hours, diarrhea and abdominal cramps develop, sometimes accompanied by fever and nausea. Symptoms usually resolve within 24 hours.

Enteritis necroticans (called Pig-bel in Papua New Guinea, where it is endemic; also called Darmbrand, which means "fire bowels" in German, after epidemics affected post-World War II Germany) is caused by *C. perfringens* type C, which is ingested in undercooked pork.[3] The bacteria multiply in the small intestine and release a clostridial beta-toxin, which is thought to be responsible for the ensuing patchy necrosis. Segmental intestinal gangrene and other severe complications necessitating surgery may occur. This very serious disease is rare in industrialized countries. People who are malnourished, alcoholic or have had pancreatic or gastric resection are at increased risk, possibly due to deficiencies in intestinal proteases, especially trypsin, which break down beta-toxin.

Gas gangrene is a rare but devastating syndrome caused by an alpha-toxin which is produced by all *C. perfringens* types. It usually occurs after traumatic injury or surgery, but spontaneous cases occasionally occur. *C. perfringens* is a common contaminant of open wounds. Factors that are thought to promote clostridial replication in a wound are foreign bodies, vascular insufficiency, and concurrent infection with other bacteria. The incubation period is 1–4 days. The first symptom is usually sudden pain at the wound site, which may be pale, edematous and tender. Crepitus may be palpated and gas from bacterial metabolism seen on radiographic studies. The skin color progresses from pale to magenta or bronze, and hemorrhagic bullae with a thin brown serosanguinous discharge, which has a characteristic offensive, sweet odor, may develop. Necrosis of muscle is an associated finding. Systemic effects in gas gangrene, thought to be due to toxin absorption, are common, and include diaphoresis, fever, tachycardia, and anxiety. Late complications include hemolytic anemia, hypotension and renal failure. In advanced cases, there may be involvement of the entire skin surface, hemoglobinuria, hemoglobinemia, and coma.

Other soft tissue infections due to *C. perfringens* include uncomplicated polymicrobial abscesses, crepitant cellulitis, suppurative myositis, emphysematous cholecystitis, anaerobic pulmonary infections (especially empyema), and, rarely, after penetrating head trauma, brain abscess.

Diagnosis

Clostridial food poisoning is diagnosed by recovery of 10^5 *C. perfringens* organisms per gram of suspected food, and recovery of 10^6 *C. perfringens* organisms per gram of stool collected within 48 hours of symptom onset. Diagnosis of enteritis necroticans is primarily clinical, and should be suspected in a patient with the risk factors outlined above, in the setting of anorexia, vomiting, abdominal pain and bloody diarrhea. Absence of colonic involvement and rapid progression of the illness to sepsis and shock favor the diagnosis. Diagnosis of gas gangrene is primarily clinical, but is supported by evidence of myonecrosis seen at surgery. Gram stained material from bullae may reveal Gram positive or Gram-variable rods with a typical "box-car" appearance, and few white blood cells. Only 15% of gas gangrene cases have bacteremia.

Treatment

No treatment is necessary for *C. perfringens* food poisoning. Treatment for enteritis necroticans includes chloramphenicol or penicillin G, supportive care, and bowel decompression. Small bowel resection may be required for persistent paralytic ileus, septicemia, peritonitis, persistent pain, or a palpable mass lesion.

Treatment of gas gangrene requires extensive surgical debridement, with wide excision for abdominal wall involvement, and usually amputation if an extremity is involved. Penicillin G is the antibiotic treatment of choice, but alternatives include chloramphenicol, metronidazole, imipenem, erythromycin, rifampin, clindamycin, and tetracycline. Use of hyperbaric oxygen in the treatment of gas gangrene is controversial, but may be considered as additional therapy after surgery and antibiotics. The mortality rate of gas gangrene treated with surgery and antibiotics still approaches 25%.

Medical surveillance

There are no recommended medical screening activities for *C. perfringens*. Outbreaks of *Clostridia* food poisoning should be reported promptly to the local health authorities in the USA.

Prevention

Foodborne clostridial disease can be prevented by adequate initial cooking temperatures and rapid cooling, and by adequate reheating of foods. Division of large batches of food into smaller units facilitates rapid cooling, which prevents germination of clostridial spores. A vaccine against *C. perfringens* beta-toxin is in use in Papua New Guinea, where it has been quite effective in preventing enteritis necroticans.[4] Cleaning grossly contaminated wounds may help prevent gas gangrene, but most cases are not easily prevented. Early detection and treatment may mitigate some of the more severe manifestations and reduce mortality.

REFERENCES

1. Hatheway CL. Toxigenic clostridia. *Clin Microbiol Rev* 1990; 3:66–98.
2. Present DA, Meislin R, Shaffer B. Gas gangrene. A review. *Orthop Rev* 1990; 19:333–41.
3. Murrell TG, Walker PD. The pigbel story of Papua New Guinea. *Trans R Soc Trop Med Hyg* 1990; 85:119–22.
4. Lawrence GW, Lehmann D, Anian G, et al. Impact of active immunization against enteritis necroticans in Papua New Guinea. *Lancet* 1990; 336:1165–7.

CLOSTRIDIUM TETANI

Common names for disease: Tetanus, lockjaw

Occupational Setting

Tetanus is caused by a toxin released by *Clostridium tetani,* a ubiquitous bacterium found in greatest numbers in soil, especially soil rich in fecal matter such as manure. It is a potential problem in any outdoor job, especially work in which minor skin trauma is frequent. The largest category of exposed workers are employed in agriculture,[1] and of the 50–100 cases reported annually in the USA, most occur in farming regions (Figure 22.4).

Exposure (route)

Spores of *C. tetani* reproduce after inoculation into traumatized skin. Although typically introduced by deep puncture wounds, in many cases no apparent wound is found, making it likely that very minor, unnoticed trauma can also be the portal of entry. Large and heavily soiled or gangrenous wounds, such as those produced by heavy equipment or artillery, are also tetanus-prone. Spores may be introduced into existing wounds through contact with manure, soil or fecal matter.

Pathobiology

Clostridium tetani is an anaerobic, Gram-positive rod that produces hardy spores found in large numbers in soil. When the spores are inoculated into a wound, they germinate and produce exotoxins, including tetanospasmin, the toxin responsible for the signs and symptoms of tetanus. The toxin travels via motor nerve axons and binds to receptors in muscle and the central nervous system, replicating the actions of neurotransmitters. The effects of the toxin last several weeks, and recovery probably requires the growth of new synapses.

The nervous system effects vary, probably depending on the size of the inoculum and degree of existing immunity. Effects can be classified into one of two categories: **central motor control effects** in which poisoning of inhibitory neuronal cells causes unopposed motor activity leading to rigidity and spasms, and **autonomic instability**, the leading cause of death in tetanus, which is manifested by increased sympathetic tone with massive catecholamine release from disinhibition of the sympathetic system neurons of the autonomic system. Occasionally, parasympathetic disinhibition is seen, with its main effect on the vagus nerve, causing bradycardia or even asystole.

Tetanus is divided clinically into four presentations, including neonatal tetanus, which will not be discussed here. In local tetanus, the muscle at the site of the injury is fixed in spasm, and progression to generalized tetanus may occur. Generalized tetanus usually starts with non-specific symptoms, including malaise, restlessness, headache, insomnia, irritability, and profuse sweating, followed by stiffness, twitching and pain at the wound site, and fever. This progresses to the classic signs of tetanus: trismus, or lockjaw, risus sardonicus, the straightening of the upper lip which has reminded some of a sardonic smile, and opisthotonic posturing, which is spasm of the back muscles such that a patient placed supine would rest on his heels and the back of his head only. There is no loss of consciousness and the condition is extremely painful. In generalized tetanus there may be airway compromise, diaphragmatic dysfunction, and autonomic dysfunction. There may be clinical progression for up to 2 weeks despite administration of antitoxin, and full recovery may take months. The cephalic form of tetanus affects the face, causing paresis.

The incubation period varies from one day to months, but the average is 10 days. The shorter the period of time between spore inoculation and symptom onset, the poorer the prognosis. Other predictors of poor prognosis are autonomic dysfunction at presenta-

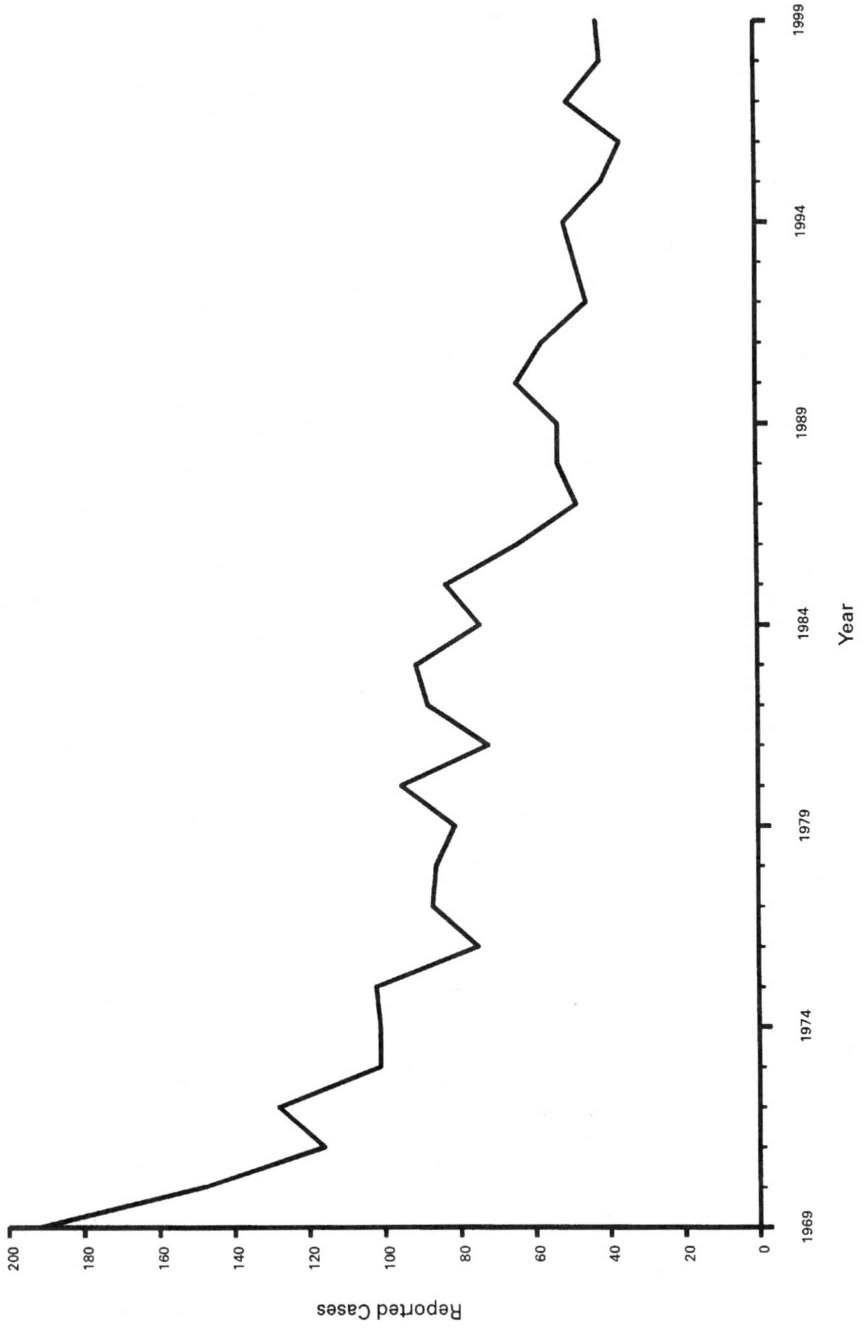

Figure 22.4 Tetanus – reported cases by year, United States, 1969–1999. Source: MMWR 2001:48(53):73.

tion, and burn or surgical site as the portal of entry.

Diagnosis

Diagnosis is by history and examination. The only condition in the differential diagnosis that closely resembles tetanus is strychnine poisoning. Laboratory measurement of antitetanus antibodies may be helpful but is unlikely to be reported in time to direct treatment. Electromyography may also help, but treatment should not be reserved pending the results of testing.

Treatment

Because tetanus may progress for several days even with appropriate therapy, all patients with tetanus should be admitted to an ICU. Passive immunotherapy with intramuscular human tetanus immune globulin is the standard therapy. The immune globulin binds only free toxin and not that already bound, which explains the progression of symptoms once therapy has begun and the long recovery period. If human tetanus immune globulin is not available, equine antitoxin may be given intravenously after testing has ruled out hypersensitivity to horse serum. Tetanus toxoid should also be administered, as clinical tetanus does not confer subsequent immunity. Wounds should be managed with debridement. Parenteral penicillin is recommended for 10–14 days, although this is of unclear benefit. Tetracycline may be used as an alternative in penicillin-allergic patients. Agents such as benzodiazepines may be used to reduce muscular spasms. Other therapy may include airway or ventilatory support, neuromuscular junction blockade, and management of autonomic dysfunction. Full recovery may take a month or longer. Complications include neuropathies, rhabdomyolysis, and myositis ossificans circumscripta (local deposits of bone in muscle).

Medical surveillance

There are no recommended medical screening activities for tetanus. Reporting of cases is required in most of the USA and in most other countries. Tetanus is a Sentinel Health Event-Occupational (SHE-O) for farmers and ranchers.[2] Case finding among all persons who receive care in emergency departments has been recommended to reduce the size of the susceptible or unvaccinated population.[3]

Prevention

Active immunization with tetanus toxoid prevents disease. Immunization usually begins in infancy with a series of injections in combination with pertussis and diphtheria vaccines. Later childhood and adult vaccines omit the pertussis component. Adults should be re-vaccinated every 10 years, or after 5 years have passed in the event of a tetanus-prone wound. Tetanus-prone wounds include those contaminated with dirt, feces or saliva, puncture wounds with unsterile needles or through unprepared skin, missile wounds, burns, frostbite, crush injuries, and avulsion injuries. Incomplete vaccination series should be completed, and patients with unknown vaccine histories should receive the full series of three injections, with the second dose 4–8 weeks after the first, and the third 6–12 months after the second, followed by revaccination every 10 years.[4] Elderly patients are at the greatest risk of contracting tetanus because of incomplete immunization or waning immunity over a lifetime.[3]

In cases of tetanus-prone injuries, passive immunization with human tetanus immune globulin should be given in patients with incomplete or unknown vaccine history and in patients with humoral immunosuppression, including human immunodeficiency virus infection.

REFERENCES

1. Benenson AS, ed. Control of communicable diseases manual 16th edn. Washington, DC: American Public Health Association, 1995.
2. Rutstein DD, Mullan RJ, Frazier TM, et al. Sentinel Health Events (Occupational): a basis

for physician recognition and public health surveillance. *Am J Public Health* 1983; 73: 1054–62.

3. Richardson JP, Knight AL. The management and prevention of tetanus. *J Emerg Med* 1993; 11:737–42.
4. Centers for Disease Control. Diphtheria, tetanus, and pertussis: recommendations for vaccine use and other preventive measures. Recommendations of the Immunization Practices Advisory Committee. MMWR 1991; 40(RR-10):6.

CORYNEBACTERIUM SPECIES

Common name for disease: Diphtheria

Occupational setting

Corynebacterium diphtheriae, the most important of the Corynebacteria, causes diphtheria. Diphtheria has been virtually eliminated in the working populations of many countries such as the USA, largely because of widespread immunization[1] (Figure 22.5). From 1980 to 1995, 41 cases of respiratory diphtheria, all of which occurred in unimmunized children, were reported in the USA.[2] Nevertheless, a large population of susceptible adults, together with a new diphtheria epidemic in countries of the former Soviet Union, have heightened concern about a possible resurgence of this disease in the USA.[3] Travelers to such locations have acquired diphtheria.[4]

C. ulcerans, frequently present in dairy cows, may also cause diphtheria in humans.[5] A variety of other species, several of which are zoonoses, have been described as sources of infection in occupational settings but are very rare. These include: *C. striatum* causing septic arthritis following a scalpel injury in a surgeon,[6] *C. aquaticum* infection of a high-pressure injection injury,[7] systemic infection from *C. ovis* in a meat-packer,[8] *C. pseudotuberculosis* causing lymphadenitis in a butcher,[9] and *C. equi* causing a lung abscess in an agricultural worker.[10]

Corynebacteria have also been cultured from the smoke plume of an operating room laser.[11]

Exposure (route)

Exposure occurs through inhalation of airborne respiratory droplets. Transmission from contact with infected skin lesions or fomites is unusual.

Pathobiology

Corynebacteria are pleomorphic, Gram-positive, aerobic bacilli. By far the most pathogenic species is *C. diphtheriae*, for which humans are the only known reservoir. Most infections are asymptomatic.

Signs and symptoms of infection develop after an incubation period of 2–5 days, locally at either mucous membranes (respiratory, ocular or genital diphtheria) or the superficial layers of the skin through pre-existing skin breaks (cutaneous diphtheria). *C. diphtheria* may or may not produce an exotoxin, depending on whether or not the bacterium has itself been infected by a bacteriophage containing the gene mediating toxin production. The toxin causes both local tissue necrosis and systemic effects with absorption. In addition to non-specific signs of shock (tachycardia, stupor), the most common systemic effects include myocarditis and neuritis.

Local infection with toxigenic diphtheria is followed by hyperemia, edema, and development of the characteristic gray exudative pseudomembrane. Although virtually any mucous membrane can be infected, the most frequent and well-known sites are the pharynx and tonsils. Fever is low-grade and a classic finding is of a "bullneck" appearance from a combination of submandibular edema and lymphadenopathy. Extensive membrane formation may lead to respiratory obstruction.

Cutaneous diphtheria is less severe than infection of the respiratory tract. Lesions usually occur in the setting of primary infection with other organisms (typically, *Staphylococcus aureus* and group A streptococci). The

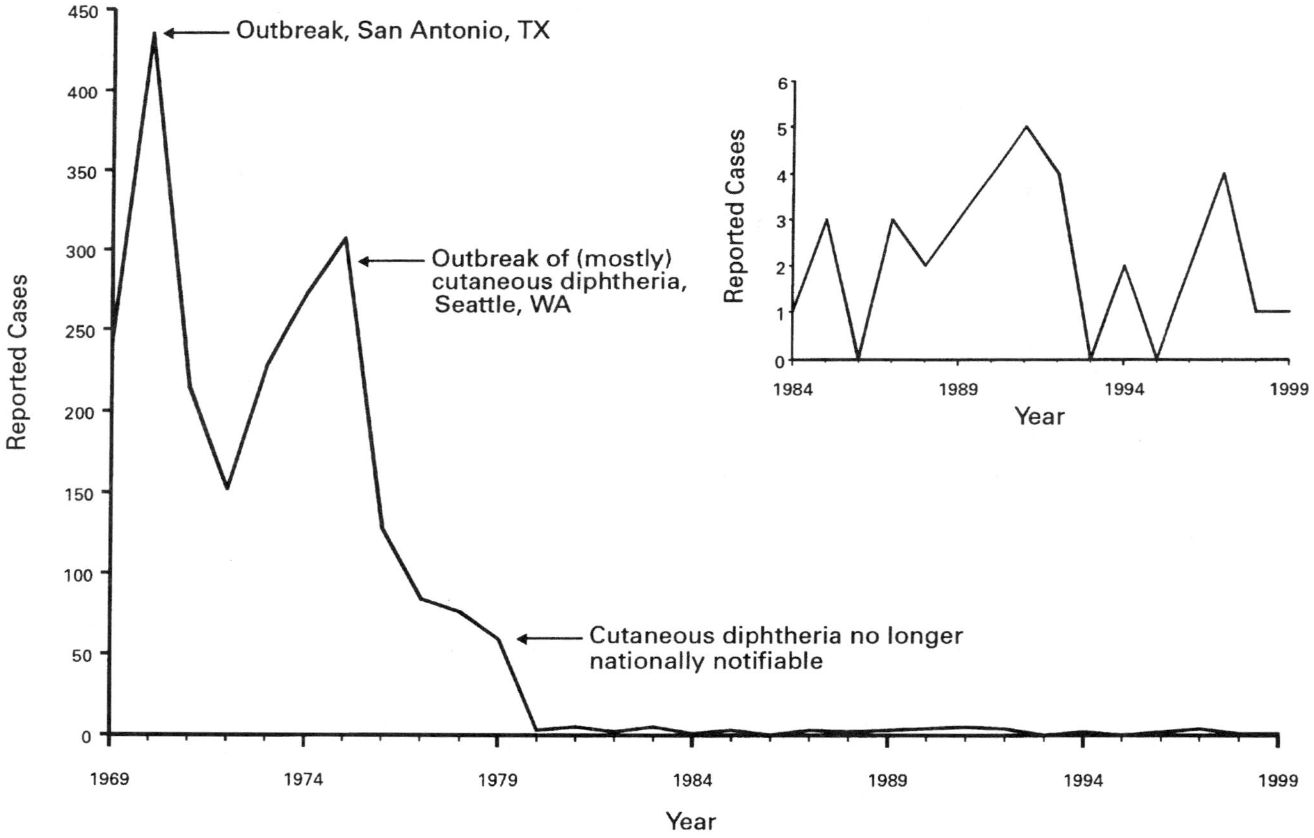

Figure 22.5 Diphtheria – reported cases by year, United States, 1969–1999. Source: MMWR 2001:48(53):33.

characteristic lesion is a non-healing ulcer with a gray membrane.

Non-toxigenic diphtheria infections are milder and confined to local effects. The usual picture is pharyngitis and tonsillitis, although endocarditis has also been described.[11]

Diagnosis

In endemic areas, toxigenic diphtheria may be diagnosed on clinical grounds alone based on the relatively specific clinical picture. Definitive diagnosis requires selective culture from nasal and throat swabs. Because this procedure may not be routinely performed in some laboratories, communication with laboratory personnel regarding a suspicion of diphtheria is advisable to ensure that appropriate isolation and identification techniques are applied. The World Health Organization has recently published guidelines for laboratories regarding the diagnosis of *C. diphtheriae* and *C. ulcerans*.[12] Direct examination through microscopy of stained samples may be unreliable because of the presence of commensals with similar appearance.

Treatment

Treatment of active diphtheria consists of diphtheria antitoxin, antibiotics directed against the organism, and supportive care. Diphtheria antitoxin is no longer licensed for use in the USA and must be obtained through an Investigational New Drug (IND) protocol by contacting the Center for Disease Control (CDC).[13] Penicillin and erythromycin are the drugs of choice, although numerous other antibiotics are also effective.

Medical surveillance

Many advocate the routine culturing of all throat swabs for *C. diphtheria*, since non-toxigenic strains have been isolated with increasing frequency and the potential exists for conversion of these organisms to toxigenic forms.[11] Such a program identified three cases of non-toxigenic *C. diphtheria* tonsillitis in British military personnel.[14] Although reporting of cases of respiratory diphtheria only is required in the USA, prompt involvement of public health authorities is advisable in any suspected case of infection with diphtheria.

Prevention

Universal vaccination is the cornerstone of prevention for diphtheria. For previously unimmunized adults, the Advisory Committee on Immunization Practices (ACIP) of the CDC recommends three doses of tetanus-diphtheria (Td) vaccine, the first two doses 1–2 months apart, and the third 6–12 months after the second dose with a booster dose every 10 years thereafter.[15] Since infection may not confer immunity, vaccination is also recommended for those with a history of the illness. Travelers should ensure that vaccinations are up to date prior to departure.

Patients with respiratory diphtheria should be strictly isolated. After 48 hours of antibiotic therapy, the disease is no longer contagious. Close contacts should be traced and receive a booster (or full series of the vaccine if unimmunized) and antibiotic prophylaxis. Antitoxin is reserved for use at early signs of illness. Two consecutive negative cultures following therapy of cases and carriers should be obtained to document elimination of the organism.

REFERENCES

1. Bisgard KM, Hardy IRB, Popovic T, Strebel PM, Wharton M, Hadler SC. Virtual elimination of respiratory diphtheria in the United States [Abstract no. G12]. In: *Abstracts of the 36th Interscience Conference on Antimicrobial Agents and Chemotherapy*. Washington, DC: American Society for Microbiology, 1995:160.

2. Golaz A, Hardy IR, Strebel P, et al. Epidemic diphtheria in the Newly Independent States of the Former Soviet Union: implications for diphtheria control in the United States. *J Infect Dis* 2000; 181 (suppl 1): S237–43.
3. Centers for Disease Control. Diphtheria acquired by US citizens in the Russian Federation and Ukraine—1994. *MMWR* 1995; 44(12):237, 243–44.
4. Public Health Laboratory Service (UK). Three cases of toxigenic Corynebacterium ulcerans infection. *Communicable Dis Rep* 2000; 10(6):49–52.
5. Cone LA, Curry N, Wuestoff MA, O'Connell SJ, Feller JF. Septic synovitis and arthritis due to Corynebacterium striatum following an accidental scalpel injury. *Clin Infect Dis* 1998; 27(6):1532–3.
6. Larsson P, Lundin O, Falsen E. "Corynebacterium aquaticum" wound infection after high-pressure water injection into the foot. *Scand J Infect Dis* 1996;28(6):635–6.
7. Battey YM, Tonge JI, Horsfall WR, McDonald IR. Human infection with Corynebacterium. *Med J Aust* 1968; 28:540–3.
8. Richards M, Hurse A. Corynebacterium pseudotuberculosis abscesses in a young butcher. *Aust NZ J Med* 1985; 15(1):85–6.
9. Golub B, Falk G, Spink WW. Lung abscess due to Corynebacterium equi. Report of first human infection. *Ann Intern Med* 1967; 66(6):1174–7.
10. Capizzi PJ, Clay RP, Battey MJ. Microbiologic activity in laser resurfacing plume and debris. *Lasers Surg Med* 1998; 23(3):172–4.
11. Wilson AP. The return of Corynebacterium diphtheriae: the rise of non-toxigenic strains. *J Hosp Infect* 1995; 30 (suppl):306–12.
12. Efstratiou A, George RC. Laboratory guidelines for the diagnosis of infections caused by Corynebacterium diphtheriae and C. ulcerans. World Health Organization. *Communicable Dis Public Health* 1999; 2(4):250–7.
13. Centers for Disease Control. Notice to readers: availability of diphtheria antitoxin through an investigational new drug protocol. MMWR 1997; 46(17):380.
14. Sloss JM, Faithfull-Davies DN. Non-toxigenic Corynebacterium diphtheriae in military personnel. *Lancet* 1993; 341(8851):1021.
15. Centers for Disease Control. Update on adult immunization: recommendations of the Immunization Practices Advisory Committee (ACIP). *MMWR* 1991; 40:16–7.

ERYSIPELOTHRIX RHUSIOPATHIAE

Common names for diseases: Erysipeloid, fish poisoning, seal finger

Occupational setting

Erysipelothrix rhusiopathiae has been termed an "occupational pathogen" since the majority of infections occur through work with animals.[1] Those at increased risk include fishermen,[2] farmers,[3] and meat processors.[4] Human infection has also occurred in a laboratory setting.[5] The organism is found in animal waste and can remain viable in the environment for long periods, thus providing an additional source of exposure in farm workers.[6]

Exposure (route)

Cutaneous inoculation of bacteria from a contaminated source is the usual mode of transmission. This can occur when fish scales or bone fragments puncture the skin surface, or from animal bites. Systemic illness has also been reported following the consumption of undercooked contaminated meat.[7]

Pathobiology

E. rhusiopathiac is a non-motile, Gram-positive rod found as a commensal or pathogen in many animal species. Swine are the primary reservoir and are also particularly susceptible to disease. The organism is harbored in the pharynx and excreted in feces. Other potential reservoirs include turkeys, chickens, ducks, sheep and fish.[1]

Infection can result in three clinical entities in humans.[1] The most common is erysipeloid, a localized skin infection at the site of contact (generally the hands). The incubation period is 2–7 days. The involved area is a clearly demarcated, edematous, violaceous lesion that fades centrally as it spreads peripherally. It is accompanied by a burning, itching or throbbing pain that can be quite severe. Fever and arthralgia occur in $\sim 10\%$ of cases, and lymphangitis and lymphadenopathy can be associated as well.[1] Cellulitis from *E. rhusiopathiae* can be distinguished from that commonly caused by streptococci or *Staphylococcus aureus* by the absence of pitting on pressure and lack of suppuration in the lesion.

The diffuse cutaneous form is an unusual presentation in which multiple skin lesions occur in a generalized pattern. The individual lesions are similar in appearance to those seen in localized presentations. Although patients may have constitutional symptoms, blood cultures are negative.

The third form is a severe systemic illness consisting of sepsis and endocarditis that, fortunately, is quite rare. Although the presentation can be subacute, with several weeks of symptoms prior to diagnosis, significant valvular damage may occur, necessitating valve replacement. This illness was uniformly fatal in the pre-antibiotic era. In a review of 45 *E. rhusiopathiae* endocarditis cases, 36% were accompanied by the characteristic skin lesion, 89% had an identifiable occupational association, and the aortic valve was preferentially affected in 61%.[8]

Diagnosis

The organism can be cultured from skin biopsies of the lesions or from blood cultures in patients with endocarditis. Identification can be difficult, since the organism resides deep in skin lesions, is difficult to culture and may be incorrectly identified.[9] Recently, PCR-based assays have been used successfully, offering a more rapid means of identification.[1]

Treatment

Penicillin is the recommended therapy for both localized and systemic infections. The bacterium is not sensitive to vancomycin, which is commonly used to treat endocarditis due to Gram positive organisms.[3] Therefore, in patients with endocarditis and a history of compatible occupational exposure, empirical antibiotic therapy should include coverage of this organism until the bacterial etiology can be established.

Medical surveillance

There are no specific medical screening or surveillance activities for this pathogen.

Prevention

Containment and control measures should be in place wherever potentially infected animals are kept, slaughtered, or processed, or where animal waste is used.[1]

Prevention efforts should also focus on avoidance of skin inoculation and worker education about the bacterium and its clinical presentations. Guards for cutting instruments and gloves with metal mesh or other reinforcement are useful in preventing skin abrasions and lacerations. Work practices designed to reduce contact with bone fragments, fish scales and knife tips should be encouraged. Handwashing after contact with infected animals and patients or their bacteriologic specimens is essential. Meat should be stored and cooked properly.

REFERENCES

1. Brooke CJ, Riley TV. Erysipelothrix rhusiopathiae: bacteriology, epidemiology and clinical manifestations of an occupational pathogen. *J Med Microbiol* 1999; 48(9):789–99.
2. Rocha MP, Fontoura PR, Azevedo SN, Fontoura AM. Erysipelothrix endocarditis with previous cutaneous lesion: report of a case and review of the literature. *Rev Inst Med Trop Sao Paulo* 1989; 31(4):286–9.

3. Venditti M, Gelfusa V, Castelli F, Brandimarte C, Serra P. Erysipelothrix rhusiopathiae endocarditis. *Eur J Clin Microbiol Infect Dis* 1990; 9(1):50–2.
4. Hill DC, Ghassemian JN. Erysipelothrix rhusiopathiae endocarditis: clinical features of an occupational disease. *South Med J* 1997; 90(11):1147–8.
5. Ajmal M. A laboratory infection with Erysipelothrix rhusiopathiae. *Vet Rec* 1969; 85(24):688.
6. Chandler DS, Craven JA. Persistence and distribution of Erysipelothrix rhusiopathiae and bacterial indicator organisms on land used for disposal of piggery effluent. *J Appl Bacteriol* 1980; 48(3):367–75.
7. Nandish S, Khardori N. Valvular and myocardial abscesses due to Erysipelothrix rhusiopathiae. *Clin Infect Dis* 1999; 29(5):1351–2.
8. Gorby GL, Peacock JE. Erysipelothrix rhuriopathiae endocarditis: microbiologic, epidemiologic, and clinical features of an occupational disease. *Rev Infect Dis* 1988; 10:317–25.
9. Dunbar SA, Clarridge JE 3. Potential errors in recognition of Erysipelothrix rhusiopathiae. *J Clin Microbiol* 2000; 38(3):1302–4.

ESCHERICHIA COLI

Common names for disease: Travelers' diarrhea, turista, food poisoning

Occupational setting

Escherichia coli is a normal commensal of the human intestinal tract. Disease is caused by novel or particularly virulent strains transmitted primarily by the fecal–oral route. Occupations with particular risk for *E. coli* diarrheal disease are day care workers and employees in chronic care facilities, where hygiene is often a problem.[1] Jobs requiring travel to other countries, especially less-developed countries, put workers (including military personnel), at risk for travelers' diarrhea due to *E. coli*.[2] The gastrointestinal tract of animal handlers will be colonized with the *E. coli* strains in the feces of the animals they handle.[3] Some studies have also found an increased incidence of acute diarrheal disease as a result of occupational agricultural exposure,[4] and outbreaks have occurred in children visiting farms.[5] This is of particular concern since the liberal use of antibiotics in animal feed has led to multiresistant bacterial strains colonizing farm workers.[6]

Numerous outbreaks of serious diarrheal disease have been caused by *E. coli* transmitted by consumption of undercooked meats, especially improperly handled and prepared ground beef. This may be a risk for workers eating in institutional cafeterias. Like agricultural workers, food processors have also been found to be colonized with potentially pathogenic *E. coli*,[7] although the risk of disease in this group is unclear. Finally, laboratory and nosocomially acquired *E. coli* infection, presumably due to failure to follow standard protective laboratory procedures, has been documented.[8]

Exposure (route)

The primary route of exposure is fecal–oral. Most cases of diarrhea in the USA are caused by the consumption of contaminated food or water. Person-to-person transmission also occurs among close contacts.

Pathobiology

E. coli is an aerobic, Gram-negative rod that has hundreds of serotypes classified by various antigens. Although any of these serotypes can cause disease, certain types have been implicated more frequently in specific diarrheal syndromes. There are five major categories of *E. coli*-induced diarrhea.

Enterotoxigenic *E. coli* (ETEC) is the class implicated in most cases of travelers' diarrhea, although enteroadherent E. coli (EAEC) are common in North Africa and Mexico. The bacteria adhere to the intestinal mucosa and produce a toxin that causes massive fluid secretion into the gut. After the victim ingests the contaminated food or water, there is usually an incubation period of a few days, followed by crampy abdominal pain and watery diarrhea. Symptoms last 3–5 days and usually resolve without specific treatment. The indigenous population is not affected, because regular exposure leads to the development of immunity to the bacterial adhesive factor.

Enteropathogenic *E. coli* (EPEC) is the major cause of childhood diarrhea in developing countries and has been implicated in institutional outbreaks, especially in nurseries.[9] These bacteria disrupt the protective mucous gel coating the intestinal cells, bind to the cells, and cause characteristic mucosal lesions. After a 2–6 day incubation period, the clinical response is watery diarrhea lasting 1–3 weeks.

Enteroinvasive *E. coli* (EIEC) invade the intestinal cells, evoking an inflammatory response that destroys the intestinal mucosa. The clinical syndrome resembles bacterial dysentery and is rare in the USA. The incubation period is usually 2–3 days, followed by fever and bloody diarrhea with numerous fecal leukocytes lasting 1–2 weeks.

Enterohemorrhagic *E. coli* (EHEC) produce verotoxins (Shiga-like toxins) and are typified by the strain O157:H7, which has caused several recent foodborne outbreaks of diarrhea in the USA (Figure 22.6). Following ingestion of contaminated food, usually undercooked ground beef, there is an incubation period of 3–5 days, followed by 7–10 days of abdominal cramping and frequently bloody diarrhea without fever or fecal leukocytes. The course may be complicated by the hemolytic–uremic syndrome with subsequent renal failure, especially in children, or by thrombotic thrombocytopenic purpura.

E. coli is also the most frequent cause of urinary tract infections and nosocomial bacteremia. It is a common cause of nosocomial pneumonia in severely ill patients.

Diagnosis

In cases of diarrhea, stool samples should be obtained for culture and sensitivity, toxin identification, and smear for fecal leukocytes. Various immunoassays, bioassays and DNA probes are used to differentiate among the serotypes. Many of these specialized tests are available through local health department laboratories. When a specific *E. coli* serotype is suspected, the laboratory should be informed so that appropriate tests may be run. Diagnosis of *E. coli* infection in other sites, such as blood or urinary tract, is made by culture and Gram stain of the appropriate samples.

Treatment

The first line of treatment for all forms of *E. coli* diarrhea should be electrolyte and fluid replacement to prevent dehydration. Severe cases of travelers' diarrhea may be treated with a short course of trimethoprim–sulfamethoxazole, doxycycline, or ciprofloxacin. Anti-motility agents are useful for symptomatic relief but are contraindicated when diarrhea is bloody. Patients with severe EIEC diarrhea may require oral or parenteral antibiotics, such as ampicillin. Antibiotics have not been shown to be effective in EHEC diarrhea, which may be complicated by hemolytic–uremic syndrome or thrombotic thrombocytopenic purpura, requiring more intensive supportive care.

The standard treatment of *E. coli* sepsis usually includes intravenous ampicillin with an aminoglycoside or a third-generation cephalosporin with or without an aminoglycoside. Intravenous therapy may be stopped after clinical improvement, but appropriate oral antibiotics should be given to complete a 10–14 day course.

Medical surveillance

There are no recommended medical screening activities for infections due to *E. coli*. Outbreaks of diarrhea due to *E. coli* should be reported to the local public health authorities.

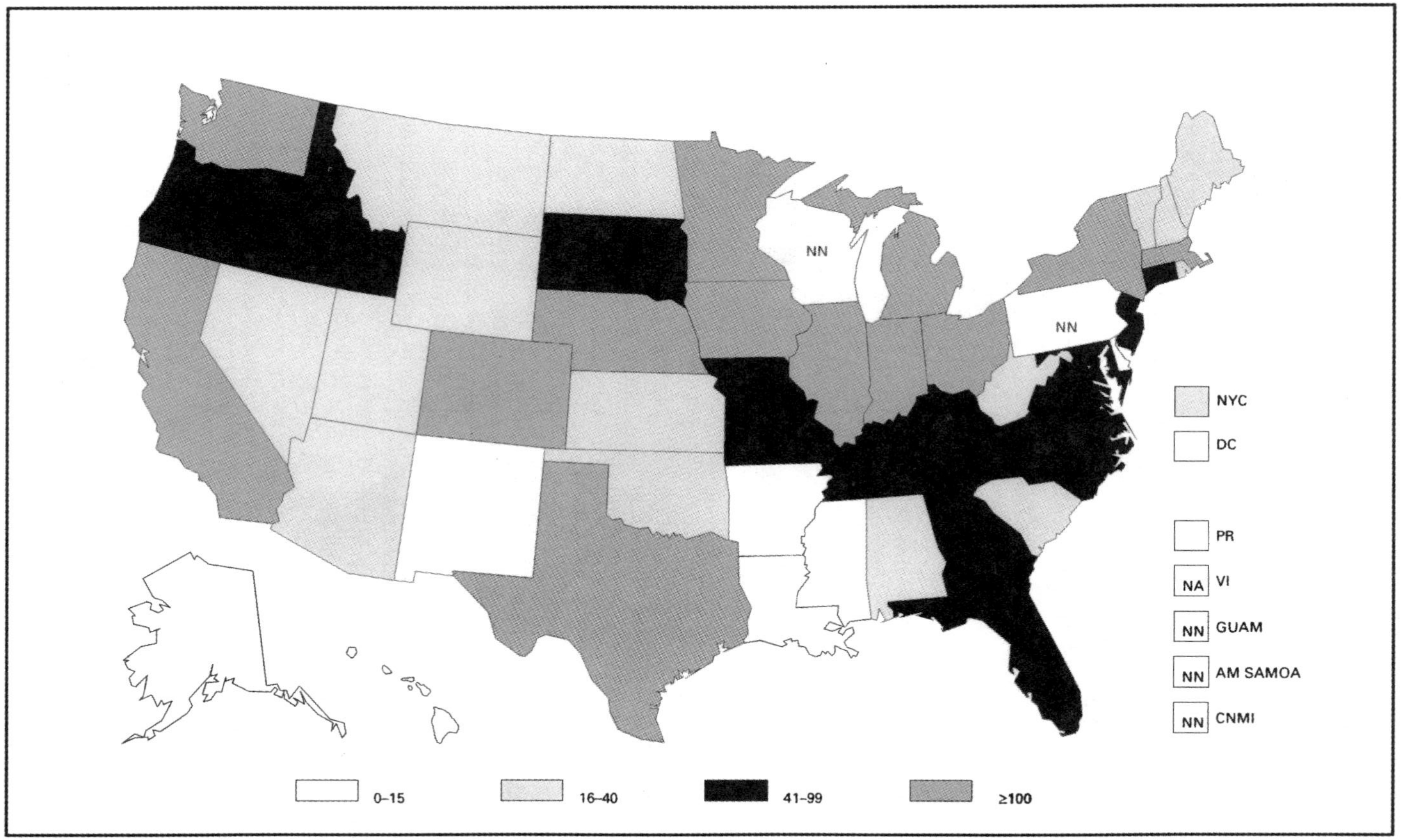

Figure 22.6 *ESCHERICHIA COLI* O157:H7 – reported cases, United States and territories, 1999. According to the CDC, since fewer than 60% of laboratories routinely test stool specimens for *E. coli* O157:H7, many cases are not recognized or reported. Source: MMWR 2001:48(53):36.

Prevention

Avoiding high-risk foods and untreated water can prevent most cases of travelers' diarrhea. Bismuth-subsalicylate has been shown to be effective in preventing travelers' diarrhea. Various antibiotics have also been shown to be effective. However, they are generally not recommended prophylactically, due to cost, the brief, self-limited nature of most cases, the increasing emergence of resistant bacteria, and the potential for serious complications such as pseudomembranous colitis.

Proper preparation and storage of food at the workplace will prevent this and other causes of infectious gastroenteritis. Work practices should emphasize good hygiene with appropriate glove use, strict handwashing after contact with potentially infected material, and disinfecting potentially contaminated surfaces. Unpasteurized dairy products and juices as well as undercooked meats should always be avoided. There is evidence that less contamination occurs in day care centers if infants and toddlers wear disposable diapers instead of cloth and wear clothing over diapers.[10]

REFERENCES

1. Pavia AT, Nichols CR, Green DP. Hemolytic–uremic syndrome during an outbreak of E. coli O157:H7 infection in institutions for mentally retarded persons: clinical and epidemiologic observations. *J Pediatr* 1990; 116:544–51.
2. Oyofo BA, Peruski LF, Ismail TF, et al. Enteropathogens associated with diarrhea among military personnel during Operation Bright Star 96, in Alexandria, Egypt. Mil Med 1997; 162(6):396–400.
3. Trevena WB, Willshaw GA, Cheasty T, Domingue G, Wray C. Transmission of Vero cytotoxin producing Escherichia coli O157 infection from farm animals to humans in Cornwall and West Devon. *Commun Dis Public Health* 1999; 2(4):263–8.
4. MacDonald IA, Gould IM, Curnow J. Epidemiology of infection due to Escherichia coli O157: a 3-year prospective study. *Epidemiol Infect* 1996; 116(3):279–84.
5. Trevena WB, Willshaw GA, Cheasty T, Domingue G, Wray C. Transmission of Vero cytotoxin producing Escherichia coli O157 infection from farm animals to humans in Cornwall and west Devon. *Communicable Dis Public Health* 1999; 2(4):263–8.
6. Al-Ghamdi MS, El-Morsy F, Al-Mustafa ZH, Al-Ramadhan M, Hanif M. Antibiotic resistance of Escherichia coli isolated from poultry workers, patients and chicken in the eastern province of Saudi Arabia. *Trop Med Int Health* 1999; 4(4):278–83.
7. Stephan R, Ragettli S, Untermann F. Prevalence and characteristics of verotoxin-producing Escherichia coli (VTEC) in stool samples from asymptomatic human carriers working in the meat processing industry in Switzerland. *J Appl Microbiol* 2000; 88(2):335–41.
8. Coia JE. Nosocomial and laboratory-acquired infection with Escherichia coli O157. *J Hosp Infect* 1998; 40(2):107–13.
9. Senerwa D, Olsvik O, Mutanda LN, Gathuma JM, Wachsmuth K. Colonization of neonates in a nursery ward with enteropathogenic Escherichia coli and correlation to the clinical histories of the children. *J Clin Microbiol* 1989; 27(11):2539–43.
10. Van R, Wun CC, Morrow AL, et al. The effect of diaper type and overclothing on fecal contamination in day-care centers. *JAMA* 1991; 265:1840–4.

FRANCISELLA TULARENSIS

Common names for disease: Tularemia, rabbit fever, deerfly fever

Occupational setting

Tularemia was first described as a zoonotic disease, resembling plague, in ground squirrels in Tulare County, California, from which the disease and species names are derived. The disease is worldwide in its distribution, and the organism can be found in over 100 species of wild animals and birds. In the USA, cases have been reported from most states, with a large proportion from Arkansas, Tennessee, Texas, Oklahoma, and Missouri. Classically, tularemia was described as a disease of small-game hunters who became infected while skinning infected rabbits, especially during the winter hunting season. Other animals commonly associated with human infections include squirrels, muskrats, beavers, woodchucks, opossums, foxes, coyotes, deer, water rats, voles, cats, sheep, occasional reptiles, and birds.[1] Approximately 150–300 cases are reported in the USA per year.[2] Most cases are sporadic, but small epidemics have occurred among muskrat trappers, outdoor workers (especially military) exposed to insect and arthropod vectors, and farmers.

Exposure (route)

The organism can be transmitted to humans from direct handling of infected animal carcasses, hides, or fur, through non-intact skin, via infected ticks and biting flies, or after bites by infected animals. Inhalation is an important route of exposure during the handling of dead animals and for laboratory technicians. Vector-borne transmission is most common in the summer months. Ticks are thought to be the most likely arthropod reservoir for the infection in nature. *Dermacentor variabilis* (dog tick), *Dermacentor andersoni* (wood tick) and *Amblyomma americanum* (Lone Star tick) are the important tick vectors; biting flies, squirrel fleas, biting gnats and mosquitoes are all thought to be possible vectors of the disease. Person-to-person transmission has not been reported.

Pathobiology

Francisella tularensis is a Gram-negative, pleomorphic, non-motile, non-spore-forming bacteria that can persist for weeks in mud and water associated with aquatic mammal dwellings.[1] Two tularemia strains (types A and B) have been identified on the basis of virulence. Type A is highly virulent, is found only in North America, produces clinical illness after exposure to as few as 10 organisms through broken skin or inhalation, is most often associated with and is often lethal for domestic rabbits, and has a case fatality ratio in humans of as high as 7% in untreated cases. Type B tularemia requires greater than 100 000 organisms for infection, is found throughout Europe, Asia, and North America, is associated with rodents and water in nature, and causes mild or even subclinical human disease.[1,2]

All forms of tularemia are commonly associated with high fever, headaches, and rigors. Skin inoculation of the organism can produce ulceroglandular tularemia, responsible for 75–85% of cases. A cutaneous ulcer gradually develops at the site of a tick bite or inoculation, with a depressed, blackened center and well-demarcated, elevated margins; proximal to this lesion, painful lymphadenopathy may ensue. The mortality rate for untreated ulceroglandular tularemia is less than 5%, but fever can last for weeks, the ulcer heals slowly over weeks to months, and lymphadenopathy can persist for months. The mode of transmission is important to the location of this lesion, with occupational inoculation most common on the hands or forearms, and insect- or arthropod-borne inoculation most common on the head, neck, legs, abdomen, groin, or axillary areas.[1,2] Inoculation by contaminated hands, splattering of infectious material or ingestion of contaminated meat can lead to infection of the eyes (oculoglandular tularemia) or pharynx (oropharyngeal tularemia), with painful lymphadenopathy in the regional lymph nodes draining these areas. Patients infected with type B strains may have mild and self-limiting disease that goes undiagnosed.

Pneumonic tularemia is a severe atypical pneumonia that has a high case fatality ratio when untreated, approaching 60% in the pre-antibiotic era. Appropriate therapy has decreased the case fatality ratio to less than 1%. Primary inoculation of the lungs by inhalation of infected aerosols or secondary pneumonia after hematogenous dissemination in septicemic tularemia can occur. Secondary pneumonia, as well as hepatitis, can develop as complications of tularemia. The septicemic, or typhoidal, form is associated with fever, prostration and weight loss without adenopathy.

Diagnosis

The diagnosis of ulceroglandular tularemia is usually made on clinical grounds, and can be assisted by serologic tests. Agglutinating antibodies are usually apparent between 10 to 14 days after the onset of illness and peak from 4–6 weeks later.[2] Culture methods are generally not recommended for diagnosis, because of the virulence and transmissibility of the organism, unless appropriate isolation laboratories are available. A skin-testing antigen, available from the Centers for Disease Control in Atlanta, but not commercially available, has proved to be useful both clinically and epidemiologically.

Treatment

Streptomycin or gentamicin is the treatment of choice and produces a rapid response. The bacteriostatic antibiotics tetracycline and chloramphenicol, while effective, have been associated with relapses in 20% of cases, and are not recommended unless there are contraindications to the use of aminoglycosides.

Medical surveillance

There are no recommended medical screening activities. Tularemia should be recognized as a Sentinel Health Event-Occupational (SHE-O) in hunters, fur handlers, sheep industry workers, cooks, veterinarians, ranchers, laboratory technicians, and veterinary pathologists.[3] The disease is reportable in selected endemic areas in the USA and in some other countries.

Prevention

A modified live intradermal vaccine for laboratory workers and others at high risk for the disease can be obtained from the United States Army Medical Research Institute of Infectious Diseases in Fort Detrick, Maryland.[1] Although not completely protective, vaccination has been shown to reduce the severity of disease. Vaccination of hunters appears unwarranted because of the low incidence of the disease in this population. Use of protective clothing and gloves is recommended during skinning or handling of potentially infected animals. Prevention of arthropod-borne disease includes strategies to decrease tick bites with protective clothing, repellent use, and careful inspection of the skin.

REFERENCES

1. Craven RB, Barnes AM. Plague and tularemia. *Infect Dis Clin North Am* 1991; 5:165–175.
2. Spach DH, Liles WC, Campbell GL, et al. Tick-borne diseases in the United States. *N Engl J Med* 1993; 329:936–47.
3. Rutstein DD, Mullan RJ, Frazier TM, et al. Sentinel Health Events (Occupational): a basis for physician recognition and public health surveillance. *Am J Public Health* 1983; 73: 1054–1062.

HAEMOPHILUS DUCREYI

Common names for disease: Chancroid, soft chancre

Occupational setting

Outbreaks of chancroid, a sexually transmitted disease, have occurred in military and civilian populations. The disease occurs often in conjunction with other sexually transmitted diseases, and appears to be a marker of high-risk behavior in some working populations. Contact with prostitutes is an important risk factor. Men, especially if uncircumcised, are at higher risk than women.[1,2]

Exposure (route)

Chancroid is sexually transmitted.

Pathobiology

Haemophilus ducreyi is a small, pleomorphic coccobacillus which causes genital and perianal ulcerative lesions. After sexual exposure to an infected person, there is a variable incubation period of a day to several weeks. The chancroidal lesion begins as a tender erythematous papule, which becomes pustular and then ulcerates; lesions are often multiple. The ulcers are usually painful and ragged in appearance, with easy bleeding upon manipulation. Tender inguinal adenopathy is common, and may progress to abscess (bubo), which often spontaneously drains.

Diagnosis

Presumptive diagnosis on clinical grounds without laboratory confirmation is often inaccurate, since other ulcerating sexually transmitted diseases, including syphilis and genital herpes simplex, may mimic chancroid. Material taken from the base of an ulcer, or aspirated from a bubo, may be gram-stained and cultured using special techniques to isolate *H. ducreyi*. If appropriate culture media are not available, presumptive diagnosis may be made on clinical grounds after ruling out (or concomitantly treating for) syphilis and *Herpes simplex*.

Treatment

The treatment of choice is oral erythromycin for at least 7 days. Single dose ceftriaxone may be as effective. Oral trimethoprim-sulfamethoxazole for 5 to 7 days is a useful alternative in areas where resistance to these drugs is uncommon. Other alternatives to multiple-dose therapy for patients in whom compliance may be a problem are single dose spectinomycin or ciprofloxacin. Fluctuant nodes should be drained by needle aspiration to prevent rupture and fistula formation. Partners should be identified and treated, even if asymptomatic.

Medical surveillance

There are no recommended medical screening activities. Reporting of cases to the local health department is obligatory in many states and countries.

Prevention

Identification and treatment of partners, who are often asymptomatic carriers, may help curtail outbreaks. Proper use of condoms is probably preventive.[2]

REFERENCES

1. Rakwar J, Jackson D, Maclean I, et al. Antibody to Haemophilus ducreyi among trucking company workers in Kenya. *Sex Transm Dis* 1997; 24:267–71.
2. Jessamine PG, Brunham RC. Rapid control of a chancroid outbreak: implications for Canada. *Can Med Assoc J* 1990; 142:1081–1085.

HAEMOPHILUS INFLUENZA

Common name for disease: None

Occupational setting

Serious invasive disease due to *Haemophilus influenzae* is most common in young children, among whom nasopharyngeal carriage rates are high. Outbreaks of disease have occurred in day care centers,[1] but the potential exists in other populations, such as those found in chronic care facilities.

Exposure (route)

The bacteria are spread by airborne droplets or by direct contact with infectious secretions.

Pathobiology

H. influenzae is a pleomorphic, Gram-negative coccobacillus that exists in both encapsulated and non-encapsulated forms. The non-encapsulated forms are commonly part of the normal human nasopharyngeal flora. In children, encapsulated *H. influenzae* type b is responsible for 95% of invasive *H. influenzae* disease, compared to about 50% of the serious disease in adults.[2,3] Non-encapsulated *H. influenzae* is responsible for many mild illnesses in adults, including sinusitis, conjunctivitis, otitis media, and bronchopneumonia. These forms may also be responsible for a greater proportion of invasive disease in adults, especially in those with underlying illnesses,[2] but invasive disease is unusual in adults.

Although previously a common cause of meningitis in children, the disease is rare in adults unless there is a history of head trauma, preceding sinusitis or otitis, or cerebrospinal fluid leak. The clinical course resembles that of other forms of purulent meningitis. Epiglottitis is also unusual in adults, but it is the only invasive *H. influenzae* disease to affect normal healthy adults without underlying or preceding illness.[2] It presents with sore throat, fever, and dyspnea, progressing to dysphagia, drooling, and upright posture with neck extended and chin protruding to maintain airflow. Death may result due to airway obstruction. Pneumonia due to *H. influenzae*, usually unencapsulated, is uncommon in children but common in adults with lung disease or alcoholism. The radiographic picture varies, but pleural effusion, usually sterile, is common. Septic arthritis occurs on occasion, usually in adults with impaired immunity or a pre-existing arthritic condition. Bacteremia may accompany invasive disease in adults.

Diagnosis

Diagnosis of clinically suspected disease is made by confirmatory Gram stain and culture of body fluid. Antigen detection, preferably by latex agglutination, in serum, cerebrospinal fluid, concentrated urine, pleural fluid, pericardial fluid, or joint fluid may also be useful. Sputum and nasopharyngeal samples are not generally helpful, due to the high normal carriage rate. All isolates should be serotyped, in view of the important distinctions between type b and other serotypes.

Treatment

Because of widespread resistance to ampicillin and chloramphenicol, infections should be treated with a cephalosporin such as cefuroxime or cefixime. For *H. influenzae* meningitis, third-generation agents with good cerebrospinal fluid penetration, such as cefotaxime or ceftriaxone, should be used. Treatment should continue for several days after resolution of signs of infection, with a typical course of 7–10 days. Treatment of pericarditis, endocarditis or osteomyelitis may require 3–6 weeks of parenteral therapy.

Medical surveillance

There are no recommended medical screening activities for diseases due to *H. influenzae*. Cases of invasive *H. influenzae* disease should be reported to the local health authorities.

Prevention

Invasive disease has almost disappeared in those countries incorporating the vaccine for *H. influenzae* type b into the primary childhood schedule.[4] Vaccination of adults against *H. influenzae* type b is not routinely indicated because of high rates of natural immunity, but should be considered in persons with underlying chronic medical conditions that may predispose to infection.

Cases of meningitis should be kept in respiratory isolation for 24 hours after starting antibiotic therapy. In cases of outbreaks, rifampin chemoprophylaxis is no longer indicated if all contacts under age 4 are fully immunized.

REFERENCES

1. Wenger JD, Harrison LH, Hightower A, et al. Day-care characteristics associated with Haemophilus influenzae disease. *Am J Public Health* 1990; 80:1455–8.
2. Takala AK, Eskola J, Van Alphen L. Spectrum of invasive Haemophilus influenzae type b disease in adults. *Arch Intern Med* 1990; 150:2573–6.
3. Farley MM, Stephens DS, Harvey RC, et al. Incidence and clinical characteristics of invasive Haemophilus influenzae disease in adults. J Infect Dis 1992; 165(suppl 1):S42–43.
4. Bisgard KM, Kao A, Leake J, Strebel PM, Perkins BA, Wharton M. Haemophilus influenzae invasive disease in the United States, 1994–1995: near disappearance of a vaccine-preventable childhood disease. Emerg Infect Dis 1998; 4(2):229–37.

HELICOBACTER PYLORI

(Formerly Campylobacter pylori.)

Common name for disease: None

Occupational setting

Several studies are now available which have examined occupational risk factors for *H. pylori* infection, the majority of which are based on comparisons between levels of seroprevalence. It should be noted that seroprevalence has been reported to vary, sometimes quite markedly, depending on numerous other factors such as cigarette smoking, alcohol consumption,[1] ethnic background, age,[2] gender, socio-economic status, and body mass index.[3] Therefore, any occupational studies must be interpreted with caution, due to the possibility of confounding factors.

The evidence for risk of infection from performing endoscopy has recently been reviewed and is contradictory.[4] Other occupations with reported increased seroprevalence include ICU nurses[5], and care-givers in an institution for the disabled.[6] One study has concluded that, while endoscopy does not pose a risk, general medical practice may slightly elevate the risk of *H. pylori* acquisition.[7] No association has been found for dentists.[8]

Whether *H. pylori* exists in animal reservoirs is also unresolved. The failure to reliably identify this organism in animals may relate to technical difficulties in detecting and isolating this organism from different environments. Investigators have reported the isolation of *H. pylori* from large domestic mammals,[9] while closely related *Helicobacter* species have been associated with gastrointestinal disease in a number of animals.[10] Shepherds have been reported to have a seroprevalence approaching 100%.[11] However, other authorities have stated that humans are likely to be the only reservoir.[12]

Exposure (route)

Transmission patterns appear consistent with fecal–oral and oral–oral spread, although further study is needed.

Pathobiology

First identified in 1982, *Helicobacter pylori* is a Gram-negative, spiral-shaped rod found within the gastric mucosal layer or adhering to the epithelium of the stomach.

H. pylori infection has been associated with atrophic gastritis, peptic ulcers, and gastric malignancy (adenocarcinoma and lymphoma).[13] There does not appear to be any interaction between *H. pylori* infection and exposure to other occupational gastric carcinogens.[14] The list of conditions which may be related to *H. pylori* continues to grow and includes functional dyspepsia, as well as a variety of disorders of the liver, skin, and immune system.[13] Although a majority of the general population is infected with *H. pylori*, it is not clear why only a small number of these people develop disease.[13]

Diagnosis

The standard approach to diagnosis includes histologic examination and culture of gastric biopsy specimens for *H. pylori*. The urea breath test is a non-invasive alternative, while new research is exploring the potential application of measuring antigen levels in stool samples.[13]

Treatment

The therapy of choice to eradicate *H. pylori* consists of a combination of a proton pump inhibitor with a triple regimen of antibiotics. Other combinations, which include ranitidine, bismuth citrate and two antibiotics, have also been used.[13]

Medical surveillance

There are no recommended medical screening or surveillance activities.

Prevention

In view of the many unresolved issues concerning transmission of *H. pylori*, specific preventive recommendations cannot be provided. General good hygiene precautions, applicable to prevent the spread of many different organisms, should be in place. Animal research currently underway offers the hope of an oral vaccine for the future.[15]

REFERENCES

1. Ogihara A, Kikuchi S, Hasegawa A, et al. Relationship between Helicobacter pylori infection and smoking and drinking habits. *J Gastroenterol Hepatol* 2000; 15(3):271–6.
2. Everhart JE, Kruszon-Moran D, Perez-Perez GI, Tralka TS, McQuillan G. Seroprevalence and ethnic differences in Helicobacter pylori infection among adults in the United States. *J Infect Dis* 2000; 181(4):1359–63.
3. Russo A, Eboli M, Pizzetti P, et al. Determinants of Helicobacter pylori seroprevalence among Italian blood donors. *Eur J Gastroenterol Hepatol* 1999; 11(8):867–73.
4. Williams CL. Helicobacter pylori and endoscopy. *J Hosp Infect* 1999; 41(4):263–8.
5. Robertson MS, Cade JF, Clancy RL. Helicobacter pylori infection in intensive care: increased prevalence and a new nosocomial infection. *Crit Care Med* 1999; 27(7):1276–80.
6. Bohmer CJ, Klinkenberg-Knol EC, Kuipers EJ, et al. The prevalence of Helicobacter pylori infection among inhabitants and healthy employees of institutes for the intellectually disabled. *Am J Gastroenterol* 1997; 92(6):1000–4.
7. Braden B, Duan LP, Caspary WF, Lembcke B. Endoscopy is not a risk factor for Helicobacter pylori infection—but medical practice is. *Gastrointest Endosc* 1997; 46(4): 305–10.
8. Lin SK, Lambert JR, Schembri MA, Nicholson L, Johnson IH. The prevalence of Helicobacter pylori in practising dental staff and dental students. *Aust Dent J* 1998; 43(1):35–9.
9. Dimola S, Caruso ML. Helicobacter pylori in animals affecting the human habitat through the food chain. *Anticancer Res* 1999; 19(5B):3889–94.
10. Fox JG. The expanding genus of Helicobacter: pathogenic and zoonotic potential. *Semin Gastrointest Dis* 1997; 8(3):124–41.
11. Dore MP, Bilotta M, Vaira D, et al. High prevalence of Helicobacter pylori infection in shepherds. *Dig Dis Sci* 1999; 44(6):1161–4.
12. Oderda G. Transmission of Helicobacter pylori infection. *Can J Gastroenterol* 1999; 13(7):595–7.
13. Chiba N, Thomson AB, Sinclair P. From bench to bedside to bug: an update of clinically relevant advances in the care of persons with Helicobacter pylori-associated diseases. *Can J Gastroenterol* 2000; 14(3):188–98.
14. Ekstrom AM, Eriksson M, Hansson LE, Lindgren A, Signorello LB, Nyren O, Hardell L.

Occupational exposures and risk of gastric cancer in a population-based case-control study. *Cancer Res* 1999; 59(23):5932–7.

15. Ruiz-Bustos E, Sierra-Beltran A, Romero MJ, Rodriguez-Jaramillo C, Ascencio F. Protection of BALB/c mice against experimental Helicobacter pylori infection by oral immunisation with H. pylori heparan sulphate-binding proteins coupled to cholera toxin beta-subunit. J Med Microbiol 2000; 49(6):535–41.

LEGIONELLA SPECIES

Common names for disease: Legionellosis, Legionnaires' disease, Pontiac fever

Occupational setting

Legionella organisms are ubiquitous in natural aquatic sources and proliferate easily in water supply systems, cooling towers, evaporative condensers, and distribution lines. Older systems, water temperature in the range 25–40°C, presence of sediment and scale, water stasis, commensal bacteria, and vertical as opposed to horizontal hot water tanks, are conditions that enhance growth of this organism.[1] Disease caused by *Legionella* species has been reported in workers cleaning steam turbine condensers[2] and cooling towers,[3] from aerosols generated by a leaking coolant system,[4] in sewage treatment workers,[5] crews aboard ships,[6] from showers in long-distance truck drivers,[7] and in well excavators.[8] Outbreaks in hospitals and long-term care facilities occur, but generally do not involve hospital staff.[9]

Although the overall rate of Legionnaire's disease in the United States in 1999 was 0.41 cases per 100 000, according to the CDC, the true rate may be approximately ten times higher (Figure 22.7).

Exposure (route)

Transmission occurs through inhalation of water mists contaminated with the organism. Person-to-person transmission is thought not to occur.

Pathobiology

Organisms in the Legionellaceae family are Gram-negative rods. *Legionella pneumophila* is responsible for the great majority of infections, but other species have also been implicated. *Legionella* organisms are particularly difficult to grow in the laboratory. Following exposure, organisms reaching the lungs are phagocytized by alveolar macrophages. They can multiply in these cells, causing lysis and release of more bacteria. *L. pneumophila* is a common cause of severe community acquired nosocomial pneumonias.

Legionella species cause two distinct clinical syndromes. The first, Pontiac fever, is a flu-like illness that occurs after a short incubation period of 1–2 days. The attack rate is over 90%, but the illness is usually mild and self-limited with a rapid recovery. Symptoms consist of fever, myalgias, headache, chills, and fatigue. Cough is a minimal symptom and the chest radiograph remains clear.

The other syndrome is Legionnaires' disease, a pneumonia with significant morbidity and mortality. The incubation period is longer, varying from 2–10 days, and the attack rate is much lower than that of Pontiac fever. The syndrome starts with a prodrome similar to that of Pontiac fever, but non-productive cough is a prominent additional feature and lung examination and chest radiograph may reveal evidence of consolidation. Chest pain and hemoptysis, which can be confused with a pulmonary embolus, may occur. Watery diarrhea is common. A number of neurologic abnormalities may be present, but altered mental status is the most common. Hyponatremia occurs more frequently in pneumonia due to *Legionella* species than in pneumonia due to other pathogens. Although there is no classic radiographic presentation, the film may worsen during the initial treatment and take several months to resolve.

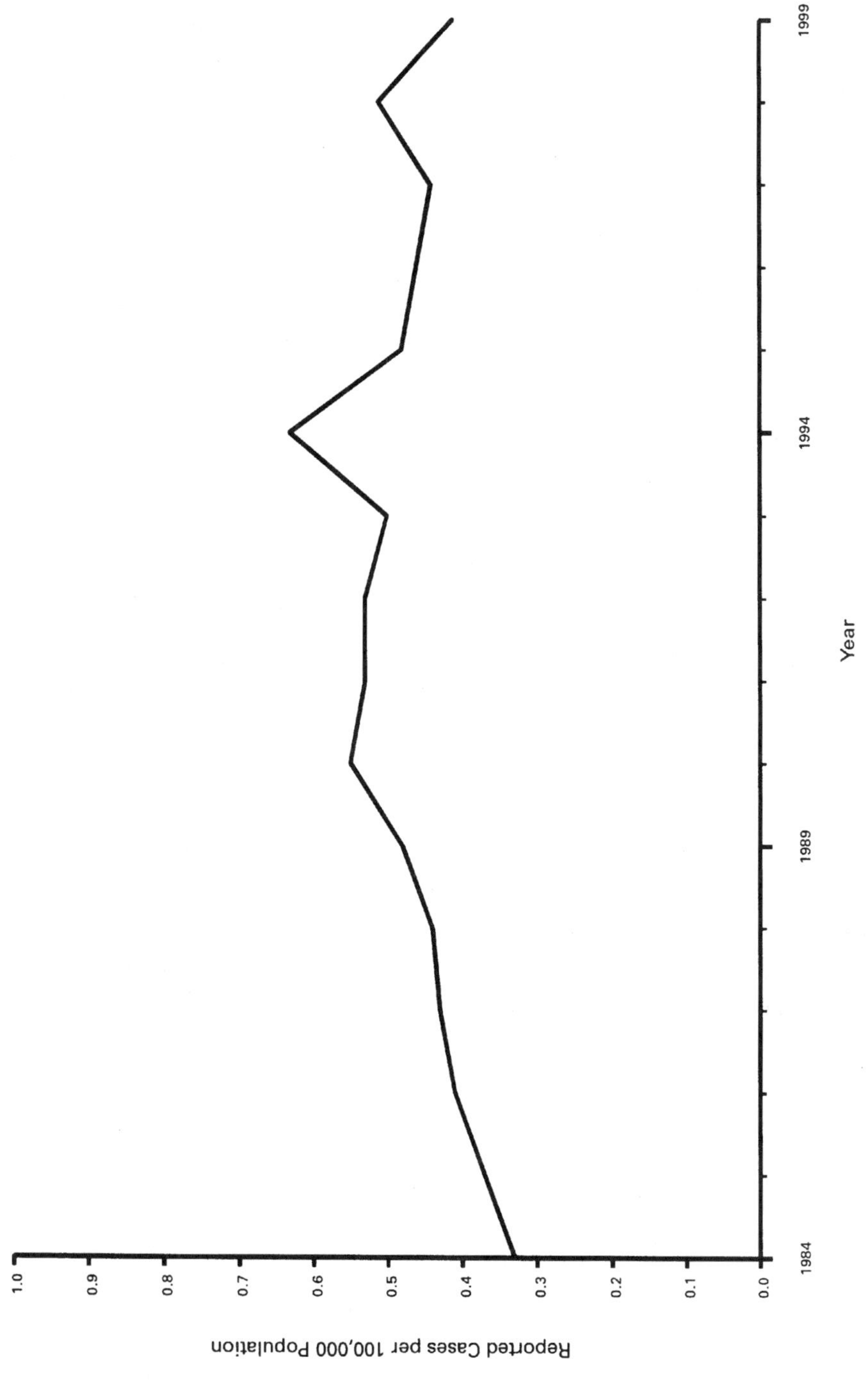

Figure 22.7 Legionnaire's disease – reported cases per 100 000 population by year, United Sates, 1984–1999. Source: MMWR 2001:48(53):48.

Diagnosis

Culture of *Legionella* from biological specimens is the standard means of diagnosis, however, this takes 3–5 days and is most sensitive for transtracheal aspirates, which are difficult to obtain. If the organism is present in sufficient quantities, a direct fluorescent antibody test on the specimen will be positive, and results will be available in a few h. Acute and convalescent antibody titers can be drawn, which requires blood samples to be taken at 3 and 6 weeks postexposure. A fourfold rise is diagnostic and, in a community with a known low prevalence of antibodies, a single titer of 1:256 may indicate exposure.[10] It must be emphasized that seroconversion indicates exposure to the organism only, which may or may not be associated with disease. Clinical correlation is therefore essential. *L. pneumophila* serogroup 1 antigen can also be measured in urine through a commercially available immunoassay.

Treatment

Pontiac fever is self-limited and requires no specific treatment. Legionnaires' disease is commonly treated with intravenous erythromycin, although the newer macrolides and quinolones (azithromycin, ciprofloxacin) are now preferable due to greater effectiveness and reduced side effects.[11]

Medical surveillance

Cases should be reported to local health authorities in selected endemic areas of the USA and in some other countries.

Prevention

Proper maintenance of water distribution and coolant systems is essential. Reducing the degree of *Legionella* growth in water systems may be achieved by raising the water temperature above 60°C and using copper–silver ionization systems. *Legionella* organisms are relatively resistant to chlorine, which is also less effective due to decomposition at higher water temperatures.

Worker education regarding this pathogen is beneficial. Respiratory protection during manual cleaning operations of water systems may be helpful. Coolant systems, particularly those using recirculated water, need regular cleaning to reduce the build up of bacteria and debris. Biocides may also be indicated. Appropriate design of buildings is essential. Water pipes should be installed to avoid areas of water stagnation and leaks repaired promptly. Materials that do not support bacterial growth should be used to connect pipes. Cooling towers should be built for ease of mechanical cleaning and should be distanced from air intake units.

REFERENCES

1. Muraca PW, Stout JE, Yu VL, et al. Legionnaires' disease in the work environment: implications for environmental health. *Am Ind Hyg Assoc J* 1988; 49:584–90.
2. Fraser DW, Deubner DC, Hill DL, et al. Nonpneumonic, short-incubation-period legionellosis (Pontiac fever) in men who cleaned a steam turbine condenser. *Science* 1979; 205:690–1.
3. Girod JC, Reichman RC, Winn WC, et al. Pneumonic and nonpneumonic forms of legionellosis: the result of a common-source exposure to Legionella pneumophila. *Arch Intern Med* 1982; 142:545–7.
4. Allen KW, Prempeh H, Osman MS. Legionella pneumonia from a novel industrial aerosol. *Communicable Dis Public Health* 1999; 2(4):294–6.
5. Gregersen P, Grunnet K, Uldum SA, Andersen BH, Madsen H. Pontiac fever at a sewage treatment plant in the food industry. *Scand J Work Environ Health* 1999; 25(3):291–5.
6. Rowbotham TJ. Legionellosis associated with ships: 1977 to 1997. *Commun Dis Public Health* 1998; 1(3):146–51.
7. Public Health Laboratory Service (UK). Legionnaires' disease in long distance lorry drivers. *Commun Dis Rep* 1998; 10:13–4.

8. Miragliotta G, Del Prete R, Sabato R, Cassano A, Carnimeo N. Legionellosis associated with artesian well excavation. *Eur J Epidemiol* 1992; 8(5):748–9.
9. Centers for Disease Control. Sustained Transmission of Nosocomial Legionnaires Disease—Arizona and Ohio. *MMWR* 1997; 46(19):416–421.
10. Yu VL. Legionella pneumophilia (Legionnaires' Disease). In: Mandell GL, Bennett JE, Dolin R, eds. *Principles and practice of infectious diseases* Fifth ed. Philadelphia: Churchill Livingstone, 2000 2424–35.
11. Klein NC, Cunha BA. Treatment of legionnaires' disease. *Semin Respir Infect* 1998; 13(2):140–6.

LEPTOSPIRA INTERROGANS

Common names for disease: Leptospirosis, Weil's disease or syndrome (name applied to severe, icteric disease), milker's fever

Occupational setting

Leptospirosis is an enzootic infection that is ubiquitous in nature. Important reservoirs include cattle, swine, dogs, rodents, and fish. Many infections in animals are not clinically apparent, and prolonged urinary shedding of the organism can occur. In an Australian study of 208 laboratory-confirmed cases of leptospirosis, 56% had a clear association with occupational exposure.[1] High risk groups include farmers,[1] sanitation and sewage workers,[2] rodent control workers,[3] laboratory animal handlers,[4] forestry workers,[5] trappers,[5] zoo workers,[6] veterinarians,[7] slaughterhouse and other meat workers,[1] and fish farmers.[8]

Dairy farmers are at very high risk during milking, because bovine urination results in an aerosol containing leptospires that can infect humans through inhalation or entry through the eyes, nose, or throat.[9]

More recent case series suggest that multiuse land development, with water from farmlands draining into recreational bodies of water, may be contributing to an increasing proportion of cases from non-occupational exposures. Water-borne disease is probably the single most important source of infection.[10] Widespread epidemics of leptospirosis have also been described following floods.[11]

Exposure (route)

Humans contract the infection from contaminated fluids, tissues or waters through direct contact with breaks in skin or mucous membranes. Urinary shedding of organisms from infected animals is the most common source of these pathogens, but meat handling is also an important route of exposure. Although person-to-person transmission is generally not thought to occur, a recent case series does raise the possibility of disease transmission to healthcare workers.[12]

Pathobiology

The organism, a spirochete that is an obligate aerobe, is easily visualized by phase contrast and dark-field microscopy but grows slowly in culture.

Leptospira consists of several species, only one of which, *L. interrogans*, is pathogenic in humans. *L. interrogans* consists of almost 300 serovars, arranged into 25 major related groups known as serogroups. This classification is important, since the animal reservoirs, clinical picture of infection and geographic distribution vary between serogroups. Active and passive immunity is also serovar-specific. The Icterohaemorrhagiae serogroup is usually carried by rats and is associated with the more severe, classic form of leptospirosis known as Weil's disease. The Australis serogroup is a common cause of infection in many parts of the world such as Australia, New Zealand and Asia. Serogroup Pomona is frequently found in pigs and cattle. The Canicola serogroup causes canine leptospirosis.

Untreated leptospirosis is most often a self-limited illness. Because of the non-specific and often mild presentation, the disease is underdiagnosed and under-reported. The incubation period is usually about 10 days, but can vary from 2 days to 4 weeks. The clinical presentation is variable, but common symptoms include the abrupt onset of fever, headache, muscular pain, nausea, vomiting, diarrhea, and conjunctival suffusion.

Rarely, the patient may appear to recover transiently only to go on to develop a more serious illness with end-organ involvement (termed Weil's disease or syndrome when icterus is present). This second phase can include liver failure, renal failure, meningitis, pulmonary involvement, and thrombocytopenia with hemorrhage. The overall case fatality rate in a recent large case series was 22.6%.[13]

Diagnosis

Leptospirosis should be considered in any patient with fever, myalgias, headache, and nausea or vomiting. Culture of leptospires from blood and cerebrospinal fluid during the first 10 days of the illness and urine beyond the first week can aid diagnosis, but the organism grows slowly and cultures may take up to 8 weeks to become positive. The only screening test approved for use in the USA is the indirect hemagglutination assay (IHA), which utilizes pooled antigens from all serogroups of leptospirosis and is broadly available. However, since the prevalence of different serovars varies geographically, this test may not be sufficiently sensitive in some regions.[14] The microscopic agglutination test (MAT) requires paired sera and is generally only available in reference laboratories. ELISA-based screening tests are also available. They have shown satisfactory sensitivity, but results should be confirmed with MAT.[15] Recently, a dipstick assay for serum has been developed which is suitable for widespread field use.[16]

Other laboratory results will show signs of systemic infection with possible renal and liver dysfunction.

Treatment

When given in the first 4 days of infection, penicillin and doxycycline are both effective in shortening the duration of illness and decreasing symptoms of fever, headache, and myalgias. Supportive therapy and careful management of renal, hepatic, hematologic and central nervous system complications are also important. Therapy has been reported to be infective when administered after day 4 of infection, and the efficacy of other antimicrobial agents has not been rigorously studied in randomized trials. The Jarisch–Herxheimer reaction, an inflammatory reaction induced during antibiotic treatment as a result of rapid release of antigen, is commonly observed during treatment.

Medical surveillance

There are no medical screening activities recommended for leptospirosis. The disease is reportable in most jurisdictions.

Prevention

Primary prevention strategies should focus on both animal reservoirs and humans. Animal preventive activities mainly consist of vaccines that are available for cattle and pigs. Vaccines are serovar-specific and are only useful where a small number of serovars are prevalent. Rodent control is important.

Strategies to prevent leptospirosis in humans have included environmental control measures, protective clothing, and antibiotic prophylaxis. A dramatic decline in *L. icterohaemorrhagiae* infections in Great Britain between 1978 and 1985 was attributed to vigorous rodent control programs, protective clothing use, attention to personal hygiene, and worker education in coal workers, sewer workers, and fish workers.[17] Personal protective equipment consisting of gloves and boots, together with careful work practices around domestic animals to avoid contact with potentially contaminated tissues and fluids (particularly urine), are recommended.

A randomized trial of chemoprophylaxis in military personnel at high risk for leptospirosis in Panama revealed that 200 mg of doxycycline administered once weekly was 95% effective in preventing the disease.[18] This strategy would seem useful only for populations at high risk for relatively brief periods, such as travelers. There are no currently available human vaccines.

REFERENCES

1. Swart KS, Wilks CR, Jackson KB, Hayman JA. Human leptospirosis in Victoria. *Med J Aust* 1983; 14:460–3.
2. De Serres G, Levesque B, Higgins R, et al. Need for vaccination of sewer workers against leptospirosis and hepatitis A. *Occup Environ Med* 1995; 52(8):505–7.
3. Demers RY, Frank R, Demers P, Clay M. Leptospiral exposure in Detroit rodent control workers. *Am J Public Health* 1985; 75(9):1090–1.
4. Natrajaseenivasan K, Ratnam S. An investigation of leptospirosis in a laboratory animal house. *J Communicable Dis* 1996; 28(3):153–7.
5. Moll van Charante AW, Groen J, Mulder PG, Rijpkema SG, Osterhaus AD. Occupational risks of zoonotic infections in Dutch forestry workers and muskrat catchers. *Eur J Epidemiol* 1998; 14(2):109–16.
6. Anderson DC, Geistfeld JG, Maetz HM, Patton CM, Kaufmann AF. Leptospirosis in zoo workers associated with bears. *Am J Trop Med Hyg* 1978; 27(1 Pt 1):210–11.
7. Kingscote BF. Leptospirosis in two veterinarians. *CMAJ* 1985 1; 133(9):879–80.
8. Gill ON, Coghlan JD, Calder IM. The risk of leptospirosis in United Kingdom fish farm workers. Results from a 1981 serological survey. *J Hyg (Lond)* 1985; 94(1):81–6.
9. Skilbeck NW, Miller GT. A serological survey of leptospirosis in Gippsland dairy farmers. *Med J Aust* 1986; 144:565–7.
10. Ciceroni L, Stepan E, Pinto A, et al. Epidemiological trend of human leptospirosis in Italy between 1994 and 1996. Eur J Epidemiol 2000; 16(1):79–86.
11. Trevejo RT, Rigau-Perez JG, Ashford DA, et al. Epidemic leptospirosis associated with pulmonary hemorrhage—Nicaragua, 1995. *J Infect Dis* 1998; 178(5):1457–63.
12. Ratnan S, Seenivasan N. Possible hospital transmission of leptospiral infection. J Commun Dis 1998;30(1):54–6.
13. Ciceroni L, Stepan E, Pinto A, et al. Epidemiological trend of human leptospirosis in Italy between 1994 and 1996. *Eur J Epidemiol* 2000; 16(1):79–86.
14. Effler PV, Domen HY, Bragg SL, Aye T, Sasaki DM. Evaluation of the indirect hemagglutination assay for diagnosis of acute lepto-spirosis in Hawaii. *J Clin Microbiol* 2000; 38(3):1081–4.
15. Winslow WE, Merry DJ, Pirc ML, Devine PL. Evaluation of a commercial enzyme-linked immunosorbent assay for detection of immunoglobulin M antibody in diagnosis of human leptospiral infection. *J Clin Microbiol* 1997; 35(8):1938–42.
16. Smits HL, Hartskeerl RA, Terpstra WJ. International multi-centre evaluation of a dipstick assay for human leptospirosis. *Trop Med Int Health* 2000; 5(2):124–8.
17. Waitkins SA. Leptospirosis as an occupational disease. *Br J Ind Med* 1986; 43:721–5.
18. Takafuji ET, Kirkpatrick JW, Miller RN, et al. An efficacy trial of doxycycline chemoprophylaxis against leptospirosis. *N Engl J Med* 1984; 310:497–500.

LISTERIA MONOCYTOGENES

Common names for disease: Listeriosis

Occupational setting

Listeria monocytogenes is a common environmental bacterium, and has been recovered from soil, dust, and water, and mammals, birds, fish, ticks and crustaceans. Most cases of listeriosis in the USA affect urban dwellers without specific occupational exposures to the bacterium; and many are linked to the ingestion of contaminated food, particularly in pregnant women, newborns, and those with compromised immunity.[1] Veterinarians handling infected calves have developed skin infections, as have laboratory personnel following accidental direct skin inoculation.[2] Mild cases of listeria conjunctivitis occur occasionally in laboratory and poultry workers.[3] Although slaughterhouse workers have been found to have five times the normal fecal carriage rate of *L. monocytogenes* (5% versus 1%), an increased risk of disease among animal handlers other than veterinarians has not been identified.

Exposure (route)

Exposure occurs by ingestion of contaminated food, by direct skin or eye inoculation, and by transplacental transmission from an infected mother to her fetus. Foods that have been implicated are unpasteurized dairy products, undercooked meats, and vegetables grown in fields fertilized with manure from infected animals.[4]

Pathobiology

L. monocytogenes is a Gram-positive, non-spore-forming, aerobic rod, which can cause a variety of clinical syndromes. Transient asymptomatic carriage in the stool is common. Serious symptomatic infection occurs almost exclusively in neonates and immunocompromised adults.[1] Infection during pregnancy (a state of relative immunodeficiency) is often unrecognized and may lead to preterm labor, intrauterine fetal demise, or a critically ill baby.

In adults, *L. monocytogenes* infection can present as sepsis of unknown origin. Symptomatic illness in adults usually occurs in those who are immunosuppressed, including those with acquired immune deficiency syndrome (AIDS) and malignancies. Bacteremia may lead to seeding of the meninges or brain. *L. monocytogenes* is the leading cause of meningitis in immunosuppressed adults, and should be considered in the differential diagnosis of any adult with meningitis. The onset is usually subacute, with low-grade fever and personality changes. Infrequently, there are focal neurologic findings; typical meningeal signs are usually absent. Cerebritis may present with headache and fever, or as a paresis resembling a cerebrovascular accident.

In cases of direct inoculation, ulcerating skin lesions have occurred, as well as purulent conjunctivitis, and, rarely, acute anterior uveitis.[2] Focal internal infections, which may arise from dissemination, include lymphadenitis, subacute bacterial endocarditis, osteomyelitis, spinal abscess, peritonitis, cholecystitis and arthritis. Disseminated listeriosis may be accompanied by hepatitis.

Diagnosis

Diagnosis is made by isolation of the organism from cultures of blood, cerebrospinal fluid, skin ulcer, conjunctival pus, or other specimens from an infected site. Presumptive diagnosis may be made pending culture results if a Gram-stained specimen reveals Gram-positive rods resembling diphtheroids (or sometimes diplococci). Large samples of infected fluid are required (at least 10 ml of cerebrospinal fluid) because the bacteria are often sparse and are difficult to isolate. A direct fluorescent antigen test is available, but is difficult to interpret and so has little practical use in most laboratories. Cerebritis is diagnosed with CT or MRI scans showing focal areas of increased uptake, without ring enhancement, and a positive blood culture; cerebrospinal fluid culture is usually negative.

Treatment

There have been no controlled studies of the efficacy of various treatment regimens, but clinical experience with penicillin and ampicillin has shown these to be usually effective, although there have been rare cases of resistance to each. Because of the refractory nature of *Listeria* to the action of most antibiotics, an aminoglycoside, usually gentamicin, is added for synergy.[5] For patients with penicillin allergy, the best alternative is probably trimethoprim–sulfamethoxazole, although erythromycin, tetracycline and chloramphenicol have all been used successfully. If gentamicin is used for central nervous system infection, it should be administered both intravenously and intrathecally. There may be progression of disease despite appropriate antibiotic therapy, and the optimum duration of therapy is unknown. Although 2 weeks of therapy is usually effective, there have been relapses in immunosuppressed patients, who may require 3–6 weeks of treatment. Effective treatment remains difficult in the immunocompromised patient, and mortality remains high (approximately 30%) despite appropriate choice of therapy.

Medical surveillance

There are no recommended medical screening activities for this disease. Cases are required to be reported to local health authorities in many parts of the USA and in some other countries. Prompt reporting of outbreaks is also required.

Prevention

Animal handlers, including veterinarians, should wear gloves and splash goggles, and should wash their hands frequently. Food borne listeriosis can be prevented by avoiding unpasteurized dairy foods and undercooked meats. Vegetables and fruits grown near the ground should be washed thoroughly before consumption. Uncooked meats should not be stored near vegetables or ready-to-eat foods. Hands, knives and cutting boards should be washed after handling uncooked foods.[4]

REFERENCES

1. Centers for Disease Control and Prevention. Update: Multistate outbreak of listeriosis—United States, 1998–1999. *MMWR* 1999; 47:1117–18.
2. McLauchlin J, Low JC. Primary cutaneous listeriosis in adults: an occupational disease of veterinarians and farmers. *Vet Rec* 1994; 135:615–17.
3. Jones D. Foodborne illness: foodborne listeriosis. *Lancet* 1990; 336:1171–74.
4. Centers for Disease Control. Update: foodborne listeriosis—United States, 1988–1990. . *MMWR* 1992; 41:251–8.
5. Jones EM, MacGowan AP. Antimicrobial chemotherapy of human infection due to Listeria monocytogenes. *Eur J Clin Microbiol Infect Dis* 1995; 14:165–75.

MYCOPLASMA PNEUMONIAE

Common names for disease: Mycoplasma pneumonia, atypical pneumonia, walking pneumonia

Occupational setting

Transmission of *Mycoplasma pneumoniae* is thought to require close, prolonged contact. Therefore, closed populations such as those in military barracks[1] or on college campuses are subject to *M. pneumoniae* infections.[2] Although epidemiologic data are limited and somewhat inconsistent, outbreaks in hospitals have been described.[3] Since many cases in such instances are quite mild, it is possible that many outbreaks go unrecognized.[3] A cluster of cases among workers in a prosthodontics laboratory was suspected to be caused by an aerosol generated during abrasive drilling on the false teeth of a patient who was diagnosed with *M. pneumoniae* 11 days later.[4] Since the organism also causes respiratory disease in animals, veterinarians, animal handlers and farmers may be exposed if contact is prolonged.[5]

Exposure (route)

Transmission occurs through contact with infected respiratory secretions.

Pathobiology

Mycoplasma are the smallest free-living organisms. They are neither true bacteria nor viruses, because, unlike the former, they lack a cell wall, and, unlike the latter, they do not require other cells to grow. *M. pneumoniae* is the most important species in this group and is a frequent cause of respiratory disease. Other pathologic mycoplasmas such as *M. hominis* and the closely related *Ureaplasma urealyticum* are common etiologic agents in infections of the urogenital tract. Infections at other sites by *M. hominis* are rare but have been described in the immunocompromised.[6] Occult infection with *M. fermentans* has been proposed as a cause of illness in Persian Gulf War veterans, but this association has not been confirmed in serologic studies.[7]

M. pneumoniae is a common cause of respiratory infections. Illness develops gradually over a period of several days following an incubation period of 1–4 weeks. Symptoms are flu-like and include fever, malaise, headache, sore throat and cough. Children and young adults are the most frequently affected age groups. Most infections result in relatively benign illness that cannot be distinguished from viral etiologies and may include tracheobronchitis, pharyngitis, and otitis. However, 3–10% of *M. pneumoniae* infections result in pneumonia.[8] Because infection with this organism is so common, its contribution to community-acquired pneumonia is significant, causing an estimated 500 000 cases per year.[8]

M. pneumoniae pneumonia classically presents with a non-productive cough, fever, and upper respiratory tract symptoms such as sore throat and rhinitis. Pleuritic chest pain, dyspnea, and rigors are less common than in other bacterial pneumonias, and the white blood cell count is often normal. Chest radiographic findings are variable, but lower lobe involvement, patchy infiltrates, and pleural effusions are common. Approximately 5–10% of patients require hospitalization.[8] These cases can be severe, resulting in respiratory insufficiency and a number of extrapulmonary complications ranging from otitis media and bullous myringitis to significant neurologic and cardiac disease. Central nervous system involvement occurs in up to 7% of hospitalized patients and includes meningitis, meningoencephalitis, and neuritis.[9] Cardiac manifestations due to myocarditis or pericarditis may result in arrhythmias and heart failure. Nausea, vomiting, or diarrhea is reported in 14–44% of patients.[9] Skin rashes are common. This infection is commonly believed to be associated with erythema multiforme; however, a systematic review of the case literature found that the association is with Stevens–Johnson syndrome and not erythema multiforme.[10]

Autoantibodies that agglutinate red blood cells at 40°C (cold agglutinins) can result in hemolytic anemia. *Mycoplasma* infections have also been associated with a variety of other autoimmune disorders, including rheumatoid arthritis and Guillain–Barré syndrome. A link between infection with *M. pneumoniae* and asthma has recently been established.[11]

Diagnosis

Three traditional methods have been used to diagnose *M. pneumoniae* infections. The organism can be cultured from biological specimens, but since growth is slow, 14–21 days may be needed before *Mycoplasma* species can be detected. The cold hemagglutinins test detects the IgM autoantibody responsible for the agglutination of red blood cells, but is neither sensitive nor specific. The complement fixation test measures antibody production to the mycoplasma lipid membrane, but is only positive late in the course of illness. Recently, rapid PCR based assays have been developed which can be performed on respiratory tract samples and are reported to be both sensitive and specific.[12]

Treatment

Erythromycin and tetracycline are traditionally used in treating *M. pneumoniae* respiratory infections. However, tetracycline resistance has been described with increasing frequency. The newer macrolides and quinolones have been shown to be effective.[13] Supportive care may be needed in the setting of complications.

Medical surveillance

Outbreaks of *Mycoplasma* infection should be promptly reported to local public health authorities. There are no recommended medical screening activities.

Prevention

Transmission of the organism generally involves prolonged close contact. Therefore, the risk is greatest in households, barracks, and dormitories. Healthcare workers and those exposed to infected animals should practice good hygiene with strict handwashing after contact. Gloves and respiratory protection may be useful if contact is prolonged.

One study has concluded that antibiotic prophylaxis of contacts with azithromycin during outbreaks may significantly reduce secondary attack rates.[14] An infected person is generally not regarded as contagious beyond 3 weeks.

REFERENCES

1. Gray GC, Callahan JD, Hawksworth AW, Fisher CA, Gaydos JC. Respiratory diseases among US military personnel: countering emerging threats. *Emerg Infect Dis* 1999; 5(3):379–85.
2. Feikin DR, Moroney JF, Talkington DF, et al. An outbreak of acute respiratory disease caused by Mycoplasma pneumoniae and adenovirus at a federal service training academy: new implications from an old scenario. *Clin Infect Dis* 1999; 29(6):1545–50.
3. Kleemola M, Jokinen C. Outbreak of Mycoplasma pneumoniae infection among hospital personnel studied by a nucleic acid hybridization test. *J Hosp Infect* 1992; 21(3):213–21.
4. Sande MA, Gadot F, Wenzel RP. Point source epidemic of Mycoplasma pneumoniae infection in a prosthodontics laboratory. *Am Rev Respir Dis* 1975; 112:213–17.
5. Jordan FT. Gordon memorial lecture: people, poultry and pathogenic mycoplasmas. *Br Poult Sci* 1985; 26(1):1–15.
6. Mattila PS, Carlson P, Sivonen A, et al. Life-threatening Mycoplasma hominis mediastinitis. *Clin Infect Dis* 1999; 29(6):1529–37.
7. Gray GC, Kaiser KS, Hawksworth AW, Watson HL. No serologic evidence of an association found between Gulf War service and Mycoplasma fermentans infection. *Am J Trop Med Hyg* 1999; 60(5):752–7.
8. Mansel JK, Rosenow EC, Smith TF, et al. Mycoplasma pneumoniae pneumonia. *Chest* 1989; 95:639–46.
9. Cassell GH, Cole BC. Mycoplasmas as agents of human disease. *N Engl J Med* 1981; 304:80–9.
10. Tay YK, Huff JC, Weston WL. Mycoplasma pneumoniae infection is associated with Stevens–Johnson syndrome, not erythema multiforme (von Hebra). *J Am Acad Dermatol* 1996; 35(5 Pt 1):757–60.
11. Daian CM, Wolff AH, Bielory L. The role of atypical organisms in asthma. *Allergy Asthma Proc* 2000; 21(2):107–11.
12. Abele-Horn M, Busch U, Nitschko H, et al. Molecular approaches to diagnosis of pulmonary diseases due to Mycoplasma pneumoniae. *J Clin Microbiol* 1998; 36(2):548–51.
13. Taylor-Robinson D, Bebear C. Antibiotic susceptibilities of mycoplasmas and treatment of mycoplasmal infections. *J Antimicrob Chemother* 1997; 40(5):622–30.
14. Klausner JD, Passaro D, Rosenberg J, et al. Enhanced control of an outbreak of Mycoplasma pneumoniae pneumonia with azithromycin prophylaxis. *J Infect Dis* 1998; 177(1):161–6.

NEISSERIA GONORRHEAE

Common name for disease: Gonorrhea, clap

Occupational setting

Gonorrhea is primarily a sexually transmitted disease. There are few occupations outside of prostitution where a true occupational risk has been demonstrated, although seafarers are known to be at increased risk of several types of sexually transmitted diseases.[1] A theoretical risk exists among dentists caring for patients with oral gonorrhea. Cutaneous infection has been reported in a laboratory worker who cut his finger on a test tube containing *Neisseria gonorrheae* prepared for lyophilization.[2]

Exposure (route)

Transmission is by direct contact of mucous membranes with the organism, usually during sexual intercourse. Oral or conjunctival splashing, as well as direct inoculation, can transmit the bacteria.

Pathobiology

N. gonorrheae is a non-motile, aerobic, Gram-negative diplococcus. The urethra, endocervix anal canal, pharynx, and conjunctiva are infected directly by contact with *N. gonorrheae*. The organism penetrates the mucosal epithelium and causes a local inflammatory response within 72 hours (although appearance of clinical symptoms may be delayed). Infection in men results in urethritis, with local extension in the urogenital tract if untreated. Infection in women results in vaginitis, cervicitis, and urethritis, can extend locally to the ducts of Skene, and Bartholin's glands, and if untreated can cause endometritis, salpingitis, and pelvic inflammatory disease. In both sexes, *N. gonorrheae* can directly infect the anorectal mucosa, pharynx, and conjunctiva, causing symptomatic disease. If untreated, infection can become systemic, resulting in bacteremia, arthritis, tenosynovitis, endocarditis, meningitis, and/or a disseminated rash.[3]

Diagnosis

Diagnosis is made by Gram stain and/or culture of the organism from infected material (discharge, synovial fluid, blood culture). There are now DNA probes and assays for gonococcal antigens available as well, although clinical judgment must be employed in their use.

Treatment

Because of widespread resistance, penicillin is no longer the recommended drug of choice for gonorrhea. The current CDC recommendations are for a single dose of ceftriaxone intramuscularly, or a single dose of oral cefixime, ciprofloxacin, or ofloxacin. This treatment should be followed by a 7-day course of doxycyline or other antibiotic to treat concomitant chlamydial infection.[4]

Medical surveillance

There are no recommended medical screening activities for gonorrhea. It is a reportable disease in the USA.

Prevention

Prevention of gonorrhea in the general population consists of treatment of infected individuals and their partners, and the use of condoms or avoidance of sexual contact with infected individuals.[4] Individuals who have direct contact with potentially infectious individuals or material, including healthcare and laboratory workers, should use appropriate personal protective equipment such as gloves.

REFERENCES

1. International Labor Organization/World Health Organization. Joint ILO/WHO committee on the hygiene of seafarers. *WHO Tech Rep Ser* 1961 No. 224 1–14.
2. Collins CH, Kennedy DA. Microbiological hazards of occupational needlestick and 'sharps' injuries. *J Appl Bacteriol* 1987; 62:385–402.
3. Cheng DSF. Gonorrhea. In: Parish LC, Gschnait F, eds. *Sexually transmitted diseases: a guide for clinicians*. New York: Springer-Verlag, 1988:59–77.
4. Centers for Disease Control and Prevention. 1993 Sexually transmitted diseases treatment guidelines. *MMWR* 1993; 42 (RR-14).

NEISSERIA MENINGITIDIS

Common names for diseases: Meningococcal meningitis, meningococcemia, cerebrospinal fever

Occupational setting

Serious disease due to *Neisseria meningitidis*, including meningitis and septicemia, usually occurs sporadically. Epidemics usually occur in institutional settings or such places as dormitories, schools, and military barracks. Occupational groups at risk during epidemics would include those who work at close quarters or with institutionalized individuals; military recruits, day care workers, prison personnel, employees of chronic care facilities, and dormitory supervisors. There have also been infections, some fatal, among laboratory personnel working with *N. meningitidis.*[1]

Exposure (route)

Spread is by direct contact with the nose, throat and upper respiratory secretions of persons infected with (or asymptomatically carrying) the bacteria or via inhalation of respiratory droplets from coughing or sneezing carriers. Transmission may be more efficient to persons already suffering with a viral upper respiratory infection.

Pathobiology

N. meningitidis is a Gram-negative diplococcus with a polysaccharide capsule. Capsular antigens form the basis for serogroup typing of the bacteria into different strains. Four serogroups are responsible for the majority of disease. Serogroups A and C cause epidemics, while serogroup B is the primary cause of sporadic cases. Serogroup Y causes sporadic pneumonia, sometimes associated with meningitis. When serogroup information was available on the 2,501 cases reported to the CDC in 1999, serogroups B, C, and Y each accounted for approximately one third of cases (Figure 22.8).

Asymptomatic carriage of meningococcus in the nose and throat is common, and may be chronic, intermittent or transient. Approximately 10% of the human population harbor meningococci in the nose or throat.[2] Nasopharyngeal carriage immunizes the host, with antibody production occurring within 2 weeks. Antibodies are serogroup-specific but confer some cross-immunity against other serogroups. There is also some cross-immunity between *Neisseria* and other species of bacteria, which probably promotes natural immunity against meningococcus in a population. Serogroup B is poorly immunogenic, possibly because antigens in the polysaccharide capsule resemble neonatal host antigens.

Cases of serious meningococcal disease occur more often in the winter and spring, and are more common in very young children. Males are affected more than females.

When transmission of meningococcus occurs, it most commonly results in asymptomatic nasopharyngeal carriage. When illness results from transmission, it usually takes the form of a mild pharyngitis. Pharyngitis may precede serious illness, but not reliably. Serious meningococcal infection can progress

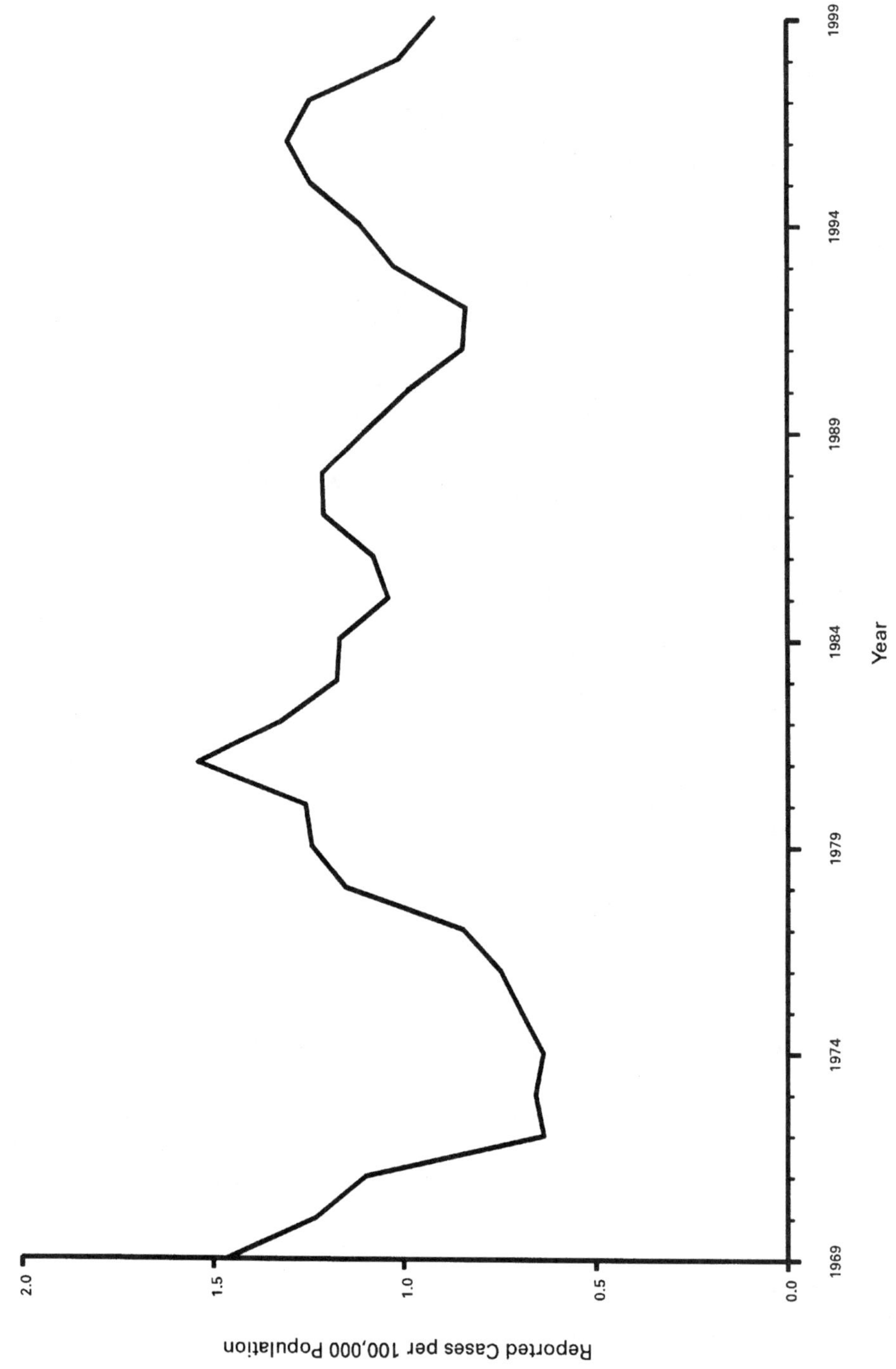

Figure 22.8 Meningococcal disease – reported cases per 100 000 population, by year, United States, 1969–1999. Source: MMWR 2001:48(53):52.

to death within a few h of symptom onset, making a high index of suspicion for the disease and a low threshold to treat essential.

The most common meningococcal syndromes are the following:

1. Bacteremia without sepsis. The patient presents with symptoms similar to those of a viral upper respiratory infection or rash, and recovers without specific therapy. Diagnosis is sometimes made when blood cultures are later found to be positive.
2. Meningococcemia without meningitis. The patient presents with signs and symptoms of bacteremia and sepsis, including fever, chills, malaise, and petechial rash. Leukocytosis and hypotension may be found on examination.
3. Meningitis with or without meningococcemia. The patient presents with headache, fever, and meningeal signs; a petechial rash may be present in meningococcemia. Mental status may range from normal to comatose. Reflexes are usually normal. Lumbar puncture will disclose cloudy cerebrospinal fluid.
4. Meningoencephalitis. The patient is obtunded, with meningeal signs, abnormal reflexes, and septic cerebrospinal fluid; pathologic reflexes may be present.

Petechial rash is common in meningococcal sepsis and meningitis. The rash is usually predominantly on the trunk and lower body, but petechiae are also often found on mucous membranes, including the palpebral conjunctiva. The lesions sometimes coalesce, especially at pressure points, to form ecchymotic-appearing lesions. A pink papular rash is an infrequent variant, and a vesicular rash is rare.

Endotoxemia may occur as a result of release of bacterial components of the Gram-negative cell wall. Endotoxemia may give rise to disseminated intravascular coagulation, septic shock with heart failure, myocarditis, pericarditis, peripheral hypoperfusion, adult respiratory distress syndrome, and adrenal hemorrhage.

Less common manifestations of meningococcal disease include chronic meningococcemia with low-grade fever, rash and arthritis, recurrent meningococcemia in persons with various complement deficiencies, meningococcal pneumonia and meningococcal urethritis.

Diagnosis

Treatment should not be withheld pending definitive diagnosis if meningococcal meningitis or meningiococcemia is suspected. Presumptive diagnosis is made by finding typical organisms (Gram-negative diplococci) on Gram-stained smear of cerebrospinal fluid or by recovering meningococcus from cerebrospinal fluid or blood culture. Occasionally, stained smears from petechiae reveal the organisms. Various antigen detection techniques can be used to identify group-specific polysaccharides. Although these techniques are rapid and specific, they are not sensitive, so they cannot be used to exclude the diagnosis of meningococcal infection. Other findings on cerebrospinal fluid examination vary, although neutrophils predominate in untreated cases. Cerebrospinal fluid chemistry usually reveals low glucose and high protein, findings seen in many infections and not specific to meningococcal meningitis.

Treatment

In adults, the treatment of choice remains intravenous penicillin G for 7–10 days. Alternatives in cases of penicillin allergy or rare penicillin resistance would include chloramphenicol or a third-generation cephalosporin. Some instances of relative penicillin resistance have been reported, and ceftriaxone has been suggested as a substitute, although its efficacy in severe cases of meningitis remains in doubt. Corticosteroids as an adjunct to antibiotic therapy have been subjected to clinical trials; their efficacy has not been demonstrated and they may impair antibiotic penetration into the cerebrospinal fluid. Patients need close monitoring early in the course of meningococcal disease for endotoxin-related complications, including disseminated intravascular coagulation (DIC) and septic shock. Because penicillin does not eradicate nasopharyngeal carriage in patients with meningococcal illness, patients should be given rifampin prior to discharge from the hospital.

Medical surveillance

Household and intimate contacts should be closely watched for signs of meningococcal infection, so that treatment can be started promptly. Surveillance of nasopharyngeal cultures is generally not indicated, since carriage is common, often transient, and not a consistent risk factor for infection. Cases must be reported immediately to local health authorities so that contact tracing can begin.

Prevention

Antibiotic prophylaxis is indicated for household and intimate contacts of cases. Household contact includes contacts in crowded quarters such as day care centers, dormitories, military barracks, prisons and chronic care facilities. Hospital personnel are not considered to be at increased risk unless there has been intimate contact such as mouth-to-mouth resuscitation.

Rifampin, given as a 600-mg dose twice daily for 2 days, is the drug of choice for chemoprophylaxis. If the isolate is appropriately sensitive to sulfadiazine, this may be used as an alternative. Other alternatives for chemoprophylaxis in adults are single-dose ceftriaxone or ciprofloxacin. Improved building ventilation may help prevent or control epidemics in close quarters.

Vaccination may be appropriate as an adjunct to antibiotic prophylaxis in an epidemic situation involving one of the serogroups for which vaccine is available, although the immunity which it confers may appear too late to be of use in an epidemic. Vaccines are currently available against single serogroups A and C, and there is also a bivalent A–C vaccine, and a quadrivalent vaccine against serogroups A, C, Y and W-135. Vaccination with the quadrivalent vaccine is recommended for patients with functional or anatomic asplenia, patients with known complement deficiencies, and for laboratory personnel working with high concentrations or large quantities of meningococcus.[3] Travelers to endemic areas, particularly the "meningitis belt" of sub-Saharan Africa extending from Mauritania to Ethiopia, should be vaccinated if their work or journey brings them into prolonged contact with the local populace. Although the CDC, as of this writing, no longer recommends meningococcal vaccine for travel to the Persian Gulf region, officials in Saudi Arabia may require that pilgrims for the Hadj produce a certificate of vaccination issued not more than 3 years and not less than 10 days before arrival in the country.

Work with high concentrations or large quantities of meningococcus should only be performed in a BSL 3 laboratory. Other laboratory personnel expected to be exposed at lower levels should be protected by use of standard microbiology laboratory procedures, including gloves and laboratory coat, and should work under a Class II biological safety cabinet when performing mechanical manipulations with the potential for aerosolization. Any incident or exposure involving meningococcus should receive prompt medical attention. In cases of percutaneous exposure, penicillin should be used for prophylaxis; for mucosal exposure, rifampin should be used.[1]

REFERENCES

1. Centers for Disease Control. Laboratory-acquired meningococcemia—California and Massachusetts. *MMWR* 1991; 40(3):46–47, 55.
2. van Deuren M, Brandtzaeg P, van der Meer JW. Update on meningococcal disease with emphasis on pathogenesis and clinical management. *Clin Microbiol Rev* 2000; 13:144–66.

3. Centers for Disease Control. Recommendation of the Immunization Practices Advisory Committee: Meningococcal vaccines. *MMWR* 1985; 34;255–9.

PASTEURELLA MULTOCIDA

Common names for disease: Pasteurellosis

Occupational setting

Increased risk of exposure is associated with work involving contact with animals. Occupations with exposure include animal laboratory personnel, veterinarians, pig, cattle and chicken breeders, zoo personnel, and abattoir workers.[1]

Exposure (route)

The bacteria are usually transmitted through animal bites or scratches, although exposure to animal secretions has been implicated in cases without a history of skin trauma. Pulmonary pasteurellosis is thought to occur via inhalation of aerosolized contaminated secretions; these may be from animals or via nosocomial spread from animals.[2,3]

Pathobiology

The organism is a small, non-motile, Gram-negative coccobacillus that frequently exhibits bipolar staining. It is present in the gastrointestinal or respiratory tracts of a number of animal species, including rodents, cats, dogs, and larger domestic and wild animals ranging from cattle to lions. Although it can cause significant disease in animals, it is commonly found in apparently healthy carriers. Approximately 70–90% of oral and nasal secretions in cats have been shown to carry the bacteria.[4] Virulence of the organism is thought to be related to a microbial capsule.

Pasteurella multocida infection results in three clinical presentations in humans, all of which are more common in older or immunocompromised individuals. The most common is local infection following a bite or scratch. Evidence of infection develops rapidly and generally can be noted within 24 hours. Local erythema, warmth and tenderness are accompanied by purulent drainage in about 40% of cases.[4] Lymphangitis and lymphadenopathy may be present. Bites or scratches from cats are the most common cause of these infections. Wound cellulitis can be complicated by osteomyelitis, from either local extension to bone or direct inoculation, and septic arthritis. The latter occurs proximal to the wound, often in a joint previously damaged by degenerative or rheumatoid arthritis.

Respiratory tract infections represent the second clinical syndrome caused by *P. multocida*. The spectrum of disease in respiratory *Pasteurella* infection ranges from sinusitis to tracheobronchitis to pneumonia with lung abscess or empyema. Pulmonary pasteurellosis is primarily a problem in individuals with underlying lung disease, including chronic obstructive pulmonary disease and malignancy. Although unusual, colonization of the respiratory tract has been reported in occupationally exposed individuals without apparent infection or underlying disease.[4]

Third, systemic infections may develop and, as with the other types of *P. multocida* illnesses, are primarily diagnosed in patients with underlying medical disease. *P. multocida* meningitis affects the very young and the elderly. Although endocarditis is rare, bacteremia can occur associated with localized sites of infection. Both respiratory and systemic infections can result without a clear history of animal-associated trauma or even documented animal exposure.

Diagnosis

Initial Gram stain of purulent material may suggest the diagnosis; culture of blood, wound, respiratory secretions or other body fluid is confirmatory. Serologic tests to detect antibody against *P. multocida* have been used in animals and as research tools in humans.

Treatment

The preferred treatment for *P. multocida* infection is penicillin; doxycycline is also effective. Other oral therapeutic agents that may be used include amoxicillin, amoxicillin–clavulanate, cefuroxime, tetracycline, and ciprofloxacin.[5] Penicillin G and its derivatives or second- and third-generation cephalosporins are effective intravenous therapy.

Medical surveillance

There are no recommended medical screening or surveillance activities.

Prevention

Those working with small animals should wear gloves, use animal handling techniques designed to avoid bites, and practice careful handwashing after contact. Exposed workers should be educated about the hazards of *Pasteurella* infection. Bites, if they occur, should be thoroughly cleansed, and attention sought if the worker becomes ill or if the injured site becomes infected. Suturing of bite wounds is controversial, and in general not recommended, especially if the hand is involved, as it may make early detection and management of infection difficult. Prophylactic antibiotics may be beneficial in high-risk wounds where infection is suspected, in hand wounds, in deep puncture wounds and in immunocompromised patients. Recent microbiological analysis of animal bite wounds found *P. multocida* to be the most common constituent isolated from cat bites.[6] Prophylaxis, if initiated, should usually be with a β-lactam antibiotic plus a β-lactamase inhibitor, such as amoxicillin–clavulinate.

REFERENCES

1. Choudat D, Le Goff C, Delemotte B, et al. Occupational exposure to animals and antibodies against *Pasteurella multocida*. *Br J Ind Med* 1987; 44: 829–33.
2. Beyt BE, Sondag J, Roosevelt TS, et al. Human pulmonary pasteurellosis. *JAMA* 1979; 242:1647–48.
3. Itoh M, Tierno PM, Milstoc M, Berger AR. A unique outbreak of Pasteurella multocida in a chronic disease hospital. *Am J Public Health* 1980; 70:1170–3.
4. Weber DJ, Wolfson JS, Swartz MN, et al. *Pasteurella multocida* infections: report of 34 cases and review of the literature. *Medicine* 1984; 63:133–54.
5. Weber DJ, Hansen AR. Infections resulting from animal bites. *Infect Dis Clin North Am* 1991; 5:663–80.
6. Talan DA, Citron DM, Abrahamian FM, et al. Bacteriologic analysis of infected dog and cat bites. *N Engl J Med* 1999; 340:85–92.

PSEUDOMONAS AND BURKHOLDERIA SPECIES

Common name for disease: Glanders (*B. mallei*), melioidosis, Whitmore's disease, pseudoglanders (*B. pseudomallei*)

Occupational setting

Pseudomonas and *Burkholderia* species are free-living, ubiquitous bacteria that are a particularly common contaminant of moist environments and microenvironments. While some species are very important causes of nosocomial infections in immunocompromised patients, these organisms are opportunistic pathogens that rarely cause infection in healthy persons. Therefore, occupationally acquired infection is uncommon.

Nevertheless, *Pseudomonas* can be a concern either through direct cutaneous infection or indirectly through reactions from exposure to pseudomonas-contaminated media. *P. aeruginosa* has been associated with skin infections in commercial divers[1] and nosocomial keratitis in a nurse.[2] *P. fluorescens* has been implicated in an outbreak of hypersensitivity pneumonitis from exposure to contaminated metalworking fluids termed "machine operator's lung".[3]

Although *B. mallei* and *B. pseudomallei* are no longer found in North America and Europe, they may be transmitted through laboratory work.[4,5] A US Army microbiologist has been reported to have acquired *B. pseudomallei* infection from working in a biological weapons defense facility.[6] *B. pseudomallei* is well known to occur in immigrants and travelers from endemic areas, including Vietnam veterans, and this infection may be a hazard to workers returning from these regions.[7] Unlike many other species in these two genera, *B. pseudomallei* has been associated with infection in otherwise healthy groups, such as military personnel.[10]

Exposure (route)

The route of exposure depends on the occupational setting. Transmission through direct contact, often because of improper handwashing, is a well-known means of propagation for nosocomial pathogens in this group.[9] Inhalational exposure applies to potential respiratory effects from media contaminated with *pseudomonas*.

Direct contact of soil onto broken skin is the principle route of exposure for *B. mallei* and *B. pseudomallei*. *B. pseudomallei* may additionally be acquired by inhalation and a recent outbreak in Australia has been attributed to contaminated water.[10]

Pathobiology

Pseudomonas and *Burkholderia* are Gram-negative, aerobic, slightly curved or straight rods. Some species previously grouped under the genus *Pseudomonas* have recently been transferred to the genus *Burkholderia*. These include *B. cepacia*, *B. mallei*, and *B. pseudomallei*.

Community-acquired *P. aeruginosa* infection is usually localized and occurs as the result of exposure to large numbers of organisms from a contaminated water source. Examples include folliculitis from swimming pools or hot tubs, otitis externa ("Swimmer's ear"), and eye infections associated with contact lens use. As noted, *P. aeruginosa* is a common and dreaded nosocomial pathogen infecting multiple sites in immunocompromised hosts.

Exposure to aerosols in contaminated environments has been associated with occupational asthma in several settings. These microenvironments usually have a mixed flora, including Pseudomonas species, and exposure to endotoxin, produced by Gram-negative bacteria, has been implicated as the agent of bronchospasm, as well as symptoms of fever, diarrhea, fatigue, headache, nausea, and eye and nasal irritation.

B. mallei produces a disease known as glanders. This is primarily an infection of horses, mules, and donkeys. Sporadic cases occur in humans in parts of South America, Africa, and Asia. Three acute clinical pictures are possible: a rapidly fatal sepsis, a pulmonary form, or an ulcerative infection of the mucosa of the nose, mouth, and conjunctiva. A chronic cutaneous form is also recognized.

B. pseudomallei causes melioidosis or pseudoglanders, a disease endemic to South East Asia and northern Australia. The presentation of melioidosis is protean with systemic and localized forms involving virtually any organ system. The diagnosis is made even more challenging by a variable time course for disease progression. There are acute, subacute and chronic forms as well as the potential for a latency period lasting several years prior to disease manifestation.

Diagnosis

Standard isolation and identification techniques can be used to diagnose infections with *Pseudomonas* and *Burkholderia* species.

Treatment

Because of the low virulence in healthy hosts, localized *pseudomonal* infections are often self-limited and require no specific management. The source of exposure obviously must be identified and disinfected. Therapy may consist of topical antibiotics with systemic antibiotics reserved for use in refractory cases or cases with complicated courses. Because of high rates of drug resistance, nosocomial infections require multiple broad-spectrum antibiotics.

For infection by *B. mallei* and *B. pseudomallei*, the choice and duration of antibiotic treatments remain to be determined. Limited experience indicates that multiple antibiotics (a quadruple regimen) may be required.

Medical surveillance

There are no recommended medical screening or surveillance activities.

Prevention

Engineering controls and work practices should be aimed at reducing microbial contamination of water and other media. Outdoor work practices in endemic regions of *B. pseudomallei* should avoid soil contact with non-intact skin.

REFERENCES

1. Ahlen C, Mandal LH, Johannessen LN, Iversen OJ. Survival of infectious Pseudomonas aeruginosa genotypes in occupational saturation diving environments and the significance of these genotypes for recurrent skin infections. *Am J Ind Med* 2000; 37(5):493-500.
2. Bowden JJ, Sutphin JE. Nosocomial Pseudomonas keratitis in a critical-care nurse. *Am J Ophthalmol* 1986; 101:612–13.
3. Bernstein DI, Lummus ZL, Santilli G, Siskosky J, Bernstein IL. Machine operator's lung. A hypersensitivity pneumonitis disorder associated with exposure to metalworking fluid aerosols. *Chest* 1995; 108(3):636–41.
4. Howe C, Miller WR. Human glanders: report of six cases. *Ann Intern Med* 1947; 26:93.
5. Schlech WF 3rd, Turchik JB, Westlake RE Jr, Klein GC, Band JD, Weaver RE. Laboratory-acquired infection with Pseudomonas pseudomallei (melioidosis). *N Engl J Med* 1981; 305(19):1133–5.
6. Laboratory-acquired human glanders – Maryland. *MMWR* 2000; 49(24):532–5.
7. Koponen MA, Zlock D, Palmer DL, Merlin TL. Melioidosis. Forgotten, but not gone! *Arch Intern Med* 1991; 151(3):605–8.
8. Lim MK, Tan EH, Soh CS, Chang TL. Burkholderia pseudomallei infection in the Singapore Armed Forces from 1987 to 1994—an epidemiological review. *Ann Acad Med Singapore* 1997; 26(1):13–17.
9. Doring G, Jansen S, Noll H, et al. Distribution and transmission of Pseudomonas aeruginosa and Burkholderia cepacia in a hospital ward. *Pediatr Pulmonol* 1996; 21(2):90–100.
10. Inglis TJ, Garrow SC, Henderson M, et al. Burkholderia pseudomallei traced to water treatment plant in Australia. *Emerg Infect Dis* 2000; 6(1):56–9.

RAT-BITE FEVER: *STREPTOBACILLUS MONILIFORMIS* AND *SPIRILLUM MINOR*

Common names for diseases: Streptobacillary fever, Haverhill fever, sodoku

Occupational setting

Rat-bite fever results from infection with *Streptobacillus moniliformis*, most frequently transmitted by the bite of a rat. Although it is rare in North America, cases continue to be reported, mostly in children.[1]

Work involving exposure to rodents, particularly rats and mice, or small animals that prey on them, such as cats and dogs, confers increased risk. Animal laboratory personnel,[2] veterinarians, animal breeders[3] and agricultural workers[4] are included in this group. Work in heavily rat-infested areas has also been implicated, either through non-bite trauma or without recognized trauma.

Exposure (route)

Bites from wild or laboratory rats,[2] whose oral cavities and upper respiratory tracts are commonly colonized with these organisms, can lead to infection, as can bites or scratches of other animal carriers. Haverhill fever occurs from ingesting food products presumably contaminated with rat excreta containing the organism. Recent outbreaks have implicated contaminated milk.[5]

Pathobiology

S. moniliformis is a pleomorphic, Gram-negative bacillus that may exhibit branching filaments. Rats are the most common reservoir, shedding the bacteria in saliva and urine.

Rat-bite fever follows an incubation period which can from 1 to 22 days but is usually less than 10 days.[1] The illness consists of a prodrome of relapsing fever, chills, headache, vomiting, myalgias, and a polyarthralgia.[1] The initial wound often appears to be healed at this point. Within 2–4 days, a rash, usually maculopapular in nature, develops over the extremities. This is followed by polyarthritis. The most frequently described complication in case reports is septic arthritis.[6] There are also reports, usually in children, of more serious complications, including septicemia,[7] endocarditis,[8] localized abscess,[9] and death.[10]

Spirillum minus is a Gram-negative spirochete responsible for rat-bite fever (sodoku) primarily in Asia, although cases are reported in the Americas.[11] It has a longer incubation period of 1–3 weeks. The symptoms are similar to those caused by *S. moniliformis*, except that ulceration at the site of the bite with associated lymphadenopathy and lymphangitis is common, while arthritis is unusual.

Diagnosis

The differential diagnosis for a patient presenting with the signs and symptoms of rat-bite fever is broad. Eliciting a history of bite or rodent exposure is essential in narrowing the possibilities but is not always present.[12] Atypical presentations occur, which may delay the diagnosis.[13]

S. moniliformis is confirmed by culture of biological specimens. The organism has fastidious growth requirements, making culture and isolation difficult. Specific agglutinins appear ~10 days after the onset of illness; a fourfold rise in titer during the following weeks or an initial titer of 1:80 is diagnostic.[2] *Spirillum minus* cannot be cultured in vitro and requires inoculation of body fluids intraperitoneally into laboratory animals with subsequent identification of the organism in peritoneal fluid by dark-field microscopy. No serologic test is available for *S. minus*.

Treatment

Penicillin is the antibiotic of choice.[4] The duration and route of therapy are dependent upon the extent of complications and response. Erythromycin does not appear to be appropriate treatment.[4]

Medical surveillance

There are no recommended medical screening or surveillance activities.

Prevention

Control of rat populations at work sites is important. Persons working with small animals should wear gloves and use handling techniques designed to avoid bites. Routine handwashing after contact is essential.

Laboratory rats should be separated from other rodents to reduce the risk of transmission between animals. Bites, if they occur, should be thoroughly cleaned. The utility of prophylactic antibiotics after a bite has not been investigated. Studies of rat bites show that only a small minority become infected,[13] but some authors recommend a short course of penicillin to reduce the potential morbidity of infection.[14]

REFERENCES

1. Centers for Disease Control. Rat-bite fever — New Mexico, 1996. *MMWR* 1998; 47(5):89.
2. Anderson LC, Leary SL, Manning PJ. Rat-bite fever in animal research laboratory personnel. *Lab Anim Sci* 1983; 33(3):292–4.
3. Wilkins EG, Millar JG, Cockcroft PM, Okubadejo OA. Rat-bite fever in a gerbil breeder. *J Infect* 1988; 16(2):177–80.
4. Hagelskjaer L, Sorensen I, Randers E. Streptobacillus moniliformis infection: 2 cases and a literature review. *Scand J Infect Dis* 1998; 30(3):309–11.
5. McEvoy MB, Noah ND, Pilsworth R. Outbreak of fever caused by Streptobacillus moniliformis. *Lancet* 1987; 2:1361–3.
6. Rumley RL, Patrone NA, White L. Rat-bite fever as a cause of septic arthritis: a diagnostic dilemma. *Ann Rheum Dis* 1986; 46(10):793–5.
7. Rygg M, Bruun CF. Rat bite fever (Streptobacillus moniliformis) with septicemia in a child. *Scand J Infect Dis* 1992; 24(4):535–40.
8. McCormack RC, Kaye D, Hook EW. Endocarditis due to streptobacillus moniliformis. *JAMA* 1967; 200(1):77–9.
9. Vasseur E, Joly P, Nouvellon M, Laplagne A, Lauret P. Cutaneous abscess: a rare complication of Streptobacillus moniliformis infection. *Br J Dermatol* 1993; 129(1):95–6.
10. Sens MA, Brown EW, Wilson LR, Crocker TP. Fatal Streptobacillus moniliformis infection in a two-month-old infant. *Am J Clin Pathol* 1989; 91(5):612–6.
11. Hinrichsen SL, Ferraz S, Romeiro M, et al. Sodoku—a case report. *Rev Soc Bras Med Trop* 1992; 25(2):135–8.
12. Fordham JN, McKay-Ferguson E, Davies A, Blyth T. Rat bite fever without the bite. *Ann Rheum Dis*; 51(3):411–72.
13. Ordog GJ, Balasubramanium S, Wasserberger J. Rat bites: fifty cases. *Ann Emerg Med* 1985; 14:126–30.
14. Weber DJ, Hansen AR. Infections resulting from animal bites. *Infect Dis Clin North Am* 1991; 5:663–80.

RELAPSING FEVER: *BORRELIA SPECIES (other than B. burgdorferi)*

Common names for disease: Relapsing fever, endemic (or sporadic) and epidemic forms

Occupational setting

Relapsing fever, with few exceptions, occurs in countries throughout the world. Louse-borne, or epidemic, disease is transmitted from person to person by the human body louse (*Pediculus humanus*). Epidemics are traditionally associated with war, famine, and other catastrophic events; migrant workers and soldiers have been particularly prone to this infection. At present, the epidemic disease is found only in Ethiopia and neighboring countries, and occurred in association with the wars there in the 1980s and 1990s.[1] Tick-borne, or endemic, disease occurs worldwide, with the largest outbreak in the western hemisphere occurring in 62 campers in Arizona in 1973.[2] As in many outbreaks, the disease appeared to be acquired through vacationing in cabins were rodents have nested. Many species of rodents and small mammals, including chipmunks, squirrels, rabbits, mice, and rats, serve as reservoirs of the infection. The vectors of the disease, ticks of the genus *Ornithodoros*, are found preferentially in forested mountain habitats, frequently above 3000 feet, especially caves, decaying wood, rodent burrows, and animal shelters.[3] Outdoor workers in selected environments, particularly those in remote natural settings, would seem to be at risk for the disease.

Exposure (route)

The disease is vector-borne, transmitted to humans by the human body louse and ticks of the genus *Ornithodoros*.

Pathobiology

The relapsing fevers (epidemic and endemic) are caused by several species of *Borrelia* spirochetes. *B. recurrentis* is the sole cause of epidemic relapsing fever and is transmitted by the human body louse. At least 15 other species of *Borrelia* are transmitted by ticks and cause the endemic, or sporadic, variety of the disease; examples include *B. turicatae*, *B. hermsii*, *B. parkeri*, and *B. duttonii*. The clinical manifestations of louse-borne and tick-borne disease are similar, characterized by acute onset of high fever with rigors, severe headache, myalgias, arthralgias, lethargy, photophobia, and cough, after an incubation period of about 7 days.[3] Fever is intermittent, with the initial episode typically lasting 3–6 days, followed by an asymptomatic period of 7–10 days. The patient is often unaware of a tick bite. In the absence of treatment, three to five relapses occur, with the duration and intensity of symptoms decreasing with each relapse of tick-borne disease. A single relapse is characteristic of louse-borne disease.

Diagnosis

Definitive diagnosis requires identification of spirochetes in peripheral blood smears by dark-field microscopy or appropriately stained specimens. Serologic tests are not standardized and are of limited diagnostic value other than to demonstrate rising titers in convalescent sera.

Treatment

As is the case in other *Borrelia* infections, relapsing fever is best treated successfully with tetracyclines, based on clinical experience. Controlled trials of therapy are lacking, but the disease will also respond to erythromycin and chloramphenicol. Penicillin has been associated with an increased rate of relapse. Treatment of tick-borne disease requires 7–10 days. Antibiotic treatment may produce a Jarisch–Herxheimer reaction characterized by fever, chills, tachycardia and hypotension. This serious complication most likely represents an inflammatory reaction induced during treatment, believed to be due to a rapid release of antigen with associated cytokine or other mediator response. Death is rare in tick-borne relapsing fever, and is limited to infants and older individuals. The case fatality ratio of untreated epidemic disease can approach 40%.

Medical surveillance

There are no recommended medical screening or surveillance activities.

Prevention

Prevention of the epidemic form of the disease requires appropriate response to natural and artificial disasters, good personal hygiene, and delousing procedures. In epidemic situations, short-term use of prophylactic antibiotics can contain the spread of infection to persons at high risk.[2] Endemic disease can be prevented by activities that limit tick exposure, including vector control, rodent control, and personal protective measures. Environmental application of insecticides and use of tick repellents can also decrease the potential for tick exposure.

REFERENCES

1. Raoult D, Roux V. The body louse as a vector of reemerging human diseases. *Clin Infect Dis* 1999; 29:888–911.
2. Centers for Disease Control. Relapsing fever. *MMWR* 1973; 22:242–6.
3. Spach DH, Liles WC, Campbell GL, et al. Tick-borne diseases in the United States. *N Engl J Med* 1993; 329:936–45.

SALMONELLA SPECIES

Common names for disease: Salmonellosis, typhoid fever, paratyphoid fever

Occupational setting

Non-typhoidal *Salmonella* species can cause infections in most animal species, including poultry, cattle, swine, cats, dogs, and turtles, and individuals in occupations with animal contact are therefore at increased risk. Furthermore, large community outbreaks have occurred in which dairy products or meat from farms have been traced as the primary source (Figure 22.9).[1] Such incidents are of particular concern because of the frequency of multiply resistant pathogens due to heavy antibiotic use in animal feeds.[1]

Food handlers are at risk through contact with contaminated animal products. Organisms causing both non-typhoidal and typhoidal illnesses represent an occupational hazard to healthcare workers. Employees in patient care and laboratory settings experience increased exposure from infected patients, resulting in documented clinical infections of both typhoid fever and enterocolitis.[2,3]

Contact with *S. typhi* in proficiency tests of clinical laboratories and training exercises for medical laboratory technician students has also resulted in typhoid fever.[4] International travelers may be exposed to *Salmonella* species due to poor sanitation.

Exposure (route)

Ingestion of bacteria is the usual route of exposure. This occurs through contamination of food or water with infected fecal material or inadequate handwashing by personnel after contact with infected humans or animals. Transmission has also been described through fomites in a report of laundry workers infected by contact with soiled linen in a nursing home.[4]

Pathobiology

Salmonellae are Gram-negative, flagellated rods. The nomenclature for this genus has changed as a result of new information gained from DNA studies. Previously distinct species are now all classified under the species *S. choleraesuis*. There are seven subgroups in this species, with subgroup I containing nearly all human pathogens. The serotypes are generally referred to in shortened form as if they were species instead of the longer, but strictly correct, designation of genus–species–serotype (i.e. *S. typhi* versus *S. choleraesuis* (group I) serotype *typhi*).

A variety of illnesses are caused by *Salmonella* species; however, they can be divided into two general groups: typhoidal and non-typhoidal. Typhoid fever is caused by *S. typhi*; the less severe illness, paratyphoid fever, is due to *S. paratyphi* A, *S. paratyphi* B (*S. schottmuelleri*), and *S. paratyphi* B (*S. hirschfeldii*). Other *Salmonella* species are occasionally responsible for this clinical presentation. Humans are the only reservoir for *S. typhi* and *S. paratyphi*.

Once ingested, *S. typhi* penetrates the intestinal wall, causing necrosis and ulceration, and eventually gains access to the bloodstream. After an incubation period of 1–3 weeks, symptoms become manifest with fever, headache, abdominal pain, and constipation or diarrhea. Respiratory symptoms may also be present. Physical examination may reveal intestinal ileus with abdominal tenderness and palpable bowel loops. Rose spots, caused by leakage from capillary endothelial cells due to bacterial infiltration, may be noted on the anterior chest and abdomen. Hepatosplenomegaly is common, and some patients may have decreased levels of consciousness. Laboratory abnormalities include anemia, leukopenia, liver function test alterations, and subclinical clotting abnormalities.

The illness is prolonged in the absence of antibiotic treatment and may be fatal. Regardless of treatment, an extended convalescence is frequently necessary. Complications are

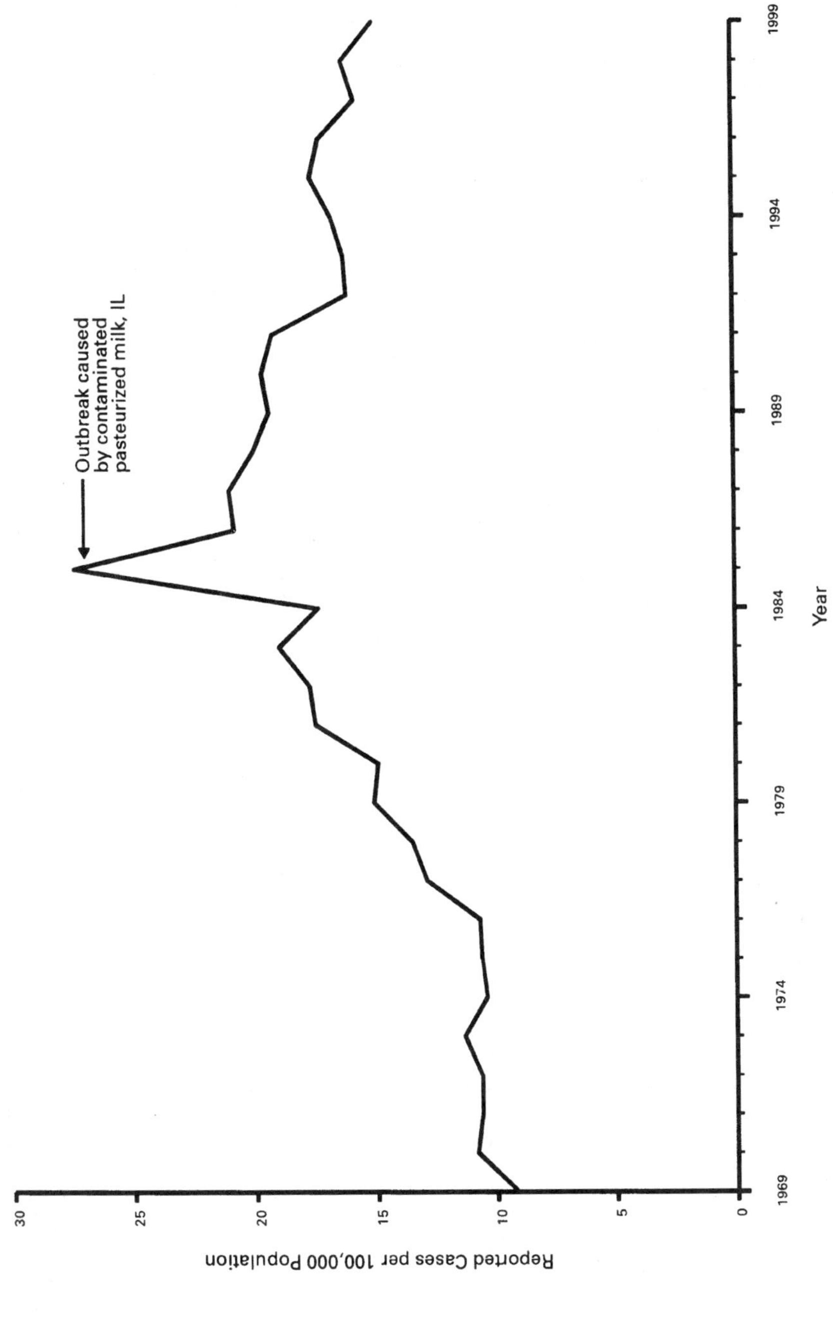

Figure 22.9 Salmonellosis – reported cases per 100 000 population by year, United States, 1969–1999. Source: MMWR 2001:48(53):62.

numerous, including gastrointestinal perforation, hemorrhage, and localized infections, such as pneumonia and meningitis. The chronic carrier state (excretion of bacteria in feces for >1 year) develops in 1–3% of those infected and may not be preceded by a serious initial illness.[5] Although typhoid fever has become quite rare in the USA, (Figure 22.10), it should be suspected in recent travelers to undeveloped countries or in exposed healthcare workers.

Non-typhoidal illness, in contrast, is an increasing problem in the USA. From a public health perspective, it is one of the most important zoonotic infections. Most infections are due to contaminated food, especially poultry and their products. The most common presentation is enterocolitis, and *S. typhimurium* and *S. enteritidis* are the most frequently responsible serotypes. The incubation period is 6–48 h after bacterial ingestion. The organisms reach the lower intestinal tract and multiply there. After mucosal invasion, the bacteria may be ingested by macrophages and multiply within these cells. Other factors contributing to bacterial virulence include: elicitation of a secretory response, which may be mediated by enterotoxin; tissue destruction, which may be cytotoxin-induced; and antimicrobial resistance, which is increasingly plasmid-related. Host defenses include: gastric acidity, which kills bacteria; presence of normal intestinal flora, which decreases bacterial multiplication; and cellular immunocompetence.

Symptoms are fever, nausea, vomiting, and headache, followed by abdominal pain and diarrhea. The diarrhea is generally of moderate volume and non-bloody; it lasts 3–7 days, although bacteria are shed in the stool for an average of 5 weeks. The abdominal pain may localize to the right lower quadrant, mimicking appendicitis. Bacteremia may accompany the illness. Approximately 1% of infected persons continue to shed bacteria in the stool for over 1 year. Bacteremia without enterocolitis or typhoid fever may occur. Localized abscesses, meningitis, pneumonia, endocarditis, arteritis, and osteomyelitis can result.

Diagnosis

Diagnosis is made by cultures of blood or stool or, in the case of localized infections, specimens from the affected area. Serologic studies are neither sensitive nor specific, although the enzyme-linked immunosorbent assay shows promise. Plasmid and phage typing of the causative organism are useful to determine the source of infection in non-typhoidal illnesses.

Treatment

Multidrug-resistant *S. typhi* is now widespread. A recent review has concluded that ciprofloxacin and ceftriaxone remain appropriate empirical therapy for suspected cases of typhoid fever, although possible resistance to these agents should be kept in mind.[6] Antibiotic therapy for most patients with enterocolitis is not necessary and, in fact, may prolong fecal excretion of bacteria and increase adverse effects.[7] However, antibiotic therapy should be considered in patients with underlying medical conditions and those with bacteremia or localized infections.

Medical surveillance

Immediate case reporting to the proper health authorities is required in the USA and most other countries.

Prevention

Work practices should emphasize good hygiene with careful handwashing after contact with infected patients, animals, or their feces. Equipment used in patient procedures is potentially infectious as well. Food handlers should wear gloves. Laboratory personnel should utilize good work practices, including, in addition to handwashing, the use of gloves, laboratory coats, and mechanical

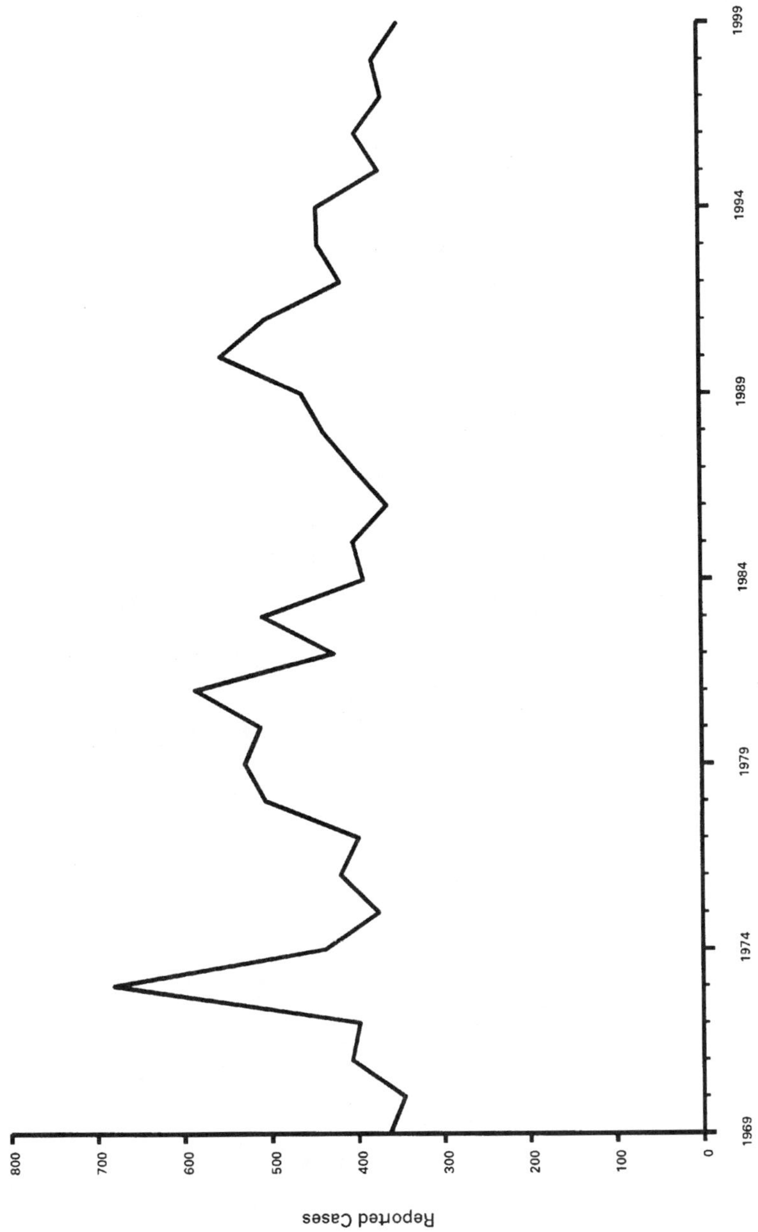

Figure 22.10 Typhoid fever – reported cases by year, United States, 1969–1999. Source: MMWR 2001:48(53):79.

pipetting devices. Eating and drinking in the workplace should be prohibited. Travelers should avoid fresh, peeled fruits and vegetables, and drink only bottled water. Individuals in sensitive jobs who develop infection, such as those in patient care and food handlers, should not return to work until their stool cultures are negative.

Worker education is essential in the management of this occupational hazard. Typhoid vaccines are available for travelers to high-risk areas, individuals persistently exposed to chronic carriers, and laboratory workers having frequent contact with *S. typhi*. New vaccines with fewer side-effects have been developed, but do not appear to provide the same level of long-term protection as the older whole-cell vaccine.[8]

The control of foodborne *Salmonella* requires a number of interventions, including the control of bacteria in animal feed, a decrease in the use of antibiotic supplements in animal feed, and proper food-preparation practices. Careful evisceration practices and physical separation of this area from the rest of the slaughterhouse is beneficial.

REFERENCES

1. Molbak K, Baggesen DL, Aarestrup FM, et al. An outbreak of multidrug-resistant, quinolone-resistant Salmonella enterica serotype typhimurium DT104. *N Engl J Med* 1999; 341(19):1420–5.
2. Grist NR, Emslie JAN. Infections in British clinical laboratories, 1984–5. *J Clin Pathol* 1987; 40:826–9.
3. Pike RM. Laboratory-associated infections: incidence, fatalities, causes, and prevention. *Annu Rev Microbiol* 1979; 33:41–66.
4. Hoerl D, Rostkowski C, Ross SL, et al. Typhoid fever acquired in a medical technology teaching laboratory. *Lab Med* 1988; 19:166–8.
5. Standaert SM, Hutcheson RH, Schaffner W. Nosocomial transmission of Salmonella gastroenteritis to laundry workers in a nursing home. *Infect Control Hosp Epidemiol* 1994; 15(1): 22–6.
6. Ackers ML, Puhr ND, Tauxe RV, Mintz ED. Laboratory-based surveillance of Salmonella serotype Typhi infections in the United States: antimicrobial resistance on the rise. *JAMA* 2000; 283(20):2668–73.
7. Sirinavin S, Garner P. Antibiotics for treating salmonella gut infections. *Cochrane Database Syst Rev* 2000; 2:CD001167.
8. Engels EA, Lau J. Vaccines for preventing typhoid fever. *Cochrane Database Syst Rev* 2000; 2:CD001261.

SHIGELLA SPECIES

Common names for disease: Shigellosis, dysentery, bacillary dysentery

Occupational setting

Shigella outbreaks have been described in laboratories,[1] child day care centers,[2] cruise ships,[3] primate animal handlers,[4] and hospitals.[5] Military personnel[6] and travelers from endemic areas[7] may also be at increased risk. The overall incidence of shigellosis has declined in the USA over the past five years, although significant outbreaks, continue to occur (Figure 22.11).

Exposure (route)

Transmission is through the fecal–oral route. This occurs most often through direct person-to-person contact. Less commonly, the organism is water- or foodborne.

Pathobiology

Shigellae are non-motile, non-encapsulated Gram-negative rods of the Enterobacteriaceae family. There are four species that can produce diarrhea. *S. sonnei* accounts for the overwhelming majority of cases in the USA.[8] *S. dysenteriae* produces a toxin (the shiga toxin)

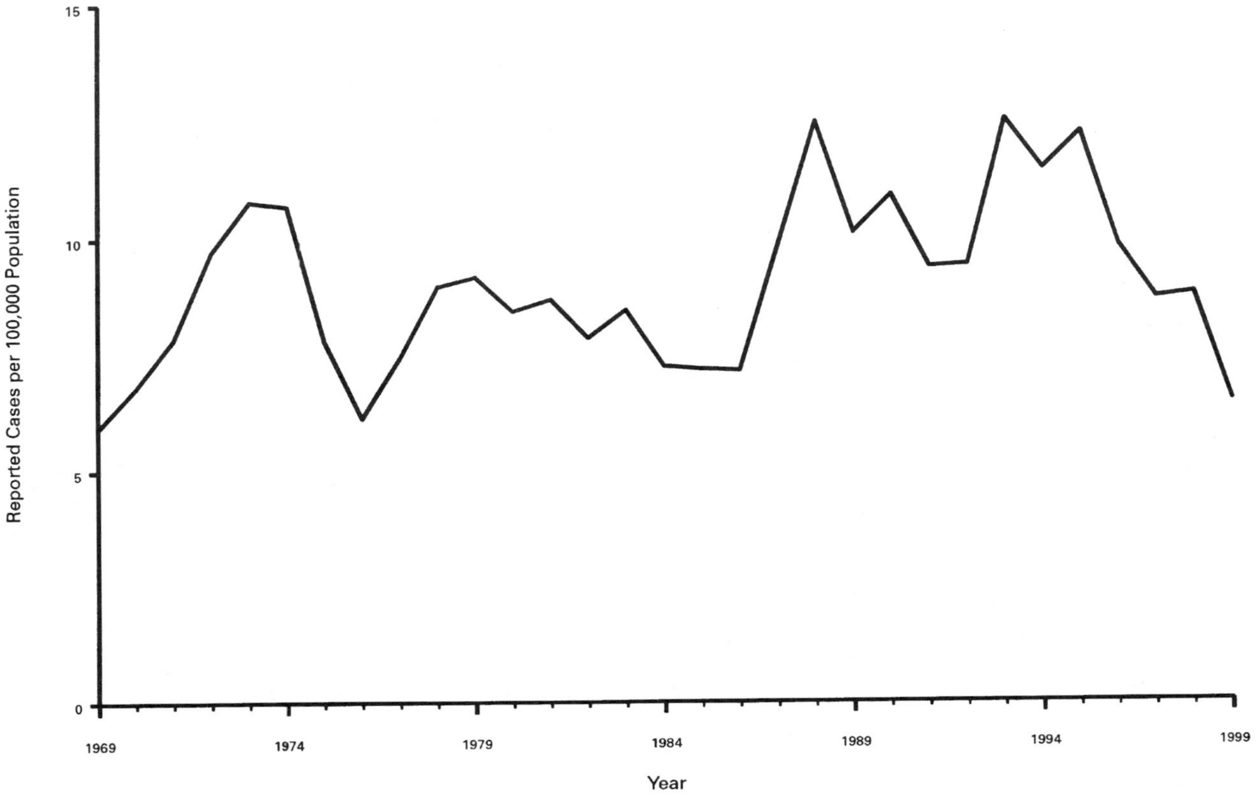

Figure 22.11 Shigellosis – reported per 100 000 population by year, United States, 1969–1999. Source: MMWR 2001:48(53):64.

and generally results in more severe disease. The shiga toxin is also produced by some serotypes of *E. coli,* such *E. coli* O157, and is associated with hemolytic–uremic syndrome in infections from either species of bacteria.[9] S. *flexneri* and *S. boydii* may also cause shigellosis.

Symptoms develop 24–48 hours after infection and include fever, abdominal cramping, tenesmus, and watery diarrhea that may contain blood, pus, and mucus. Mild cases are self-limited and resolve within a few days. Complications are rare and generally occur in children or immunocompromised adults. They include intestinal perforation,[10] sepsis,[11] and hemolytic–uremic syndrome.[9] Delayed sequelae include reactive arthritis and Reiter's syndrome.[12]

Shigella demonstrates a high degree of pathogenicity; disease may result from an inoculation with as few as 10–100 bacteria. After ingestion and passage through the stomach, the organism invades the mucosa of the colon. Infection is thought to involve primarily local cell invasion and destruction, although the bacilli also produce several toxins. The superficial mucosal layer of the colon then ulcerates and sloughs. Toxins may also contribute to a moderate secretory diarrhea.

Diagnosis

Since bleeding may not be clinically apparent, *Shigella* should be included within the differential diagnosis of any compatible acute diarrhea. Fecal leukocytes are a frequent finding. A definitive diagnosis is made by culturing the organism from stool samples.

Treatment

Treatment is primarily supportive and consists of oral or, if necessary, intravenous rehydration. Antibiotic therapy can shorten the course of disease and prevent transmission of infection. Ampicillin or trimethoprim–sulfamethoxazole have been traditional empirical choices. However, since antibiotic resistance to both of these agents is now widespread, either a quinolone or cefixime should be selected when antibiotic therapy is indicated.[13]

Medical surveillance

There are no recommended medical screening activities for infections due to *Shigella.* Reporting of cases to local public health authorities is required in most jurisdictions.

Prevention

Work practices common to prevent transmission of any enteric pathogen should be in place. These measures include handwashing, glove use, and appropriate preparation and storage of food. Chlorination of water supplies and provision of sanitary facilities are also important.

Persons with active infections can easily transmit the disease, so adequate precautions are essential. These include enteric (or universal) precautions for healthcare workers caring for the infected patient and verification of noninfectious status (with two successive cultures) before allowing the individual to prepare food or resume child or patient care.

Current research offers the hope of a vaccine for *Shigella* in the future.[14]

REFERENCES

1. Mermel LA, Josephson SL, Dempsey J, Parenteau S, Perry C, Magill N. Outbreak of Shigella sonnei in a clinical microbiology laboratory. *J Clin Microbiol* 1997; 35(12):3163–5.
2. Mohle-Boetani JC, Stapleton M, Finger R, et al. Community-wide shigellosis: control of an outbreak and risk factors in child day-care centers. *Am J Public Health* 1995; 85(6):812–16.
3. Centers for Disease Control. Outbreak of Shigella flexneri 2a infections on a cruise ship. *MMWR* 1994; 43(35):657.
4. Kennedy FM, Astbury J, Needham JR, Cheasty T. Shigellosis due to occupational contact with non-human primates. *Epidemiol Infect* 1993; 110(2):247–51.

5. Hunter PR, Hutchings PG. Outbreak of Shigella sonnei dysentery on a long stay psychogeriatric ward. *J Hosp Infect* 1987; 10(1):73–6.
6. Mikhail MM, Mansour MM, Oyofo BA, Malone JD. Immune response to Shigella sonnei in US Marines. *Infect Immun* 1996; 64(9):3942–5.
7. Aleksic S, Bockemuhl J, Degner I. Imported shigellosis: aerogenic Shigella boydii 74 (Sachs A 12) in a traveller followed by two cases of laboratory-associated infections. *Tropenmed Parasitol* 1981; 32(1):61–4.
8. Centers for Disease Control. Outbreaks of Shigella sonnei infection associated with eating fresh parsley—United States and Canada, July–August 1998. *MMWR* 1999; 48(14):285–9.
9. Bhimma R, Rollins NC, Coovadia HM, Adhikari M. Post-dysenteric hemolytic uremic syndrome in children during an epidemic of Shigella dysentery in Kwazulu/Natal. *Pediatr Nephrol* 1997; 11(5):560–4.
10. Upadhyay AK, Neely JA. Toxic megacolon and perforation caused by Shigella. *Br J Surg* 1989; 76(11):1217.
11. Trevett AJ, Ogunbanjo BO, Naraqi S, Igo JD. Shigella bacteraemia in adults. *Postgrad Med J* 1993; 69(812):466–8.
12. Finch M, Rodey G, Lawrence D, Blake P. Epidemic Reiter's syndrome following an outbreak of shigellosis. *Eur J Epidemiol* 1986; 2(1):26–30.
13. Replogle ML, Fleming DW, Cieslak PR. Emergence of antimicrobial-resistant shigellosis in Oregon. *Clin Infect Dis* 2000; 30(3):515–19.
14. Shata MT, Stevceva L, Agwale S, Lewis GK, Hone DM. Recent advances with recombinant bacterial vaccine vectors. *Mol Med Today* 2000; 6(2):66–71.

STAPHYLOCOCCUS SPECIES

Common name for diseases: Common names for some staphylococcal illnesses include impetigo, toxic shock syndrome, scalded skin syndrome, and staphylococcal food poisoning

Occupational setting

Staphylococcal skin infections can occur in any setting where trauma to the skin occurs, including agricultural workers,[1] construction workers, poultry process workers, and meat packers.[2,3] Outbreaks of skin infection outside of healthcare settings usually occur as a result of direct physical trauma, but have occurred in some unusual outdoor occupations such as river rafting guides.[4] Staphylococcal infections and asymptomatic carriage of *S. aureus* in healthcare employees, either of which can cause outbreaks of staphylococcal infection in patients, are of particular concern in hospitals.[5] Food-poisoning outbreaks have been caused by food handlers or preparers with staphylococcal skin infections,[6] and much current food-handling regulation is directed at reducing this risk to the general public. Toxic shock syndrome has been linked in at least one case to a possible occupational exposure, involving skin exposure to blood-contaminated water.[7]

Exposure (route)

Humans are the reservoir for *Staphylococcus* species. The coagulase-negative staphylococci (*S. epidermidis* and *S. saprophyticus*) are ubiquitous inhabitants of the skin; *S. aureus* may colonize the skin in individuals who come into contact with it through work or other activities. Carriage of the bacteria may be either chronic or transient, and it can be transmitted from carriers to other individuals or objects. *S. aureus* may be cultured from 30–50% of healthy individuals from the nasopharynx, skin, gastrointestinal and urogenital tracts, and perineum; persistent colonization occurs in 10–20% of cases.[8] Infection occurs by direct invasion of a tissue, usually through traumatic breaks in the skin or through conditions such as eczema that disrupt the skin's integrity as a barrier.

Pathobiology

Staphylococcus is a Gram-positive, non-motile coccus that grows in clusters. It is a facultative anaerobe that colonizes human skin. The most important human pathogen, *S. aureus*, is the only coagulase-positive staphylococcus species. *S. epidermidis* and *S. saprophyticus* are clinically the most important of the coagulase-negative staphylococci.

S. aureus is the most common staphylococcal pathogen in the occupational setting. Factors which predispose to infection with *S. aureus* include injury to normal skin, prior viral infections, immunologic compromise, indwelling foreign bodies including sutures and catheters, prior antibiotic treatment, and pre-existing illness, including diabetes, alcoholism, or cancer.[8] *S. aureus* infection occurs when the organism finds a suitable environment for growth, usually through a break in the skin. The bacteria produce a wide variety of enzymes and toxins that enhance their virulence and pathogenicity. Non-specific toxins, including hemolysins and leukocidins, and those specific to some strains, including epidermolytic toxins (which cause bullous impetigo and the scalded skin syndrome), increase the invasive properties of staphylococci. Additional strains produce very specific exotoxins, which include toxic shock toxin and the enterotoxins responsible for food poisoning.[9]

The spectrum of disease arising from *S. aureus* depends on the specific organism, the location of invasion, and host characteristics. **Superficial infections** include folliculitis, furunculosis, skin abscesses, impetigo, hydradenitis suppurativa, mastitis (in nursing mothers), wound infections, and spreading pyodermas. **Systemic infections** include superficial scalded skin syndrome, a serious condition which can result in desquamation of the entire skin, and toxic shock syndrome, a severe illness characterized by fever, hypotension, rash with subsequent desquamation, and involvement of several organ systems.[8] Toxic shock syndrome is caused by strains of *S. aureus* that produce a unique toxin, TSST-1 or toxic shock toxin. The disease may arise from local staphylococcal infections such as blisters,[10] may be introduced from a site of colonization rather than infection, and despite its familiar association with tampon use, has been reported in both sexes. **Organ infections** caused by *S. aureus* include endocarditis, pericarditis, pneumonia, osteomyelitis, septic arthritis, septic bursitis, and pyomyositis. Staphylococcal bacteremia and endocarditis have occurred as a result of accidental needlesticks.

Food poisoning is caused by an *S. aureus* enterotoxin that produces vomiting, diarrhea, fever, and abdominal pain. The source of the organism can be direct contact of food with the skin or infectious discharge of someone harboring the organism, with subsequent incubation of the organisms and production of enterotoxin while in storage, or from an animal source, such as meat or milk that has been inadequately processed or stored. The onset of symptoms is typically only a few h after ingestion; the enterotoxin is heat-stable and is not inactivated by subsequent cooking of food after contamination.

S. epidermidis and *S. saprophyticus* are not of major concern in occupational settings. *S. epidermidis* is typically associated with infections in patients with prosthetic or intravenous access devices. *S. saprophyticus* is usually associated with urinary tract infections.

Diagnosis

Staphylococcal disease can be diagnosed by demonstrating the organism on Gram stain and culture. Specific diagnostic criteria have been established for the diagnosis of toxic shock syndrome, which must be distinguished from other diseases with similar clinical presentations, including Rocky Mountain spotted fever, leptospirosis, scarlet fever, and measles.[8]

Treatment

Treatment of all staphylococcal infections includes appropriate antibiotic therapy. Selection of antibiotics depends on the type of infection, the host, and the likely resistance patterns of the infecting organism, but choices include penicillins with β-lactamase inhibitors in combination (such as amoxicillin/clavula-clavulanate), certain cephalosporins, erythromycin and macrolide antibiotics (clarithromycin or azithromycin) or quinolones, such as ciprofloxacin. Methicillin-resistant *S. aureus* (MRSA) is of concern primarily in healthcare settings because of selection pressures for bacterial antibiotic resistance in these locations. Treatment of MRSA infections often includes intravenous vancomycin. Resistance to quinolone antibiotics has also rapidly developed within the past decade, and their use in staphylococcal infections may be limited. In particular types of infection (such as food poisoning), additional supportive therapy, such as intravenous hydration, may be required.

Medical surveillance

Reporting to local health authorities is required for cases or outbreaks of staphylococcal food poisoning, for community outbreaks (especially in schools or camps) of other staphylococcal infections, for epidemics in hospitals, and for cases of toxic shock syndrome in most of the USA and in other countries. Medical screening activities, such as culture of nares swabs for staphylococcal diseases or asymptomatic carriage, may be performed in hospitals and other settings where nosocomial transmission of *S. aureus* to patients is suspected.

Prevention

Food handlers should use strict personal hygiene, including handwashing and gloves. Additional precautions, including temporary work removal, should be taken if they have purulent lesions of the hands, nose, or face. Food itself should be appropriately cooked, processed, and rapidly refrigerated when not used. Meat packers should be provided with appropriate tools and personal protective equipment to minimize the risk of exposure and skin trauma, such as cut-resistant gloves. Equipment should also be cleaned regularly.

Healthcare workers are more likely to be asymptomatic carriers of *S. aureus* than the general population,[11,12] and must be scrupulous in the use of handwashing techniques and personal protective equipment to prevent transmission to susceptible individuals. For some healthcare workers who are carriers of *S. aureus*, particularly MRSA, it may be desirable to remove them from direct patient contact until they are culture negative. Clearance of colonizing bacteria may occur spontaneously, or colonization may be treated using topical mupirocin or oral antibiotics (trimethoprim–sulfamethoxazole or rifampin).[8,13,14] It is important to reduce the opportunity for needlesticks and other trauma through the use of work practices, equipment redesign, and personal protective equipment, and to sterilize equipment and other fomites such as microscopes and ocular eyepieces that may be contaminated to eliminate potential reservoirs for infection.

REFERENCES

1. Pardo-Castello V. Common dermatoses in agricultural workers in the Caribbean area. *Indust Med Surg* 1962; 31:305–7.
2. Barnham M, Kerby J. A profile of skin sepsis in meat handlers. *J Infect* 1984; 9:43–50.
3. Fehrs LJ, Flanagan K, Kline S, et al. Group A beta-hemolytic streptococcal skin infections in a US meat-packing plant. *JAMA* 1987; 258:3131–4.
4. Decker MD, Lybarger JA, Vaughn WK, et al. An outbreak of staphylococcal skin infections among river rafting guides. *Am J Epidemiol* 1986; 124:969–76.
5. Patterson WB, Craven DE, Schwartz DA, et al. Occupational hazards to hospital personnel. *Ann Intern Med* 1985; 102:658–80.
6. Eisenberg MS, Gaarslev K, Brown W, et al. Staphylococcal food poisoning aboard a commercial aircraft. *Lancet* 1975; ii:595–9.

7. Tack KJ. Possible tampon-associated toxic shock syndrome in a man. *Lancet* 1981; ii:1354.
8. Lowy FD. Staphylococcus aureus infections. *N Engl J Med* 1998; 339:520–32.
9. Noble WC. Skin bacteriology and the role of Staphylococcus aureus in infection. *Br J Dermatol* 1998; 139(Suppl 53):9–12.
10. Berkeley SF, McNeil JG, Hightower AW, et al. A cluster of blister-associated toxic shock syndrome in male military trainees and a study of staphylococcal carriage patterns. *Military Med* 1989; 154:496–9.
11. Godfrey ME, Smith IM. Hospital hazards of staphylococcal sepsis. *JAMA* 1958; 166:1197–201.
12. Ballou WR, Cross AS, Williams DY, et al. Colonization of newly arrived house staff by virulent staphylococcal phage types endemic to a hospital environment. *J Clin Microbiol* 1986; 23:1030–3.
13. Yu VL, Goetz A, Wagener M, et al. *Staphylococcus aureus* nasal carriage and infection in patients on hemodialysis. Efficacy of antibiotic prophylaxis. *N Engl J Med* 1986; 315:91–6.
14. Goetz MB, Mulligan ME, Kwok R, et al. Management and epidemiologic analyses of an outbreak due to methicillin-resistant Staphylococcus aureus. *Am J Med* 1992; 92:607–14.

STREPTOCOCCUS SPECIES

Common names for diseases: Impetigo, erysipelas, strep throat, scarlet fever, rheumatic fever

Occupational setting

Like staphylococcal infections, the most common occupational streptococcal infections are skin infections, which are prominent among workers with frequently traumatized or abraded skin, including construction workers, foresters, and farmers. Group A β-hemolytic streptococcal and *S. pyogenes* infections have been reported in slaughterhouse workers, meat-packers and poultry handlers.[1–3] Streptococcal septicemia from *S. pyogenes* has also been reported in a mortuary technician who punctured himself while conducting a post-mortem examination.[4] Healthcare workers can be asymptomatic carriers of streptococci (group A or B) and transmit nosocomial infection to patients.[5] There are several other species of *Streptococcus*, especially those found in several animal species, that rarely cause skin infections or systemic disease in humans. *S. agalactiae* (group B), *S. milleri* (α-hemolytic) and *S. equisimilis* (group C) have been associated with local and systemic infections in persons with exposure to pigs and after pig bites.[6] An excess of pneumococcal pneumonia has been noted in welders, although these workers were considered to have an increased susceptibility to infection from unspecified welding fumes rather than direct exposure to *S. pneumoniae*.[7]

Streptococcus suis type II (group R β-hemolytic streptococci) was first noted in 1968 as the cause of a syndrome of meningitis and sepsis in both pigs and humans.[8] Groups at risk include pig farmers and handlers of raw pork, such as meat-packers and butchers.[9]

Exposure (route)

Streptococci are ubiquitous human pathogens. Skin infection probably occurs through breaks in the skin arising from laceration, trauma, surgery or skin breakdown, although in some cases the breaks may be unnoticed. Respiratory exposure occurs through inhalation of droplets. *S. suis* colonizes the snout and pharynx of healthy pigs; diseased pigs may exhibit a bacteremia. Transmission to humans occurs from exposure to work with pigs or raw pork products, most likely through minor skin breaks, although transmission may also take place through respiratory exposure.

Pathobiology

Streptococci are Gram-positive, non-spore-forming bacteria that typically grow in pairs or chains of spherical cells. They are classified as **β-hemolytic** if a clear zone of hemolysis surrounding bacterial colonies is seen on sheep blood agar medium. If the zone is only partly clear (usually noted as a greenish tint, giving rise to the term "viridans" streptococci), they are considered **α-hemolytic**. Another classification scheme, developed by Lancefield, designates groups (A to D, G, R) based on antigenic differences.

Group A streptococcus (*Streptococcus pyogenes*) is the most important human pathogen, causing both streptococcal skin infections and pharyngitis. Streptococcal **skin infections** include: erysipelas, a rapidly progressive skin infection accompanied by fever and, in some cases, bacteremia; pyoderma or impetigo, a localized purulent skin infection; and cellulitis.[10] Lymphangitis and lymphadenitis may also occur. The most severe form of group A streptococcal infection is necrotizing fasciitis, an infection of subcutaneous tissue with relative sparing of overlying skin and underlying muscle. Symptoms include severe local pain and tenderness, fever and systemic toxicity. There is rapid progression to tissue gangrene and death unless the infection is quickly treated.[10,11] Elaboration of pyrogenic exotoxins in strains of *S. pyogenes* is responsible for the streptococcal toxic shock syndrome, characterized by tachycardia, tachypnea, fever, chills, and diarrhea, progressing to septic shock and organ failure. Approximately half the cases are seen in association with necrotizing fasciitis. Skin or vaginal mucosa appears to be the portal of entry in 60% of cases, with the rest probably arising from bacteremia originating in the pharynx.[10]

Streptococcal **pharyngitis**, along with skin infection, is one of the most common streptococcal illnesses; its importance lies in it potential sequelae. Although nasopharyngeal carriage of group A streptococci declines somewhat from childhood to adulthood, it is not uncommon to see pharyngitis in adults, and outbreaks in crowded conditions (such as military barracks) can occur. Outbreaks may also arise from foodborne or waterborne transmission. Clinically, an incubation period of several days is followed by fever and sore throat. The posterior pharynx is red and edematous, the tonsils are enlarged and frequently have a patchy white exudate, and the cervical lymph nodes are swollen and tender. Although very early treatment may shorten the duration of symptoms and the period of communicability, the primary purpose and importance of treatment is to prevent complications of the infection.

Complications and sequelae of streptococcal pharyngitis include local head and neck infections, including otitis media, sinusitis, peritonsillar cellulitis or abscess, suppurative lymphadenitis, and bacteremia. Scarlet fever, which is caused by a toxin (erythrogenic toxin) produced by some strains, is characterized by a distinctive rash and, in more severe cases, systemic toxicity. Acute rheumatic fever and acute post-streptococcal glomerulonephritis are two delayed complications of streptococcal infection. Both are inflammatory diseases which develop after the streptococcal infection itself has resolved. Acute rheumatic fever is a systemic illness that involves connective tissue, primarily in the joints, manifesting as acute arthritis. Other organs may be affected, including the heart, skin and blood vessels, leading to carditis, erythema marginatum and subcutaneous nodules, and chorea. Acute post-streptococcal glomerulonephritis can follow either pharyngitis or pyoderma, and consists of a proliferative glomerular disease that is manifested clinically by edema, proteinuria, hematuria, and hypertension.

S. suis infection is usually manifested by a flu-like prodromal illness, followed by fever and meningismus. Hearing loss and vestibular dysfunction with ataxia occur in approximately 50% of cases. Other manifestations of infection may include endocarditis, arthritis, septicemia with shock and disseminated intravascular coagulation, and rhabdomyolysis.[12]

Diagnosis

Diagnosis in most cases of infection is made by Gram stain and culture of the organism. Kits are available to detect group A antigen for rapid diagnosis of streptococcal pharyngitis. Throat culture is the preferred method of diagnosis, however, to distinguish streptococcal pharyngitis from other causes of exudative pharyngitis such as *C. diphtheriae*, *C. haemolyticum*, *N. gonorrhoea*, *M. pneumoniae*, *Yersinia enterocolitica*, and several species of viruses. Mononucleosis should be suspected in cases of exudative pharyngitis, particularly in adolescents and younger adults. Detection of specific streptococcal antibodies such as anti-streptolysin O is not useful in the diagnosis of acute infections.

S. suis infection should be considered in cases of systemic illness where a history of exposure to pigs or raw pork products is obtained or suspected. Signs of meningitis or septicemia with characteristic findings of hearing loss and vestibular dysfunction are nonspecific, but may lead to a higher index of suspicion for the infection. The bacteria can be cultured from both blood and cerebrospinal fluid.

Treatment

Treatment of streptococcal infections requires antibiotic therapy. Penicillin is the antibiotic of choice, followed by erythromycin. Other β-lactam antibiotics and clindamycin are also effective; addition of the latter may be advisable in cases of streptococcal toxic shock. Pharyngitis should be treated for 10 days to prevent post-streptococcal complications, unless a single dose of a long-acting penicillin is used intramuscularly. Most strains of *S. suis* are penicillin-sensitive, though consideration should be given to addition of a second antibiotic until culture and sensitivity results are obtained. Aggressive treatment and surgical debridement is essential in necrotizing fasciitis to reduce the mortality from this condition. Additional supportive measures, including fluid resuscitation and mechanical ventilation, may be necessary for serious infections and streptococcal toxic shock.

Medical surveillance

There are no recommended medical screening activities for streptococcal diseases. Community or school screening programs for identifying group A streptococcal carriers have not been shown to be effective. Epidemics must be reported to local health authorities in the USA.

Prevention

Prevention of streptococcal disease requires good hygiene practices. Food handlers should use strict personal hygiene, and food handlers with active streptococcal respiratory infections should be considered for job reassignment. Food itself should be appropriately cooked, processed, and refrigerated. Meat-packers and meat-handlers should be provided with appropriate tools and personal protective equipment to minimize the risk of exposure and skin trauma, such as cut-resistant gloves. Equipment should also be cleaned regularly. Antibiotic prophylaxis of asymptomatic swine herds and changes in breeding conditions have not been well studied.[9] Pneumococcal vaccine (the polyvalent vaccine for *S. pneumoniae*) is indicated for those at risk of pneumococcal disease on the basis of underlying health status, not occupational exposure.[13]

REFERENCES

1. Fehrs LJ, Flanagan K, Kline S, et al. Group A beta-hemolytic streptococcal skin infections in a US meat-packing plant. *JAMA* 1987; 258:3131–4.
2. Phillips G, Efstratiou A, Tanna A, et al. An outbreak of skin sepsis in abattoir workers caused by an "unusual" strain of Streptococcus pyogenes. *J Med Microbiol* 2000; 49:371–4.
3. Barnham M, Kerby J, Skillin J. An outbreak of streptococcal infection in a chicken factory. *J Hyg* 1980; 84:71–5.

4. Hawky PM, Pedler SJ, Southall PJ. *Streptococcus pyogenes*: a forgotten occupational hazard in the mortuary. *Br Med J* 1980; 281:1058.
5. Patterson WB, Craven DE, Schwartz DA, et al. Occupational hazards to hospital personnel. *Ann Intern Med* 1985; 102:658–80.
6. Barnham M. Pig bite injuries and infection: report of seven human cases. *Epidemiol Infect* 1988; 101:641–5.
7. Coggon D, Inskip H, Winter P, Pannett B. Lobar pneumonia: an occupational disease in welders. *Lancet* 1994; 344:41–3.
8. Zanen HC, Engel HW. Porcine streptococci causing meningitis and septicaemia in man. *Lancet* 1975; 1(7919):1286–8.
9. Dupas D, Vignon M, Geraut C. *Streptococcus suis* meningitis: a severe noncompensated occupational disease. *J Occup Med* 1992; 34:1102–5.
10. Bison AL, Stevens DL. Streptococcal infections of skin and soft tissue. *N Engl J Med* 1996; 334:240–5.
11. Green RJ, Dafoe DC, Raffin TA. Necrotizing fasciitis. *Chest* 1996; 110:219–29.
12. Tambyah PA, Kumarasinghe G, Chan HL, Lee KO. Streptococcus suis infection complicated by purpura fulminans and rhabdomyolysis: case report and review. *Clin Infect Dis* 1997; 24:710–12.
13. Centers for Disease Control. Update on adult immunization: recommendations of the Immunization Practices Advisory Committee (ACIP). *MMWR* 1991; 40(RR-12):43–4.

TREPONEMA PALLIDUM

Common name for disease: Syphilis

Occupational setting

As with other infections whose primary route of transmission is sexual, occupational groups at increased risk of syphilis are those who have direct contact with infectious lesions, such as prostitutes, and workers who have enforced separation from their families, such as seafarers and migrant laborers. In the early part of the century, infection was reported among some laboratory workers who were handling animals infected with strains of *T. pallidum*. Transmission may have occurred through either scratches, bites, or self-inoculation.[1]

Exposure (route)

Transmission of the organism takes place through direct contact of mucous membrane or skin with an infectious lesion, and almost always involves sexual contact. Infectious lesions include chancres, skin rashes, mucous patches, or condylomata lata. Syphilis can also be acquired congenitally, and by contact with infected human blood through transfusion or auto-inoculation.

Pathobiology

After an incubation period of 9–90 days (median of 21 days), a primary lesion develops at the site of infection. The chancre of primary syphilis is typically painless and indurated with a well-defined border. It usually heals within 2 months. Secondary syphilis, a systemic disease, typically occurs 6–8 weeks after the primary lesion is healed. A wide variety of clinical manifestations may be present, including involvement of the skin, mucous membranes and lymphatic, renal, gastrointestinal and skeletal systems. The most characteristic lesion at this stage is the punctate or pox-like rash, which appears most often on the palms and soles.[2] Late or tertiary syphilis occurs years after primary infection. Consequences of syphilis at this stage include neurosyphilis, with numerous clinical manifestations including ataxia and ocular lesions, cardiovascular syphilis, which may be seen as aortitis and valvular disease, and a "benign" form in which the characteristic syphilitic gumma may be found in organs including liver and bone systems.

Diagnosis

As *T. pallidum* is extremely difficult to grow in culture, diagnosis is made either by direct examination or serologic testing. Identification of the organism is made by dark-field microscopic examination of material (scrapings or exudate) from an active lesion or lymph node. Serologic tests fall into two categories. Those that detect antigenic indicators of host tissue damage (the Venereal Disease Research Laboratory (VDRL), rapid plasma reagin (RPR) tests) are non-specific but inexpensive and can be used for initial testing of individuals suspected of having disease, or for population screening. Specific treponemal antigen testing is used to confirm the diagnosis in individual patients. These tests include microhemagglutinin assays for *T. pallidum* antibody (MHA-TP) and the fluorescent treponemal antibody absorption (FTA-ABS) test.[3]

Treatment

All stages of syphilis can be treated with penicillin.

Medical surveillance

There are no recommended occupational screening activities for syphilis. Syphilis is a reportable disease in the USA and many other countries.

Prevention

Infection with *T. pallidum* is prevented by treatment of infected individuals, and the use of condoms and antiseptic prophylactic agents. In addition, healthcare workers should use appropriate personal protective equipment in cases where the organism is being handled when examining patients in whom clinical syphilis is suspected.

REFERENCES

1. Collins CH, Kennedy DA. Microbiological hazards of occupational needlestick and "sharps" injuries. *J Appl Bacteriol* 1987; 62:385–402.
2. Brown TJ, Yen-Moore A, Tyring SK. An overview of sexually transmitted diseases. *J Am Acad Dermatol* 1999; 41:511–32.
3. Larsen SA, Steiner BM, Rudolph AH. Laboratory diagnosis and interpretation of tests for syphilis. *Clin Microbiol Rev* 1995; 8:1–21.

VIBRIO CHOLERAE

Common name for disease: Cholera

Occupational setting

Persons at risk of cholera include travelers to areas where cholera is endemic[1] and those in occupations where exposure to contaminated sea water or food is possible.[2] Those who handle or consume undercooked shellfish may also be at risk.[3] Healthy commercial divers have been found to become colonized following dives at contaminated sites.[4] Outbreaks have occurred in hospitals.[5] In 1999, five states reported cases of cholera to the CDC, primarily from international travelers returning from endemic areas (Figure 22.12).

Exposure (route)

Exposure is through ingestion of food or water containing live organisms, or through direct contact with water bearing the organisms.

Pathobiology

Vibrios are curved, flagellated Gram-negative rods. The organism is found in surface waters (both fresh and salt water) all over the world. *V. cholerae* is a diverse species, consisting of numerous different strains. Strains are primarily grouped according to the type of cell wall O antigen present. Most cases of epidemic cholera are due to serogroup O1. Within this serogroup there are also subdivisions into three serotypes, Ogawa, Inaba, and Hikojima,

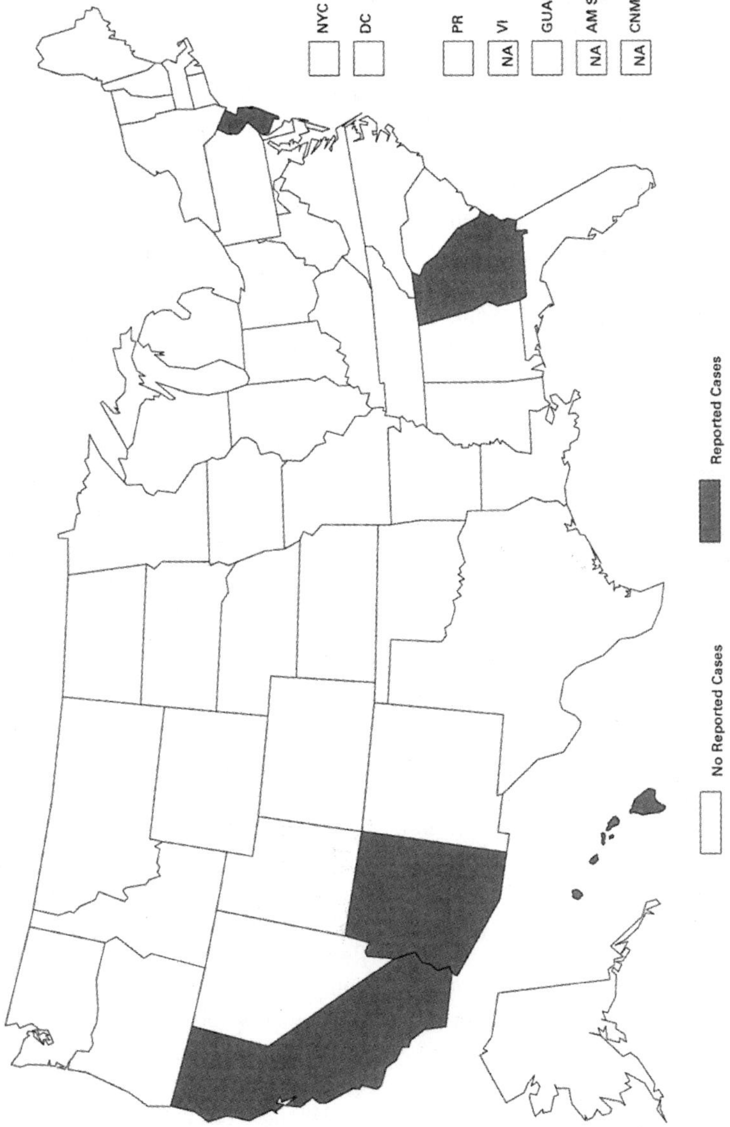

Figure 22.12 Cholera – reported cases, United States and territories, 1999. Source: MMWR 2001:48(53):31.

as well as two biotypes, classical and El Tor. However, not all O1 strains are pathogenic. While some members of serogroups O2 to O138 have the potential to cause isolated cases of cholera,[6] they are not thought to result in epidemics. *V. cholerae* O139 (Bengal strain) has been isolated from a large proportion of cases in the recent cholera epidemic in Asia.[7] New serotypes are likely to be implicated in future epidemics.[8]

When ingested with water or food, *V. cholerae* passes through the stomach into the small bowel, where it adheres to the mucosal lining. There, it secretes an enterotoxin that acts on intestinal cell receptors, mediated by cyclic adenosine monophosphate, to cause the active secretion of sodium chloride into the gut lumen. This, in turn, causes a voluminous watery diarrhea, which may occur at a rate of 1 liter per hour.

The incubation period after exposure ranges from 6 hours to 5 days. Prodromal symptoms consist of abdominal discomfort and anorexia. This is followed by diarrhea that progresses from brown to a 'rice water' appearance (due to mucus secretion). The complications of cholera all derive from the rapid loss of fluids and electrolytes and include hypotension, hypoglycemia, electrolyte imbalance, renal failure, and acidosis. Rarely, an ileus may occur. Death is most often due to the loss of glucose, electrolytes, and fluid.

It should be noted that in most studies of institutional outbreaks, over 75% of those infected are asymptomatic.[5]

Diagnosis

Diagnosis of cholera can be made by dark-field examination of stool or by culture. A history of travel to areas with endemic cholera is also helpful.

Treatment

The cornerstone of cholera treatment is fluid and electrolyte replacement, either by oral or intravenous rehydration. With adequate rehydration, the case fatality rate is low, 2% in a case series from the USA.[9] Oral rehydration solutions can be made from salt and sugar (5 g sodium chloride and either 20 g glucose or 40 g sucrose per liter of water), or, according to the World Health Organization formulation, salt, sugar, potassium chloride and bicarbonate (3.5 g sodium chloride, 2.5 g sodium bicarbonate, 1.5 g potassium chloride, and 20 g glucose per liter of water). Since this formulation does not decrease (and may even increase) the duration and volume of diarrhea, formulations which add starch from a variety of sources have been advocated.[10] Most cases can be successfully treated with oral rehydration alone, however, more severe diarrhea may require intravenous rehydration.

Antibiotics are useful in shortening the duration of infection. Tetracycline, doxycyline, erthromycin and trimethoprim–sulfamethoxazole have all been used successfully. However, multiple antibiotic resistance has emerged among strains of *V. cholerae*. Currently, the quinolones, such as ciprofloxacin and norfloxacin, have shown excellent results and offer the advantages of efficacy with a short course or even a single dose.[11]

Medical surveillance

Reporting of cholera cases and epidemics to local health authorities is mandated in virtually all countries. There are no recommended medical screening activities for this disease.

Prevention

Prevention of cholera requires avoiding contaminated food and water. The absence of municipal water chlorination has been identified as a major contributor to the re-emergence of cholera in South America.[12]

An inactivated bacteria cholera vaccine is available, but it is not normally recommended. The vaccine is about 50% effective in preventing illness for 3–6 months, does not prevent transmission of infection, and may not offer cross-protection against diarrhea caused by *V.*

choleraee O139.[13] Newer oral live attenuated vaccines have been developed which appear to offer protection for up to 2 years.[14]

Persons with cholera infections should not return to work until clearance of the organism is complete as documented by negative stool cultures. In outbreaks, asymptomatic carriers should be identified and treated to eradicate the organism. Healthcare workers caring for patients with cholera should apply appropriate enteric precautions.

REFERENCES

1. Cooper G, Hadler JL, Barth S, et al. Cholera associated with international travel, 1992. *MMWR* 1992; 41:664–7.
2. Hunt MD, Woodward WE, Keswick BH, Dupont HL. Seroepidemiology of cholera in Gulf coastal Texas. *Appl Environ Microbiol* 1988; 54(7):1673–7.
3. Weber JT, Mintz ED, Canizares R, et al. Epidemic cholera in Ecuador: multidrug-resistance and transmission by water and seafood. *Epidemiol Infect* 1994; 112(1):1–11.
4. Huq A, Hasan JA, Losonsky G, Diomin V, Colwell RR. Colonization of professional divers by toxigenic Vibrio cholerae O1 and V. cholerae non-O1 at dive sites in the United States, Ukraine and Russia. *FEMS Microbiol Lett* 1994; 120(1–2):137–42.
5. Goh KT, Teo SH, Lam S, Ling MK. Person-to-person transmission of cholera in a psychiatric hospital. *J Infect* 1990; 20(3):193–200.
6. Holmes FF, Wells B, Dees DJ, Lindsey NJ, Godwin JR, Montgomery JC. A Vibrio cholerae infection in a transient teamster. *Am J Trop Med Hyg* 1981; 30(6):1277–80.
7. Bhattacharya SK, Bhattacharya MK, Nair GB, et al. Clinical profile of acute diarrhoea cases infected with the new epidemic strain of Vibrio cholerae O139: designation of the disease as cholera. *J Infect* 1993; 27(1):11–15.
8. Dalsgaard A, Forslund A, Bodhidatta L, et al. A high proportion of Vibrio cholerae strains isolated from children with diarrhoea in Bangkok, Thailand are multiple antibiotic resistant and belong to heterogenous non-O1, non-O139 O-serotypes. *Epidemiol Infect* 1999; 122(2):217–26.
9. Weber JT, Levine WC, Hopkins DP, Tauxe RV. Cholera in the United States, 1965–1991. Risks at home and abroad. *Arch Intern Med* 1994; 14:551–6.
10. Rabbani GH. The search for a better oral rehydration solution for cholera. *N Engl J Med* 2000; 342(5):345–7.
11. Usubutun S, Agalar C, Diri C, Turkyilmaz R. Single dose ciprofloxacin in cholera. *Eur J Emerg Med* 1997; 4(3):145–9.
12. Ries AA, Vugia DJ, Beingolea L, et al. Cholera in Piura, Peru: a modern urban epidemic. *J Infect Dis* 1992; 166(6):1429–33.
13. Albert MJ, Alam K, Ansaruzzaman M, Qadri F, Sack RB. Lack of cross-protection against diarrhea due to Vibrio cholerae O139 (Bengal strain) after oral immunization of rabbits with V. cholerae O1 vaccine strain CVD103-HgR. *J Infect Dis* 1994; 169(1):230–1.
14. Graves P, Deeks J, Demicheli V, Pratt M, Jefferson T. Vaccines for preventing cholera. *Cochrane Database Syst Rev* 2000; 2:CD000974.

VIBRIO SPECIES OTHER THAN V. CHOLERAE (V. PARAHEMOLYTICUS, V. VULNIFICUS)

Common name for disease: None

Occupational setting

Individuals at risk for non-cholera *Vibrio* infection are those in close contact with both aquatic environments such as commercial divers,[1] fishermen,[2] fish farmers,[3] and workers who handle seafood or shellfish.[4] Outbreaks have also occurred aboard cruise ships.[5] Bacteria in this group have been described as "occupational pathogens", since asymptomatic carriage rates approaching 4% have been documented in high-risk worker groups which result in sporadic outbreaks.[6]

Exposure (route)

Exposure is through ingestion or direct contact with marine organisms or water containing live organisms.

Pathobiology

Vibrios are curved, flagellated, Gram-negative rods. These organisms are found in surface waters (fresh and salt water) around the world. A number of *Vibrio* species have been identified which may cause disease in humans. Most, like *V. cholerae*, cause toxic gastrointestinal disease. Several species, however, also have varying degrees of predilection for causing soft tissue infections, usually at the site of pre-existing skin breaks or wounds. Sepsis and death are well-known complications of either presentation, but typically only in those with chronic disease.[7] In cases complicated by sepsis, there may be lower extremity edema and bullae.[8] The most important species to consider are *V. parahaemolyticus* and *V. vulnificus*.

V. parahaemolyticus is a common cause of diarrheal disease throughout the world. The most frequent route of exposure is consumption of raw shellfish, usually oysters.[8] Four such outbreaks have occurred in the USA since 1997.[8] Other *Vibrio* species primarily associated with gastroenteritis include *V. mimicus*,[9] *V. hollisae*,[10] and *V. fluvialis*.[11]

V. vulnificus causes local wound infections which may follow an aggressive course with rapid spread and necrosis of surrounding tissue.[12] Other conditions associated with this organism include corneal ulcers,[4] and a fulminant systemic illness characterized by a hemorrhagic rash, fever, gastroenteritis, and hypotension.[13] Other vibrios that cause local wound infections are *V. alginolyticus*,[14] *V. damsela*,[15] and *V. metschnikovii*.[16] The infections occur frequently in pre-existing wounds or skin breaks, or they develop as an acute otitis.[1] The most common route of exposure is direct contact with open sea water.[12]

Diagnosis

All *Vibrio* infections are diagnosed by culturing the organism from stool, blood or wound samples. A history of exposure to sea water or raw shellfish, or ingestion of raw or undercooked shellfish, is helpful in the diagnosis. Development of a severe cellulitis of the extremities after exposure to sea water, especially if the cellulitis does not respond to aminoglycosides, should raise the suspicion of infection with *Vibrio* species.

Treatment

Most of the gastrointestinal diseases caused by non-cholera vibrios are self-limited and do not require specific therapy other than rehydration.

Local wound infections generally respond to antibiotics such as tetracycline, with chloramphenicol or penicillin as a second choice. Surgical debridement is frequently required when soft tissue necrosis is present.[12] Systemic infection requires parenteral antibiotic therapy.

Medical surveillance

There are no recommended medical screening or surveillance activities for non-cholera *Vibrio* infections.

Prevention

There is no effective vaccine for any of the non-cholera *Vibrio* infections. Prevention of foodborne disease consists primarily of proper handling, cooking, and refrigeration. Immunocompromised individuals, particularly those with chronic liver disease, should avoid ingestion of uncooked shellfish.[12] The concentration of *Vibrio* species in sea water increases with increasing water temperature, increasing the risk of contamination for seafood harvested during the summer.[17]

Individuals with pre-existing wounds should take precautions to avoid exposure to sea water or other potentially contaminated material.

REFERENCES

1. Tsakris A, Psifidis A, Douboyas J. Complicated suppurative otitis media in a Greek diver due to a marine halophilic Vibrio sp. *J Laryngol Otol* 1995; 109(11):1082–4.
2. Hoi L, Dalsgaard A, Larsen JL, Warner JM, Oliver JD. Comparison of ribotyping and randomly amplified polymorphic DNA PCR for characterization of Vibrio vulnificus. *Appl Environ Microbiol* 1997; 63(5):1674–8.
3. Bisharat N, Agmon V, Finkelstein R, et al. Clinical, epidemiological, and microbiological features of Vibrio vulnificus biogroup 3 causing outbreaks of wound infection and bacteraemia in Israel. Israel Vibrio Study Group. *Lancet.* 1999; 354(9188):1421–4.
4. Massey EL, Weston BC. Vibrio vulnificus corneal ulcer: rapid resolution of a virulent pathogen. *Cornea* 2000; 19(1):108–9.
5. Centers for Disease Control. Gastroenteritis caused by Vibrio parahaemolyticus aboard a cruise ship. *MMWR* 1978; 27:65–6.
6. Morris JG Jr. Non-O group 1 Vibrio cholerae: a look at the epidemiology of an occasional pathogen. *Epidemiol Rev* 1990; 12:179–91.
7. Klontz KC. Fatalities associated with Vibrio parahaemolyticus and Vibrio cholerae non-O1 infections in Florida (1981 to 1988). *South Med J* 1990; 83(5):500–2.
8. Centers for Disease Control. Outbreak of Vibrio parahaemolyticus infection associated with eating raw oysters and clams harvested from Long Island Sound—Connecticut, New Jersey, and New York, 1998. *MMWR* 1999; 48(3):48–51.
9. Campos E, Bolanos H, Acuna MT, et al. Vibrio mimicus diarrhea following ingestion of raw turtle eggs. *Appl Environ Microbiol* 1996; 62(4):1141–4.
10. Carnahan AM, Harding J, Watsky D, Hansman S. Identification of Vibrio hollisae associated with severe gastroenteritis after consumption of raw oysters. *J Clin Microbiol* 1994; 32(7):1805–6.
11. Klontz KC, Cover DE, Hyman FN, Mullen RC. Fatal gastroenteritis due to Vibrio fluvialis and nonfatal bacteremia due to Vibrio mimicus: unusual vibrio infections in two patients. *Clin Infect Dis* 1994; 19(3):541–2.
12. Howard RJ, Bennett NT. Infections caused by halophilic marine Vibrio bacteria. *Ann Surg* 1993; 217(5):525–30.
13. Serrano-Jaen L, Vega-Lopez F. Fulminating septicaemia caused by Vibrio vulnificus. *Br J Dermatol* 2000; 142(2):386–7.
14. Mukherji A, Schroeder S, Deyling C, Procop GW. An unusual source of Vibrio alginolyticus-associated otitis: prolonged colonization or freshwater exposure? *Arch Otolaryngol Head Neck Surg* 2000; 126(6):790–1.
15. Tang WM, Wong JW. Necrotizing fasciitis caused by Vibrio damsela. *Orthopedics* 1999; 22(4):443–4.
16. Hansen W, Freney J, Benyagoub H, Letouzey MN, Gigi J, Wauters G. Severe human infections caused by Vibrio metschnikovii. *J Clin Microbiol* 1993; 31(9):2529–30.
17. Shapiro RL, Altekruse S, Hutwagner L, et al. The role of Gulf Coast oysters harvested in warmer months in Vibrio vulnificus infections in the United States, 1988–1996. Vibrio Working Group. *J Infect Dis* 1998; 178(3):752–9.

YERSINIA PESTIS

Common names for disease: Plague

Occupational setting

Human plague arises from infection with *Yersinia pestis*, a bacterium that is maintained in a natural reservoir of small rodents and the fleas that infest them. Since 1925, plague in the USA has been associated with exposures to wild rodents only in the southwestern states, primarily around the Four Corners region where New Mexico, Arizona, Utah, and Colorado join, and in California.[1] The disease is also endemic in many areas of South America, Africa, and South East Asia. Historically, throughout the world, urban and domestic rats have been the most important reservoirs for epidemic plague. As control measures have reduced the proximity of rats to humans in urban areas, foci of the sporadic illness have shifted to rural areas where burrowing mammals make their habitat. The oriental rat flea (*Xenopsylla cheopis*) is the most important vector. In the USA, plague is maintained in well-established enzootic foci among wild rodents, including rock squirrels, the California ground squirrel, prairie dogs, chipmunks, and woodrats. Although they are not part of the enzootic cycle, are only incidentally infected, and rarely develop overt illness, rabbits and hares, deer, antelope, gray fox, badger, bobcat and coyote have been occasionally associated with human plague in hunters and trappers. Plague associated with rock squirrels is the most important cause of human disease in North America, as housing developments have been introduced into habitats where plague was enzootic. From 1947 to 1996, 390 cases of plague were reported in the USA, with 60 deaths (Figure 22.13).[2] Hunters, trappers, foresters, rangers, and others working in remote locations where contact with rodent habitats might be expected are at risk for contracting plague, as are veterinarians in enzootic areas. Military personnel and individuals who might be stationed in areas where rats or other animal reservoirs are present would also be at increased risk.

Exposure (route)

Plague is transmitted from infected animals to humans by several species of rodent fleas, including in the USA the ground squirrel fleas *Diamana montana* and *Thrassis bacchi*. Domestic cats can also transmit the infection after consuming plague-infected animals, or after bites from infected rodent fleas, to humans by bites or scratches.[3] Human-to-human transmission can occur in pneumonic plague, and such transmission constitutes a public health emergency. Human-to-human pneumonic plague transmission has not been reported in the USA since a 1925 epidemic in Los Angeles. Human pneumonic plague can be acquired from domestic cats with secondary pneumonic plague; this mode of transmission has been associated with disease in veterinarians. *Yersinia pestis* can also be acquired through infectious fluids or tissues entering through cuts or abrasions in the skin, a mode of transmission most commonly seen in hunters and trappers who skin infected animals.

Pathobiology

Y. pestis is a Gram-negative, non-motile coccobacillus. Human infection is associated with four common clinical presentations: bubonic plague, septicemic plague, pneumonic plague, and meningitis. From 2 to 6 days after the bite of an infected flea, the patient develops a febrile illness, followed by the development of very painful suppurating lymphadenopathy (buboes) proximal to the bite. These are seen most commonly in the groin, axillae, or cervical region.[1] Septicemic plague is usually secondary to untreated bubonic plague, and is usually rapidly fatal without treatment. Hematogenous dissemination can affect any organ system, but most commonly involves lungs, eyes, meninges, joints, and skin. Occlusion of small cutaneous blood vessels can result in necrosis and gangrene of the fingers and skin (which gave rise to the term "Black Death" in the 14th century). Meningitis is a rarer complication that follows inadequately treated bubonic plague, and is more commonly associated with axillary buboes. Case fatality rates for untreated bubonic and septicemic plague exceed 50% in most reports.[4]

Pneumonic plague can be secondary to septicemia or arise as a primary infection after the inhalation of infectious droplet nuclei. Primary pneumonic plague has a short incubation period and can spread rapidly in close contacts; its presence is therefore a public health emergency, as the infected case

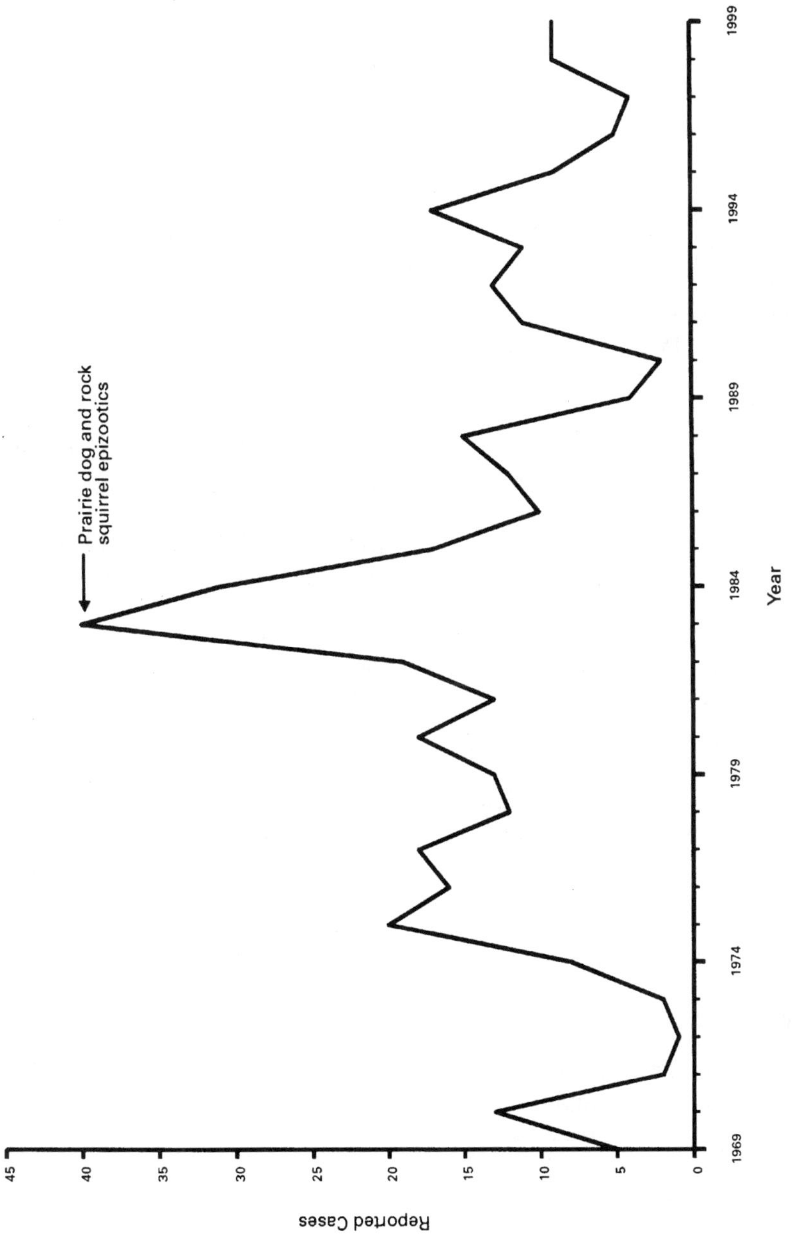

Figure 22.13 Plague – reported cases among humans, United States, 1969–1999. Source: MMWR 2001:48(53):56.

will generate infective aerosols. In the USA, the few reported human cases of primary pneumonic plague have been acquired from domestic cats that developed secondary pneumonic infection. Survival is unlikely unless treatment is initiated within 18 hours after onset of respiratory symptoms.[4]

Diagnosis

Infectious clinical material can be examined with light microscopy after appropriate staining. Buboes and blood can also provide specimens for culture. Large quantities of bacteria in lesions and blood typically make bacteriologic diagnosis relatively easy. A rapid fluorescent antibody test is available at certain reference laboratories.

Treatment

Streptomycin has been the traditional treatment of choice, but other aminoglycoside antibiotics, including gentamicin, are also effective. Quinolone antibiotics, including ofloxacin, show promise of greater in vitro efficacy, and may be used in place of tetracycline, the usual alternative in patients in whom streptomycin is contraindicated. Chloramphenicol is the preferred treatment for meningitis and endophthalmitis.

Medical surveillance

There are no recommended human medical screening activities for plague. Suspected and confirmed cases of plague are required to be immediately reported to local health authorities worldwide. Monitoring of rodent populations for increased die-offs and evidence of epizootics should be performed in endemic areas. Plague should be recognized as a Sentinel Health Event-Occupational (SHE-O) in shepherds, farmers, ranchers, hunters, and field geologists.[5]

Prevention

Discussion of the prevention of human plague can be grouped into three categories: (1) very high-risk populations such as laboratory workers or medically underserved endemic or epidemic areas; (2) exposure to human plague cases or epizootic foci; and (3) environmental management of hyperendemic residential or recreational foci.[1] An inactivated plague vaccine is available in the USA, consisting of a primary immunization series of two injections and boosters every 6 months, and is recommended for laboratory workers with frequent exposure to *Y. pestis* and persons such as mammalogists, ecologists and other field workers who have regular contact with wild rodents or their fleas in areas in which plague is enzootic or epizootic.[4]

Management of contacts of human plague cases and of exposures to epizootic plague are potential public health emergencies. Contacts should be carefully identified, and appropriate follow-up, including consideration of prophylactic antibiotic therapy, should be ensured. Available environmental control measures for management of epizootic plague include: application of pesticides to control fleas; closure of recreational areas to humans; and judicious use of rodenticides. Control may also be exercised through management of the environment near the interface of human and rodent activities, such as appropriate handling of trash, and removal of rock piles and dilapidated buildings that provide habitats for rodents. Catastrophic events, such as war or natural disasters, that disrupt normal sanitary activities may lead to spread of plague from rural foci into urban centers; control measures should be reinstituted as rapidly as possible to prevent this scenario.

REFERENCES

1. Craven RB, Barnes AM. Plague and tularemia. *Infect Dis Clin North Am* 1991; 5:165–75.

2. Centers for Disease Control and Prevention. Fatal Human Plague—Arizona and Colorado, 1996. *MMWR* 1997; 46:617–20.
3. Doll JM, Zeitz PS, Ettestad P, et al. Cat-transmitted fatal pneumonic plague in a person who traveled from Colorado to Arizona. *Am J Trop Med Hyg* 1994; 51:109–14.
4. Centers for Disease Control and Prevention. Prevention of Plague: Recommendations of the Advisory Committee on Immunization Practices (ACIP). *MMWR* 1996; 45(RR-14):1–15.
5. Rutstein DD, Mullan RJ, Frazier TM, et al. Sentinel Health Events (Occupational): a basis for physician recognition and public health surveillance. *Am J Public Health* 1983; 73:1054–62.

YERSINIA PSEUDOTUBERCULOSIS AND ENTEROCOLITICA

Common names for disease: None

Occupational setting

Yersinia enterocolitica and *Y. pseudotuberculosis* are widespread in natural settings, and have been isolated from wild and domestic animals, foods, water and soil. *Y. enterocolitica* is a frequent cause of diarrhea and gastroenteritis in European countries and North America. Swine and pigs are common asymptomatic carriers of the bacteria, with Yersiniae frequently isolated from pigs' tongues and tonsils, and abattoir workers have been found to have an elevated risk of infection.[1] Other animal reservoirs of the infection include rodents, rabbits, sheep, cattle, horses, dogs, and cats. Workers who come in contact with refrigerated meat after slaughter, including butchers, meat-handlers and meat-packers, are also at risk, as Yersiniae can propagate at low temperatures. *Y. pseudotuberculosis* is a zoonosis of the aforementioned animals and also several species of birds (turkeys, ducks, geese, pigeons). It is an uncommon disease, more frequent in children and during the winter months, and is associated with exposure to animals or common-source outbreaks from contaminated food or water.

Exposure (route)

Transmission occurs via ingestion of contaminated food or water and by direct contact with infected animals, possibly by a fecal–oral route. Fecal–oral transmission between humans has not been documented.

Pathobiology

These Yersiniae are facultatively anaerobic Gram-negative bacilli that are motile when grown at 25°C. Asymptomatic infection with either *Yersinia* species is common. The prevalence of antibodies to Yersiniae is up to 10% in the general population and up to 40% in slaughterhouse workers. The incubation period for *Y. enterocolitica* enterocolitis ranges from 1 to 11 days.[2] Symptomatic infection most commonly is a diarrheal illness with fever and severe abdominal pain that can mimic acute appendicitis. Septicemia is less common, has an untreated case fatality ratio of 50%, and is usually associated with moderate to severe underlying medical problems such as diabetes or cancer. Interestingly, in one study, abattoir workers were reported to have almost a fourfold increased risk, and pig farmers a twofold increased risk, of appendectomy compared to grain or berry farmers. The authors hypothesized that the severe abdominal pain associated with *Yersinia* infections in these workers could have accounted for the increased risk.[3] Postinfection complications include reactive polyarthritis, erythema nodosum, and eye inflammation (e.g. iridocyclitis).[1] Person-to-person transmission of *Y. enterocolitica* has also been reported as a cause of septicemia arising from blood transfusions.[4] Patients developed the abrupt onset of fever and hypotension within 50 min after transfusion had begun; four of six patients reported between 1989 and 1991 died from the infection. Transient bacteremia in donors with proliferation of the bacterium under cold storage conditions was considered responsible for the contamination of blood products. *Y. pseudotuberculosis* most commonly causes mesenteric adenitis in adults, mimicking acute appendicitis, which is often self-limiting.

Diagnosis

Culture of appropriate clinical specimens (usually stool samples) often yields yersiniae. Cold enrichment increases the yield of cultures by selectively favoring the growth of *Eurasian*, although, because of asymptomatic colonization, care must be taken with culture results that become positive only after prolonged culture. Serologic tests using adsorption methods (to remove cross-reacting antibodies) are also useful in diagnosis. If transfusion-associated bacteremia is suspected, the residual blood in the bag should be examined by Wright–Giemsa or other hematologic stain, and cultured.[4]

Treatment

Enterocolitis and mesenteric adenitis are usually self-limited, and the need for antibiotic therapy in these conditions is unclear. Doxycycline and trimethoprim–sulfamethoxazole are effective in complicated gastrointestinal infection or focal extraintestinal infection.[2] Septicemia should be treated with a combination of doxycycline and an aminoglycoside. Laparotomy for suspected appendicitis should be avoided if *Yersinia* infection is a likely diagnosis.

Medical surveillance

There are no recommended medical screening activities. Reporting of cases is mandatory in many areas in the USA and in many other countries.

Prevention

Prevention should focus on the animal reservoirs of the infection. Institution of work practices to minimize contamination of meat, such as altered methods of slaughter of pigs and avoidance of prolonged refrigeration of meat before consumption, is advised. Careful handwashing and cleaning of surfaces after food preparation is essential to prevent bacterial spread to other foods. Personal protective equipment use may also afford some protection from infection, but its effectiveness has not been evaluated.

REFERENCES

1. Merilahti-Palo R, Lahesmaa R, Granfors K, et al. Risk of *Yersinia* infection among butchers. *Scand J Infect Dis* 1991; 23:55–61.
2. Cover TL, Aber RC. Yersinia enterocolitica. *N Engl J Med* 1989; 321:16–22.
3. Seuri M. Risk of appendicectomy in occupations entailing contact with pigs. *Br Med J* 1991; 301:345–6.
4. Centers for Disease Control and Prevention. Epidemiologic Notes and Reports Update: Yersinia enterocolitica bacteremia and endotoxin shock associated with red blood cell transfusions—United States, 1991. *MMWR* 1991; 40:176–8.

23

MYCOBACTERIA

Linda M. Frazier, M.D., M.P.H.

MYCOBACTERIUM TUBERCULOSIS

Common names for disease: Tuberculosis, consumption

Occupational setting

Tuberculosis (TB) exposure may occur in healthcare facilities, including hospitals, dental clinics and nursing homes, and in clinical or research laboratories processing TB cultures or infected specimens. Exposure may occur in funeral homes and is a significant risk in drug treatment centers, correctional institutions and facilities for the homeless, alcoholics, or persons with AIDS. Animal care-takers can contract TB from primates even though the animal may not appear ill. Maintenance and construction workers may be exposed while manipulating ventilation systems for patient care isolation rooms or for biological safety cabinets in which infectious samples are handled.

Exposure (route)

TB is contracted after inhalational exposure via droplet nuclei that are 1–5 μm in size and thus remain airborne for long periods of time. TB is not contracted by skin contact with surfaces such as hospital room furniture, equipment, or walls. Infection by gastrointestinal exposure is not a significant risk.

Pathobiology

There are more than 30 members of the genus *Mycobacterium*, many of which are saprophytes that cause no human disease. *Mycobacterium tuberculosis* is the organism that causes TB. The surface lipids of mycobacteria cause them to be resistant to decolorization by acid alcohol during staining procedures. This property gives rise to the name acid-fast bacilli.

Physical and Biological Hazards of the Workplace, Second Edition, Edited by Peter H. Wald and Gregg M. Stave
ISBN 0-471-38647-2

Mycobacteria will not grow in common culture media but require techniques and reagents found in specialized laboratories.

TB may be insidious in onset, causing symptoms that the affected individual may ignore. These non-specific symptoms include anorexia, weight loss, low-grade fever, fatigue, and cough. Pulmonary and pleural TB is the most common acute manifestation, presenting with pleuritic chest pain, cough productive of bloody sputum, high fever, and profuse sweating. Multidrug-resistant TB has a high case fatality rate.

TB can affect organs other than the lungs. Tuberculous infection in the genitourinary system can cause ureteral obstruction or irregular menses. Lymphatic infection can cause the swelling of the lymph nodes, known as scrofula. If vertebral bodies are affected, pain and compression fractures may occur. Meningeal TB is associated with abnormal behavior, headaches, or seizures. Tuberculous peritonitis causes abdominal pain and ascites. TB can affect the pericardium, causing heart failure, and the larynx, causing persistent hoarseness. Adrenal involvement may cause Addison's disease, and tuberculous skin infiltration has been reported. If infection is overwhelming and disseminated (miliary TB), the presentation can mimic acute leukemia.[1,2]

Eliminating occupational exposure to TB has become crucial, because large numbers of strains have become resistant to several of the commonly used anti-tuberculous agents (termed multidrug-resistant, or MDR). Intermittent compliance in taking anti-tuberculous drugs increases the risk that initially susceptible strains will develop resistance. After a century of decline, the incidence of TB began rising in the 1980s, especially in inner cities and certain states (Figure 23.1, 23.2). The resurgence of TB has been attributed to economic deprivation, homelessness, alcoholism, drug use, and the rising incidence of AIDS. Public health workers and clinicians subsequently increased their efforts to trace contacts and ensure completion of therapy among individuals with TB. After reaching a peak of >26 000 confirmed cases of TB reported per year in the USA, incidence rates fell by the late 1990s to about 14 000 cases per year.[3]

Nosocomial spread of TB has been documented; some outbreaks have been associated with tuberculin skin test conversions in >50% of healthcare workers in a single year. These significant occupational exposures have been associated with failure to comply with basic personal protective practices, delayed diagnosis of infected patients, and inadequate hospital room ventilation.[4–6] In recent years, healthcare facilities have made improvements in the availability of isolation rooms for potentially infectious patients, in use of proper respiratory protective devices, and in training healthcare workers; even so, optimal practices are not always followed.[7–10]

Diagnosis

Skin testing coupled with clinical evaluation serve to distinguish between infection with the tubercle bacillus and disease caused by the infection. When first exposure results in the entry of tubercle bacilli into the body, bacilli that are not immediately cleared can reside in lymph nodes and other tissues for long periods. A cell-mediated immune response develops that will cause a tuberculin skin test to register positive in ~2–10 weeks. Most individuals never develop disease manifestations. An estimated 5% of newly infected individuals develop clinical illness within 1 year. Later in life, particularly during advanced age, another 5% of infected people develop "reactivation" TB. In these cases, dormant tubercle bacilli begin multiplying and clinical TB develops, often in the lungs.[1,2,11]

When symptoms are present, organ system evaluations coupled with cultures for acid-fast bacilli are needed to confirm the diagnosis. Tuberculous lung infection typically causes lesions in the upper lung fields, such as granulomatous infiltrates. Radiographs may also show a diffuse miliary pattern or sterile pleural effusions. Urinary system TB can cause white blood cells in the urine with negative cultures

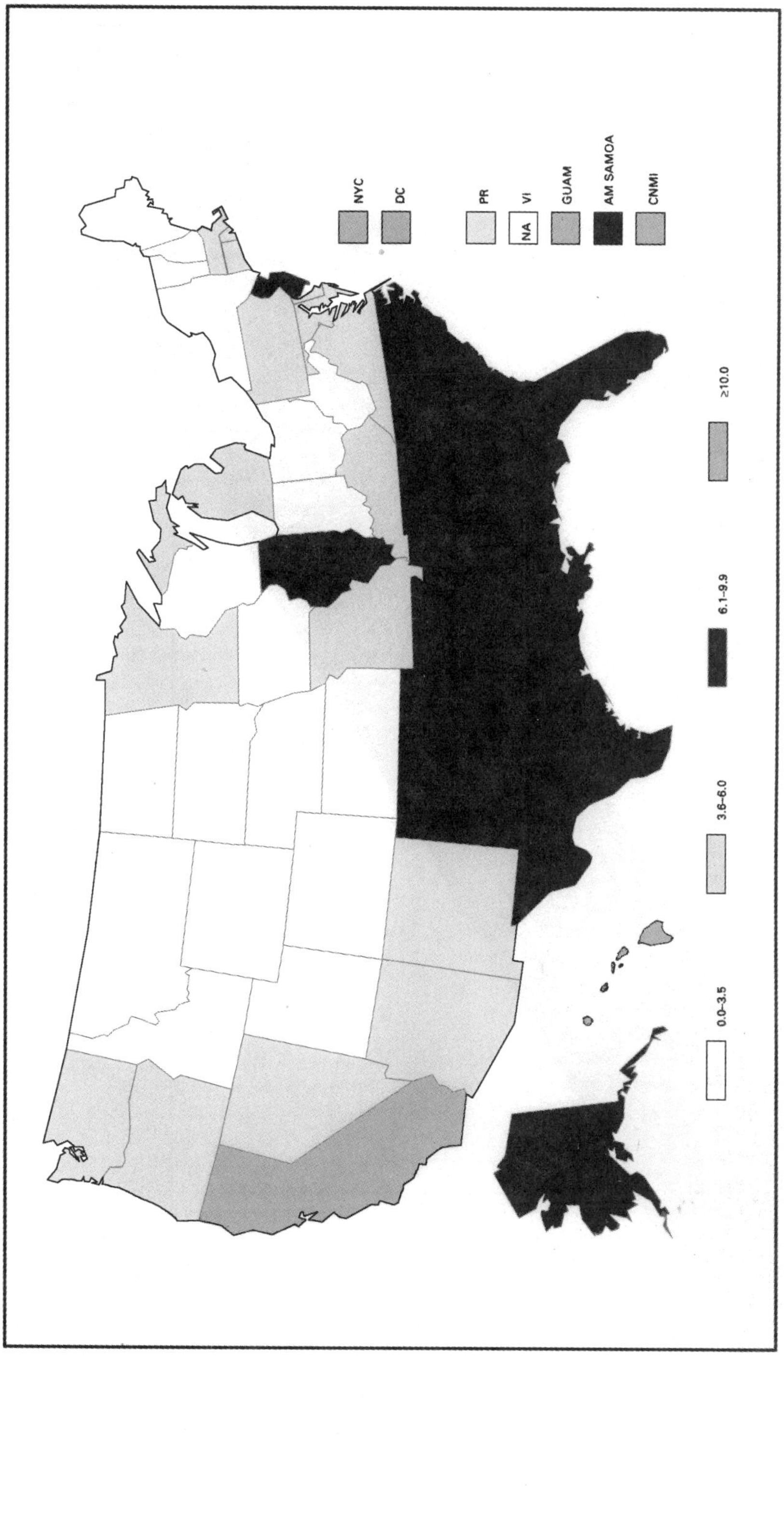

Figure 23.1. Tuberculosis—rates by state, USA (including Puerto Rico and the US Virgin Islands), 1992. Source: MMWR 1992:41(55):59.

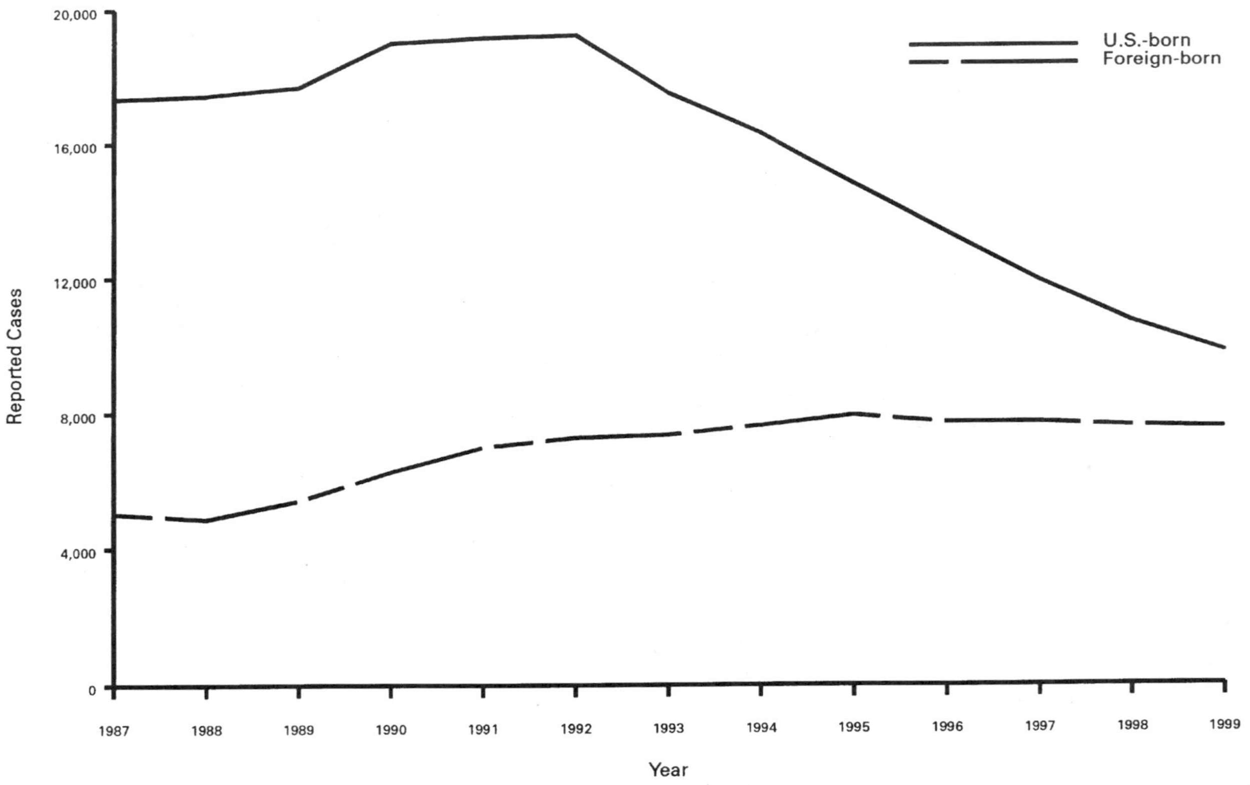

Figure 23.2 Tuberculosis (TB)—reported cases among US-born and foreign-born persons* by year, US, 1987–1999. *In 1999, place of birth was unknown for 169 case-patients. The number of TB cases among foreign-born persons in the US increased from 22% (4925 cases) of the total in 1986 to 43% (7553 cases) of the total in 1999. Source: MMWR 2001:48(53):78.

for routine urinary pathogens. Skin tests can be used to help diagnose TB, but individuals with extensive TB or other immunosuppressing conditions may have false-negative skin tests.[1,2,11]

Skin testing by the Mantoux method using purified protein derivative (PPD) is preferred over other skin test methods because of its greater sensitivity and reproducibility. Contact with non-pathogenic mycobacteria can cause small cross-reactions to tuberculin skin tests, so minimum skin induration diameters have been established to define a tuberculin skin test as positive (Table 23.1). Using the Mantoux method, 5 tuberculin units (5 TU) are injected intradermally by syringe according to the manufacturer's instructions. The 1-TU and the 250-TU products have limited clinical applications. Improper administration, such as failure to administer the skin injection soon after drawing the solution from the vial, can lead to false-negative results. Other causes of false-negative PPD results are immunosuppression, including AIDS, hematogenous malignancy, concurrent serious infection, recent live virus vaccination (measles, mumps, polio), and improper storage or denaturation of the reagent. One can differentiate between general anergy and a truly negative PPD by simultaneously skin testing for reaction to other common exposures, such as *Candida albicans*.[1,2] Diagnostic methods and other information about treatment and prevention of TB can be found on the web site of the Centers for Disease Control and Prevention (www.cdc.gov).

Prior vaccination with bacillus calumet guerin (BCG) does not completely protect against TB infection and is not a likely cause of a positive skin test. Among individuals who have had infections with the tubercle bacillus, the PPD skin response may wane after age 55. To assess for this phenomenon, PPD boosting can be performed by administering two 5-TU tests 1–2 weeks apart. Immunocompetent individuals who carry the tubercle bacillus but who have experienced a wane in PPD response will react to the second PPD. This is known as "the booster phenomenon". Repeated skin testing will not cause a positive PPD among individuals who are uninfected.[1,11]

Sputum, urine or other tissues can be evaluated rapidly by Ziehl–Nielsen (acid-fast) staining for tubercle bacilli. Biopsy of infected tissue may show caseating granulomas as well as acid-fast organisms. Acid-fast organisms

Table 23.1 Criteria for positive Mantoux test.

Clinical situation	Minimum size of positive Mantoux test
No sign of illness, no risk factors for tuberculosis, no contact with tuberculosis	15 mm
Employees in facilities where a person with disease would pose a hazard to large numbers of susceptible persons (such as healthcare facilities, schools, child care facilities, certain food services)	10 mm
Individuals with potential occupational exposure to tuberculosis	
Employees with silicosis	
People with other non-occupational risk factors for tuberculosis (foreign-born persons from endemic area; medically underserved low-income populations; residents of long-term care facilities; persons with certain medical conditions, such as gastrectomy, severe underweight, renal failure, diabetes mellitus, immunosuppressive diseases other than HIV infection, corticosteroid or other immunosuppressive therapy, etc.)	
HIV infection or unknown HIV status with risk factors for HIV infection	$\geq$5 mm
Close recent contact with infectious tuberculosis case	
Chest radiography consistent with old healed tuberculosis	

can be detected and further classified by culture. A proficient and experienced microbiology laboratory should be used. Specimens must be prepared correctly and cultures may require 1 or more months to grow. New techniques, such as DNA polymerase reactions and immunoassays, are under development to allow more rapid identification of *Mycobacterium* species and to determine drug sensitivity.[1] DNA fingerprinting of TB isolates using IS6110 RFLP and spoligotyping has been used to study how TB is transmitted through a population.[12] The technique was able to determine that contact with a patient who had TB was not the source of a healthcare worker's infection in many cases, and to confirm in other cases that unsuspected transmission in occupational, nosocomial and school settings had occurred.

Active TB is a reportable condition. Local health departments can provide information about this procedure and will assist with tracing contacts who might also be infected.

Treatment

A physician familiar with the most up-to-date treatment recommendations should always be consulted when an employee needs anti-tuberculous drug therapy. When therapy is initiated, it is crucial that the individual comply with therapy, both to combat his or her disease and to reduce the likelihood that multidrug resistance will develop. Initial therapy should follow established guidelines; pregnant women may also be safely treated.[13,14]

If an individual who has had a skin test conversion does not have evidence of active TB, prophylactic treatment should be considered. The purpose of prophylaxis is to reduce the likelihood of tuberculous infection progressing to active disease. The usual prophylaxis regimen is isoniazid 300 mg per day for 6 months; however, certain groups, such as immunosuppressed or pregnant individuals, should be treated for 9 months.[5,13,14] Pyridoxine (vitamin B_6) should be taken with isoniazid to prevent peripheral neuropathy, especially in individuals with poor nutrition or other diseases that may cause peripheral neuropathy.[5,13–15]

Isoniazid can cause hepatic toxicity, particularly among individuals over age 35 or those exposed to other hepatotoxicants. Isoniazid can rarely cause liver failure and death, so patients should receive clear instructions to report promptly malaise, nausea, abdominal pain or jaundice and to discontinue isoniazid until cleared by a physician.

Baseline liver function tests and periodic surveillance for symptoms should be performed. Surveillance should include a contact within the first month of beginning isoniazid. At baseline, providers should take a complete medication history to assess concurrent use of hepatotoxic drugs, including acetaminophen. Because alcohol abuse can increase the risk of hepatic toxicity, abstention should be recommended. Close follow-up of alcoholics is warranted.[15]

Active TB is treated with a combination of drugs. The most commonly used drugs are isoniazid, rifampin, pyrazinamide, ethambutol, and streptomycin. An accepted protocol must be used. Microbiological tests for drug resistance are crucial.[16] An effective duration of therapy can be as short as 6 months for low-risk, compliant individuals with no evidence of drug resistance. Each drug has potential side-effects that should be monitored. To prevent peripheral neuropathy, concomitant use of pyridoxine (vitamin B_6) should be considered when administering isoniazid. Persons at greatest risk include individuals with poor nutrition or other diseases, such as diabetes and alcoholism, that may cause peripheral neuropathy.[5,13–15] Patients with active TB should be evaluated clinically at least monthly. They should also have sputum cultures and sensitivities performed monthly for the first 3 months. At the conclusion of therapy for pulmonary TB, sputum cultures should be repeated if sputum can be obtained and a chest radiograph performed. Monthly chest radiographs are not necessary if the patient is progressing well.

Individuals with high likelihood of poor compliance should be placed on observed therapy—i.e. they should report to an outpatient healthcare facility for each dose of medication. Any individual may be non-compliant with therapy, but most at risk are homeless people, migrants, alcoholics, and drug abusers. Local health departments can assist with providing observed therapy.

TB is common among individuals with AIDS. A smaller skin test response is positive for people with HIV infection (Table 23.1) than for individuals without immunosuppression. Even if a skin test is negative, the index of suspicion should be high for any AIDS patient with symptoms. Treatment, which should be managed by a physician expert in AIDS, may require more aggressive drug therapy than is needed for individuals without HIV infection.

Employees with active pulmonary TB should be restricted from work until they are documented to be non-infectious. Coworkers of employees with active TB should be assessed for skin test conversion if there was a likelihood of exposure to aerosolized bacilli. Employees with skin test conversion in the absence of active pulmonary TB do not need to be restricted from work or from patient care activities (Table 23.2).

Medical surveillance

Medical surveillance should be conducted among selected employee groups. Skin testing using the Mantoux technique is the preferred surveillance test. Chest radiographs should not be used for routine surveillance, but they should be performed if skin test conversions are documented or if symptoms suspicious of TB develop.

Workers who have direct contact with patients, clients, animals or tissues known to have a high risk of infectiousness require surveillance. These workers should have skin testing annually. The frequency of skin testing can be increased based on the work site's recent past history of skin test conversions. For workers at very high risk, skin testing every 6 months has been recommended. Many individuals who work in healthcare facilities, such as secretaries or medical record clerks, do not require TB surveillance because they do not have patient contact. Guidelines from the TB branch of the Centers for Disease Control and Prevention can help determine which worker groups should be included in surveillance.[17]

The size of skin reaction considered to be a positive response among populations that have various pretest probabilities of tuberculous infection are shown in Table 23.1. Many clinicians, however, feel uncomfortable with reading as negative an 11–14-mm PPD, even if the patient is not at increased risk for TB. When skin test surveillance is indicated in occupational settings, employees by definition have some increased risk of TB exposure or transmission, so a 10-mm reaction is the correct definition of positive. In addition, employees with skin test reactions of 5–9 mm should be considered positive if they are immunosuppressed or have had close recent contact with an infectious TB case. Courses of action for typical scenarios when TB surveillance is conducted are listed in Table 23.2.

Prevention

Methods to prevent TB transmission in occupational settings focus on using a combination of tactics involving exposure control. Reducing individual susceptibility by means of vaccination with BCG has not been recommended in the USA as a general public health measure, and nor has BCG vaccination been recommended as a general form of worker protection, although its use could be considered in settings where the likelihood is high that transmission and subsequent infection with *M. tuberculosis* strains resistant to isoniazid and rifampin may occur.[18,19] BCG vaccination should not be used, however, in lieu of infection control precautions.

OSHA has developed a TB standard (29 CFR 1910.1035) that is expected to become

Table 23.2 Courses of action for typical scenarios when occupational tuberculosis surveillance is conducted.

Surveillance scenario	Course of action
Previously had negative skin test; now skin test is also negative	No action unless false-negative reaction is suspected
False-negative skin test is suspected	If age 55 or over, repeat 5-TU PPD in 2 weeks If immunosuppressive condition is suspected, refer to experienced personal physician for evaluation Restrict from work pending evaluation only if active pulmonary tuberculosis is suspected, regardless of skin test results
Previously had negative skin test; now has positive skin test	Consider booster phenomenon Refer to experienced personal physician to evaluate for active tuberculosis Consider prescribing prophylactic therapy with isoniazid if active tuberculosis is ruled out (consider need for pyridoxine with isoniazid) Restrict from work pending evaluation only if active pulmonary tuberculosis is suspected If skin test conversion is likely to have been caused by work, check infection control practices and improve them where needed
Untoward reaction to PPD (large, ulcerated reaction, lymphangitis, regional adenopathy, fever)	Refer to experienced personal physician to evaluate for active tuberculosis; restrict from work pending this evaluation if index of suspicion for active pulmonary tuberculosis is high Exclude from further routine PPD skin testing
History of vaccination with bacillus calumet guerin (BCG)	OK to skin test If positive skin test, evaluate for active tuberculosis If active tuberculosis excluded, consider prophylaxis with isoniazid
Signs of possible tuberculosis	Refer to experienced personal physician Exclude from work pending this evaluation if index of suspicion is high
Active pulmonary tuberculosis	Refer to experienced physician for treatment and follow-up May not return to work until acid-fast bacteria have been cleared from the sputum; this usually requires 2 or more weeks of anti-tuberculous therapy and follow-up sputum examination by Ziehl–Nielsen staining for confirmation If high likelihood of non-compliance with drug therapy, observed treatment may be warranted to prevent relapse and multidrug-resistant tuberculosis Evaluate coworkers for skin test conversion if history of exposure to aerosolized bacilli from the index case is likely If tuberculosis is likely to have been caused by work, check infection control practices and improve them where needed
Active extrapulmonary tuberculosis	Experienced personal physician should treat and follow-up If well enough to tolerate working, may work unless there is a draining skin lesion If high likelihood of non-compliance with drug therapy, observed treatment may be warranted to prevent relapse and multidrug-resistant tuberculosis
Recent significant exposure to infectious tubercle bacilli from a person, specimen, culture, primate, or other source	Skin test immediately to establish baseline reactivity status Retest in 6–12 weeks to determine if PPD converts to positive In the interim, if any significant new symptoms develop, refer to personal physician for evaluation Check infection control practices and improve them where needed

final in the near future. The TB standard is based on a series of recommendations made by the Centers for Disease Control and Prevention, OSHA's experience with its 1993 mandatory guidelines that have been enforced under the general duty clause, public comment on the proposed standard and other information.[17,20,21] The proposed OSHA standard is different in a few ways from the recommendations made by the Centers for Disease Control and Prevention. Workplaces that have developed exposure control plans prior to the issuance of the standard will need to review their procedures to determine if they require modification to meet regulatory requirements. Topics that have been modified in the proposed OSHA regulation include risk assessment, medical surveillance and respiratory protection.

Control of occupational exposure to TB should begin with an assessment of past occupational exposure to TB in the facility, and the frequency with which active TB occurs in the surrounding community. Under the proposed standard, certain workplaces may be able to implement an abbreviated exposure control program if they do not admit or provide services to individuals with TB, they have no confirmed cases of TB in the past 12 months, and they are located in a county with a negligible incidence of TB. Otherwise, the full exposure control program should be implemented. In addition, the proposed standard defines the responsibilities of the employer for prevention of TB exposure among contract workers.

The following preventive practices are important for worker protection and regulatory compliance: a written exposure control plan, training for employees, a medical surveillance program, procedures for prompt identification of individuals with suspected or confirmed infectious TB, use of isolation rooms employing engineering controls for such individuals, placement of warning signs at the entrance to high-risk areas, use of specific work practices including respiratory protection, procedures to evaluate employee exposures, medical removal protection for workers who contract TB occupationally, and a record-keeping system that protects confidential medical information appropriately.

Employee training should be provided on hire and annually for those workers with occupational exposure. The training should include the signs and symptoms of TB, hazards of transmission, the purpose and nature of medical surveillance, and site-specific controls. The primary method of transmission of TB through aerosols should be reviewed. Methods to ensure early identification of suspected cases of TB should be in place at the facility, and should be emphasized during training. Medical surveillance should include preplacement evaluation, periodic Mantoux testing and management of persons with positive test results. The proposed OSHA standard states that skin tests must be read by a qualified person, instead of self-reading by an untrained employee.

Isolation rooms with a negative pressure gradient should be available. This will prevent tubercle bacilli from being blown into halls, neighboring patient care rooms, and other work areas. Contaminated air from isolation rooms should not be exhausted near air intake vents, or public walkways. At least six air exchanges per hour without recirculation are recommended for isolation rooms, although the proposed OSHA standard does not specify a required air exchange rate and will allow recirculation if the ventilation system employs HEPA filtration meeting certain criteria. Ultraviolet light may be used to supplement the engineering controls in a facility, but OSHA does not consider ultraviolet light to be a substitute for negative pressure, exhaust ventilation or HEPA filtration. Isolation rooms that are engineered correctly need regular maintenance to ensure that clogged filters do not prevent negative pressure from being achieved. Maintenance and construction employees who are potentially exposed to TB require training, and need to use personal protective equipment.

Respirator use must comply with the provisions of the OSHA respiratory protection standard, as well as specific requirements for prevention of occupational TB. Surgical masks are not respirators, as they are designed to prevent the healthcare provider from contaminating the patient, but they do not prevent the wearer from being exposed to ambient organisms.[5,6,20] Although the Centers for Disease Control and Prevention recommended use of HEPA filter respirators in all circumstances, use of an N95 respirator approved by the National Institute for Occupational Safety and Health (NIOSH) is now acceptable in most exposure settings. Workers must receive medical clearance prior to using the N95 respirator or other respirators. Training should include instructions on the correct way to wear the respirator, how to self-fit test each time a disposable respirator is worn, and how to store a respirator properly. A worker with a full beard cannot achieve an adequate face-to-respirator seal. The proposed OSHA standard lists detailed protocols for employee fit testing using either a qualitative or a quantitative method.

Individuals with active pulmonary TB whose sputum contains acid-fast bacilli should wear a fitted mask (a surgical mask is acceptable) when they leave the isolation room to be transported. They should be instructed to cover their mouth and nose with a tissue when coughing. When these measures are impossible to achieve, e.g. when a patient is combative, the worker should wear respirator. Attendants should also wear respirators when transporting a person with active pulmonary TB in an enclosed vehicle, or when working in the home of such a person, even if the infected individual is wearing a mask.

Sputum induction, administration of aerosolized medications, bronchoscopy and other procedures likely to produce airborne bacilli should be performed in a properly ventilated setting, such as a sputum induction booth, and attendants must wear proper respiratory protection. In the proposed standard, posting a specific sign at the entrance to high-risk areas will be required. The proposed sign will be red, in the shape of a stop sign, and will contain this text: "No admittance without wearing a Type N95 or more protective respirator."

In work settings where tissues or cultures are the source of infectious organisms, engineering controls such as biological safety cabinets, instead of respirators, are the preferred method to prevent exposure. To ensure that the correct air balance occurs in a biological safety cabinet, the HEPA-filtered exhaust air must be discharged either directly to the outside, or by means of proper connections with the building exhaust system. Equipment such as continuous flow centrifuges used to process specimens containing tubercle bacilli may produce aerosols and so these should be exhausted in such a way as to prevent contamination of the laboratory.

In all work settings where there is potential exposure to TB, workers need to follow standard infection control practices, especially since infectious agents other than *M. tuberculosis* may be present. These practices include handwashing and use of protective clothing. Specimen containers should be transported in a plastic bag using gloves. Biohazardous waste should be stored in labeled containers and disposed of properly. Work areas should be cleaned with disinfectant. Although extensive activities to prevent exposure to fomites are not necessary, the manipulation of biological materials containing high concentrations of infectious tubercle bacilli can create hazardous aerosols. Therefore, sputum, body fluids or tissues containing tubercle bacilli should be cleaned up by trained individuals who use respirators and other appropriate personal protective equipment.

Engineering controls, administrative controls, patient care procedures and other work practices should be evaluated periodically for areas that need improvement, with attention to results from medical surveillance such as skin test conversion rates. In the event that an employee contracts TB at work, the proposed OSHA standard stipulates that medical removal procedures should include protection of the person's job with full medical earnings and all other rights and benefits until the worker becomes non-infectious and can return to work, or for 18 months, whichever comes first.

REFERENCES

1. American Thoracic Society. Diagnostic Standards and Classification of Tuberculosis, 1990. *Am Rev Respir Dis* 1990; 142:725–35.
2. Ravigilione MC, O'Brien RJ. Tuberculosis. In: Braunwald E, Fauci AS, Kasper DL, et al, eds. *Harrison's principles of internal medicine*, 15th edn. New York: McGraw-Hill, 2000.
3. Centers for Disease Control and Prevention. Provisional cases of selected notifiable diseases, United States, week ending January 1, 2000. *MMWR*, Available at http:/wonder.cdc/gov/mmwr, accessed 22 September 2000.
4. Centers for Disease Control. Outbreak of multidrug-resistant tuberculosis at a hospital-New York City, 1991. *MMWR* 1993; 42(427):433–4.
5. Mahmoudi A, Iseman MD. Pitfalls in the care of patients with tuberculosis: common errors and their association with the acquisition of drug resistance. *JAMA* 1993; 270:65–8.
6. Dooley SW, Villarino ME, Lawrence M, et al. Nosocomial transmission of tuberculosis in a hospital unit for HIV-infected patients. *JAMA* 1992; 267:2632–5.
7. Manangan LP, Simonds DN, Pugliese G, et al. Are US hospitals making progress in implementing guidelines for prevention of *Mycobacterium tuberculosis* transmission? *Arch Intern Med* 1998; 158:1440–4.
8. Sutton PM, Nicas M, Harrison RJ. Tuberculosis isolation: comparison of written procedures and actual practices in three California hospitals. *Infect Control Hosp Epidemiol* 2000; 21:28–32.
9. Porteous NB, Brown JP. Tuberculin skin test conversion rate in dental health care workers—results of a prospective study. *Am J Infect Control* 1999; 27:385–7.
10. Steenland K, Levine J, Sieber K, et al. Incidence of tuberculosis infection among New York State prison employees. *Am J Public Health* 1997; 87:2012–17.
11. Centers for Disease Control, American Thoracic Society. *Core curriculum on tuberculosis,* 2nd edn. Publication no. 00-5763. Washington, DC: US Government Printing Office, 1991.
12. Bauer J, Kok-Jensen A, Faurschou P, et al. A prospective evaluation of the clinical value of nation-wide DNA fingerprinting of tuberculosis isolates in Denmark. *Int J Tuberc Lung Dis* 2000; 4:295–9.
13. Centers for Disease Control. Initial therapy for tuberculosis in the era of multidrug resistance: recommendations of the Advisory Council for the Elimination of Tuberculosis. *MMWR* 1993; 42(No. RR-7):1–8.
14. Centers for Disease Control. Tuberculosis among pregnant women—New York City, 1985–1992. *MMWR* 1993; 605:611–12.
15. Centers for Disease Control. Severe isoniazid-associated hepatitis—New York, 1991–1993. *MMWR* 1993; 42:545–7.
16. Bloch NB, Cauthen GM, Onorato IM, et al. Nationwide survey of drug-resistant tuberculosis in the United States. *JAMA* 1994; 271:665–71.
17. Centers for Disease Control and Prevention. Guidelines for preventing transmission of tuberculosis in health-care facilities, 1994. *MMWR* 1994; 43(RR13):1–132.
18. Centers for Disease and Prevention. The role of BCG vaccine in the prevention and control of tuberculosis in the United States: a joint statement by the Advisory Council for the Elimination of Tuberculosis and the Advisory Committee on Immunization Practices. *MMWR* 1996; 45(RR4):1–18.
19. Marcus AM, Rose DN, Sacks HS, Schecter CB. BCG vaccination to prevent tuberculosis in health care workers: a decision analysis. *Preventive Med* 1997; 26:201–7.
20. Centers for Disease Control. Guidelines for preventing transmission of tuberculosis in health-care settings, with special focus on HIV-related issues. *MMWR* 1990; 39(No. RR-17):1–29.
21. Occupational Safety and Health Administration. Occupational exposure to tuberculosis; proposed rule—62-54159-54309, http://www.osha-slc.gov/FedReg_osha_data/FED199710-17.html, accessed 22 September 2000.

MYCOBACTERIA other than *Mycobacterium tuberculosis*

Occupational setting

Several mycobacteria other than *M. tuberculosis* can cause disease in humans. Significant occupational exposure is much less common for these bacilli than for *M. tuberculosis*.

Potential contact with *M. kansasii* and *M. avium intracellulare* could occur in patient care settings, especially during care of immunosuppressed patients with pulmonary infections from these organisms. However, person-to-person transmission may not be the only route of exposure.[1] The organisms are also found in the environment and in animal reservoirs. *M. kansasii* outbreaks have occurred among miners, particularly when the miners are in poor health, water used during mining is contaminated with the organism and the miners have evidence of dust-related pulmonary disease or silicosis.[2,3]

Occupational contact with *M. bovis* may occur from working with dairy cattle. Infected beef cattle and game animals such as deer can also transmit the disease.[4,5] Tuberculin testing of cattle and pasteurization of milk have greatly reduced the incidence of *M. bovis* infection, but in some developing countries infection rates remain high in cattle herds.[6]

M. marinum can be acquired via skin exposure to infected fish or to fresh water or salt water aquatic environments. Individuals who are occupationally infected often give a history of minor skin trauma to the hands or other skin immersed in water.[7]

Occupational contact with *M. leprae* is rare in the USA. In the 1980s, more than 900 immigrants and refugees from South East Asia had leprosy, but there was no evidence that these individuals transmitted the disease to others after their arrival[8] (Figure 23.3). The organism is difficult to transmit, although nasal colonization can occur occupationally. In one study, a polymerase chain reaction test revealed *M. leprae* genetic material in nasal swab specimens among 55% of untreated patients, 19% of occupational contacts, and 12% of controls from an endemic region.[9] less than 10% of close family members of affected patients contract the disease.[2,10]

If infected specimens are handled improperly, laboratory workers can be exposed. Contact during travel to underdeveloped nations, where the incidence of mycobacterial disease may be high, is a theoretical possibility, but it is unlikely unless the traveler has prolonged close physical contact with infected, untreated individuals.

Exposure (route)

Routes of exposure have not been entirely elucidated; however, person-to-person contact is not believed to be as high a risk as with *M. tuberculosis*.[1] Most pathogenic *Mycobacterium* species are thought to be absorbed from the upper respiratory tract. Infection with some species can occur after skin contact (*M. marinum, M. ulcerans, M. fortuitum, and M. leprae*). The gastroenterologic route of exposure is important for *M. bovis* (contaminated milk), as is respiratory exposure to aerosols produced during the slaughtering process.[1,4,5,10]

Pathobiology

M. avium intracellulare, *M. kansasii* and other species can cause a potentially serious pulmonary infection similar to that caused by *M. tuberculosis*. Disseminated disease is more common among immunosuppressed individuals, especially those with AIDS or organ transplants. *M. bovis* can cause pulmonary disease. Localized skin infection or lymphadenitis can occur after a percutaneous exposure from *M. fortuitum, M. marinum or M. ulcerans*. Lymphadenopathy, primarily cervical, appears after oral exposure to *M. scrofulaceum*.[1,11]

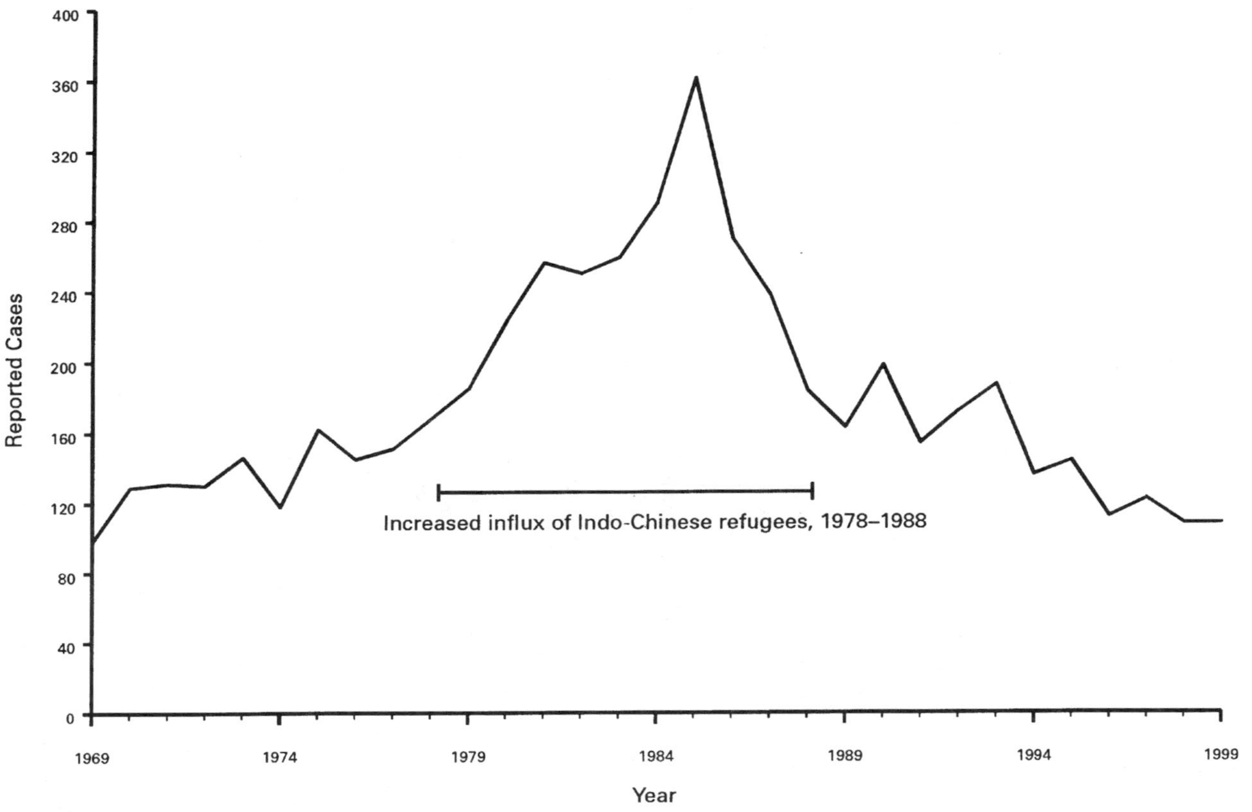

Figure 23.3 Hansen disease (leprosy)—reported cases by year, US, 1969–1999. In 1999, a total of 108 cases of Hansen disease was reported in the US. The number of cases peaked at 361 in 1985, and since 1988, has remained relatively stable. Source: MMWR 2001:48(53):43.

M. leprae affects the skin, soft tissues, and peripheral nerves. The average incubation period is 2–4 years. It is common in some underdeveloped nations but rare in the USA. Anesthetic or paresthetic skin lesions with a typical appearance are usually the first clinical sign of infection with *M. leprae*. In advanced cases, destruction of the soft tissues of the nose and face, fingers, toes and other structures leads to cosmetic deformities. The granulomatous reaction of the host, rather than a neurotoxic effect of the bacilli themselves, is thought to cause nerve damage.[10,11]

Diagnosis

Pulmonary infection is diagnosed in the same manner as *M. tuberculosis*—using microscopic examination of sputum (Ziehl–Nielsen staining), sputum cultures, and chest radiographs. Sputum cultures should be sent to a qualified laboratory. Biopsy material from skin or lymph node lesions suspected of harboring *Mycobacterium* species should be evaluated similarly with staining and culture. Skin tests with PPD can be positive, but tissue diagnosis is more definitive, especially since many infected patients are immunosuppressed.[1,11]

Hansen's disease (leprosy) is diagnosed among individuals with suggestive symptoms by demonstrating *M. leprae* organisms on biopsy of skin or peripheral nerves. Skin testing is of no use, but a serodiagnostic test to a specific phenolic glycolipid from the surface of the bacillus has a high diagnostic specificity. The organism can be cultured in footpads of mice and in the armadillo. The Gillis W. Long Hansen's Disease Center (Baton Rouge, LA) specializes in the care of this disorder.[10,11]

Treatment

With the exception of Hansen's disease, infections with the most common species are treated with the many of the same drugs used with *M. tuberculosis*. *M. avium intracellulare* is commonly multidrug resistant; therefore, two to six drugs are usually chosen from among isoniazid, ethambutol, rifampin, ethionamide, pyrazinamide, cyclosterine, amikacin, and clofazamine. Surgical excision is sometimes used for lymphadenitis and some other localized infections. A physician who is experienced with infectious diseases should direct the patient's care.[1]

Three drugs have been traditionally used to treat the more severe forms of Hansen's disease—dapsone, rifampin, and clofazimine. At least 2 years of drug therapy are required, and indefinite treatment may be necessary. Cosmetic deformities can often be surgically corrected. For less severe cases, dapsone and rifampin are given for at least 6 months.[10] In 1998, the US Food and Drug Administration approved the use of thalidomide to treat skin lesions in leprosy known as erythema nodosum leprosum.

Medical surveillance

Many workers at risk for exposure to these *Mycobacterium* species will already be in a medical surveillance program for *M. tuberculosis*, which should be adequate. The Centers for Disease Control can be used as a resource for program planning. Recommended medical surveillance for the rare worker who may have significant occupational contact with *M. leprae* should consist at least of a clinical examination focusing on detection of typical skin lesions. Serologic testing may be a useful adjunct.[10]

Prevention

A prudent strategy for prevention of possible exposure to bioaerosols of *Mycobacterium* species that cause pulmonary infection follows the same principles as the program for *M. tuberculosis*. Because of concomitant risk of exposure to *M. tuberculosis*, control measures should already be in place in most targeted workplaces. Control measures include rapid diagnosis of infectious patients, respiratory isolation, negative-pressure room ventilation, and maintenance of ventilation systems. Worker training should focus on the use of gloves, protective clothing, and an approved respirator. Biological wastes should be disposed of properly. Patient care procedures likely to produce bioaerosols should be conducted in a properly ventilated area, such as a sputum induction booth. Laboratory workers should use biological safety cabinets. Workers in the fishing industry and in agriculture should use protective clothing and gloves.

To prevent exposure to *M. leprae*, standard infection control measures should be followed. Prophylactic postexposure chemotherapy under the direction of a physician experienced in caring for patients with Hansen's disease should be considered for individuals who have had a significant occupational exposure.[10]

REFERENCES

1. Hirschel B. Other mycobacterial infections. In: Braunwald E, Fauci AS, Kasper DL, et al, eds. *Harrison's principles of internal medicine*, 15th edn. New York: McGraw-Hill, 2000.
2. Corbett EL, Blumberg L, Churchyard GJ, et al. Nontuberculous mycobacteria: defining disease in a prospective cohort of South African miners. *Am J Respir Crit Care Med* 1999; 160:15–21.
3. Chobot S, Malis J, Sebakova H, et al. Endemic incidence of infections caused by Mycobacterium kansasii in the Karvina district in 1968–1995 (analysis of epidemiological data & review). *Cent Eur J Public Health* 1997; 5:164–73.
4. Cousins DV, Williams SN, Dawson DJ. Tuberculosis due to mycobacterium bovis in the Australian population: DNA typing of isolates, 1970–1994. *Int J Tuberc Lung Dis* 1999; 3:722–31.
5. Liss GM, Wong L, Kittle DC, et al. Occupational exposure to Mycobacterium bovis infection in deer and elk in Ontario. *Can J Public Health* 1994; 85:326–9.
6. de Kantor IN, Ritacco V. Bovine tuberculosis in Latin American and the Caribbean: current status, control and eradication programs. *Vet Microbiol* 1994; 40:5–14.
7. Iredell J, Whitby M, Blacklock Z. Mycobacterium marinum infection: epidemiology and presentation in Queensland 1971–1990. *Med J Aust* 1992; 157:596–8.
8. Mastro TD, Redd SC, Breiman RF. Imported leprosy in the United States, 1978 through 1988: an epidemic without secondary transmission. *Am J Public Health* 1992; 82:1127–80.
9. de Wit MY, Douglas JT, McFadden J, et al. Polymerase chain reaction for detection of Mycobacterium leprae in nasal swab specimens. *J Clin Microbiol* 1993; 31:502–6.
10. Geiber RH. Leprosy (Hansen's disease). In: Braunwald E, Fauci AS, Kasper DL, et al, eds. *Harrison's principles of internal medicine*, 15th edn. New York: McGraw-Hill, 2000.
11. Anonymous. Mycobacteria. In: Joklik WK, Willet HP, Amos DB, Wilfert CM, eds. *Zinsser microbiology*, 20th edn. Norwalk: Appleton & Lange, 1992:497–525.

24

FUNGI

Craig S. Glazer, M.D., M.S.P.H., and Cecile S. Rose, M.D., M.P.H.

ALTERNARIA SPECIES

Common name for disease: Wood pulp worker's disease

Occupational setting

Wood-workers exposed via inhalation to wood dusts contaminated with the mold *Alternaria* may develop hypersensitivity pneumonitis, asthma and allergic rhinitis.

Exposure (route)

Inhalation of the pyriform-shaped spores with an average length of 18 µm and diameter of 5 µm can cause hypersensitivity lung disease. Infections of the nails and cornea in an immunocompromised wood-pulp worker has also been reported.[1]

Pathobiology

In 1930, Hopkins described a 37-year-old man whose asthma was associated with exposure to damp, musty areas contaminated with *Alternaria* spp. and confirmed by bronchial challenge.[2] Bronchial challenge with both fungal extracts and whole spores of *Alternaria* has produced both immediate and late asthmatic reactions in sensitized asthmatics.[3] More recently, immunologic sensitization to *Alternaria* has been linked to increased bronchial hyperractivity and to life-threatening asthma.[4,5]

Progressive hypersensitivity pneumonitis leading to chronic interstitial fibrosis documented in two workers with prolonged exposure to *Alternaria* during the manufacture of wood pulp.[6] Eosinophilic pneumonia related to *Alternaria* exposure in a water-damaged home has been reported.[7]

Diagnosis

Diagnosis of sensitizing occupational asthma relies on the symptom and exposure histories,

Physical and Biological Hazards of the Workplace, Second Edition, Edited by Peter H. Wald and Gregg M. Stave
ISBN 0-471-38647-2

findings on pulmonary function testing (including positive methacholine or histamine challenge), and results of peak flow monitoring at and away from work. Allergy skin prick testing to *Alternaria* is often but not always positive. Diagnosis of hypersensitivity pneumonitis relies on a constellation of clinical findings, including a careful symptom and occupational history.[8] Physical examination may be normal or show basilar crackles; the chest radiography may be normal or show diffuse alveolar or interstitial opacities; pulmonary function tests show restriction, obstruction, or a mixed picture. Fiber optic bronchoscopy with bronchoalveolar lavage and transbronchial biopsies may be helpful when the clinical suspicion is strong but routine tests are normal. Exercise physiolgy testing may be helpful in patients with dyspnea but normal resting pulmonary function. Serum-precipitating antibodies to *Alternaria* are often found in asymptomatic exposed workers and may be negative if the wrong antigen preparation is used.

Treatment

The Treatment of hypersensitivity lung diseases begins with prompt removal from antigen exposure. Inhaled corticosteroids are useful as first-line pharmacotherapy for asthma, often in combination with inhaled bronchodilators. Oral corticosteroids may be indicated in patients with hypersensitivity pneumonitis or eosinophilic pneumonia manifested by severe symptoms and radiographic or functional abnormalities.

Prevention

Adequate ventilation and process enclosure, appropriate respiratory protection and work practices that reduce airborne dust levels are recommended. Workers in at-risk industries should be educated regarding exposure risks and encouraged to seek early medical attention for persistent respiratory and systemic symptoms.[8]

REFERENCES

1. Arrese JE, Pierard-Franchimont C, Pierard GE. Onychomycosis and keratomycosis caused by *Alternaria* spp. A bipolar opportunistic infection in a wood-pulp worker on chronic steriod therapy. *Am J Dermatopathol* 1996; 18:611–13.
2. Hopkins J, Benham R, Kesten B. Asthma due to a fungus—*Alternaria. JAMA* 1930; 94:6–11.
3. Licorish K, Novey H, Kozak P. Role of *Alternaria* and *Penicillium* spores in the pathogenesis of asthma. *J Allergy Clin Immunol* 1985; 76:819–25.
4. Nelson HS, Szefler SJ, Jacobs J, Huss K, Shapiro G, Sternberg AL. The relationships among environmental allergen sensitization, allergen exposure, pulmonary function, and bronchial hyperresponsiveness in the Childhood Asthma Management Program. *J Allgery Clin Immunol* 1999; 104:775–85.
5. Black PN, Udy AA, Brodie SM. Sensitivity to fungal allergens is a risk factor for life-threatening asthma. *Allergy* 2000; 501–4.
6. Schlueter D, Fionk J, Hensley G. Wood pulp workers' disease: a hypersensitivity pneumonitis caused by *Alternaria. Ann Intern Med* 1972; 77:907–14.
7. Ogawa H, Fujimura M, Amaike S, Matsumoto Y, Kitagawa M, Matsuda T. Eosinophilic pneumonia caused by *Alternaria alternata. Allergy* 1997; 52:1005–8.
8. Rose C. Hypersensitivity pneumonitis. In: Murray and Nadel, eds. *Textbook of respiratory medicine.* 3rd edn. Philadelphia, 2000: 1867–84.

ASPERGILLUS SPECIES

Common names for disease: Hypersensitivity diseases: Extrinsic allergic alveolitis, farmer's lung, malt worker's lung (*A. clavatus*), allergic bronchopulmonary aspergillosis (ABPA), allergic *Aspergillus* sinusitis, allergic fungal sinusitis (AFS), baker's asthma, mushroom grower's asthma. Infections: Invasive aspergillosis, aspergilloma. Mycotoxins: Mycotoxicosis, aflatoxin-induced liver cancer.

Occupational setting

There are over 600 species in the genus *Aspergillus*. Most *Aspergillus* species are found in soil. Many species are found on a wide variety of substrates, including forage products, food products, cotton, and other organic debris. *A. fumigatus,* the most common species, accounts for most disease, both allergic and infectious. Farmers, sawmill workers, mushroom workers, greenhouse workers, tobacco workers and bird hobbyists are among the many groups at risk from this fungal exposure.[1–10] Workers who deal with compost piles, decomposing haystacks or moldy grains may develop hypersensitivity responses and are at increased risk of developing liver cancer.[11–14] Allergy to *Aspergillus*-derived enzymes has been associated with baker's asthma.[15]

Exposure (route)

Aspergillus species produce conidia in chains measuring between 2–5 μm in diameter that are readily airborne and easily respirable. *Aspergillus* aflatoxin may become airborne as well.[16] Most *Aspergillus* diseases are acquired from inhalation of spores, but airborne spores probably also infect tissues exposed during surgery. Hospital renovations may increase the risk of aspergillosis in immunocompromised hosts due to release of spore-bearing dust. This risk can be attenuated by the addition of either high-efficiency particulate air filtration or laminar airflow.[17] Contaminated ventilation systems and compost sites have been associated with case clusters.[18]

Other portals of entry (including the eye, paranasal sinuses, skin, gastrointestinal (GI) tract, and via intravenous drug use) have been associated with invasive aspergillosis.

Pathobiology

There are four main categories of disease involving *Aspergillus* spp.—allergic aspergillosis, colonizing aspergillosis, invasive diseases (both pulmonary and non-pulmonary), and aflatoxin-induced malignancy.

ALLERGIC ASPERGILLOSIS The four major allergic manifestations of *Aspergillus* sensitization are asthma, hypersensitivity pneumonitis, allergic bronchopulmonary aspergillosis, and allergic sinusitis.

The clinical manifestations of *Aspergillus*-related asthma are no different from those of other forms of extrinsic asthma. The symptoms are: cough, wheezing, chest tightness, and dyspnea, wheezing on examination; and obstructive changes on pulmonary function testing during acute exacerbations. Peripheral eosinophilia is common, and *Aspergillus*-specific IgE antibodies are detectable with serum RAST or ELISA.

Hypersensitivity pneumonitis (extrinsic allergic alveolitis) can occur in individuals with repeated exposure to organic dusts containing *Aspergillus* species. Symptoms may resemble an acute, self-limited flu-like illness occurring 6–12 hours after exposure. Signs include crackles on lung examination, peripheral leukocytosis, infiltrates on chest radiographs, and normal, restrictive or obstructive changes on lung function testing. Symptoms of hypersensitivity pneumonitis may also be subacute or chronic. They include myalgias, weight loss, fatigue, chest tightness, cough, and subtle, progressive dyspnea on exertion. The chest radiograph and pulmonary function tests are often normal in early disease. More sensitive diagnostic testing such as fiberoptic bronchoscopy, may be needed to detect the characteristic lymphocytic alveolitis and granulomatous lung disease. Continued antigen exposure typically leads to worsening symptoms, restrictive physiologic changes, and permanent interstitial fibrosis.

Allergic bronchopulmonary aspergillosis (ABPA) is an inflammatory disease caused by an immunologic response to *Aspergillus fumigatus* and other *Aspergillus* species growing in the bronchi of patients with asthma and cystic fibrosis.[19] Classically, patients present with chest radiographic infiltrates and peripheral blood eosinophilia. ABPA can range from mild asthma to end-stage fibrotic lung disease. Major ABPA dignostic criteria include: (1) asthma or cystic fibrosis; (2) immediate cutaneous reactivity to *Aspergillus*; (3) precipitating (IgG) antibodies to *Aspergillus*; (4) elevated total serum IgE (>1000 ng/ml); and (5) elevated serum IgE antibodies to *Aspergillus*. Minor diagnostic criteria include: (1) proximal bronchiectasis; (2) infiltrates on chest radiograph; and (3) peripheral eosinophilia coincident with radiographic infiltrates.[19]

Allergic fungal sinusitis (AFS) due to *Aspergillus* species typically occurs in immunocompetent atopic patients. Most patients have asthma; 85% have nasal polyposis.[20] Patients present with chronic recurrent sinusitis unresponsive to conventional therapy. Many describe blowing dark, rubbery plugs out of their nose. Fungal hyphae are identified with Gomori methenamine silver stain of the allergic mucin. Diagnostic criteria are similar to those for ABPA; they include peripheral eosinophilia, increased serum IgE levels, and precipitating antibodies to fungal extracts.

COLONIZING ASPERGILLOSIS Saprophytic colonization of air spaces by *Aspergillus* species can result in a number of outcomes, including otomycosis, fungus ball of the paranasal sinuses,[21] endobronchial colonization, and fungus ball of the lung (aspergilloma). Fungus balls of the lung typically occur in patients with pre-existing cavitary lung diseases or chronic allergic aspergillosis, with an aspergilloma forming in an ectatic bronchus. The condition is associated with eosinophilic pneumonia and bronchiectasis. Primary aspergillomas may also occur in patients without allergic disease; they present as fuzzy infiltrates that progress to rounded cavities on the chest radiograph. The major symptom of pulmonary aspergilloma is recurrent hemoptysis. Masses of fungal elements are found on surgical resection. Resolution with residual fibrosis is uncommon. Aspergillomas may occur in association with other cavitary diseases, including tuberculosis and sarcoidosis. Fungus balls occur rarely in the urinary bladder, gallbladder, and bile ducts.

INVASIVE ASPERGILLOSIS This is an uncommon form of *Aspergillus*-related disease. It is also the most serious. It typically occurs in immunocompromised patients, most notably those with leukemia and lymphoma. Fatal invasive microgranulomatous aspergillosis occurred after shoveling moldy cedar wood chips in a patient with chronic granulomatous disease.[22] An unusual case of fatal invasive pulmonary aspergillosis occurred in a non-immunocompromised gardener exposed to heavy environmental concentrations of *A. fumigatus*.[3] The disease presents as pneumonia with fever, cough, leukocytosis, and respiratory distress. Radiographically, there may be diffuse, patchy, alveolar infiltrates or consolidation with mass effect (lung abscess). The necrotizing pneumonia that develops is usually fulminant, and death occurs in 1–2 weeks. Diagnosis is difficult, because *Aspergillus* antibodies are often not detectable in the immunocompromised patient. Depending on the patient's clinical status, invasive diagnostic procedures, such as bronchoscopy or open lung biopsy, early in the course of disease may enable early diagnosis and treatment with intravenous amphotericin B.

Extrapulmonary forms of invasive aspergillosis may involve the eye (keratitis), ear, paranasal sinuses, skin, central nervous system (CNS), heart and GI tract. Intravenous drug injection and iatrogenic procedures (i.e. dialysis, cardiac surgery, intrathecal medication) can create portals of entry for fungal dissemination.

MYCOTOXIN DISEASE Aflatoxin, a mycotoxin produced by many aspergillus species, is an IARC class I carcinogen.[24] Specifically, aflatoxin increases the risk for developing liver cancer in exposed subjects. Increased standardized mortality ratios (SMRs) for liver cancer have been seen in Swedish grain millers, animal feed production plant workers and other agricultural occupations in both Denmark and the USA where aflatoxin exposures probably occur.[12–14]

Treatment

Management of *Aspergillus*-related asthma includes treatment with inhaled steroids and bronchodilators and avoidance of exposure to high airborne concentrations of conidia. Similarly, the mainstay of management of hypersensitivity pneumonitis is removal from further antigen exposure. Treatment with oral corticosteroids may be indicated in persistently or severely symptomatic individuals with abnormal physiology. For patients with ABPA, treatment with systemic corticosteroids is indicated to prevent or minimize bronchiectasis that occurs with each episode of pneumonia. Antifungal therapy with itraconazole may allow a reduction in corticosteroid dose and may be useful if oral corticosteroids are contraindicated.[23] Therapy for AFS usually includes surgical debridement followed by systemic corticosteroids; however, evidence of efficacy is largely anecdotal. Early diagnosis and treatment with amphotericin B is necessary for treatment of invasive aspergillosis.[25]

Prevention

Although *Aspergillus* spp. are ubiquitous, immunocompromised patients should be protected from exposure to high concentrations of fungal conidia and should avoid occupational settings where high spore concentrations are generated. High-efficiency filters remove *Aspergillus* from the air of operating rooms and laminar flow rooms. Isolating immunosuppressed patients from dusty hospital renovation and construction appears useful, as does keeping potted plants out of their rooms. High-efficiency particulate air filtration has been used during renovations to minimize exposure risk.[17]

Occupational groups such as agricultural workers, bird breeders, sawmill workers and brewery workers should be provided with adequate ventilation or respiratory protection in circumstances where exposure concentrations are likely to be high. These occupational groups should be educated regarding the exposure risks and instructed to seek early medical attention for recurrent or persistent respiratory and systemic symptoms. Aggressive diagnostic evaluation of suspect hypersensitivity pneumonitis (including early bronchoscopy) is essential, as is removal from exposure if disease is confirmed.

REFERENCES

1. Yoshida K, Ueda A, Sato K, et al. Hypersensitivity pneumonitis resulting from *Aspergillus fumigatus* in a greenhouse. *Arch Environ Health* 1993, 48:260–2.
2. Yoshida K, Ando M, Ito K, et al. Hypersensitivity pneumonitis of a mushroom worker due to *Aspergillus glaucus*. *Arch Environ Health* 1990; 45:245–7.
3. Zuk J, King D, Zakhour HD, et al. Locally invasive pulmonary aspergillosis occurring in a gardener; an occupational hazard? *Thorax* 1989; 44:678–9.
4. Land C, Hult K, Fuchs R, et al. Tremorigenic mycotoxins from *Aspergillus fumigatus* as a possible occupational health problem in sawmills. *Appl Environ Microbiol* 1987; 58: 787–90.
5. Reijula K, Sutinen S. Immunohistochemical identification of *Aspergillus fumigatus* in farmer's lung. *Acta Histochem* 1984; 75:211–13.
6. Mehta S, Sandhu R. Immunological significance of *Aspergillus fumigatus* in cane-sugar mills. *Arch Environ Health* 1983, 38:41–6.
7. Palmas F, Cosentino S, Cardia P. Fungal airborne spores as health risk factors among workers in alimentary industries. *Eur J Epidemiol* 1989; 5:239–43.
8. Anonymous. Fatal pulmonary aspergillosis following a farm accident. *Chest* 1984; 85: 448–9.
9. Riddle H, Channell S, Blyth W, et al. Allergic alveolitis in a malt worker. *Thorax* 1968; 23:271–80.

10. Huuskonen M, Husman K, Jarvisalo J, et al. Extrinsic allergic alveolitis in the tobacco industry. *Br J Ind Med* 1984; 41:77–83.
11. Vincken W, Roels P. Hypersensitivity pneumonitis due to *Aspergillus fumigatus* in compost. *Thorax* 1984; 39:74–5.
12. Alvanja MC, Malker H, Hayes RB. Occupational cancer risk associated with the storage and bulk handling of agricultural foodstuff. *J Toxicol Environ Health* 1987; 22:247–54.
13. Autrup JL, Schmidt J, Autrup H. Exposure to aflatoxin B1 in animal-feed production plant workers. *Environ Health Perspect* 1993; 99:195–7.
14. Peraica M, Radic B, Lucic A, Pavlovic M. Toxic effects of mycotoxins in humans. *Bull WHO* 1999; 77:756–66.
15. Quirce S, Cuevas M, Diez-Gomez, et al. Respiratory allergy to *Aspergillus*-derived enzymes in bakers' asthma. *J Allergy Clin Immunol* 1992; 90:970–8.
16. Selim MI, Juchems AM, Popendorf W. Assessing airborne aflatoxin B1 during on-farm grain handling activities. *Am Ind Hyg Assoc J* 1998; 59:252–6.
17. Cornet M, Levy V, Fleury L, et al. Efficacy of prevention by high-efficiency particulate air filtration or laminar airflow against *Aspergillus* airborne contamination during hospital renovation. *Infect Control Hosp Epidemiol* 1999; 20:508–13.
18. Topping MD, Scarisbrick D, Luczynska C, et al. Clinical and immunological reactions to *Aspergillus niger* among workers at a biotechnology plant. *Br J Ind Med* 1985; 42:312–18.
19. Cockrill BA, Hales CA. Allergic bronchopulmonary aspergillosis *Annu Rev Med* 1999; 50:303–16.
20. Schwietz L, Gourley D. Allergic fungal sinusitis. *Allergy Proc* 1992; 13:2–6.
21. Robb P. Aspergillosis of the paranasal sinuses: a case report and historical perspective. *J Laryngol Otol* 1986; 100:1071–7.
22. Conrad DJ, Warnock M, Blanc P, et al. Microgranulomatous aspergillosis after shoveling wood chips: report of a fatal outcome in a patient with chronic granulomatous disease. *Am J Ind Med* 1992; 22:411–18.
23. Leon EE, Craig TJ. Antifungals in the treatment of allergic bronchopulmonary aspergillosis. *Ann Allergy Asthma Immunol* 1999; 82:511–17.
24. Anonymous. Aflatoxins. *IARC Monogr Eval Carcinog Risks Hum* 1993; 56:245–395.
25. Stevens DA, Kan VL, Judson MA, et al. Practice Guidelines for Diseases Caused by *Aspergillus*. Infectious Diease Society of America. *Clin Infect Dis* 2000; 30:696–709.

BASIDIOMYCETES (INCLUDING *MERULIUS LACRYMANS*, *LYCOPERDON*, AND MUSHROOMS)

Common names for disease: Mushroom spore asthma, hypersensitivity pneumonitis, extrinsic allergic alveolitis, mushroom worker's lung, mushroom compost worker's lung, lycoperdonosis

Occupational setting

Basidiomycetes are common in nature but are rarely associated with human disease. Allergic reactions (asthma, rhinoconjunctivitis, and hypersensitivity pneumonitis) to inhaled spores have been described for a number of species. Contact dermatitis has also been described but infection is very rare.[1] Certain basidiospores (e.g. *Merulius lacrymans*) contaminate wood with 20–25% water content, and mycelia typically extend in sheets over timber and adjacent brickwork.[2,3] Mushroom workers may develop asthma, allergic rhinitis or hypersensitivity pneumonitis from inhalation of several species of mushroom spores.[4–8] Mushroom soup workers have been known to develop asthma and allergic rhinoconjunctivitis from inhaled mushroom dusts.[9]

Exposure (route)

Inhalation of basidiospores causes hypersensitivity lung disease. Contact dermatitis after skin exposure to *Hericium erinaceum* has been reported.[1]

Pathobiology

The *Basidiomycetes* are known as club fungi because, after mycelial growth, a fruiting body is formed and club-shaped structures called basidia develop, where basidiospores are produced. Many diverse forms are included in this class, including puffballs, common rusts and smuts, mushrooms, and certain yeasts. Spores may be discharged in bursts during times of high humidity.

Atopic asthmatics may demonstrate IgE-mediated reactivity to Basidiomycete aeroallergens. *M. lacrymans,* a basidiomycete found in buildings in cool temperate climates, has been associated with both asthma and hypersensitivity pneumonitis.[2,3,10] In one report, a teacher developed insidious onset of symptoms of dyspnea, cough, malaise, fever and weight loss. He also had rales on physical examination. His chest radiograph showed diffuse micronodular infiltrates. Pulmonary function tests showed low-diffusing capacity for carbon monoxide. Serum precipitins to *M. lacrymans* present in his home were positive, as was inhalation challenge with the fungus. Clinical recovery progressed slowly following cessation of antigen exposure.

Inhalation of dried powder from the fleshy basidiomycete, *Lycoperdon* (puffball) has been associated with lycoperdonosis, a form of hypersensitivity pneumonitis that develops following treatment of epistaxis.[11]

In another report, eight workers exposed to dried mushroom soup developed symptoms of asthma and rhinoconjunctivitis.[6] A type I hypersensitivity reaction was suggested by clinical history, positive immediate skin prick test reactivity to mushroom extracts, and immediate response to inhalation challenge.

Mushroom worker's lung typically occurs during the first few months of employment, though sensitization after many years of exposure is also described.[6] Seven workers at a mushroom farm in Florida developed episodic dyspnea, cough, fever, chills and myalgias. Pulmonary function tests and chest radiographs showed evidence of hypersensitivity pneumonitis. The workers were subsequently removed from exposure but a specific causative antigen was not identified.[4] In a similar case series from Japan, specific precipitating antibodies were found to mushroom spore extracts.[8]

Diagnosis

Diagnosis of sensitizing occupational asthma relies on the symptom and exposure histories, findings on pulmonary function testing (including positive methacholine or histamine challenge), and results of peak flow monitoring at and away from work. Allergy skin prick testing to mushroom antigen is often (but not always) positive. Diagnosis of hypersensitivity pneumonitis relies on a constellation of clinical findings, including a careful symptom and occupational history.[12] Physical examination may be normal or show basilar crackles; the chest radiography may be normal or show diffuse alveolar or interstitial opacities; pulmonary function tests show restriction, obstruction, or a mixed picture. Fiber optic bronchoscopy with bronchoalveolar lavage and transbronchial biopsies may be helpful when the clinical suspicion is strong but routine tests are normal. Exercise physiology testing may be helpful in patients who have dyspnea but normal resting pulmonary function. Serum-precipitating antibodies to mushroom extracts may be found in asymptomatic exposed workers, and may be negative if the wrong antigen preparation is used.

Treatment

The treatment of hypersensitivity lung diseases begins with early recognition and prompt removal from antigen exposure. Following removal, inhaled corticosteroids are useful as first-line pharmacotherapy for asthma. The efficacy of oral corticosteroids in the treatment of hypersensitivity pneumonitis is untested, but treatment is probably indicated in patients with severe symptoms, or radiographic or functional abnormalities.

Prevention

Once sensitization has occurred, the prognosis for recovery from occupational asthma is directly related to the duration of exposure. If removal from exposure is delayed, individuals with hypersensitivity pneumonitis are at risk for disease progression. A high index of clinical suspicion for workers in at-risk environments is therefore crucial. Reduction of workplace exposure to high fungal concentrations through engineering and process controls reduces the risk of disease.

REFERENCES

1. Maes MF, van Baar HM, van Ginkel CJ. Occupational allergic contact dermatitis from the mushroom White Pom Pom (*Hericium erinaceum*). *Contact Dermatitis* 1999; 40: 285–90.
2. Herxheimer H, Hyde H, Williams D. Allergic asthma caused by basidiospores. *Lancet* 1969; 131–3.
3. O'Brien I, Bull J, Creamer B, et al. Asthma and extrinsic allergic alveolitis due to *Merulius lacrymans*. *Clin Allergy* 1978; 8:535–42.
4. Sanderson W, Kullman G, Sastre J, et al. Outbreak of hypersensitivity pneumonitis among mushroom farm workers. *Am J Ind Med* 1992; 22:859–72.
5. Michils A, DeVuyst P, Nolard J, et al. Occupational asthma to spores of *Pleurotus cornucopiae*. *Eur Respir J* 1991; 4:143–7.
6. Mori S, Nakagawa-Yoshida K, Tsuchihashi H, et al. Mushroom worker's lung resulting from indoor cultivation of *Pleurotus osteatus*. *Occup Med* 1998; 48:465–8.
7. Helbling A, Gayer F, Brander KA. Respiratory allergy to mushroom spores: not well recognized, but relevant. *Ann Allergy Asthma Immunol* 1999; 83:17–19.
8. Akizuki N, Inase N, Ishiwata N, et al. Hypersenstivity pneumonitis among workers cultivating *Tricholoma conglobatum* (Shimeji). *Respiration* 1999; 66:273–8.
9. Symington I, Kerr J, McLean D. Type I allergy in mushroom soup processors. *Clin Allergy* 1981; 11:43–7.
10. Horner WE, Helbling A, Lehrer SB. Basidiomycete allergens. *Allergy* 1998; 53:1114–21.
11. Strand R, Neuhauser E. Lycoperdonosis. *N Engl J Med* 1967; 277:89–91.
12. Rose C. Hypersensitivity pneumonitis. In: Murray and Nadel, eds. *Textbook of respiratory medicine*. 3rd edn. Philadelphia, 2000:1867–84.

BLASTOMYCES DERMATITIDIS

Common names for disease: Blastomycosis, North American blastomycosis, Gilchrist's disease, Chicago disease, Namekagon fever

Occupational setting

Exposure to the fungus *Blastomyces dermatitidis* can cause the infection blastomycosis. Blastomycosis is most prevalent in the southeastern USA and the Ohio–Mississippi River Valley area. However, the geographic range may be broader than previously believed, as blastomycosis has recently been reported in Colorado following prairie dog relocation.[1,2] An African form of blastomycosis has been reported. Disease is much more common in males than in females (9 : 1).

Epidemiologic studies suggest that patients often work outdoors and have intimate contact with soil. A horticulturist developed progressive blastomycosis from exposure to contaminated fertilizer.[3] A tobacco worker in Switzerland and a packing material handler in England developed the illness after handling fungal fomites.[4] There have been occasional reports of small clusters or disease outbreaks in many areas of the USA and Canada.[5] In Virginia, four hunters were infected while raccoon hunting at night in swampy wooded areas. In a Minnesota outbreak, four cases developed in three families constructing a cabin in a wooded area near a lake.[6] In a 1979 Wisconsin outbreak, seven of eight individuals camping near a river developed acute pneumonia.[7] In a larger outbreak in Wisconsin in 1984, numerous elementary school children and several adults who visited a beaver pond at a campground developed symptomatic blastomycosis with an incubation period between 21–106 days after exposure to the presumed point source.[8,9] A technician working for several years in a small, dusty, wooden petroleum-filtering shed in southwest Ontario developed systemic blastomycosis with meningeal involvement; *B dermatitidis* was isolated from the earthen floor of the shed.[10]

These data suggest that *B. dermatitidis* survives in wet soil of acidic pH with a high organic content and probably exists in point sources close to rivers, streams, or swamps. Environmental conditions during cool months may be more favorable for the saprophytic growth and survival of the fungus; disturbance of these sites either through occupational or avocational activities may lead to airborne dispersal of the spores.

Several cases of laboratory-acquired disease have been reported; the majority resulting from finger inoculation with the yeast form during autopsy by pathologists who developed primary cutaneous blastomycosis.[11–13] Primary pulmonary blastomycosis can be a laboratory-acquired infection.[14]

Exposure (route)

The most important route of exposure leading to infection from *B. dermatitidis* is disturbance of contaminated point sources leading to airborne dispersal of spores. Accidental skin inoculation of the yeast form has been reported, as has transmission via dog bites. Venereal transmission of genitourinary infection and in utero transmission are rare.

Pathobiology

B. dermatitidis is a dimorphic fungus that grows as a mycelial form at room temperature and as a yeast form at 37°C.

The lung is the organ most commonly infected by *B. dermatitidis;* the resulting illness is usually indolent in onset. Symptoms may be present for weeks, months or even years before diagnosis. Symptoms typically include cough, weight loss, chest pain, skin lesions, fever, hemoptysis, and localized swelling. In almost half the patients with pulmonary infection, respiratory symptoms are mild or absent. It is usually systemic symptoms or extrapulmonary lesions that lead to medical attention.

A number of chest radiographic patterns have been described, including a patchy alveolar airspace process with air bronchograms, fibronodular densities, miliary nodules, linear interstitial infiltrates, and cavitation. Pleural effusions and hilar adenopathy are uncommon. Laryngeal, tracheal or endobronchial lesions are seen occasionally. The rate of progression of indolent disease may be gradual or sudden and rapid. Occasionally, patients present with acute symptoms of fever, productive cough, and pleuritic chest pain. The chest radiograph typically shows single or multiple nodular or patchy infiltrates. Spontaneous improvement of acute blastomycotic pneumonia may occur after 2–12 weeks of symptoms.

Blastomycotic skin lesions involving the face, extremities, neck and scalp are common and provide ready access for biopsy and culture. Most lesions arise from hematogenous seeding from the lung, although local inoculation may occur in researchers, pathologists and morticians handling infected tissue.

Osteomyelitis involving vertebrae, skull, ribs, and distal extremities is found in 14–60% of cases. Osseous lesions may produce symptoms from abscess development in adjacent soft tissue, by spread to contiguous joints, or by vertebral collapse. Radiography shows a sharply defined area of osteolysis.

Blastomycotic arthritis, typically monarticular, is not rare, and first appears as swelling, pain, and limited range of motion in an elbow, knee, or ankle. Infection of the prostate, epididymis, or kidney can be documented in cultured urine in one-fourth of cases. Hematogenous spread to the brain occurs in 3–10% of cases and may present as meningitis, brain abscess, spinal epidural lesions, or cranial lesions. Blastomycotic lymphadenitis and intraocular infection have been reported.

Diagnosis

Diagnosis of blastomycosis requires isolation of the fungus in culture or the demonstration of characteristic yeast-like cells in pus, sputum, or tissue. Mycelial growth is usually evident within 3–14 weeks of incubation at 25–30°C, but cultures should be kept for at least 4 weeks before recording them as negative.

Treatment

Oral itraconazole is the treatment of choice for patients with indolent extracranial blastomycosis. Amphotericin B remains the treatment of choice for patients with severe, rapidly progressive or CNS infection.[15]

Prevention

Since environmental point sources in soils close to rivers and swamps are difficult to identify, and since occupational inhalational exposures are rare, few preventive methods have been identified. Careful handling of laboratory specimens using BSL2 practices and proceedures,[16] and the use of impermeable gloves will minimize the risk of aerosol exposure and hand inoculation.

REFERENCES

1. De Groote MA, Bjerke R, Smith H, et al. Expanding epidemiology of blastomycosis: clinical features and investigation in Colorado. *Clin Infect Dis* 2000; 30:582–4.
2. Anonymous. From the Centres for Disease Control and Prevention. Blastomycosis acquired occupationally during prairie dog relocation—Colorado, 1998. *JAMA* 1999; 282:21–2.
3. Sarosi G, Serstock D. Isolation of *Blastomyces dermatitidis* from pigeon manure. *Am Rev Respir Dis* 1976; 114:1179–93.
4. Anonymous. Blastomycosis. In: Rippon JW, ed. *Medical mycology: the pathogenic fungi and the pathogenic actinomycetes*. WB Saunders, 1982:428–58.
5. Dwight PJ, Naus M, Sarsfield P, et al. An outbreak of human blastomycosis: the epidemiology of blastomycosis in the Kenora catchment region of Ontario, Canada. *Can Communicable Dis Rep* 2000; 26:82–91.
6. Vaaler A, Bradsher R, Davies S. Evidence of subclinical blastomycosis in forestry workers in northern Minnesota and northern Wisconsin. *Am J Med* 1990; 89:470.
7. Cockerill F, Roberts G, Rosenblatt J, et al. Epidemic of pulmonary blastomycosis (Namekagon fever) in Wisconsin canoeists. *Chest* 1984; 86:688–92.
8. Klein B, Vergeront J, Weeks R, et al. Isolation of *Blastomyces dermatitidis* in soil associated with a large outbreak of blastomycosis in Wisconsin. *N Engl J Med* 1986; 314:529–34.
9. Klein B, Vergeront J, DiSalvo A, et al. Two outbreaks of blastomycosis along rivers in Wisconsin: isolation of *Blastomyces dermatitidis* from riverbank soil and evidence of its transmission along waterways. *Am Rev Respir Dis* 1987; 136:1333–8.
10. Bakerspigel A, Kane J, Schaus D. Isolation of *Blastomyces dermatitidis* from an earthen floor in southwestern Ontario, Canada. *J Clin Microbiol* 1986; 24:890–1.
11. Larson D, Eckman M, Alber R, et al. Primary cutaneous (inoculation) blastomycosis: an occupational hazard to pathologists. *Am J Clin Pathol* 1983; 79:523–5.
12. Larsh H, Scharz J. Accidental inoculation blastomycosis. *Cutis* 1977; 19:334–7.
13. Kantor G, Roenigk R, Mailin P, et al. Cutaneous blastomycosis. Report of a case presumably acquired by direct inoculation with carbon dioxide laser vaporization. *Cleve Clin J Med* 1987; 54:121–4.
14. Baum G, Lerner P. Primary pulmonary blastomycosis: a laboratory-acquired infection. *Ann Intern Med* 1970, 73:263–9.
15. Chapman SW, Bradsher, RW Jr, Douglas G, et al. Practice guidelines for the management of patients with Blastomycosis. Infectious Disease Society of America. *Clin Infect Dis* 2000; 30:679–83.

16. Richmond JY, McKinney RW, eds. *Biosafety in Microbiological and biomedical laboratories*, 4th edn. HHS publication no. (CDC) 93–8395. Washington, DC: US Government Printing Office, 1999:118.

CANDIDA SPECIES

Common names for disease: Candidiasis, Candidosis, thrush, moniliasis

Occupational setting

Candida species are found in soils, especially those with heavy organic debris, and have been recovered from hospital environments and inanimate objects. The intact integument is the most important defense against cutaneous candidiasis. Environmental factors that lead to increased moisture, such as prolonged immersion of hands in water or tight clothing worn in hot climates, increase the risk for cutaneous candidiasis by compromising the tissue.[1] *Candida* paronychia often arises after continued immersion and mechanical irritation of the hands. Homemakers, dishwashers, bartenders, cannery workers and nurses are at risk for cutaneous candidiasis.[2] Non-occupational iatrogenic factors (antibiotics, immunosuppressants, barrier breaks, prostheses) and chronic disease such as diabetes are the most common causes of systemic candidiasis.

Exposure (route)

Organisms are normal commensals, and the vast majority of human infections are of endogenous origin. Person to person transmission has been described.

Pathobiology

C. albicans and the other *Candida* species are budding yeasts that produce mycelium with continued growth. *Candida* invasion of the moist areas of the skin produces a red, "scalded skin" lesion with a scalloped border. Satellite pustular lesions surround the primary lesion, and dry scaly lesions may also occur.

Diagnosis

Skin scrapings examined microscopically in potassium hydroxide (KOH) show budding yeast and mycelial hyphae.

Treatment

Nystatin ointment, topical amphotericin, gentian violet and a number of other topical treatments are effective in the treatment of *Candida* paronychia and intertriginous candidiasis.

Prevention

Avoidance of tight clothing in tropical climates, of tight boots which macerate the skin and of prolonged immersion of hands will prevent tissue compromise and diminish the risk of cutaneous candidiasis.

REFERENCES

1. Campbell M, Stewart J. *The medical mycology handbook*. New York: Wiley, 1980; 244–52.
2. Hunter P, Harrison G, Fraser C. Cross-infection and diversity of *Candida albicans* strain carriage in patients and nursing staff on an intensive care unit. *J Med Vet Mycol* 1990; 28:317–25.

CLADOSPORIUM SPECIES

Common names for disease: Asthma, allergic rhinoconjunctivitis, hypersensitivity pneumonitis

Occupational setting

Cladosporium spp. are ubiquitous in nature. Peak ambient air levels generally occur in summer and early fall.[1,2] Farmers, agricultural workers, and occupants of water-damaged buildings are at risk for exposure and associated hypersensitivity lung disease.[3] Rarely, skin, corneal, CNS, and pulmonary infections may occur.

Exposure (route)

Exposure occurs primarily through inhalation of airborne conidia or spores. Skin infection may occur with direct inoculation.

Pathobiology

Cladosporium sensitization occurs with a prevalence of about 3%,[4] and is associated with allergic rhinitis and eczema.[5] *Cladosporium* sensitization has also been associated with increased bronchial hyperresponsiveness and is a risk factor for both the development of asthma and fatal asthma attacks.[4,6,7] Occupational asthma related to *Cladosporium* sensitization is rare. Other hypersensitivity reactions to *Cladosporium* include allergic bronchopulmonary mycosis and hypersensitivity pneumonitis (HP).[8,9] HP developed in a 48-year-old woman afer exposure to a contaminated indoor hot tub. *Cladosporium* was confirmed as the cause by positive specific challenge.[9]

There are approximately 30 reports of brain abscess caused by *C. trichoides* in the literature, primarily in immunocompromised hosts.[10] Skin and pulmonary infections also occur but are uncommon.

Diagnosis

Diagnosis of asthma relies on symptom and exposure histories, findings on pulmonary function testing (including positive non-specific bronchial challenge), and the results of peak flow monitoring. Allergy skin prick testing and specific IgE to *Cladosporium* are often but not always positive. Diagnosis of HP relies on a constellation of clinical findings, including a careful symptom and exposure history.[11] Physical examination may be normal or show basilar crackles; the chest radiography may be normal or show diffuse alveolar or interstitial opacities; and pulmonary function tests show restriction, obstruction, or a mixed picture. Fiber optic bronchoscopy with bronchoalveolar lavage and transbronchial biopsies may be helpful when the clinical suspicion is strong but routine tests are normal. Exercise physiology testing may be helpful in patients who have dyspnea but normal resulting pulmonary function.

Diagnosis of infectious disease related to *Cladosporium* requires positive culture or the demonstration of fungal forms in histologic specimens.

Treatment

As with all hypersensitivity lung diseases, prompt removal from exposure is essential. Inhaled corticosteroids are useful as first-line pharmacotherapy for asthma. Oral corticosteroids may be indicated in patients with HP manifested by severe symptoms and radiographic or functional abnormalities.

Fluconazole at a dose of 400 mg/day in combination with surgery has successfully treated CNS infection.[12] Progressive disease is generally treated with intravenous amphotericin B; however, this has not been shown to alter outcome.[10]

Prevention

Rapid remediation of water damage in homes and office buildings will prevent fungal growth and thus exposure. Proper attention to building practices during new construction will help prevent subsequent leaks and water damage. In agricultural settings, engineering and process controls can reduce exposure to high fungal concentrations. Workers in at-risk industries should be educated regarding exposure risks and encouraged to seek early medical attention for persistent respiratory and systemic symptoms.[11]

REFERENCES

1. Mediavilla MA, Angulo RJ, Dominguez VE, et al. Annual and diurnal incidence of *Cladosporium* conidia in the atmosphere of Cordoba, Spain. *J Invest Allerg Clin Immunol* 1997; 7:179–82.
2. Ren P, Nakun TM, Leaderer BP. Comparisons of seasonal fungal prevalence in indoor and outdoor air in house dusts of dwellings in one Northeast American country. *J Expos Anal Environ Epidemiol* 1999; 9:560–8.
3. Kotimaa MH, Terho EO, Husman K. Airborne moulds and actinomycetes in work environment of farmers. *Eur J Respir Dis Suppl* 1987; 152:91–100.
4. Chinn S, Jarvis D, Luczynska C, et al. Indivicual allergens as risk factors for bronchial responsiveness in young adults. *Thorax* 1998; 53:662–7.
5. Bundgaard A, Boudet L. Reproducibility of early asthmatic response to *Cladosporium herbarum. Eur J Respir Dis Suppl* 1986; 143:37–40.
6. Abramson M, Kutin JJ, Raven J, et al. Risk factors for asthma among young adults in Melbourne, Australia. *Respirology* 1996; 1:29–7.
7. Black PN, Udy AA, Brodie SM. Sensitivity to fungal allergens is a risk factor for life-threatening asthma. *Allergy* 2000; 501–4.
8. Moreno-Ancillo A, Diaz-Pena JM, Ferrer A, et al. Allergic bronchopulmonary cladosporiosis in a child. *J Allergy Clin Immunol* 1996; 97:714–5.
9. Jacobs RL, Thorner RE, Holcomb JR, et al. Hypersensitivity pneumonitis caused by *Cladosporium* in an enclosed hot-tub area. *Ann Intern Med* 1986; 105:204–6.
10. Dixon DM, Walsh TJ, Merz WG, et al. Infections due to *Xylohypha bantiana* (*Cladosporium, trichoides*), *Rev Infect Dis* 1989; 11: 515–25.
11. Rose C. Hypersensitivity pneumonitis. In: Murray & Nadel, eds. *Textbook of respiratory medicine*, 3rd edn. Philadelphia, 2000:1867–84.
12. Turker A, Altinors N, Aciduman A, et al. MRI findings and encouraging fulconazole treatment results of intracranial *Cladosporium trichoides* infection. *Infection* 1995; 23: 60–2.

COCCIDIODES IMMITIS

Common names for disease: Coccidioidomycosis, valley fever, desert rheumatism, valley bumps, Calfornia diseases, Posada's mycosis

Occupational setting

Coccidioides immitis is a soil fungus that is endemic in the semiarid or desert-like regions of the USA, Mexico, Guatemala, Honduras, Colombia, Venezuela, Bolivia, Paraguay, and Argentina.[1] *C. immitis* may be dispersed by wind or by disruptions from construction work, farming, or archeological digs. Outbreaks after natural disasters (e.g. earthquakes) have been reported.[2] Occupations at risk for developing coccidioimycosis include agricultural workers, construction crews, telephone post diggers, archeologists, and military personnel traveling to endemic areas.[3–5] Laboratory workers are at risk for infection from inhalation of the arthroconidia.[6] Although there are no racial, gender, or age differences in susceptibility to primary coccidioidomycosis, dark-skinned people are more prone to severe primary illness and disseminated disease.

Cases of occupational person-to-person transmission are very rare. Six medical staff members were infected after inhaling arthrospores that had grown on the plaster cast of a patient with coccidioidal osteomyelitis. An embalmer developed disease after accidentally piercing his skin with a needle during preparation of the body of a victim of disseminated coccidioidomycosis.[7,8] There have been occasional reports of coccidioidomycosis in

Georgia, Virginia and North Carolina among textile workers who inhaled dust particles from wool or cotton shipped from the San Joaquin Valley.

Exposure (route)

The primary route of exposures is inhalation. Skin inoculation and transplacental infection from mothers with disseminated disease have rarely been reported.[9]

Pathobiology

C. immitis exists in the mycelial phase in soil, where it matures to form arthroconidia that can be inhaled. In the host, these spores swell to form thick-walled, non-budding, round cells that contain endospores.

In most cases, coccidioidomycosis is a mild respiratory infection or is completely asymptomatic. Primary coccidioidomycosis occurs in $\sim$40% of patients with positive coccidioidin skin tests. Symptoms typically begin 10–16 days after exposure; they include fever, pleuritic or dull chest pain, cough, white or blood-streaked sputum, and constitutional symptoms of malaise, headache, anorexia, myalgia, and fever. A fine diffuse, erythematous skin rash often occurs within the first few days of illness. Erythema nodosum and erythema multiforme are more common in Caucasian women with primary coccidioidomycosis; they are accompanied by arthralgias of the knee or ankle in a third of cases. The rash, arthralgias and mild conjuctivitis that often occur are probably all manifestations of exuberant delayed-type hypersensitivity reactions to *C. immitis* antigens. The chest radiograph typically shows alveolar infiltrates, with or without hilar adenopathy. Paratracheal or mediastinal adenopathy suggests that infection may be spreading beyond the lung. Small pleural effusions may occur, but large effusions are uncommon. Laboratory studies often show a mild leukocytosis, elevated sedimentation rate, and eosinophilia. Conversion of the coccidioidin skin test to positive occurs 2–21 days after oneset of symptoms. The appearance of complement-fixing antibodies in serum is often delayed.

Coccidioidal pneumonia may resolve by forming dense spherical nodules (coccidioidomas) in the area of infiltrate. The mass may cavitate, leaving a single, thin-walled cavity; approximately half of these cavities close spontaneously within 2 years. Though hemoptysis can occur, most are asymptomatic. Extension of the cavity to the pleura can cause bronchopleural fistula, pneumothorax, and coccidioidal empyema. Acute coccidioidal pneumonia may disseminate rapidly; it is potentially fatal. Some individuals develop a chronic progressive form of pulmonary coccidioidomycosis that mimics tuberculosis, with apical fibronodular lesions and cough, weight loss, fever, and chest pain of many months duration.

Extrapulmonary dissemination occurs in <5% of cases, most often in dark-skinned men. Pregnancy also appears to increase the risk of dissemination. Infection in later stages of pregnancy results in increasing morbidity and mortality for the mother. In disseminated infection, skin and subcutaneous lesions are the most common manifestations. Bone lesions occur in 20% of patients with disseminated disease. Meningitis occurs in one-third to one-half of cases of disseminated coccidioidomycosis, usually with a subacute or chronic presentation including headache, lethargy, confusion, or decreased memory. Anorexia, nausea, weight loss, and ataxia may occur. The most valuable diagnostic test for meningitis is the complement fixation test for cerebrospinal fluid (CSF) antibody to *C. immitis*, which is positive in 75–95% of cases. Multiple organ systems may be involved in disseminated coccidioidomycosis. Patients are at risk for anterior and posterior uveitis, lymphadenitis, cystitis, renal abscess, orchitis, epididymitis, peritonitis, urethroscrotal fistula, laryngitis, and otitis.

Diagnosis

Together with the typical clinical manifestations, a positive coccidioidin skin test is suggestive for disease in someone who has recently traveled to an endemic area. Coccidioidal serology on acute and convalescent sera is a reliable means of diagnosis. Newer antibody tests may assist in diagnosing progressive disease.[10] Recovery of *C. immitis* from sputum, urine, or bronchial washing is definitive, but a negative culture does not exclude the diagnosis. Microscopic identification of *C. immitis* spherules on wet smear can aid in diagnosis. In disseminated coccidioidomycosis with meningitis, CNS serologic tests are helpful. Demonstration of *C. immitis* by culture, smear, or biopsy is the most definitive diagnostic test in disseminated disease. Fungemia is detected in ~12% of disseminated cases; it is an extremely grave prognostic sign.

Treatment

Because spontaneous cure is common, treatment is not usually necessary for acute pulmonary coccidioidomycosis.[11] Intravenous amphotericin B may be indicated in some very ill patients with primary illness without proof of dissemination in an effort to prevent extrapulmonary foci. The only effective treatment for cavitary disease is surgical resection, with intravenous amphotericin B serving an adjunctive role. Repeated bacterial superinfection and the presence of an expanding cavity near the pleural surface are indications for resection of the cavities. Treatment of patients with chronic, indolent, apical coccidioidal pneumonia is difficult. Initial treatment with azoles is generally preferred. For those patients who fail initial therapy, options include switching to an alternative azole or to intravenous amphotericin B. When infection is confined to one lobe, combination treatment with intravenous amphotericin B and resection may be useful.

In the treatment of disseminated coccidioidomycosis, therapy with an oral azole is begun unless the patient is severely ill, in which case intravenous amphotericin B is used. When amphotericin B is used initially, an oral azole is usually substituted once the patient has stabilized.[11] Oral ketoconazole, itraconazole, or fluconazole following a course of treatment with intravenous amphotericin B can help in management of skin lesions, subcutaneous abscesses, and joint effusions. Complete cure is elusive, and improvement often takes many months. Oral fluconazole is the preferred treatment for coccidioidal meningitis, often beginning at high doses of 800 mg/day. Intrathecal amphotericin B may be added initially.[11] For those patients who respond, lifelong therapy with oral azoles is required.[11,12]

Prevention

Given the considerable danger of laboratory infection by *C. immitis*, precautionary measures in handling cultures should be emphasized. The organism should be handled using BSL2 procedures and practices in clinical laboratories.[8] Petri dishes should not be used for isolation of the organism from clinical specimens. Subculturing and harvesting of the arthrospores should be carried out under a laminar flow hood or other isolation hood using BSL3 practices and facilities.[8] Viable plate cultures should never be discarded or sent through the mail. For outdoor work in endemic areas, dust control procedures should be followed.

REFERENCES

1. Pappagianis D. Epidemiology of coccidioidomycosis. *Curr Top Med Mycol* 1988; 2:199–238.
2. Schneider E, Hajjeh RA, Spiegel RA, et al. A coccidioidomycosis outbreak following the Northridge, California, earthquake. *JAMA* 1997; 277:904–8.
3. Johnson W. Occupational factors in coccidioidomycosis. *J Occup Med* 1981; 23:67–74.
4. El-Ani A, Elwood C. A case of coccidioidomycosis with unique clinical features. *Arch Intern Med* 1978; 138:1421–2.
5. Stander SM, Schooner W, Galgiani JN, et al. Coccidioidomycosis among visitors to a *Coccidioides immitis*-endemic area: an out-break in a

military reserve unit. *J Infect Dis* 1995; 171:1672–5.

6. Kohn G, Linne S, Smith C. Acquisition of coccidioidomycosis at necropsy by inhalation of coccidioidal endospores. *Diagn Microbial Infect Dis* 1992; 15:527–30.
7. Canoil F, Haley K. Brown J. Primary cutaneous coccidioidomycosis: a review of the literature and a report of a new case. *Arch Dermatol* 1977; 113:933–6.
8. Richard JY, McKinney RW, eds. *Biosafety in microbiological and biomedical laboratories*, 4th edn. HHS publication no. (CDC) 93–8395 Washington. DC: US Government Printing Office, 1999:118–19.
9. Charlton V, Ramsdell K, Sehring S. Intrauterine transmission of coccidioidomycosis. *Pediatr Infect Dis J* 1999; 18:561–3.
10. Orsborn KI, Galgiani JN. Detecting serum antibodies to a purified recombinant proline-rich antigen of *Coccidioides immitis* in patients with coccidioidomycosis. *Clin Infect Dis* 1998; 27:1475–8.
11. Galgiani JN, Ampel NM, Catanzaro A, et al. Practice guidelines for the treatment of coccidioidomycosis. Infectious Disease Society of America. *Clin Infect Dis* 2000; 30:658–61.
12. Dewsnup DH, Galgiani JN, Graybill JR, et al. Is it ever safe to stop azole therapy for *Coccidioides immitis* meningitis? *Ann Intern Med* 1996; 124:305–10.

CRYPTOCOCCUS NEOFORMANS

Common names for disease: Cryptococcosis, torulosis, European blastomycosis

Occupational setting

The most important natural source of *Cryptococcus neoformans* is weathered droppings from pigeons and soil contaminated with avian droppings. The organism is most likely to be found in old pigeon droppings that have accumulated over years in roosting sites such as towers, window ledges, hay mows of barns, and upper floors of old buildings, *C. neoformans* has also been isolated from the droppings of other birds, dairy products, soil, wood, rotting vegetables and fruits, and swallows' nests.[1] Cryptococcal antibodies are detected more commonly in pigeon breeders than in other occupational groups,[2] but the infection rate is no greater because the disease mainly affects immunocompromised hosts (including patients with AIDS, sarcoidosis, lymphoma, and those requiring chronic steroids). The organism was cultured from bagpipes played by a patient with leukemia who developed pulmonary cryptococcosis.[3] Cases of cryptococcosis rarely occur in clusters, and there is no clear occupational predisposition. Histories of exposure to pigeons or dust are usually unhelpful.

Exposure (route)

Inhalation of fungal spores is the major route of entry.

Pathobiology

C. neoformans is an encapsulated, yeast-like fungus that reproduces by budding into 4–6-μm-diameter cells that can cause disease when they are aerosolized and inhaled.

The most common clinical manifestation of cryptococcosis is infection of the cerebral cortex, brainstem, cerebellum, or meninges.[4] Symptom onset may be insidious (with headache, dizziness, irritability, subtle altered mental status, personality change, and visual symptoms) or explosive (with rapid deterioration and death within 2 weeks of onset).

Pulmonary cryptococcosis has a variety of clinical manifestations and an unpredictable course. A self-limited pneumonia with indolent onset and symptoms of dry cough, chest pain and little or no fever can occur. The chest radiograph typically shows one or more well-circumscribed areas of pneumonitis, occasionally with central cavitation. Pleural effusions, hilar adenopathy, and calcification are rare. Resolution often requires months of treatment and occasionally progresses to chronic pneumonia. The most serious outcome in cryptococcal pneumonia is silent dissemination to the CNS. Patients with underlying lung disease may develop asymptomatic colonization of the bronchial tree with *C. neoformans*.

Papular skin lesions from hematogenous dissemination occurs in ~10% of patients with cryptococcosis, and is more common in immunocompromised patients. Local cutaneous cryptococcosis as a result of direct inoculation may occur in immunocompetent individuals.[5] Bone and joint involvement may also occur following hematogenous dissemination, with vertebral lesions the most common sites. Ocular lesions of cryptococcosis include chorioretinitis, papilledema, optic atrophy, scotomata, and ocular motor palsies. Rarely, cryptococcosis may involve the genitourinary tract, heart valves, liver, sinuses, and GI tract.

Diagnosis

The most important procedure in the diagnosis of cryptococcal meningitis is lumbar puncture, which characteristically shows pleocytosis, elevated protein, and hypoglycorrhachia. India ink smear of CSF shows the encapsulated yeast in 50% of cases, and cryptococcal antigen is detected in 94% of cases. Diagnosis of cryptococcal pneumonia is often challenging, since cryptococci are scanty in sputum except in cases of cavitary lung disease or widely disseminated infection. Cultures of sputum and bronchoalveolar lavage (BAL) are occasionally helpful. Cryptococcal antigen can be measured in BAL fluid. Early studies showed high diagnostic sensitivity of BAL cryptococcal antigen at 98%; however, follow-up investigations indicate a sensitivity closer to 70%.[6,7] Positive serum cryptococcal antigen is suggestive, but it occurs only in cases with extensive pulmonary infiltrates or extrapulmonary dissemination. Transbronchial biopsy is occasionally helpful, but surgical lung biopsy is often necessary to confirm the diagnosis. Central punch biopsy of cutaneous cryptococcosis with culture and histology has a high diagnostic yield.

Treatment

Aggressive search for disseminated disease (including CSF culture, multiple urine cultures, and blood cultures) is necessary in patients with pulmonary cryptococcosis. If results are negative, chemotherapy can be withheld except in patients at risk for dissemination, such as those with steroid therapy, diabetes, HIV infection, or underlying malignancy. However, careful follow-up of immunocompetent patients with suspected illness is required. Therapy for pulmonary cryptococcosis is recommended if the serum level of cryptococcal antigen is greater than 1:8.[8] Fluconazole at a dose of 200–400 mg/day for 3–6 months is the recommended therapy for pulmonary disease. Itraconazole is an acceptable alternative. For CNS or severe disseminated disease, combination therapy with intravenous amphotericin B may be required. Meningitis is usually treated with 2 weeks of amphotericin B plus flucytosine followed by at least 10 weeks of fluconazole at 400 mg/day, and then 6–12 months of fluconazde at 200 mg/day.[8] Intrathecal amphotericin B has been used for refractory CNS disease. All patients with meningitis should be evaluated for elevated intracranial pressure. Daily large volume lumbar puncture to reduce intracranial pressure until a normal opening pressure is achieved on several consecutive days is recommended. Symptomatic hydrocephalus should be treated by ventriculoperitoneal shunt even if viable cryptococci are still present in CSF. Treatment for cryptococcosis in AIDS patients is beyond the scope of this discussion.

Prevention

Since cryptococcosis is typically a disease of the immunocompromised host, and since occupational cases are rare, few preventive strategies have been identified. In the laboratory, BSL2 procedures and practices should be followed.[9] In endemic areas, pigeon dropping control procedures should be utilized.

REFERENCES

1. Gordon M. Cryptococcosis: a ubiquitous hazard. *Occup Health Safety* 1980; 49:61–3.
2. Tanphaichitra D, Sahaphongs S, Srirnuang S. Cryptococcal antigen survey among racing pigeon workers and patients with cryptococcosis, pythiosis, histoplasmosis and penicilliosis. *Int J Clin Pharmacol Res* 1988; 8:433–9.
3. Cobcroft R, Kronenberg H, Wilkinson T. *Cryptococcus* in bagpipes. *Lancet* 1978; 1:1368–9.
4. White P, Kaufman L, Weeks R, et al. Cryptococcal meningitis: a case report and epidemiologic study. *J Med Assoc Ga* 1982; 71: 539–42.
5. Micalizzi C, Persi A, Parodi A. Primary cutaneous cryptococcosis in an immunocompetent pigeon keeper. *Clin Exp Dermatol* 1997; 22:195–7.
6. Baughman RP, Rhodes JC, Dohn MN, et al. Detection of cryptococcal antigen in bronchoalveolar lavage fluid: a prospective study of diagnostic utility. *Am Rev Respir Dis* 1992; 145:1226–9.
7. Kralovic SM, Rhodes JC. Utility of routine testing of bronchoalveolar lavage fluid for cryptococcal antigen. *J Clin Microbiol* 1998; 36:3088–9.
8. Saag MS, Graybill RJ, Larsen RA, et al. Practice Guidelines for the Management of Cryptococcal Disease. Infectious Disease Society of America. *Clin Infect Dis* 2000; 30:710–18.
9. Richmond JY, McKinney RW, eds. *Biosafety in microbiological and biomedical laboratories*, 4th edn. HHS publication no. (CDC) 93–8395. Washington, DC: US Government Printing Office, 1999:119.

CRYPTOSTROMA CORTICALE

Common names for disease: Maple bark disease, maple bark stripper's disease

Occupational setting

Wood and sawmill workers engaged in the debarking of logs prior to cutting are at risk for developing HP or asthma from a variety of fungi that contaminate wood, including *Cryptostroma corticale*.[1,2]

Exposure (route)

Inhalation of respirable spores can cause sensitization and subsequent occupational lung disease.

Pathobiology

Maple bark disease is a rare disorder that can affect both trees and humans. Both sensitizing asthma and HP from the fungus contaminating maple bark have been described.[1,2] In one report five workers in the wood room of a paper mill where logs were de-barked and cut developed cough, dyspnea, chest tightness, fever and weight loss during the winter months when workplace ventilation was minimal.[3] Physical examination showed pulmonary crackles; chest radiographs showed a diffuse reticulonodular infiltrate, occasionally with patchy alveolar infiltrates; and arterial oxygen saturation was reduced in three of the five patients. Lung histology demonstrated granulomas and scattered fibrosis; fungal spores were detected by methenamine silver staining of lung tissue in four individuals.

Diagnosis

Diagnosis of sensitizing occupational asthma relies on the symptom and exposure histories, findings on pulmonary function testing (including positive non-specific bronchial challenge), and results of peak flow monitoring at and away from work. Diagnosis of HP relies on a constellation of clinical findings, including a careful symptom history.[4] Physical examination may be normal or show basilar crackles; the chest radiograph may be normal or show diffuse alveolar or interstitial opacities; and pulmonary function testing may be normal or show restriction, obstruction, or a mixed picture. Fiber optic bronchoscopy with bronchoal-veolar lavage (BAC) and transbronchial biopsies may be helpful when the clinical suspicion is strong but routine tests are normal. Exercise physiology testing may be helpful in patients with dyspnea but normal resting pulmonary function. Serum precipitins are often found in asymptomatic exposed workers, and may be negative if the wrong antigen preparation is used.

Treatment

A strong index of clinical suspicion and prompt removal of a sensitized worker from the antigen-containing environment are the mainstays of therapy. Treatment with inhaled steroids and a long acting β-agonist for asthma and with oral corticosteroids for severe HP may be indicated in some patients.

Prevention

Removal of symptomatic individuals from the spore-laden environment and changes in the manufacturing process to reduce spore concentrations should lead to eradication of the disease.

REFERENCES

1. Towey J, Sweany H, Huron W. Severe bronchial asthma apparently due to fungus spores found in maple bark. *JAMA* 1932; 99:453–9.
2. Emanuel D, Lawton B, Wenzel F. Maple-bark disease; pneumonitis due to *Cryptostroma corticale*. *N Engl J Med* 1962; 266:333–7.
3. Emanuel D, Wenzel J, Lawton B. Pneumonitis due to *Cryptostroma corticale* (maple-bark disease). *N Engl J Med* 1966; 274: 1413–18.
4. Rose C. Hypersensitivity pneumonitis. In: Murray, Nadel eds. *Textbook of respiratory medicine*, 3rd edn. Philadelphia, 2000:1867–84.

FONSECAEA AND OTHER AGENTS OF CHROMOMYCOSIS

Common names for disease: Chromomycosis, chromoblastomycosis

Occupational setting

Chromomycosis is a chronic cutaneous and subcutaneous fungal infection which occurs most commonly in the tropics and subtropics among barefoot agricultural workers.[1] Corneal infections may also occur.[2] These opportunistic fungi are common in soil, decayed vegetation, and rotting wood. Handling lumber and sitting on wooden planks in Finnish saunas are additional documented sources of infection.[3]

Exposure (route)

Traumatic inoculation of fungi into the skin is the main mode of infection. Person-to-person transmission has not been documented.

Pathobiology

Fonsecaea pedrosi is the most commonly isolated agent of chromomycosis, accounting for the majority of cases in Brazil and 61% of the cases in Madagascar,[1,4] *Cladosporium carrioni* is the major pathogen in South Africa, Venezuela, and Australia.[5] All species produce slow-growing, 4–12-µm round, brown, thick-walled cells, often in clumps; hyphae may be seen in crusts from lesions.

The typical verrucous cutaneous infection occurs primarily below the knee in workers in contact with soil, especially when shoes are not worn.[6] Lesions often remain localized for years. Early ulcerated nodules develop into cauliflower-like masses. Small ulcerations ("black dots") are seen on the warty surface, they may be pruritic but are rarely painful. Some infected patients may instead develop an annular, flattened lesion with a raised border.

Scarring can cause lymphostasis and lymphedema of the involved extremity. Hematogenous spread to brain, lymph nodes and other organs is rare.

Diagnosis

Characteristic pigmented sclerotic bodies ("copper pennies" when seen microscopically) are present in tissue and exudate in all types of chromomycosis. Fungi appear as long, septate, branched hyphal forms in crusts and exudate. Serologic testing is unhelpful, and culture is only positive in a minority of cases.[4]

Treatment

In early stages, when lesions are small, wide and deep surgical excision is the treatment of choice. Medical therapy for chromomycosis has been disappointing. Topical antifungals, potassium iodide, amphotericin B, S-fluorocytosine, ketoconazole, and local thermotherapy, alone or in combination, have all had varying degrees of success. Recently, treatment with terbinafine has shown promise, with cure rates of 82% at 1 year in open trials; however, randomized controlled trials are still needed.[7]

Prevention

Since close contact with soil is the most prevalent predisposing condition, appropriate protective clothing is recommended to prevent cutaneous inoculation.

REFERENCES

1. Silva JP, de Souza W, Rozental S. Chromoblastomycosis: a retrospective study of 325 cases in Amazonic region. *Mycopathologia* 1998–99; 143:171–5.
2. Barton K, Miller D, Pflugfelder SC. Corneal chromoblastomycosis. *Cornea* 1997; 16:235–9.
3. Sonck CE. Chromomycosis in Finland. *Dermatologia* 1975; 19:189–93.
4. Esterre P, Andriantsimahavandy A, Ramarcel ER, et al. Forty years of chromoblastomycosis in Madagascar: a review. *Am J Trop Med Hyg* 1996; 55:45–7.
5. McGinnis M. Chromoblastomycosis and phaeophyphomycosis: new concepts, diagnosis and mycology. *J Am Acad Dermatol* 1983; 8:1–16.
6. Fader R, McGinnis M. Infections caused by dermatiaceous fungi: chromoblastomycosis and phaeophyphomycosis. *Infect Dis Clin North Am* 1988; 2:925–38.
7. Esterre P, Inzan CK, Ramarcel ER, et al. Treatment of chromomycosis with terbinafine: preliminary results of an open pilot study. *Br J Dermatol* 1996;134(suppl 46):33–6.

HISTOPLASMA CAPSULATUM

Common names for disease: Histoplasmosis, Ohio Valley disease, cave disease, Tingo Maria fever, Darling's disease, reticuloendotheliosis

Occupational setting

Working on or under structures that have been habitats for birds or bats can lead to histoplasmosis.[1,2] Epidemics of acute pneumonia due to *Histoplasma capsulatum* result from group exposures to inhaled particulates containing high concentrations of the fungus.[3] Common sites of outbreaks are bat-infested caves, starling roosts, and old chicken houses with dirt floors.[4] Outbreaks related to disposal of bird droppings also occur.[5] The endemic areas with the highest concentrations of disease are located in the eastern USA (Ohio River Valley) and Latin America. Activities such as exploring bat-infested caves, clearing bird roosts or cleaning chicken houses are associated with the disease.[6] Clean-up, construction, or demolition activities in urban areas may be associated with inhalation of airborne conidia.[7,8] Cases of laboratory-acquired pulmonary histoplasmosis have been reported. Accidental inoculation in a hospital worker assisting on an autopsy and in a laboratory worker who accidentally pricked his thumb with a contaminated needle have led to primary cutaneous histoplasmosis.

Infection caused by *H. capsulatum* variant *duboisii* occurs in tropical Africa. Human cases of *duboisii* infection have been associated with bat-infested caves and chicken roosts, suggesting that the var. *duboisii* shares the same ecological niche as the var. *capsulatum*.

Exposure (route)

The major route of exposure is inhalation of airborne conidia. Cutaneous inoculation has been reported.

Pathobiology

H. capsulatum is a yeast-like fungus with oval, budding uninucleate cells measuring 1.5–2.0 × 3.0–3.5 μm that are often found within macrophages in viable tissue. The mycelial form is found in soil and bears infectious spores called microconidia (2–6 μm) and macroconidia (8–14 μm).

Most infections are mild or subclinical. They are diagnosed in retrospect by a positive skin test or small, scattered pulmonary calcifications. Acute pulmonary histoplasmosis typically presents with symptoms of cough, pleuritic chest pain, fever, chills, myalgias, malaise, nausea, anorexia, and weight loss. The chest radiograph shows pulmonary infiltrates, usually patchy, finely nodular, and involving both lungs, often with hilar adenopathy. There are often slight elevations in peripheral blood leukocyte count and erythrocyte sedimentation rate. Illness may be mild or severe, accompanied by hypoxemia. Pleural effusions resolve over several weeks. Scattered calcification throughout the lung fields is a hallmark of healed acute pulmonary histoplasmosis. Microcalcifications of the spleen may be seen on the chest radiograph. Healing of a localized pulmonary infiltrate may lead to formation of a pulmonary nodule called a histoplasmoma.

Acute pulmonary histoplasmosis may result in lymphatic spread, leading to hilar or mediastinal lymphadenitis. In severe cases, granulomatous inflammation with central areas of caseation may replace entire mediastinal structures, causing fibrosing mediastinitis. Massive adenopathy typically appears in the hilar or right paratracheal area on chest radiograph. Obstruction of pulmonary veins causes hemoptysis and dyspnea. Heart failure and tracheobronchial hemorrhage are lethal complications of pulmonary venous obstruction. Histoplasmosis is the most common nonmalignant cause of superior vena cava syndrome, another potential sequellae. Compression of the recurrent laryngeal nerve can cause hoarseness. Pericarditis may also occur, leading to potentially lethal complications such as tamponade and constrictive pericarditis. Esophageal complications of fibrosing mediastinitis may include ulceration, tracheoesophageal fistula, or traction diverticulum.

Chronic pulmonary histoplasmosis occurs most commonly in middle-aged men with underlying emphysema or chronic bronchitis. Symptoms include cough and sputum production, chest pain, dyspnea, malaise, weakness, fever, weight loss, and easy fatigability. Hemoptysis may also occur. The chest radiograph shows interstitial infiltrates, predominantly in the upper lobes. Nodular areas may slowly shrink or cavitate and expand. The course is marked by progressive hypoxemia, dyspnea, hemoptysis, bacterial pneumonia, and cor pulmonale. Laboratory abnormalities may include anemia of chronic disease, mild leukocytosis, and elevated alkaline phosphatase.

Hematogenously disseminated histoplasmosis is a rare but often lethal complication which occurs most commonly in immunosuppressed patients (e.g. those with hematologic malignancies, with AIDS, or on high-dose corticosteroids). The clinical presentation may vary from acute to indolent. A variety of organ systems may be affected, causing meningitis, endocarditis, ulcerated lesions of the intestinal tract, hepatitis, and mucous membrane lesions of the oropharynx, face, and external genitalia. Ocular histoplasmosis occurs rarely.

Diagnosis

A point-source exposure to *H. capsulatum* should be suspected when several individuals develop respiratory illness 2 weeks following a common outdoor exposure. Serodiagnosis is presumptive but should be sought by immunodiffusion testing early in the disease course and by fourfold elevation of complement fixation titer between acute and convalescent sera. Cultures of blood, bone marrow, and urine should be obtained in hospitalized patients to rule out disseminated infection. Urinary antigen testing has a sensitivity approaching 90% for disseminated disease. Unfortunately, sensitivity is less than 50% for localized pulmonary infection. Specificity of urinary antigen testing is excellent, although false-positive reactions in patients with paracoccidioides can occur.[9]

In mediastinitis and pericarditis, Gomori methenamine silver (GMS) staining of biopsied lymph node sections has the best chance of demonstrating organisms. Sputum culture is recommended for diagnosis of chronic pulmonary histoplasmosis. If sputum is inadequate, bronchoal-veolar lavage (BAL) may be indicated. In disseminated histoplasmosis, taking a smear or biopsy or uninary antigen tesing leads to a faster diagnosis than awaiting culture results. Transbronchial biopsies may be useful in patients with diffuse radiographic infiltrates. In AIDS patients, GMS staining of BAL fluid has a high yield. Liver biopsy is useful in the setting of hepatic enlargement or abnormal liver function tests. Cultures of blood, bone marrow and focal lesions have the highest yield for positive results.

Treatment

In most patients with acute pulmonary histoplasmosis, spontaneous improvement has begun at or before diagnosis, so no therapy is necessary. However, the presence of persistent symptoms for more than 1 month is an indication for therapy.[10] Some immunocompetent patients will develop severe diffuse pneumonia and respiratory failure after high spore exposures, requiring therapy with amphotericin B. Corticosteroids are sometimes used as adjuvant therapy in acute fulminant histoplasmosis to attenuate the inflammatory response.[10] Early surgery to relieve pericardial tamponade and to confirm the diagnosis may be useful in pericarditis; however, antifungal therapy is often not required in this setting.[10] The role of surgery in early mediastinal infection has not been adequately evaluated, but late in the course of fibrosing mediastinitis, complications, including massive hemoptysis from venous obstruction, may require surgical management. Neither high-dose steroid therapy nor antifungal chemotherapy appears to be beneficial in fibrosing mediastinitis.

For chronic pulmonary histoplasmosis, itraconazole is the drug of choice at a dose of 200 mg/day. The duration of therapy is typically 6–12 months. Ketoconazole is an effective alternative but fluconazole has a higher rate of treatment failure. In patients where compliance is a problem or in those who have contraindications to azole therapy, intravenous amphotericin B may be useful.

In patients with disseminated histoplasmosis, amphotericin B is recommended for hospitalized patients. For outpatient therapy, itraconazole is the treatment of choice. Hospitalized patients may be switched to itraconazole once stable. Amphotericin B followed by at least 9–12 months of fluconazole at a dose of 800 mg/day is recommended for patients with meningitis. Fluconazole is used because of the poor CNS penetration of itraconazole. Weekly maintenance therapy with high-dose fluconazole is often necessary to prevent relapse in patients with AIDS or in patients who have a first relapse after successful treatment.[10]

Prevention

Avoidance of circumstances in which *H. capsulatum* is likely to be found in high concentrations is the best approach to prevention. Although there are no data on efficacy, fit-tested negative-pressure respirators with HEPA filters or powered air-purifying respirators probably decrease the risk for inhalation exposure when cleaning or bulldozing bird roosts or chicken houses, activities that tend to increase the number of airborne spores. BSL2 practices and facilities are recommended for handling and processing clinical specimens, and for identifying cultures and isolates in diagnostic laboratories. BSL3 practices and procedures should be used for manipulating identified cultures and for processing soil or other environmental materials that contain infectious spores.[11]

REFERENCES

1. Sorley D, Levin Warren J, et al. Bat-associated histoplasmosis in Maryland bridge workers. *Am J Med* 1979; 67:623–6.
2. Taylor ML, Perez-Mejia A, Yamamoto-Furusho JK, et al. Immunologic, genetic and social human risk factors associated to histoplasmosis: studies in the State of Guerrero, Mexico. *Mycopathologia* 1997; 138:137–42.
3. Goodwin R, Loyd J, Des Prez R. Histoplasmosis in normal hosts. *Medicine (Baltimore)* 1981; 60:231–66.
4. Stobierski MG, Hospedales CJ, Hall WN, et al. Outbreak of histoplasmosis among employees in a paper factory—Michigan, 1993, *J Clin Microbiol* 1996; 34:1220–3.
5. Tosh F, Doto I, Beecher S, et al. Relationship of starling–blackbird roosts and endemic histoplasmosis. *Am Rev Respir Dis* 1970; 101:283–6.
6. Furcolow M. Environmental aspects of histoplasmosis. *Arch Environ Health* 1975; 10:4–8.
7. Dean A, Bates J, Sorrels C, et al. An outbreak of histoplasmosis at an Arkansas courthouse with five cases of probable reinfection. *Am J Epidemiol* 1978; 108:36–46.
8. Anonymous. Case records of the Massachusetts General Hospital. *N Engl J Med* 1991; 325:949–56.
9. Durkin MM, Connolly PA, Wheat LJ. Comparison of radioimmunoassay and enzyme-linked immunoassay methods for detection of *Histoplasma capsulatum* var. *capsulatum* antigen. *J Clin Microbiol* 1997; 35:2252–5.
10. Wheat J, Sarosi G, McKinsey D, et al. Practice guidelines for the management of patients with histoplasmosis. Infectious Disease Society of America. *Clin Infect Dis* 2000; 30:688–95.
11. Richmond JY, McKinney RW, eds. *Biosafety in microbiological and biomedical laboratories*, 4th edn. HHS publication no. (CDC) 93–8395. Washington, DC: US Government Printing Office, 1999:120.

MADURELLA SPECIES AND OTHER AGENTS OF MYCETOMA

Common names of disease: Mycetoma, Madura foot, maduromycetoma, maduromycosis

Occupational setting

Cases of this chronic indolent infection are most often seen in tropical and subtropical countries, such as India, Mexico, sub-Saharan Africa, and Venezuela. In some cases, saprophytic soil fungi enter the hands or feet after local trauma such as a thorn prick, or they enter the chest wall and back from soil-contaminated sacks carried on the shoulders; in other cases, mycetomas form on the head and neck as a result of carrying bundles of wood.[1] Mycetomas occur most frequently in male farmers and other rural laborers exposed to penetrating wounds from thorns and splinters. Inadequate nutrition and hygiene are probably contributory factors.

Exposure (route)

The route of infection is through skin inoculation. Hematogeous spread has been rarely reported.

Pathobiology

Causal fungi include *Pseudallescheria (Petriellidium) boydii*, *Madurella mycetomati*, *Madurella grisea*, *Acremonium (Cephalosporium)* species, *Fusarium* species, *Exophiala (Phialophora) jeanselmei*, and a number of others.

A triad of signs—indurated swelling, multiple sinus tracts draining grain-filled pus, and localization to the foot (the most common site of infection)—characterize mycetomas.[2] The primary lesion is a locally invasive, indolent, painless, subcutaneous swelling that slowly enlarges, causing subsequent distortion, pain, and disability. The radiographic findings include necrosis, osteolysis, and bone fusion.

Diagnosis

In addition to the classic clinical triad, characteristic grains in draining sinuses can be seen on hematoxylin–eosin staining. GMS or periodic acid-Schiff staining will detect hyphae in tissue. Speciation requires culture of the grain and isolation of the organism.

Treatment

Surgical resection of a localized mycetoma may be necessary. Medical therapy is often unsuccessful, but ketoconazole may be helpful.

Prevention

Since the major predisposing factor is inoculation through bare feet in contact with soil, adequate shoes and clothing are recommended. Improvements in nutrition and hygiene would undoubtedly diminish the risk for infection as well.

REFERENCES

1. Sugar AM. Agents of mucormycosis and related species. In: Mandell GL, Douglas RG, Bennett JE, eds. *Principles and practice of infectious disease*, 5th edn. New York: Churchill Livingstone, 2000.
2. Butz W, Ajello L. Black grain mycetoma. *Arch Dermatol* 1971; 104:197–201.

PARACOCCIDIOIDES BRASILIENSIS

Common names for disease: Paracoccidioidomycosis, South American blastomycosis, Brazilian blastomycosis, paracoccidioidal granuloma, Lutz's disease

Occupational setting

Paracoccidioidomycosis is a chronic granulomatous disease that is geographically restricted to areas of Central and South America. The etiologic agent, *Paracoccidioides brasiliensis*, is found in soil in humid mountain forests. Over 5000 cases have been reported, the majority from Brazil.[1] Disease is much more common in men than in women (7–70 : 1) and typically occurs between the ages of 20 and 50. Women appear to acquire the disease at a younger age than men.[2] Most cases occur in rural workers such as tree cutters, (46% of all cases in one region of Brazil occurred in rural occupations[2]), and most patients are at least moderately malnourished.

Exposure (route)

The primary route of entry is by fungal inhalation into the lungs. Local trauma with inoculation of the organism is less common.

Pathobiology

P. brasiliensis is a dimorphic fungus that forms 2–30-μm round, budding cells that are released when small.

The pulmonary infection is usually subclinical. It then disseminates to form ulcerative granulomata of the buccal, nasal and, occasionally, GI mucosa. In clinically evident lung disease, the alveolitis is manifested as patchy bilateral radiographic infiltrates with hilar adenopathy. Occasionally, chronic progressive pulmonary disease can occur, leading to diffuse cavitary and alveolar involvement. Symptoms and signs include dyspnea, productive cough, chest pain, fever, and rales. Hematogenous and lymphatic dissemination without lung involvement is more common, especially to mucous membranes and mucocutaneous junctions. Papules first become vesicles, then granulomatous ulcers. The spleen, liver, CNS, bones, lymph nodes and intestine may be involved.

Diagnosis

Sputum, crusts, material from the granulomatous bases of ulcers, biopsies of lesions and pus from draining lymph nodes contain fungal elements; typically budding yeast forms. Serologic studies (complement fixation, immunodiffusion) are usually positive.[3] In addition, an antigen test may soon become available.[4]

Treatment

Most cases are self-limiting. Itraconazole is the drug of choice for cases requiring treatment.[4] Amphotericin B may be required for extensive pulmonary and severe disseminated forms of infection.

Prevention

Disease is limited to endemic areas in Central and South America, where prevention of malnutrition and other diseases (such as Chagas' disease and tuberculosis) may decrease the risk for paracoccidioidomycosis in rural workers.

REFERENCES

1. Franco M, Montenegro M, Mendes R, et al. Paracoccidioidomycosis: a recently proposed classification of its clinical forms. *Rev Soc Brasil Med Trop* 1987; 20:129–32.
2. Blotta MH, Mamoni RL, Oliveira SJ, et al. Endemic regions of paracoccidioidomycosis in Brazil: a clinical and epidemiologic study of 584 cases in the southeast region. *Am J Trop Med Hyg* 1999; 61:390–4.
3. Restrepo A, Robledo M, Giraldo R, et al. The gamut of paracoccidioidomycosis. *Am J Med* 1976; 61:33–42.
4. Brummer E, Castaneda E, Restrepo A. Paracoccidioidomycosis: an update. *Clin Microbiol Rev* 1993; 6:89–117.

PENICILLIUM SPECIES

Common names for disease: Humidifier lung, Suberosis (*Penicillium frequentans*), cheese washer's disease (*P. caseii*, *P. rouefortii*); cheese worker's lung, woodman's disease, allergic bronchopulmonary mycosis (ABPM), penicilliosis (*Penicillium marneffei*), peat moss worker's lung (*Penicillium citreonigrum*)

Occupational setting

Because the blue–green *Penicillium* molds are ubiquitous in nature, they are common contaminants of indoor environments. Exposure to *Penicillium* spp. has been associated with hypersensitivity lung disease in cork workers,[1] cheese workers,[2,3] laboratory workers, farmers,[4] tree cutters,[5] sawmill workers, other handlers of mold-contaminated wood,[6] peat moss workers,[7] and salami factory workers.[8] Contaminated humidifier water and moldy HVAC (heating, ventilation and air-conditioning) systems have been associated with *Penicillium*-induced hypersensitivity pneumonitis.[9,10]

Exposure (route)

Inhalation of airborne spores is the major route of entry.

Pathobiology

A variety of *Penicillium* species have been associated with hypersensitivity lung diseases; the most common is HP, although asthma has also been described. *Penicillium frequentans* spores in the air of factories where cork bark is processed can cause suberosis, a form of HP. *P. caseii* and *P. roquefortii* have been associated with HP in cheese workers exposed to *Penicillium*-contaminated cheese. Several *Penicillium* species isolated from contaminated humidifier water were shown to induce a precipitating antibody response in an entomologist exposed to mists generated by a reservoir type of humidifier. Wood-workers, including those exposed to mold-contaminated fuel chips, de-barking of live trees, and sawmill particulates, are at risk for *Penicillium*-induced HP. Workers exposed to *Penicillium*-contaminated peat moss have also developed HP.

The clinical presentation of HP is variable, ranging from severe, acute respiratory and systemic symptoms to subtle, chronic symptoms.[11,12] Acute illness is manifested by fevers, chills, cough, dyspnea, abnormal chest radiograph, and leukocytosis; improvement is seen within a few days following removal from exposure. The more subacute or chronic illness is manifested by insidious onset of cough and progressive dyspnea on exertion. Pulmonary physiology may be normal, show isolated obstruction, or show the more classic restrictive or mixed restrictive and obstructive pattern. Exercise physiology often demonstrates gas exchange abnormalities. The chest radiograph may be normal or show inhomogeneous, patchy alveolar infiltrates or interstitial opacities. High-resolution CT scans typically show diffuse, fine, poorly defined centrilobular micronodules. Serum-precipitating antibodies to *Penicillium* species are often positive.

Rare cases have been reported of allergic bronchopulmonary penicilliosis causing intermittent airways obstruction, transient pulmonary infiltrates, blood and sputum eosinophilia, a positive dual skin test (types 1 and 3), and precipitating antibodies in serum. Proximal saccular bronchiectasis is often found in the segment of lung containing the infiltrate. Bronchial hygiene alone or in combination with inhaled or oral corticosteroids is usually effective treatment.

Penicillium infections of clinical importance are very rare. There have been case reports of *Penicillium* infection of the ear, foot, urinary bladder, and lung. *Penicillium* presenting as a solitary pulmonary nodule in a non-immunocompromised host has also been reported. *P. marneffei* is endemic in South East Asia, and is also found in Africa. It has been associated with recurrent episodes of hemoptysis attributed to bronchitis and bronchiectasis. Histopathologically, lung tissue shows granulomata with central areas of necrosis and neutrophilic infiltration with many yeast-like tissue-forming cells of *P. marneffei*. In addition, *P. marneffei* has caused disseminated infection manifested by fever, weight loss, anemia and skin lesions (penicilliosis) in immunocompromised hosts.[13,14] *P. marneffei* can cause peritonitis in peritoneal dialysis patients.[15]

Treatment

Treatment of *Penicillium*-induced HP relies on removal of the affected individual from exposure to the contaminated environment. Systemic steroids have been used in severely ill patients with interstitial pneumonitis, but controlled clinical trials are lacking. Treatment of *Penicillium*-induced asthma involves elimination of antigen exposure and use of inhaled steroids and bronchodilators. Allergic bronchopulmonary penicilliosis is rare, but treatment should be similar to that for ABPA, with inhaled or oral corticosteroids to prevent recurrent pneumonitis and subsequent bronchiectasis. Treatment of *P. marneffei* injection generally includes induction with amphotericin B followed by itraconazole therapy.[14]

Prevention

Attention should be paid to safe handling of all types of solid fuel (wood, chips, and peat) and other materials in which mold may grow. Dry storage, prevention of mold growth and use of appropriate respiratory protection are important.

Avoidance or elimination of water damage to HVAC systems is important in preventing significant indoor mold contamination. Regular HVAC maintenance and inspection, appropriate filtration of outside air, and provision of indoor environment free from water intrusion, are crucial to prevent fungal amplification and dissemination. Hard surfaces supporting fungal growth in indoor environments should be cleaned with dilute bleach (1:10–1:50 dilution); then, the surface should be rinsed with clean water and dried. Mold-contaminated materials such as furniture, draperies and insulation material should be discarded.

A variety of measures have been introduced in the cheese production industry to reduce contamination with airborne molds. These measures include: wrapping cheese in foil or plastic film before entering the aging room, thus preventing surface mold formation; careful temperature and humidity control in aging rooms; and removal of surface mold contamination before it becomes an aerosolized dust.

REFERENCES

1. Avila R, Lacey T. The role of *Penicillium frequentans* in suberosis (respiratory disease in workers in the cork industry). *Clin Allergy* 1974; 4:109–17.
2. Campbell J, Kryda M, Treuhaft M, et al. Cheese worker's hypersensitivity pneumonitis. *Am Rev Respir Dis* 1983; 127:495–6.
3. Schlueter D. "Cheesewasher's disease": a new occupational hazard? *Ann Intern Med* 1973; 78:606.
4. Nakagawa-Yoshida K, Ando M, Etches RI, et al. Fatal cases of farmer's lung in a Canadian family. Probable new antigens, *Penicillium brevicompactum* and *P. olivicolor*. *Chest* 1997; 111:245–8.
5. Dykewicz M, Laufer P, Patterson R, et al. Woodman's disease: hypersensitivity pneumonitis from cutting live trees. *J Allergy Clin Immunol* 1988; 81:455–60.
6. Van Assendelft A, Raitio M, Turkia V. Fuel chip-induced hypersensitivity pneumonitis caused by *Penicillium* species. *Chest* 1985; 87:394–6.
7. Cormier Y, Israel-Assayag I, Bedard G, et al. Hypersensitivity pneumonitis in peat moss workers. *Am J Respir Crit Care Med* 1998; 158:412–77.
8. Marchisio VF, Sulotto F, Botta GC, et al. Aerobiological analysis in a salami factory: a possible case of extrinsic allergic alveolitis by *Penicillium camembertii. Med Mycol* 1999; 37:285–9.
9. Baur X, Behr J, Dewair M, et al. Humidifier lung and humidifier fever. *Lung* 1988; 166:113–24.
10. Bernstein R, Sorenson W, Garabrant D, et al. Exposures to respirable, airborne *Penicillium* from a contaminated ventilation system: clinical, environmental and epidemiological aspects. *Am Ind Hyg Assoc J* 1983; 44: 161–9.
11. Richerson H, Bernstein I, Fink J, et al. Guidelines for the clinical evaluation of hypersensitivity pneumonitis. *J Allergy Clin Immunol* 1989; 84:839–44.
12. Rose C. Hypersensitivity pneumonitis. In: Murray, Nadel, eds. *Textbook of respiratory medicine*, 3rd edn. Philadelphia, 2000:1867–84.
13. Lo Y, Tintelnot K, Lippert U, et al. Disseminated *Penicillium marneffei* infection in an African AIDS patient. *Trans R Soc Trop Med Hyg* 2000; 94:187.
14. Kurup A, Leo YS, Tan Al, et al. Disseminated *Penicillium marneffei* infection: a report of five cases in Singapore. *Ann Acad Med Singapore* 1999; 28:605–9.
15. Chang HR, Shu KH, Cheng CH, et al. Peritoneal-dialysis-associated *Penicillium* peritonitis. *Am J Nephrol* 2000; 20:250–2.

SPOROTHRIX SCHENCKII

Common names for disease: Sporotrichosis

Occupational setting

Although sporotrichosis occurs worldwide, it is found mainly in temperate, warm, and tropical areas. *Sporothrix schenckii* is isolated most often from soil and living plants or plant debris. Cutaneous infection develops where the organism is introduced to sites of skin injury. Subsequent nodular lymphangitic spread is common. Pulmonary sporotrichosis is rare condition caused by inhalation of fungal spores.

Occupations that predispose to infection include gardening, farming, masonry, outdoor work, floral work, and other activities with exposure to contaminated soil or vegetation such as sphagnum moss, prairie hay, salt marsh hay, or roses.[1–3] Outbreaks have occurred among miners, nursery workers, and forestry workers who handle contaminated timbers, seedlings, mulch, hay, or other plant materials.[4–7] In one study, risk for infection was related to exposure to moss and to seedlings from a particular nursery.[7] Arm and finger infections have been reported in laboratory workers through contact with experimentally infected animals or contaminated material.[8] Two laboratory workers developed conjunctival and eyelid infections after mycelial elements were accidentally splattered into the eyes. Transmission to humans from infected cats has also been reported.[9] Armadillo hunting in Uruguay has been associated with sporotrichosis, presumably from exposure to the fungus isolated from the dry grass used by armadillos and rodents to prepare their nests. An outbreak of the illness in South African gold miners was traced to contaminated mine timbers.[10]

Exposure (route)

S. schenckii usually enters the body through traumatic implantation,[11] but inhalation of fungal conidia is occasionally associated with pulmonary infection.

Pathobiology

S. schenkii is a dimorphic fungus with 4–6-μm round, oval, or cigar-shaped cells.

Cutaneous disease arising at sites of minor trauma begins as a small, erythematous papule which enlarges over days or weeks. It usually remains painless, but it may secrete a clear discharge. Typically, discrete nodular lesions spread along lymphatic channels. Skin lesions may also result from hematogenous dissemination. Such lesions may herald the onset of osteoarticular sporotrichosis, which is manifested by stiffness and pain in a large joint, particularly the knee, elbow, ankle, or wrist. Radiologic evidence of osteomyelitis develops slowly, and additional joints may become involved in untreated patients. An indolent tenosynovitis of the wrist or ankle with pain, limitation of motion, and nerve entrapment can occur. Endophthalmitis, brain abscess, chronic meningitis, and other manifestations of disseminated disease are rare. Patients with a history of alcoholism or immunosuppression are at incresed risk for dissemination.[12]

Pulmonary sporotrichosis typically presents with cough, low-grade fever, weight loss, hemoptysis, and an upper lobe single cavitary lesion with or without surrounding parenchymal infiltrate. Pleural effusion, hilar adenopathy and calcification are rare. In the absence of treatment, the lung lesion gradually progresses to death. Coronary obstructive pulmonary disease (COPD) is a risk factor for pulmonary infection.[12]

Diagnosis

Accurate diagnosis requires detection of the fungus in clinical specimens, including skin biopsy, joint aspirate, or sputum, either through culture or by fluorescent antibody staining.

Treatment

The key to appropriate diagnosis and treatment of sporotrichosis is a high index of clinical suspicion combined with culture to confirm results. Itraconazole is the drug of choice for lymphocutaneous infection: 200 mg/day for 3–6 months results in a 90% cure rate. Fluconazole at a dose of 400 mg/day is second-line treatment; ketoconazole should not be used. Local hyperthermia is an alternative for isolated cutaneous lesions.[12] Intravenous amphotericin B is reserved for treatment failures.

Itraconazole at a dose of 200 mg twice daily can be used for mild pulmonary disease. However, severe pulmonary disease or progression despite itraconazole therapy is best managed with a combination of surgery and amphotericin B.[12]

In osteoarticular sporotrichosis, itraconazole at a dose of 200 mg twice daily is recommended, as accompanying systemic illness is unusual. Cure rates of 60–80% have been achieved with this regimen. Amphotericin B is indicated for severely ill patients or for those in whom itraconazole fails. Amphotericin B is also the drug of choice for disseminated or meningeal sporotrichosis.[12]

Prevention

Use of heavy, impermeable gloves and long-sleeved shirts in at-risk occupations has been shown to limit traumatic fungal implantation.[7] The use of alternative packing materials for plant products, such as shredded paper or cedar wood chips, has been recommended. Laboratory workers who handle contaminated material should work under appropriate hoods and utilize BSL2 work practices and facilities which minimize inhalation of conidia.[13] Early recognition and prompt treatment of disease limit morbidity and mortality.

REFERENCES

1. Center for Disease Control. Sporotrichosis among hay-mulching workers—Oklahoma, New Mexico. *MMWR* 1984, 33:682–3.
2. Dixon D, Salkin I, Duncan R, et al. Isolation and characterization of *Sporothrix schenckii* from clinical and environmental sources associated with the largest US epidemic of sporotrichosis. *J Clin Microbiol* 1991; 29:1106–13.
3. Cox R, Reller L. Auricular sporotrichosis in a brickmason. *Arch Dermatol* 1979; 115:1229–30.
4. Grotte M, Younger B. Sporotrichosis associated with sphagnum moss exposure. *Arch Pathol Lab Med* 1981; 105:50–1.
5. Powell K, Taylor A, Phillips B, et al. Cutaneous sporotrichosis in forestry workers. Epidemic due to contaminated sphagnum moss. *JAMA* 1978; 240:232–5.
6. Dooley DP, Bostic PS, Beckius ML. Spook house sporotrichosis. A point-source outbreak of sporotrichosis associated with hay bale props in a Halloween haunted-house. *Arch Intern Med* 1997; 157:1885–7.
7. Hajjeh R, McDonnell S, Reef S, et al. Outbreak of sporotrichosis among tree nursery workers. *J Infect Dis* 1997; 176:499–504.
8. Cooper C, Dixon D, Salkin I. Laboratory-acquired sporotrichosis. *J Med Vet Mycol* 1992; 30:169–71.
9. Reed K, Moore F, Geiger G, et al. Zoonotic transmission of sporotrichosis: case report and review. *Clin Infect Dis* 1993; 16:384–7.
10. Einstein H. ACCP Committee Report: occupational aspects of deep mycoses. *Chest* 1978; 73:115.
11. Tan T, Field C, Faust B. Cutaneous sporotrichosis. *J La State Med Soc* 1988; 140:41–5.
12. Kauffman CA, Hajjeh R, Chapman SW. Practice guidelines for the management of patients with sporotrichosis. *Clin Infect Dis* 2000; 30:684–7.
13. Richmond JY, McKinney RW, eds. *Biosafety in microbiological and biomedical Laboratories,* 4th edn. HHS publication no. (CDC) 93–8395. Washington, DC: US Government Printing Office, 1999:121.

STACHYBOTRYS CHARTARUM

Common name for disease: None

Occupational setting

Stachybotrys is a rare fungal contaminant of water-damaged cellulose materials. High exposure levels have been reported among farmers, wood-workers, and composting workers exposed to moldy plant products.[1–4] High concentrations have also been described in water-damaged office buildings, courthouses and homes.[5–8]

Exposure (route)

Exposure occurs through inhalation of mycotoxin-containing spores. Disease in animals has been reported after ingestion of contaminated plant feeds.

Pathobiology

Stachybotrys chartarum is a rare contaminant of nitrogen-poor straw and cellulose-based water-damaged materials.[6] Pathogenic effects were first described in the veterinary literature of the 1920s, when cattle and horses developed anemia and GI hemorrhage after ingesting contaminated grain.[9]

Stachybotrys is believed to cause disease through the production of powerful trichothecene mycotoxins. The mechanism of trichothecene toxicity is potent inhibition of DNA, RNA and protein synthesis. These mycotoxins act as powerful immunosuppressants and can induce hemolysis and bleeding in animal models.[10,11]

Many possible human health effects have been described, but none definitively proven. Occupational exposure to *Stachybotrys* in farmers and agricultural workers handling contaminated straw has been associated with epistaxis, hemoptysis, skin irritation, and alterations in white blood cell counts.[12] *Stachybotrys* has been implicated (along with many other toxigenic fungi) in sick-building syndrome.[5,6,8,13,14] A possible association has been described between *Stachybotrys* contamination of homes and idiopathic pulmonary hemorrhage/hemosiderosis in infants.[7,15] A cluster of 10 infants suffering from idiopathic pulmonary hemorrhage/hemosiderosis occurred in Cleveland between January 1993 and December 1994. Only three cases had been reported in the prior 10 years. Symptoms began with a prodrome featuring abrupt cessation of crying, limpness, and pallor. The prodrome was followed by hemoptysis, grunting, and respiratory failure. Chest radiographs revealed diffuse, bilateral alveolar infiltrates, and laboratory examination revealed decreased hematocrits and hemolysis on blood smears. Fifty per cent of the cases recurred after returning home, and one infant died. Initial epidemiologic investigations implicated water-damaged homes as a significant risk factor.[16] Follow-up industrial hygiene studies revealed increased levels of *Stachybotrys* in case versus control homes.[17] However, re-examination by the CDC concluded that the data were not strong enough to definitively support an association between pulmonary hemorrhage in infants and exposure to *S. chartarum*.

Diagnosis

Clinical evaluation requires careful occupational and environmental histories to elicit circumstances of water damage, and environmental sampling to confirm the exposure.

Treatment

Removal of contaminated materials and prevention of circumstances leading to moisture intrusion into indoor environments are key to management of *Stachybotrys* exposure.

Prevention

Rapid remediation of water damage in homes and office buildings will diminish exposure risks. Proper attention to building practices during new construction will help prevent subsequent leaks and water damage. In agricultural settings, engineering and process controls can reduce exposure to high fungal concentrations.

REFERENCES

1. Croft WA, Jarvis BB, Yatawara CS. Airborne outbreak of trichothecene toxicosis. *Atmos Environ* 1986; 20:549–52.
2. Mainville C, Auger PL, Smoagiewica W, et al. Mycotoxins and chronic fatigue syndrome in a hospital. In: Anderson K, ed. Healthy buildings conference. Stockholm: Swedish Council of Building Research, 1988:1–10.
3. Auger PL, Gourdeau P, Miller JD. Clinical experience with patients suffering from a chronic fatigue-like syndrome and repeated upper respiratory infections in relation to airborne molds. *Am J Ind Med* 1994; 25:41–2.
4. Johanning E, Auger PL, Reijula K. Building-related illnesses. *N Engl J Med* 1998; 338:1070.
5. Johanning E, Landsbergis P, Gareis M, et al. Clinical experience and results of a sentinel health investigation related to indoor fungal exposure. *Environ Health Perspect* 1999; 107(suppl 3):489–94.
6. Hodgson MJ, Morey P, Leung WY, et al. Building-associated pulmonary disease from exposure to *Stachybotrys chartarum* and *Aspergillus versicolor. J Occup Environ Med* 1998; 40:241–9.
7. Etzel RA, Montana E, Sorenson WG, et al. Acute pulmonary hemorrhage in infants associated with exposure to *Stachybotrys atra* and other fungi. *Arch Pediatr Adolesc Med* 1998; 152:757–62.
8. Sudakin DL. Toxigenic fungi in a water-damaged building: an intervention study. *Am J Ind Med* 1998; 34:183–90.
9. Hinitikka EL, Stachybotryotoxicosis as a veterinary problem. In: Rodricks JV, Hesseltine CW, Mehlman MA, eds. *Mycotoxins in human and animal health*. Park Forest South: Pathotox Publishers, 1977:277–84.
10. Pang VF, Lambert RJ, Felsburg PJ, et al. Experimental T-2 toxicosis in swine following inhalation exposure: clinical signs and effects on hematology, serum biochemistry, and immune response. *Fundam Appl Toxicol* 1988; 11:100–9.
11. Ueno Y. Trichothecene mycotoxins—mycology, chemistry and toxicology. *Adv Nutr Res* 1980; 3:301–53.
12. Hintikka EL. Human Stachybotrytoxicosis. In: Wyllie TD, Morehouse LG, eds. *Mycotoxigenic fungi, mycotoxins, mycotoxicoses*. New York: Marcel Dekker, 1987:87–9.
13. Johanning E, Biagini R, Hull D, et al. Health and immunology study following exposure to toxigenic fungi (*Stachybotrys chartarum*) in a water-damaged office environment. *Int Arch Occup Environ Health* 1996; 68:207–18.
14. Mahmoudi M, Gershwin ME. Sick building syndrome. III. *Stachybotyrs chartarum. J Asthma* 2000; 37:191–8.
15. CDC. Update: pulmonary hemorrhage/ hemosiderosis among infants—Cleveland, Ohio, 1993–1996. *MMWR* 1997; 46:33–5.
16. Montana E, Etzel RA, Allan T, et al. Environmental risk factors associated with pediatric idiopathic pulmonary hemorrhage and hemosiderosis in a Cleveland community. *Pediatrics* 1997; 99:117–24.
17. CDC. Update: pulmonary hemorrhage/ hemosiderosis among infants—Cleveland, Ohio, 1993–1996. *MMWR* 2000; 49:180–4.

TRICHOPHYTON AND OTHER DERMATOPHYTES

Common names for diseases: Dermatophytosis, tinea corporis (ringworm), tinea glabrosa, Majocchi's granuloma, tinea cruris ("jock itch"), tinea pedis ("athlete's foot")

Occupational setting

Cutaneous infections occur from exposure to the fungal dermatophytes, most commonly species of *Trichophyton*. The primary sources of dermatophytes are animals (zoophilic), humans (anthropophilic), and soil (geophilic). Dermatophytes originating from soil are occasionally responsible for outbreaks of human disease in exposed occupational groups such as gardeners and farm workers. Zoophilic outbreaks among cattle workers have also been reported.[1] Spread of the anthropophilic organisms that infect glabrous skin is typically through contact with infected desquamated skin scales, such as in bathing or shower facilities in military barracks or factories.[2] As many as 30–35% of British coal miners have dermatophyte infections of the feet.[3] *Trichophyton tonsurans* was responsible for an outbreak of dermatophytosis in hospital personnel exposed to an infected patient.[4] *T. verrucosum,* the cause of cattle ringworm, and *Microsporum canis* in cats and dogs, are the most common zoophilic dermatophytes that cause human infection in temperate countries.[5]

Podiatrists exposed to toenail dust generated when drills and burrs are used to reduce the thickness of hyperkeratotic nails can develop hypersensitivity reactions to *T. rubrum* (including nasal and eye symptoms, restrictive changes on pulmonary function tests, and specific IgG-precipitating antibodies).[6,7]

Exposure (route)

Dermatophytes invade the stratum corneum of the skin, most commonly of the feet, groin, scalp, and nails. Hypersensitivity reactions of the nasal mucus membranes and lungs can occur from exposure to *Trichophyton* dusts.

Pathobiology

The three genera of pathogenic dermatophyte fungi are *Trichophyton*, *Microsporum*, and *Epidermophyton.* The classic lesion of dermatophytosis is an annular scaling patch with a raised edge and a less inflamed central area.

Tinea refers to dermatophyte infection and is followed by the Latin word for the affected site. *T. rubrum* is the most common cause of tinea cruris, the dermatophyte infection of the groin. Scaling and irritation are the usual presenting findings. The disease is most common in young adult males. The leading edge extending onto the thighs is prominent and may contain follicular papules and pustules. Tinea pedis usually begins in the lateral interdigital spaces of the foot. The main symptom is itching; the skin usually cracks and may macerate. Tinea corporis usually involves the trunk or legs.

Diagnosis

Fungal hyphae are easily observed as chains of arthrospores in wet mount preparations from skin scrapings. It is important to sample the edge of skin lesions and to allow the material to soften in potassium hydroxide before microscopic examination.

Treatment

Topical treatment with keratolytics and compounds with specific antifungal activity is usually successful. Nail, hair and widespread skin infections may require oral agents such as griseofulvin. *T. tonsurans* infection may respond to oral ketoconazole.

Prevention

To prevent tinea cruris, cool and loose-fitting clothing should be worn in hot and humid environments, where perspiration and irritation of skin are contributing factors. Avoidance of contact with contaminated clothing and towels is helpful. The floors of locker rooms or showers contaminated with dermatophytes should also be avoided.

REFERENCES

1. Lehenkari E, Silvennoinen-Kassinen S. Dermatophytes in northern Finland in 1982–90. *Mycoses* 1995; 38:411–14.
2. Korting H, Zienicke H. Dermatophytoses as occupational dermatoses in industrialized countries. Report on two cases from Munich. *Mycoses* 1990; 33:8609.
3. Gugnani H, Oyeka C. Foot infections due to *Hendersonula toruloidea* and *Scytalidium hyaline* in coal miners. *J Med Vet Mycol* 1989; 27:167–79.
4. Arnow P, Houchins S, Pugliese G. An outbreak of tinea corporis in hospital personnel caused by a

patient with *Trichophyton tonsurans* infection. *Pediatr Infect Dis J* 1991; 10:355–9.

5. Chmel L, Buchvald J, Valentova M. Ringworm infection among agricultural workers. *Int J Epidemiol* 1976; 5:291–5.
6. Davies R, Ganderton M, Savage M. Human nail dust and precipitating antibodies to *Trichophyton rubrum* in chiropodists. *Clin Allergy* 1983; 13:309–15.
7. Abramson C, Wilton J. Nail dust aerosols from onychomycotic toenails. Part II. Clinical and serologic aspects. *J Am Podiatr Med Assoc* 1992; 82:116–23.

ZYGOMYCETES

Including the order Mucorales, and *Hyphomycetes (Verticillium, Fusarium* and *Neurospora*)

Common names for disease: Hypersensitivity diseases: Paprika splitter's disease (*M. stolonifer*), wood trimmer's disease (rhizopus, mucor), tomato grower's asthma (*Verticillium albo-atrum*), sinus fusariosis (*Fusarium*)
Infections: Mucormycosis, phycomycosis, hyphomycosis, zygomycosis

Occupational setting

Hypersensitivity pneumonitis can occur in workers exposed to respirable *Mucorales* in paprika and from contaminated wood bark in sawmills.[1,2] In addition to occupational hypersensitivity lung disease, various genera and species of the class Zygomycete, order Mucorales, can cause infectious mucormycosis. The Zygomycetes grow in the environment and in tissue as hyphae. They are thermotolerant fungi that are commonly found in decaying organic debris. Despite their ubiquity, human infection is infrequent. Typically, it is associated with severe immunocompromise, malnutrition, iron chelation therapy, diabetes mellitus, or trauma.[3,4] The hyaline hyphomycete *Verticillium albo-atrum* has been associated with cases of occupational asthma in tomato and tobacco growers exposed to crop outbreaks.[4–6] *Fusarium* species are very common soil organisms. Maxillary sinus fusariosis has been described in agricultural workers exposed to *Fusarium*.[7] Occupational asthma from immune sensitization to *Neurospora* spp. occurred in a plywood factory worker exposed to moldy wood products.[8] Occupational sensitization was demonstrated by allergy prick skin testing, the presence of specific serum IgE antibodies, and inhalation challenge with the *Neurospora* mold growing on plywood. Suggested preventive strategies included sealing of the wood drying machine to minimize dust concentrations and shorter outdoor storage times for the plywood to prevent fungal growth.

Exposure (Route)

For the hypersensitivity diseases (asthma and hypersensitivity pneumonitis), the route of exposure is inhalation of respirable airborne hyphae. Cutaneous and subcutaneous mucormycosis can occur by direct implantation from "barrier breaks" or by hematogenous dissemination. Several cases have been associated with contaminated bandages and surgical dressings. Rarely, gastrointestinal transmission may occur.[1]

Pathobiology

All of the Zygomycetes grow as 4–15 μm wide hyphae in the environment and tissue, and are identified microscopically by their morphology. As with other forms of hypersensitivity pneumonitis (HP), symptoms may be acute and flu-like or subtle, chronic and predominantly respiratory in nature. Acute illness is manifested by fevers, chills, cough, dyspnea, abnormal chest radiograph, leukocytosis; improvement follows within a few days after removal from exposure. The more subacute or chronic forms of HP are typically manifested by insidious onset of cough, progressive dyspnea on exertion, and weight loss. Pulmonary physiology may be normal, show isolated obstruction, or show the more classic restrictive or mixed restrictive and obstructive pattern. Exercise physiology often shows gas exchange abnormalities. The chest radiograph may be normal or show inhomogeneous, patchy alveolar infiltrates or interstitial opacities. Serum precipitating antibodies to *Zygomycete* antigens species are often positive.

Many different species in the order *Mucorales* have been implicated in disease, including *Rhizopus, Mucor, Mortierella,* and *Absidia* species. Mucormycosis is the most acute, fulminant fungal infection known. Organisms invade arterial vessels and may infect the face and cranium, lungs, GI tract or skin. Rhinocerebral disease typically occurs in the acidemic patient with uncontrolled diabetes, beginning in the nasal turbinates, paranasal sinuses, palate, pharynx or ears and spreading to the central nervous system. Renal dialysis patients treated with deferoxamine may also be at increased risk. Presenting signs usually include a thick, dark, blood-tinged discharge which often reveals hyphal strands on KOH mount. Rapid invasion of surrounding tissues with sloughing and cranial nerve dysfunction follows, and death usually occurs in a few days. Pulmonary mucormycosis occurs from inhalation of spores or from aspiration of nasopharyngeal secretions, leading to bronchitis and pneumonia with subsequent arterial invasion, often massive cavitation, and rapid death. Primary gastrointestinal and pelvic mucormycosis are less common, and malnourished patients or those with hematologic malignancies are most at risk. Cerebral zygomycosis has been described in intravenous drug abusers.[9] This infection is most likely spread by the bloodstream following intravenous injection of infectious organisms.

Treatment

Treatment of hypersensitivity lung disease involves, most importantly, early disease recognition and elimination of further exposure to the offending antigen. Oral corticosteroids may be useful in patients with severe symptoms or physiologic abnormalities.

Treatment of the underlying disease (e.g. diabetes) is the most effective method to control infectious mucormycosis, and the prognosis is poor (80–90% mortality rate) even with institution of therapy with amphotericin B. Surgical debridement of necrosed tissue may be necessary.

Prevention

Prevention of occupational sensitization to these organisms involves the provision of adequate ventilation, process controls, and/or respiratory protection to limit exposure to high fungal bioaerosol levels.

REFERENCES

1. Eduard W, Sandven P, Levy F. Relationships between exposure to spores from *Rhizopus microsporus* and *Paecilomyces variotii* and serum IgG antibodies in wood trimmers. *Int Arch Allergy Appl Immunol* 1992; 97(4):274–82.
2. Hedenstierna G, Alexandrsson R, Belin L, et al. Lung function and rhizopus antibodies in wood trimmers. A cross-sectional and longitudinal study. *Int Arch Occup Environ Health* 1986; 58(3):167–77.
3. Gordon G, Indeck M, Bross J, et al. Injury from silage wagon accident complicated by mucormycosis. *J Trauma* 1988; 28(6):866–7.
4. Eaton K, Hannessy T, Snodin D, et al. *Verticillium lecanii.* Allergological and toxicological studies on work exposed personnel. *Ann Occup Hyg* 1986; 30(2):209–17.
5. Anonymous. Occupational asthma in tomato growers. *Occup Health (Lond)* 1989; 41(3):70–1.
6. Davies P, Jacobs R, Mullins J, et al. Occupational asthma in tomato growers following an outbreak of the fungus *Verticillium albo-atrum* in the crop. *J Soc Occup Med* 1988; 38(1–2):13–17.
7. Kurien M, Anandi V, Raman R, et al. Maxillary sinus fusariosis in immunocompetent hosts. *J Laryngol Otol* 1992; 106(8):733–6.

8. Cote J, Chan H, Brochu G, et al. Occupational asthma caused by exposure to neurospora in a plywood factory worker. *Br J Ind Med* 1991; 48(4):279–82.

9. Stave GM, Heimberger T, Kerkering T. Cerebral zygomycosis in intravenous drug abuse. *Am J Med* 1989; 86:115–7.

Table 24.1 Common fungi associated with hypersensitivity diseases.

Fungus	Exposure	Syndrome
Alternaria sp	Wood pulp	Wood pulp worker's lung mold allergy, asthma
Aspergillus sp	Moldy hay	Farmer's lung
	Water	Ventilation pneumonitis
Aspergillus fumigatus	General enviroment	Allergic bronchopulmonary aspergillosis
Aspergillus clavatus	Barley	Malt worker's lung
Aureobasidium pullulans	Water	Humidifier lung
Cladosporium sp	Hot-tub mists	Hot tub HP
	General environment	Mold allergy, asthma
Cryptostroma corticale	Wood bark	Maple bark stripper's lung
Graphium, Aureobasidium pullulans	Wood dust	Sequoiosis
Merulius lacrymans	Rotten wood	Dry rot lung
Penicillium sp	Fuel chips, sawmills, tree cutting, Shiitake mushroom manufacturing, humidifier water	Hypersensitivity pneumonitis, asthma
Penicillium frequentans	Cork dust	Suberosis
Penicillium casei, P. roqueforti	Cheese	Cheese washer's lung
Rhisopus, Mucor	Damp basements	Asthma, mold allergy
Trichosporon cutaneum	Damp wood and mats	Japanese summer-type HP

Table 24.2 Common toxigenic fungi.

Toxin	Fungal Source	Occupational Exposure
Aflatoxin	*A. flavus*	Farmers
	A. parasiticus	Peanut handlers
Fumitoxin	*A. fumigatus*	Compost workers
Satratoxin	*Stachybotrys atra*	Maintenance workers (on insulated pipes)
Sterigmatocyslin	*A. versicolor*	Housekeepers
T-2 toxin	*Fusarium*	Machinists
		Farmers

Table 24.3 Occupational fungal infections.

Fungus	Source	Exposures
Blastomyces dermatitidis	Acid soil near rivers, streams, and swamps	Hunters Campers
Coccidioides immitis	Semi-arid or desert soils	Construction workers Farmers Archaeologists Labortory workers Textile workers
Cryptococcus neoformans	Pigeon/avian droppings	Pigeon breeders
Histoplasma capsulatum	Bat-infested caves, starling/chicken roosts	Spelunkers Construction workers Labortory workers
Sporothrix schenkii	Contaminated soil and vegetation in warm or tropical areas	Gardeners Farmers Florists Hunters Gold miners Laboratory workers

25

RICKETTSIA AND CHLAMYDIA

Dennis J. Darcey, M.D., M.S.P.H., and Ricky L. Langley, M.D., M.P.H.

CHLAMYDIA PSITTACI

Common name for disease: Psittacosis, ornithosis, parrot fever

Occupational setting

Bird handlers, pet shop workers, zoo attendants, poultry workers, laboratory personnel and veterinarians and technicians are at risk from exposure.

Exposure (route)

Inhalation of discharges or excrement of infected domestic birds (parakeets, parrots, pigeons, turkeys, chickens, ducks, etc.) is the most common route of exposure. Person-to-person transmission is rare.

Pathobiology

Chlamydia psittaci is an obligate intracellular parasite whose genome contains both DNA and RNA. The organism may be seen as a large cytoplasmic inclusion (0.3–1.0 μm in diameter) that is glycogen negative upon staining. It has a cell wall, like Gram-negative bacteria. These organisms cannot synthesize ATP and are considered energy parasites.[1,2]

Once inhaled, *Chlamydia psittaci* rapidly enters the bloodstream and is transported to the reticuloendothelial cells of the liver and spleen. It replicates in these sites and then invades the lungs and other organs by hematogenous spread. The incubation period ranges from 5 to 15 days. Disease often starts abruptly with chills and fever.[1] Headache is a constant symptom. Malaise, anorexia, myalgias and arthralgias are common. The pulse rate may be slow relative to the temperature elevation. Respiratory symptoms are typically mild compared to the extensive changes

Physical and Biological Hazards of the Workplace, Second Edition, Edited by Peter H. Wald and Gregg M. Stave
ISBN 0-471-38647-2

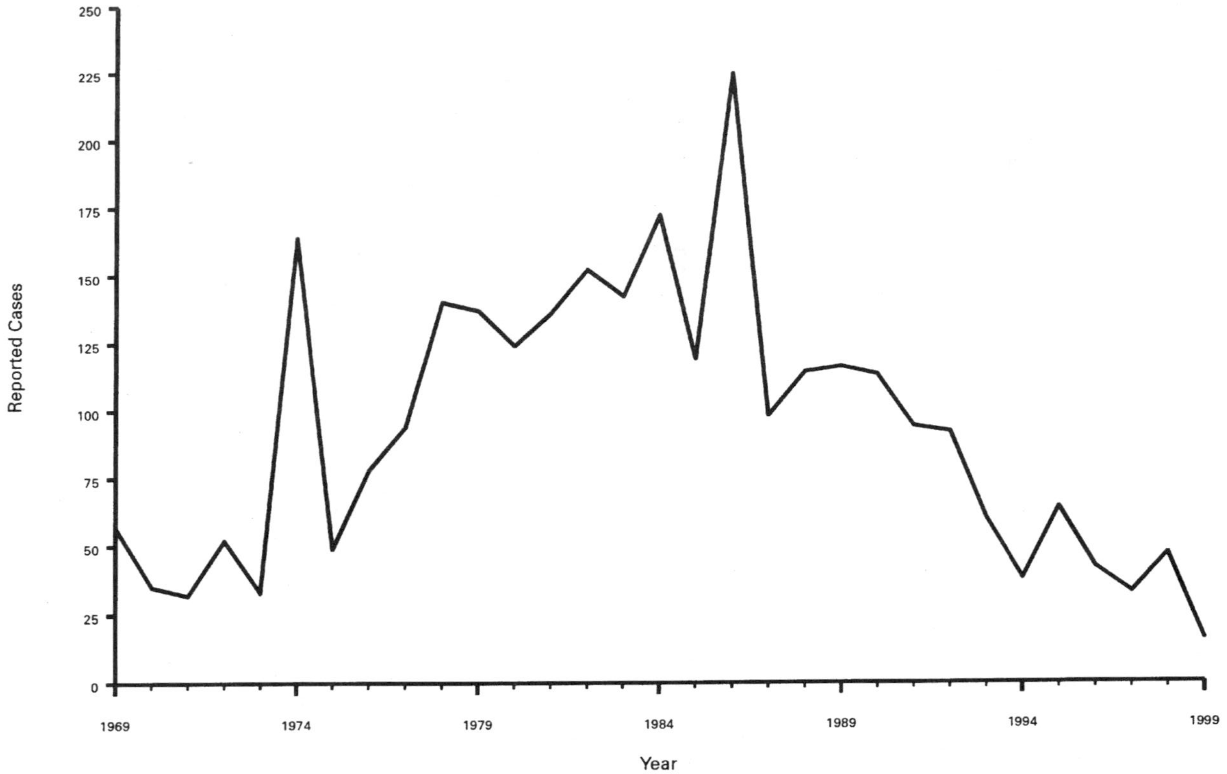

Figure 25.1 Psittacosis—reported cases by years, US, 1969–1999. During the 1990s, the number of reported psittacosis cases steadily declined. This decline could reflect both improved diagnostic testing to distinguish *Chlamydia psittaci* from *C. pneumoniae* infections, as well as improved control measures for psittacosis among birds. Source: MMWR 2001:48(53):58.

present on chest X-ray. A persistent dry hacking cough is prominent. Occasionally, changes in mentation may be noted. Hepatosplenomegaly may also occur in a significant number of patients. Myocarditis and encephalitis complications are rare but may occur. There are no characteristic laboratory or roentgenographic changes.

Diagnosis

The diagnosis can be confirmed by either isolation of the organism in culture or by serologic studies.[1,3,5] Isolation of the organism should be attempted only in specialized laboratories. A fourfold rise between acute and convalescent antibody titers by complement fixation is diagnostic. A single titer of 1 : 32 in a patient with a compatible illness is presumptive evidence of ornithosis.

Treatment

Tetracycline and doxycycline are the antibiotics of choice.[1,2] Erythromycin can be used as an alternative when tetracycline is contraindicated (e.g. pregnancy, young children).

Medical surveillance

No medical surveillance practices are routinely recommended. If cases can be linked to a source of exposure, then surveillance of the pet shop, aviary or farm should be undertaken. Infected birds should be treated or destroyed. The area where they are housed should be cleaned and disinfected.

Prevention

Education of the public as to the potential danger of household and occupational exposure to birds is needed. Regulation of the importation and proper quarantine of exotic birds is recommended.[5] Prophylactic treatment of psittacine birds with antibiotic-supplemented feed reduces the risk of disease in bird handlers. Occupational health personnel responsible for employee health in poultry-processing plants should be aware that headaches and pneumonia among workers may indicate psittacosis.

REFERENCES

1. Schlossberg D. Chlamydia psittaci (Psittacosis). In: Mandell GL, Dolin R, Bennett JE, eds. *Principles and practice of infectious disease,* 5th edn. Philadelphia: Churchill Livingstone; 2000:2004–7.
2. Wyriuck PB, Gutman LT, Hodinka RL. Chlamidae. In: Joklick WK, Willet HP, Amos DB, Wilfert CM, eds. *Zinsser microbiology,* 20th edn. Norwalk: Appleton and Lange, 1992:727–8.
3. Barnes RC. Laboratory diagnosis of human chlamydial infections. *Clin Microbiol Rev* 1989; 2:119–36.
4. Kuritsky JN, Schmid GP, Potter ME, et al. Psittacosis: a diagnostic challenge. *J Occup Med* 1984; 26:731–3.
5. Chin J, ed. *Control of communicable diseases manual,* 17th edn. Washington DC: American Public Health Association, 2000.

COXIELLA BURNETII

Common names for disease: Q fever

Occupational setting

Occupations at risk include abattoir and livestock workers, dairy workers, veterinarians, veterinary and laboratory technicians, laboratory animal handlers, farmers, ranchers, and hide and wool handlers. Farmers who are in contact with cattle, sheep and goats, assisting in the birthing of lambs, or exposed to birth by-products of other animals such as dogs and cats, are at particularly high risk.[1,2]

Exposure (route)

Exposure occurs through inhalation of aerosolized particles or direct contact with infected animals, primarily cattle, sheep or goats. Placentas from infected sheep are extremely infectious. In addition to domesticated livestock, a broad range of domestic and wild animals are natural hosts for *Coxiella burnetii*, including horses, dogs, swine, pigeons, ducks, geese, turkeys, squirrels, deer, mice, cats, and rabbits. Exposure to *C. burnetii* in research laboratories and veterinary hospitals has resulted in large outbreaks of Q fever.[3–5] Transmission has occurred following exposure at autopsy, but has not been documented during the clinical care of infected patients.[6] Person-to-person transmission of Q fever is unusual but has been reported.[7] Although organisms have been found in non-pasteurized milk and cheese, studies of *C. burnetii* infections in persons who consume unpasteurized dairy products have found that most or all of the serum-positive persons recall no acute illness.[8]

Although many species of ticks have been found to be infected, and there may be tick transmission among animals, ticks are not considered a source of infection in humans. Organisms can live in water and milk for prolonged periods of time.

Pathobiology

C. burnetii is a member of the family Rickettsiaceae and, like other rickettsia, is an obligate intracellular parasite that appears as a short pleomorphic Gram-negative rod. Unlike other Rickettsiaceae, *C. burnetii* grows in the phagosomes of the cell rather than the cytoplasm or the nucleus. It is very resistant to inactivation and can survive in the environment for long periods of time in the spore stage. *C. burnetii* is extremely infectious, and a single organism is enough to cause infection.[9]

Q fever usually presents as a mild respiratory illness and is often described as one of the atypical pneumonias. The incubation period varies, ranging from 9 to 39 days, but is usually 2–3 weeks. Clinical signs and symptoms include fever, malaise, headache, weakness, and transient pneumonitis with cough, chest pain, myalgias, and arthralgias. Physical examination is often unremarkable, and the most common physical finding is inspiratory crackles. Q fever does not usually present with a rash, and the acute illness resolves within 2–4 weeks. The majority of infections are mild and self-limiting. With a prolonged course, granulomatous hepatitis with jaundice can occur. A chronic form often associated with an infective endocarditis can develop as late as 1–20 years following an acute illness. Over 100 well-documented cases of endocarditis have been reported in the literature, but for most there was a history of pre-existing valvular disease or presence of a prosthetic valve.[10] Rare complications of Q fever include acute pleuropericarditis, hemolytic anemia, bone marrow necrosis, inflammatory pseudotumor of the lung, meningitis, optic neuritis, arthritis, cerebella ataxia, myocarditis, thrombophlebitis, nephritis, cystitis and epididymitis. The case fatality rate in untreated cases may be as high as 2.4%.

Diagnosis

The diagnosis of Q fever can be made by isolating the organism in the laboratory or by serologic demonstration of infection. The most common serologic tests include complement fixation (CF) and indirect fluorescent antibody (IFA) procedures. Antibodies to phase II antigen become detectable during the second week following the onset of illness and peak around week 8 for IFA titers or week 12 for CF titers. Confirmation of the diagnosis rests on demonstrating a fourfold or greater rise in paired serum specimens or a titer of IgM antibody greater than or equal to 1 : 20 by IFA. A phase II CF titer of 1 : 8 is considered significant. In chronic Q fever, the phase I antibody level rises and a phase I CF antibody titer greater than 1 : 200 is considered diagnostic.

Treatment

The treatment of choice for Q fever pneumonia is doxycycline or tetracycline for 2 weeks. There have been no controlled trials to assess the treatment of chronic Q fever, but therapy with tetracycline in combination with trimethoprim–sulfamethoxazole or rifampin or oflaxacin for several years has been recommended.[11,12]

Prevention

C. burnetii organisms are widespread in the environment and resistant to inactivation. Control of major animal reservoirs of the organism is impractical. Personal protective equipment when handling affected animals, bedding and their byproducts and respiratory protection when working in dusty environments contaminated with organisms are recommended. Other preventive measures include pasteurization of milk to reduce the potential risk for transmission through milk and cheese products. Workers with prosthetic heart valves and liver disease are at particularly high risk for the sequelae of infection and are best restricted from high-risk environments. Biosafety level 2 (BSL2) practices, containment equipment and facilities are recommended for non-propagative laboratory procedures, including serologic examinations and staining of impression smears. BSL3 practices and facilities are recommended for activities involving the inoculation, incubation and harvesting of embryonated eggs or tissue cultures, the necropsy of infected animals and manipulation of infected tissue. Experimentally infected rodents should also be maintained under Animal biosafety level 3 (ABSL3).[13] An investigational phase I Q fever vaccine is available from the US Army Medical Research Institute for Infectious Diseases. The use of the vaccine has been limited to high-risk groups who have no demonstrated sensitivity to Q fever antigen.[12]

REFERENCES

1. Buhariwalla F, Cann B, Marrie TJ. A dog-related outbreak of Q Fever. *Clin Infect Dis* 1996; 23:753–5.
2. Marrie TJ, Durant H, Williams JC, et al. Exposure to parturient cats: a risk factor for acquisition of Q fever in Maritime Canada. *J Infect Dis* 1988; 158:101–8.
3. Johnson JE II, Kadull PJ. Laboratory-acquired Q fever. *Am J Med* 1966; 41:391–403.
4. Hall CJ, Richmond SJ, Caul EO, et al. Laboratory outbreak of Q fever acquired from sheep. *Lancet* 1982; 1:1004–6.
5. Meiklejohn G, Reimer LG, Graves PS, et al. Cryptic epidemic of Q fever in a medical School. *J Infect Dis* 1984; 144:107–14.
6. Marrie TJ, Coxiella burnetii (Q fever). In: Mandel GL, Douglas GR, Bennett JE. eds. *Principles and practice of infectious diseases,* 5th edn. New York: Churchill Livingstone, 2000.
7. Mann JS, Douglas JG, Inglis JM, et al. Q fever: person to person transmission within a family. *Thorax* 1986; 41:974–5.
8. Fishbein DB, Raoult D. A cluster of Coxiella burnetii infections associated with exposure to vaccinated goats and their unpasteurized dairy products. *Am J Trop Med Hyg* 1992; 47(1):35–40.
9. Leedom JM. Q fever: an update. *Curr Clin Top Infect Dis* 1980; 1:304.
10. Sawyer LA, Fishbein DB, McDade JE. Q fever: Current Concepts. *Rev Infect Dis* 1987; 9:935–46.
11. Holtom PD, Leedom JM. Coxiella burnettii (Q Fever). In: Gorbach SL, Bartlett JG, Blacklow NR, eds. *Infectious diseases*. Philadelphia: WB Saunders, 1992:1657–9.
12. Chin J, ed. Control of communicable diseases manual, 17th edn. Washington DC: American Public Health Association, 2000:407–10.
13. Richmond JY, McKinney RW, eds. *Biosafety in microbiological and biomedical laboratories,* 4th edn. Washington DC: US Government Printing Office, HHS Publication No. (CDC) 93-8395. 1999:148–9.

EHRLICHIA SPECIES

Common name for disease: Ehrlichiosis

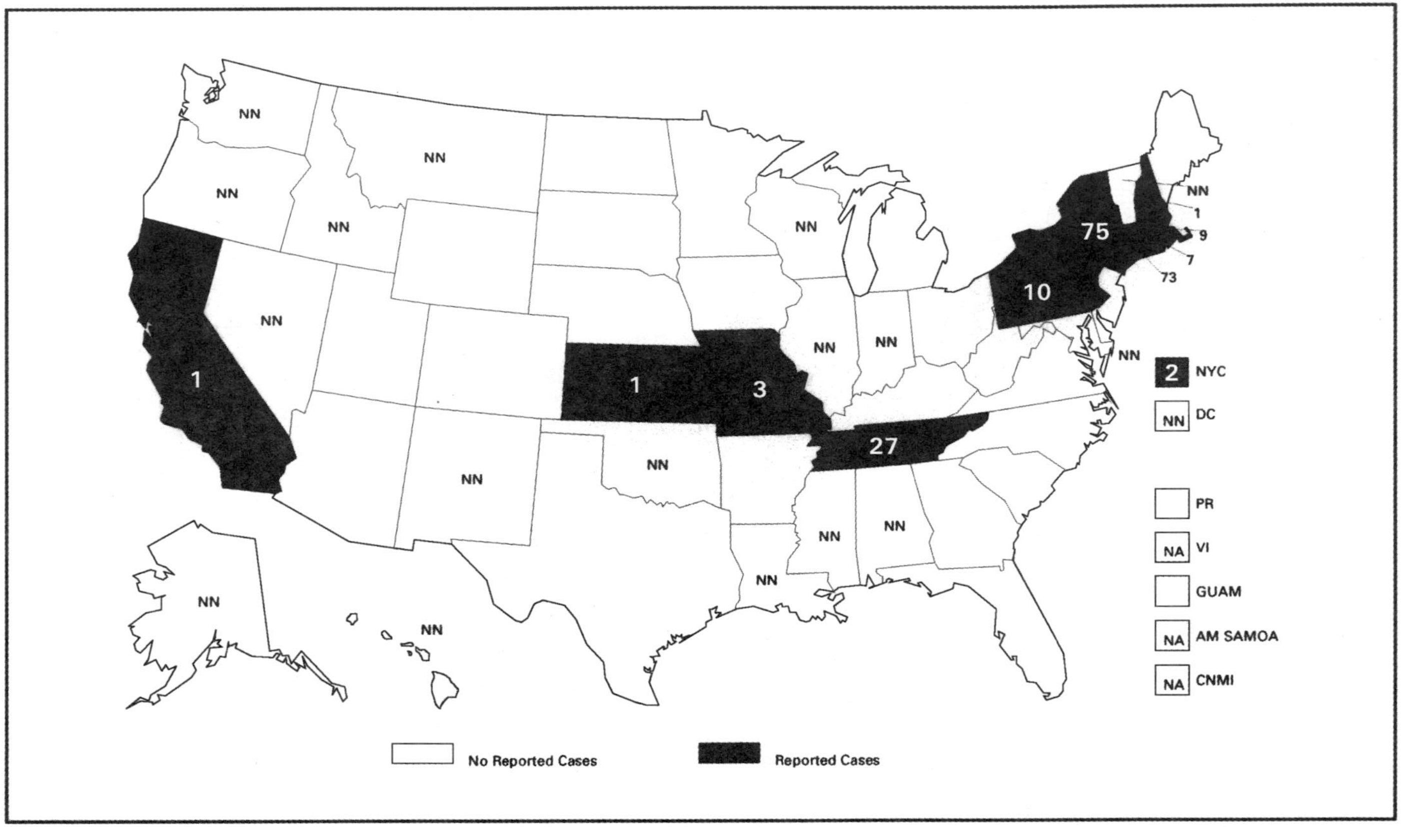

Figure 25.2 Ehrlichiosis, human granulocytic–reported cases, US and territories, 1999. Human ehrlichiosis is an emerging infectious disease that became nationally notifiable in 1999. Identification and reporting of human ehrlichiosis are incomplete, and the number of cases reported here do not represent the overall distribution or regional prevalence of disease. Source: MMWR 2001:48(53):34.

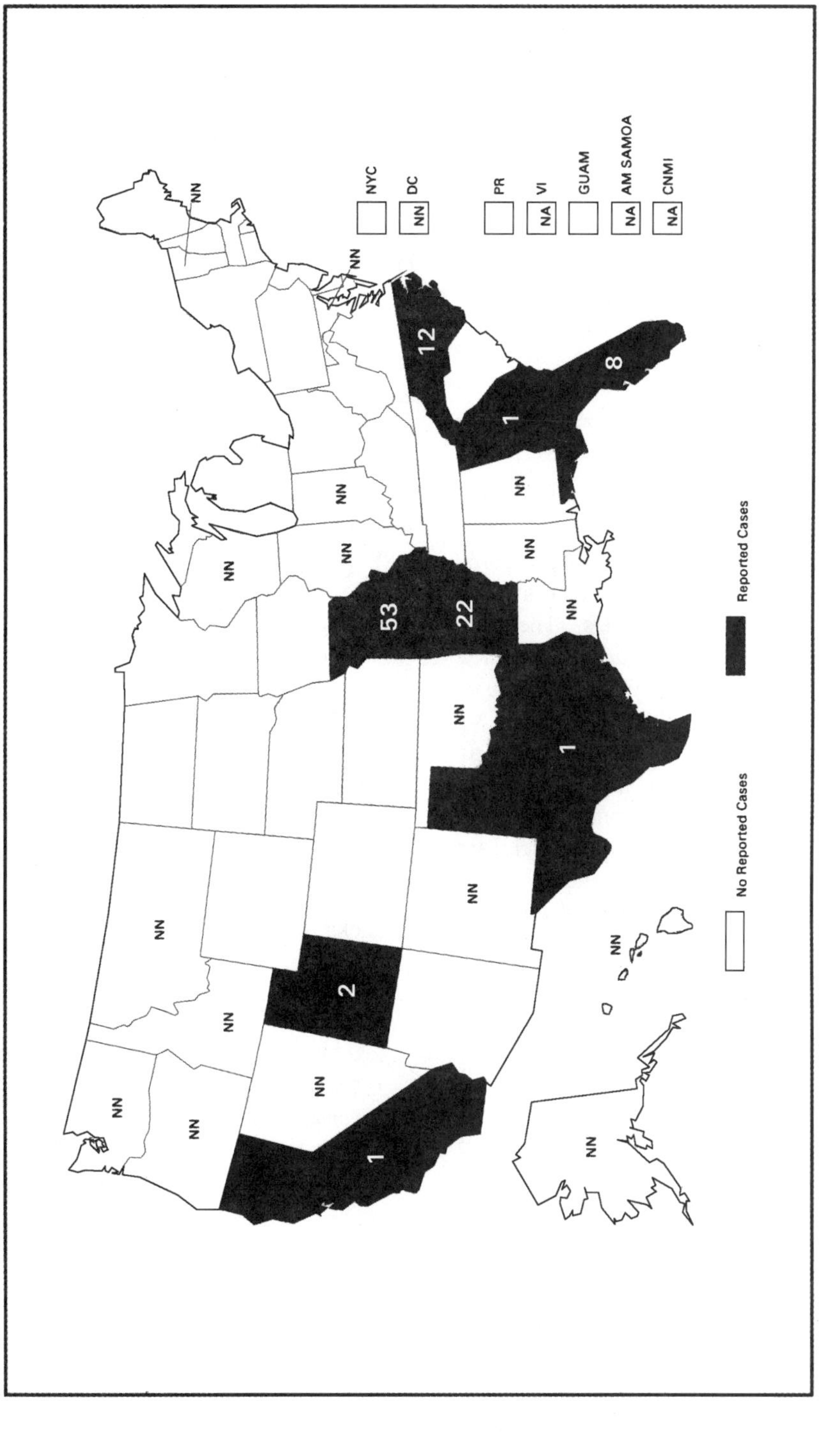

Figure 25.3 Ehrlichiosis, human monocytic—reported cases, US and territories, 1999. Human ehrlichiosis is an emerging infectious disease that became nationally notifiable in 1999. Identification and reporting of human ehrlichiosis is incomplete, and the number of cases reported here do not represent the overall distribution or regional prevalence of disease. Source: MMWR 2001:48(53):35.

Occupational setting

Ranchers, farmers, foresters, rangers, hunters and lumberjacks are at increased risk of exposure.

Exposure (route)

Infection is most likely caused by the bite of an infected tick. The species of tick transmitting infection to humans are probably *Amblyomma americanum* for human monocytic ehrlichiosis and *Ixodes scapularis* for human granulocytic ehrlichiosis.[1]

Pathobiology

Ehrlichia species are obligate intracellular Gram-negative coccobacilli and are members of the Rickettsiaceae family.[1,2] They form intracellular inclusion bodies measuring 2–5 μm in size within leukocytes. The first proven human infection with *Ehrlichia* species was described in Japan in 1953 and was due to *Ehrlichia sennetsu*. This illness was similar to infectious mononucleosis. The first case of human ehrlichiosis in the USA was reported in 1987. Two similar but distinct organisms cause illness. Human monocytic ehrlichiosis (HME) affects primarily mononuclear cells. The causative agent of HME is *E. chaffeensis*. Most cases of HME in the USA occur in the southeast and south central areas of the country.[3,4] Human granulocytic ehrlichiosis (HGE) affects the granulocytes and occurs primarily in the upper midwestern and northeastern USA. The causative agent of HGE is identical to or closely related to *E. phagocytophila* and *E. equi*.[1] The incubation period ranges from 1 to 21 days. The typical illness for American ehrlichiosis begins abruptly with prostration, severe headaches, fever, malaise, myalgias and occasionally nausea and vomiting. Other signs and symptoms may include interstitial pneumonia, abdominal pain, arthralgias, jaundice, diarrhea, encephalopathy, and lymphadenopathy. A transient rash occurs in 10–50% of cases. In contrast to Rocky Mountain spotted fever, the rash is usually not petechial. Laboratory findings may include hyponatremia, thrombocytopenia, leucopenia (especially lymphopenia), anemia and elevated aminotransferase levels. Deaths do occur but are rare.

Diagnosis

Disease usually occurs from April through October.[5] Besides the clinical and laboratory manifestations noted above, serologic diagnosis is confirmed with a four-fold rise in antibody titer or an antibody titer greater than or equal to 1 : 80 by indirect immunofluorescence.[3,5] Rarely, cytoplasmic inclusion bodies (morulae) are seen and can aid in early diagnosis of disease. Other diagnostic methods include immunohistochemistry, PCR, and culture.[1]

Treatment

The treatment of choice is one of the tetracyclines, although chloramphenicol is also effective.[1,2,5]

Prevention

Use of tick repellents may be helpful. After being outdoors, the body should be examined for ticks.

REFERENCES

1. Chin J, ed. *Control of communicable diseases manual*, 17th edn. Washington DC: American Public Health Association, 2000:181–2.
2. Sexton DJ, Willet HP, Rickettsiae. In: Joklik W, Willet H, Amos D, Wilfert C, eds. *Zinsser microbiology*, 20th edn. Norwalk: CT Appleton and Lange, 1992:700.
3. Anderson BE, Dawson JE, Jones DC, Wilson KH. Ehrlichia chaffeensis, A new species associated with human ehrlichiosis. *J Clin Microbiol* 1991; 29:2838–42.
4. Rathore MH. Infection due to Ehrlichia canis in Children. *Southern Med J* 1992; 85:703–5.

5. Armstrong RW. Ehrlichiosis in a visitor to Virginia. *Western J Med* 1992; 157:182–4.

RICKETTSIA RICKETTSII

Common name for disease: Rocky Mountain spotted fever

Occupational setting

Persons in outdoor occupations, including farm, forestry logging and construction workers, telephone linemen, environmental technicians and some laboratory workers are at risk from exposure.

Exposure (route)

Transmission of infection occurs primarily following bites from infected ticks or from skin contamination with tick tissue or feces when removing ticks from humans or animals. Infections have also been associated with needlestick injuries, blood transfusions, and laboratory handling of rickettsii. Laboratory-acquired infection has also occurred following exposure to infectious aerosols.[1] The tick is both the vector and main reservoir.

Pathobiology

Rickettsia rickettsii belongs to the spotted fever group of Rickettsiae, which are genetically related but different in their surface antigenic proteins. The rickettsia are small obligate intracellular bacteria measuring approximately 0.3–1.0 μm in size. The cell wall has the ultrastructural appearance of a Gram-negative bacterium and contains lipopolysaccharide (LPS).[2] The LPS contains immunogenic antigens that are shared among the rickettsia and cross-react with *Proteus* and *Legionella*. Cross-reactivity with *Proteus* is the basis for the Weil–Felix test.

The American dog tick, *Dermacentor variabilis*, is the most prevalent vector in the Eastern USA. The Rocky Mountain wood tick, *Dermacentor andersoni*, is the prevalent vector in the western USA. Of the three tick stages, larvae, nymph, and adult, only the adult ticks feed on humans. Transmission of infection occurs after about 6–10 hours of feeding. Exposure to infected tick hemolymph can occur during the removal of ticks from humans or domestic animals, especially when the tick is crushed between the fingers. Infection spreads from skin inoculum through the lymphatics and small blood vessels, causing widespread endothelial injury leading to increased vascular permeability.

First recognized in the Rocky Mountain region, cases have been reported in almost every state. The incidence of disease varies, with between 600 and 1000 cases reported to the CDC each year. The disease is most prevalent in the South and Midwest, with the highest rates reported in southeastern North Carolina and southwestern Oklahoma (Figure 25.1). The disease is more prevalent in the spring and summer months but has been reported throughout the year in the USA.

The incubation period ranges from 2 to 14 days. Onset of symptoms is abrupt with fever, headache, chills, myalgia and malaise. Frequently this is accompanied by gastrointestinal symptoms of nausea, vomiting, abdominal pain and diarrhea, which often lead to confusion and delay in the diagnosis. The characteristic rash usually appears 3–5 days after the onset of fever. In many cases, it appears first on the ankles and wrist and then becomes generalized. Involvement of the palms and soles is considered characteristic, but often appears late in the course of the acute illness. The rash, which is initially maculopapular, becomes petechial and hemorrhagic as the illness progresses. Vasculitis involving the brain, heart, liver and kidneys may cause complications from seizures to congestive heart failure and acute renal failure.

Fatality rates have decreased over the years with improvements in early detection and

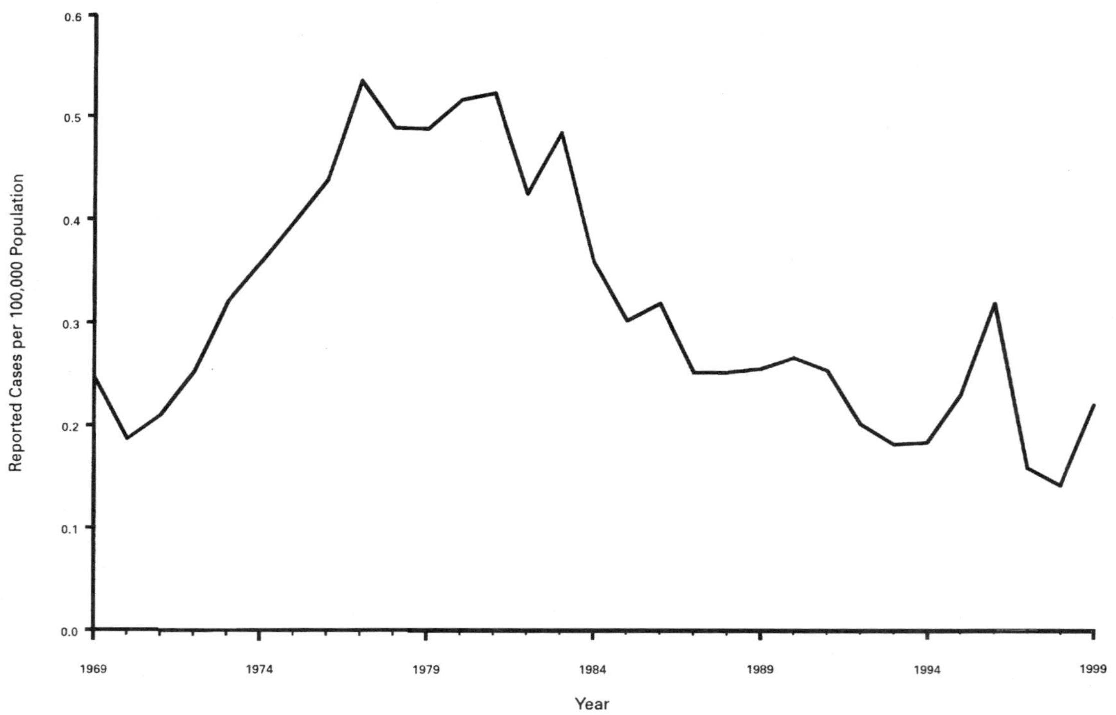

Figure 25.4 Rocky Mountain spotted fever—reported cases per 100 000 population by year, US, 1969–1999. Changes in the number of reported cases of Rocky Mountain spotted fever could reflect alterations to surveillance algorithms for this and other tickborne diseases. Biological factors (e.g., changes in tick populations resulting from fluctuating enviromental conditions) also could be involved. Source: MMWR 2001: 48(53):60.

treatment. Risk factors associated with a higher mortality rate include increasing age, delay in initiation of chemotherapy, male gender and glucose-6-phosphate dehydrogenase deficiency.

Diagnosis

The initial diagnosis of Rocky Mountain spotted fever is largely based upon clinical signs and symptoms (fever, headache, myalgia, rash) and epidemiology (geographic and seasonal variability). Treatment should be initiated before laboratory confirmation of the diagnosis. In endemic regions, an appropriate constellation of symptoms and signs is sufficient justification to begin treatment, even in those with no discernable history of tick bite.

Punch biopsies of skin lesions stained with fluorescent conjugates of anti-rickettsial serum are useful in diagnosing Rocky Mountain spotted fever in its early stages. However, this test lacks sensitivity and may fail to accurately diagnose Rocky Mountain spotted fever in as many as 50% of cases.[2]

The laboratory diagnosis of Rocky Mountain spotted fever can be achieved with isolation of the organism from blood, demonstration of positive immunofluorescence of skin lesion (biopsy), or organ tissue (autopsy), and positive polymerase chain reaction assay to *R. rickettsii*. A fourfold or greater rise in antibody titer to *R. rickettsii* antigen by immunofluorescence antibody (IFA), complement fixation (CF), latex agglutination (LA), microagglutination (MA), or indirect hemagglutination antibody (IHA) test in acute- and convalescent-phase specimens, ideally taken greater than or equal to 3 weeks apart, meet the CDC laboratory criteria for diagnosis. The microimmunofluorescent antibody test is the most widely used serologic test to diagnose Rocky Mountain spotted fever in the USA. A fourfold rise in titer is evidence of recent *Rickettsia* infection. A single convalescent titer of 1 : 64 or greater is also considered to be diagnostic but weaker evidence of acute infection.[3]

Treatment

Since the introduction of tetracycline antibiotics, the lethality of the disease has declined dramatically, but remains significant at around 5%.[2] Starting treatment early in the course of the disease before the rash is apparent is critical in reducing mortality. Rocky Mountain spotted fever responds to treatment with oral tetracycline, doxycycline or chloramphenicol. Antibiotic treatment is usually given for at least 7 days, continuing until 2 days after the patient has become febrile. Doxycycline or another tetracycline is considered the drug of choice, but chloramphenicol is preferred during pregnancy because of tetracycline's effects on fetal bones and teeth. Although tetracyclines have been avoided in young children in the past because of concerns about staining the teeth, it is recommended that doxycycline be used for suspected Rocky Mountain Spotted Fever because of the life-threatening nature of the disease and the minimal risk for staining the teeth after one course of treatment.

Medical surveillance

It is essential that laboratories working with *R. rickettsii* have an effective system for reporting febrile illness in laboratory personnel. A medical surveillance program should include evaluation of potential cases, and, when indicated, institution of appropriate antibiotic therapy.

Prevention

At present, there is no commercially available vaccine for Rocky Mountain spotted fever. The best means of prevention remains avoidance of contact with ticks, wearing protective clothing, and using insect repellents. Body checks, with particular attention to the scalp, pubic and axillary hair, should be conducted daily for at-risk workers. Ticks can be removed using forceps and gentle traction, being careful to remove all of the mouthparts from the skin. Tick bites should be cleansed, and care taken during removal to prevent crushing the tick and contaminating fingers with tick tissue and feces, which are potentially infectious. In endemic areas, physicians caring for injured and ill workers should be periodically reminded of the importance of early diagnosis and treatment to prevent serious sequelae. In the laboratory, BSL2 practices, containment equipment and facilities are recommended for all non-propagation laboratory procedures, including serologic and fluorescent antibody tests, and staining of impression smears. BSL3 practices and facilities are recommended for all other manipulations of known or potentially infectious materials, including necropsy of experimentally infected animals and trituration of their tissues, and inoculation, incubation, and harvesting of embryonated eggs or tissue cultures.[4]

REFERENCES

1. Johnson JE, Kadull PJ. Rocky Mountain spotted fever acquired in a laboratory. *N Engl J Med* 1967; 227:842–7.
2. Walker DH, Raoult D. Rickettsia rickettsii and other spotted fever group Rickettsiae. In: Mandell GL, Dolin R, Bennett JE, eds. *Principles and practice of infectious diseases*, 5th edn. Philadelphia: Churchill Livingstone, 2000:2035–241.
3. Centers for Disease Control and Prevention Epidemiology Program Office. *Case definitions for infectious conditions under public health surveillance Rocky Mountain spotted fever—1996.*
4. Richmond JY, McKinney, RW, eds. *Biosafety in microbiological and biomedical laboratories,* 4th edn. HHS Publication No. (CDC) 93-8395, Washington DC: US Government Printing Office, 1999:149–52.

26

PARASITES

William N. Yang, M.D., M.P.H.

CRYPTOSPORIDIUM PARVUM

Common name for disease: Cryptosporidiosis

Occupational setting

Farmers, animal handlers, veterinarians, laboratory personnel, healthcare workers and day care workers are at risk from exposure.[1–3]

Exposure (route)

Occupational disease results most commonly through ingestion following exposure to infected calves and other farm animals. The prevalence of infection is higher in young animals, such as calves and lambs.[4] Other sources of infection are water supplies and swimming and wading pools, that have been contaminated by animal or human sewage. Person-to-person (fecal–oral) infection may also occur in healthcare and day care settings.[3–7] The 1993 cryptosporidiosis waterborne outbreak in Milwaukee, Wisconsin was the largest waterborne outbreak in US history, affecting over 400 000 people. The outbreak was estimated to cost over 53 million dollars in lost wages, lost productivity, and medical costs.[8]

Pathobiology

Cryptosporidium parvum is an intracellular, but extracytoplasmic protozoan parasite. When a human ingests the oocyst that has been passed in the feces of an infected host, the oocyst wall dissolves and sporozoite forms invade the host gastrointestinal (GI) epithelial cells. They pass through a trophozoite stage and an asexual multiplication stage that results in schizonts. These schizonts can reinvade the host or continue to a sexual multiplication stage that leads to new oocysts that are passed outside the body. The sporulated oocyst is the only developmental stage that

Physical and Biological Hazards of the Workplace, Second Edition, Edited by Peter H. Wald and Gregg M. Stave
ISBN 0-471-38647-2

occurs extracellularly. Since this part of the cycle is completed before the oocyst is excreted, the oocysts are immediately infectious when passed in feces.[7] In immunocompromised persons, the ability to reinfect the host can lead to continuing infection.[5]

Cryptosporidiosis is the intestinal infection caused by *C. parvum*. It occurs in mammals, birds, reptiles and fish worldwide. *Cryptosporidium* is a major cause of diarrhea in infants and children, particularly in developing countries. In immunocompetent hosts, it is usually an asymptomatic or self-limited diarrheal disease. However, in an immunosuppressed person, it can be a life-threatening condition that causes malabsorption and severe weight loss (Figure 26.1). Although the jejunum seems to be the most heavily infected area, in some immunocompromised persons *Cryptosporidium* has been found throughout the GI tract, gallbladder, liver and pancreas and even in the lungs of AIDS patients.[7] The prevalence of *Cryptosporidium* infection in HIV-positive people is 14% (range 6–70%) in the developed countries and 24% (range 8.7–48%) in developing countries.[7]

The immune status of the person determines the severity and length of the symptoms. In immunocompetent persons, the symptoms begin quickly, after a 2–12 day incubation period, and then last from 2 to 55 days, with a mean duration of 9 days. The illness is usually self-limited. Symptoms of disease include watery diarrhea, cramping abdominal pain, weight loss, and flatulence. Less common symptoms are nausea, vomiting, myalgias, and fever. While cryptosporidial cysts can be found in the stool, blood and white blood cells are usually absent. Oocysts may remain in the stool for 8–50 days (mean 12–14 days) after the clinical symptoms have ended.[7]

In immunocompromised persons including AIDS patients, the infection may start slowly and become chronic and progressively worse as the immunosuppression worsens. The diarrhea can last for months and lead to dehydration, electrolyte imbalance, malnutrition, and weight loss. Losses of more than 20 liters/day have been reported. Persons with AIDS can have biliary cryptosporidiosis that presents with the signs and symptoms of cholecystitis and elevated alkaline phosphatase levels with normal serum transaminases and bilirubin levels.[8] Symptoms in patients with reversible causes of immunodeficiencies usually resolve quickly when the cause of the immunosuppression is eliminated.[7]

Diagnosis

The diagnosis of cryptosporidiosis can be made by microscopic examination of small bowel biopsy specimens or by staining the parasite in stool specimens. With a modified acid-fast technique, the oocysts are red-stained and can be differentiated from similar-appearing yeast forms that do not take up the acid-fast stain. Strict morphologic criteria must be applied to the diagnosis to avoid confusion with other oocysts, such as *Cyclopsora* oocysts, which are much larger (10 μm).[9] Enzyme immunoassays and immunofluorescence assay methods are also available and have the advantage of high sensitivity and the disadvantage of high cost.[9] No serologic tests are currently available commercially.[7]

Treatment

There is no known effective treatment for cryptosporidiosis. The inability to cultivate the organism in vitro and the lack of an animal model for the disease have hampered finding effective agents.

In the immunocompetent host, the infection is self-limited; however, the gastroenteritis can be severe enough to require anti-motility drugs and oral/intravenous hydration. Those on corticosteroids or cytotoxic drugs will recover if the agents are stopped.[5,6]

In the immunocompromised person, fluid and electrolyte balance and anti-diarrheal agents are used as supportive therapy. No chemotherapy is currently recommended as treatment. Paromomycin, clarithromycin, nita-

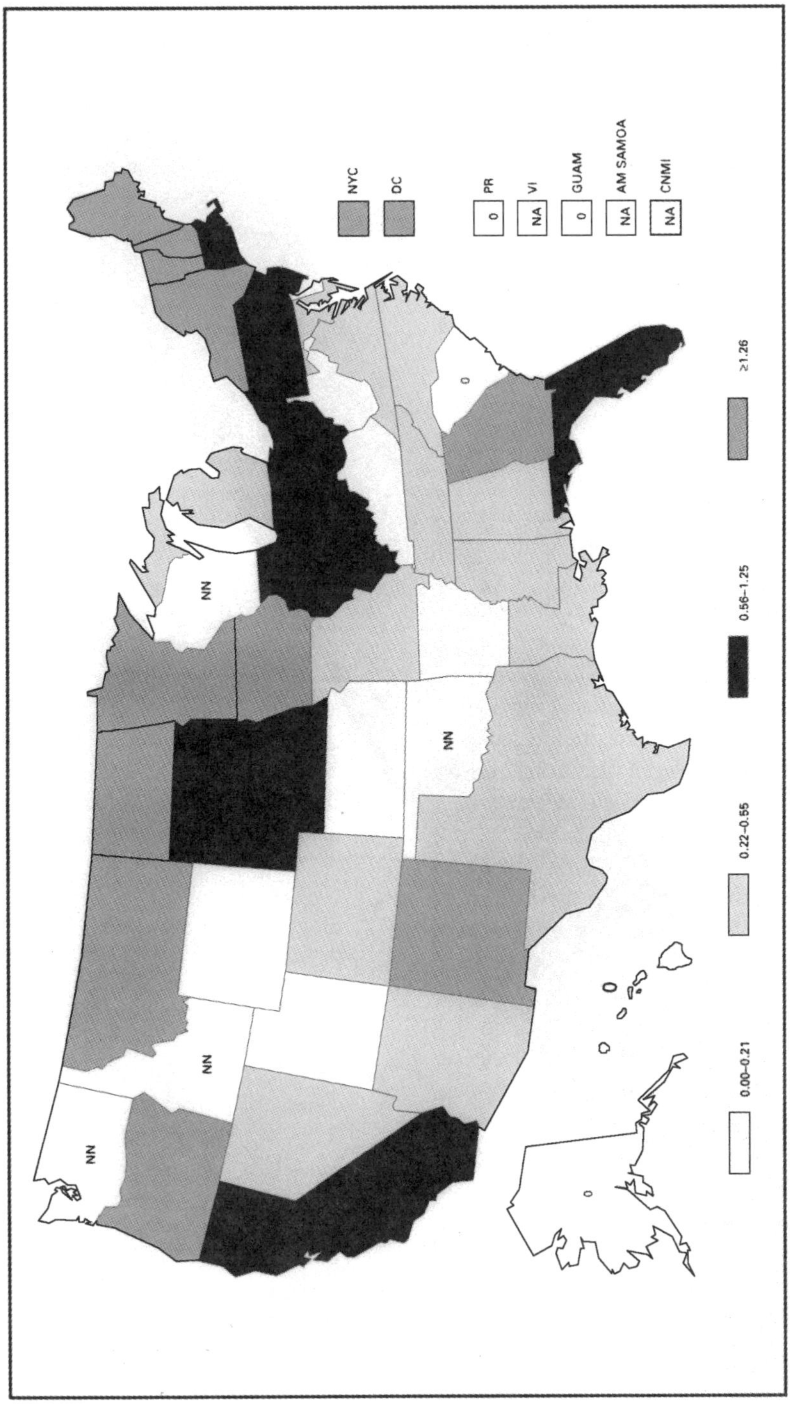

Figure 26.1 Cryptosporidiosis-reported cases per 100,000, United States and Territories, 1999. Source: MMWR 2001: 48(53):32.

zoxanide and hyperimmune bovine colostrum are being evaluated.[7] Highly active anti-retroviral therapy (HAART) has been shown to decrease the prevalence of cryptosporidiosis in HIV-infected patients.[9]

Medical surveillance

Occupationally exposed or at-risk workers should be monitored for signs or symptoms of diarrhea.

Prevention

No effective vaccine and no prophylactic drugs exist. Eliminating the disease in farm animals decreases the risk for farmers and veterinary personnel. In farm and institutional settings, personal hygiene practices, especially good handwashing, and environmental sanitation should be emphasized.[7] Strict enteric precautions should be practiced around infected animals or humans. Cooking kills *Cryptosporidium* in food, but vegetables and fruits to be consumed uncooked must be thoroughly washed with potable water before eating. No safe and effective chemical disinfectants have been found to decontaminate produce.[7]

Chlorination of water does not kill *Cryptosporidium*. Drinking water sources must be protected from human and animal fecal material from surface runoff.[7] Maintaining proper filtration procedures for drinking water supplies will minimize infection but may not prevent it, due to the small size of the oocysts.[10] Travelers should minimize their exposure by avoiding untreated water and uncooked fruits and vegetables. The oocyst will survive 3% hypochlorite solution, iodophors, benzalkonium chloride, and 5% formaldehyde. Reverse osmosis filters can remove all oocysts but require low-turbidity source water to prevent plugging. Contaminated water can be purified for personal use by boiling for 1 min[7] and prolonged treatment with bleach or 10% formalin with either bleach or 5% ammonia.

Immunocompromised persons should not have contact with any animal that has diarrhea.[7] They should also realize that there is risk of infection from the accidental ingestion of recreational lake, river or swimming pool water due to intermittent contamination with cryptosporidia from human or animal waste.[7] One approach to prevent cryptosporidiosis in HIV-infected persons is to maintain the immune system function by using HAART.[9]

REFERENCES

1. Current VVL, Reese NC, Ernst JY, Bailey WS, Heyman MB, Weinstein WM. Human cryptosporidiosis in immunocompetent and immunodeficient persons. Studies of an outbreak and experimental transmission. *N Engl J Med* 1983; 308:1252–7.
2. Pohjola S, Oksanen H, Jokipii L, Jokipii AMM. Outbreak of cryptosporidiosis among veterinary students. *Scand J Infect Dis* 1986; 18:173–8.
3. Guerrant RL. Cryptosporidiosis: an emerging highly infectious threat. *Emerg Infect Dis* 1997; 3:51–7.
4. Wolfson JS, Richter JM, Waldron MA, Weber DJ, McMarthy DM, Hopkins CC. Cryptosporidiosis in immunocompetent patients. *N Engl J Med* 1985; 312:1278–82.
5. Soave R. Human coccidial infections: cryptosporidiosis and isosporiasis. In: Warren KS, Mahmoud AA, eds. *Tropical and geographical medicine*, 2nd edn. New York: McGraw-Hill, 1990, 352–8.
6. Naven TR, Juranek DD. Cryptosporidiosis: clinical, epidemiologic, and parasitologic review. *Rev Infect Dis* 1984; 6:313–27.
7. Juranek DD. Cryptosporidiosis. In: Strickland GT, ed. *Hunter's tropical medicine and emerging infectious diseases*, 8th edn. Philadelphia: WB Saunders Company, 2000.
8. Franzen C, Muller A. Cryptosporidia and microsporidia–waterborne diseases in the immunocompromised host. *Diagn Microbiol Infect Dis* 1999; 34:245–62.
9. Clark DP. New insights into human cryptosporidiosis. *Clin Microbiol Rev* 1999; 12:554–63.

10. Soave R. Treatment strategies for cryptosporidiosis. *Ann 2NV Acad Sci* 1990; 616:442–51.

CYCLOSPORIASIS

Common names for disease: None

Cyclosporiasis is an intestinal infection caused by the coccidian *Cyclospora cayetanensis*. Prior to 1996, cyclosporiasis in North America was associated with overseas travelers. In 1996, a large outbreak in North America occurred in the spring and summer and was associated with eating Guatemalan raspberries.[1] The clinical presentation of cyclosporiasis comprises diarrhea, fatigue, and anorexia. The clinical course can be prolonged without treatment.[2]

Occupational setting

Cyclosporiasis can be an important cause of diarrhea both domestically and in travelers. *Cyclospora* has been found throughout the world and in both immunocompetent and immunosuppressed persons.[2] *Cyclospora* infected 7% of the Nepal American Embassy community during the 1992 *Cyclospora* season and was the cause of 11% of the cases of diarrhea seen at the Canadian International Water and Energy Consultants' Clinic in Katmandu.[2]

Although travelers do not seem to be at high risk for acquiring cyclosporiasis, the Indian subcontinent and Indonesia have been reported as high-risk areas.[3] *C. cayetanensis* should be considered as a possible cause of diarrhea in an individual with prolonged watery diarrhea with a history of travel to a developing country.[3]

Exposure (route)

The transmission of *Cyclospora* appears to be fecal–oral, but there is no definite evidence of person-to-person spread. Water-borne transmission and food-borne transmission evidence exists.[2] A water supply from a hospital building at a Chicago hospital was implicated as the source of a 1990 summer outbreak in the hospital staff and water was associated with a case–control study of cyclosporiasis in Katmandu.[2,4] An airline pilot developed *Cyclospora* diarrhea after eating airline food prepared in Haiti.[2] The large outbreak that occurred in 1996 was related to the consumption of raspberries imported from Guatemala.[2] In 1997, several other outbreaks in Canada and the USA were also linked to raspberries from Guatemala and to basil and mesclun lettuce.[5]

The mode of contamination of the raspberries is unclear. However, following the 1996 outbreak, berry growers and exporters in Guatemala, with assistance from the Food and Drug Administration (FDA) and Centers for Disease Control, began voluntary water quality and sanitary control measures.[1,5] After another outbreak in spring 1997, Guatemala began classifying the farms, and only "low-risk" farms were allowed to export to North America.[5] In 1998, the FDA banned the importation of fresh raspberries into the USA, but Canada continued to import them until June 1998.

Two surveys of indigenous Peruvian children under age 30 months found 6% and 18% with Cyclospora cysts in their feces.[2,4] Of those infected, 11% and 28% respectively presented with diarrhea. In Cyclospora endemic areas, the risk is higher in the late spring and summer.[2]

Pathobiology

In retrospect, cases of cyclosporiasis were first noted in the 1980s and the organism was referred to as an unsporulated coccidian and alga-like organism and an atypical *Cryptosporidium*. In 1993, Ortega correctly classified it as *Cyclospora cayetanensis*.[4] The oocysts of *C. cayetanensis* are immature when evacuated in stool but mature in about 5 days when observed in the laboratory.[6] The ingestion of mature oocysts causes infection in the duodenum.[6]

The incubation period is 2–11 days after exposure, and the clinical onset can be abrupt. The acute symptoms that occur with invasion of enterocytes of the small intestine are fever, vomiting and frequent watery diarrhea that usually last for several days.[2] After the acute symptoms, affected individuals may note fatigue, anorexia, nausea, belching, bloating and further diarrhea that can be daily or intermittent.[2] Weight loss generally occurs in untreated cases.[2]

If untreated, clinical manifestations can last for an average of 6–7 weeks in immunecompetent hosts, and in HIV-positive individuals the diarrhea can last for months. The organism can be detected in stools up to 8 weeks after the onset of symptoms.[2] Asymptomatic individuals who are excreting cysts have been documented.[6]

Diagnosis

Identifying the typical oocysts in stools after using concentration methods or flotation methods makes the diagnosis. Staining stool smears with acid-fast stain can result in variable results, particularly if small numbers of oocysts are present. Fluorescence microscopy can assist in the diagnosis, since the oocysts autofluoresce. The diagnosis can also be made by observing the oocysts in wet preparation from jejunal aspirates.[6]

Current CDC Guidelines for the confirmation of *C. cayetanensis* food-borne disease outbreaks require the demonstration of the organism in stools of two or more ill persons.[7]

Treatment

Trimethoprim–sulfamethoxazole (160/800 mg) given twice daily for 7 days is an effective treatment. AIDS patients have also responded to the same medication given for 10 days. Norfloxacin, metronidazole, tinidazole, quinacrine and azithromycin have not been effective.[2]

Medical surveillance

Currently, cyclosporiasis is not a nationally notifiable disease in the USA. However, clinicians and laboratories that identify cases of cyclosporiasis unrelated to travel outside the USA are encouraged to contact the CDC's Division of Parasitic Diseases, National Center for Infectious Diseases at (770) 488–7760.[5]

Since 1997, the CDC has been collecting data on cyclosporiasis cases from selected cities through its Emerging Infections Program Foodborne Diseases Active Surveillance Network FoodNet. With the voluntary controls in Guatemala followed by the 1998 import ban on Guatemalan raspberries, the rate of *Cyclospora* infections has decreased by 70%, from 0.3% to <0.1% per 100 000 population.[8]

Prevention

Cyclospora is killed by boiling water but not by chlorination. In countries where there is risk of exposure to *Cyclospora*, the risk is minimized by boiling drinking water and drinking tea and coffee. Fresh fruits, especially raspberries and salads, should be thoroughly washed before eating.[2]

REFERENCES

1. Herwaldt BL, Ackers M. The Cyclospora Working Group. An outbreak in 1996 of cyclosporiasis associated with imported raspberries. *N Engl J Med* 1997; 336:1548–56.
2. Conner BA, Shlim DR. Cyclosporiasis. In: Strickland GT, ed. *Hunter's tropical medicine and emerging infectious diseases*, 8th edn. Philadelphia: WB Saunders Company, 2000.
3. Jelling T, Lots M, Eichenlaub S, Loscher T, Nothdurft HD. Prevalence of infection with cryptosporidium parvum and cyclospora cayetanensis. *Gut* 1997; 41:801–4.
4. Ortega YR, Sterling CR, Gilman RH, Cama VA, Diaz F. Cyclospora species—a new protozoan

pathogen of humans. *N Engl J Med* 1993; 328:1308–12.
5. CDC. Outbreak of cyclosporiaris—Ontario, Canada, May 1998. *MMWR* 1998; 47:806–9.
6. Gutierrez Y. The protists. In: *Diagnostic pathology of parasitic infections with clinical correlations*, 2nd edn. New York: Oxford University Press, 2000.
7. CDC. Surveillance for foodborne-disease outbreaks—United States, 1993–1997. *MMWR CDC Surveillance Summaries* 2000; 49(SS-1):61.
8. CDC. Preliminary FoodNet data on the incidence of foodborne illness–selected sites, United States, 1999. *MMWR* 2000; 49:201–5.

CUTANEOUS AND MUCOCUTANEOUS LEISHMANIASIS

Common names for disease: Baghdad or Delhi boil, Oriental sore, espundia, in the Old World; uta, chichero ulcer, or forest yaws in the New World

Occupational setting

In the western hemisphere, workers in forested areas or workers living adjacent to or working next to forested areas, loggers, road builders, agricultural workers, hunters, explorers, scientists, missionaries and military personnel are at risk.[1] Workers who spend months in the forests of southern Mexico collecting chewing gum latex, "chicleros", have a high incidence of infection—30% during the first year of employment.[1] In the Americas, leishmaniasis areas include southern Texas, Central America, and South America, with the exception of Chile and Uruguay.[2] Urban populations and workers in Asia, Africa and Europe may also be at risk.[1]

Exposure (route)

Humans acquire cutaneous leishmaniasis (CL) through the bite of an infected female phlebotomine sandfly. Other reservoirs for the organisms include rodents, edentates (sloths), marsupials, and carnivores (Canidae), including the domestic dog.[2]

Pathobiology

Cutaneous and mucocutanous leishmaniasis are infections that affect the skin and mucous membranes, respectively; they are caused by the vector-borne intracellular protozoan *Leishmania.* In Asia, Africa, and southern Europe, the agents are *L. tropica*, *L. major*, and *L. aethiopica*. In the western hemisphere, the *L. braziliensis* complex and *L. mexicana* cause cutaneous and mucocutanous lesions. The *L. donavani* complex can cause single cutaneous lesions in both hemispheres; it also can cause visceral disease.[1]

After the sandfly feeds on an infected host, flagellated forms develop and multiply in the sandfly gut. After 8–20 days, infective parasites are present and can be transmitted to another host during a blood meal. After the parasites are injected, they are taken up by macrophages, where they can become amastigote forms. The amastigotes multiply, causing macrophage rupture and leading to further spread to other macrophages.

When an infected sandfly bites and feeds on exposed skin, a small erythematous papule appears after an incubation period of 1–2 weeks to 1–2 months. Eventually, the papule becomes a nodule, and then an ulcer with characteristic firm, raised and reddened edges. The ulcer can be dry with a central crust or it may weep seropurulent fluid. It is not usually painful. Subcutaneous nodules may develop along lymphatics, but they represent collections of infected macrophages, not lymph nodes. The cutaneous lesions heal spontaneously, with scarring, after several weeks to several months or as long as 1 year later.[1–3]

The site and appearance of the lesion may be specific for an occupational work group and a geographic region. Chiclero ulcer occurs in Central American forest workers who harvest chicle gum from plants and characteristically develop CL from *L. mexicana* on the pinna of the ear.[1]

Mucocutaneous leishmaniasis (MCL), also known as espundia, occurs when parasites from cutaneous lesions metastasize and cause destructive lesions in the oronasopharynx. MCL is found in Brazil, Bolivia, Ecuador, Peru, and other countries of northern and central South America. The most common cause is the *L. braziliensis* complex.[3]

MCL usually occurs after the initial cutaneous lesion has healed, often several years later. The nose is commonly involved, and initial symptoms include stuffiness and intermittent nosebleed. Tissue destruction can involve just the nasal septum, or it can destroy the nose. The upper lip, soft and hard palate and larynx can also be involved. The parasites in MCL may be difficult to find in spite of the extensive tissue involvement.[1,3]

Diagnosis

No single diagnostic test will give a definitive answer in all clinical settings. The diagnosis is made by finding *Leishmania* (amastigotes) in tissue or by culturing the flagellates on suitable media from tissue biopsy specimens or aspirates.[1] Since cultures may take 4 weeks to grow, it is important to specifically request that the laboratory culture for it.

The Montenegro test is an intradermal test using antigen from flagellated forms (promastigotes). It is positive in established disease, and, once positive, the Montenegro test remains positive. It is not helpful in early disease. Serologic tests, such as immunofluorscence assay (IFA) or ELISA, are available, but the antibody levels are low and they should not be the sole basis for a definitive diagnosis of leishmaniasis.[1]

Treatment

The majority of Old World CL spontaneously heals over months to years. Systemic chemotherapy is usually indicated for New World and for Old World patients with large or multiple lesions. Treatment with sodium stibogluconate (Pentostam), a pentavalent antimonial, speeds the time to healing.[4] In the USA, Pentostam is considered an investigational drug and is available only from the CDC. By contrast, MCL does not heal spontaneously. It should be treated with sodium stibogluconate and may require several courses of treatment for cure.[3] Other drugs used outside the USA for CL include pentamidine, and ketoconazole. In South America, amphotericin B is used when sodium stibogluconate fails to cure MCL. Topical paromycin has been used to treat New World and Old World CL, but dosages and formulations vary by location.[1] Since *L. tropica* species do not survive temperatures above 37°C, local heat therapy has been used to treat unresponsive lesions.[1]

Prevention

Workers and travelers with the potential for exposure should be educated about the transmission and clinical manifestations of leishmaniasis, as well as control methods for the vector phlebotomines (sandflies). Insecticides with residual activity can be used to control sandfly populations. Screens and insecticide-treated bednetting with fine mesh screen (10–12 holes/linear cm) should be used, since sandflies are about one-third smaller than mosquitoes. Sandfly breeding sites should be eliminated, and control of principal animal reservoirs and burrows should be exercised. No vaccine against leishmaniasis is available at this time.[1–3]

REFERENCES

1. Magill AJ. Leishmaniasis. In: Strickland GT, ed. Hunter's tropical medicine and emerging infectious diseases, 8th edn. Philadelphia: WB Saunders Company, 2000, 665–87.
2. Herwaldt BL, Stokes SL, Juranek DD. American cutaneous leishmaniasis in US travelers. *Ann Intern Med* 1993; 118:779–84.
3. Neva F, Sacks D. Leishmaniasis. In: Warren KS, Mahmoud AAF, eds. *Tropical and geographical medicine,* 2nd edn. New York: McGraw-Hill, 1990, 296–308.

4. Herwaldt BL, Berman JD. Recommendations for treating leishmaniasis with sodium stibogluconate (Pentostam) and review of pertinent clinical studies. *Am J Trop Med Hyg* 1992; 46:296–306.

VISCERAL LEISHMANIASIS

Common name for disease: Kala-azar

Occupational setting

Visceral leishmaniasis (VL) occurs in rural tropical and subtropical areas, including India, Bangladesh, Pakistan, China, the southern part of the former USSR, the Middle East (including Turkey), the Mediterranean basin, Mexico, and Central and South America (primarily Brazil). It is also found in Sudan, Kenya, Ethiopia, and the sub-Saharan savanna parts of Africa. Over 90% of cases are found in the following three regions: Sudan/Ethiopia/Kenya, India/Bangladesh/Nepal, and Brazil.[1] Outdoor workers in these locations are at risk. VL has recently been reported in US veterans of the Gulf War (Operation Desert Storm).[2]

Exposure (route)

Humans acquire VL through the bite of an infected phlebotomine sandfly. Humans and dogs (wild and domesticated) serve as reservoirs. Humans can serve as reservoir hosts and remain infectious even after clinical symptoms have resolved. In India, Nepal, and Bangladesh, humans are the only known reservoir. Transmission by blood transfusions, occupational exposure, congenital transmission and sexual contact has been reported.[1,2]

Pathobiology

VL is a chronic systemic disease caused by the vector-borne intracellular protozoans *Leishmania donovani*, *L. infantum*, and *L. chagasi*. The recent cases in Operation Desert Storm veterans were due to *L. tropica*, usually associated with CL.[3,4]

After the sandfly feeds on an infected host, flagellated forms develop and multiply in the sandfly gut. After 8–20 days, infective parasites are present and can be transmitted to another host during a blood meal. After the parasites are injected, they are taken up by macrophages, where they can become amastigote forms. The amastigotes multiply, causing macrophage rupture and leading to spread to other macrophages. The incubation period can range from several weeks to 8 months. Although infection can be subclinical, symptomatic disease can be chronic and systemic, leading to death if left untreated.

Malnutrition appears to be a risk factor for development of disease and reactivation of latent infection. Death can result from secondary bacterial infections, tuberculosis, or dysentery. Latent infections can also become active when immunosuppression occurs.[1,2] The symptoms of fever, leukopenia, weight loss and weakness can be mistaken for a malignancy. The fever can be gradual, sudden, continued, or irregular.[1]

In VL, the liver and spleen are enlarged, often to the point of abdominal discomfort, by increased numbers of cells containing parasitized macrophages.[2] Lymph nodes may also show hypertrophy. Hypopigmentation in dark-skinned persons or hyperpigmentation in lighter-skinned persons can occur along with petechiae or ecchymoses from the thrombocytopenia. Small, parasite-containing maculopapular skin lesions, known as post-kala-azar dermal leishmaniasis, may appear after recovery.

Leishmaniasis is a significant opportunistic infection in patients with HIV or immunosuppression for other reasons. The combination of HIV and VL is a problem in France, Italy, Portugal, and Spain. Up to 70% of adult VL cases in Mediterranean countries are associated with HIV, and 70% of AIDS patients have newly acquired or reactivated VL.[1,2]

Diagnosis

Diagnosis is made by culture or demonstration of leishmania (amastigotes) from tissue biopsy specimens or aspirates of the spleen, liver, bone marrow, lymph node, or blood. Since cultures may take 4 weeks to grow, it is important to specifically request that the laboratory culture for it.

The Montenegro test is an intradermal test using antigen from flagellated forms (promastigotes). It is positive in established disease, and, once positive, the test remains positive. It is not helpful in early disease. Serologic tests, such as IFA or ELISA, are available, but the antibody levels are low and they should not be the sole basis for a definitive diagnosis of leishmaniasis.[1,2]

Treatment

VL is treated with sodium stibogluconate (Pentostam), a pentavalent antimonial.[22] In the USA, Pentostam is considered an investigational drug and is available only from the CDC in Atlanta. Two or three 20-day courses of treatment with sodium stibogluconate may be required for cure.[1,2,5] In Kenya, up to 30% of the cases treated with Pentostam recur in 6 months.[1,2] Amphotericin B deoxycholate is also used as a treatment and is effective in cases that are resistant to other medications.[1] An oral agent, miltefosine, is currently being evaluated.[4]

Prevention

Workers and travelers with the potential for exposure should be educated about transmission and clinical manifestations of leishmaniasis, as well as control methods for the vector phlebotomines (sandflies). Insecticides with residual activity can be used to control sandfly populations. Screens and insecticide-treated bednetting with fine mesh screen (10–12 holes/linear cm) should be used, since sandflies are about one-third smaller than mosquitoes. Sandfly breeding sites should be eliminated, and control of principal animal reservoirs and burrows should be exercised. No vaccine against leishmaniasis is available at this time.[1,2]

REFERENCES

1. Magill AJ. Leishmaniasis. In: Strickland GT, ed. *Hunter's tropical medicine and emerging infectious diseases*, 8th edn. Philadelphia: WB Saunders Company, 2000, 665–87.
2. Davidson RN. Visceral leishmaniasis in clinical practice. *J Infect* 1999; 39:112–16.
3. Magill AJ, Grogl M, Gasser RA, Sun XV, Oster CN. Visceral infection caused by Leishmania tropica in veterans of Operation Desert Storm. *N Engl J Med* 1993; 328:1383–7.
4. CDC. Viscerotropic leishmaniasis in persons returning from operation desert storm 1990–1991. *MMWR* 1992; 41:131–134.
5. Jha TK, Suncar S, Thakur CP, et al. Miltefosine, an oral agent, for the treatment of indian visceral leishmaniasis. *N Engl J Med* 1999; 341:1795–1800.

NANOPHYETUS

Common name for disease: Human nanophyetiasis

Occupational setting

Fish handlers working with salmon and trout are at risk from exposure.[1]

Exposure (route)

Most cases of human intestinal infection with *Nanophyetus salmincola salmincola* are caused by eating raw or incompletely cooked, smoked or salted salmon or steelhead trout. Infections have also been reported from ingestion of raw steelhead trout eggs, as well as from handling infected salmonid fish. Infection has been reported in the Pacific northwest from *Nanophyetus salmincola salmincola* and in Siberia from *Nanophyetus salmincola schikhobalowi*.[2,3]

Pathobiology

This zoonotic disease is caused by the trematode *N. salmincola salmincola*. Disease usually results from ingestion of raw, undercooked or undersmoked salmonid fishes. Recently, a case was reported in a biological technician due to hand contamination from handling fresh-killed, infected juvenile coho salmon.[4]

N. salmin*cola salmincola* can also infect dogs through the ingestion of infected raw fish. Although it does not cause clinical disease in dogs, it can be the vector of a rickettsial organism, *Neorickettsia helminthoeca*, which causes a systemic infection called salmon poisoning of dogs. This infection can be fatal in dogs; however, it does not cause disease in humans.[2]

When *N. salmincola* eggs from the adult worm are shed in the feces of fish-eating animals such as raccoons, otters, spotted skunks, coyotes, foxes, herons, and diving ducks, miracidia hatch which then penetrate an intermediate snail host. In the snail, the parasite grows; it is shed from the snail as xiphidiocercaria that can penetrate and encyst in 34 species of fish.[4] Salmonid fishes seem to be more susceptible.[4] The cycle is completed when fish containing the encysted metacercaria are ingested by another animal, allowing the fluke to mature in the intestine. If humans ingest infected fish, they become definitive hosts.[2,4]

In humans, the clinical findings can range from no symptoms to abdominal pain, bloating, diarrhea, nausea and vomiting, and fatigue. Symptoms seem to be related to worm burdens.[3] Eosinophilia can be present and may be significant.[2] Eggs appear in the stool 1 week after eating infected fish. The number of eggs in the stool is related to the number of worms causing the infection.[2]

In the case of the biological technician who was handling infected coho salmon and removing the posterior one-third of the kidney of each fish, infection occurred by accidental hand-to-mouth ingestion of infectious metacercariae.[4]

Diagnosis

The diagnosis is made in patients with GI symptoms or unexplained eosinophilia by examining the stool for eggs or mature flukes.

Treatment

Bithionol (50 mg/kg orally on alternate days, for a total of two doses), niclosamide (2 g orally on alternate days, for a total of three doses) and praziquantel (20 mg/kg three times daily for 1 day) are effective treatments for *Nanophyetus* infection.[3,4] In the series of patients treated with praziquantel, stool examinations done 2–12 weeks after treatment were negative for eggs.[3]

Medical surveillance

Fish handlers who clean and eviscerate infected salmonid fishes should be monitored for symptoms of diarrhea and eosinophilia.[4]

Prevention

Workers at risk for exposure should wear gloves and practice regular handwashing and good personal hygiene. Fish viscera should be disposed of safely. Thorough cooking, or freezing at $-20°C$ for 24 hours, inactivates metacercarial cysts. Individuals should be advised to avoid eating incompletely cooked, salted or smoked, or raw, salmon or steelhead trout.

REFERENCES

1. Dieckhaus KD, Garibaldi RA. Occupational infections. In: Rom WN, ed. *Environmental occupational medicine*, 3rd edn. Philadelphia: Lippincott-Raven, 1998, 768.
2. Eastburn RL, Fritsche TR, Terhune CA Jr. Human intestinal infection with *Nanophyetus salmincola* from salmonid fishes. *Am J Trop Med Hyg* 1987; 36:586–91.
3. Fritsche TR, Eastbum RL, Wiggins LH, Terhune CA Jr. Praziquantel for treatment of human *Nanophyetus* salmincola (*Troglotrerna salmincola*) infection. *J Infect Dis* 1989; 160:896–9.

4. Harrel LW, Deardorff TL. Human nanophyetiasis: transmission by handling naturally infected coho salmon (*Oncorhynchuskisutch*). *J Infect Dis* 1990; 161:146–8.

PFIESTERIA PISCICIDA

Common names for disease: Possible estuary-associated syndrome (PEAS)

Pfiesteria piscicida is a one-celled dinoflagellate (alga) that has been associated with flesh ulcers in fish, with fish kills in estuaries along the eastern seaboard, and possibly with human health effects. *P. piscicida* is usually found only between April and October. It is not found in fresh water or in the open ocean.[1]

In the autumn of 1996, watermen reported seeing fish with "punched-out" skin ulcers and erratic swimming behavior in the Pocomoke and neighboring estuaries on the eastern shore of the Chesapeake Bay, Maryland. Sightings continued and increased in the spring and summer of 1997.[2]

People in Maryland who had environmental exposure to water from the affected waterways began reporting learning and memory difficulties. Other complaints included headache, skin lesions, and skin burning on contact with water. A study of 24 exposed individuals showed a dose-related reversible clinical syndrome consisting of difficulties with learning and memory. The symptoms resolved by 3–6 months after stopping exposure.[2]

No other reports of clusters of disease attributed to *P. piscicida* or other *Pfiesteria*-like organisms (PLOs) have been reported.[3] A team of North Carolina medical specialists investigated 67 persons exposed to fish kill waters in North Carolina and found no evidence of adverse health effects from their exposure.

Since no *P. piscicida* toxin has been isolated, there is no biomarker of exposure and the relationship of *P. piscicida* and *P. piscicida*-related illness in humans remains speculative. *P. piscicida* has been found in waters where there were no reports of symptoms or findings in fish in the waters or in persons exposed to the waters.[4]

In January 1998, a CDC-sponsored work group suggested using the term possible estuary-associated syndrome, (PEAS) and surveillance criteria were developed.[3,4]

Occupational setting

Fisherman and crabbers (watermen), environmental workers and laboratory workers who come in contact with *P. pisicicida* toxins in water from river or estuary waters during periods of "fish kills" or when fish with *Pfiesteria*-like lesions are present may be at risk.

Exposure (route)

Skin contact with water in affected waterways and exposure to aerosolized spray from affected waters are the routes of exposure.

Pathobiology

The relationship of PEAS to the dinoflagellate species and the toxins has not been fully characterized. In fact, it is not clear that the "clinical neurotoxic" symptoms are directly due to the toxins produced by *P. piscicida* or *Pfiesteria*-like dinoflagellates.[2] It is unknown whether individuals exposed to *P. piscicida* while swimming, boating or engaging in other types of recreational activities in coastal waters are at risk for developing PEAS. PEAS does not appear to be infectious, since there is no association with the consumption of fish or shellfish caught in waters containing *P. piscicida*.[4]

The evidence that suggests a relationship between PEAS and toxins produced by *P. piscicida* is based on the reports that individuals exposed to estuary water in Maryland prior to and during fish kills associated with *P. piscicida* toxin developed symptomatic neurocognitive deficits. All deficits resolved by 3–6 months after stopping exposure.[2,3] There is a report of learning difficulties in laboratory rats associated with exposure to water from aquaria containing *P. piscicida* toxins.[2,3]

A study designed and conducted under the guidance of an independent expert Task Force on Health Risk from Exposure to Fish Kill Waters by a team of medical specialists from north Carolina found no evidence of severe, chronic, or widespread adverse health effects from exposures to fish kill waters in North Carolina.[3]

Diagnosis

The presence of *P. piscicida* does not indicate risk to fish or humans. Fish lesions can also result from a variety of biological physical and environmental factors. The CDC, along with other federal, state and local government agencies, and academic institutions has set up organized surveillance for PEAS.[4] The surveillance system tracks PEAS rather than *P. piscicida*-related illness, since a *P. piscicida* toxin(s) has not been identified and cause and effect between *P. piscicida* and PEAS has not been established.[5]

PEAS criteria:

1. Developing symptoms within 2 weeks after exposure to estuarine water.
2. Reporting memory loss or confusion of any duration and/or three or more selected symptoms (i.e. headache, skin rash at the site of water contact, sensation of burning skin, eye irritation, upper respiratory irritation, muscle cramps, and GI symptoms) that—with the exception of skin rash at the site of water contact and sensation of burning skin—persist for 2 or more weeks.
3. A healthcare provider cannot identify another cause for the symptoms.[4]

Treatment

There is no current treatment other than avoiding further exposure.

Medical surveillance

Individuals should be screened for skin diseases or conditions that will place them at increased risk for exposure. Individuals who will wear respirators and/or protective clothing while working in *P. piscicida* toxin areas should be certified as medically fit to wear a respirator and be screened for the risk of heat stress.

Individuals who experience health problems or providers who see patients with symptoms after exposure to estuarine water, a fish disease event or a fish kill site should contact their healthcare provider and their state or local public health agency.

The following states have established PEAS information lines: Delaware (800) 523-3336; Florida, (888) 232-8635: Maryland, (888) 584-3110; North Carolina, (888) 823-6915; South Carolina, (888) 481-0125; and Virginia, (800) 238-6154(4).

From 1 June, 1998 to 31 December 1999, the six state systems received 1984 calls about *P. piscicida*, PLOs and PEAS. Seventy-eight calls reported a symptomatic person, 54 (69%) of whom had a possible exposure to estuarine water. Forty-four of the 54 individuals were seen by or referred to a healthcare provider. Twenty-four of the 44 did not meet PEAS symptom or exposure criteria, and 15 had another identified cause for their symptoms. Five individuals have environmental and medical results pending. As of May 2000, none of the illnesses met the PEAS criteria.[5]

Prevention

Individuals should avoid areas with diseased, dying or dead fish or shellfish. Persons should not go in or near water in areas that have been closed by governmental agencies and should not harvest or eat fish or shellfish from such areas.[3,4] People who come in contact with water from fish kill areas should wash well with plain soap and water and wash any clothes that come in contact with such water. Use waterproof gloves when handling items or objects contaminated with fish kill waters.

Individuals who are occupationally exposed to fish kill waters should consider wearing personal protective equipment (PPE) to minimize exposure to toxin by protecting the respiratory tract and skin. In addition to PPE, workers should be educated and trained in *P. piscicida* toxin risk prevention.

REFERENCES

1. CDC. Results of the Public Health Response to Pfiesterisa Workshop—Atlanta, Georgia, September 29–30, 1997. *MMWR* 1997; 46: 951–2.
2. Grattan LM, Oldach D, Tracy JK, et al. Learning and memory difficulties after environmental exposure to waterways containing toxic-producing Pfiesteria or Pfiesteria-like dinoflagellates. *Lancet* 1998; 352:532–9.
3. Smith CG, Music SI. Pfiesteria in North Carolina: the medical inquiry continues. *North Carolina Med J* 1998; 59:216–9.
4. CDC. Possible estuary-associated syndrome. *MMWR* 1999; 48:381–2.
5. CDC. Surveillance for possible estuary-associated syndrome (Pfiesteria), 6 states. *MMWR* 2000; 49;372–3.

PLASMODIUM SPECIES

Common name for disease: Malaria

Occupational setting

Workers or travelers in endemic areas are at risk from exposure. Occupational factors that increase the risk include working in rural areas where mosquito control methods are less cost-effective or non-existent, working during mosquito biting hours (dusk to dawn for *Anopheles sp.*), and working without protective clothing or insect repellents. Airport ground personnel who work at airports that receive airplanes from malarious countries that have not complied with international regulations requiring a malaria vector-free perimeter of 400 m around the airport may be at risk when cabin and cargo hold doors are opened.[1,2]

Human population movement is a significant factor in the re-emergence of malaria worldwide. Besides the intercontinental movement of infected people from endemic areas to non-malarious areas through modern transportation, population movement may lead to environmental changes that increase the risk of malaria by creating better habitats for *Anopheles* mosquitoes.[1] Examples are deforestation, the creation of irrigation systems, and rapid, unregulated urbanization accompanied by poor housing and sanitation.[1]

Malaria was eradicated from the USA in the 1940s. However, 1000–1400 cases are reported to the CDC each year. The majority are acquired during international travel, and 75% are associated with failure to use recommended prophylaxis. Rarely, cases in the USA occur through exposure to infected blood products, by congenital transmission, or by local mosquito-borne transmission.[3] In the USA, prevention of transfusion cases of malaria depends on collecting a travel history from presenting donors and deferring those who have traveled to malarious areas in the last year and deferring for 3 years donors who have lived in malaria-endemic countries or who have a history of malaria.[4]

Imported cases of malaria occur in the USA in migrant workers, in foreign citizens traveling to the USA, and in permanent US citizens who acquired malaria from endemic areas while traveling (including military personnel returning from duty in endemic areas). In addition to purely imported cases of malaria, there have been cases of transmitted malaria in the USA since the end of World War II. These cases were transmitted by *Anopheles* sp. mosquitoes. Cases have been reported in California, New York, Michigan, and New Jersey.[3,5] The number of cases per 100 000 population is shown in Figure 26.2.

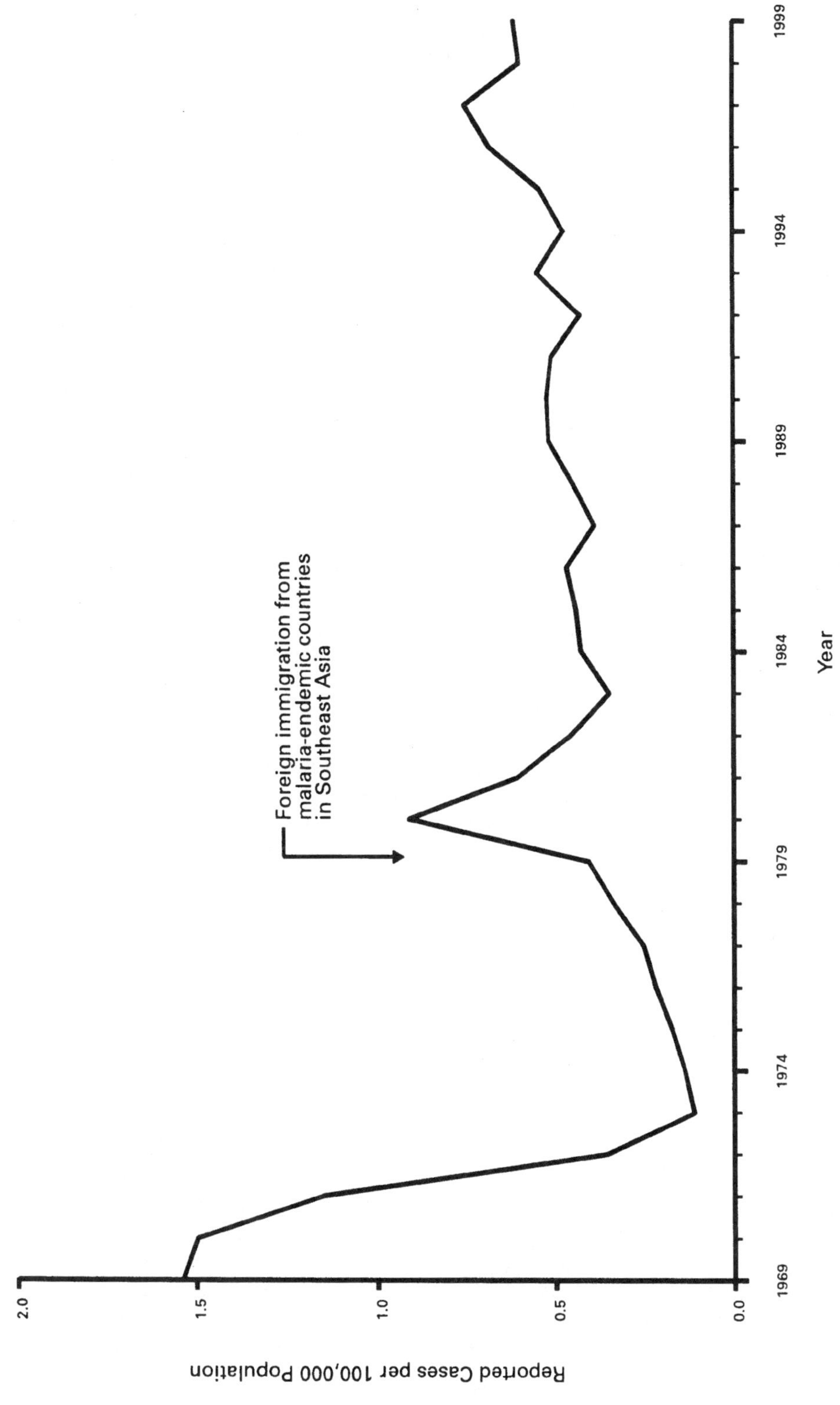

Figure 26.2 Malaria—reported cases per 100 000 population, United States, 1969–1999. Source: MMWR 2001:48(53):50.

Exposure (route)

Malaria is transmitted by infected female *Anopheles* sp. mosquitoes. The malaria sporozoite is introduced into humans when the mosquito punctures the skin to feed.

Pathobiology

The spread of disease from person to person depends on the availability of an appropriate mosquito vector in which the parasite can develop. It then feeds on humans, transmitting the parasite. Human malaria can result from four distinct species of the parasite of the genus *Plasmodium*: *P. falciparum*, *P. vivax*, *P. malariae*, and *P. ovale*. *P. falciparum* causes 40–60% of the cases of malaria worldwide; it is responsible for the most severe form of malaria, causing 95% of all malaria deaths. *P. vivax* causes 30–40% of the cases of malaria; the rest are due to *P. malariae* and *P. ovale*. Cases of malaria in US travelers are split about 50 : 50 between *P. falciparum* and *P. vivax*.

The life cycle of the *Plasmodium sp.* that cause human malaria involves humans and the *Anopheles sp.* mosquito. The mosquito becomes infected after feeding on an infected human. If *Plasmodium* gametocytes of both sexes are present, the sexual cycle leads to the creation of an ookinete (a mobile, fertilized egg). The ookinete penetrates through the gut, where an oocyst forms. After ~2 weeks, this oocyst ruptures, releasing sporozoites. Many of these make their way into the mosquito's salivary glands. An infected female *Anopheles sp.* mosquito is then capable of inoculating malaria sporozoite forms into humans while she feeds. The sporozoites rapidly enter the bloodstream and within hours enter liver cells (hepatocytes) and develop into liver-stage schizonts. After an asexual multiplication stage, each shizont ruptures, releasing 10 000–40 000 uninucleate merozoites, which can invade an erythrocyte. Inside the red blood cell, each merozoite develops asexually before rupturing the red blood cell, releasing more merozoites and continuing the erythrocyte cycle. The erythrocytic cycle takes 48–72 hours, depending on the species. *P. falciparum*, *P. vivax* and *P. ovale* take 48 hours, and *P. malariae* takes 72 hours. The erythrocyte stages of the malaria parasite are primarily asexual. Occasional merozoites become male or female gametocytes, permitting the life cycle to continue.[5]

Both *P. vivax* and *P. ovale* can have dormant forms of the parasites, hypnozoites, that can remain in the liver and become liver schizonts months or years after the initial inoculation. The liver schizonts can start new cycles (relapses) of erythrocytic infections.[5–7]

Malaria is a serious worldwide disease that the World Health Organization (WHO) estimates results in 100 million episodes of clinical disease per year. The WHO also estimates that >270 million people are infected at any one time. Over 100 countries are classified as sites where endemic transmission occurs. The distribution includes sub-Saharan Africa, Central and South America, the Dominican Republic, Haiti, the Middle East, the Indian subcontinent, Asia, and Oceania.[5] *P. falciparum* is found in tropical and subtropical areas, because the development of the parasite is greatly impaired in areas where the temperature is below 20°C.

However, all countries, including developed countries, can have cases of malaria due to the increased mobility of today's population. Malaria may occur in immigrants, refugees, and travelers on both personal and work-related trips. As many as 30 000 American and European travelers develop malaria every year.[6]

Even though most symptoms occur within 2 weeks of exposure and 95% of cases occur within 6 weeks of inoculation, an occasional case of *P. vivax* may not develop symptoms for years after exposure.[6] In a person with no prior exposure to malaria, the presenting symptoms go through four successive stages of chills, fever, sweats, and remission of fever. Nonspecific symptoms associated with malaria include headache, back pain, muscle pain, and malaise. When partial immunity has developed due to repeated exposure, the symptoms of infection may be much milder or, on occasion, non-existent.[5]

Malaria is classically described in terms of cyclical episodes of fever every 3 days (tertian) for *P. falciparum*, *P. vivax,* and *P. ovale*, and every 4 days (quarten) for *P. malariae*. Fever follows the release of blood merozoites from rupture of erythrocytes after shizogony. The cyclical fevers are only evident when a population of parasites predominates and the majority of the schizogony ends cyclically and in synchronization. *P. falciparum* causes the most severe complications. It can cause neurologic symptoms, from headaches to seizures and cerebral edema.

P. falciparum can also cause pulmonary edema, glomerulonephritis and renal failure, liver dysfunction, anemia, and hypoglycemia. *P. vivax*, *P. ovale*, and *P. malariae* can lead to anemia, splenomegaly, and headache; however, they generally do not cause the severe symptoms and mortality associated with *P. falciparum*.[5]

One interesting, although uncommon, occupational exposure is airport malaria. Airport malaria refers to malaria case reports of individuals who never traveled to malarious areas and who also lacked other risk factors for malaria, such as history of blood transfusions or intravenous drug abuse. In several reported cases, the victims worked near or at an international airport and were thought to have been infected by the bite of an infected tropical *Anopheles sp.* mosquito. Airport malaria does not include cases in persons who became infected during brief stops at airports in malaria endemic areas; nor does it include those who may have acquired the disease from an infected *Anopheles sp.* mosquito during a flight.[8]

During the summer of 1994 six cases of airport malaria were found around Roissy-Charles-de-Gaulle Airport.[1] Four of the cases were in airport workers and two other cases resided in Villeparisis about 7.5 km away. The mechanism of infection is thought to be anopheline mosquitoes that were carried in the cars of airport workers who lived next door to the two cases. Hot humid summer weather is thought to be a factor that allows the survival of infected *Anopheles* mosquitoes brought by airplanes.[1] International sanitary regulations requiring airports and a 400-m perimeter to be vector-free and WHO disinfection guidelines prior to opening doors on arrival are not always followed.[2]

Diagnosis

An important aspect of the diagnosis of malaria is the necessity for suspicion of malaria in a febrile patient who has been in an endemic area, no matter what preventive measures were taken. Suspicion should lead to a search for parasites in Giemsa-stained blood smears. Thick blood smears contain 20–40 times more blood than thin smears and are more sensitive. Thin blood smears are better for species identification once malaria parasite forms have been seen.[6] If the first smear is negative, repeat smears should be done every 12 hours for 36–48 hours, since parasitemia will vary in intensity due to sequestration in the microcirculation of the deep organs.[6] Parasites may also be undetectable in asymptomatic semi-immune people and in people who have taken antimalarial drugs.

Once parasites have been seen, the thin smear is used to identify the species. At this point, it is important to determine if the infection is due to *P. falciparum*, because this species causes the greatest morbidity and mortality and is often resistant to choroquine. Identifying *P. vivax* and *P. ovale* is also important, because they require treatment of the hypnozoites in the liver to prevent relapses, as well as treatment of the blood-stage infection.[6] Mixed cases of malaria can also occur.

Treatment

Treatment should rapidly reduce or eliminate the parasitemia and should prevent the complications of severe malaria (cerebral malaria; >3% of red blood cells parasitemia; hematocrit

<20%, hypoglycemia; renal, cardiovascular, liver, or pulmonary dysfunction; disseminated intravascular coagulation; prolonged hyperthermia; or severe vomiting or diarrhea.[6]

Drug therapy must reduce or eliminate the asexual blood stage. Drugs that are used include chloroquine, quinine, quinidine, the combination of pyrimethamine and sulfadoxine (Fansidar™), tetracycline, and mefloquine. A fixed-dose combination of atovaquone and proguanil (Malarone™) has been used in other countries and was approved for use in the USA in July 2000. Specific therapy will depend on the area where the infection was acquired and whether the malaria is uncomplicated or severe.[6] In addition, patients with *P. vivax* or *P. ovale* must also receive a drug such as primaquine, which eliminates the liver stage, to prevent relapses.[6] Consult an infectious disease/tropical disease expert or contact the CDC for specific treatment regimens.

P. falciparum resistance to the antimalarial chloroquine was first seen in South America in 1961, in South East Asia in 1962, and in Africa in 1978. Chloroquine-resistant *P. falciparum* has been seen in all malarious areas except some parts of the Middle East, North Africa, Central America, Haiti, and the Dominican Republic. *P. falciparum* resistant to quinine, mefloquine and pyrmethamine–sulfadione has also been reported. *P. vivax* resistant to chloroquine has been reported in South America, South East Asia, and Oceania. Specific country-by-country information is available at the CDC Travel Website—http://www.cdc.gov/travel/.

Prevention

Prevention strategies include avoidance and control of the mosquito vector, as well as chemoprophylaxis for travel to endemic areas. Individuals can minimize their chance of contact with the *Anopheles sp.* mosquito by wearing clothing that covers the arms and legs, and using mosquito repellant, bednets, and anti-mosquito spray. They should also avoid malarious areas from dusk to dawn, the time *Anopheles sp.* mosquito feeds.[5,6] Repellants containing up to 35% *N,N*-diethyl-metatoluamide (DEET), spraying rooms with pyrethrim-containing spray, and spraying clothing with permethrin, are specific methods to minimize mosquito contact. Sleeping in rooms with window screens and sleeping under insecticide (permethrin)-treated bednets are also helpful.[5–7]

Vector control methods include reducing the breeding areas of mosquitoes by draining or eliminating standing water and by killing the adult and (larval) stages of the mosquito. Antimalarial chemoprophylaxis should be based on the areas to be visited, length of stay, age, pregnancy status, interaction with other medications and vaccines, general medical history, and side-effects.[9]

The 1993 chemoprophylaxis recommendation for visitors to Central America, Haiti, the Dominican Republic and malarious areas of the Middle East is chloroquine (300 mg base) weekly. For the rest of the malarious areas of the world where chloroquine-resistant *P. falciparum* is present, mefloquine 250 mg weekly is the first choice and doxycycline 100 mg daily is the second choice.[6] They should be started 1–2 weeks before entering the malarious area and continued during the time in the endemic area and for 4 weeks after leaving the area. Primaquine, 15 mg base (26.3 mg salt), is given once a day for 14 days after leaving the endemic area to eliminate hypnozoites.[6] Since primaquine can cause severe hemolysis in persons with low activity of glucose-6-phosphate dehydrogenase (G6PD), patients should undergo G6PD testing before receiving the medication. Primaquine should not be given to pregnant women, because the drug may cross the placenta to a G6PD-deficient fetus and lead to hemolysis in utero.[6] Because resistance patterns and chemoprophylaxis recommendations may change, the physician should obtain current information from the CDC or another source before prescribing medication for chemoprophylaxis.

A final aspect of malaria prevention is the necessity for prompt diagnosis and treatment of malarial disease. Ninety per cent of all travelers who develop malaria do not show symptoms until they return home. The number of cases of *P. falciparum* infections from travel to sub-Saharan Africa continues to increase, even though it represents less than 2% of all travel. The increase in cases is due to the high risk of malaria to travelers in both urban and rural areas of sub-Saharan Africa. Malaria can still occur in persons who use personal protective measures and correctly take chemoprophylaxis.

REFERENCES

1. Martens P, Hall L. Malaria on the move: human population movement and malaria transmission. *Emerg Infect Dis* 2000; 6:103–9.
2. Guillet P, Germain MC, Chandre F, et al. Origin and prevention of airport malaria in France. *Trop Med Int Health* 1998; 3:700–5.
3. Malaria Surveillance—United States, 1995. *MMWR* 1999; 48(SS-1):1–21.
4. Dodd RY. Transmission of parasites by blood transfusion. *Vox Sanguinis* 1998; 74(Suppl 2): 161–3.
5. Taylor TE, Strickland GT. Malaria. In: Strickland GT, ed. Hunter's tropical medicine and emerging infectious diseases, 8th edn. Philadelphia: WB Saunders Company, 2000, 614–43.
6. Hoffman SL. Diagnosis, treatment, and prevention of malaria. *Med Clin North Am* 1992; 76:1327–55.
7. Kain KC, Keystone JS. Malaria in travelers epidemiology, disease, and prevention. *Infect Dis Clin North Am* 1998; 12:267–84.
8. Isaacson M. Airport malaria: a review. *Bull WHO* 1989; 67:737–43.
9. Baird JK, Hoffmand SL. Prevention of malaria in travelers. *Med Clin North Am* 1999; 83:923–44.

TOXOPLASMA GONDII

Common name for disease: Toxoplasmosis

Occupational setting

Butchers, slaughterhouse workers, laboratory workers, gardeners, farmers, veterinarians, pet store owners and cat breeders are at risk from exposure.[1]

Exposure (route)

Human infection is through contact with and ingestion of oocysts found in cat feces or in "contaminated" soil, ingestion of cysts in raw or undercooked infected meats (mutton, pork, hamburgers, steak tartare) or congenital transmission to a fetus born to a woman first infected during her pregnancy.[2] Water supplies contaminated with oocysts have been implicated as sources of infection.[2–4] A cat that has eaten a single infected bird or mouse can shed millions of infectious oocysts that can persist in moist soil for up to 18 months.

Pathobiology

Toxoplasma gondii is a protozoan parasite found in the cat family (Filadae). The cat serves as the definitive host and does not require an intermediate host to become infected. Intermediate hosts are pigs, sheep, deer, rabbits, rats, mice birds, cattle, and humans.[2]

T. gondii can exist in three forms: the tachyzoite, the trophozoite form of acute infections; the bradyzoite; found in the tissue cyst, and the oocyst, shed in cat feces. *Toxoplasma* infection is the presence of the trophozoite form or the cyst form in tissues whether there is clinical disease or not.[2] The tachyzoite is present in the host tissues during the acute stage of the infection. Tissue cysts containing the bradyzoites develop in host tissues and are the main mode of transmission of the organism to carnivores, including humans.

The oocyst is produced only through an enterepithelial cycle in the intestines of the cat family. Cats are important in the spread of toxoplasmosis, because they can periodically deposit massive numbers of oocysts that remain infective for months. Cats develop a primary infection during their first year of life, when they begin hunting and ingest tissue cysts in rodents or oocysts from other cat feces. Following primary infection, a cat may release up to 20 million oocysts per day for a period of 10–15 days. The oocysts sporulate 1–5 days after excretion. Oocysts are easily spread by wind and rain or on fomites. Oocysts tend to survive longer in moist, shaded soil. Their survival is decreased under temperature extremes, low humidity, and exposure to direct sunlight. The common chemical disinfectants are not effective in inactivating oocysts.[2]

Once sporulated oocysts or tissue cysts are ingested, the extraintestinal cycle can occur. Trophozoite forms are released from the oocysts. They invade the GI mucosa and reach all organs by means of the bloodstream. Within the organs, the organisms invade and kill cells. When the immune system gains control of the infection, the organism encysts and remains as a tissue cyst in the host.

The prevalence of toxoplasmosis infection varies from 20–30% of the population in the USA to 81% in French Parisians.[2] The high antibody prevalence rates in France are probably due to the local habits of eating undercooked meet, especially mutton.[2] Most immunocompetent persons who become infected by sporulated oocysts are asymptomatic. When symptoms develop, the most common are fever and transient head and neck lymphadenopathy, most commonly of the posterior cervical nodes. Other systemic, non-specific symptoms can include fatigue, sore throat, myalgias, and headache. The usual time course for symptoms is 2–4 weeks. The disease is usually self-limited. The differential diagnosis includes infectious mononucleosis and lymphoma. Occasionally, patients develop myocarditis, splenomegaly, encephalitis, hepatomegaly, and retinochoroiditis.[2]

After parasitemia, antibody appears, usually leading to lifelong immunity against reinfection.

Tissue cysts are not eradicated by antibody, and maintenance of an antibody titer can be an indicator of the continuing presence of live organisms in the cysts.[2]

Immunosuppressed hosts are at risk for relapses of chronic infections or primary infection from natural routes, organ transplant or leukocyte transfusion.[2] About 25% of AIDS cases with *Toxoplasma* antibody will relapse. The most common manifestation is encephalitis. When primary infection occurs in an immunosuppressed patient, the disease is generalized, involving multiple organs similar to acute acquired toxoplasmosis and often more severe.[2]

Diagnosis

The diagnosis of acute toxoplasmosis can be made by the serologic pattern, a finding of trophozoites in tissue, isolation of trophozoites from blood or body fluids, lymph node histology, or a finding of cysts in tissue.[2,5]

The Sabin–Feldman dye test is very specific and sensitive, but it creates a potential occupational hazard because it requires the use of live parasites. Because so many cases are asymptomatic, the dye test is not helpful in distinguishing recent from past infection, since the titer peaks early and remain elevated.

IgG and IgM antibody (Ab) tests by IFA or by ELISA are available. The IgG Ab test is specific and sensitive, with levels peaking in 1–2 months and persisting for several months or years. The IgM antibodies are the first to be detected and are useful in diagnosing acute infection.[2]

IgG Ab levels are helpful in determining the prevalence of infection, the risk of reactivation in the immunocompromised, and the risk of acquiring *Toxoplasma* infection. Diagnosis of acute acquired infection can be based on seroconversion, a fourfold rise, two-tube rise in IgG antibody titer, or the presence of a high level of IgM antibody. Because they do not show a rise in titer, diagnosis of reactivated infection in immunocompromised patients is more difficult.

Treatment

Most acquired cases of toxoplasmosis in immunocompetent persons are minimally symptomatic and therefore go unrecognized. Treatment is indicated for diagnosed clinical illness, active lesions such as eye disease, congenital infection, and symptoms and signs consistent with toxoplasmosis in immunosuppressed patients.[2]

The recommended treatment regimen for the immunocompetent patient is the combination of pyrimethamine and sulfadiazine or trisulphapyrimidines with folinic acid for >2–4 weeks. Additional therapy is based on clinical data after the initial treatment. Spiramycin, which is not available in the USA, is used often in the treatment of acute acquired toxoplasmosis in pregnant women because it is potentially less toxic to the fetus. Clindamycin has been used to treat toxoplasmosis, but does not penetrate well into the central nervous system, except when encephalitis is present.[2,6] Treatment recommendations and regimens for immunodeficient patients, AIDS patients, and transplant recipients are more complex.[2,5]

Medical surveillance

Surveillance may be appropriate for workers in specific high-risk settings, particularly for women of childbearing age. Initial screening may include IgG and IgM antibodies. Periodic screening can be done annually. Seronegative women at high risk of acquiring toxoplasmosis who become pregnant should be screened early in pregnancy, again at 20–22 weeks, and near or at term to find those who may have acquired primary infections during pregnancy.[5] Another alternative would be to offer removal from potential exposure to all seronegative pregnant women for the duration of their pregnancy.

Prevention

Persons at risk should be educated about the organism, the symptoms of infection, and control practices. Emphasis should be placed on thorough handwashing prior to eating or touching the face and the avoidance of raw or undercooked meat. Avoiding contact with cat feces is prudent but not sufficient, since infection may also result following exposure to oocysts in the environment.[5,7]

For those working outdoors, gloves should be worn in environments where fecal contamination may be found. Workers should avoid touching facial mucous membranes with gloves. All river and lake water should be boiled prior to drinking.[5,7]

On farms, pregnant women should not take part in lambing activities. Cats should not be used to control rodents; traps or rodenticides should be used instead. Cats should not be allowed inside feed-storage facilities, and all cat feces should be removed from buildings and stalls. Farm cats should be fed palatable dry or canned food, not forced to forage and hunt. Pigs or cats should never be permitted to eat uncooked scraps of pork or sheep. Dead animals should be removed promptly to prevent cannibalism.[5]

REFERENCES

1. Dieckhaus KD, Garibaldi RA. Occupational infections. In: Rom WN, ed. *Environmental occupational medicine*, 3rd edn. Philadelphia: Lippincott-Raven, 1998, 768.
2. Frenkel JK, Fishback JL. Toxoplasmosis. In: Strickland GT, ed. *Hunter's tropical medicine and emerging infectious diseases,* 8th edn. Philadelphia: WB Saunders Company, 2000, 691–701.
3. Benenson MW, Takafuju ET, Lemon SM, Greenup RE, Sulzer AJ. Oocyst-transmitted toxoplasmosis associated with ingestion of contaminated water. *N Engl J Med* 1982; 307:66–9.
4. Bowie WR, King AS, Werker DH, et al. Outbreak of toxoplasmosis associated with municipal drinking water. *Lancet* 2000; 350: 173–7.
5. August JR, Chase TM. Toxoplasmosis. *Vet Clin North Am* 1987; 17:55–71.

6. McCabe RE, Oster S. Current recommendations and future prospects in the treatment of toxoplasmosis. *Drugs* 1989; 38:973–87.

7. Weigel RM, Dubey D, Siegel AM. Risk factors for infection with toxoplasma gondii for residents and workers on swine farms in Illinois. *Am J Trop Med Hyg* 1999; 60:793–8.

27

ENVENOMATIONS

James A. Palmier, M.D., M.P.H., M.B.A., and Catherine E. Palmier, M.D.

ARTHROPOD ENVENOMATIONS

HYMENOPTERA

Common names: *Apoidea* (bumblebee, honeybee), *Vespoidea* (wasp, hornet, yellow jacket), *Formicoidea* (fire ants)

Occupational setting

Gardeners, orchard workers, farmers and agricultural workers, florists and workers in the flower industry, forestry workers and sanitation workers are potentially at risk.

Exposure (route)

Exposure occurs through physical trauma to the skin and occasionally the oral mucosa as a result of a bite or sting.

Pathobiology

Bees have a barbed stinger that becomes attached to the human skin after a sting, when it deposits its venom. In honeybee stings, the stinger apparatus becomes disengaged from the bee with its venom sac intact. Wasps also have a stinging apparatus but it does not contain barbs; therefore, they are able to withdraw the stinger and sting again. Ants have powerful jaws that grasp the skin and cause the release of venom locally; they are capable of inflicting multiple bites.

The Hymenoptera venom of the bee (*Apoidea*) consists of histamine, dopamine, enzymes (phospholipase A_1 and hyaluronidase), and peptides, including neurotoxins and hemolysins.[1] The *Vespoidea* venom is very similar to the bee venom, except that, both wasp venom and hornet venom contain serotonin, and the venom of the hornet also contains acetylcholine.[1] The *Formicoidea* venom contains alkaloids and a small amount of proteins. Histamine is released when these

Physical and Biological Hazards of the Workplace, Second Edition, Edited by Peter H. Wald and Gregg M. Stave
ISBN 0-471-38647-2

alkaloids come in contact with the mast cells, causing a local reaction. The hypersensitivity-type reaction appears to be caused by the low-molecular-weight proteins found in the venom.[2,3] Hymenoptera stings most often cause pain characterized as sharp stabbing, or burning; they are accompanied by a local reaction consisting of mild erythema, edema, and puritis.

Stings can cause severe local reactions, systemic or anaphylactic reactions, as well as delayed and unusual reactions. Severe local reactions consist of increased swelling (edema) around the sting site. The severity depends on the sting location. Stings in the mouth, upper airway or esophagus may cause obstruction of the airway or dysphagia.[4] Stings that occur on the eye structures or around the eye may cause local complications, including cataract formation and glaucoma. Severe local reactions on an extremity may include swelling of one or more joints without systemic toxicity.

Systemic or anaphylactic reactions may consist of mild symptoms and signs, including malaise, urticaria, pturitis, and anxiety. A moderate response may present with tightness in the throat or chest, generalized angioedema, nausea, vomiting, and abdominal discomfort. A severe reaction consists of cyanosis, hypotension, loss of consciousness, and other signs of shock, as well as dyspnea, hoarseness, increased malaise, and incontinence. A systemic or anaphylactic reaction may occur from single or multiple stings; its severity may increase rapidly within minutes to cardiovascular collapse and death.

Delayed-hypersensitivity reactions may occur within the first 2 weeks after an envenomation. Symptoms often include fever, urticaria, arthralgias, malaise, headache, lymphadenopathy, and myalgias. Unusual reactions include neurologic complications of encephalopathy, neuritis, Guillain–Barré-type reaction, myasthenia gravis, multiple sclerosis exacerbations, extrapyramidal disorders,[5–7] hemolytic anemia,[8] and the nephrotic syndrome.[9]

Diagnosis

The specific identification of the organism causing the sting is difficult and is usually dependent on history. The history should include the geographic location where the sting took place, and the location of any nests should be carefully noted. Wasps usually make nests under eaves, hornets choose bushes or tree limbs, and yellow jackets usually make nests in the ground. Pictures or photographs of the various Hymenoptera species often help the patient to identify the insect.

The site of the sting is usually not helpful, except in the stings of the honeybee and the fire ant. The honeybee almost always leaves its stinger at the site of the sting with its venom sac attached. The fire ant usually inflicts a distinctive pattern of multiple bites, which subsequently produce pustules. Laboratory tests are not usually helpful in the diagnosis of Hymenoptera envenomations, except in the few unusual cases of hemolytic anemia subsequent to a wasp bite.

Treatment

Most local reactions can be treated with local wound care. If a stinger is in place at the envenomation site, it should be removed immediately by gently scraping or teasing the affected area, thereby minimizing the spread of venom into the wound. All sites should be washed thoroughly with an antiseptic, and the tetanus status of the patient should be assessed. All potentially constrictive jewelry should be removed. Ice packs at the sting site decrease venom absorption and edema. Oral analgesics and antihistamines (e.g. diphenhydramine) usually relieve the pain and subsequent puritis associated with a mild reaction. Elevation and rest are indicated if there is increased edema of a limb. Antibiotics are indicated only if a secondary infection occurs.

Anaphylactic reactions are medical emergencies. Initial mild systemic symptoms may progress very rapidly to respiratory obstruction, cardiovascular compromise, and death. Any patient who displays signs of a systemic reaction should immediately be placed on a cardiac monitor, an intravenous line should be started, and epinephrine should be administered. In patients who are not in shock, 0.3–0.5 ml for an adult, or 0.01 ml/kg (~0.3 ml) for a child, of 1 : 1000 epinephrine hydrochloride should be administered subcutaneously and the site of administration massaged. A second injection of epinephrine may be needed in 10–15 minutes if symptoms do not improve or have become more severe. For severe anaphylactic reactions, epinephrine should be given slowly intravenously in a 1 : 10 000 solution and the patient observed and monitored closely. A repeat dose may be needed in 20–30 minutes.

Antihistamines such as diphenhydramine may be given intramuscularly or intravenously, depending on the severity of the anaphylactic reaction. Corticosteroids may also be helpful; but, as with antihistamines, their effect may be delayed and they should be used only as adjuncts to epinephrine therapy. Bronchospasm usually responds well to intravenous aminophylline after a loading dose (5–6 mg/kg over 20 minutes). The use of nebulized β-agonists may also be considered for patients with significant bronchospasm. Vasopressors are indicated in situations where hypotension is unresponsive to intravenous fluids and epinephrine.

Individuals who have moderate to severe systemic reactions to Hymenoptera venom with positive skin reactions should undergo desensitization with venom immunotherapy.[10,11] Immunotherapy usually consists of extracts containing purified insect venom rather than whole-body extracts, which appears to limit substantially the number of systemic reactions to the therapy.[12,13]

Prevention

Preventive measures, especially in allergic individuals, include the avoidance of bright or flowered clothing and the avoidance of scented deodorants, perfumes, and shampoos. Individuals should avoid going outdoors barefoot. Nests around the home or frequented dwellings should be destroyed, preferably with professional assistance. Sensitized individuals should carry commercial epinephrine kits for self-administration; they should be available for use in their home and car and carried along with other personal belongings (e.g. in golf bags). Sensitized patients should wear a medical alert bracelet or tag. Potential victims should stand still or retreat slowly when a Hymenoptera species is encountered. They should be instructed to brush the insect off their skin but avoid swatting or moving quickly. Sensitized individuals should avoid gardening. They should use caution when visiting places with open garbage cans, such as picnic areas, or other locations where insects are likely to be found.

REFERENCES

1. Haberman E. Bee and wasp venom. *Science* 1972; 177:314–21.
2. Lund N. Mechanisms of action of fire venoms: I. Lytic release of histamine from mast cells. *Toxicon* 1982; 20:831–40.
3. Kemp SF. Expanding habitat of the imported fire ant (Solenopsis invicta): a public health concern. *J Allergy Clin Immunol* 2000; 105(4):683–91.
4. Farivar M. Bee sting of the esophagus. *N Eng J Med* 1981; 305:1020.
5. Behan PG. Insect sting encephalopathy. *Br Med J* 1982; 284:504–5.
6. Incorvaia C. Clinical aspects of Hymenoptera venom allergy. *Allergy* 1999; 54 (Suppl 58):50–2.
7. Dionne AJ. Exacerbation of multiple sclerosis following wasp stings. *Mayo Clin Proc* 2000; 75(3):317–8.
8. Schulte KL, Kochen MM. Haemolytic anemia in an adult after a wasp sting. *Lancet* 1981; 2:478.
9. Tareyeva JE, Nikolaev AJ, Janushkevitch TN. Nephrotic syndrome induced by insect sting. *Lancet* 1982; 2:825.

10. Schwartz HJ. Appropriate evaluation and therapy of sting insect hypersensitivity. *Arch Intern Med* 1984; 144:1560–1.
11. Rieger-Ziegler V. Hymenoptera venom allergy: time course of specific IgE concentrations during the first weeks after a sting. *Int Arch Allergy Immunol* 1999; 120(2):166–8.
12. Hunt KJ, Valentine MD, Sobotka AK, et al. A controlled trial of immunotherapy in insect hypersensitivity. *N Engl J Med* 1978; 299: 157–61.
13. Ebner C. Immunological mechanisms operative in allergen-specific immunotherapy. *Int Arch Allergy Immunol* 1999; 119(1):1–5.

LATRODECTUS SPECIES

Common names: Black widow spider, brown widow spider, red-legged spider, hourglass spider, poison lady spider, deadly spider, red-bottom spider, T-spider, gray lady spider

Occupational setting

Farmers, gardeners and forestry, construction and sanitation workers are at risk for exposure.

Exposure (route)

Exposure occurs through physical trauma (bite) to the human skin with injection of venom.

Pathobiology

The female black widow spider is almost twice the size of its male counterpart. Although both are considered venomous, only the female spider is able to bite and envenomate humans. During the summer months, the female black widow spider is the most venomous. The body of the female can range from 8 to 15 mm long, with a leg span of 5 cm. The spider undergoes multiple moltings throughout the year and often changes color. The female is most often shiny black in color and has a rounded abdomen with a red distinctive hourglass on its ventral surface. Occasionally, two red spots may be seen instead of the hourglass configuration.

The five species in the USA are *Latrodectus mactans*, *L. hesperus*, *L. variolus*, *L. bishopi*, and *L. geometricus*. Although variations occur between species, most *Latrodectus* venoms contain multiple proteins and a potent neurotoxin. The proteins include hyaluronidase, phosphodiesterase, GABA, and 5-hydroxytryptamine. This neurotoxin is one of the most potent human neurotoxins. Neurotransmitters (GABA, acetylcholine, and norepinephrine) appear to be released secondary to the *Latrodectus* venom.[1,2] The clinical presentation usually begins with a pinprick sensation, followed by the appearance of mild swelling and erythema around the bite wound. It is not unusual for the patient to be unaware of the bite until a local reaction has occurred. Close evaluation of the site may reveal two erythematous fang marks. In the first hour after the bite, pain often increases around the area of the bite and spreads to the entire body. Symptoms peak at 3–4 hours. Upper extremity bites often lead to spasm of the upper trunk muscles; bites of the lower extremity often lead to abdominal spasms. These muscle spasms can become very intense and painful and have been misdiagnosed as an acute surgical abdomen. However, cases of abdominal spasms secondary to the black widow bite do not present with associated fever or a leukocytosis.

Other common symptoms include paresthesias and hyperesthesia, especially of the lower extremities and feet. Increased deep tendon reflexes, headache, anxiety, nausea, vomiting, diaphoresis, tremor, restlessness and seizures may also be seen. Symptoms usually resolve within 24–48 hours. Pregnancy,[3] infancy, increased age, and chronic debilitating illness appear to increase the risk of complications.

Diagnosis

Laboratory studies are usually not helpful in diagnosis. A description of the spider's appearance, identification of the geographic area where the bite took place, and the clinical presentation and evaluation are the most helpful factors in diagnosing a black widow spider bite. The bite area usually consists of two closely approximated erythematous fang marks. In *L. hesperus* bites, there may be a target or halo lesion.[4] Severe chest and back spasms must be carefully distinguished from myocardial infarction or an acute surgical abdomen.

Treatment

In general, treatment is supportive and includes local wound care, tetanus prophylaxis, and pain control (i.e. salicylates, opioids). All victims should be monitored for 6–8 hours for progression of symptoms. Airways, breathing and circulation should be closely monitored and assisted, if necessary. The pain of severe generalized spasm may be treated with intravenous morphine. Intravenous calcium gluconate (10 ml of 10% solution) can be used to decrease cramping and spasm in select individuals. Methocarbamol and diazepam have also been used as muscle relaxants and for their anxiolytic effect. The use of intravenous dantrolene sodium for muscle relaxation is controversial and needs further study.[5]

Most cases can be appropriately managed without the equine-derived antivenom. However, in cases where there is severe hypertension, or in the case of a pregnant woman at risk for premature labor or spontaneous abortion because of severe muscle spasms, the lyophilized antivenom may be indicated. Other indications for antivenom include respiratory distress, protracted symptoms despite treatment, cardiovascular symptoms, and symptomatic patients younger than 16 years and older than 65 years.[2,6–8] The *Latrodectus* antivenom is effective for all species. Unfortunately, it has the potential to cause immediate hypersensitivity, including anaphylaxis, as well as delayed onset serum sickness within 2 weeks. One vial of antivenom is usually necessary for envenomations. All patients should be skin tested for horse serum sensitivity prior to administration.

Prevention

Prevention measures should focus on the use of personal protective equipment such as gloves, heavy garments that are fully buttoned, and protective footwear when working in endemic areas. These spiders generally inhabit places that are dark and protected, such as woodpiles, barns, stables, garages, homes, outdoor sanitation facilities, and walls made of rock. Areas suspected of harboring black widows should be subjected to professional pest control.

REFERENCES

1. Grishin E. Polypeptide neurotoxins from spider venoms. *Eur J Biochem* 1999; 264(2):276–80.
2. Akhunov AA. Kininase of the Latrodectus tredecimguttatus venom: a study of its enzyme substrate specificity. *Immunopharmacology* 1996; 32(1–3):160–2.
3. Russell FE. Black widow spider envenomation during pregnancy: report of a case. *Toxicon* 1979; 2:620–3.
4. Rauber A. Black widow spider bites. *J Toxicol Clin Toxicol* 1984; 21:473–85.
5. Ryan PJ. Preliminary report: experience with the use of Dantrolene sodium in the treatment of bites by the black widow spider Latrodectus hesperus. *J Toxicol Clin Toxicol* 1984; 21:487.
6. Pennell T, Babu S, Meredith J. The management of snake and spider bites in the southeastern United States. *Am Surg* 1987; 53:201.
7. Heard K. Antivenom therapy in the Americas. *Drugs* 1999; 58(1):5–15.
8. Banner W. Bites and stings in the pediatric patient. *Curr Prob Pediatr* 1988; 18:1–69.

LOXOSCELES SPECIES

Common names: Brown recluse spider, violin spider, Arizona brown spider, fiddle spider, necrotizing spider

Occupational setting

Gardening, housekeeping, forestry, farming and agricultural workers are at risk of exposure.

Exposure (route)

Exposure occurs through physical trauma (bite) to the human skin with injection of venom.

Pathobiology

The brown recluse spider is ~1 cm in body length, with a leg span of up to 2.5 cm. The color of these spiders is usually tan to brown. In the USA, there are 13 species of *Loxosceles*, but *L. reclusa* is the most commonly found. The female is more dangerous than the male. These spiders contain a very distinctive violin-shaped, dark brown to yellow marking on the dorsal cephalothorax.

The *Loxosceles* venom contains a number of enzymes, including sphingomyelinase D, the agent believed to be responsible for skin necrosis and hemolysis by its direct effect on the cell plasma membrane.[1–3] The venom also contains hyaluronidase, which does not appear to cause skin necrosis. *Loxosceles* envenomation is initially painless for most victims. Within the first few hours, pain and erythema occur at the bite site. A central bleb or vesicle forms. It is surrounded by a blanched, whitish ring of ecchymosis and ischemia. The appearance resembles a bull's-eye and it is most often 1–5 cm in diameter. Over the next few days, the central vesicle ulcerates, necroses, spreads in diameter and involves the skin and subcutaneous fat. In ~1 week, an eschar develops that eventually sloughs and can require many months to heal.

Systemic reactions to *Loxosceles* envenomation, while uncommon, can occur. Systemic symptoms and signs usually occur within the first 2 days after envenomation; they include fever, chills, generalized rash, nausea, vomiting, and hemolysis with possible renal failure. Disseminated intravascular coagulation is rare but has been reported.[4] The size of the necrotic ulcer does not seem to correlate with the frequency or severity of the systemic symptoms.[5]

Diagnosis

Laboratory studies are not usually helpful in diagnosing a *Loxosceles* envenomation. Description of the spider's appearance, identification of the geographic area where the bite took place and the clinical presentation and evaluation are the most helpful factors in diagnosing a brown recluse spider envenomation. Moreover, it should be noted that there are other species that may reproduce the symptoms of the brown recluse spider.[6]

Treatment

In general, treatment should begin with local wound care, tetanus prophylaxis, immobilization, elevation, observation, and antipruritic agents, as needed. Local wound care has been controversial. Some authorities advocate early surgical excision of the wound; others recommend allowing complete delineation of the necrotic area before surgical intervention.[7] The use of dapsone in *Loxosceles* envenomations is controversial[6] and may cause methemoglobinemia in some cases.[8] The use of hyperbaric oxygen to decrease the extent of the necrosis has been successful in a number of cases.[9]

All patients with systemic symptoms must be hospitalized and followed closely for hemolysis, coagulopathy, and renal failure. Treatment for systemic symptoms is supportive. There is no commercially available antivenom.

Prevention

Prevention measures should focus on the use of personal protective equipment like gloves, fully buttoned heavy garments, and protective footwear when working in endemic areas. The spider is usually not aggressive. Its habitat includes woodpiles, barns, the underside of rocks, closets, attics, clothing, carpets, linen, and other dry, dark places. Areas suspected of harboring these spiders should undergo professional pest control.

REFERENCES

1. Rees R, Phillips A, Lynch J, et al. Brown recluse spider bites: a comparison of early surgical excision versus dapsone and delayed surgical excision. *Ann Surg* 1985; 202:659.
2. Gomez HF. Loxosceles spider venom induces the production of alpha and beta chemokines: implications for the pathogenesis of dermonecrotic arachnidism. *Inflammation* 1999; 23(3):207–15.
3. Wright RP, Elgert KD, Campbell BJ, et al. Hyaluronidase and esterase activities of the venom of the poisonous brown recluse spider. *Arch Biochem Biophys* 1973; 159:415–26.
4. Vorse IT, Seccareccio P, Woodruff K, et al. Disseminated intravascular coagulopathy following fatal brown spider bite (necrotic arachnidism). *Pediatrics* 1972; 80:1035–7.
5. Wasserman CS, Siegel C. Loxoscelism (brown recluse spider bite): a review of the literature. *Clin Toxicol* 1979; 14:353–8.
6. Wong R, Hughes S, Voorhees J. Spider bites. *Arch Dermatol* 1987; 123:99.
7. Rees R, Shack B, Withers E, et al. Management of the brown recluse spider bite. *Plast Reconstr Surg* 1981; 68:768–73.
8. Iserson KV Methemoglobinemia from dapsone therapy for a suspected brown spider bite. *J Emerg Med* 1985; 3:285–8.
9. Maynor ML. Brown recluse spider envenomation: a prospective trial of hyperbaric oxygen therapy. *Acad Emerg Med* 1997; 4(3):184–92.

SCORPIONIDA (SCORPIONS)

Common names: Scorpion, Centruroides

Occupational setting

Persons employed in farming, agriculture, construction, forestry and other outdoor occupations in the southwestern part of the USA are at risk of exposure.

Exposure (route)

Exposure occurs through physical trauma (sting) to the human skin with injection of the venom.

Pathobiology

The scorpion consists of anterior pincers, a pseudoabdomen, and a tail that ends in a bulbous structure called the telson. Within the telson is the stinger and venom. The scorpion usually grasps the victim using its pincers, then arches its body and strikes the victim with the stinger located in its tail, injecting the venom. It ranges in size from 1 to 7 cm. A nocturnal creature, it prefers to take shelter during the day under rocks, debris, or objects outside the house. It generally climbs rather than burrows and therefore is often found among trees.

The venom consists of enzymes (hyaluronidase, acetylcholinesterase, and phospholipase), proteins, serotonin, and neurotoxins, which appear to affect the sodium channel.[1] It results in increased stimulation at the neuromuscular junction and increased axonal discharges.[2]

The clinical presentation usually consists of immediate, severe burning pain at the site of the sting. Paresthesias around the area of the sting may occur. There may be some local inflammation around the site, but most often there is no erythema, swelling, or evidence of ecchymosis.

Severe systemic symptoms may occur in some victims, with a greater incidence in victims under the age of 10 years. Systemic symptoms may occur within minutes or hours after the envenomation. Symptoms result from excitation of the sympathetic and parasympathetic nervous systems, as well as the neuromuscular junction. Symptoms include hyperexcitability, diplopia, nystagmus, restlessness, diaphoresis, muscle fasciculations, tachycardia, hypertension, opisthotonus, and convulsions. Cardiovascular collapse, respiratory arrest, disseminated intravascular coagulation, renal failure and death are rare.

A clinical grading system for *Centruroides sculpaturatus* envenomation uses the following 4-point scale: I (local pain), II (pain or paresthesia remote from sting site), III (either cranial nerve or somatic skeletal neuromuscular dysfunction), and IV (both cranial nerve and somatic skeletal neuromuscular dysfunction).[3] There is a case report of a scorpion sting that appeared to relieve the symptoms of multiple sclerosis.[4]

Diagnosis

Laboratory studies are usually not helpful in diagnosis. Description of the scorpion's appearance, identification of the geographic area where the sting took place, and the clinical presentation are the most helpful factors in making the diagnosis. In *Centruroides* stings, tapping or pressure on the site of the wound ("tap test") usually produces severe discomfort.

Treatment

Most *Centruroides* stings can be managed with local wound care, tetanus prophylaxis, and the application of ice to the affected area. Systemic reactions should be treated as necessary, with advanced life support and supportive measures, and should be monitored very closely in the hospital. Antivenom has been used successfully in many severe cases.[5,6] Complications of the use of antivenom include immediate hypersensitivity (anaphylaxis) and delayed hypersensitivity (serum sickness) reactions.

Prevention

Prevention measures should focus on the use of personal protective equipment like gloves, heavy garments that are fully buttoned, and protective footwear when working in areas that the scorpion inhabits. Areas suspected of harboring scorpions should undergo professional pest control.

REFERENCES

1. Nencioni AL. Neurotoxic effects of three fractions isolated from Tityus serrulatus scorpion venom. *Pharmacol Toxicol* 2000; 86(4): 149–55.
2. Rachesky IJ, Banner XV, Dansky J, et al. Treatments for Centruroides exilicauda envenomation. *Am J Dis Child* 1984; 138:1136–9.
3. Curry SC, Vance MV, Ryan PJ, et al. Envenomation by the scorpion Centruroides sculpturatus. *J Toxicol Clin Toxicol* 1983–1984; 21:417–49.
4. Breland AE, Currier RD. Scorpion venom and multiple sclerosis. *Lancet* 1983; 2:1021.
5. Amitai Y. Clinical manifestations and management of scorpion envenomation. *Public Health Rev* 1998; 26(3):257–63.
6. Krifi MN. Evaluation of antivenom therapy in children severely envenomed by Androctonus australis garzonii (Aag) and Buthus occitanus tunetanus (Bot) scorpions. *Toxicon* 1999; 37(11):1627–34.

MARINE ENVENOMATIONS

CATFISH

Common names: Catfish (channel, blue, bullhead, brown) and Carolina mudtorn catfish[1]

Occupational setting

Occupations with potential exposure include commercial fishing, diving, seafaring, and fish handling.

Exposure (route)

Exposure occurs through physical trauma (sting) to the skin with release of venom.

Pathobiology

The catfish has dorsal and pectoral fins with venom-containing spines. The most common affected sites are the hands of fishermen and fish handlers.[2–5] The venom contains agents that can cause skin necrosis and vasoconstriction, as well as other heat-labile agents.

The clinical presentation includes burning and throbbing at the envenomation site that usually resolves within a few hours but can last for days.[4] The wound site may contain fragmented spines and is often edematous or may appear ischemic. Gangrene is possible secondary to severe envenomation.

Diagnosis

There are no known laboratory tests that are helpful in diagnosing these envenomations. Positive identification of the offending organism is helpful. Barring positive identification, the clinical presentation and geographic location of the envenomation are used to determine the appropriate treatment modality. Since most catfish spines are radiopaque, a radiograph of the affected area may be useful in identifying retained spines.

Treatment

Local wound care, including careful removal of the spines and tetanus prophylaxis, is extremely important. Immediate immersion in hot water (110–115°F) gives adequate pain relief.[2–5] Analgesics are a helpful adjunct in pain relief. Close observation of the wound site for infection is indicated.

Prevention

Preventive measures include wearing gloves and long-sleeved garments when handling organisms or specimens, avoiding any contact with reefs, and never contacting or touching an unknown organism. Persons at risk should always dive with a skilled diver (never alone) in unknown waters. Before working in unknown waters, they should become familiar with the potential marine envenomations and the local medical facilities in case treatment is needed.

REFERENCES

1. Eastaugh J, Shepherd S. Infectious and toxic syndromes from fish and shellfish consumption: a review. *Arch Intern Med* 1989; 149:1735–40.
2. Kizer KW. Marine envenomations. *J Toxicol Clin Toxicol* 1983–1984; 21:527–55.
3. Blomkalns AL. Catfish spine envenomation: a case report and literature review. *Wilderness Environ Med* 1999; 10(4):242–6.
4. Auerbach PS. Stings of the deep. *Emerg Med* 1989; 21:27–31.
5. Schwartz S. Venomous marine animals of Florida: morphology, behavior, health hazards. *J Fla Med Assoc* 1997; 84(7):433–40.

COELENTERATE—ANTHOZOA

Common names: Soft coral, true coral, sea anemones

Occupational setting

Workers with potential exposure include sailors, divers, fishermen and seafood producers.

Exposure (route)

Exposure occurs through physical trauma (sting) to the skin caused by nematocysts that release venom.

Pathobiology

Sea anemones contain stinging structures, called *nematocysts*, that contain venom. They are abundant, flower-like structures that are usually sessile and have the ability to envenomate with their tentacles but rarely do so.[1] The soft and true corals do not contain nematocysts and therefore cannot envenomate. The venom of coelenterates is not well characterized, but serotonin (5-hydroxytryptamine (5HT)) appears to be a major component.[2,3] It produces pain, vasoconstriction, and histamine release.

The clinical presentation of sea anemone envenomation is stinging, pain, and a burning rash. Systemic reactions are rare and the envenomations are infrequent. The coral usually causes injury secondary to trauma sustained when the victim brushes against the extremely sharp, calcified exoskeleton, suffering lacerations.

Diagnosis

There are no known laboratory tests that are helpful in diagnosing these envenomations. Positive identification of the offending organism is helpful. Barring positive identification, the clinical presentation and geographic location of the envenomation are very useful in determining the appropriate treatment modality.

Treatment

Treatment for the sea anemone envenomation includes inactivating the nematocysts by applying alcohol (i.e. isopropyl) or salt water. Alternatively, dilute vinegar, papain or baking soda can be used. Fresh water should be avoided because it can cause increased venom release from the nematocysts. Immobilization of the affected area is also useful. Baking soda and abrasion of the area with a sharp knife can help to remove any remaining tentacles. Soaking the affected area in hot water (110–115°F) will alleviate severe pain.[4] Narcotics, such as codeine, may also be given for pain control. Tetanus prophylaxis and local wound care should be administered. Oral antihistamines are often given routinely and can be especially effective with symptoms of pruritis. Tissue trauma caused by the corals should be treated with appropriate wound care and tetanus prophylaxis.

Prevention

Preventive measures include wearing gloves and long-sleeved garments when handling organisms or specimens, avoiding contact with reefs, and never contacting or touching an unknown organism. At-risk workers should always dive with a skilled diver (never alone) in unknown waters. Before working in unknown waters, they should become familiar with the potential marine envenomations and the local medical facilities, in case treatment is needed.

REFERENCES

1. Kizer KW. Marine envenomations. *J Toxicol Clin Toxicol* 1983–4; 21:527–55.
2. Grotendorst GR. Enzymatic characterization of the major phospholipase A2 component of sea anemone (Aiptasia pallida) nematocyst venom. *Toxicon* 2000; 38(7):931–43.
3. Burnett JVV, Carleton GJ. The chemistry and toxicology of some venomous pelagic coelenterates. *Toxicon* 1977; 15:177–96.
4. Schwartz S. Venomous marine animals of Florida: morphology, behavior, health hazards. *J Fla Med Assoc* 1997; 84(7):433–40.

COELENTERATE—HYDROZOA

Common names: Portuguese man-of-war (*Physalia*), fire coral, hydroids

Occupational setting

Sailors, divers, fishermen and seafood producers are at risk for exposure.

Exposure (route)

Exposure results from physical trauma (sting) to the skin inflicted by a stinging structure (nematocyst) containing the venom.

Pathobiology

All members capable of inflicting stings possess stinging structures, called nematocysts, that contain venom. The Portuguese man-of-war consists of a body and tentacles. On these tentacles, there are millions of nematocysts that can release venom upon contact. The tentacles of the *Physalia* can be up to 10 feet long. This creature can be found in all oceans. The hydroids and fire coral, also members of the Coelenterata, possess calcareous growths and inhabit coral reefs in tropical oceans. They also have venom-containing nematocysts.

The venom of coelenterates is not well characterized, but serotonin(5-hydroxytryptamine) appears to be a major component.[1,2] It produces pain, vasoconstriction, and histamine release. In particular, the venom of the Portuguese man-of-war contains a neurotoxin that produces severe pain and possible systemic symptoms.[3]

The clinical presentation of the Portuguese man-of-war envenomation is severe pain with multiple, linear, erythematous macules and papules, which may progress to vesicular lesions. Systemic reaction to envenomation may cause nausea, vomiting, chills, myalgias, and respiratory and cardiovascular depression. The clinical presentation of the envenomation caused by the fire coral and hydroids consists of erythema, pruritis, and edema of the affected skin area. Vesicular lesions and ulcer formation can occur. Systemic reactions of fever, chills, fatigue, myalgias, and abdominal discomfort occur infrequently.[4]

Diagnosis

There are no known laboratory tests that are helpful in diagnosing these envenomations. Positive identification of the offending organism is useful. Barring positive identification, the clinical presentation and geographic location of the envenomation are very helpful in determining the appropriate treatment modality.

Treatment

Treatment includes inactivating the nematocysts by applying alcohol (i.e. isopropyl) or salt water. Alternatively, dilute vinegar, papain or baking soda can be used. Vinegar appears to be especially effective in Portuguese man-of-war envenomations.[5,6] Fresh water should be avoided because it can cause increased venom release from the nematocysts. Immobilization of the affected area is also helpful. Baking soda and abrasion of the area with a sharp knife can help to remove any remaining tentacles. Soaking the affected area in hot water (110–115°F) will alleviate severe pain. Narcotics, such as codeine, can be given for pain control. Tetanus prophylaxis and local wound care should be administered. Oral antihistamines are often given routinely and can be especially helpful with symptoms of pruritis.

Prevention

Preventive measures include wearing gloves and long-sleeved garments when handling organisms or specimens, avoiding contact with reefs, and never contacting or touching an unknown organism. Persons at risk should always dive with a skilled diver (never alone) in unknown waters. Before working in unknown waters, they should become familiar with the potential marine envenomations and the local medical facilities, in case treatment is needed.

REFERENCES

1. Edwards L. Portuguese man-of-war (Physalia physalis) venom induces calcium influx into cells by permeabilizing plasma membranes. *Toxicon* 2000; 38(8):1015–28.
2. Burnett JXV, Carlton GJ. The chemistry and toxicology of some venomous pelagic colenterates. *Toxicon* 1977; 15:177–96.
3. Joannides C, Davis JH. Portuguese man-of-war stinging. *Arch Dermatol* 1965; 91:448–51.
4. Kizer KW. Marine envenomations. *J Toxicol Clin Toxicol* 1983–4; 21:527–55.

5. Burnett JXV, Carlton GJ, Rubinstein M. First aid for jellyfish envenomation. *South Med J* 1983; 76:870–2.
6. Fenner PJ. Dangers in the ocean: the traveler and marine envenomation. I. Jellyfish. *J Travel Med* 1998; 5(3):135–41.

COELENTERATA—SCYPHOZOA

Common names: True jellyfish, sea nettles, sea wasp (box jellyfish)

Occupational setting

Sailors, divers, fishermen and seafood producers are at risk for exposure.

Exposure (route)

Exposure occurs through physical trauma (sting) to the skin inflicted by a stinging structure (nematocyst) containing venom.

Pathobiology

All members contain stinging structures called nematocysts that contain venom. Jellyfish are found in many shapes, sizes, and colors. They can ride the waves or swim freely on their own. Their tentacles contain numerous nematocysts. Sea wasps are found in Australian waters; the sea nettle is a type of jellyfish found in waters off the Middle Atlantic States.

The venom of coelenterates is not well characterized, but serotonin(5-hydroxytryptamine) appears to be a major component.[1] It produces pain, vasoconstriction, and histamine release.

Jellyfish envenomations usually cause erythematous papular lesions followed by vesicles within the first day. The victim experiences pain, regional lymphadenopathy, and burning. Anaphylactic reactions can occur in some envenomations. The sea wasp, which is found only in Australian waters, causes the most poisonous of all marine envenomations. It can result in death within minutes.[2] Envenomation by this creature can cause pain, nausea, vomiting, erythematous lesions, headache, chills, and occasionally immediate cardiovascular collapse.

Diagnosis

There are no known laboratory tests that are helpful in diagnosing these envenomations. Positive identification of the offending organism is helpful. Without positive identification, the clinical presentation and geographic location of the envenomation are very helpful in determining the appropriate treatment.

Treatment

Treatment includes inactivating the nematocysts by applying alcohol (i.e. isopropyl) or salt water. Alternatively, dilute vinegar, papain or baking soda can be used. Fresh water should be avoided because it can cause increased venom release from the nematocysts. Immobilization of the affected area is also helpful. Baking soda and abrasion of the affected area can be effective in removing any remaining tentacles. Soaking the affected area in hot water (110–115°F) will alleviate severe pain. Narcotics, such as codeine, may be given for pain control. Tetanus prophylaxis and local wound care should be administered. Oral antihistamines are given routinely and can be especially helpful with symptoms of pruritis.

A sheep antivenom is available for sea wasp envenomations. It should be given immediately after skin testing in all severe envenomations. Cardiovascular support (e.g. fluids, vasopressors) may also be needed.[3]

Prevention

Preventive measures include wearing gloves and long-sleeved garments when handling organisms or specimens, avoiding contact with reefs, and never contacting or touching an unknown organism. Persons at risk should always dive with a skilled diver (never alone) in unknown waters. Before working in unknown waters, they should become familiar with the potential marine envenomations and the local medical facilities, in case treatment is needed.

REFERENCES

1. Welsh JH. 5-Hydroxytryptamine in coelenterates. *Nature* 1960; 186:811–72.
2. Flecker H. Fatal stings to North Queensland bathers. *Med J Aust* 1952; 1:35–8.
3. Tibballs J. The effects of antivenom and verapamil on the haemodynamic actions of Chironex fleckeri (box jellyfish) venom. *Anaesth Intens Care* 1998; 26(1):40–5.

DASYATIS (STINGRAY)

Common name: Stingray

Occupational setting

Commercial fishermen, divers and sailors are at risk of exposure.

Exposure (route)

Exposure occurs through physical trauma to the skin with the release of venom.

Pathobiology

The stingray is a bottom dweller that usually envenomates by raising its barbed, stinger-like tail in a defensive posture after it has been touched or stepped on. The tail apparatus penetrates the skin, usually fracturing the creature's spine in the wound, and envenomates the victim. The laceration and wound are usually deep and irregular.[1] The venom appears to contain cardiotoxins, convulsants, and respiratory depressants.[1]

The stingray envenomation is followed by almost immediate severe pain. The pain's intensity is grossly out of proportion to the degree of trauma visualized; thus, it is helpful in making the diagnosis when the organism was not seen. Edema, erythema and cyanosis often appear around the wound site. The fragmented spine is also often seen. Systemic symptoms include nausea, vomiting, muscle cramps, syncope, tachycardia, abdominal pain, cardiac arrhythmias, convulsions, hypotension, and (rarely) death.[2]

Diagnosis

There are no known laboratory tests that are helpful in diagnosing these envenomations. Positive identification of the offending organism is useful. Barring positive identification, the clinical presentation and geographic location of the envenomation are very helpful in determining the appropriate treatment.

Treatment

Initial treatment includes local wound care, including removal of any of the fragmented spines, tetanus prophylaxis, and immersion in hot water (110–115°F) for 1 hour.[3,4] Analgesics may be an important adjunct for pain control. Prophylactic antibiotics are not indicated, but these wounds must be observed closely for evidence of infection. Systemic symptoms should receive prompt supportive care and close observation in a hospital setting.

Prevention

Preventive measures include wearing gloves and long-sleeved garments when handling organisms or specimens, avoiding contact with reefs, and never contacting or touching an unknown organism. Persons at risk should always dive with a skilled diver (never alone) in unknown waters. Before working in unknown waters, they should become familiar with the potential marine envenomations and the local medical facilities, in case treatment is needed.

REFERENCES

1. Kizer KW. Marine envenomations. *J Toxicol Clin Toxicol* 1983–4; 21:527–55.

2. Rathjin WE, Halstead BW Report on two fatalities due to stingrays. *Toxicon* 1969; 6:30–2.
3. Evans RJ. Stingray injury. *J Accid Emerg Med* 1996.
4. Baldinger PJ. Treatment of stingray injury with topical becaplermin gel. *J Am Podiatr Med Assoc* 1999; 89(10):531–3.

ECHINODERMATA

Common names: Starfishes, sea urchins, sea cucumbers

Occupational setting

Sailor, divers, fishermen and seafood producers are at risk from exposure.

Exposure (route)

Exposure occurs through physical trauma to the skin with the release of venom.

Pathobiology

Sea urchins are bottom-dwelling organisms that have calcareous exoskeletons. They are covered with spines that become embedded in the victim's skin upon contact. Some species contain venom in their spines. The venom contains steroid glycosides, acetylcholine-like substances, and serotonin.[1]

The clinical manifestation of sea urchin envenomation includes both an immediate and a delayed reaction. The immediate reaction consists of severe pain and localized edema and erythema. Dye within the sea urchin spine (usually black or purple) may stain the surrounding skin. Paresthesias around the untreated wound site are common. Occasionally, systemic symptoms of nausea, myalgias, fatigue, syncope and respiratory difficulties occur.[2,3] The delayed reaction consists of diffuse inflammation or local granuloma formation a few months after the initial injury.[2] Bone destruction is a rare occurrence. Sea urchins can also cause direct injury to joints, nerves and soft tissue secondary to their sharp spines.[2]

Starfish have sharp spines and secrete a slimy venom. Envenomation can cause nausea, vomiting, paresthesias, and muscle paralysis.[4] Sea cucumbers are sausage-shaped organisms that produce a venom called holothurin, which is a cardiac glycoside. The venom causes a macular and papular rash. Contact with the eyes can cause severe conjunctivitis, opacities, and blindness. The sea cucumber also ingests nematocysts from other species and can secrete them later in intact form as a self-defense measure, causing symptoms of coelenterate envenomation.

Diagnosis

There are no known laboratory tests that are helpful in diagnosing these envenomations. Positive identification of the offending organism is useful. Barring positive identification, the clinical presentation and geographic location of the envenomation are very helpful in determining the appropriate treatment.

Treatment

Treatment should begin with local wound care and tetanus prophylaxis. Immersion in hot water (110–115°F) should be done and analgesics should be administered. Spines should be removed carefully in order to avoid fragmentation. Granulomas secondary to delayed reaction may need specific surgical or dermatologic procedures. Steroids, ammonia, antibiotics and acetone have all been used with limited success.[5]

Prevention

Preventive measures include wearing gloves and long-sleeved garments when handling organisms or specimens, avoiding contact with reefs, and never contacting or touching an unknown organism. Persons at risk should always dive with a skilled diver (never alone) in unknown waters. Before working in unknown waters, they should become familiar with the potential marine envenomations and the local medical facilities, in case treatment is needed.

REFERENCES

1. Ritchie KB. A tetrodotoxin-producing marine pathogen. *Nature* 2000 23; 404(6776):354.
2. Schwartz S. Venomous marine animals of Florida: morphology, behavior, health hazards. *J Fla Med Assoc* 1997; 84(7):433–40.
3. Kizer KW. Marine envenomations. *J Toxicol Clin Toxicol* 1983–4; 21:527–55.
4. Liram N. Sea urchin puncture resulting in PIP joint synovial arthritis: case report and MRI study. *J Travel Med* 2000; 7(1):43–5.
5. Millar JS. Clinical curio: treatment of sea urchin stings. *Br Med J* 1984; 288:390.

MOLLUSCA

Common names: Cone shell (*Conidae*), blue-ringed octopus

Occupational setting

Sailors, divers, fishermen and seafood producers are at risk from exposure.

Exposure (route)

Exposure occurs through physical trauma to the skin with injection of venom.

Pathobiology

Conidae, which are found mainly in tropical waters, are cone-shaped, univalve organisms surrounded by a shell. They contain a radula tooth apparatus that envenomates its victims. The blue-ringed octopus is found mainly in Australian waters. It injects its venom through a beak-like apparatus. The venom appears to contain a number of neurotoxins.

The initial symptoms of envenomation include pain, stinging, burning, numbness, and paresthesias around the wound site. In severe envenomations, local ischemia and cyanosis can occur. Systemic reactions include fatigue, dysphagia, diplopia, paresthesias, dyspnea, coma, and cardiovascular collapse.[1] Systemic reactions to severe envenomations include paresthesias, slurred speech, weakness, dysphagia, and respiratory depression. Allergic reactions to the envenomation have been reported.[2,3]

Diagnosis

There are no known laboratory tests that are helpful in diagnosing these envenomations. Positive identification of the offending organism is useful. Barring positive identification, the clinical presentation and geographic location of the envenomation are very helpful in determining the appropriate treatment.

Treatment

Conidae envenomations respond to hot water immersion (110–115°F) as well as analgesics. All patients should be observed for 6 hours for systemic reactions. Blue-ringed octopus envenomations do not respond to hot water immersion. These victims also need to be observed for systemic symptoms.

Prevention

Preventive measures include wearing gloves and long-sleeved garments when handling organisms or specimens, avoiding contact with reefs, and never contacting or touching an unknown organism. At-risk persons should always dive with a skilled diver (never alone) in unknown waters. Before working in unknown waters, they should become familiar with the potential marine envenomations and the local medical facilities, in case treatment is needed.

REFERENCES

1. Kizer KW. Marine envenomations. *J Toxicol Clin Toxicol* 1983–4; 21:527–55.
2. Edmonds C. A nonfatal case of blue-ringed octopus bite. *Med J Aust* 1969; 2:601.
3. Bonnet MS. The toxicology of Octopus maculosa: the blue-ringed octopus. *Br Homeopath J* 1999; 88(4):166–71.

PORIFERA

Common names: Caribbean and Hawaiian fire sponge (*Tedania ignis*), poison burn sponge (*Fihula nolitangere*), red sponge (*Microionamprolifera*)

Occupational setting

Sailors, divers, fishermen and seafood producers are at risk from exposure.

Exposure (route)

Exposure occurs through physical and chemical irritation of the skin.

Pathobiology

Sponges contain calcareous spines that can cause direct trauma to the skin. The components of the sponge's venom are unknown, but it does cause symptoms similar to those of *Rhus* dermatitis.[1,2] Initially, there is pruritis and erythema in the area of contact, followed by progressive swelling and vesicle formation within a few hours. Lymphadenopathy can also occur. Anaphylactoid reactions and erythema multiforme have been reported.[3]

Diagnosis

There are no known laboratory tests that are helpful in diagnosing these envenomations. Positive identification of the offending organism is useful. Barring positive identification, the clinical presentation and geographic location of the envenomation are very helpful in determining the appropriate treatment.

Treatment

Local wound care and tetanus prophylaxis should be administered. Topical steroids and antihistamines may be helpful. Tape can be used to remove spines. Dilute vinegar (30 ml in 1 liter) or 5% acetic acid applications may be helpful with the acute dermatitis.

Prevention

Preventive measures include wearing gloves and long-sleeved garments when handling organisms or specimens, avoiding contact with reefs, and never contacting or touching an unknown organism. Persons at risk should always dive with a skilled diver (never alone) in unknown waters. Before working in unknown waters, they should become familiar with the potential marine envenomations and the local medical facilities, in case treatment is needed.

REFERENCES

1. Kizer KW. Marine envenomations. *J Toxicol Clin Toxicol* 1983–4; 21:527–55.
2. Kobayashi M. Marine spongean cytotoxins. *J Nat Toxins* 1999; 8(2):249–58.
3. Yaffee S, Stargardter F. Erythema multiforme from Tedania ignis: report of a case and an experimental study of the mechanism of cutaneous irritation from the fire sponge. *Arch Dermato* 1963; 87:601–4.

SCORPAENIDAE

Common names: Lionfish, turkeyfish (*S. pterosis*), stonefish (*S. synaceja*), scorpionfish (*S. scorpaena*)

Occupational setting

Workers with potential exposure include commercial fishermen, divers, sailors, aquarists, and fish handlers.

Exposure (route)

Exposure occurs through physical trauma (sting) to the skin, with the potential release of venom.

Pathobiology

The Scorpaenidae are usually bottom dwellers with fins that are supported by venomous spines. They live in shallow waters around coral reefs and rocky formations. Envenomation can occur when the fish is stepped on or touched by the hand. A fragmented spine is often left in the wound site. The venom is a heat-labile mixture that can cause hypotension, weakness, vasodilation, and respiratory depression.[1,2]

The clinical presentation of envenomation includes local effects such as edema, erythema, and ecchymosis at the wound site with intense, severe pain that can last up to 2 days if not treated. Paresthesias of the affected extremity are common. Neuropathy is rarely seen. Severe envenomation can cause hypotension, cardiac arrhythmias, syncope, nausea, vomiting, diaphoresis, convulsions, and paralysis. These systemic symptoms can develop very quickly.

Diagnosis

There are no known laboratory tests that are helpful in diagnosing these envenomations. Positive identification of the offending organism is useful. Barring positive identification, the clinical presentation and geographic location of the envenomation are very helpful in determining the appropriate treatment. Laboratory studies such as CBC (Complete Blood Count), electrolytes, glucose, BUN (Blood Urea Nitrogen), creatinine phosphokinase, creatinine, ECG, chest radiograph and urinalysis will not be diagnostic, but they are helpful adjuncts to care of the patient.

Treatment

The local wound site should be examined, and local wound care and tetanus prophylaxis should be given. The extremity should be immediately immersed in hot water at 110–115°F for at least 1 hour. This treatment usually decreases the severe pain of the sting.[3,4] Analgesics can also be used as an adjunct. The wound site should be examined carefully for any fragmented spines, which should be removed with care. Radiograph of the affected site may be helpful in revealing these fragmented spines; however, a negative radiograph does not rule out the presence of spines, because not all spines are radiopaque.

In severe envenomations, *Scorpaenidae* antivenom may be indicated. Assistance can be obtained from the local regional poison control center. Immediate supportive care is needed for patients with severe envenomations who present with hypotension, arrhythmias, convulsions, or paralysis.

Prevention

Preventive measures include wearing gloves and long-sleeved garments when handling organisms or specimens, avoiding contact with reefs, and never contacting or touching an unknown organism. At-risk persons should always dive with a skilled diver (never alone) in unknown waters. Before working in unknown waters, they should become familiar with the potential marine envenomations and the local medical facilities, in case treatment is needed.

REFERENCES

1. Kizer KW, McKinney HE, Auerbach PS. Scorpaenidae envenomation. A five year poison center experience. *JAMA* 1985; 253:807–10.
2. Schaeffer RC, Carlson RW, Russell FE. Some chemical properties of the venom of the scorpion fish, Scorpaena guttata. *Toxicon* 1971; 9:69–78.
3. Patel MR. Lionfish envenomation of the hand. *J Hand Surg (Am)* 1993; 18(3):523–5.
4. Kizer KW. Marine envenomations. *J Toxicol Clin Toxicol* 1983–4; 21:527–55.

SNAKE ENVENOMATIONS

COLUBRIDAE

Common names: Garter snake (*Thamnophis elegans vagrans*), king snake (*Lampropeltis*), hognose snake (*Heterodon*), boomslang snake (*Dispholidus*), racer snake (*Coluber*), and red-neck keelback snake (*Rhabdophis*)

Occupational setting

Farming and agricultural workers, gardeners and orchard workers, forestry workers, construction workers, veterinary office workers, pet store workers and zoo personnel are at risk of exposure.

Exposure (route)

Exposure occurs through physical trauma (bite) to the human skin, with injection of venom.

Pathobiology

There are ~1400 species in the Colubridae family. The majority contain teeth and no venom-injecting apparatus and are therefore non-poisonous. However, a small percentage of colubrids, like the hognose snake, have rear fangs with a venom-injecting apparatus.[1] Because of its small size and the posterior location of its fangs, these snakes do not usually envenomate their victims, but they are capable of doing so, given the right circumstances. All colubrids, whether rear-fanged or not, contain saliva that has been shown to be toxic.[2]

The colubrid bite usually presents as a minor abrasion or an imprint of teeth on the skin surface. Occasionally, there is evidence of a puncture wound inflicted by rear-fanged colubrids, but this would take prolonged skin contact time and can usually be elicited with the history. Occasionally, erythema, pruritis and swelling are seen around the bite site; however, most bites cause only minor discomfort for the victim secondary to the abrasion.

Rare envenomations usually produce low-grade, mild symptoms. Fibrinolysis was reported secondary to a bite of a red-neck keelback snake that was purchased in a pet store.[3] Fibrinolysis has also been reported with other colubrid bites.[1,4,5] In one report, a garter snake produced local hemorrhage, pain, and swelling of an upper extremity without systemic symptoms.[6]

Diagnosis

The identification of the snake that inflicted the bite is extremely helpful in the management of all snake bites. The actual snake should be brought to the treating physician for identification, if possible, or the snake should be identified using well-defined photographs of different varieties of snakes. A herpetologist or regional poison control center may need to be consulted. Envenomation can occur hours after a snake's death; therefore, these snakes should be handled carefully.

The victim should be evaluated carefully for fang marks, swelling, erythema, pain, local reaction (i.e. blebs, vesicles, petechiae), and any systemic reactions. Most colubrid bites present with minor abrasions or the imprint of teeth on the skin surface. Any deviation from this minor abrasion, without positive identification of the offending snake, should make the treating physician suspicious of a more venomous species. Laboratory data are usually not helpful in making the diagnosis. Coagulation abnormalities suggest a severe envenomation.

Treatment

Treatment usually consists of local wound care and tetanus prophylaxis. Antibiotics may be indicated if there is any suspicion of infection. After an uncomplicated colubrid bite, observation for at least 4 hours is indicated to rule out systemic symptoms. If fang marks are present and the snake has been identified as a colubrid, then at least 6 hours of observation is prudent to rule out systemic symptoms and a significant envenomation. Fibrinolysis and hemorrhage should be treated with conventional therapy. There is no approved antivenom currently available. The vast majority of bites need only local wound care. Any systemic symptoms respond well to supportive measures.

Prevention

Whether dead or alive, snakes should always be handled carefully, especially when the treating personnel are attempting to identify the snake. Snakes can still envenomate hours after their death. Snakes are ubiquitous in all outdoor areas. Areas like rock crevices, caves and unusual terrain should be avoided. Trousers, gloves, boots and long-sleeved shirts should be worn in areas where snakes may be present. Pest control professionals should be consulted when work must be performed in high-risk areas. One should never awaken or try to molest a snake. Venomous snakes should not be sold or kept as pets.

REFERENCES

1. Mandell F, Bates J, Mittleman M, et al. Major coagulopathy and "non-poisonous" snake bites. *Pediatrics* 1980; 65:314–17.
2. McKinstry DM. Evidence of toxic saliva in some colubrid snakes of the United States. *Toxicon* 1978; 16:523–34.
3. Cable D, MeGehee XV, Wingert WA, et al. Prolonged defibination after a bite from a "non-venomous" snake. *JAMA* 1984; 25(1):925–6.
4. Hill RE. Venom yields from several species of colubrid snakes and differential effects of ketamine. *Toxicon* 1997; 35(5):671–8.
5. Minton SA. Beware: nonpoisonous snakes. *Clin Toxicol* 1979; 15:259–65.
6. Vest DK. Envenomation following the bite of a wandering garter snake (Thamnophis elegans vagrans). *Clin Toxicol* 1981; 18:573–9.

CROTALIDAE

Common names: Rattlesnake (*Crotalus*), copper-head, cottonmouth, water moccasin (*Agkistrodon*), fer-de-lance (*Bothrops*)

Occupational setting

Farming and agricultural workers, gardeners and orchard workers, construction workers, veterinary and ancillary staff workers, pet store personnel and zoo workers are at risk of exposure.

Exposure (route)

Exposure occurs through physical trauma (bite) to the human skin, with injection of venom.

Pathobiology

All pit vipers (*Crotalidae*) have: a vertical, elliptical pupil; facial pits located between the nostril and the eye, which are used to sense heat and vibration; a triangular head; and a single row of subcaudal plates (in contrast to the double row found in non-venomous snakes). Most true rattlesnakes also contain a terminal segment called a rattle, but not all species contain this rattle. In some cases, trauma or congenital anomaly can also contribute to the lack of the rattle apparatus. Cottonmouths (*Agkistrodon piscivorus*) have a white buccal mucosa and usually a dark green to black appearance; they are often found in swampy marshes, lakes, or other aquatic environments. Copperheads have a head that is copper to brown in color and a distinctive band-like appearance on their bodies. Copperheads can be found in any outdoor area. Neither copperheads not cottonmouths have rattles.

Pit vipers have anterior fangs on the roof of their mouths that can be up to 20 mm in length, allowing for easy envenomation and injection of venom. The venom contains numerous components, with a predominance of enzymes and neurotoxins that are used to immobilize and kill the animal prey.[1] The potency and amount of venom injected varies within the individual species. In general, the potency of the venom of the true rattlesnake is greater than that of the cottonmouth, which is, in turn, greater than that of the copperhead.

After a Crotalidae bite, there are typically two or more fang or puncture marks on the victim's skin. In minutes, the victim experiences pain, edema, and erythema around the site of envenomation. Over the next few hours, the edema progresses proximally, the erythema increases, and there is evidence of hemorrhagic blebs or vesicles and ecchymosis. An affected extremity can become quite large, secondary to edema. Hypovolemic shock may occur in some cases. If progressive swelling has not occurred in the first few hours, then envenomation probably has not occurred.[2] The exception is the Mojave rattlesnake (*Crotalus scutulatus scutulatus*), whose envenomation produces little or no local symptoms but may cause the delayed systemic symptoms of muscle fatigue, ptosis, and respiratory depression.

Systemic effects of Crotalidae envenomation include fatigue, muscle fasciculations, lightheadedness, minty or metallic taste, nausea, vomiting, and paresthesias (both perioral and peripheral). Thrombocytopenia and coagulopathy can occur.[3,4] Hypotension, pulmonary edema and renal failure may also be seen.[5]

Diagnosis

The identification of the snake that inflicted the bite is critical in the management of all snake bites. The snake should be brought to the treating physician for identification, if possible, or it should be identified using well-defined photographs of different varieties of snakes. A herpetologist located at a local zoo, park or aquarium may be consulted. The regional poison control center or the Antivenom Index of the Arizona Poison Control Center may be helpful. Snakes should be handled carefully even when dead, since envenomation can occur hours after a snake's death.

The victim should be evaluated for fang marks, swelling, erythema, pain, local reactions (i.e. blebs, vesicles, petechiae), and any systemic reactions. The classic symptoms and signs of Crotalidae envenomation include puncture or fang wounds, pain, erythema, and edema around the bite site. The evaluation of the site may reveal one or more puncture wounds, which could represent the number of fangs the snake had or the number of strikes that occurred.[6]

There is no specific assay or laboratory test that can reliably identify and thereby assist in the diagnosis of Crotalidae envenomation. However, cell blood count, coagulation studies, disseminated intravascular coagulopathy (DIC) panel, blood type and screen, platelet count, creatinine phosphokinase, and urine for myoglobin can be helpful.

Treatment

In the field, prior to definitive care by medical personnel, the affected bite area should be immobilized in a dependent position. Ice should not be applied. Tourniquets as well as incision and suction are controversial and can be deleterious.[7]

Medical personnel should monitor all Crotalidae envenomation for at least 4 h for local and systemic effects. In the case of the Mojave rattlesnake, observation should be for 12–24 hours because of the delayed effects of the venom in this species. The affected limb should be closely monitored for local swelling, ecchymosis, and circulatory compromise. Tetanus prophylaxis and local wound care should be administered. Any complication regarding the wound, especially a compartment syndrome, should be referred to an experienced surgeon.

Systemic effects should be monitored closely and treated immediately. Respiratory depression requires immediate airway maintenance and assisted ventilation. Coagulopathy, hypotension and rhabdomyolysis also require immediate treatment.

If documented envenomation has occurred, antivenom should be administered. Most local and systemic symptoms improve with the antivenom. The Crotalidae antivenom is equine-based and is therefore contraindicated in individuals with known hypersensitivity to the antivenom or horse serum. Antivenom may need to be given cautiously in severe envenomations despite a positive allergic history. All patients should be initially skin-tested. Both immediate hypersensitivity (anaphylaxis) and delayed hypersensitivity (serum sickness) can occur.

The amount of antivenom to administer is determined by the degree of envenomation.[8,9] It is wise to consult the regional poison control center prior to its use. For bites that produce very minor local symptoms and no systemic symptoms after 6 hours, no antivenom is recommended. In moderate envenomations where there is progression of local symptoms beyond the bite site and minor systemic symptoms such as nausea, anxiety, or paresthesias, 4–10 vials of Crotalidae antivenom should be administered, depending on the severity of the symptoms. In severe envenomations, where there is progressive local pathology and severe systemic symptoms,
10–40 vials may be needed. All patients receiving antivenom should be closely observed and monitored in a hospital setting.

Prevention

Whether dead or alive, snakes should be handled carefully, especially when the treating personnel are attempting to identify the snake. Dead snakes can still envenomate hours after their death. Snakes are ubiquitous in all outdoor areas. Areas like rock crevices, caves and unusual terrain should be avoided. Trousers gloves, boots and long-sleeved shirts should be worn in areas where snakes may be present. Pest control professionals should be consulted when work must be performed in high-risk areas. One should never awaken or try to molest a snake. Venomous snakes should not be sold or kept as pets.

REFERENCES

1. Russell FE, Picchioni AL. Snake venom poisoning. *Clin Toxicol Consult* 1983; 5:73–87.
2. Russell FE, Carlson RXV, Wainschel J, et al. Snake venom poisoning in the United States: experience with 550 cases. *JAMA* 1975; 233:341–4.
3. Simon TL, Grace TG. Envenomation coagulopathy in wounds from pit vipers. *N Engl J Med* 1981; 305:443–7.
4. Gibly RL. Intravascular hemolysis associated with North American crotalid envenomation. *J Toxicol Clin Toxicol* 1998; 36(4):337–43.
5. Danzig LE, Abells GH. Hemodialysis of acute renal failure following rattlesnake bite with recovery. *JAMA* 1961; 175:160–1.
6. Snyder CC, Knowles RP. Snakebites; guidelines for practical management. *Postgrad Med* 1988; 83:53–75.
7. Wasserman GS. Wound care of spider and snake envenomation. *Ann Emerg Med* 1988; 17:12.
8. Holstege CP. Crotalid snake envenomation. *Crit Care Clin* 1997; 13(4):889–921.
9. Seifert SA. Relationship of venom effects to venom antigen and antivenom serum concentrations in a patient with Crotalus atrox envenomation treated with a Fab antivenom. *Ann Emerg Med* 1997; 30(1):49–53.

ELAPIDAE

Common names: Coral snake (*Micrurus*), cobra (*Naja*), krait (*Bungarus*), mamba (*Dendroaspis*)

Occupational setting

Farming and agricultural workers, gardeners and orchard workers, forestry workers, construction workers, veterinary and ancillary staff workers, pet store personnel, zoo workers and research personnel are at risk from exposure.

Exposure (route)

Exposure occurs through physical trauma (bite) to the human skin with injection of venom.

Pathobiology

Coral snakes, the only members of the Elapidae family native to the USA, are found in the southeastern and south central states. They include the Sonoran or Arizona coral snake (*Micruroides euryxanthus*), the Eastern coral snake (*Micrurus fulvius fulvius*), and the Texas coral snake (*Micrurus fulvius tenere*). The eyes are round and the mouth small, with a row of teeth and rear fangs. This anatomy forces the coral snake to hold on and "chew" its victim in order to get its rear fangs into the skin and cause envenomation. Because of its anatomy, the coral snake is unable to envenomate its victim in most circumstances. The venom is a complex mixture of many components, including a neurotoxin.[1] Coral snake venom causes very little local destruction, but it can cause systemic reactions that are primarily neurologic.

Coral snakes have a distinct pattern of red and black bands that are interspaced with narrower yellow bands. This is often remembered as: "Red on yellow, kill a fellow"—coral snake; "Red on black, venom lack"—non-venomous snake.

There is usually very mild (if any) swelling, erythema and pain around the bite site. Occasionally, paresthesias occur around the site. A few hours after envenomation, the victim experiences nausea, vomiting, confusion, euphoria, or drowsiness. The systemic effects, which are predominantly neurologic, can progress to diplopia, fasciculations, slurred speech, dysphagia, paralysis, and respiratory depression. It is important to realize that the systemic symptoms of a coral snake envenomation may be delayed for up to 10 hours.[2] The Arizona coral snake appears to cause greater neurologic symptomatology than the Texas or Eastern coral snakes.[3]

Diagnosis

The identification of the snake that inflicted the bite is critical in the management of all snake bites. The actual snake should be brought to the treating physician for identification, or the snake should be identified using well-defined photographs of different varieties of snakes. A herpetologist located at a local zoo, park or aquarium may need to be consulted. The regional poison control center or the Antivenom Index of the Arizona Poison Control Center may be helpful. Because envenomation can occur even hours after a snake's death, snakes should be carefully handled even when dead.

The victim should be evaluated for fang marks, swelling, erythema, pain, local reaction (i.e. blebs, vesicles, petechiae), and any systemic reaction. Most victims of coral snake bites present with minor abrasion or the imprint of teeth on the skin surface. Fang or puncture marks reveal possible severe envenomation. The victim should be observed for up to 12 hours for systemic symptoms, which are predominantly neurologic and would be consistent with a coral snake envenomation.

Treatment

Most coral snake envenomations can be treated with local wound care and tetanus prophylaxis. The presence of fang or puncture marks is a sign of a potentially serious envenomation and necessitates observation for at least 12–24 hours. The affected area should be immobilized in a dependent position to decrease the spread of the venom. Cryotherapy should not be used, and incision and suction are usually not warranted.

In severe envenomations, consultation with a regional poison center and local herpetologist is recommended before using the *Micrurus fulvius* antivenom. In general, any victim with severe envenomation from a Eastern coral snake or Texas coral snake, which causes systemic neurologic symptomatology or coagulopathy should receive the antivenom. It is not effective for envenomation by the Arizona or Sonoran coral snake. Hypersensitivity to horse serum or the antivenom is a relative contraindication, but it may be given with extreme caution in severe envenomations. All victims should be skin-tested with the antivenom prior to administration. Adverse effects include immediate hypersensitivity and possible anaphylaxis as well as delayed hypersensitivity (serum sickness).

Prevention

Whether dead or alive, snakes should be handled carefully, especially when the treating personnel are attempting to identify the snake. Dead snakes can still envenomate hours after their death.

Snakes are ubiquitous in all outdoor areas. Areas like rock crevices, caves and unusual terrain should be avoided. Trousers, gloves, boots and long-sleeved shirts should be worn in areas where snakes may be present. Pest control professionals should be consulted when work must be performed in high-risk areas. One should never awaken or try to molest a snake. Any snake that has the potential of envenomation should not be sold or kept as a pet.

REFERENCES

1. Silveira de Oliveira J. Cloning and characterization of an alpha-neurotoxin-type protein specific for the coral snake Micrurus corallinus. *Biochem Biophys Res Commun* 2000; 267(3): 887–91.
2. Deer PJ. Elapid envenomation: a medical emergency. *J Emerg Nurs* 1997; 23(6):574–7.
3. Russell FE. Bites by the Sonoran coral snake Micruroides euryxanthus. *Toxicon* 1967; 5:39–42.
4. Kitchens CS, Van Mierop LH. Envenomation by the eastern coral snake (Micrurus fulvius fulvius): a study of 39 victims. *JAMA* 1987; 258:1615.

HYDROPHIDAE

Common name: Sea snake

Occupational setting

Fishermen and diving personnel, especially those in the Pacific and Indian oceans, are at risk from exposure. There do not appear to be any sea snakes in the Atlantic Ocean or the Caribbean.

Exposure (route)

Exposure occurs through physical trauma (bite) to the human skin, with potential envenomation.

Pathobiology

Sea snakes have a compressed, fin-like tail that allows them to propel themselves in the water. They have two to four fangs, which are attached to venomous glands, but the fangs are short; therefore, most bites result in no envenomation. There are at least 52 species and all are venomous. *Enhydrina schistosa* is the most dangerous. The venom is a potent neurotoxin[1,2] that contains myotoxins, such as phospholipase A, which can cause striated muscle necrosis.[3] Many other substances are present in this complex venom.

Most often, the initial bite is painless and the victim asymptomatic. If envenomation did occur, symptoms can develop within 10 minutes and will usually be present within an 8-hour period. In actual envenomations, there are puncture wounds on the victim's skin. The classic envenomation results in muscle pain, difficulty with muscle movement, and myoglobinuria.[1] Ophthalmoplegia, trismus, ptosis, paralysis (flaccid or spastic) of the lower extremities and respiratory failure can occur.[2]

Diagnosis

The identification of the snake that inflicted the bite is critical in the management of all snake bites. The snake should be brought to the treating physician for identification, if possible, or it should be identified using well-defined photographs of different varieties of snakes. Specifically, with sea snakes, the bite must have been inflicted in the Pacific or Indian ocean. The evaluation of the bite site should reveal two or more puncture wounds secondary to the fang marks. The clinical presentation of a painless bite followed by trismus, ptosis, myalgias, difficulty with muscle movement or lower extremity paralysis should make the treating personnel suspicious of a sea snake envenomation.

Treatment

Treatment consists of immobilizing the victim with the affected site in a dependent position to decrease the spread of the venom. Cryotherapy is not indicated, and nor is incision and drainage of the affected site in most cases. Most victims can be treated with supportive care.

For severe envenomations, there is antivenom available from Australia, as well as a polyvalent tiger snake antivenom.[3,4] As with all antivenom, sensitivity testing should be performed prior to administration and the victim observed for signs of anaphylaxis and serum sickness. In all systemic reactions to the sea snake bite, the victim should be closely monitored in a hospital setting, since respiratory failure is possible.

Prevention

Trousers, gloves, boots and long-sleeved shirts should be worn in areas where the snakes may be present. Sea snakes are often found in fishing nets. Fishermen should be extremely careful when removing entangled sea snakes from these nets. Whether dead or alive, sea snakes should be handled carefully. Dead snakes can still envenomate hours after their death.

REFERENCES

1. Kizer KW. Marine envenomations. *J Toxicol Clin Toxicol* 1983–4; 21:527–55.
2. Tu AT. Biotoxicology of sea snake venoms. *Ann Emerg Med* 1987; 16:1023–8.
3. Soppe GG. Marine envenomation and aquatic dermatology. *Am Fam Physician* 1989; 40:97–106.
4. Amarasekera N. Bite of a sea snake (Hydrophis spiralis): a case report from Sri Lanka. *J Trop Med Hyg* 1994; 97(4):195–8.

28

ALLERGENS

Gwendolyn S. Powell, M.D., M.P.H.

ENZYMES

Common names: Detergents, digestive aids, dough improvers, papain, others

Occupational setting

Enzymes are used in the chemical, pharmaceutical, cosmetic, textile, medical, detergent and food and beverage industries. The paper and pulp industries use enzymes to break down wastes. Cellulase is used as a digestive aid. *Aspergillus*-derived alpha-amylase and cellulase are added to flour as dough improvers. Enzymes derived from *Bacillus subtilis*, such as alcalase (subtilisin A), are used in the detergent industry as an aid in removing stains from clothing. Papain and other enzymes are used in pharmaceutical products where the manufacturing process may involve sieving, blending and compressing powders. Papain also has uses as a meat tenderizer (which use has resulted in illness in industrial kitchens), in the treatment of wool and silk for textiles, and in clarifying beer. Trypsin, a pancreatic enzyme, is used in the rubber industry.

Exposure (route)

Enzyme powders tend to be fine and easily airborne. Exposure to enzymes may occur through inhalation of enzyme dusts or through skin contact with liquid enzyme preparations or airborne dust. Dust may impact the upper airways or be carried directly into the lungs.

Pathobiology

Enzymes are proteins that catalyze chemical reactions. They are usually high-molecular-weight proteins and are effective in very small quantities. Cellulase catalyzes the cellulose-to-glucose reaction, xylanase catalyzes the breakdown of xylan, and alpha-amylase also has carbohydrate-cleaving activity. Papain

Physical and Biological Hazards of the Workplace, Second Edition, Edited by Peter H. Wald and Gregg M. Stave
ISBN 0-471-38647-2

and bromelain are proteolytic enzymes derived from the latex of ripe fruit of the pawpaw tree and the pineapple, respectively. Alcalase is a proteolytic enzyme derived from *Bacillus subtilis* bacteria. Enzyme preparations may be contaminated with byproducts of their production, such as growth media material, microorganisms, and preservatives.

Enzymes are potent respiratory sensitizers that may cause immediate, delayed or dual allergy consisting of rhinoconjunctivitis and several patterns of asthma. In addition to respiratory allergy, immediate and delayed types of skin allergy have been described from enzyme exposure. Respiratory allergy may occur from exposure to cellulase, xylanase, papain, flaviastase, trypsin, bromelain, pepsin, alpha-amylase, and other enzymes. Asthmatic illnesses occur in a portion of sensitized workers who all typically have rhinitis and conjunctivitis along with nasal congestion. The pattern of asthma symptoms may be immediate (type I, IgE-mediated), biphasic with both early and late reactions, or delayed. Hypersensitivity pneumonitis may also occur. The occurrence of systemic symptoms along with delayed resolution of symptoms after removal from exposure and the presence of precipitins provides evidence for type III hypersensitivity. In papain asthma, the pathology seems to involve small airways, and alveolar and interstitial lung tissue, in an inflammatory manner. Higher doses of enzyme seem to cause more intense pulmonary symptoms. Atopic individuals are thought to be at higher risk of developing enzyme-related allergies, such as that to papain, and to develop antibody and symptoms sooner than non-atopics.

Skin disorders from enzyme exposure may be irritant, digestive (commonly), or allergic. Skin allergy may be of the immediate (type I, IgE-mediated) variety, manifested by urticaria following skin contact with enzymes in liquid or powder form. Papain, cellulase and xylanase are among the enzymes that cause contact urticaria.

Contact dermatitis, a delayed (type IV) allergy that can be verified by positive patch tests, may also occur following exposure to enzymes. Contact dermatitis consists of an itchy, raised, red skin rash that may be slow to resolve. Although alcalase skin disorders may be irritant or digestive in nature, the combination of detergent with alcalase may be more allergenic than alcalase alone. Some enzymes, such as cellulase and xylanase, have been found to cause both urticaria and contact dermatitis. Patients with enzyme-related contact dermatitis may or may not be able to eat related products without problems.

Allergy to enzymes may develop after months to years of exposure. Allergic individuals exhibit rhinorrhea, conjunctivitis, and often shortness of breath, cough, chest tightness and wheezing that occurs minutes to hours following exposure to enzymes. The oculonasal complaints typically precede the development of chest symptoms.

Anaphylaxis has followed the ingestion of papain in sensitized individuals. Cellulase and xylanase exhibit some cross-reactivity, as do chymopapain, papain, and bromelain. There is conflicting evidence regarding cross-reactivity between alcalase, sarinase, and esperase.[1]

Diagnosis

Medical and occupational history can be used to diagnose this disorder when the temporal relationship to exposure is elucidated in association with allergic symptoms. Symptoms that appear following exposure may resolve on weekends and holidays. Physical examination may reveal mucous membrane inflammation and respiratory wheeze. In patients with skin allergy, an itchy, red skin rash may be observed on exposed surfaces.

Pre- and post-shift spirometry can be used to document the association of bronchoconstriction with exposure. One case report documented a 48–50% decrement in peak expiratory flow rate (PEFR) in a patient following entry into rooms where xylanase was handled.[2] Confirmation of the diagnosis of enzyme allergy may be obtained through specific immunologic testing. RAST, skin prick and skin patch testing may demonstrate specific IgE and hyperreactivity, respectively. Skin prick tests have been used to document reactivity of allergic patients to enzymes, including cellulase, xylanase, papain, bromelain, and trypsin. In some cases, skin prick testing has resulted in a systemic reaction.[2] Specific IgE measurements using RAST have been done for antibodies to cellulase, xylanase, alpha-amylase, alcalase, and papain.

ELISA has been reported to measure specific IgE antibody to alcalase in exposed detergent workers. In this study, ELISA was more sensitive than RAST, detecting 85% versus 68.4% of skin-test-positive workers.[1] IgG antibodies to papain have been documented. Bronchial provocation has been used to diagnose alpha-amylase, cellulase,[3] trypsin and papain asthma.

Treatment

The preferred treatment for enzyme allergies is removal from exposure. In severe cases, there is no alternative to a job change. Job change should be the ultimate goal for any sensitized worker. An anaphylactic death has been reported in a papain-allergic worker who accidentally received a large-dose exposure.[4]

In workers who remain in exposure situations, conventional allergy treatments may be tried. Cromolyn sodium reportedly prevented immediate papain-related symptoms in allergic workers, although steroids were needed to control the late response.[5] Long-acting bronchodilators may abate late symptoms.

Medical Surveillance

In the detergent industry, workers are monitored by prick testing for specific IgE antibodies to enzymes. Total IgE and enzyme-specific IgE antibodies are also measured, typically by RAST.[1] It has been suggested that persons with IgE antibodies to enzymes are at higher risk of developing symptoms during prolonged or repeated exposure, or during high-dose exposures such as accidental spills.

Prevention

The problem of respiratory sensitization in the detergent industry led to the implementation of engineering controls that reduced exposures and the incidence of disease. Fume hoods or other containment devices should be used to prevent powder from becoming airborne and being inhaled. When possible, enzymes should be handled in liquid preparations rather than powders to prevent airborne exposures to the respiratory tract and skin. Personal protective equipment may be used to further reduce the risk of exposure.

REFERENCES

1. Sarlo K, Clark ED, Ryan CA, Bernstein DI. ELISA for human IgE antibody to subtilisin A (alcalase): correlation with RAST and skin test results with occupationally exposed individuals. *J Allergy Clin Immunol* 1990; 86:393–9.
2. Tarvoinen K, Kanerva L, Tupasela O, et al. Allergy from cellulase and xylanase enzymes. *Clin Exp Allergy* 1991; 21:609–15.
3. Quirce S, Caevas M, Diez-Gome ML, et al. Respiratory allergy to Aspergillus-derived enzymes in bakers asthma. *J Allergy Clin Immunol* 1992; 90:970–8.
4. Flindt MLH. Respiratory hazards from papain. *Lancet* 1978; 1:430–2.
5. Novey HS, Keenan WJ, Fairshter RD, Wells ID, Wilson AF, Culver BD. Pulmonary disease in workers exposed to papain: clinico-physiological and immunological studies. *Clin Allergy* 1980; 10:721–31.

FARM ANIMALS

Common names: Cows, horses, pigs, reindeer, sheep, chickens

Occupational setting

The raising of livestock occurs in various settings ranging from indoor to outdoor and small, family-run enterprises to large, commercial facilities. Reindeer herding is an important and prevalent occupation in Scandinavia. Horse exposure occurs in mounted law enforcement personnel and racetrack workers as well as agricultural workers. Livestock exposure also poses risk to veterinary surgeons.

Exposure (route)

Exposure to farm animal antigens may be through inhalation of airborne particles or through skin contact with animals. Housekeeping tasks cause increased airborne dust concentrations, which may result in inhalation exposure.

Pathobiology

Animal proteins from farm animals may cause occupational allergy. Farm animals include various species of domesticated animals. Animals whose antigenic dander is associated with occupational allergy include horses, cows, sheep, pigs, and reindeer. Cat and dog allergies will not be discussed here.

Skin contact with farm animal antigens may cause immediate or delayed skin allergy that produces dermatitis.[1] Delayed dermatitis can be predicted to be an IgG-mediated process following skin exposure. One case report described delayed hypersensitivity in a piggery worker to pig epithelium that resulted in hand and body eczema. This contact dermatitis was substantiated by positive patch test to pig epithelium.

Allergic respiratory disease has been associated with exposure to farm animals, including cows, horses, and reindeer. Dander from these animals is allergenic. Respiratory disorders in hog farmers are not thought to be IgE or IgG allergic illnesses.[2] Poultry workers have been found to have rhinitis and asthma in association with poultry-related antigens.[3] This IgE antibody-mediated allergy may result from chicken antigens (to which antibodies to serum and droppings have been described) or from related antigens such as northern fowl mite antigen (see section on Mites). Bird fanciers and pigeon breeders have been shown to have bird-related immediate respiratory symptoms associated with immediate wheal and flare reactions. Although 8–19% of allergic patients reportedly react to horse allergens by intracutaneous testing, clinical disorders from horse allergy seem to be relatively infrequent.[4]

Diagnosis

Contact dermatitis from farm animal exposure may be diagnosed through a thorough medical and occupational history and physical examination along with patch skin testing. Workers with occupational dermatitis develop red, raised, itchy, often lichenified patches of skin inflammation in areas where direct skin contact with an allergen has occurred. Hand dermatitis is more common, but the disorder can occur on other exposed areas, such as limbs, face, and thorax. The dermatitis may be chronic, with clearing only during vacations.

In immediate-hypersensitivity disorders, rhinitis and asthma may develop following months to years of exposure. Workplace challenge can be used to demonstrate signs of allergy, such as bronchoconstriction as measured by pre- and post-shift peak flow measurements. Skin prick test and specific IgE antibody measurements with reindeer epithelium have been used to document reindeer allergy in herders.[5] Skin prick test antigens from cow, sheep, goat and horse are also available. Nasal provocation has been used to confirm cow-dander allergy.[6] Skin prick test and RAST have also been used to demonstrate antibody to poultry-related antigens in poultry workers and cow dander in farmers.

Treatment

Avoidance of antigen exposure should be the ultimate goal in the treatment of occupational animal allergies. Since this solution is normally not practical, conventional treatments for dermatitis and asthma may be tried.

Medical surveillance

Farm workers tend to be self-employed or employed in relatively small numbers per facility. Therefore, surveillance has not been addressed.

Prevention

Farmers need to be educated about the risks of allergy in association with farm work. Informed workers are more likely to avoid antigen exposure. Gloves, protective clothing and respirators may decrease exposure. Work practice controls that minimize the generation of airborne dust and the need for direct skin contact should be instituted. An increased risk of symptoms in winter associated with closed quarters may necessitate increased vigilance in protective measures.

REFERENCES

1. Malanin G, Kalimo K. Occupational contact dermatitis due to delayed allergy to pig epithelia. *Contact Dermatitis* 1992; 26:134–5.
2. Matson SC, Swanson MC, Reed CE, et al. IgE and IgG-immune mechanisms do not mediate occupation-related respiratory or systemic symptoms in hog farmers. *Allergy Clin Immunol* 1983; 72:299.
3. Bar-Sela S,Teichtahl H, Lutskvl. Occupational asthma in poultry workers. *J Allergy Clin Immunol* 1984; 73:271–5.
4. Bardana EJ Jr. Occupational asthma and related conditions in animal workers. In: Bardana EJ Jr, Montanaro A, O'Hollaren MT, eds. *Occupational asthma*. Philadelphia: Hanley & Belfus, 1992:225–35.
5. Reijula K, Halmepuro L, Hannaksela M, Larmi E, Hassi J. Specific IgE to reindeer epithelium in Finnish reindeer herders. *Allergy* 1991; 46:577–81.
6. Valero Santiago AL, Rosell Vives E, Lluch Perez M, Sancho Gomez J, Piulats Xanco J, Malet Casajuana A. Occupational allergy caused by cow dander: detection and identification of the allergenic fractions. *Allergol Immunopathol* 1997: 25:259–65.

GRAIN DUST

Common names: Rye grass, soybeans, buckwheat, oat grass, barley

Occupational setting

Grain dust allergy occurs in any occupation where grains are handled, stored, or used. Farming, cereal manufacturing and grain loading and unloading operations are examples of settings associated with occupational grain dust allergy. Rye flour is used in agglomerate board manufacturing glue and therefore may be an allergen in wood dust. Buckwheat hulls are sometimes used as fillers in pillow and cushion manufacturing.[1] *Aspergillus*-derived enzymes are used to enhance baked products.[2]

Exposure (route)

Inhalation exposure to respirable dust is the route of exposure in the development of occupational respiratory sensitization to grain dust. Grain dust results from the abrasion of kernels during handling; it forms at an estimated rate of 3–4 lb per ton of grain handled.[3] Airborne dust measurements made during wheat and oat loading in Canada showed a mean particle size of 1.7–3.1 μm, which is respirable.[4] Some grain-associated allergens, such as cellulase and buckwheat, may also cause skin reactions from skin contact.

Pathobiology

Grain dust is a product of various grass species used primarily in food manufacturing. It contains fractured grain kernels as well as various contaminants and additives, including molds, fungi, *Aspergillus*-derived enzymes (high-molecular-weight proteins with catalytic activity), bacteria, insects such as grain weevil, storage mites, pollens, fractured weed seeds, and mineral particles.[2,5] Wheat dust and storage mites will be considered separately.

Because there are many allergenic components of grain dust, the etiology of grain-related allergy varies. Potential allergens range from contaminants such as insects, bacteria and fungi to components of the grains themselves, such as soybean flour (which contains a number of antigens) and buckwheat flour. Wheat, rye and triticale grasses are closely related species that have significant cross-antigenicity. Rye, barley and oat are less closely related but still exhibit some cross-reactivity.[5] Allergic individuals may be sensitized to more than one grain-related antigen. Most grain-associated antigens cause an immediate-type IgE-mediated hypersensitivity that may result in grain-associated rhinoconjunctivitis and asthma. Symptoms typically occur within minutes of exposure. A late-phase aggravation of symptoms after the workday may occur. Buckwheat and cellulase have also been associated with contact urticaria. One patient with prior symptoms related to occupational buckwheat and kapok exposure experienced an anaphylactic episode after buckwheat ingestion. Community asthma epidemics have been associated with soybean loading and unloading in Barcelona, Spain.[6] A case of occupational allergy to oilseed rape dust has been documented.[7]

Among grain workers, grain dust exposure is a well-documented cause of non-allergic respiratory symptoms, including cough, shortness of breath, and decrease in lung function. Grain elevator workers have been found to have lower forced expiratory volume in 1 s (FEV_1) and lower forced vital capacity (FVC) than non-exposed controls on a chronic basis.[8] Severity of symptoms tends to be related to duration of exposure. Non-allergic acute grain-related respiratory disorders also occur, including an asthma-like acute syndrome and grain fever. These types of pathology of the bronchial airways must be taken into account in assessing allergic disease from grain dust.

Diagnosis

The presence of rhinitis, conjunctivitis, sneezing, coughing, wheezing and shortness of breath upon exposure to grain dust provides evidence for grain dust allergy. However, a medical history should be taken to obviate the existence of an underlying respiratory illness. Grain-exposed workers are known to have a high prevalence of respiratory symptoms, including productive cough, wheezing, and shortness of breath, from non-allergic grain-related pathology (e.g., exposure to toxic gases, pesticides, fertilizers).

Conjunctivitis, rhinitis, edematous mucosa and wheezing may be present on physical examination. Serial peak flow measurements may be used to establish reversibility of airflow obstruction and the temporality of symptoms in relation to exposure. Also, a trial of removal from the workplace to watch for resolution of symptoms may be useful. Demonstration of atopic status by history or skin prick test to common allergens is useful, since atopic individuals are at increased risk of some types of grain dust allergy.

Demonstration of positive skin prick test or RAST or positive bronchial challenge to grain-related antigens can support the diagnosis of grain dust hypersensitivity. Skin prick tests and reverse-enzyme immunoassay have been used to demonstrate hypersensitivity to fungal alpha-amylase and cellulase in bakers who were also sensitized to wheat flour.[2] Specific IgE antibodies to some cereal grains, including wheat and rye, can be measured through commercially available tests. Skin prick test methods and specific IgE tests have been developed for a number of research applications in grain dust allergy. Skin prick tests and RAST have been used for barley and rye flour. Bronchial challenge has been used to demonstrate reactions in individuals sensitized to alpha-amylase, cellulase, and wheat flour. Because of coincident exposures in the grain industry, non-allergenic causes of respiratory pathology should be ruled out.

Treatment

Treatment for grain dust allergy focuses on the avoidance of exposure. If workers are unable or unwilling to change occupations, various standard preventive and symptomatic therapies for allergies and asthma may be instituted. However, when the allergic disorder is severe, avoidance of exposure is prudent. If specific antigens are identified to which the patient is allergic, immunotherapy could be considered; however, this is not standard treatment.

Medical surveillance

An occupational history, concentrating on recent tasks, exposures, and respiratory symptoms, may be taken as part of routine medical surveillance. Periodic spirometry is indicated in workers exposed to grain dust. Although other screening tests for surveillance purposes have not been well studied, skin prick testing or RAST should be considered.

Prevention

Occupational allergy develops following exposure to antigenic materials. Therefore, avoidance of inhalation of and skin contact with these substances is the best way to prevent the development of hypersensitivity. The American Conference of Governmental Industrial Hygienists (ACGIH) has recommended $4\,mg/m^3$ 8-hour TWA (time-weighted average) as an exposure limit for oat, wheat, and barley dust. For buckwheat, however, allergic illness has been documented among workers exposed to as little as $1–2\,mg/m^3$ of airborne dust.[9] Work practice, engineering, housekeeping and personal protective equipment controls can help to minimize dust exposure in the food and other grain-handling industries.

REFERENCES

1. Davidson AE, Passero MA, Settipane GA. Buckwheat-induced anaphylaxis: a case report. *Ann Allergy* 1992; 69:158–9.
2. Quirce S, et al. Respiratory allergy to *Aspergillus*-derived enzymes in baker's asthma. *J Allergy Clin Immunol* 1992; 90:970–8.
3. Chan Yeung M, Enarson D, Kennedy S. The impact of grain dust on respiratory health. *Am Rev Respir Dis* 1992; 145:476–87.
4. Williams N, Skoulas A, Merriman JE. Exposure to grain dust. I. A survey of the effects. *J Occup Med* 1964; 6:319–29.
5. O'Holleran MT. Bakers' asthma and reactions secondary to soybean and grain dust. In: Bardana EJ Jr, Montanaro A, O'Holleran MT, eds. *Occupational asthma*. Philadelphia: Hanley & Belfus, 1992:107–16.
6. Anto JM, Sunyer J, Rodrigues-Roisin R, Suarez-Cervera M, Vasquez L. Community outbreaks of asthma associated with inhalation of soybean dust. *N Engl J Med* 1989; 320:1097–102.
7. Suh CH, Park HS, Nahm, DH, Kim HY. Oilseed rape allergy presented as occupational asthma in the grain industry. *Clin Exp Allergy* 1997; 28:1159–63.
8. Chan-Yeung M, Dimich-Word H, Enarson DA, Kennedy SM. Five cross-sectional studies of grain elevator workers. *Am J Epidemiol* 1992; 136:1269–79.
9. Goehte CJ, Wieslander G, Ancker K, Forsbeck M. Buckwheat allergy: health food, an inhalation health risk. *Allergy* 1983; 38:155–9.

INSECTS

Common names: Moths, bees, beetles, cockroaches, others

Occupational setting

Exposure to insects occurs in a variety of occupational settings. Workers at risk of exposure include those who work directly with insects, those who work with materials that may be contaminated by insects, and those who work in outdoor environments where insects coincidentally live.[1] Workers who have direct contact with insects include entomologists, lepidopterists, ecologists, aquarists, toxicologists, beekeepers, spice or dye factory workers, organic farmers, and pest control researchers. Workers who have contact with potentially contaminated materials include bakers, process and warehouse workers (e.g., honey processors, grain mill workers), silk weavers, dock loaders, and sewage treatment workers. People who are in occupations where insects coincidentally occur include farmers, fishermen, gardeners, firefighters, forestry workers, environmental researchers, hydroelectric plant workers, fruit pickers, and those involving waterside work.

Exposure (route)

Insect-related organic materials readily become airborne. Shedded insect exoskeletons and scales are thin, readily dried, and very friable. In environments where these and related materials are found, dust may be raised through nearly any activity. Particulates from insects may be inhaled, resulting in respiratory illnesses.

Skin contact with certain insect parts may result in irritant or allergic disorders. Such contact can arise through direct handling of insects or by contact with contaminated objects or surfaces. Bites, such as from beetles, can cause urticaria. Insect stings may also cause allergy.

Pathobiology

Insects are small, invertebrate animals that have an adult stage characterized by three pairs of legs, a segmented body with three major divisions, and usually two pairs of wings. Insect parts and byproducts are the source of allergenic proteins that cause insect-related allergies and asthma.

Allergenic particles can arise from shed skeleton, scales, excretions, and secretions of insects. Caterpillar hairs and moth scales are known skin irritants that, upon repeated contact, can cause dermal sensitization. Insect hairs that are not irritants may also give rise to allergic disorders following repeated exposure. Larvae may contain the same antigenic material found in adult insects.[2]

Inhalation of insect fragments may result in respiratory sensitization, which may occur after weeks to years of exposure. Intensity and duration of exposure are important determinants of allergy development. Although the underlying mechanism of sensitization has not been established for all insect allergies, these disorders are generally immediate-type IgE-related hypersensitivities. In one survey of entomologists, 33% of workers directly exposed to insects had allergic conditions.[3] In evaluating other documents that estimate incidence, 30% is a representative figure. In moth- and butterfly-rearing laboratories, an incidence of allergy of 53–75% has been documented.[4]

Many insect species have been implicated in occupational allergic disorders. Some examples include moths of various species, grasshoppers, locusts, screwworms, blowflies, beetles, parasitic wasps, *Drosophila*, yellow jackets, honeybees, bumble bees, houseflies, caddis flies, cochineal insect (source of carmine dye), red midges and larvae, and bloodworms, as well as cockroaches.[5] Reactions that occur upon development of insect-related occupational allergy include eye symptoms, rhinitis, nasal congestion, urticaria, and often cough, wheezing, and shortness of breath. Once sensitized, workers develop illness within minutes of exposure to insect-related allergens.

The exoskeleton of many insect species is allergenic. The scales of butterflies and moths produce allergy. Hemoglobins are major allergens of red midge larvae and adults.[2] In species where feces cause sensitization, the allergen arises from gut-derived cellular material, possibly the peritrophic membrane.[6] Other insect byproducts, such as "bee dust" and cocoons, are also allergenic.

Diagnosis

Diagnosis of insect allergy can be made presumptively by a thorough medical and occupational history along with physical examination. Immunologic testing can be used to confirm that the suspected allergenic material will cause the signs and symptoms of the illness.

For example, skin prick tests, RAST and bronchial provocation with honeybee whole-body extract have been used to document IgE-mediated occupational asthma in a honey processor.[7] Intracutaneous tests produce reactions in bumblebee-allergic patients.[8] ELISA has been applied in the measurement of specific IgE and IgG antibodies to insect extracts from locust, mealworms, cockroaches, spring stick insects, and mulberry moon moths.[6] Specific IgE can be documented by immunoblot. Skin prick tests and bronchial challenge have been used to document occupational asthma from beetles in a museum curator[9] and from blowflies in researchers.[10] Nasal provocation in cockroach-sensitized workers demonstrated decreased nasal flow rates.[5] Ophthalmic challenge produced conjunctivitis in a chironomid-sensitized researcher and a housefly-sensitized patient who worked in a barn.[11,12] Caution must be used in interpretation of skin prick test results in chironomid-exposed workers, since cross-sensitization with other insect or crustacean species may occur.[13]

Treatment

Definitive treatment of occupational insect allergies is removal from exposure. Treatment by conventional therapies such as antihistamines, bronchodilators and inhaled steroids may be used. In one survey, 28% of individuals reporting allergy indicated that job transfer or discontinuation was necessary.[4] Desensitization injections have been successfully used to treat occupational insect allergies. Patients with anaphylactic reactions to bumble bees have been desensitized using honeybee venom.[14] Injectable adrenaline kits are indicated for workers with history of anaphylaxis.

Medical surveillance

No literature is available on the application of surveillance methods to insect-handling workers. In theory, a program including a respiratory questionnaire and examination along with pulmonary function testing would be useful. Research programs could be conducted utilizing skin prick testing or RAST methods.

Prevention

Prevention of inhalation of or skin contact with insects and insect byproducts will prevent insect-related allergies. Engineering and work practice controls may reduce the amount of airborne dust in occupational settings where insects are used. Laboratories should be designed with appropriate air circulation, segregation of insects and ease of maintenance and housekeeping in mind. Good hygiene is imperative.

Personal protective equipment may be useful in avoiding exposure and preventing symptoms. However, one survey found that individuals at institutions that reported no insect allergies were less likely to use protective equipment on either routine or as-needed bases than individuals at institutions that reported allergies.[6] Protective equipment may include gloves, respirators (including face masks), head nets, and laboratory coats.

Educating workers about the risk of insect allergy is important in preventing these disorders, particularly since the illnesses are not well known. Workers may be at risk of non-occupational exposure to insects and may have illness because of antigenic cross-reactivity. In one case report, a spice factory worker sensitized to carmine developed food allergy from ingestion of the dye which had initially produced respiratory symptoms from inhalation.[15] Workers who are aware of the possibility of developing respiratory allergy are more likely to protect themselves from exposure.

REFERENCES

1. Bellar TE. Occupational inhalant allergy to arthropods. *Clin Rev Allergy* 1990; 8:15–29.
2. Galindo PA, Feo F, Gomez E, et al. Hypersensitivity to chironomid larvae. *Invest Allergol Clin Immunol* 1998; 8(4):219–225.
3. Bauer M, Patnode R. NIOSH, GHETA, West Virginia, 1983:81–121.
4. Wirtz RA. Occupational allergies to arthropods—documentation and prevention. *Bull Entomol Soc Am* 1980; 26:356–60.
5. Steinberg DR, Bernstein DI, Gallagher JS, Arlian L, Bernstein IL. Cockroach sensitization in laboratory workers. *J Allergy Clin Immunol* 1987; 80:586–90.
6. Edge G, Burge PS. Immunological aspects of allergy to locusts and other insects. *Clin Allergy* 1980; 10:347.
7. Ostrom NK, Swanson MC, Agarwal MK, Yunginger JM. Occupational allergy to honey bee-body dust in a honey-processing plant. *J Allergy Clin Immunol* 1986; 77:736–40.
8. deGroot H, de Graafin T, Veld C, Gerth van Wijk R. Allergy to bumblebee venom. I. Occupational anaphylaxis to bumblebee venom: diagnosis and treatment. *Allergy* 1995; 50:581–4.
9. Sheldon JM, Johnston JH. Hypersensitivity to beetles (Coleoptera): report of a case. *J Allergy* 1941; 493–4.
10. Kaufman GL, Baldo BA, Tovey ER, Bellas TE, Gandevia BH. Inhalant allergy following occupational exposure to blowflies. *Clin Allergy* 1986; 16:65–71.
11. Teranishi H, Kawai K, Murakami G, Miyao M, Kasayu M. Occupational allergy to adult chironomid midges among environmental workers. *Int Arch Allergy Immunol* 1995; 106:271–7.
12. Wahl R, Fraedrich J. Occupational allergy to the housefly (Musca domestica). *Allergy* 1997; 52:236–8.
13. Galindo PA, Lombardero M, Mur P, Feo F, Gomez E, Borja J. Patterns of immunoglobulin E sensitization to chironomids in exposed and unexposed subjects. *Invest Allergol Clin Immunol* 1999; 9(2):117–22.
14. Kochuyt AM, Van Hoeyveld E, Stevens EAM. Occupational allergy to bumble bee venom. *Clin Exp Allergy* 1993; 23:190–5.
15. Acero S, Tabar AI, Alvarez MJ, Garcia BE, Olaguibel JM, Moneo I. Occupational asthma and food allergy due to carmine. *Allergy* 1998; 53:897–901.

LABORATORY ANIMALS

Common names: Rats, mice, guinea pigs, rabbits, hamsters, monkeys

Occupational setting

Laboratory animal allergy (LAA) is an important work-related illness that occurs in 11–33% of exposed workers. It occurs in workers who have laboratory animal contact ranging from casual, indirect exposure, through infrequent animal handling, to full-time daily care of animals and their housing facilities. Laboratory animals are housed and handled at many types of research institutions, including academic centers, medical schools, and private sector research facilities such as pharmaceutical companies. LAA may also result from work in other settings where rodents are present, such as pet shops, vivariums, and veterinary offices.

Exposure (route)

LAA may result from skin exposure or respiratory exposure. Percutaneous exposures may result from animal bites or allergen contamination of wounds. Different routes of exposure result in different disorders with different mechanisms.

Exposure to allergens may result from direct contact through handling contaminated animal bedding and laboratory animals as well as from indirect contact through airborne dust.

Inhalation exposure may result from general contamination of air within the facilities or from airborne dust created through animal handling or care. Allergens, such as those from rodent urine, may contaminate animal cage bedding and other materials, and these allergens may become airborne through any type of disturbance. Additionally, antigen is carried on animal fur and easily becomes airborne when the animal is handled.

Any factors that increase the concentration of allergens in the workplace also increase potential exposure. These factors include increased numbers of animals, decreased frequency of cage changing, increased manipulation of animals, and accumulation of dust.

Pathobiology

Laboratory animals reared for use in research consist mainly of rodents. Those laboratory animals best studied regarding occupational allergy include mice, rats, guinea pigs, hamsters, and rabbits. In addition, frogs and monkeys that are laboratory-reared may also cause occupational allergy.

Exposure to laboratory animals and associated materials may result in development of immediate-type IgE-mediated respiratory hypersensitivity. Immediate skin reactions including wheal and flare from skin contact (contact urticaria) may also occur.

The prevalence of allergy among laboratory animal handling workers is $\sim$15–30%. About 10% of these workers develop asthma. The risk of allergy in relation to increasing intensity and duration of exposure is unclear.[1]

Evidence that exposure–response relationships exist has been reported with respect to rat allergy.[2] Although many cases have been reported that developed more than 20 years after initial exposure, most LAAs develop within the first few years of exposure. It has been suggested that 50% of those who develop LAAs do so within 2 years.[3]

Allergy typically presents as rhinitis and conjunctivitis, which progresses into asthma in a minority of patients. Symptoms typically occur within 5–30 minutes after exposure. Asthma may occur immediately or may exhibit dual or delayed patterns. Skin reactions include contact urticaria from direct contact of the animal with exposed skin and maculopapular pruritic rash on exposed skin in association with airborne exposure and respiratory symptoms.

The antigens responsible for LAA come from urine, dander, saliva, and serum. These antigens are typically small acidic glycoproteins with molecular weights of 15–30 000.[4] For mice, rats, rabbits and guinea pigs, urinary proteins and saliva are thought to cause the majority of hypersensitivity problems. The main allergen in mice is a urinary protein, possibly prealbumin,[1] whereas the main allergens in rats are serum albumin, urinary α_2-globulins and prealbumin.[3] Monkey dander has caused occupational asthma in researchers.[5]

IgE antibody is produced in response to allergen exposure. IgE antibodies bind to mast cells. Adsorption of the corresponding allergen triggers the release of histamine and other mediators from the sensitized mast cell.[3] Other antigens, such as those derived from molds or mites, cause laboratory animal-related allergies and are found in the same settings.

The presence of several general allergic symptoms and several historical indicators of atopy is moderately predictive of the new onset of LAA.[6] Development of asthma from laboratory animals may predispose sensitized workers to the development of other hypersensitivities.

Diagnosis

A clinical diagnosis of LAA can be made based on a careful medical and occupational history along with a physical examination. A detailed respiratory and dermatologic history will reveal symptoms of allergic disorders such as wheal and flare reaction to contact, e.g., a rat's tail wrapping around the hand. Respiratory symptoms may include rhinitis, conjunctivitis, sneezing spells, nasal congestion, cough, shortness of breath, and wheezing.

It is critical to document the fact that the symptoms are temporally associated with exposure to laboratory animal antigens and that they resolve when the individual is away from work. Objective evidence of temporality may be obtained by doing a physical examination or peak flow measurements prior to exposure and again at the end of the workshift. The finding of objective evidence of bronchoconstriction, such as wheezes or a decrement in FEV_1 and FVC, supports the diagnosis of occupational asthma.

Immunologic testing may be used to confirm a diagnosis of LAA. Skin prick testing with relevant allergens or RAST may be used to demonstrate IgE antibody. RAST to detect mouse- and rat-specific IgE are commercially available and have been developed for other animal allergens. Non-laboratory animal antigens, such as house dust mite or molds, may be used to rule out disease from these animal-associated allergens or to establish coincidental allergies. Bronchial challenge testing with animal antigens can be used to confirm reversible airway constriction related to exposure.

Treatment

Avoidance of exposure to animal antigens should be the goal of treatment. Sensitized workers who have rhinitis should be counseled on their risk of developing asthma and the risk of eventual intolerance of exposure.

Symptomatic care may include antihistamines or corticosteroids for upper tract symptoms and bronchodilators, corticosteroids and theophylline for asthma. Cromolyn may prevent asthma from allergen exposure. Long-acting bronchodilators may be tried as preventive therapy. Immunotherapy is not likely to be helpful for most sensitized individuals who remain in an environment where regular exposure occurs.[7]

Medical surveillance

Whether atopic status is a risk factor for the development of LAA is controversial. Recent literature supports the notion that atopy predisposes to LAA.[8,9] Conversely, it has been suggested that HLA-B16 may confer a protective effect against the development of LAA.[10] Skin prick testing and RAST have been suggested as surveillance methods that may prove useful in detecting preclinical allergy to laboratory animals. The utility of these methods has yet to be assessed experimentally.

Respiratory history may be used to document early symptoms of LAA. If nasal symptoms are detected early and further exposure is eliminated, subsequent progression to asthma may be prevented.

Prevention

Containment of the source protein is the cornerstone of animal allergy prevention measures. A comprehensive program including education, engineering controls, administrative controls, use of personal protective equipment and medical surveillance can prevent the development of LAA.[11] Efforts should be made to limit airborne dust, such as by housing and transporting animals in filter-top cages and wetting them prior to handling. Engineering controls such as construction of clean corridors, provision of high ventilation rates with limited recirculation of air, use of downdraft (ceiling to floor) ventilation and installation of HEPA filters are helpful.

Although no exposure level is without risk, one study showed smaller mediator changes in sensitized workers exposed to antigen concentrations of $<10\,\mathrm{mg/m^3}$ than in those more heavily exposed.[37] Airborne antigen levels have been found to be related to litter type and stock density. In one study, significant reductions in rat allergen concentrations were achieved by replacing wood-based (sawdust) contact litter with non-contact absorbent pads.[4]

The concentration of airborne animal allergen is dependent on the activities in the contaminated areas. For example, rat allergen concentrations in air have been found to be 10–100 times higher during animal handling or disturbance of bedding than at quiet times.[2]

Because allergens from laboratory animals adhere to all types of particulate matter, contamination of facilities occurs easily, as evidenced by the presence of antigen in all types of dust. Therefore, a reservoir of allergens exists outside the immediate animal housing areas.[7] Removal of animals from work areas will not eliminate exposure. In homes where pets are removed, it takes 4–6 months before allergen levels are at clinically insignificant levels.

Personal protective equipment can be used to supplement engineering and hygiene controls. Gloves and laboratory coats should be worn routinely for animal handling. Respiratory protection should be used in exposure situations because of the high incidence of LAA.

Many animal-allergic workers also have household pets, which may complicate treatment. Whenever possible, pets in the home should be removed. In sensitized individuals, the use of cromolyn and possibly long-acting bronchodilators prior to exposure may prevent asthma.

REFERENCES

1. Kibby T, Powell G, Cromer J. Allergy to laboratory animals: a prospective and cross-sectional study. *J Occup Med* 1989; 31: 842–6.
2. Hollander A, Heederick D, Doekes G. Respiratory allergy to rats: exposure–response relationships in laboratory animal workers. *Am J Respir Crit Care Med* 1997; 155:562–7.
3. Hunskaar S, Fosse RT. Allergy to laboratory mice and rats: a review of the pathophysiology, epidemiology and clinical aspects. *Lab Anim* 1990; 24:358–74.
4. Gordon S, Tee RD, Lowson D, Wallace J, Newman-Taylor AJ. Reduction of airborne allergenic urinary proteins from laboratory rats. *Br J Ind Med* 1992; 49:416–22.
5. Petry RW, Voss MJ, Kroutil LA, Crowley W, Bush RK, Busse WW. Monkey dander asthma. *J Allergy Clin Immunol* 1985; 75:268–71.
6. Eggleston PA, Newill CA, Ansan AA, et al. Task-related variation in airborne concentrations of laboratory animal allergens. Studies with rat n1. *J Allergy Clin Immunol* 1988; 84:347–52.
7. Eggleston PA, Wood KA. Management of allergies to animals. *Allergy Proc* 1992; 13:289–92.
8. Bryant DH, Boscato LM, Mboloi PN, Stuart MC. Allergy to laboratory animals among animal handlers. *Med J Aust* 1995; 163:415–8.
9. Hollander A, Doekes G, Heederik D. Cat and dog allergy and total IgE as risk factors of laboratory animal allergy. *J Allergy Clin Immunol* 1996; 98:545–54.
10. Sjostedt L, Willers S, Orbaek P. Human leukocyte antigens in occupational allergy: a possible protective effect of HLA-B16 in laboratory animal allergy. *Am J Indust Med* 1996; 30:415–20.
11. Fisher R, Saunders B, Murray SJ, Stave GM. Prevention of laboratory animal allergy. *J Occup Environ Med* 1998; 7:609–13.

MITES

Common names: Dust mites, storage mites, red spider mites, citrus red mites

Occupational setting

Occupational illness from exposure to mites may occur in a variety of settings. Workplaces associated with mite-related illnesses include warehouses, barns, poultry houses, greenhouses, flower farms, fruit orchards, livestock farms, grain storage facilities, bakeries, and animal-housing facilities.

Exposure (route)

Occupational allergy to mites results from inhalation exposure or skin contact. Because mites are tiny creatures, entire animals can be airborne and readily inhaled. In addition, skin exposure to mite-infested materials may result in contact allergy: occupational dermatitis or contact urticaria.

Pathobiology

Mites are a class of arthropods. Storage mites are wingless, translucent, microscopic invertebrates. Dust mites, or pyroglyphid mites, are more abundant than non-pyroglyphid, or storage mites. In optimal conditions, mites can self-multiply at rates of 4–10-fold weekly.[1]

Various mite species are ubiquitous in our environment. Dust mites infest organic particles in dusts such as mattress, bedding, pillow, carpet, or floor dust. Storage mites infest stored food and vegetable products as well as hay and straw. Red spider mites, *Tetranychus* species, parasitize flowers such as carnations in greenhouse cultivation as well as in open fields. Mites, especially storage types, thrive in damp environments. The major species in infestations vary by season.

Inhalation of mites or their byproducts may result in immediate-type IgE-mediated respiratory hypersensitivity. Mites as well as their feces are allergenic.[2] Mites that have been implicated in occupational allergy include storage mites, grain mites, dust mites, red spider mites, citrus red mites, fowl mites, and hay itch mites. Spider silks have also been implicated as workplace allergens. Exposure to antigens from these arachnids may cause rhinoconjunctivitis and asthma upon exposure in sensitized individuals. With *Lepidoglyphus destructor*, late-phase asthma (occurring hours after exposure) has been described and may be related to leukotriene release.[3] Contact allergy has been reported with two spotted spider mites and red spider mites.[4] Case reports have described occupational allergy to cheese mites (*Blomia kulagini*), chorizo mites (*Euroglyphus maynei*), and salty ham mites (*Tyrophagus putrescens*).[5]

Storage mites contain the most important antigens that have been related to asthma and allergic rhinoconjunctivitis among grain farmers.[6] A strong association between storage mite allergy and house dust allergy has been noted in farmers.[7] In one survey, >80% of patients thought to be at risk of occupational exposure had storage mite allergy. Also, there is a high degree of cosensitization to storage mites among wheat flour allergic bakers.[8]

In one study, *Dermatophagoides pteronyssinus* was the most allergenically potent of four mite species evaluated. The *Dermatophagoides* genus provides the major allergens of house dust. Up to three-quarters of serum IgE to mites is directed against antigen P_1, which is associated with fecal particles.[9] Cross-reactivity may occur between several species of house dust mite. There may also be shared antigens among storage mite species.

Diagnosis

Medical history may reveal symptoms of nasal and ocular itching, rhinorrhea, sneezing, shortness of breath, wheezing, and dry or productive cough.[6] Mite-induced skin disorders include contact urticaria and dermatitis. These symptoms in association with exposure to mite-infested material suggest the diagnosis of allergy. Immunologic testing may provide supportive evidence for IgE-mediated hypersensitivity, or, in the case of contact dermatitis, delayed-type allergy.

Skin prick test materials are available for house dust mite (*Dermatophagoides farinae* and *D. pteronyssinus*), storage mite (*Acarus siro*, *Tyrophagus putrescentiae*, *Glycyphagus domesticus*, and *Lepidoglyphus destructor*) and northern fowl mite. Skin prick testing may be the most sensitive immunologic test in confirming a diagnosis of occupational mite allergy. Skin patch testing (finn chamber, urticaria-inducing open test, and acute eczema-inducing open test) is used to detect allergic contact dermatitis or overlapping contact cutaneous syndrome to red spider mite.[10] Specific IgE can be measured by RAST for house dust mites, storage mites, and northern fowl mites. ELISA and EAST have been used to document citrus mite and red spider mite allergy, respectively. Correlation between skin prick testing with RAST and skin prick/RAST testing with symptoms is variable. Conjunctival provocation can be used to substantiate rhinoconjunctivitis from specific allergens. Bronchial challenge with mite antigen may cause immediate, dual-type or delayed reactions.[6] Methacholine challenge, documenting bronchial hyperreactivity, may be useful.

Treatment

Avoidance of further exposure to allergen will prevent symptoms of mite allergy. Since changing vocations is often difficult, medical management of symptoms may be tried, such as conventional therapies including antihistamines and inhaled β-agonists. Immunotherapy has been used with some success. Pretreatment with cromolyn sodium can prevent the occurrence of asthma from exposure to mite allergens.[3] Because of the abundance of various mites in the environment, attempts to avoid exposure must extend to non-occupational settings.

Medical surveillance

Respiratory symptoms can be assessed periodically in mite-exposed workers. Skin prick testing and RAST have no proven utility in surveillance for occupational arachnid allergy but may be considered on an experimental basis.

Prevention

Ninety-one per cent of 46 spider mite-allergic farm workers had a family history of allergy, hay fever, or respiratory problems.[11] This finding suggests that atopic individuals may be at increased risk of mite allergy. Also, sensitization to house dust mites may indicate increased susceptibility to spider mite allergy.[11] Preventive measures should be aimed at reducing the mite population in the workplace. Intensification in such fields as poultry husbandry may have actually increased the concentration of mites in some workplaces.[12] Open field cultivation may be associated with a lower prevalence of clinical spider mite allergy than greenhouse environments, because of ventilation.[13] To decrease mite populations, decreasing reservoirs, decreasing humidity and using miticides may be helpful. HEPA filters on air ducts and vacuum cleaners can be used. Once the population has been minimized through housekeeping measures and other types of dust control, secondary means of preventing exposure can be undertaken. Secondary exposure control should aim at minimizing the opportunities for mites and related dusts to become airborne.

Educating workers about the risk of allergy to mites is important in preventing the illness. Workers who are aware of the risks are better prepared to prevent exposure. Home-related exposures should be assessed in workers with occupational allergic symptoms to mites.

REFERENCES

1. Wraith DC, Cunningham AM, Seymour WM. The role and allergenic importance of storage mites in house dust and other environments. *Clin Allergy* 1979; 9:545–62.
2. Warren CPW, Holford-Strevens V, Sinha RN. Sensitization in a grain handler to the storage mite *Lepidoglyphus destructor* (Schrank). *Ann Allergy* 1983; 50:30–3.
3. Gerces Sotillos M, Blanco Carmona J, Juste Picon S. Late asthma caused by inhalation of *Lepidoglyphus destructor*. *Ann Allergy* 1991; 67:126–8.
4. Wirtz RA. Occupational allergies to arthropods—documentation and prevention. *Bull Entermol Soc Am* 1980; 26:356–60.
5. Armentia A, Fernandez A, Perez-Santos C, et al. Occupational allergy to mites in salty ham, chorizo, and cheese. *Allergol Immunopathol* 1994; 22:152–4.

6. Armentia A, Tapias J, Bowber D, et al. Sensitization to the storage mite *Lepidoglyphus destructor* in wheat flour respiratory allergy. *Ann Allergy* 1992; 68:398–406.
7. Iversen M, Korsgaard J, Hallas T, Dahi R. Mite allergy and exposure to storage mites and house dust mites in farmers. *Clin Exp Allergy* 1990; 20:211–9.
8. Revsbech P, Dueholm M. Storage mite allergy among bakers. *Allergy* 1990; 45:204–8.
9. Tovey ER, Chapman MD, Platt Mills TXE. Mite faeces are a major source of house dust allergens. *Nature* 1981; 289:592–3.
10. Astarita C, Di Martino P, Scala G, Franzese AA, Sproviero S. Contact allergy: another occupational risk to *Tetranychus urticae*. *J Allergy Clin Immunol* 1996; 98:732–8.
11. Astarita C, Franzese A, Scala G, Sproviero S, Raucci G. Farm workers' occupational allergy to *Tetranychus urticae*: clinical and immunologic aspects. *Allergy* 1994; 49: 466–71.
12. Lutsky I, Teichtahl H, Bar-Sela S. Occupational asthma due to poultry mites. *J Allergy Clin Immunol* 1984; 73:56–60.
13. Burches E, Pelaez A, Morales C, et al. Occupational allergy due to spider mites: *Tetranychus urticae* (Koch) and *Panonychus citri* (Koch). *Clin Exp Allergy* 1996; 26:1262–67.

PLANTS

Common names: Poison ivy (*Toxicodendron radicans*), eastern poison oak (*Toxicodendron quericifolium*), western poison oak (*Toxicodendron diversilobum*), poison sumac (*Toxicodendron vernix*), coffee beans, tobacco, psyllium, tea leaves, ipecac, colophony, others (natural rubber latex is covered in Chapter 29)

Occupational setting

Plant allergies may result from exposure in agricultural settings such as outdoor farming areas. In addition, facilities that process plant materials may provide a setting for the development of occupational allergy. Tobacco and garlic farmers, coffee bean processors, tea blenders, plant-leasing farm workers, plant wholesalers, woodwork teachers, spice processors, paper and rubber workers, florists, gardeners and beekeepers may have exposure to plant-derived antigens. Exposure to plant products occurs in the pharmaceutical industry and in healthcare settings. For example, ispaghula husks (psyllium) and senna pods are used as bulk laxatives. In addition to these settings, occupations where workers may be exposed to natural vegetation also provide an opportunity for plant-derived antigen exposure. Examples of outdoor occupations where workers may be exposed to poison ivy or poison oak include forest rangers (and other forest workers), surveyors, and utility company field workers.

Exposure (route)

Plant allergy results from dermal or respiratory exposure. For those substances that cause skin allergy, exposure occurs via direct contact of the plant material with exposed skin. Alternatively, contamination of objects such as clothing and tools with antigen such as that from *Rhus* plants (of the family Anacardiascae) may result in exposure when these objects come into contact with skin. Inhalation of plant-derived allergens is necessary for the development of respiratory sensitization to plants.

Pathobiology

The plant kingdom contains organisms with a vast array of characteristics, and many plants have sensitizing capabilities. The plants considered here are ones that have been documented to be capable of inducing occupational allergy. Included are tobacco, tea, poison oak and ivy, colophony from *Pinus* species, ispaghula, senna, ipecacuanhae, and others. Many other plants have been implicated in occupational allergy, including spathe flower, saffron, Compositae species (sesquiterpene lactone allergen, (e.g., chicory, camomile), weeping fig, christmas cactus, carnation, umbrella tree, mugwort, alstroemeria, and narcissus.[1,2] Cross-sensitization to other flowering plants is common.[3]

Exposure to plant materials may cause skin or respiratory allergy. There may be some overlap of respiratory allergy and skin symptoms; for example, contact or generalized urticaria can occur along with respiratory symptoms in IgE-mediated disorders.

Contact dermatitis is caused by a delayed IgG-mediated hypersensitivity. It is manifested by an itchy, red, raised skin rash that results from repeated contact with an allergen. Once allergy has developed, contact dermatitis occurs one to several days after exposure. Repeated allergen exposure may cause a chronic dermatitis characterized by thickening and lichenification of the skin. Occupational contact dermatitis typically involves the hands and other exposed skin such as arms and face.

Toxicodendron (*Rhus*) dermatitis from poison ivy, poison oak and poison sumac may occur in outdoor workers. *Rhus* rash occurs within 48 hours after exposure in sensitized individuals. It is characterized by intense itching followed by inflammation and grouped vesicles resulting from contact with plant oil. No spread of the dermatitis occurs after the oil has been removed; however, new patches may occur because of the delayed nature of this disorder. Attacks usually last 2–3 weeks, with patches becoming crusted and dry. The antigen involved in *Rhus* dermatitis is pentadecylcatechol, a component of the oleoresin urushiol.[4]

Colophony is resin that is derived from *Pinus* and other species of trees. Rosin may be obtained from living trees or may be a byproduct of paper pulp manufacturing.[5] It is used as a resin in solder. Colophony is an important cause of contact allergy, especially hand eczema. Positive patch test results are frequent.

Propolis, a sticky, resinous material collected by bees from the bud scales of plants and trees, is a well-known cause of allergic eczematous contact dermatitis in beekeepers.[6] The major allergens in propolis seem to come from poplar species, including 1,1-dimethylallyl caffenic acid ester from poplar bud extracts.

In one case report, a saxophonist developed cheilitis due to a musical reed. In this patient, a skin prick test to cane reed scraping was positive, indicating an IgE-mediated dermatitis. Contact skin allergy to reed has been noted in other types of musicians, as well as workers who handle the reed *Arundo donax*.[7]

Although reports of cutaneous reactions are rare, olive oil has been reported to cause contact allergy, including two cases of hand eczema in pedicurists.[8] The sensitizers in olive oil are unknown. Coffee bean dust may cause occupational contact dermatitis. Dandelion, a member of the daisy family, has been reported to cause allergic contact dermatitis in gardeners.[9] Cross-reactivity between dandelions and other allergenic *Compositae* may occur.

Occupational asthma has been found to occur from tea leaves, ficus, ispaghula, senna, green tobacco, green coffee beans, baby's breath (*Gypsophila*), and many other plants. These hypersensitivities are in general of the type I IgE-mediated type.

Asthma from inhalation exposure to green tobacco leaf has been documented in tobacco workers. In one study, green tobacco leaf asthma was found to result from an immediate, IgE-mediated hypersensitivity, as evidenced by positive RAST, nasal and bronchial provocation, and histamine release assay.[10] Because this illness is related to green tobacco exposure, it is postulated that the antigen degrades in the curing process. Other allergens found in association with tobacco plants—e.g., microfungi—may also cause asthma. Asthma associated with green tobacco is a distinct entity from green tobacco sickness (GTS). Green tobacco sickness is a form of nicotine poisoning resulting from dermal absorption during the handling of wet plant leaves.

Respiratory sensitization (IgE-mediated) from green coffee beans may result in upper respiratory symptoms as well as asthma. As with tobacco, the processing of coffee beans destroys some antigen, although roasted coffee allergy also occurs. An estimated 10% of exposed workers develop allergy, a small proportion of who also develop asthma. Immediate- and delayed-type allergy may coexist in coffee bean-sensitized workers.[11]

Inhalation allergy to pharmaceutical products such as psyllium (from ispaghula husks) and senna pods seems to cause asthma less frequently than other respiratory allergies. One cross-sectional study found a prevalence of asthma of 3.2% in a population of whom 7.6% were allergic to ispaghula.[12] Anaphylaxis and eczema have also been reported following ispaghula husk exposure. Asthma from inhalation of ipecac powder has also been reported.[13]

Descriptions in the literature of respiratory allergy and asthma from plants are varied and may be documented in only an article or two for each plant. For example, a few studies have described mushroom workers' lung, a hypersensitivity pneumonitis that may result from mushroom spores or microorganisms.[14] Farm workers who harvest garlic bulbs and spice factory workers have been shown to have garlic allergy. Pectin has been found to cause IgG_4-mediated occupational asthma in a candy maker.[15] Pectin is a high-molecular-weight product of fruits and fruit rinds that contains methyl-esterified galacturonan, galactan, and araban. Tea leaf allergy typically occurs at facilities where tea leaves are mixed together or blended, generating a fine dust.

Diagnosis

Medical and occupational history coupled with physical examination may be used to make a presumptive diagnosis of contact dermatitis from plants. The presence of an itchy, red skin rash occurring hours or days after contact with allergenic material should lead to a suspicion of allergic dermatitis. Acute eruptions are usually vesicular and edematous. *Rhus* contact occurs in a linear fashion, such as from brushing against twigs or leaves. Rashes associated with these plants are usually linear.

Patch skin testing should usually be done to confirm a diagnosis of allergic contact dermatitis when the causative agent is in question, or if the consequences of the diagnosis may affect decisions about the risk of further exposure. Usually, the allergen is applied to the skin for 24–48 hours, after which a delayed rash appears in positive cases. Care must be taken to distinguish between irritant and allergic causes. Other allergenic exposures such as insects and mites should be considered in appropriate settings. Castor bean allergy may complicate the diagnosis of green coffee bean allergy; onion or pollen allergy may be comorbid in garlic-allergic patients.[16]

Respiratory sensitization to plants results in similar symptoms to those found with other inhalant allergens. These symptoms, which must be temporally associated with exposure, include rhinitis, conjunctivitis, nasal congestion, cough, wheezing, and shortness of breath.

Skin prick tests and RAST, among other immunologic studies, may be used to confirm plant allergies of the immediate, respiratory type. Pulmonary function testing is also helpful in diagnosing inhalant allergy. Bronchial provocation has been used to demonstrate coffee bean, garlic and tea leaf allergy, among others. Significant differences have been documented in peak expiratory flow rates between tobacco workers and unexposed controls.[17] Coffee bean-exposed workers may have a higher than average incidence of chronic respiratory problems.

Treatment

For allergic contact dermatitis, treatment with local topical corticosteroid creams or ointments is recommended. For more extensive and severe cases, systemic steroids should be used. For *Rhus* dermatitis, standard steroid protocols usually begin with 40–60 mg of prednisone orally in a single daily dose with a 3-week taper.[4] Calamine lotion may ease the symptoms. Local antihistamine and anesthetic ointments should be avoided because of the possibility of contact sensitization.

Allergic respiratory disorders from plant exposure may be treated with conventional allergy therapies. These include antihistamines, bronchodilators, and steroids. Avoidance of exposure is the only definitive treatment.

Prevention

Atopy may be a risk factor in green coffee bean allergy.[18] Protecting workers from skin and respiratory exposure can prevent occupational plant allergies. Engineering controls and personal protective equipment as well as sound work practices may be helpful.

Use of mechanical blending processes with extraction ventilation rather than hand mixing reduces personnel exposure to tea leaf dust. Wearing cotton glove liners reduces contact with skin sensitizers and therefore prevents allergy. Personal protective equipment to prevent inhalation and skin exposure to airborne materials, such as respirators and barrier clothing, should also be considered.

REFERENCES

1. Sanchez-Guerrero IM, Escudero AI, Bartolome B, Palacios R. Occupational allergy caused by carnation (*Dianthus caryophyllus*). *J Allergy Clin Immunol* 1999; 104:181–5.
2. Grob M, Wuthrich B. Occupational allergy to the umbrella tree (*Schefflera*). *Allergy* 1998; 53:1008–9.
3. deJong NW, Vermeulen AM, Gerth van Wijk R, de Groot H. Occupational allergy caused by flowers. *Allergy* 1998; 53:204–9.
4. Ellenhorn MJ, Barceloux DG. *Medical toxicology*. New York: Elsevier, 1988:1299–304.
5. Karlberg AT, Linden C. Colophony (rosin) in newspapers may contribute to hand eczema. *Br J Dermatol* 1992; 126:161–5.
6. Hay KD, Greig DE. Propolis allergy: a cause of oral mucositis with ulceration. *Oral Surg Oral Med Oral Pathol* 1990; 70:584–6.
7. Van der Wegen-Keijser MH, Bruynzeel DP. Allergy to cane reed in a saxophonist. *Contact Dermatitis* 1991; 25:268.
8. Padoan SM, Petterson A, Svensson A. Olive oil as a cause of contact allergy in patients with venous eczema, and occupationally. *Contact Dermatitis* 1990; 23:73–6.
9. Lovell CR, Rowan M. Dandelion dermatitis. *Contact Dermatitis* 1991; 25:185–8.
10. Gleich GJ, Welsh PXV, Yunginger JW, et al. Allergy to tobacco: an occupational allergy. *N Engl J Med* 1980; 302:617–9.
11. Treudler R, Tebbe B, Orfanos CE. Coexistence of type I and type IV sensitization in occupational coffee allergy. *Contact Dermatitis* 1997; 36:109.
12. Marks GB, Salome SM, Woodcock AJ. Asthma and allergy associated with occupational exposure to ispaghula and senna products in a pharmaceutical work force. *Am Rev Respir Dis* 1991; R4:1065–9.
13. Poshkin MM. Ipecac sensitization and bronchial asthma: report of a case. *JAMA* 1920; 76:1133.
14. O'Neil C. Occupational respiratory diseases resulting from exposure to eggs, honey, spices, and mushrooms. *Allergy Proc* 1990; 11:69–70.
15. Kraut A, Peng Z, Becker NB, Warren CPW. Christmas candy maker's asthma. IgG5-mediated pectin allergy. *Chest* 1992; 102:1605–7.
16. Anibarro B, Fontela JL, De La Hoz F. Occupational asthma induced by garlic dust. *J Allergy Clin Immunol* 1997; 100:73–738.
17. O'Holleran MT. Byssinoses and tobacco related asthma. In: Bardana EJ Jr Montanaro A, O'Holleran MT, eds. *Occupational asthma*. Philadelphia: Hanley & Belfus, 1992: 77–85.
18. Larese F, Fiorito A, Casasola F, et al. Sensitization to green coffee beans and work-related allergic symptoms in coffee workers. *Am J Ind Med* 1998; 34:623–7.

SHELLFISH AND OTHER MARINE INVERTEBRATES

Common names: Crustacean, crab, lobster, shellfish, prawn

Occupational setting

Food-processing facilities that handle prawns, lobster, crabs and other shellfish are a potential source of exposure to crustacean-derived allergens. Oyster farming may also cause exposure to invertebrate allergens. Because of the diversity of sea animals associated with occupational and non-occupational allergy, any setting where marine animals are handled may pose risks from exposure. Exposure to horseshoe crab-derived antigen may occur in laboratory settings where assays for bacterial endotoxins are performed.

Exposure (route)

Inhalation of allergenic components of shellfish may result in respiratory sensitization. Allergens become airborne in food-processing facilities by aerosolization or by being carried along with steam and water vapor. Dust from the processing of dried products may also cause inhalant allergy.

Pathobiology

Shellfish, such as molluscs and crustaceans, are aquatic invertebrate animals with a shell or exoskeleton. Crustaceans such as crabs and lobsters are arthropods. A variety of other marine invertebrates may cause illnesses similar to those seen with shellfish.

Allergy to marine invertebrates typically results from an IgE-mediated immediate hypersensitivity. As with other IgE-mediated respiratory allergies, these disorders are manifested by watery, itchy eyes and nose, and sneezing; less frequently, they produce cough and wheezing, with shortness of breath.

Many marine species have allergenic components. Occupational allergy has been reported from exposure to marine sponge,[1] clam,[2] abalone,[3] brine shrimp, and daphnia. Sea squirt-induced asthma has been associated with oyster farming.[4] *Limulus* amoebocyte lysate (LAL), a horseshoe crab-derived product, used in a laboratory assay, has been reported to cause occupational allergy.[5] Cross-reactivity between shrimp, crab, lobster and crayfish antigens has been documented.[6]

Diagnosis

The presence of respiratory allergic symptoms in temporal association with exposure to marine animal products suggests the diagnosis of occupational allergy. A thorough history, including qualitative assessment of exposure and the relationship of symptoms to exposure, is critical. Physical examination may support the diagnosis of occupational allergy if mucosal change or wheezing is found. Removal from the workplace with resolution of symptoms and rechallenge with work can provide sufficient evidence for a working diagnosis.

Immunologic studies such as skin prick testing may be employed. RAST is available for crab, shrimp, chironomids, lobster,[5] and sea squirt. Skin prick testing has been used with horseshoe crab[5] and others. Bronchial provocation may be used to demonstrate asthma in the presence of the suspected allergen.

Treatment

There is no specific treatment for marine animal allergy. Definitive treatment involves removal of the patient from exposure to the allergen. Conventional allergy therapies such as antihistamines and bronchodilators may be used. Immunotherapy has been used in treating sea squirt asthma with reported success.[4]

Medical surveillance

Surveillance methods should be directed toward detecting respiratory symptoms. History and pulmonary function testing may be useful.

Prevention

Inhalation of dusts and vapors containing marine animal allergen should be avoided. Engineering and administrative controls can help prevent personnel exposure. Personal protective equipment, including respirators, may be useful.

REFERENCES

1. Baldo BA, Krils S, Taylor KM. IgE mediated acute asthma following inhalation of a powdered marine sponge. *Clin Allergy* 1982; 12:171–86.
2. Karlin JM. Occupational asthma to clam's liver extract. *J Allergy Clin Immunol* 1979; 63:197.
3. Clarke PS. Immediate respiratory hypersensitivity to abalone. *Med J Aust* 1979; 1:623.
4. Montanaro A. Asthma in the food industry. In: Bardana EJ Jr, Montanaro A, O'Hollarcn MT, eds. *Occupational asthma*. Philadelphia: Hanley & Belfus, 1992:125–30.
5. Ebner C, Kraft D, Prasch F, Steiner R, Ebner H. Type I allergy induced by limulus amoebocyte lysate (LAL). *Clin Exp Allergy* 1992; 22: 417–9.
6. Lehrer SB. Hypersensitivity reactions in seafood workers. *Allergy Proc* 1990; 11:69–70.

WHEAT FLOUR AND EGG

Common names for disease: Baker's asthma, wheat flour asthma, egg allergy

Occupational setting

Baker's asthma has been associated with occupational wheat flour exposure in bakers since the 1700s. Other settings in which occupational grain dust allergy occurs include flour-mill work, grain handling (including loading, unloading, and storage operations), pastry factories, cereal factories, and animal feed facilities. Confectionary and baking industry workers also have exposure to egg products.

Exposure (route)

Flour and grain dust, and powdered egg, readily become airborne. Inhalation exposure may result in occupational allergic disorders, including asthma. Extrinsic allergic alveolitis may result from egg allergy.[1]

Pathobiology

Wheat flour is a grass product that contains albumin, globulin, gliadin, and glutenin proteins. Contaminants of and additives to flour are considered separately in the section on grain dust. Egg proteins such as ovalbumin are allergenic.[1]

Repeated inhalation of wheat flour may result in IgE-mediated, immediate-type allergic respiratory disorders in susceptible individuals. These allergic disorders, which may become manifest after months or years of exposure, occur in 10–30% of workers and include rhinoconjunctivitis and asthma.[2] Asthma is typically immediate but may also be delayed. Although >40 wheat antigens have been documented, many of which produce disease, albumin in wheat flour is most closely linked to baker's asthma.[2] Many wheat antigens are insoluble and therefore of questionable significance in baker's asthma. Other allergens associated with flour dust, including egg, molds, fungi, insects, storage mites, pollens, and bacteria, may also cause asthma in flour dust workers. Workers who are skin-tested with combinations of flour and related allergens rather than wheat flour alone have been shown to have a high prevalence of positivity.[3] In addition, cross-reactivity has been reported among wheat, rye, and barley.[4]

The risk of development of wheat flour allergy is related to the intensity and duration of exposure. Working conditions important to the development of allergy include dust concentration, ventilation, lack of engineering controls on machinery, and varying levels of allergen.[5] Rhinitis and conjunctivitis typically precede the development of asthma. The incidence of baker's asthma increases with longer duration of wheat flour exposure.[6] Exposure to wheat flour at levels of 1–2 mg/m^3 results in a significant risk of allergy development.[7] Individuals with atopic characteristics are at increased risk.

Diagnosis

Baker's asthma may be diagnosed through a thorough history and physical examination, along with demonstration of reversible airway obstruction in association with flour dust exposure. Rhinitis, conjunctivitis, cough, wheezing and shortness of breath are typical. Symptoms develop after varying periods of exposure; rhinitis usually precedes asthma. Once sensitized, workers become ill within minutes to hours after starting the workday and are generally well during weekends and holidays. Late responses often occur, typically 4–8 hours after exposure; antecedent rhinorrhea, congestion and sneezing are typical. Atopic individuals are at higher risk of developing disease.

Workplace physical examination and serial peak flow testing may be useful in documenting reactive airways. Diagnostic tests include skin prick testing with commercially available wheat flour standardized extract, specific IgE RAST, leukocyte histamine release, and inhalation challenge with wheat flour extract. In one study, 25% of wheat flour workers with occupational rhinitis or asthma had positive skin prick tests and positive RAST to wheat flour. It has been suggested that RAST may be more reliable than skin testing for flour allergy,[8] although this conclusion warrants further study. Those with positive prick tests are more likely to have reactive airways. Because bakers may have hypersensitivity to other flour-associated antigens, care must be taken not to miss multiple hypersensitivities. In one study, five patients with wheat flour hypersensitivity also had allergy to alpha-amylase or cellulase.[9] Nasal provocation has been used to reproduce symptoms from egg allergy.[1] Nasal challenge test has been validated for use in flour-allergic subjects through the measurement of increases in eosinophil and basophil numbers, albumin/total protein ratio, eosinophil cationic protein, and tryptase levels.[10]

Treatment

Definitive treatment of wheat flour-related allergic disorders requires prevention of further exposure. Since changing occupations is often not possible, affected workers often rely on palliative therapy. Cromolyn sodium or salmeterol pretreatment and albuterol treatment of symptoms may allow patients to continue with work. Respirator use may reduce work-related symptoms. Immunotherapy with weekly wheat flour extract injections has been shown to produce a decrease in hyperresponsiveness to methacholine, skin sensitivity, and specific IgE to wheat flour, as well as subjective improvement in symptoms. When considering immunotherapy, all potential hypersensitivities, such as enzymes, mites, and multiple grains, should be investigated and ruled out or included in therapy.

Medical surveillance

Surveillance for baker's asthma should include a periodic detailed respiratory history that may be supplemented by pulmonary function testing. Although the relationship of positive skin prick tests to the development of asthma is not known, consideration should be given to the use of skin prick tests as a marker of exposure. An increase in the prevalence of positive skin tests among bakers from 9% initially to >30% by the 5th year has been documented.[11] Some of these workers reverted to negative within 12 months.

Prevention

The demonstration of exposure–response relationships in baker's asthma indicates that preventive efforts are worthwhile.[12] Avoiding inhalation of dusts can prevent occupational allergy in flour workers. Engineering controls can help to ensure proper dust containment and exhaust ventilation. As with any inhalation exposure, respirator use will decrease the dose of antigen received. Tobacco use may reduce the latency for allergy development and should be avoided.[13]

REFERENCES

1. Valero A, Lluch M, Amat P, Serra E, Malet A. Occupational egg allergy in confectionary workers. *Allergy* 1996; 51: 588–92.
2. O'Holleran MT. Baker's asthma and reactions secondary to soybean and grain dust. In: Bardana Jr., EJ, Montanaro A, O'Holleran MT, eds., *Occupational Asthma*. Philadelphia: Hanley and Belfus. 1992:107–16.
3. Bakhri ZI. Causes of hypersensitivity reactions in flour mill workers in Sudan. *Occup Med* 1992; 42:149–54.
4. Baldo B, Krilis S, Wrigley CS. Hypersensitivity to inhaled flour allergens. *Allergy* 1980; 35:45.
5. Thiel H, Ulmer WT. Bakers' asthma: development and possibility for treatment. *Chest* 1980; 78:400–4.
6. Prichard MG, et al. Skin test and RAST responses to wheat and common allergens and respiratory disease in bakers. *Clin Allergy* 1985; 15:203–10.
7. Goehte CJ, Wieslander G, Ancker K, Forscheck M. Buckwheat allergy: health food, an inhalation health risk. *Allergy* 1983; 38:155–9.
8. Sutton R, Skerritt JH, Baldo BA, Wrigley CW. The diversity of allergens involved in bakers' asthma. *Clin Allergy* 1984; 14:93–107.
9. Quirce S, et al. Respiratory allergy to *Aspergillus*-derived enzymes in bakers' asthma. *J Allergy Clin Immunol* 1992; 90:970–8.
10. Gorski P, Krakowiak A, Pazdrak K, Palczynski C, Ruta U, Walusiak J. Nasal challenge test in the diagnosis of allergic respiratory diseases in subjects occupationally exposed to a high molecular allergen (flour). *Occup Med* 1998; 48:91–7.
11. Herxheimer H. The skin sensitivity to flour of bakers' apprentices. *Acta Allergol* 1973; 28:42.
12. Houba R, Doekes G, Heederik D. Occupational respiratory allergy in bakery workers: a review of the literature. *Am J Ind Med* 1998; 34:529–46.
13. Armentia A, et al. Bakers' asthma: prevalence and evaluation of immunotherapy with a wheat flour extract. *Ann Allergy* 1990; 65: 265–72.

29

LATEX

Charles C. Goodno, M.D., M.P.H., and Carol A. Epling, M.D., M.S.P.H.

OCCUPATIONAL SETTING

Contemporary awareness of the problem of occupational latex hypersensitivity (LH) began in 1979 with Nutter's report of a case of contact urticaria produced by latex gloves.[1] Over the next 10 years there were a few literature reports of occupational latex allergies, but the problem remained a minor curiosity to occupational physicians. About 1990, however, the trickle of reports expanded into a steady stream, and LH has become recognized as a significant occupational problem, particularly for healthcare workers.

Various theories have been advanced to explain this phenomenon. Perhaps the most plausible is that occupational exposure to latex antigens was low until the implementation of universal precautions in the late 1980s prompted an increase of more than 10-fold in the use of natural rubber latex gloves. As the usage increased, manufacturers reduced quality control in an attempt to keep pace with demand, and new manufacturers with little experience entered the market. This led to a marked rise in the antigen content of latex gloves. The juxtaposition of elevated antigen content and increased wear-time has probably increased the latex antigen exposure of the typical healthcare worker by more than an order of magnitude. Meanwhile, the apparent prevalence of latex-related problems has risen markedly. While case reporting may have increased somewhat due to increasing physician awareness, there can be little doubt that there have been dramatic increases in both the prevalence and severity of latex-related occupational illness.

The occupational population at risk for latex allergy comprises primarily end-users of latex gloves, predominantly healthcare workers. As a group, healthcare workers account for virtually all the occupational epidemiology studies of latex allergy. However, numerous other groups have significant occupational exposure to latex. These include morticians,

Physical and Biological Hazards of the Workplace, Second Edition, Edited by Peter H. Wald and Gregg M. Stave
ISBN 0-471-38647-2

hairdressers, greenhouse workers, and a variety of manufacturing workers. Interestingly, the literature contains few reports of latex allergy associated with the manufacture of latex products. This may be related to the movement of much of latex product manufacturing to overseas locations and to relatively low exposure to latex aerosols during production. As a result, this chapter will focus on latex allergy and latex glove dermatitis in healthcare workers. It should also be noted that patients with spina bifida also have a significant risk of latex allergy, presumably because of repeated mucous membrane exposures to latex from operations early in life.

EXPOSURE (ROUTE)

Exposure to latex antigens typically occurs through skin contact and inhalation. Although ingestion is a theoretical possibility, there have been no reports of occupational exposure by this route. Skin exposure occurs primarily by direct contact with latex products such as gloves. Latex gloves contain highly variable amounts of extractable protein, which acts as the antigenic material. The wearer's skin moisture provides an effective mechanism for extraction of these antigens. Glove powder also provides an effective vehicle for the transfer of latex antigen from the glove matrix to the wearer's skin. Although the proteins that comprise the latex antigen are non-volatile, they bind to glove powder. When the glove is donned or removed, a puff of glove powder with bound antigen is released into the air as an aerosol which can be inhaled by the wearer or nearby coworkers. When this process is extrapolated to an area such as an operating room or a laboratory, where dozens of workers may don and remove dozens of pairs of gloves each day, the potential for inhalation exposure can be magnified. Thus, glove powder can act as a classical vector. Latex antigen exposure can be estimated by the latex content of gloves, the duration of wear, the route of exposure (with mucosal and parenteral routes thought to be most significant), and the average daily number of glove pairs worn by the worker and coworkers. Considering these variables, the prevalence of LH in subsets of the population seems to correlate roughly with the intensity of latex antigen exposure (Table 29.1)

Table 29.1 Prevalence of latex hypersensitivity according to exposure group (note that reported prevalence varies with measurement technique as well as population).

Population	Prevalence (%)
General population	0.5–1.0
Healthcare workers	3–15
Clinical laboratory workers	10–20
Spina bifida patients	35–65

A recent prevalence study by Page et al,[2] however, raises some questions about the association between latex exposure and the prevalence of LH. These investigators found equal prevalence rates for LH in exposed healthcare workers and non-exposed controls (about 6% each). Misclassification bias and dropout rates are a potential concern in this study. Workers classified as non-exposed might have had exposure as a result of airborne latex antigens from either occupational or environmental sources. Workers classified as exposed might have had low-dose exposure as a result of using powder-free low-protein latex gloves. Moreover, we know that some workers leave the healthcare field as a result of LH. Since these effects might combine to reduce the power of a study, this type of negative result seems less compelling than the positive results in the literature.

PATHOBIOLOGY

The raw material for natural rubber latex is the milky sap of the rubber tree *Hevea brasiliensis*. This sap is an aqueous emulsion of isoprene droplets (which polymerize to form natural rubber) and a variety of other substances, including over 250 distinct proteins, some 60 of which have been shown to be antigenic. As the sap is processed, various stabilizers and vulcanizing agents are added. These include thiurams, mercaptobenzothiazoles, carbamates, and phenol derivatives. When the emulsion is deposited on hand-shaped ceramic forms and polymerized to form a glove, some of the proteins are adsorbed and trapped in the matrix. Depending on the manufacturing process and the subsequent washing steps, the protein content of gloves can range from undetectable to over 1000 micrograms per gram of finished glove.[3]

Latex gloves can produce three specific types of problems: irritant contact dermatitis, allergic contact dermatitis, and LH. In addition, all types of gloves can produce nonspecific irritation in the form of hyperhidrosis and maceration of the stratum corneum from accumulation of heat and moisture inside the glove. The vast majority of latex glove problems fall into the category of irritant contact dermatitis (ICD), which is produced by direct local tissue irritation without activation of the immune system. ICD may be immediate or slowly progressive and is characterized by erythema limited to skin areas in direct contact with gloves. It does not spread to non-contact areas or produce more serious sequelae.

Allergic contact dermatitis (ACD) is a type IV (cell-mediated) immune reaction and is typically characterized by a delay of one to several days between exposure and development of local erythema. With chronic glove wear, however, the delay may be less apparent. Although latex proteins may occasionally incite ACD, the antigens for this reaction are usually the low-molecular-weight additives such as thiurams and carbamates. ACD is usually limited to areas in direct contact with gloves, but the rash can occasionally generalize. Frequently, ACD and ICD are difficult to distinguish clinically.

LH is a type I (IgE-mediated) immune reaction and typically occurs as an immediate reaction (within minutes of exposure). LH may occur in a spectrum of presentations (listed in rough order of increasing severity):

- positive skin prick test in an asymptomatic patient
- contact urticaria
- rhinitis and conjunctivitis
- angioedema
- asthma
- anaphylaxis.

Since it is somewhat unusual to see serious reactions such as asthma and anaphylaxis as the initial presentations of LH, some investigators believe that susceptible individuals who experience ongoing exposure to latex will tend to show a progression toward serious manifestations with time. In this respect, LH appears to act like other IgE-mediated responses to protein antigens.

While LH is the least common of latex glove problems, it is potentially the most serious and expensive. When LH is defined by positive skin prick testing, which is the current gold standard, the prevalence is between 3% and 15% for healthcare workers, in general, and possibly higher for laboratory workers. Unfortunately, skin prick testing simply identifies workers with sufficient latex-specific IgE to produce a skin reaction but does not provide a clear estimate of the prevalence of symptomatic disease. In an attempt to circumvent this problem, Horwitz and Arvey[4] used workers' compensation claims as an index of latex-induced disease. Their results showed a strikingly lower rate of latex-related claims than might be expected from the reported prevalence of LH. Unfortunately, these results seem to suffer from potentially serious limitations of information bias. At present, insufficient studies have been done to define the extent of symptomatic disease in workers with positive latex skin prick tests.

The natural history of LH is poorly understood. The only incidence study published to date found an annual incidence of 1% among hospital-based healthcare workers (LH defined as positive latex skin prick test).[5] This annual incidence would be sufficient to account for the current prevalence of 3–15%, assuming low attrition of workers and significantly increased exposure around the time of promulgation of the Bloodborne Pathogens Standard. Although we do not know how the incidence of LH will ultimately translate into symptomatic latex allergy, there is a possibility that current levels of latex-related symptoms are the beginning of an epidemic.

Several studies have detected a correlation between airborne latex antigen levels and measures of LH.[6–8] The corollary of this is that the indices of LH appear to decrease when airborne latex antigen is reduced. These indices include severity of rhinoconjunctivitis, latex-specific IgE levels and frequency of asthma attacks. Equally importantly, the study of Allmers et al showed that removal of powdered latex gloves led to dramatic reduction of airborne latex antigen levels within a few days.[8]

An association between LH and allergies to certain foods has been observed by a number of investigators. The most prominent of these associations are listed in Table 29.2.[9] Until recently, there has not been a clear biochemical basis for these associations. The studies of Mikkola et al and Posch et al, however, provide a model for such cross-reactions.[10,11]

Table 29.2 Major food allergies associated with latex hypersensitivity and their prevalence among latex-hypersensitive patients. Adapted from Kim and Hussain.[9]

Food	Prevalence (%)
Banana	18
Avocado	16
Kiwi fruit	12
Shellfish	12
Fish	8
Tomato	6
Watermelon, peach, carrot (each)	4
Apple, chestnut, cherry, coconut, apricot, strawberry, loquat (each)	2

Using sera from banana- and avocado-sensitized latex-hypersensitive patients, respectively, they identified 31–33 kDa proteins from banana and avocado that cross-react with hevein (one of the major latex antigens). Amino acid sequencing revealed that these proteins (class I endochitinases) contain a domain of sequence homology with hevein, which explains the IgE cross-reactivity. Since endochitinases are common plant proteins, this type of cross-reactivity may offer a general explanation for associations between LH and allergies to fruits and vegetables. Other associations with LH include an atopic history, multiple operations, mucosal exposure to latex, and spina bifida.

DIAGNOSIS

Non-specific glove irritation is generally characterized by mild local itching or burning in the distribution of glove contact that resolves rapidly with cessation of glove wear. The skin appears normal or pale with possible mild edema and mild fissuring. Excoriation and lichenification are minimal. Usually this can be distinguished clinically from ICD or ACD.

ICD and ACD both present with local itching, burning, and erythema in the glove distribution. Marked erythema, vesicles and weeping may be present in severe cases. In chronic cases, fissuring and lichenification may be prominent. ICD and ACD can sometimes be distinguished on the basis of the history of an early reaction in the case of ICD and a delayed reaction for ACD. However, most workers are unable to give a sufficiently detailed history to allow the distinction to be made with confidence. Rashes that begin in the glove distribution but then generalize are not ICD. Both ACD and LH should be considered in such cases. Patch testing with a kit of typical latex additives may clarify the diagnosis by confirming ACD.

LH, like all allergic reactions, is primarily a clinical diagnosis. This is illustrated by analysis of the results of the Multi-Center Latex Skin Testing Task Force.[12] These results show that a clinical evaluation provides about 90% sensitivity and specificity for diagnosis of LH (defined by subsequent positive skin prick test). In the evaluation, historical features of exposure, proximate symptoms, nature of symptoms and family history are most important. Reported symptoms should fit into a recognizable pattern of type I reactions (rhinitis/conjunctivitis, urticaria, angioedema, asthma, or anaphylaxis). A useful early sign of LH is development of a papular rash consistent with contact urticaria. Extension of the papular rash beyond the area of the glove is further evidence of LH. Physical examination should be used to provide objective confirmation of signs and symptoms but may be unremarkable in the asymptomatic patient. In the occupational setting, consistent provocation of typical symptoms by known work exposure to latex is frequently sufficient to make the diagnosis. In the case of latex-induced asthma, serial peak expiratory flow measurements are sometimes useful.

Laboratory testing is useful but primarily serves to confirm the clinical evaluation. Probably the ultimate diagnostic test is the wear and puff test.[13] This involves starting with a finger cut from a high-antigen-content latex glove and wearing progressively larger portions until the entire glove is worn. Development of symptoms is assessed at each stage. If the glove can be worn without symptoms, a glove is inflated like a balloon, and the air inside (with suspended glove powder) is exhausted into the patient's breathing zone. While highly sensitive, this test can provoke severe asthma or anaphylaxis in susceptible individuals and is not recommended except at specialized centers where occupational aerosol challenge testing is routinely performed.

Skin prick testing is a more practical confirmatory test. A standardized latex skin testing antigen manufactured by Greer Laboratories was evaluated by the Multi-Center Latex Skin Testing Task Force and found to be both sensitive (sensitivity = 95%), specific (specificity = about 99%), and safe (no epinephrine-requiring reactions).[12] This antigen preparation promises to become the first standardized latex preparation licensed by the FDA. Since its licensure is still pending, skin testing for LH in the USA is presently limited to either experimental or foreign (e.g. Canadian) antigen preparations.

In vitro testing for latex-specific IgE has also been used in evaluation of LH. Currently, three FDA-licensed test kits are commercially available in the USA. The characteristics of two of these kits (Diagnostic Products Corp. AlaSTAT and Pharmacia-Upjohn CAP) are comparable, with sensitivity in the range of 75% and specificity in the range of 97%.[14] The third test (Hycor HYTEC) has a sensitivity of about 92% but a specificity of only 73%. As a result, we are confronted with a peculiar situation where the laboratory test may have a lower sensitivity and specificity than the clinical evaluation it is intended to confirm. Accordingly, the role of in vitro latex IgE testing in clinical decision-making is murky. The interested reader is referred to Hamilton et al, for a discussion of the performance of in vitro latex testing.[14]

TREATMENT

Non-specific glove irritation usually responds to a glove holiday and/or use of glove liners made from materials such as white cotton or Dermapore®. Liners should be changed whenever they become damp. ICD and ACD will sometimes respond to the same regimen, but the ultimate treatment is substituting a different glove material for the one that provokes the problem. In the acute phase, a glove holiday combined with treatment using a low-to medium-potency topical corticosteroid is beneficial. This may be combined with astringent soaks in the case of weeping lesions or emollient preparations in the case of dryness. The use of barrier creams is not a suitable alternative to removal from latex gloves, and some authors speculate that creams might act as a vehicle for extraction of antigens from latex gloves.

Symptomatic LH poses a special problem, since the definitive treatment is complete removal from exposure to latex antigens. Few medical institutions have achieved truly "latex-safe" status at this point. Thus, sending a latex-hypersensitive worker back to work in a medical institution usually involves a degree of risk. As mentioned in the 1997 NIOSH Alert and confirmed by recent research, one of the most critical issues is whether the workplace is free of powdered latex gloves.[15] Institution-wide substitution of powder-free latex gloves for powdered ones has been reported to reduce the level of airborne latex antigen to undetectable levels. Thus, it seems reasonably safe to send workers with LH back to work in institutions that have made this change. Care must still be exercised with activities that tend to release latex particles from reservoirs such as HVAC filters and plenums. The remaining issue is one of helping the worker avoid direct contact with latex. Given the variety of alternative glove materials, selecting alternative gloves need not be a problem. Avoiding contact with latex-containing supplies and equipment (Table 29.3), however, may require detailed examination of specifications for these items. In addition to these measures, LH workers should be educated to follow the recommendations in Table 29.4.

Table 29.3 Common items containing latex.

Hospital supplies:
Gloves, barium enema catheters, adhesive tape, EKG electrodes, blood pressure cuffs, stethoscopes, anesthesia masks, surgical masks, goggles, catheters, injection ports
Office supplies:
Rubber bands, erasers, adhesives
Household objects:
Automobile tires, elastic waistbands, dishwashing gloves, condoms, diaphragms, balloons, pacifiers, baby bottle nipples

Table 29.4 Recommendations for healthcare workers with latex hypersensitivity.

1. Inform employer, healthcare providers and family members of latex allergy
2. Identify latex-containing products and avoid contact
3. Wear medical alert bracelet with latex allergy notation
4. Carry and understand use of an epinephrine injector (e.g. Epi-Pen)
5. Notify employer's occupational health service of changes in latex-related symptoms

MEDICAL SURVEILLANCE

Since the prevalence of LH is likely to rise, and early diagnosis is almost certainly beneficial, LH is suitable for a medical surveillance program. Screening can be implemented with a simple symptom inventory for exposed workers, since history is the key feature in LH diagnosis. Workers classified as high risk by screening might undergo subsequent clinical history and physical examination, serial peak expiratory flow measurements, or nonspecific inhalation challenge testing for confirmation. Assuming the existence of an appropriate index of suspicion, clinical evaluation alone is probably sufficient to make the diagnosis in many cases. When a standardized skin-testing antigen becomes available, skin prick testing will be the confirmatory test of choice in view of its sensitivity, specificity, and safety. For latex-hypersensitive workers who return to work with some degree of ongoing latex exposure, periodic follow-up in a medical surveillance program is indicated to detect development of new symptoms. If new symptoms occur, further intervention to eliminate latex exposure is vital.

PREVENTION

The operating principle of LH prevention is that prolonged exposure produces sensitization and that ongoing exposure after sensitization increases the likelihood of development of serious symptoms. Assuming a 1% annual incidence of LH in healthcare workers, the current prevalence of serious latex problems might be expected to double in 5–10 years (unless latex exposure is reduced). Thus, primary prevention should be aimed at avoidance of high-level and long-term latex exposure in all workers. Careful selection of barrier protection according to risk for contact with infectious materials may reduce the use of latex gloves among workers who are unlikely to encounter infectious substances (i.e. food services, maintenance and housekeeping). Strict avoidance would mean eventually converting from latex to non-latex gloves and perhaps making substitutions for latex articles such as Foley catheters and tourniquets that are routinely handled by workers. At present, the cost of substituting non-latex gloves could represent a doubling of glove cost for medical institutions. Many institutions find this too great a financial sacrifice. In addition, wearer satisfaction and barrier properties of non-latex gloves are significant issues for workers who wear gloves for prolonged periods during direct contact with blood. A compromise solution is to convert to low-antigen powder-free latex gloves as suggested by the NIOSH Alert.[15] This produces a more modest increase in glove cost but provides a significant measure of protection from inhalational exposure for latex-hypersensitive individuals. The study of Phillips et al provides an economic analysis of the increased glove cost versus disability cost associated with continued use of powdered latex gloves.[16]

A powder-free latex workplace also affords a measure of secondary prevention for those who already have LH. In this situation, workers with LH, once identified, can return to work wearing non-latex gloves while exercising caution in tasks that involve handling latex-containing supplies and equipment. One can be reasonably confident that these affected workers will have minimal latex aerosol exposure in this situation. Close medical follow-up, however, is indicated to monitor for possible progression of symptoms.

REFERENCES

1. Nutter AF. Contact urticaria to rubber. *Br J Dermatol* 1979; 101:597–8.
2. Page EH, Esswein EJ, Peterson MR, Lewis DM, Bledsoe TA. Natural rubber latex: Glove use, sensitization, and airborne and latent dust concentrations at a Denver hospital. *J Occup Environ Med* 2000; 42:613–20.
3. Williams PB, Halsey JF. Endotoxin as a factor in adverse reactions to latex gloves. *Ann Allergy Asthma Immunol* 1997; 79:303–10.
4. Horwitz IB, Arvey RD. Workers' compensation claims from latex glove use: a longitudinal analysis of Minnesota data from 1988 to 1997. *J Occup Environ Med* 2000; 42:932–8.
5. Sussman GL, Liss GM, Deal K, et al. Incidence of latex sensitization among latex glove users. *J Allergy Clin Immunol* 1998; 101:171–8.
6. Swanson MC, Bubak ME, Hunt LW, Yunginger JW, Warner MA, Reed CE. Quantification of occupational latex aeroallergens in a medical center. *J Allergy Clin Immunol* 1994; 94:445–51.
7. Tarlo SM, Sussman G, Contala A, Swanson MC. Control of airborne latex by use of powder-free latex gloves. *J Allergy Clin Immunol* 1994; 93:985–9.
8. Allmers H, Brehler R, Chen Z, Raulf-Heimsoth M, Fels H, Bair X. Reduction of latex aeroallergens and latex-specific IgE antibodies in sensitized workers after removal of powdered natural rubber latex gloves in a hospital. *J Allergy Clin Immunol* 1998; 102: 841–5.
9. Kim KT, Hussain H. Prevalence of food allergy in 137 latex-allergic patients. *Allergy Asthma Proc* 1999; 20:95–7.
10. Mikkola JH, Alenius H, Kalkkinen N, Turjanmaa K, Palosuo T, Reunala T. Hevein-like protein domains as a possible cause for allergen cross-reactivity between latex and banana. *J Allergy Clin Immunol* 1998; 102:1005–12.
11. Posch A, Wheeler CH, Chen Z, et al. Class 1 endochitinase containing a hevein domain is the causative allergen in latex-associated

avocado allergy. *Clin Exp Allergy* 1999; 29:667–72.

12. Hamilton RG, Adkinson NF Jr. The Multi-Center Latex Skin Testing Study Task Force. Diagnosis of natural rubber latex allergy: multi-center latex skin testing efficacy study. *J Allergy Clin Immunol* 1998; 102:482–90.
13. Hamilton RG, Adkinson NF Jr. Validation of the latex glove provocation procedure in latex-allergic subjects. *Ann Allergy Asthma Immunol* 1997; 79:266–72.
14. Hamilton RG, Biagini RE, Krieg EF. Diagnostic performance of food and drug administration-cleared serologic assays for natural rubber latex-specific IgE antibody. The multi-center latex skin testing study task force. *J Allergy Clin Immunol* 1999; 103:925–30.
15. National Institute for Occupational Safety and Health. *Preventing allergic reactions to natural rubber latex in the workplace*. NIOSH publication no. 97–135. Cincinnati: Government Printing Office, 1997.
16. Phillips VL, Goodrich MA, Sullivan TJ. Health care worker disability due to latex allergy and asthma: A cost analysis. *Am J Pub Health* 1999; 89:1024–8.

FURTHER READING

Goodno LE. *Prevalence of latex allergy and latex sensitivity in healthcare workers.* Master's thesis. University of North Carolina, 1999.

Laoprasert N, Swanson MC, Jones RT, Schroeder DR, Yunginger JW. Inhalation challenge testing of latex-sensitive health care workers and the effectiveness of laminar flow HEPA-filtered helmets in reducing rhinoconjunctival and asthmatic reactions. *J Allergy Clin Immunol* 1998; 102: 998–1004.

SMTL glove protein information sheet. Available at URL: http://www.smtl.co.uk/MDRC/Gloves/smtl-latex-protein-summary.html.

Sussman GL, Beezhold DH. Allergy to latex rubber. *Ann Int Med* 1995; 122:43–6.

Vandenplas O, Delwiche J-P, Evrard G, et al. Prevalence of occupational asthma due to latex among hospital personnel. *Am J Respir Crit Care Med* 1995; 151:54–60.

Warshaw EM. Latex allergy. *J Am Acad Dermatol* 1998; 39:1–24.

30

MALIGNANT CELLS

Aubrey K. Miller, M.D., M.P.H.

OCCUPATIONAL SETTING

Workers at risk include: (1) laboratory workers who handle malignant cells during the performance of in vitro and in vivo research, (2) histology and pathology workers involved in the preparation and processing of neoplastic tissues, (3) medical and nursing staff involved in surgical procedures (i.e. aspirates, biopsies, resections) on cancer patients; (4) surgical scrub personnel handling sharp instruments contaminated with malignant cells; and (5) housekeeping workers (especially those in cancer research areas) exposed to sharp objects contaminated with malignant cells.

EXPOSURE (ROUTE)

The occupational risk of cancer occurring in humans from exposure to malignant cells is not well recognized, owing to the few cases reported in the medical literature. Based on these reports, the most likely route of occupational transmission involves needlestick or sharp object injuries whereby malignant cells are cutaneously injected or possibly implanted into an open wound. This risk is best understood by the well-described occurrence of occupational transmission of infectious diseases such as HIV, hepatitis B and hepatitis C via needlesticks or other sharp objects (i.e. surgical instruments, histologic tissue cutters, broken capillary tubes and pipettes).

There are currently more than 8 million healthcare workers in the USA in hospitals and other healthcare settings. While precise data are not available with respect to the actual number of annual needlestick injuries or other percutaneous injuries among healthcare workers, it is estimated that 600 000–800 000 occur annually and that half of these go unreported.[1] It is estimated that as many as 2800 injuries may occur each year from handling glass capillary tubes.[2]

Physical and Biological Hazards of the Workplace, Second Edition, Edited by Peter H. Wald and Gregg M. Stave
ISBN 0-471-38647-2

A review of studies reporting needlestick injuries found that 34–50% of health care workers were injured and that 10–70% of those injuries were due to recapping of needles.[3] Studies of hospital workers have shown that the highest incidence of needlestick injuries occurs in housekeeping personnel (during trash disposal) and laboratory and nursing personnel (during needle disposal or recapping).[4,5] Pathologists and surgeons have also been shown to be at increased risk for cutaneous injuries from sharp instruments and needlesticks (especially involving the distal fingers of the non-dominant hand) during operative procedures.[6,7] Although the incidence of cutaneous injuries resulting from sharps contaminated with viable cancer cells is unknown, it probably represents only a small fraction of the cutaneous injuries incurred by potentially exposed workers.

PATHOBIOLOGY

Transplantation of foreign human tissue to a healthy recipient normally leads to an immune response resulting in destruction of the transplanted tissue (rejection).[8] Southam et al,[9,10] showed that normal recipients given subcutaneous injections of human cancer cells responded with a marked local inflammatory reaction and a rapid complete regression of the cancer implants within 3–4 weeks. In contrast, cancer cell injections given to advanced cancer patients showed little or no acute inflammatory reaction; the cancer cells typically grew for 3 weeks or longer before regression, and in some recipients growth continued beyond 6 weeks.[10] One recipient exhibited local recurrence of tumor growth even after three excisional biopsies, and another recipient had lymph node metastasis.[9] In another study, local cancer growth occurred in two patients who received small allogeneic tumor implants as part of an immunotherapy protocol for advanced cancer.[11]

Scanlon et al[12] reported that some patients with advanced cancer have even tolerated tissue grafts from other animal species. Growth of transplanted cancer cells has also been reported to occur in healthy immunocompetent individuals. In one case, death from metastatic disease was reported in a woman who received a small melanoma graft taken from her daughter as part of an immunotherapy protocol.[12] In another case, a healthy 19-year-old laboratory worker developed an actively growing adenocarcinoma of colonic origin on her hand following a needlestick injury. At the time of the injury, only a small superficial wound was noted, with no apparent injection of the cancer cell suspension. The tumor, which was widely excised after 19 days, showed no evidence of an inflammatory response or necrosis. The worker was noted to be free from recurrence 4 years after the injury.[13]

The occurrence of transplanted cancer cell growth and metastasis in some individuals appears to be related to alterations of immune functioning. Rejection of foreign tissue depends upon recognition of major cell surface histocompatibility (HLA) antigens and involves both cell-mediated and humoral immunities.[8] The most important cell-mediated reactions involve both $CD4^+$ T-helper cells and $CD8^+$ cytotoxic T-cells, which play a crucial role in the recognition of foreign tissue cells and regulation of the immune response.[8] Humoral immune reactions to transplantation antigens appear to be mediated by antibodies formed against foreign class I and class II HLA antigens.[8] Therefore, HIV/AIDS patients, other immunocompromised individuals and those on immunosuppressive medications (i.e. steroids, cyclosporine, azathioprine) may be at increased risk for developing viable neoplasms when exposed to malignant cells. The occurrence of cancerous growth in two apparently healthy immunocompetent adults is not well understood. Immune tolerance, lack of an immune response to specific antigens, under these conditions may be due to certain mechanisms which allow the tumor cells to escape immunosurveillance, such as loss or reduced expression of histocompatibility antigens, shedding or modulation of tumor antigens, and production of immunosuppressive factors.[8]

TREATMENT

Injuries should be medically treated as with other needlestick or cutaneous injuries. In addition, the injury site should be periodically evaluated for any tumor growth for at least the ensuing 3–4 weeks (immunocompromised individuals may require longer follow-up). If tumor growth occurs, wide excision of the tumor with close follow-up should be considered. This treatment was apparently effective in at least one of the reported cases.[13]

MEDICAL SURVEILLANCE

All percutaneous injuries should be handled in accordance with the OSHA Bloodborne Pathogens Standard, which covers workers occupationally exposed to unfixed human tissues or blood products.[14]

PREVENTION

All non-essential sharps should be eliminated where possible, especially in laboratory situations. The risk of needlestick injury can be reduced by eliminating all needle recapping and non-essential unprotected needle use; needleless or protected needle devices should be used where possible.[15] Where the use of needles or sharp instruments is indicated, workers should be trained in the safe techniques for handling and disposal (i.e. using puncture-resistant containers) of these objects. Additionally, workers should be encouraged to report all needlesticks and contaminated cutaneous injuries so that appropriate postexposure treatment can be given and so that the incident can be studied to prevent similar accidents in the future. Further, worker education and training, needle handling and sharps disposal should be conducted in accordance with the OSHA Bloodborne Pathogens Standard and NIOSH recommendations to reduce the likelihood of worker injuries.[1,14–16]

REFERENCES

1. NIOSH. *Preventing needlestick injuries in health care settings*. DHHS (NIOSH) Publication No. 2000–108. Department of Health and Human Services, Public Health Servicc, Centers for Disease Control and Prevention, National Institute for Occupational Safety and Health, 2000.
2. Jagger J, Bentley M, Perry J. Glass capillary tubes: eliminating an unnecessary risk to healthcare workers. *Adv Exp Prev* 1998; 3(5):49–55.
3. Martin LS, Hudson CA, Strine PW. Continued need for strategies to prevent needle-stick injuries and occupational exposures to bloodborne pathogens. *Scand J Work Environ Health* 1992; 18:94–6.
4. McCormick JD, Maki DG. Epidemiology of needles-stick injuries in hospital personnel. *Am J Med* 1981; 70:928–32.
5. Neuberger JS, Harris JA, et al. Incidence of needle-stick injuries in hospital personnel: implications for prevention. *Am J Infect Control* 1984; 12:171–6.
6. O'Brian DS. Patterns of occupational hand injury in pathology. *Arch Pathol Lab Med* 1991; 115:610–3.
7. Tokars JI, Bell DM, et al. Percutaneous injuries during surgical procedures. *JAMA* 1992; 267:2899–904.
8. Cotran RS, Kumar V, Robbins SL. *Robbins pathologic basis of disease*, 5th edn. Philadelphia: WB Saunders Company, 1994, 175–7, 190–7.
9. Southam CM. Homotransplantation of human cell lines. *Bull NY Acad Med* 1958; 34:416–23.
10. Southam CM, Moore AE. Induced immunity to cancer cell homografts in man. *An NY Acad Sci* 1958; 73:635–53.

11. Nadler SH, Moore GE. Immunotherapy of malignant disease. *Arch Surg* 1969; 99:376–81.
12. Scanlon EF, Hawkins RA, et al. Fatal homotransplantated melanoma: a case report. *Cancer* 1965; 18:782–9.
13. Gugal EA, Sanders ME. Needle-stick transmission of human colonic adenocarcinoma. *N Engl J Med* 1986; 315:1487.
14. OSHA Bloodborne Pathogens Standard. Title 29 Code of Federal Regulations, Part 1910.30; 56 Federal Register 64004:1991.
15. NIOSH. *What every worker should know: how to protect yourself from needlestick injuries.* DHHS (NIOSH) Publication No. 2000-135. Department of Health and Human Services, Public Health Service, Centers for Disease Control and Prevention, National Institute for Occupational Safety and Health, 2000.
16. NIOSH. *Guidelines for protecting the safety and health of health care workers.* DHHS (NIOSH) Publication No. 88–119. Department of Health and Human Services, Public Health Service, Centers for Disease Control and Prevention, National Institute for Occupational Safety and Health, 1988

31

RECOMBINANT ORGANISMS

Jessica Herzstein, M.D., M.P.H., Ed Fritsch, Ph.D., and John L. Ryan, Ph.D., M.D.

OCCUPATIONAL SETTING

Recombinant organisms are used routinely in the biotechnology industry and in academic laboratories. They are the source of many of the most innovative biopharmaceuticals that have contributed to medical science in the past 20 years. Because of the relative effectiveness of production techniques, certain processes may move from laboratory to pilot plant (often still a laboratory) to production with very little apparent change. Occasionally, relatively small facilities can produce large quantities of complex and previously rare or unattainable products. Increasingly, though, the scale of commercial manufacturing has grown significantly in order to support these valuable products.

The workforce at these facilities may be small but is very highly trained. Most biotechnology research and pilot production personnel have advanced degrees. Commercial facilities tend to be staffed with a mixture of individuals with high school or technical education and managers with advanced degrees. As a group, biotechnology workers are highly invested, both emotionally and economically, in the success of their new enterprises. Long work hours, secrecy concerning processes and product development, an emphasis on rapid progress toward production and very high economic stakes increase the potential for worker hazards.

EXPOSURE (ROUTE)

Exposure to three types of hazards must be considered in biotechnology research and production: recombinant organisms, biological (human- or animal-derived) reagents used for recombinant organism growth and non-biological hazards such as chemicals, radiation and recombinant products. Of these, biological

Physical and Biological Hazards of the Workplace, Second Edition, Edited by Peter H. Wald and Gregg M. Stave
ISBN 0-471-38647-2

reagents and chemicals/radiation pose the greatest concern. Exposure may theoretically occur through ingestion, inhalation, skin and mucous membrane contact, or skin penetration by a contaminated needle or other sharp object.

PATHOBIOLOGY

Recombinant biology is the ability to insert specific pieces of DNA into selected organisms for the purpose of creating desired products. Recombinant organisms are the resulting genetically modified bacteria, fungi, and cells. Humans have selectively bred animal and plant species for desired traits for thousands of years. Recombinant biology extends that activity to the molecular level and can produce very precise outcomes.

More than 80 biopharmaceutical products have now been approved in the USA, EU or both.[1] These fall into multiple broad categories including blood factors, hormones, hematopoietic growth factors, vaccines, monoclonal antibodies, enzymes and chimeric molecules (Table 31.1). Many more products are currently in development, and the human genome and bioinformatics efforts are likely to result in an explosion of new possibilities. In addition, new approaches to the production of biopharmaceutical products such as transgenic animals and gene therapy are aggressively being developed, extending the scope of technologies, facility design and environmental impact. Non-medical products include pesticide-resistant and pest-resistant plants,

Table 31.1 Selected products from recombinant organisms.

Biopharmaceutical class	Examples	Date first approved (EU or USA)
Blood factors	Factor VIII	1993
	Factor IX	1997
	Hirudin	1997
Hormones	Insulin	1982
	Growth hormone	1985
	Glucagon	1998
	Follicle stimulating hormone	1995
Hematopoieic growth factors	Erythropoietin	1989
	Granulocyte colony stimulating factor	1991
	Granulocyte–macrophage colony stimulating	1991
	factor	1995
	Alpha and beta interferon	1992
	Interleukin-2	
Vaccines	Hepatitis B surface antigen	1986
Monoclonal antibodies	Anti-CD-3	1986
	Anti-GII_bIII_a	1994
	Anti-HER 2	1998
	Anti-CD20	1997
Enzymes	DNase	1993
	β-Glucocerebrosidase	1994
Chimeric molecules	IL-2 Diptheria toxin	1999
	TNF Receptor-IgG	1998

Data are from the following sources: http://www.fda.gov, http://www.eudra.org/en_home.htm, http://www.phrma.org

industrial or food-processing enzymes, chemicals, fuels, and foods. In the future, specialty products may appear in clothing, construction, or transportation materials. Even before the public became aware of commercial recombinant technology in the mid-1980s, hundreds of start-up enterprises existed in the USA (and elsewhere). Continued progression from laboratory to pilot plant to commercial products is inevitable for a wide new array of products. As in every new industry, the potential also exists for health and safety hazards.

One of the fastest-growing areas in the biotechnology industry is gene therapy. Gene therapy employs viral and bacterial vectors to deliver specific human genes to patients with hereditary and acquired diseases. Recombinant vectors are often replication-deficient viruses, e.g. retrovirus, adenovirus, or herpes virus. These are used in gene therapy both for protein or enzyme replacement and for oncology. There are serious concerns that these viruses may become replication-competent in the host. Recognizing that large numbers (10^{10} or more) of viruses are often delivered emphasizes the need for caution, because there may be risks to the care-givers in preparation of the therapy or to the patient from the vector carrying the intended therapeutic gene.

Simple infection is the most significant biological hazard for workers. At the research stage, hazardous organisms such as drug-resistant microorganisms, hepatitis viruses and oncogenic viruses are frequently used. Potentially infectious organisms such as adenovirus have specialty uses in the field of gene therapy. Obviously, careful attention to biological containment and good microbiological practice is critical. Commonly, tissue culture work is carried out with non-pathogenic *E. coli*, *Bacillus subtilis*, *Saccharomyces cerevisiae*, *Aspergillus niger* or mammalian cells. Exposure to the relatively non-pathogenic *E. coli* used for much recombinant technology work has not been associated with diarrheal or other gastrointestinal illness. Several studies have demonstrated the lack of colonization of workers with recombinant organisms.[2]

For production of recombinant proteins, the most common infection risk comes from exposure to reagents used to support growth and productivity of the recombinant organism itself. Mammalian cells typically used for production are not dangerous themselves, as non-pathogenic recombinant organisms are used exclusively and they have been extensively tested for a variety of known and possible infectious agents. However, material of animal or human origin, such as raw serum, partially purified blood components (human or bovine serum albumin) or even highly purified reagents from animals (bovine or porcine insulin) are frequently required. The risk of known or previously unknown infectious agents in the source animals or humans must be carefully assessed. The devastating impact of unknown HIV infection in the blood supply on the hemophilia population is ample demonstration of the concern becoming reality. More recently, the apparent emergence of a new variant of a human transmissible spongiform encephalopathy agent, new variant Creutzfeld–Jacob Disease (nvCJD), following from the epidemic of bovine spongiform encephalopathy in the UK, has again highlighted the possibility of infectious risk. Although the exposure of workers to such infectious agents is probably lower than for those receiving the products, the need for appropriate precautions is clear, especially when the volume of materials handled or stored may be significant.

Another potential biological hazard involves special applications of hybridoma technology. Typically, antibodies are produced from two fused murine cells to produce specific antibodies, creating an immortal cell line. Atypically, a theoretical hazard of this activity exists when one of the two cell lines is human, which carries the risk of latent or unapparent viral infection. In vitro human cells, such as lymphocytes, are susceptible to infection with tumorigenic virus. In addition, the murine fusion partner can carry mouse type C leukemia and sarcoma. Although these viruses are typically not infectious of human cells, it is not known whether residence in hybrid cell lines could alter the host range. If so, this may create a new hazard whereby pathogenic animal viruses extend their host range to humans (and vice versa). Studies to date have shown no evidence of clusters of cancer or other diseases or elevated rates of disease in workers.

The pharmaceutical industry creates final recombinant products as well as intermediate biological products that are highly concentrated and physiologically active. Normal physiologic responses to inhaled or absorbed substances may include hormonal and allergic responses. This is the same problem encountered in the conventional pharmaceutical industry.[3] Workers exposed to estrogens, growth hormone or other physiologically active chemicals can be expected to have unintended but predictable responses.

Allergy to the concentrated product has been seen in biotechnologists. Products and intermediate products may elicit immune responses. In addition, product development involves laboratory animal handling, which has its own set of allergic issues. Normal engineering controls, with occasional use of personal protective devices, represent the best means of preventing sensitization or other unintended physiologic responses.

In addition to these concerns surrounding recombinant organisms and their products, other important non-biological hazards exist in the biotechnology industry. These may be of a substantially greater magnitude than in traditional research laboratories. For example, many laboratory research operations use radioisotopes, including tritium (^{3}H), carbon-14 (^{14}C), sulfur-35 (^{35}S), calcium-45 (^{45}Ca), phosphorus-32 (^{32}P), chromium-51 (^{51}Cr), and iodine-125 (^{125}I). Several of these are volatile and therefore quite dangerous. Safe handling and dose reduction techniques should be collaborative efforts between a laboratory radiation safety liaison and a consulting certified health physicist. Given the relatively young age of the biotechnology workforce and the reproductive implications of hazards such as ^{125}I, additional emphasis should be placed on workplace safety. Most monitoring is done by radiation badge, a process comparable to industrial hygiene monitoring. Health physicists should perform thyroid and whole-body scanning when needed.

A surprising number of physical hazards are associated with both research and product preparation in this highly technical industry. Biotechnologists must deal with compressed gas, cryogenics, humidified atmospheres (with associated building biological hazards), high-voltage electricity, as well as some flammable and corrosive materials. Heat and cryogenic burns are a common hazard of laboratory and start-up production facilities, especially if laboratory personnel transport laboratory-scale hoses for production use. Ultraviolet light and associated skin hazard may be encountered in manufacturing suites, research biological safety cabinets, and electrophoresis operations. Laboratory and even manufacturing operations can be hands on, with lots of lifting and frequent exposure to glassware. In industry operations, glassware handling (and cleaning) present opportunities for sharps exposure. For one biotechnology company, lacerations comprised 41% of injuries and back injuries accounted for 93% of lost time.

One largely unappreciated aspect of biotechnology is shiftwork. Relatively small labor forces are engaged in both research and production, yet the cell cycle is continuous in both settings. Round-the-clock care of cultures is dictated by biological realities. Shiftwork may be informal in the laboratory or formal in production facilities, but problems of shifting human sleep cycles are inherent in the production of cell cultures. Young research workers cheerfully accept poorly designed work cycles and do not think of their tasks in terms of shiftwork.

Biotechnology processes require significant chemical use, with the potential for exposure.[4] Classes of chemicals common in biotechnology include culture additives and antibiotics, solvents and extractants (including those used in chromatography and sequencing), and the corrosive additives needed to maintain a working chilled-water facility. Ethidium bromide is a genotoxic compound used for fluorescent staining in sequencing operations. Cell culture additives are frequently mutagenic. The most common acute chemical overexposures have been to acetonitrile, which is used as an extractant. Patients with mild cyanide-like toxicity of acetonitrile have been reported anecdotally in several emergency situations following spills and inappropriate handling during clean-up operations. One potentially hazardous repetitive process involves the use of acrylamide in gel preparations. The purchase of preformed gels can reduce the hazard of handling the neurotoxic bis-acrylamide powder. Finally, many production processes, especially those using *E. coli*, result in a fusion protein as the primary product. The fusion protein requires peptide bond cleavage with a strongly reactive and specific chemical in order to prepare the final product. This cleavage is carried out at large volume with highly reactive chemicals, including hydrofluoric acid. Precautions to prevent both exposure of workers and inadvertent release into the environment must be in place.

MEDICAL SURVEILLANCE

Reduction of health risks in biotechnology is attained through control of work hazard and exposure, which requires in-depth focus on work processes and the methods of containment. Medical surveillance is a strategy for disease prevention that is based on a risk assessment of the jobs performed by the worker.[5] Medical surveillance is usually targeted to specific chronic risks.

Medical surveillance is not routinely indicated for workers simply because they work with recombinant organisms. There is no prescribed medical surveillance related directly to work with recombinant organisms. Studies of workers in the biotechnology industry have concluded that there is no evidence of adverse health effects related to the unique aspects of recombinant DNA technology.[6,7] Medical surveillance data have documented neither the occurrence of clusters of disease nor an increased incidence of diseases in workers who are frequently exposed to recombinant organisms.[5] Medical surveillance for health effects related to exposure is best conceived when certain criteria are fulfilled:

1. The exposure can potentially cause an identifiable health effect.
2. It is reasonably likely that the disease or effect may occur (and it is related to work).
3. There is an acceptable and scientifically sound methodology for diagnosing the condition or disease.
4. Early diagnosis has the potential to reduce morbidity or mortality.

When data collection and understanding the prevalence of a work-related condition are the goals, surveillance may target specific conditions even if therapeutic approaches are not yet available.

Some common exposures that may warrant medical surveillance in biotechnology laboratories and industrial settings include: infectious agents, sensitizing chemicals, radio-isotopes, and animal handling (a potential for zoonotic infections and allergic responses). The goal is detection of early health effects and prevention of long-term morbidity related to agents such as animal dander, enzymes, and endotoxin.

Work with recombinant organisms that present no identifiable risk to human health (group I in the European Hazard Classification scheme) would not entail routine medical surveillance. For work with highly pathogenic organisms or biologically active substances such as enzymes that are expressed by genetically modified microorganisms, periodic medical evaluation may be desirable. Medical surveillance for infectious disease endpoints emphasizes the identification of workers at risk and the early diagnosis of an infectious process.[8]

The goals of periodic surveillance are primarily to (1) detect early signs and symptoms of disease and (2) detect changes in the health of employees indicating a need for changes in job functions and/or work process. The genetically modified organism does not typically have human health effects different from those associated with the unmodified organism. However, the genetically modified organism may cause allergenic responses or the expressed products may have toxic effects and these may warrant medical surveillance. The periodic evaluation usually includes a health questionnaire. Other components of the evaluation depend on the type of exposure and health effect and should be targeted to specific exposures and health risks.[9] For example, a lung function assessment is appropriate if there is potential exposure to asthmagen(s) or endotoxin.

In general, implementation of the hazard control plan at work protects the workers from a significant risk of health effects related to work. However, removal from work with certain pathogenic microorganisms or biologically active products may be necessary in the case of an allergic or a susceptible individual.

Is medical surveillance also indicated for a special subpopulation of workers who are more susceptible to health effects from recombinant DNA technology? Who is potentially susceptible to health effects related to work with genetically modified microorganisms? Similar to any work with pathogenic microorganisms, individuals with reduced immunocompetence (including steroid treatment) and individuals with less effective barriers to infection (usually related to disease of the respiratory tract or gastrointestinal tract or illness or injury of the skin) represent a potentially susceptible population. Preplacement evaluations are therefore recommended to identify persons with medical conditions that may increase risk of adverse health effects in work with infectious organisms, including recombinant microorganisms. The focus of the preplacement evaluation is the potential for altered host defenses.

PREVENTION

Guidelines for laboratory animal use were formulated in the 1960s, and with the expanding use of recombinant organisms in research laboratories, the NIH has developed guidelines for research involving recombinant DNA, which have been revised annually.[10,11] The CDC/NIH revised biosafety guidelines were issued in 1999.[12] There are four levels of biosafety containment (BSL).[12,13] Level 1 is for well-characterized agents not known to cause disease in healthy adults and of minimal hazard to personnel. Level 2 is for agents with moderate potential hazard and requires special training for personnel. Level 3 is for potentially lethal infectious agents and requires specific containment, protective clothing, and special safeguards, such that many laboratories are not able to handle this work. Level 4 is for exotic highly pathogenic, poorly understood pathogens and is extremely rare in either the academic or industrial setting. Very recently, a new European Council Directive on contained use of genetically modified agents was approved.[14] Member states have the option of requiring even more stringent safety measures.[15] The European Directive focuses on appropriate training and supervision in the workplace where genetically modified organisms are used. In the USA, the NIH Recombinant DNA Advisory Committee (RAC) has set up strict guidelines for most recombinant DNA technology including gene therapy. Thus it is clear that biological safety has become a paramount issue on a global scale and that recombinant organisms have provided the impetus for these developments.

REFERENCES

1. Walsh G. Biopharmaceutical benchmarks. *Nature Biotechnology* 2000; 18:831–3.
2. Cohen R, Hoerner CL. Recombinant DNA technology: a 20-year occupational health retrospective. *Rev Environ Health* 1996; 11(3):149–65.

3. Klees JE, Joines R. Occupational health issues in the pharmaceutical research and development process. *Occup Med: State of the Art Rev* 1997; 12(2):5–27.
4. Ducatman Alan M, Coumbis John J. Chemical hazards in the biotechnology industry. *Occup Med: State of the Art Rev,* 1991; 6(2):193–208.
5. Liberman DF, Israeli E, Fink R. Risk assessment of biological hazards in the biotechnology industry. *Occup Med: State of the Art Rev* 1991; 6:2.
6. Vidal DR, Paucod JC, Thibault F, Isoard P. Biological safety in the laboratory. Biological risk, standardization and practice. *Ann Pharmaceut Francaises* 1993; 51(3): 154–66.
7. Finn AM, Scott AJ, Stave GM. Genetic modification and biotechnology. In: Baxter PJ, Adams PH, Aw T-C, Cockcroft A, Harrington JM, eds. *Hunter's diseases of occupations,* 9th edn. London, 2000, 521–35.
8. Rosenberg J, Clever LH. Medical surveillance of infectious disease endpoints. *Occup Med: State of the Art Rev* 1990; 5(3):583–605.
9. Goldman RH., Medical surveillance in the biotechnology industry. *Occup Med: State of the Art Rev,* 1991; 6(2):209–225.
10. Department of Health and Human Services, National Institutes of Health. *Guidelines for research involving recombinant DNA molecules*; October 2000. 65 Federal Register 60328. Available at: http://www.nih.gov/od/orda/
11. McGarrity GJ, Hoerner C. Biological safety in the biotechnology industry. In: *Laboratory safety: principles and practices,* 2nd edn. 1995, 119–131.
12. Centers for Disease Control and Prevention, National Institutes of Health. *Biosafety in microbiological and biomedical laboratories,* 4th edn. HHS Publication no. (CDC) 93–8395 1999.
13. Gilpin Richard W. Research activities including pathogens, recombinant DNA, and animal handling. In: McCunney RJ, ed. *Medical Center Occupational Health and Safety* 1999; 115–136.
14. Vranch S. New directive on biosafety: the contained use of genetically modified microorganisms. *Pharmaceut Technol Eur* 2000; 12(5):42.
15. Health and Safety Executive. *Compendium of guidance from the Health and Safety Commission's Advisory Committee on Genetic Modification.* London: HMSO, 1997.

32

PRIONS: CREUTZFELDT–JAKOB DISEASE (CJD) AND RELATED TRANSMISSIBLE SPONGIFORM ENCEPHALOPATHIES (TSE)

Dennis J. Darcey, M.D., M.S.P.H.

Common names for disease: Sporadic Creutzfeldt–Jacob disease, new variant Creutzfeldt–Jakob disease, fatal familial insomnia, Kuru, Gerstmann–Straussler–Scheinker disease

Classification: Prion diseases

OCCUPATIONAL SETTING

There are theoretical but as yet unproven risks to healthcare workers, including physicians, surgeons, pathologists, nurses, dentists, laboratory technicians, veterinarians, veterinary technicians, agriculture workers, farmers, meat processors, butchers, abattoir workers and cooks exposed to blood and uncooked animal products. Travelers to the UK during 1990–96 may have consumed meat from cattle with bovine spongiform encephalopathy.

EXPOSURE (ROUTE)

Transmission of Creutzfeldt–Jakob disease (CJD) from human to human has been reported for patients receiving corneal transplants, dural grafts, and human growth hormone and gonadotropins derived from pooled human cadaver pituitary glands. Transmission has also been linked to inadequately sterilized instruments and stereotactic electrodes. In addition, transmission of Kuru by ritual

Physical and Biological Hazards of the Workplace, Second Edition, Edited by Peter H. Wald and Gregg M. Stave
ISBN 0-471-38647-2

cannibalism has been documented. Recent reports from the UK suggest that a new variant Creutzfeldt–Jakob disease may be associated with consumption of brain and spinal cord in sausages, hamburger and other processed meats from cattle with bovine spongiform encephalopathy (BSE), "mad cow disease". Transmission of CJD through human blood or blood products has not been documented but is of some concern.

In August 1999 the FDA issued guidelines to reduce the possible risk of transmission of CJD and new variant CJD (nvCJD) by blood and blood products.[1] As a result, the Red Cross has implemented restrictions on blood donors who have been diagnosed with CJD or who have relatives with CJD. Blood donors who spent a total time of 6 months during 1980–96 in the UK during the mad cow epidemic are also restricted from blood donation.

Although there have been case reports of healthcare personnel, veterinarians and farmers developing CJD, there is no firm epidemiologic evidence to date to suggest that work in these occupations increases the risk for developing CJD.

PATHOBIOLOGY

A number of transmissible spongiform encephalopathies or prion diseases have been described in animals and humans. Animal diseases include scrapie in sheep and goats, transmissible mink encephalopathy, wasting disease of deer and elk, bovine spongiform encephalopathy, and feline spongiform encephalopathy. Human diseases include sporadic CJD, nvCJD, Kuru, genetically transmitted familial CJD, Gerstmann–Straussler–Scheinker disease and fatal familial insomnia.

The term "prion protein" (PrP) was coined by Prusiner in 1982 to describe a novel host membrane sialoglycoprotein that caused scrapie in sheep.[2] This PrP was apparently able to "replicate" without DNA or RNA. The PrP was resistant to treatments that inactivate nucleic acids and viruses, including ionizing radiation, alcohol, formalin, proteases and nucleases. It was inactivated by treatments that disrupt proteins, including phenol, detergents, autoclaving, and extremes of pH. The protease-resistant protein associated with disease proved to be an isoform of a protease-sensitive normal host cellular protein.

In the pathologic process, the normal host cellular protein appears to undergo a post-translational conformational change to the abnormal protein. This conformational change converts the PrP from a predominantly alpha-helical structure into a structure with a large beta-sheet content. This difference in tertiary structure causes a major difference in the molecule, with a high tendency of the disease producing PrP to form insoluble deposits. These deposits accumulate in neural cells, disrupting function and leading to vacuolization and cell death.[3] The "replication" of the PrP agent is thought to result in the continuous conversion of normal PrP into confirmationally changed PrP that contributes to successive changes in the molecule.

There appears to be some species barrier due to differences between species PrP structure. In addition, genetic prion diseases have been associated with mutations to the PrP gene, with different mutations associated with different phenotypes. However, some controversy remains as to the etiology of the transmissible spongiform encephalopathies, and a competing viral hypothesis has recently been resurrected.[4]

All of the transmissible spongiform encephalopathies have incubation periods of months to years and all gradually increase in severity and lead to death over a period of months. None seem to invoke an immune response and cause a non-inflammatory degenerative process in the central nervous system. In all infected species, infectivity appears to be greatest in brain tissue and in limited peripheral tissues, but generally has not been seen in body fluids except for cerebrospinal fluid. Prions can remain "infectious" for years even in the dried state.

DIAGNOSIS

CJD was first described in 1920 and occurs worldwide with an incidence of 0.5–1.5 cases per million population per year. CJD generally occurs between the ages of 50 and 70. The incidence of the disease has been constant over many decades and there is no apparent geographic or seasonal clustering of the sporadic form. A genetically determined familial variant has also been reported. To date, no environmental risk factors for the disease have been identified, although dietary factors, exposure to animals and occupational exposures have been evaluated in epidemiologic studies. Surgeons and pathologists handling infected human brain tissue, as well as abattoir workers and cooks exposed to blood and uncooked animal products, do not appear to have an increased risk for disease.[5,6]

Patients with CJD complain of fatigue, disordered sleep, decreased appetite, memory loss, and confusion. Focal neurologic signs, such as ataxia, aphasia, visual loss, hemiparesis and amyotrophy, are reported. A diagnosis of CJD is suggested by a progressive loss of cognitive abilities, the development of mild chronic jerking, and pyramidal tract, cerebellar and extrapyramidal signs. During the latter stages of disease, the patient becomes mute and akinetic. The mean survival time is 5 months, and 80% of patients with sporadic disease die within 1 year.[7]

There appear to be no diagnostic peripheral blood abnormalities or abnormalities in the cerebrospinal fluid that are helpful in the diagnosis. The electroencephalogram can be normal early in the disease but has been shown to be abnormal in 90% of patients if repeated during the course of the illness. As the disease progresses, computed tomography may show progressive generalized atrophy, and magnetic resonance imaging may show hyperintense signals in the basal ganglion on T2-weighted images.

Abnormal protein patterns in the cerebrospinal fluid of patients with CJD were found with two-dimensional electrophoresis, but the method was not practical for routine use.[8] Histologic examination of the brain and immunostaining for the abnormal PrP are the gold standards for diagnosis. Spongiform changes accompanied by neuronal loss and gliosis are common features. Amyloid plaques are found in 10% of the brains in the sporadic form of CJD. In contrast, plaques are common in Kuru, some of the familial spongiform encephalopathies and nvCJD. Tonsillar biopsy has been suggested as a possible pathologic diagnostic technique in the living patient, particularly for nvCJD.[9]

Because of concern about cross-species transmission in the UK, a national surveillance unit for CJD was established in 1990. Between 1994 and 1999, 38 cases of what is now called nvCJD were reported.[10,11] Recent laboratory studies provide evidence that the causative agents of nvCJD disease and BSE may have a common origin.[12] However, the common origin of the PrP does not prove that humans were actually infected by eating bovine nervous tissue or other animal products. New variant CJD appears to occur in younger patients, with a mean age at onset of 29. The duration of disease is more than twice as long (14 months) as that of the sporadic form. In contrast to the dementia and myoclonus seen in the sporadic form of CJD, nvCJD patients tend to have more psychiatric and behavioral manifestations with persistent paresthesias, dysesthesias and cerebellar ataxia. The electroencephalogram does not show the typical periodic complexes seen in the sporadic form, and the pathologic examination shows more prominent and diffuse plaques similar to Kuru. No human disease resembling nvCJD disease has been found in North America, and there has been no increase in the incidence of CJD in persons under 45 years of age.

TREATMENT

CJD and the related transmissible spongiform encephalopathies are universally fatal. There is no effective treatment or vaccine.

PREVENTION

Although iatrogenic transmission of CJD has been reported and possible transmission of BSE to humans has been proposed, the cause of sporadic CJD, which comprises 90% of the cases, remains unknown. To date, there does not appear to be an increased risk of transmission from CJD patients or tissues to surgeons, pathologists, dentists or other healthcare professionals. Nor does there appear to be an increased risk to veterinarians, meat packers or abattoir or agricultural workers handling infected BSE animals or tissues.

Unlike most other pathogens, prions resist ordinary disinfection and sterilization techniques and can remain infectious for years even in a dried state. Thus, infected spinal fluid and tissue from the brain, spine and eyes are theoretically of concern. Healthcare personnel caring for patients with CJD should follow universal precautions for blood-borne pathogens. Gloves should be worn when in contact with central nervous system tissue, contaminated instruments and surfaces and when handling blood and body fluids, particularly cerebrospinal fluid. Gowns and plastic aprons are indicated if splattering of blood or tissue is anticipated. Masks or protective goggles should be worn if aerosol generation or splattering is likely to occur, such as in dental or surgical procedures, wound irrigations, postmortem examinations, bronchoscopy, or endoscopy. Needles and other sharp objects should never be clipped, recapped or removed from disposable syringes. Needles should be disposed of in a puncture-proof, leakproof, rigid plastic biohazard container.

All precautions should be taken when handling cerebrospinal fluid and brain tissue. Where practicable, disposable instruments are preferable. Controversy continues concerning the best procedures for fully sterilizing instruments, tissues, or other materials known to contain prions. The Committee on Health Care Issues of the American Neurological Association has suggested either steam autoclaving (1 hour at 132°C) or immersion in 1 M NaOH (1 hour at room temperature).[13] Biopsy material should be soaked in concentrated formic acid for 1 h, followed by formalin for 48 hours minimum, or treated with a mixture of formalin in 0.1% hypochlorite (1:1) for at least 48 hours. After that, the infectivity is destroyed, and the material can be processed. Instruments and surfaces can be cleaned with a hypochlorite solution or 2 M sodium hydroxide solution. Sodium hydroxide should not be used on aluminum material.[14] Recent studies have suggested that concentrated (3 M) guanidine thiocyanate solutions may be highly effective as disinfectants.[15] Specific guidelines and precautions have also been suggested for special situations such as the performance of neuropathologic autopsies in suspected cases of CJD,[16] and for dental practice.[17]

REFERENCES

1. Revised precautionary measures to reduce the possible risk of Creutzfeldt–Jacob disease (CJD) and the new variant Creutzfeldt–Jacob disease (nvCJD) by blood and blood products. FDA Guidance for Industry, 7 August, 1999.
2. Prusiner SB. Novel proteinaceous infectious particles cause scrapie. *Science* 1982; 216:136–44.
3. Johnson RT, Gibbs C. Creutzfeldt–Jakob disease and related transmissible spongiform encephalopathies. *N Engl J Med* 1998; 339: 1994–2004.
4. Manuelidis L, Fritch W, Xi YG. Evolution of a strain of CJD that induces BSE-like plaques. *Science* 1997; 277:94–9.
5. Harris-Jones R, Knight R, Will RG, Cousens S, Smith PG, Matthews WB. Creutzfeldt Jakob disease in England and Wales, 1980–1984: a case–control study of potential risk factors. *J Neurol Neurosurg Psychiatry* 1988; 51:1113–19.
6. Wientjens DPWM, Davanipour Z, et al. Risk factors for Creutzfeldt–Jacob disease: a reanalysis of the case control studies. *Neurobiology* 1996; 46:1287–91.
7. Brown P, Gibbs CJ Jr, Rodgers-Johnson P, et al. Human spongiform encephalopathy; the National Institutes of Health series of Health

series of 300 cases of experimentally transmitted disease. *Ann Neurol* 1994; 35:513–29.

8. Harrington MG, Merril CR, Asher DM, Gajdusek DC. Abnormal proteins in the cerebrospinal fluid of patients with Creutzfeldt–Jakob disease. *N Engl J Med* 1986; 315:279–83.
9. Hill AF, Zeidler M, Ironside J, Collinge J. Diagnosis of new variant Creutzfeldt–Jakob disease by tonsil biopsy. *Lancet* 1997; 349:99–100.
10. Alperovitch A. Commentary: uncertainty over length of incubation tempers optimism. *Br Med J* 1999: 318(7190):1045.
11. Collinge J. Variant Creutzfeldt–Jakob disease. *Lancet* 1999: 351:317–23.
12. Hill AF, Desbruislais M, Joiner S, et al. The same prion strain causes nvCJD and BSE. *Nature* 1997; 389:448–50.
13. Committee on Health Care Issues ANA. Precautions in handling tissues, fluids and other contaminated materials from patients with documented or suspected Creutzfeldt–Jakob disease. *Ann Neurol* 1986; 19:75–7.
14. van der Valk. Prion diseases: what will be next? *J Clin Pathol* 1988; 51:265–9.
15. Manuelidis L. Decontamination of Creutzfeldt–Jakob disease and other transmissible agents. *J Neurovirol* 1997; 3:62–5.
16. Ironside J, Bell JE. The "high risk" neuropathological autopsy in AIDS and Creutzfeldt–Jakob disease: principles and practice. *Neuropathol Appl Neurobiol* 1996; 22:388–93.
17. Porter S, Scully C, Ridgway GL, Bell J. The human transmissible spongiform encephalopathies (TSEs): implications for dental practitioners. *Br Dental J* 2000; 188:432–6.

33

ENDOTOXINS

Brian A. Boehlecke, M.D., M.S.P.H., and Robert Jacobs, Ph.D.

Common names for disease: Cotton dust-byssinosis and brown lung. Additionally, an active febrile response to endotoxin has been given many names depending on the source of exposure such as mattress maker's fever, mill fever, and card room fever.

OCCUPATIONAL SETTING

Agricultural workers and processors of vegetable fibers are most likely to be at risk. Workplaces with potentially high airborne concentrations of endotoxins include cotton and flax mills, grain storage and handling operations, poultry houses and processing plants, saw and paper mills, sewage treatment plants, and swine confinement buildings. Workers may also be exposed during animal handling in various facilities or during composting operations. Contamination of cutting fluids used in machining operations or of workplace humidifying systems can also result in significant exposure.

EXPOSURE (ROUTE)

Occupational exposures are predominantly by inhalation. Significant exposure can occur wherever aerosols of materials contaminated with Gram-negative bacteria are generated.

PATHOBIOLOGY

Endotoxin refers to the lipopolysaccharide (LPS) complex and associated proteins found in the outer layer of the cell wall of Gram-negative bacteria. Although the lipid component (lipid A) is responsible for most of the toxic effects, variability of the biological responses to endotoxin exposures in natural

Physical and Biological Hazards of the Workplace, Second Edition, Edited by Peter H. Wald and Gregg M. Stave
ISBN 0-471-38647-2

settings may be due in part to influences from associated cell wall components, which differ among bacterial species.

Most medical attention has focused on the toxicity of endotoxins in reaching the bloodstream from endogenous Gram-negative bacterial infections of the host or via contaminated parenteral products.[1,2] However, little endotoxin is detectable in the blood after inhalational exposure. Local uptake in the respiratory system appears to account for the manifestations observed in the workplace.

An acute febrile response may follow inhalation of aerosols containing high concentrations of endotoxin.[3] This is especially true for individuals with no prior occupational exposure or after a hiatus in exposure for workers with chronic lower-level exposures. Historically, process- or material-specific descriptors such as mattress maker's fever, card room fever and grain fever were used. Monday fever is named for its occurrence on the first workday after a weekend break. Recently, the generic term organic toxic dust syndrome (OTDS) has been proposed to emphasize the commonality of this response to many agents and the non-immunologic nature of its pathobiology.

The fever usually begins several hours after exposure. Other symptoms include chills, myalgias, malaise, anorexia, headache, cough, and chest tightness. The worker may consider these symptoms to be manifestations of a viral flu. The condition is self-limited and usually lasts only a day or two, but it may be more prolonged if the exposure is massive. Tolerance develops with repeated exposures, and symptoms diminish despite similar doses of endotoxin. The underlying mechanism appears to be endotoxin-induced release of interleukin-1 from alveolar macrophages.

A syndrome commonly described in association with exposure to cotton dust consists of recurrent chest symptoms that begin several hours into the workshift, often on the first workday after a break. Chest tightness with or without cough or shortness of breath develops gradually and may persist for several hours after the exposure ceases. Symptoms generally remit by the following day and do not recur or are markedly diminished on subsequent days, despite continued exposure. Some, but not all, workers with these symptoms also show a significant decline in ventilatory function over the workshift, with a decrease in the forced expiratory volume in the first second (FEV_1) and the forced vital capacity (FVC) measured on spirometry. These values generally return to their pre-exposure levels by the following day. The term byssinosis has been used when this syndrome is associated with exposure to cotton dust. However, this term is also used in reference to chronic respiratory symptoms and ventilatory impairment in textile workers who probably have a form of chronic bronchitis and who may never have experienced the recurrent chest symptoms.

The exact mechanism for the symptoms and ventilatory changes has not been established. Animal and human exposure studies have shown that endotoxin inhalation stimulates production or release of various mediators from alveolar macrophages. For example, in vitro production of tumor necrosis factor α by alveolar macrophages in response to exposure to urban air particles is amplified by priming the cells with LPS.[4] Chemotactic agents recruit polymorphonuclear leukocytes to the airways and lung interstitium. Various products of arachadonic acid metabolism mediate bronchoconstriction.

Thus, endotoxin inhalation can produce airway inflammation and bronchoconstriction, which could account for the symptoms. Some epidemiologic studies of textile workers have found better correlation between chest symptoms and estimated doses of inhaled endotoxin than with crude measures of dust exposure. Controlled exposures of subjects preselected for acute ventilatory responses to cotton dust showed a strong correlation between drop in FEV_1 during exposure and airborne endotoxin concentration, but no association with respirable dust concentration uncorrected for endotoxin content.[5] These findings suggest that the endotoxin content of organic dusts and other aerosols is likely to be an important factor in producing recurrent chest symptoms in exposed persons. However, other studies have shown biological activity of endotoxin-free extracts of organic dusts. Therefore, the mechanism for responses to natural exposures may be complex and dependent on interactions of multiple agents.

Although not extensively studied, inhalation of organic dusts or purified endotoxin has been shown to transiently increase non-specific airway reactivity, as measured by methacholine or histamine bronchoprovocation. Both healthy people and persons with mild atopy or frank asthma have shown bronchial hyperresponsiveness lasting from a few hours to 24 hours after exposure. The clinical significance of these findings is uncertain. Acute airway inflammation may be a contributing factor to these changes.[6] Numerous epidemiologic studies have shown that workers exposed to cotton dust are at increased risk for symptoms of chronic bronchitis and for airway obstruction detectable by spirometry. Retrospective autopsy studies have confirmed the increased prevalence of pathologic changes of chronic bronchitis, even in non-smoking textile workers. Prospective studies have now confirmed decrements in ventilatory function in textile workers that exceed the expected rate and are correlated with dust exposure. Cigarette smoking is clearly an important factor; however, accelerated loss in FEV_1 also occurs in non-smokers. Although this condition has been referred to as byssinosis, or brown lung, when associated with cotton dust exposure, workers exposed to other organic dusts are also at risk.

The role of endotoxin in the pathogenesis of this condition is not clear.[7] In animals, the inflammatory response seen acutely after endotoxin inhalation appears to diminish with repeated exposures. Also, chronic high-level exposures to organic dusts containing endotoxin do not reproduce the pathologic changes of chronic bronchitis in animal models. Tolerance to the effects of endotoxin given intravenously is well documented in humans as well as animals. In healthy young volunteers, the Th1 type of lymphocyte cytokine response is decreased after intravenous LPS challenge.[8] Therefore, the effects of chronic organic dust exposure on the airways may be mediated by a complex interaction of the agents present, including endotoxin.

Respiratory effects similar to those seen in textile workers have also been reported in workers exposed to non-cotton organic dusts. Poultry house workers had drops in FEV_1 over the workshift that correlated with endotoxin exposure.[9] However, the correlation of lung function decrement with total dust exposure was stronger than that with the endotoxin component. Sewage treatment plant workers had increased non-specific airway responsiveness measured by methacholine challenge compared to controls, but the difference was small and no difference in lung function was found.[10] Thus, a causal role of endotoxin for the respiratory findings in these groups has not been clearly established.

DIAGNOSIS

The non-specific nature of these syndromes makes definitive diagnosis difficult. A careful history is critical to establish a pattern consistent with an association between symptoms and the putative causal exposure. For clinical evaluations, self-measurement and recording of peak flows using portable devices several times per day for 2 weeks may be helpful. Consistent acute worsening of ventilatory function during exposure or a progressive decline during the workweek with improvement on weekends suggests an association with exposure. However, lack of significant findings does not exclude an association between the symptoms and the exposure. Auscultation of the chest will generally not reveal wheezing or rales unless another underlying condition is present, which accounts for the finding. Persons with underlying hyper-reactive airways may develop symptomatic bronchospasm with wheezing in response to organic aerosol exposures without detectable sensitization to any components of the aerosol.

No specific diagnostic tests are available for the syndrome of recurrent chest symptoms. No evidence for a direct immunologic mechanism has been found in textile workers with either underlying recurrent chest symptoms (byssinosis) or acute declines in FEV_1 over the workshift. RAST testing for specific IgE is therefore of no value. Chest radiographs will be normal unless there is other pulmonary disease. Clinical judgment is, of course, necessary to determine if other causes for chest symptoms, such as coronary artery disease, should be pursued. Longer-term trends in ventilatory function may show losses in FEV_1 greater than expected from aging alone. Normal variability and the influence of non-work-related exposures, including cigarette smoking, must be considered when interpreting these findings. However, an accelerated decline in function associated with recurrent work-related chest symptoms or acute decrements in function during exposure merits attention and consideration of reduction in exposure.

Demonstration of environmental airborne concentrations of endotoxin of the magnitude that has been associated with these findings in other settings would support a causal relationship. Threshold concentrations of airborne endotoxin for the syndromes described have been postulated for cotton dust environments.[3] They vary from 0.5–1 $\mu g/m^3$ for the febrile reaction to 0.1–0.5 $\mu g/m^3$ for chest tightness and acute declines in ventilatory function in chronically exposed individuals. Previously non-exposed persons may require higher concentrations (2–3 $\mu g/m^3$) to develop acute symptoms of chest tightness. Endotoxin concentrations associated with increased airway reactivity and bronchitis are postulated to be lower, but they are more speculative, given the interaction of the many potential agents to which a worker may be exposed over the course of many years.

These endotoxin levels are based on measurement of endotoxin by the *Limulus amoebocyte* lysate (LAL) method. However, bioavailability of endotoxin in aerosols may vary depending on the nature of the materials. Therefore, concentrations measured in extracts by LAL may not be accurate predictors of the toxic potential in the natural setting. Based on experimental exposures, cell-bound endotoxin has been estimated to be three times as biologically active as isolated endotoxin in equal amounts, as measured by LAL. Also, interlaboratory variation in measurement of endotoxin is large. A recent comparative study showed a sixfold range in reported endotoxin content of the same cotton dust by 11 laboratories experienced with this measurement, despite use of a common extraction protocol.[11]

This is probably due to an underestimation of the amount present in the former preparation. Analyses should be done by a laboratory familiar with measurement of endotoxin in the type of material sampled. Caution should be exercised when comparisons are made between observed concentrations and proposed effect threshold levels. The acute febrile response may mimic an infectious illness. For example, findings may include an elevated peripheral blood leukocyte count. However, chest radiographs and cultures are usually negative. The self-limited nature of the illness coupled with a history of exposure should be sufficient to reach a presumptive diagnosis.

Established chronic bronchitis and airway obstruction associated with organic dust exposure has no inherent pathophysiologic features that allow differentiation from that caused by cigarette smoking. A history of prior recurrent chest symptoms or acute ventilatory declines associated with the exposure is helpful, but its absence does not eliminate the possibility that exposure is a causal factor. The magnitude and duration of exposure must always be considered; the greater the cumulative dose, the more likely it is the exposure contributed to the development of the condition. When a history of significant cigarette consumption is also present, the contribution of exposure to the pathogenesis of the condition is difficult to assess.

TREATMENT

Reduction in exposure to the organic aerosol is the treatment of choice. The acute febrile reaction is usually self-limited and can be treated symptomatically if necessary. Recurrent chest symptoms may improve with pre-exposure use of an inhaled bronchodilator or regular use of inhaled steroid medication, but these methods should not be used in lieu of reduction in exposure.

Established chronic bronchitis and airway obstruction may be treated with inhaled bronchodilators and steroids. Courses of antibiotics may be given when bacterial superinfections occur. Cessation of smoking and avoidance of other respiratory irritants are also indicated. Vaccination against pneumococcal pneumonia and yearly influenza vaccination is recommended for persons with moderate or severe airway obstruction.

MEDICAL SURVEILLANCE

Periodic medical questionnaires focusing on respiratory symptoms are helpful. The association of symptoms with exposure should be explored carefully with questions referring to the intensity of symptoms at home compared to at work, improvement during days off or vacations, and changes in severity related to specific job activities or use of personal protective equipment.

Periodic measurement of pre- and post-shift spirometry may show significant reductions in ventilatory capacity after exposure. Normal variability in spirometry must be considered when interpreting these results. The acute response to inhalation of organic dust may vary significantly on different occasions. Therefore, repeated observations are most helpful to determine if a meaningful pattern is present. Traditionally, yearly testing has been recommended, with modification of frequency if symptoms or abnormalities in lung function are noted.

Traditionally, most surveillance examinations have been done on a yearly basis. However, modification of the interval based on the intensity of exposure and the presence or absence of symptoms is justifiable. If work-related symptoms are absent and exposure is relatively low, ventilatory function testing every 2 years is probably adequate to detect important long-term trends. Changes in intensity of exposure or the onset of symptoms are indications for consideration of more frequent monitoring.

PREVENTION

Currently, there are no official health standards limiting exposure to endoxins. No observable effect levels (NOELs) have been proposed for several disease endpoints,[12] and the Health Council of the Netherlands proposed a recommended health-based exposure limit.[13] However, until a reliable and reproducible method for measuring endotoxin is developed, the most effective primary preventative measure is to reduce exposure by decreasing work-place concentrations of endotoxin-containing material. Traditional methods of reducing airborne contaminants include enclosure of aerosol-generating processes and increasing general ventilation. Decreasing the endotoxin content of organic dusts by selective removal has not yet reached commercial feasibility. Water washing of cotton prior to processing has been shown to reduce both the endotoxin content of the dust generated during processing and the acute dilatory function decline associated with exposure. Other methods of detoxification, such as heat treatment, have not been successful due to the relative stability of endotoxin and the toxicity of cell wall components of non-viable bacteria.

Use of personal protective equipment to reduce the dose of inhaled aerosol can be effective. However, because the highest concentration of endotoxin appears to be associated with the smallest particulates in an organic dust-contaminated environment and workers dislike using respirators, due to their discomfort and interference with job activities, reliance on this technique is questionable. Nevertheless, a respirator program with adequate training and monitoring of use can be a valuable adjunct to other measures and may be necessary for adequate protection of especially sensitive workers. There is both in vitro and in vivo evidence that atopic persons have an increased risk for adverse effects. Atopic persons showed an increase in nonspecific airway reactivity when exposed experimentally to cotton dust.[14] Atopic individuals also had increased nasal airway eosinophils when challenged with LPS compared to after saline challenge.[15] Lymphoctyes from persons allergic to birch pollen had increased expression of CD154, a component of the signaling pathway for allergic inflammatory responses, after incubation with pollen and LPS compared to incubation with pollen alone.[16] However, despite these findings, at present there is not sufficient justification for excluding atopic persons from jobs with potential exposure for endotoxin. If work-related symptoms and acute decrements in ventilatory function are found, reduction in exposure is clearly indicated.

REFERENCES

1. Brigham KL, Meyrick B. Endotoxin and lung injury. *Am Rev Respir Dis* 1986; 133:913–27.
2. Chosh S, Latimer RD, Gray BM, Harwood RJ, Oduro A. Endotoxin-induced organ injury. *Crit Care Med* 1993; 21:519–24.
3. Rylander R. Health effects of cotton dust exposures. *Am J Ind Med* 1990; 17:30–45.
4. Imrich A, Yu Ning Y, Koziel H, Coull B, Kobzik L. Lipopolysaccharide priming amplifies lung macrophage tumor necrosis factor production in response to air particles. *Toxicol Applied Pharmacol* 1999; 159:117–24.
5. Castellan RM, Olenchock SA, Kinsley KB, Hankinson JL. Inhaled endotoxin and decreased spirometric values. *N Engl J Med* 1987; 317:605–10.
6. Michel O, Ginanni R, LeBon B, Content J, Duchateau J, Sergysels R. Inflammatory response to acute inhalation of endotoxin in asthmatic patients. *Am Rev Respir Dis* 1992; 146:352–7.
7. Rylander R.The role of endotoxin for reactions after exposure to cotton dust. *Am J Ind Med* 1987; 12:687–97.
8. Lauw FN, Ten Hove T, Dekkers P, De Jonge E, Van Deventer S, Van Der Poll T. Reduced Th1 but not Th2 cytokine production by lymphocytes after in vivo exposure of healthy subjects to endotoxin. *Infect Immun* 2000; 68:1014–18.
9. Donham KJ, Cumro D, Reynolds SJ, Merchant JA. Dose–response relationships between occupational aerosol exposures and cross-shift declines of lung function in poultry workers: recommendations for exposure limits. *J Occup Environ Med* 2000; 42:260–9.
10. Rylander R. Health effects among workers in sewage treatment plants. *Occup Environ Med* 1999; 56:354–7.
11. Chun DTW, Chew V, Bartlett K, et al. Preliminary report on the results of the second phase of a round-robin endotoxin assay study using cotton dust. *Appl Occup Environ Hyg* 2000; 15:152–7.
12. Rylander R. Evaluation of the risk of exposures. *Intern J Occup Environ Health* Jan/March 1997; V3(1):532–536.
13. Health Council of the Netherlands. Dutch Expert Committee on Occupational Standards (DECOM). Endotoxins. Rijswijk: Health Council of the Netherlands, 1998; publication no. 1998/O3WDG.
14. Jacobs R, Boehlecke B, Van Hage-Hamsten M, Rylander R. Bronchial reactivity, atopy and airway response to cotton dust. *Am Rev Respir Dis* 1993; 148:19–24.
15. Peden DB, Tucker K, Murphy P, et al. Eosinophil influx to the nasal airway after local, low level LPS challenge in humans. *J Allergy Clin Immunol* 1999; 104:388–94.
16. Nakstad B, Kahler H, Lyberg T. Allergen-stimulated expression of CD154 (CD40 ligand) on CD3+ lymphocytes in atopic but not in nonatopic individuals. Modulation by bacterial lipopolysaccharide. *Allergy* 1999; 54:722–9.

34

WOOD DUST

Harold R. Imbus, M.D., M.Sc.D.

OCCUPATIONAL SETTING

Exposure to wood dust occurs in many industries, including logging and sawmill operations, furniture manufacturing, paper manufacturing and construction of residential and commercial buildings, especially carpentry and cabinet making. Workers are exposed when wood is sawed, chipped, routed, or sanded.

Wood may also contain biological or chemical contaminants. Biological contaminants include molds and fungi, which often grow on the bark of wood. Exogenous chemicals include those used in treating the wood. Common wood preservatives are arsenic, chromium, copper, creosote, and pentachlorophenol. Wood also contains many endogenous chemicals that are responsible for its biological actions.

EXPOSURE (ROUTE)

Wood dust exposure occurs through inhalation. In general, the finer particles of wood dust are more biologically active due to their greater surface area and their ability to penetrate and adhere to the respiratory mucosa. Furniture manufacturing and cabinet making are operations that produce the finer particles. Contact with skin or mucous membranes may also have health consequences.

PATHOBIOLOGY

Wood dust is composed of wood particles generated by the processing or handling of wood. Hardwoods, such as maple, oak, and cherry, come from deciduous trees with broad leaves. Softwoods come from evergreen trees such as pine, spruce, and fir. The terms are somewhat misleading in that some of the hardwoods may be soft and some of the softwoods relatively hard.

Health effects of wood dust may be classified primarily as irritation, sensitization, and cancer. In the case of irritation, these effects

Physical and Biological Hazards of the Workplace, Second Edition, Edited by Peter H. Wald and Gregg M. Stave
ISBN 0-471-38647-2

can involve the skin, eyes, or respiratory tract. Allergic manifestations can involve skin or respiratory tract, and cancer associated with wood dust exposure involves the sinonasal tract.

Skin irritation caused by wood dust is often mechanical. Splinters or tiny particles of wood can get under the skin and cause irritation or infection. Soaps and chemicals can add to the irritation. Particles may lodge between the folds of skin, and sweat and rubbing can result in inflammation. Good personal hygiene practices and protection of exposed areas can obviate this type of problem.

Some woods, mostly foreign and exotic species, contain chemicals that are irritants. These can cause dermatitis, resulting in redness and blistering. Eyes may also become irritated. Teak, mansonia and radiata pine have been reported to cause such reactions.

Allergic contact dermatitis can result from some species, again mostly foreign woods such as teak and African mahogany. However, some of the North American woods, such as Douglas fir, western red cedar, poplar, airborne pine dust due to colophony,[1] and rosewood, may cause allergic contact dermatitis.[2] Specific chemicals used in glues and resin binders, such as urea or phenol-formaldehyde, potassium dichromate, ethylene glycol, and propylene glycol, can also be responsible for allergic contact dermatitis.

Wood dust is a particulate that causes irritation of the eyes and the upper and lower respiratory tracts. As with all particulates, the magnitude of the effect depends on the size of the particulate and the amount of exposure. Some woods with strong chemicals are more irritating, and most wood is more irritating than inert dust. For example, wood dust has been found to be almost four times as irritating as plastic dust in the same concentration.[3]

Clinical epidemiologic studies of woodworkers have found frequent symptoms and physical findings of nasal irritation. Symptoms often consist of continued colds, nosebleeds, sneezing, and sinus inflammation. Unfortunately, most of the studies lack adequate controls and dose–response information. In one study, the employees in a dusty furniture plant, including office workers, had decreases in 1-second forced expiratory volume (FEV_1) and forced vital capacity (FVC) during the workshift.[4] A large Vermont study that also measured pulmonary function found an inverse association between pulmonary flow (FEV_1/FVC ratio) and indexes of wood dust exposure.[5] Mucociliary clearance in the nose has also been measured in wood-workers. Andersen et al found a higher percentage of individuals with mucostasis among workers exposed to higher wood dust levels.[6] Mucociliary clearance is important in cleansing the nasal passages. Individuals with impaired mucous flow may be more susceptible to infection and other problems.

Western red cedar is well known for its potential to cause asthma. Workers in British Columbia, Canada, have been studied extensively.[7–18] Plicatic acid, an extract of western red cedar, was shown to be the cause of bronchial asthma.[18] Other domestic species capable of causing asthma include oak, ash, redwood, mahogany, and eastern white cedar.[19–21] A number of foreign exotic species are capable of causing allergic asthma.

Microorganisms on wood bark can cause type III allergic reactions, or extrinsic allergic alveolitis.[22–24] Maple bark disease, sequoiosis and suberosis have been shown to be caused by the inhalation of fungal spores associated with maple, redwood, and cork, respectively.

Wood dust may also cause other lung diseases. High levels of small dust particles that penetrate into the bronchial tubes or smaller airways may produce irritation and bronchitis. Whether they produce irreversible changes is not known. Most reports of serious or permanent lung disease associated with wood dust exposure, until recently, have been anecdotal. However, recent studies showed lower pulmonary function in wood mill workers,[25] higher prevalence of respiratory impairment,[26] and significant cross-shift declines of pulmonary function in wood dust-exposed workers.[27]

An unusually high incidence of nasal cancer was first noted by an English otolaryngologist in the chair-making and furniture industry of High Wycombe, Buckinghamshire, England.[28] This finding was confirmed by Acheson et al,[29,30] who found an incidence of nasal adenocarcinoma from 1956 to 1965 of 0.7 cases per 1000 per year, an incidence ~1000 times greater than that seen in the general population. Dust depositions were noted on the nasal mucosa of furniture workers at the anterior part of the nasal septum and at the anterior ends of the middle turbinates.[31] A biopsy of these areas revealed squamous metaplasia. Excesses of nasal adenocarcinoma have also been described in France, Australia, Finland, Italy, Holland, Denmark, and Belgium.[32–40] An early study in Canada showed no excess of nasal cancer in woodworkers; however, few furniture workers were involved in the study.[41] A more recent Canadian study showed an odds ratio of 2.5 for occupations involving exposure to wood dust when compared to the general population.[42] In 1995, the International Agency for Research on Cancer (IARC) classified wood dust: "Group I, carcinogenic to humans".[43]

In the USA, a case–control study showed an approximate fourfold increase of nasal cancer in furniture workers in North Carolina.[44] Also, a Connecticut study showed an odds ratio of 4.0 for nasal cancer among persons exposed to wood dust.[45] In a follow-up study, Brinton et al found an overall relative risk of nasal cancer among furniture workers of 0.74 in men and 0.91 in women; however, the relative risk of the rarer adenocarcinoma in male furniture workers was 5.68.[46] These excesses of adenocarcinoma, a very rare disease, are far lower than those noted in the UK and other countries. A review of nasal cancer in furniture manufacturing and woodworking in North Carolina suggested that there was some excess of nasal cancer in the industry prior to World War II, but it found little or no evidence of continuing excess risk,[47] while another review of US wood-workers confirmed this disparity between the USA and the UK.[48] Another pooled reanalysis observed excesses of nasal cancer only among workers in British furniture manufacturers as compared with those in the USA.[49] Likewise, in the UK, Acheson postulated that the factor in furniture manufacturing that gave rise to the nasal adenocarcinomas was present only between 1920 and 1940, since cases did not occur among workers who entered the industry after World War II.[50]

It is unlikely that exogenous chemicals were responsible for the excess nasal cancer, since they were not widely used prior to World War II. It may well be that differences in processing, including the use of tools that create more heat, or dust, or different particle sizes, account for the increased disease of the earlier years.[47] Certain woods, such as beech, may contain genotoxic chemicals.[51–53] Also, wood that has been combusted may contain mutagenic compounds.[54] Though the cause of nasal cancer excesses is not known, it appears that the risk has declined markedly in industry.

DIAGNOSIS

A clinical diagnosis of allergy or asthma can be made based on a careful medical and occupational history along with a physical examination. A detailed respiratory and dermatologic history will reveal symptoms of allergic disorders. Respiratory symptoms may include rhinitis, conjunctivitis, sneezing spells, nasal congestion, cough, shortness of breath, and wheezing.

Documenting the occurrence of these symptoms as temporally associated with exposure to wood dust and the resolution of symptoms while away from work is critical to the diagnosis. Objective evidence of temporality may be obtained by doing a physical examination or peak flow measurements prior to exposure and again at the end of the workshift. The finding of objective evidence of bronchoconstriction, such as wheezes or a decrement in FEV_1 and FVC, supports the diagnosis of occupational asthma. However, asthma due to allergenic wood dust may be delayed, occurring during evening or night after the workshift.[55]

To evaluate suspected contact dermatitis, a patch test can be done with the sawdust of the wood itself or, in some cases, with an extract of woods such as teak. Patch testing can also be performed with exogenous chemicals if they are suspected of being the causative agent.

TREATMENT

Avoidance of exposure to wood dust should be the goal of treatment of allergy. Sensitized workers who have rhinitis should be counseled on their risk of developing asthma and the risk of eventual intolerance of exposure.

Symptomatic care may include antihistamines or corticosteroids for upper tract symptoms and bronchodilators, corticosteroids and theophylline for asthma. Cromolyn may prevent asthma from allergen exposure. Immunotherapy is not likely to be helpful for sensitized individuals who remain in an environment where regular exposure occurs.

Skin disease can be treated with topical or, in severe cases, systemic corticosteroids. Work practices and controls should be modified to prevent further exposure.

MEDICAL SURVEILLANCE

Respiratory history may be used to document early symptoms of allergy. If nasal symptoms are detected early and further exposure is eliminated, subsequent progression to asthma may be prevented. Preplacement, periodic and possibly cross-shift pulmonary function testing may be useful, depending upon the type and quantity of wood dust, and any indicators of employee health problems related to their exposure.

PREVENTION

Containment of wood dust should be the cornerstone of allergy and irritation prevention measures. Efforts should be made to limit airborne dust through engineering controls such as ventilation systems. Personal protective equipment, including skin coverings and respirators, should be considered where significant exposure cannot be engineered out. The concentration of airborne dust is dependent on the activities in the contaminated areas.

Wood dust can be measured by pumping air into a preweighed filter and weighing the filter after exposure. Particle size determinations can be made with a cascade impactor. The American Conference of Governmental Industrial Hygienists (ACGIH) standard for hardwood dust such as beech and oak is 1 mg/m^3; for softwood dust, it is 5 mg/m^3. However in its 2001 "TLVs and BEIs,"[55] the ACGIH published a notice of intended changes as follows:

- Hardwoods and softwoods (non-allergenic), 2 mg/m^3
- Beech and oak, 1 mg/m^3
- Birch, mahogany, teak, walnut, 1 mg/m^3
- Softwoods and other hardwoods (allergenic), 1 mg/m^3
- Western red cedar, 0.5 mg/m^3

When these changes may go into effect is unknown. Until 19 January 1989, the OSHA regulated wood dust as a nuisance dust with a permissible exposure level (PEL) of 15 mg/m^3. After that date, the OSHA adopted the PEL standard, which regulated wood dust at 5 mg/m^3 for hardwood and softwood; because of its allergic effects, the PEL is 2.5 mg/m^3 for western red cedar. On 7 July 1992, the US Court of Appeals vacated the PEL standard; therefore, as of this writing, wood dust is regulated by the OSHA as a nuisance dust at 15 mg/m^3. However, use of the lower exposure levels seems prudent to reduce the risk of illness.

REFERENCES

1. Watsky KL. Airborne allergic contact dermatitis from pine dust. *Am J Contact Dermat* 1997; 8:118–20.
2. Weber LE. Dermatitis veneata due to native woods. *AMA Arch Dermatol Syphilol* 1953; 67:388–94.
3. Andersen I. Effects of airborne substances on nasal function in human volunteers. In: Carrow CS, ed. *Toxicology of the nasal passages.* Washington, DC: Hemisphere, 1986, 143–54.
4. Zuhair YS, Whitaker CJ, Cinkotai EF. Ventilatory function in workers exposed to tea and wood dust. *Br J Ind Med* 1981; 38:339–45.
5. Whitehead LW, Asbikaga T, Vacek P. Pulmonary function status of workers exposed to hardwood or pine dust. *Ann Ind Hyg Assoc* 1981; 42:1780–6.
6. Andersen HC, Solgaard J, Andersen I. Nasal cancer and nasal mucus transport rates in woodworkers. *Acta Otolayngol* 1976; 82:263–5.
7. Brooks SM, Edwards JJ, Edwards FE. An epidemiologic study of workers exposed to western red cedar and other wood dusts. *Chest* 1981; 80(suppl 1):30–2.
8. Chan-Yeung M. Maximal expiratory flow and airway resistance during induced bronchoconstriction in patients with asthma due to western red cedar (Thuja plicata). *Am Rev Respir Dis* 1973; 108:1103–10.
9. Chan-Yeung M. Fate of occupational asthma. A follow-up study of patients with occupational asthma due to western red cedar (Thuja plicata). *Am Rev Respir Dis* 1977; 116:1023–9.
10. Chan-Yeung M. Immunologic and nonimmunologic mechanisms in asthma due to western red cedar (Thuja plicata). *J Allergy Clin Immunol* 1982; 70:32–7.
11. Chan-Yeung M, Abboud R. Occupational asthma due to California redwood (Sequoia sempervirens) dusts. *Am Rev Respir Dis* 1976; 114:1027–31.
12. Chan-Yeung J, Ashley MJ, Corey P, Willson G, Dorken E, Grzybowski S. A respiratory survey of cedar mill workers. I. Prevalence of symptoms and pulmonary function abnormalities. *J Occup Med* 1978; 20:323–7.
13. Chan-Yeung M, Barton GM, MacLean L, Grzybowski S. Bronchial reactions to western red cedar (Thuja plicata). *Can Med Assoc J* 1961; 105:56–8.
14. Chan-Yeung M, Barton GM, MacLean L, Grzybowski S. Occupational asthma and rhinitis due to western red cedar (Thuja plicata). *Am Rev Respir Dis* 1973; 108:1094–102.
15. Chan-Yeung M, Lam S, Koener S. Clinical features and natural history of occupational asthma due to western red cedar (Thuja plicata). *Am J Med* 1982; 72:411–5.
16. Chan-Yeung M, Veda S, Kus J, MacLean L, Enarson D, Tse KS. Symptoms, pulmonary function, and bronchial hyperreactivity in western red cedar workers compared with those in office workers. *Am Rev Respir Dis* 1984; 130:1038–41.
17. Vedal S, Chan-Yeung M, Enarson D, et al. Symptoms and pulmonary function in western red cedar workers related to duration of employment and dust exposure. *Arch Environ Health* 1986; 41:179–83.
18. Chan-Yeung M, Gicias PC, Henson PM. Activation of complement by plicatic acid, the chemical responsible for asthma due to western red cedar (Thuja plicata). *J Allergy Clin Immunol* 1980; 65:333–7.
19. Malo JL, Cartier A, Desjardins A, Van de Weyer R, Vandenplas O. Occupational asthma caused by oak wood dust. *Chest* 1995; 108:856–8.
20. Fernandez-Rivas M, Perez-Carral C, Senent CJ. Occupational asthma and rhinitis caused by ash (Fraxinus excelsior) wood dust. *Allergy* 1997; 52:196–9.
21. Malo JL, Cartier A, L'Archeveque J, Trudeau C, Courteau JP, Bherer L. Prevalence of occupational asthma among workers exposed to eastern white cedar. *Am J Respir Crit Care Med* 1994; 150:1697–701.
22. Rask-Andersen A, Land CJ, Enlund K, Lundin A. Inhalation fever and respiratory symptoms in the trimming department of Swedish sawmills. *Am J Ind Med* 1994; 25:65–7.
23. Dahlqvist M, Johard U, Alexandersson R, et al. Lung function and precipitating antibodies in low exposed wood trimmers in Sweden. *Am J Ind Med* 1992; 21:549–59.
24. Dahlqvist M, Ulfvarson U. Acute effects on

forced expiratory volume in one second and longitudinal change in pulmonary function among wood trimmers. *Am J Ind Med* 1994; 25:551–8.

25. Rastogi SK, Gupta BN, Husain T, Mathur N. Respiratory health effects from occupational exposure to wood dust in sawmills. *Am Ind Hyg Assoc J* 1989; 50:574–8.
26. Liou SH, Cheng SY, Lai FM, Yang JL. Respiratory symptoms and pulmonary function in mill workers exposed to wood dust. *Am J Ind Med* 1996; 30:293–9.
27. Mandryk J, Alwis KU, Hocking AD. Work related symptoms and dose-response relationships for personal exposures and pulmonary function among woodworkers. *Am J Ind Med* 1999; 35:481–90.
28. Macbeth R. Malignant disease of the paranasal sinuses. *J Laryngol Otol* 1965; 79:592–612.
29. Acheson ED, Hadfleld EH, Macbeth RG. Carcinoma of the nasal cavity and accessory sinuses in wood workers. *Lancet* 1967; 1:311–2.
30. Acheson ED, Cowdell RH, Hadfield F, et al. Nasal cancer in woodworkers in the furniture industry. *Br Med J* 1968; 2:587–96.
31. Hadfield EH. A study of adenocarcinoma of the paranasal sinuses in woodworkers in the furniture industry. *Ann R Coll Surg Engl* 1970; 46:301–19.
32. Cignoux M, Bernard P. Malignant ethmoid bone tumors in woodworkers. *J Med Lyon* 1969; 25:92–3.
33. Ironside P, Matthews J. Adenocarcinoma of the nose and paranasal sinuses in woodworkers in the state of Victoria, Australia. *Cancer* 1975; 36:1115–21.
34. Klintenberg C, Olofsson J, Hellquist H, et al. Adenocarcinoma of the ethmoid sinuses: a review of 28 cases with special reference to wood dust exposure. *Cancer* 1984; 54:482–8.
35. Cecchi F, Buiatti E, Kriebel D, et al. Adenocarcinoma of the nose and paranasal sinuses in shoemakers and woodworkers in the province of Florence, Italy (1963–77). *Br J Ind Med* 1980; 37:222–5.
36. Battista G, Cavallucci F, Coinba P, et al. A case-referent study on nasal cancer and exposure to wood dust in the province of Sienna Italy. *Scand J Work Environ Health* 1983; X:9–29.
37. Debois JM. Tumors of the nasal cavity in woodworkers. *Tjidschr Diergeneeskd* 1969; 25:92–3.
38. Mosbech J, Acheson ED. Nasal cancer in furniture makers in Denmark. *Dan Med Bull* 1970; 18:34–5.
39. Van den Oever R. Occupational exposure to dust and sinonasal cancer. An analysis of 386 cases reported to the NCCSF Cancer Registry. *Acta Otorhinolaryngol Belg* 1996; 50:19–24.
40. Demers PA, Kogevinas M, Boffetta P, et al. Wood dust and sino-nasal cancer: pooled reanalysis of twelve case–control studies. *Am J Ind Med* 1995; 28:151–66.
41. IARC Working Group. Wood dust and formaldehyde. In: IARC Monographs on the evaluation of carcinogenic risks to humans, Vol. 62. Lyon: WHO, 1995.
42. Ball MJ. Nasal cancer and occupation. *Lancet* 1967; 2:1089–90.
43. Elwood JM. Wood exposure and smoking: association with cancer of the nasal cavity and paranasal sinuses in British Columbia. *Can Med Assoc J* 1981; 124:1573–7.
44. Brinton LA, Blot WJ, Stone BJ. A death certificate analysis of nasal cancer among furniture workers in North Carolina. *Cancer Res* 1977; 37:3473–4.
45. Roush GC, Meigs JW, Kelly J. Sinonasal cancer and occupation: a case control study. *Am J Epidemiol* 1980; 111:183–93.
46. Brinton LA, Blot WJ, Becker JA. A case–control study of cancers of the nasal cavity and paranasal sinuses. *Am J Epidemiol* 1984; 119:896–906.
47. Imbus HR, Dyson WE. A review of nasal cancer in furniture manufacturing and woodworking in North Carolina, the United States, and other countries. *J Occup Med* 1987; 29:734–40.
48. Blot WJ, Chow WH, McLaughlin JK. Wood dust and nasal cancer risk. A review of the evidence from North America. *J Occup Environ Med* 1997; 39:148–56.
49. Demers PA, Boffetta P, Kogevinas M, et al. Pooled reanalysis of cancer mortality among five cohorts of workers in wood-related industries. *Scand J Work Environ Health* 1995; 21:179–90.

50. Acheson ED. Nasal cancer in the furniture and boot and shoe manufacturing industries. *Prezí Med* 1976; 5:295–315.

51. Nelson E, Zhou Z, Carmichael PL, Norpoth K, Fu J. Genotoxic effects of subacute treatments with wood dust extracts on the nasal epithelium of rats: assessment by the micronucleus and 32P-postlabelling. *Arch Toxicol* 1993; 67:586–9.

52. Mohtashamipur E, Norpoth K, Ernst H, Mohr U. The mouse-skin carcinogenicity of a mutagenic fraction from beech wood dusts. *Carcinogenisis* 1989; 10:483–7.

53. Mohtashamipur E, Norpoth K, Hallerberg B. A fraction of beech wood mutagenic in the Salmonella/mammalian microsome assay. *Int Arch Occup Environ Health* 1986; 58:227–34.

54. Mohtashamipur F, Norpoth K. Non-mutagenicity of some wood-related compounds in the bacterial/microsoine plate incorporation and microsuspension assays. *Int Arch Occup Environ Health* 1984; 54:83–90.

55. American Conference of Governmental Industrial Hygienists. *2001 TLVs and BEIs*. Cincinnati: 2001.

INDEX

ABOUT THE AUTHORS

Peter H. Wald, M.D., M.P.H., F.A.C.P., F.A.C.O.E.M., F.A.C.M.T., is Medical Director and Principal at WorkCare, Orange, CA, and Clinical Assistant Professor of Occupational Medicine at the University of California, Los Angeles and University of California, Irvine.

Dr. Wald received his undergraduate education at Harvard before attending Tufts Medical School. He also holds an M.P.H. degree from the University of California, Berkeley. Dr. Wald is board certified in internal medicine, occupational medicine, and medical toxicology. Before joining WorkCare, he was Corporate Medical Director for ARCO, and prior to that worked in the medical departments at Mobil Oil and the Lawrence Livermore National Laboratory.

Dr. Wald is a Fellow of the American College of Physicians, the American College of Occupational and Environmental Medicine, and the American College of Medical Toxicology. He lives in Los Angeles with his wife and two sons.

Gregg M. Stave, M.D., J.D., M.P.H., F.A.C.P., F.A.C.O.E.M., is Director, Strategic Health Planning, GlaxoSmithKline, Research Triangle Park, NC, and Consulting Assistant Professor in the Division of Occupational and Environmental Medicine at Duke University Medical Center.

Dr. Stave received his undergraduate education at the Massachusetts Institute of Technology before attending the dual-degree M.D./J.D. program at Duke University. He received his M.P.H. degree in Epidemiology from the University of North Carolina, Chapel Hill. Dr. Stave is board certified in internal medicine and preventive medicine (occupational medicine).

A Fellow of the American College of Physicians, American College of Occupational and Environmental Medicine, and the American College of Preventive Medicine, Dr. Stave is also a member of the Bar in North Carolina and the District of Columbia. He lives in Chapel Hill, North Carolina with his wife and daughter.